Advances in Intelligent Systems and Computing

Volume 556

Series editor
Janusz Kacprzyk, Polish Academy of Sciences, Warsaw, Poland
e-mail: kacprzyk@ibspan.waw.pl

About this Series

The series "Advances in Intelligent Systems and Computing" contains publications on theory, applications, and design methods of Intelligent Systems and Intelligent Computing. Virtually all disciplines such as engineering, natural sciences, computer and information science, ICT, economics, business, e-commerce, environment, healthcare, life science are covered. The list of topics spans all the areas of modern intelligent systems and computing.

The publications within "Advances in Intelligent Systems and Computing" are primarily textbooks and proceedings of important conferences, symposia and congresses. They cover significant recent developments in the field, both of a foundational and applicable character. An important characteristic feature of the series is the short publication time and world-wide distribution. This permits a rapid and broad dissemination of research results.

More information about this series at http://www.springer.com/series/11156

Himansu Sekhar Behera
Durga Prasad Mohapatra
Editors

Computational Intelligence in Data Mining

Proceedings of the International Conference on CIDM, 10–11 December 2016

Springer

Editors
Himansu Sekhar Behera
Department of Computer Science and Engineering & Information Technology
Veer Surendra Sai University of Technology
Sambalpur, Odisha
India

Durga Prasad Mohapatra
Department of CSE
National Institute of Technology (NIT)
Rourkela, Odisha
India

ISSN 2194-5357 ISSN 2194-5365 (electronic)
Advances in Intelligent Systems and Computing
ISBN 978-981-10-3873-0 ISBN 978-981-10-3874-7 (eBook)
DOI 10.1007/978-981-10-3874-7

Library of Congress Control Number: 2017931531

Printed on acid-free paper

This Springer imprint is published by Springer Nature
The registered company is Springer Nature Singapore Pte Ltd.
The registered company address is: 152 Beach Road, #21-01/04 Gateway East, Singapore 189721, Singapore

Preface

In this decade, with advancements in data mining, many new models and tools in discovering knowledge and extracting intelligence brought forth revolutionary developments with the help of computational intelligence techniques. The present scenario of storage of the amount of data is quite huge in the modern database due to the availability and popularity of Internet. Thus, information needs to be summarized and structured in order to maintain effective decision-making. When the quantity of data, dimensionality, and complexity of the relations in the database are beyond human capacities, there is a requirement for intelligent data analysis techniques, which could discover useful knowledge from data. While data mining evolves with innovative learning algorithms and knowledge discovery techniques, computational intelligence harness the results of data mining for becoming more intelligent than ever. In the present scenario of computing, computational intelligence is playing a major role in solving many real world complex problems. The 3rd International Conference on "Computational Intelligence in Data Mining (ICCIDM 2016)" organized by Kalinga Institute of Industrial Technology (KIIT), Bhubaneswar, Odisha, India on 10 and 11 December 2016. ICCIDM is an international forum for representation of research and developments in the fields of data mining and computational intelligence. More than 300 prospective authors had submitted their research papers to the conference. This time the editors have selected 79 papers after double-blind peer-review process by experienced subject experts chosen from the country and abroad. The proceedings of ICCIDM are a mix of papers from some latest findings and research of the authors. It is a great honor for us to edit the proceedings. We have enjoyed considerably working in cooperation with the International Advisory, Program, and Technical Committees to call for papers, review papers, and finalize papers that are included in these proceedings.

This International Conference ICCIDM aims at encompassing a new breed of engineers, technologists making it a crest of global success. All papers are focused on thematic presentation areas of the conference and they have provided ample opportunity for presentation in different sessions. This year's program includes exciting collections of contributions resulting from a successful call for papers. The selected papers have been divided into thematic areas including both review and research papers and which highlight the current focus of computational intelligence techniques in data mining. The conference aims at creating a forum for further discussion for an integrated information field, incorporating a series of technical issues in the frontier analysis and design aspects of different alliances in the related field of intelligent computing and others. Therefore, the call for paper was on three major themes such as methods, algorithms, and models in data mining and machine learning, advance computing and applications. Further the papers discussing the issues and applications related to the theme of the conference were also welcomed in ICCIDM.

We hope readers will enjoy the collection papers published in this volume.

Sambalpur, India Himansu Sekhar Behera
Rourkela, India Durga Prasad Mohapatra

Acknowledgements

The 2016 edition of ICCIDM has drawn nearly eighty research articles authored by numerous academicians, researchers, and practitioners throughout the world. We thank all of them for sharing their knowledge and research findings on an international platform like ICCIDM and thus contributing towards producing such a comprehensive conference proceedings of ICCIDM.

The level of enthusiasm displayed by the members of organizing committee right from day one is commendable. The extraordinary spirit and dedication shown by the organizing committee in every phase throughout the conference deserves sincere thanks from the bottom of my heart.

After two successful versions of ICCIDM, it is an honor for us to edit the proceedings of this third series of ICCIDM. We are fortunate to work in cooperation with a brilliant international as well as national Advisory, Program, and Technical Committees, comprising eminent academicians, to call, review, and finalize the papers for the proceedings.

We would like to express our heartfelt gratitude and obligations to the benign reviewers for sparing their valuable time and putting in effort to review the papers in a stipulated time and providing their valuable suggestions and appreciation in improving presentation, quality, and content of these proceedings. The eminence of these papers is an accolade not only to the authors but also to the reviewers who have guided towards perfection.

Last but not least, the editorial members of Springer Publishing deserve a special mention and we offer our sincere thanks to them not only for making our dream come true in the shape of these proceedings, but also for its hassle-free and in-time publication in the reputed Advances in Intelligent Systems and Computing Series.

The ICCIDM conference and proceedings are a credit to a large group of people and everyone should be proud of the outcome.

Acknowledgements

About the Conference

The International Conference on "Computational Intelligence in Data Mining" (ICCIDM) has become one of the reputed conferences in data mining and its applications amongst the researchers across the globe after its two successful versions in 2014 and 2015. ICCIDM 2016 aims to facilitate cross-cooperation across diversified regional research communities within India as well as with other international regional research programs and partners. Such active discussions and brainstorming sessions among national and international research communities are the need of the hour as new trends, challenges, and applications of computational intelligence in the field of science, engineering and technology are cropping up by each passing moment. The 2016 edition of ICCIDM is an opportune platform for researchers, academicians, scientists, and practitioners to share their innovative ideas and research findings, which will go a long way in finding solutions to confronting issues in related fields.

The conference aims to:

- Provide a sneak preview into the strengths and weakness of trending applications and research findings in the field of computational intelligence and data mining.
- Enhance the exchange of ideas and achieve coherence between the various computational intelligence methods.
- Enrich the relevance and exploitation experience in the field of data mining for seasoned and naïve data scientists.
- Bridge the gap between research and academics so as to create a pioneering platform for academicians and practitioners.
- Promote novel high-quality research findings and innovative solutions to the challenging problems in intelligent computing.
- Make a fruitful and effective contribution towards the advancements in the field of data mining.
- Provide research recommendations for future assessment reports.

At the end, we hope the participants will enrich their knowledge with new perspectives and views on current research topics from leading scientists, researchers and academicians around the globe, contribute their own ideas on important research topics like data mining and computational intelligence, as well as will collaborate with their international counterparts.

Contents

About the Editors

Prof. Himansu Sekhar Behera is working as Associate Professor in the Department of Computer Science and Engineering & Information Technology, Veer Surendra Sai University of Technology (VSSUT)—an unitary technical university established by the Government of Odisha. He has received M.Tech. in Computer Science and Engineering from NIT, Rourkela (formerly R.E.C, Rourkela) and Doctor of Philosophy in Engineering (Ph.D.) from Biju Pattnaik University of Technology (BPUT), Rourkela, Government of Odisha, respectively. He has published more than 100 research papers in various international journals and conferences, edited 11 books and is acting as a member of the editorial/reviewer board of various international journals. He is an expert in the field of computer science engineering and served in the capacity of program chair, tutorial chair and as an advisory member of committees of many national and international conferences. His research interest includes data mining and intelligent computing. He is associated with various educational and research societies like OITS, ISTE, IE, ISTD, CSI, OMS, AIAER, SMIAENG, SMCSTA, etc. He is currently guiding eight Ph.D. scholars.

Prof. Durga Prasad Mohapatra received his Ph.D. from Indian Institute of Technology Kharagpur and is presently serving as Associate Professor in NIT Rourkela, Odisha. His research interests include software engineering, real-time systems, discrete mathematics, and distributed computing. He has published more than 30 research papers in these fields in various international journals and conferences. He has received several project grants from DST and UGC, Government of India. He has received the Young Scientist Award for the year 2006 from Orissa Bigyan Academy. He has also received the Prof. K. Arumugam National Award and the Maharashtra State National Award for outstanding research work in Software Engineering for the years 2009 and 2010, respectively, from the Indian Society for Technical Education (ISTE), New Delhi. He is going to receive the Bharat Sikshya Ratan Award, for his significant contribution in academics, awarded by the Global Society for Health and Educational Growth, Delhi.

Safety and Crime Assistance System for a Fast Track Response on Mobile Devices in Bhubaneswar

Debabrata Singh, Abhijeet Das, Abhijit Mishra and Binod Kumar Pattanayak

Abstract We have developed an android application which will locate the user using GPS system and recommend the nearest emergency point along with their route direction and contact details. We will try to solve the various issues in the area of emergency prevention and mitigation. In this technology, the system is highly effective for local residents and a boon for the tourists. Service for providing highway emergency number, which is different for different highways sections, is in pipeline. There is also an issue of non-availability of information about emergency numbers in many areas or the credibility of information is questioned so we are working to provide the facility at Bhubaneswar, updating the information through crowd computing, i.e., through people of that concerned area.

Keywords Location awareness · Emergency handling mechanism · Tracking crime · Safety index · Mobile crime assistance

1 Introduction

India is a developing nation with 1.28 billion of population that has scattered emergency rescue handling mechanism which leads to higher death rates and criminal index as well. India still requires a one-stop emergency handling mechanism, like 911 phone call in USA, which will cater all the emergency need of

D. Singh (✉) · A. Das · A. Mishra · B.K. Pattanayak
Institute of Technical Education and Research, S'O'A University,
Bhubaneswar, Odisha, India
e-mail: debabratasingh@soauniversity.ac.in

A. Das
e-mail: dasabhijeet01@gmail.com

A. Mishra
e-mail: mishra.abhijit94@gmail.com

B.K. Pattanayak
e-mail: binodpattanayak@soauniversity.ac.in

H.S. Behera and D.P. Mohapatra (eds.), *Computational Intelligence in Data Mining*, Advances in Intelligent Systems and Computing 556,
DOI 10.1007/978-981-10-3874-7_1

civilian thereby mitigating the affect rate of such emergency. On the other hand, there is an increasing level of penetration of Internet and android-based smart phones even in rural areas. Integration of these two technologies gives a powerful opportunity to streamline the emergency handling procedure in India.

There is need of time to bring together all the emergency service providers as well as family and friends of victim under one umbrella by use of advancement in mobile technology so that the victim saves the precious time during such events which will also reduce the crime index. India is a 'mobile first' country, we want to build an ecosystem with victim and his smart phone as center point. It will advance the digital India initiative. Our projects target on two main aspects of emergency handling procedure, one is emergency prevention and another is emergency mitigation. We are also working on advancement of basic app which includes, one-touch information forwarding about victim's whereabouts over the Internet to the location wise nearest available friend or registered *volunteer* in that particular area so that the victim can be assisted in sooner time. Proposal of creating *safety index,* i.e., frequency of crime occurrence in an area through use of crowd computing will be useful for security agency to develop strategies to curb crime rate in that area as well as to prompt passersby in that area to take adequate security measures.

This system can be further enhanced if federal government adopts this technology and includes police, ambulance, trauma care centers where the app will instantly send instant notification to the nearest police personnel, ambulance, volunteer regarding the whereabouts of the victim through his smart phone which will be concurrent message passing than linear message passing, and will drastically reduce the time taken to reach the victim and rescue the victim.

2 Problem Statement

India being a developed country is still struggling to provide basic necessities to its citizens. Though India is progressing in the way to provide such services but till today a billion+ population country has no one-stop dedicated emergency handling mechanism. As a result, many people find a great difficulty to locate their emergency service providers.

There is no one-stop app for emergency service provider's contact numbers and their location details. The project aims to develop an android application which will simplify the process of locating and contacting the nearest emergency service providers, e.g., Police Station, Hospital Services, Trauma Services, Fire Services, Blood banks, etc.

It will save precious time during a panic situation by guiding the user to navigate as well as to contact the emergency providers in real time primarily targeted toward the citizens of the country. We already have android apps that provide police station, fire, hospital ambulance services, blood bank, but this facility is scattered. Currently you need to have one application for locating a particular emergency service provider subject to availability of such apps. No one really knows when, how, and what kind of emergency (S) he will face and when cases arises so it will be useful for people to locate such services. Particularly travelers, highway commentators will like it as it will provide them the information in an unknown location.

3 Current Mechanism of Response to Crimes

1. **Accidents and crimes are ever rising**. Due to increasing growth of civilization, the rise in vehicles as well as safety of citizens is now becoming a serious problem to our nation. But to handle the ever-increasing need of security concerns our security personnel's are not adequate. They are unable to tackle the issues in the conventional issues. And it is not about just security issue there is also need of attending any crime [1, 2] or accident victim in the shortest possible time as every second counts in a typical accident scenarios. So often it is observed that people lose precious time in locating the victim which drastically increases the seriousness of the victim.
2. **Typically the victim or surrounding call the police control room, ambulance, hospital, victim's family in a sequential manner in time line**. When an accident occurs in real world, people often call the police ambulance in linearly which takes a lot of valuable time. The police personnel's or ambulance people waste a lot of time to pin pointedly locate the victim which loses time which cannot be compensated against patient safety. This linear approach to message passing to the friend's relatives and security personnel's takes a lot of time.
3. **Increases the risk for victim for any kind of crime scene due to loss of time**. As people loose precious time in sending the patient to hospital and in worst case if the people do not know where the hospital is (In case of foreign travelers) the matter gets worse; the patient sometimes die due to no medical attention.
4. **If the victim is all alone there the risk is much higher**. In case the victim is traveling alone and meets some unfortunate events then the matter gets worst. He is not in situation to call any of the police family or friends. This makes the matter worse and a life is lost.
5. **Often it is much tougher to locate the victim pin-pointedly by the rescue personnel which increases the risk of victim health safety**. As we already discussed about the issues that how important it is that the person gets immediately and accurately located to give him/her best possible attention to rescue him/her from that situation. In conventional matter we cannot locate the victim accurately and immediately.
6. **Rescue personnel often get late information about incidents which consequences either death of victim or escape of crime mastermind**. People often do not know the details of nearest police station or trauma center and there is no scientific mechanism to pass the accurate information about incidents to the

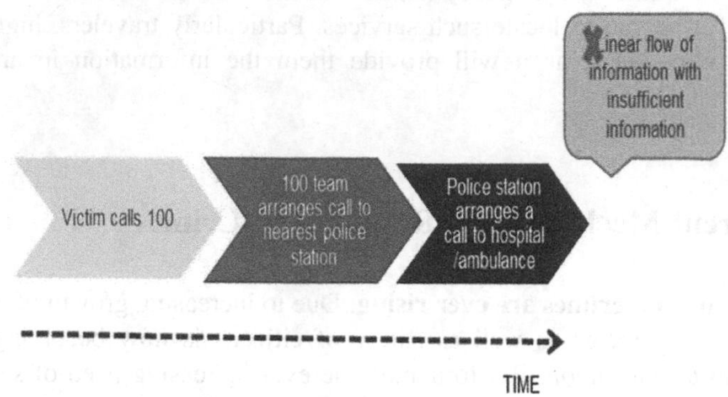

Fig. 1 Indicating loss of time in linear flow of message

nearest security or ambulance personals which create problems in survival of patient and escape of criminals.

7. **Information passing in conventional mode**. As depicted in Fig. 1 when an accident occurs in real world people often call the police ambulance in linearly which takes a lot of valuable time. The police personnel's or ambulance people waste a lot of time to pin-pointedly locate the victim which loses time which cannot be compensated against patient safety. This linear approach to message passing to the friend's relatives and security personnel's takes a lot of time.

We are currently experiencing some problems in Bhubaneswar to meet the ever-increasing demand of security and safety. But these issues cannot be tackled in the conventional mode without the use of mobility technologies. So to provide better security and safety service to citizens, we need to heavily rely on mobility technologies which will solve some serious problems of our country. We are working to build ecosystem of victim, victim's friend and relatives, ambulance system, hospitals, and police system.

4 How the Situation can be Tackled?

Location information can be used for reporting accident and crime [3], so that a quick and fast response can be achieved [4]. This paper argues that people having smart phones can be easily traced using cellular technology which will help in fastest possible tracking of victim. But we argue that with use of GPS and cellular tracking technology, we can pin pointedly and accurately locate the user. Moreover we can also pass this information to other stake holders to help the victim in fastest possible way and accurately. Smart phones with Internet (generally android apps) [5] are penetrating to the even rural areas.

Today smart phones are available at the rates of as low as $100 and there is heavily penetration of 3G and 4G data connectivity even in rural areas. Even in worst case, the operators are providing 2G services in rural areas. And the highways or heavily populated areas are connected with 3G or 4G services. So it is safe to say that smart phones along with internet are capable of transmitting the vital information at such unfortunate incidents. It will also act as a medium to create an eco-system of victim friend's family, police and hospitals.

Today's smart phones being capable of sharing location based data through various technologies can be a boon to handle these scenes. As all android phones have GPS and track over Wi-Fi and cellular data is available by default so we can safely argue that any android phonc will be capable of availing the facility of this system. Later we can extend the service to different mobile operating system. We can bring together all the stakeholders of a crime/accident scene under one roof for fast track response with victim at center approach. As of now all the stakeholders work differently with any synchronism. They all work individually to address a crime or accident victim. But it should be understood that this job is collective work. All the stakeholders must work in synchronous manner in a coordinated way to get fruitful results from noble intention of saving lives.

So there is "ZERO time loss" due to communicational delay [6, 7] as the accurate information about victim can be shared by all the stakeholders and can actively monitor the rescue process. Right action at Right time can be taken and we can save thousands of peoples by optimizing assisting personnel. This model assumes that the victim as well as all the stakeholders have a smart phone with necessary apps and data connectivity.

4.1 Victim at Center Approach

With victim at center approach, everyone's primary attention will be locating the victim and rescuing him. With this approach, we assume victim being the central point and all the stakeholders are surrounded to the victim so that they can proactively help the victim with zero time loss as depicted in Fig. 2. With all the stake holders in one system, i.e., Mobile Crime Assistance System [8, 9].

4.2 Information Passing in Victim at Center Approach

With victim at center approach the information reaches to all stakeholders in seconds through mobility technology conventional to linear approach in information passing which takes sequential time which loses valuable time as depicted in Fig. 3.

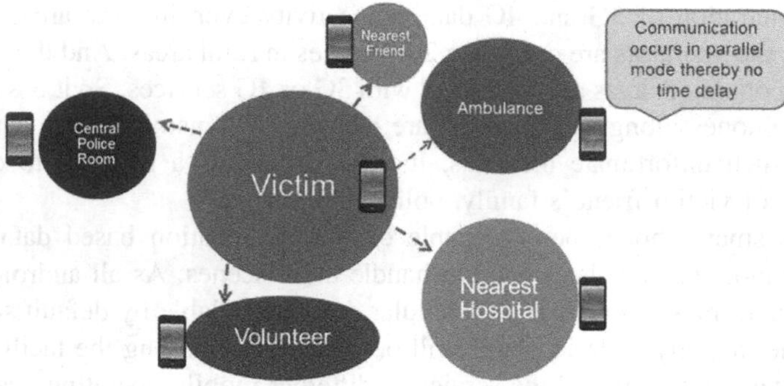

Fig. 2 Communication in victim at centers

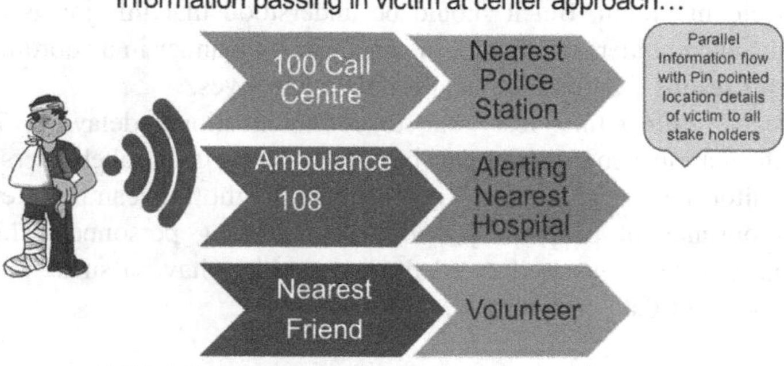

Fig. 3 Information passing in victim at center approach

5 Safety and Crime Assistance System

This system can be achieved through Safety and Crime Assistance System. It will be multi-tier, multi-view, multi-user, distributed architecture with no central control point for faster response. Being a multi-tier, it will be accessible through multi-devices and multi-view refers to web view as well mobile view. Here the control will be peer to peer, i.e., there will be no central authority to pass the message [10]. Once the user sends a message it will be available to all the stakeholders of the system in that geographical area. This will allow ground rescuing authority to get into the victim as fast as possible as they do not need to wait from there call center, e.g., 108 Ambulance call center and the driver need not to be manually notified about the crime or accident spot. It will be synchronously available to the ground rescuing authority. And once the rescuing people reach there they do the needful and update the same in app. This again notifies all the stakeholders about the current status about the victim [11, 12] and alerts the next required rescuing personnel to attain the victim which describes in Chap. 6.

The whole process being peer to peer will drastically reducing the time gap due to communicational issues. The rider taps a button requesting a car and the driver accepts the ride and traces the user in realtime and reaches him. This process will be same in SCAS as depicted in Fig. 5, the technical architecture is depicted on Fig. 4 and process workflow is also given on Fig. 7. Figure 6 describes about the safety and crime assistance system. Once the rescuing personnel collects the victim they can inform the nearest hospital and police for their schedule arrival in the hospital which will keep the trauma care team at ready state for providing necessary assistance.

5.1 SCAS Architecture

See Fig. 4

5.2 SCAS Advantages in Accidental Scenes

1. No loss of time due to concurrent message passing. Message passes to stakeholders in real-time with parallel flow of information which reduces the time gap. 2.

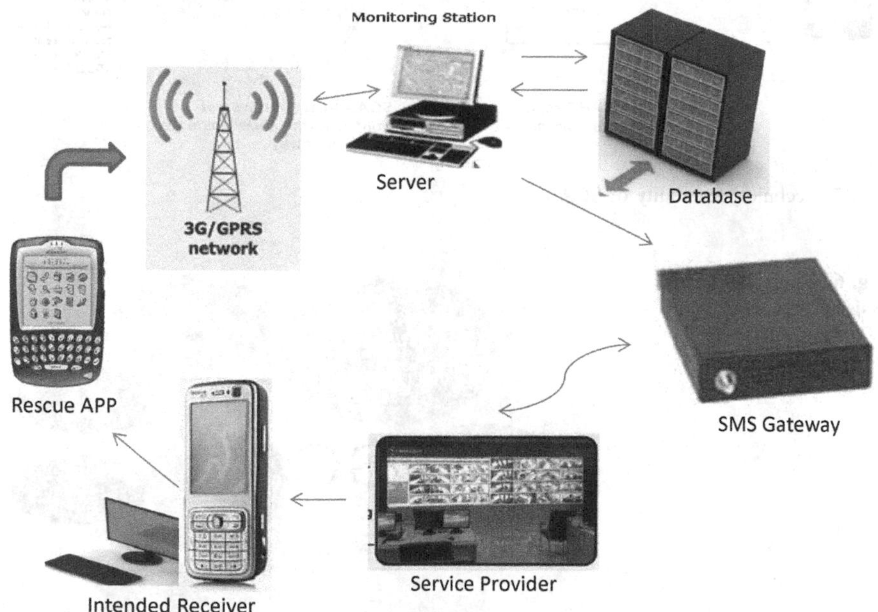

Fig. 4 Technical architecture of our application

Pin pointed location details of victim to everyone, for faster victim locating process. All the rescuing authority gets pin pointed real time location of victim there by reducing time loss in locating the victim. 3. No need of remembering emergency numbers. 4. Hospital is alerted in advance about victim's arrival thereby prepares its clinical arrangements, e.g., concerned doctor, medicine, etc. 5. The system does all the work of informing at that crucial time. 6. Very useful if victim is not in condition to pass the info verbally, e.g., being kidnapped.

5.3 Technical Architecture and Feasibility of SCAS

See Figs. 5 and 6.

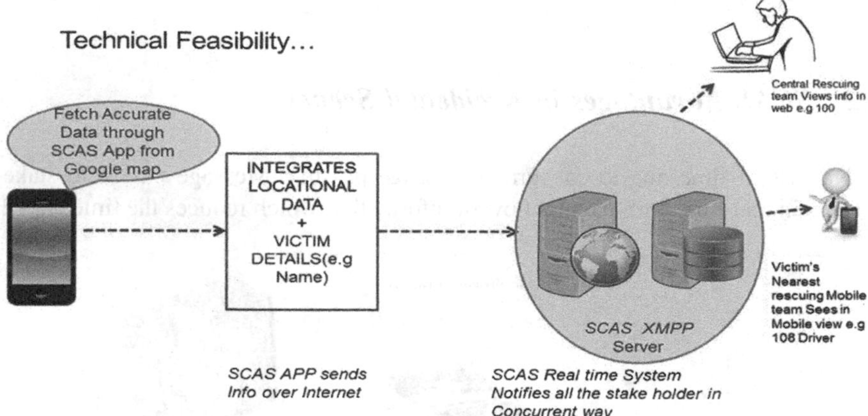

Fig. 5 Technical feasibility of SCAS

Fig. 6 Safety and crime assistance system

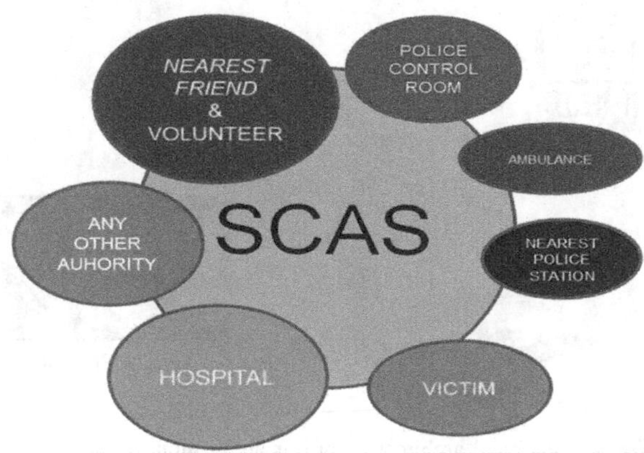

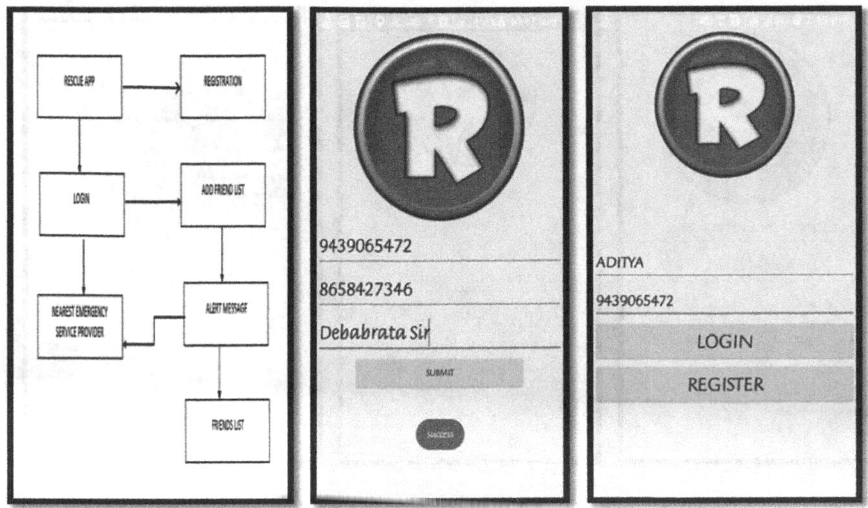

Fig. 7 Process workflow, login and registration page

6 Issues

We tried to develop a safety and crime assistance system but we are unable to build a complete and fully reliable system due to various technical issues. We tried to implement the message passing through Google Cloud Messaging using Push notification technology then the system would have more reliable failed on which we implanted the same using SMS technology through SMS gateway. Another point where we failed to implement one of the features, where we could locate the victims nearest friend in real time basis which would have been possible by constantly tracking each friend and registered volunteer on real time is our next research. We tried to represent a near replica model for parallel flow of information with one single touch rather than multiple information passing which saves and provided that model using alternative technology which is less reliable and cost expensive.

7 Process Workflow, Login Page, and Registration Page

The above and below workflow sequences detailing how activities are launched and in what sequence the correspondence to reach the goal, which are given from Figs. 8, 9 and 10. Figure 8 describes about the login and registration page and adds the friends name with phone numbers. At the receiver end, the message of victim's location is received at registered friend's mobile. Finally Fig. 9 shows the list of nearby fire stations and blood banks and Fig. 10 shows the list of nearby hospitals and police Stations.

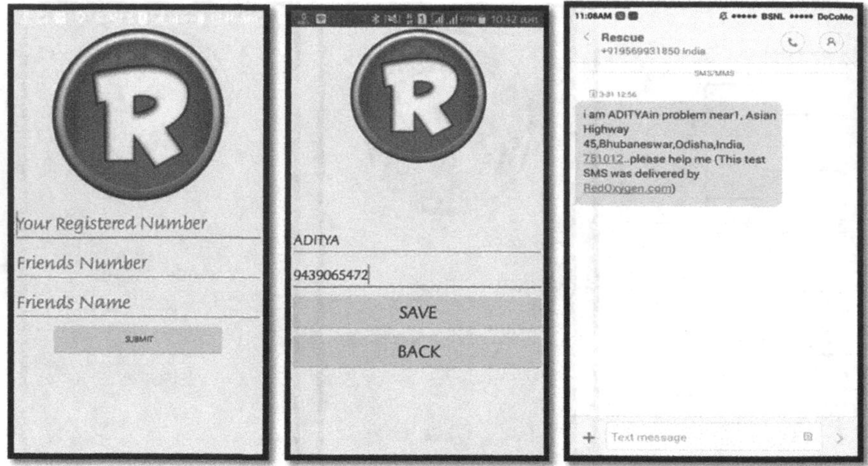

Fig. 8 Add friend name mobile numbers and message received at registered friend's mobile

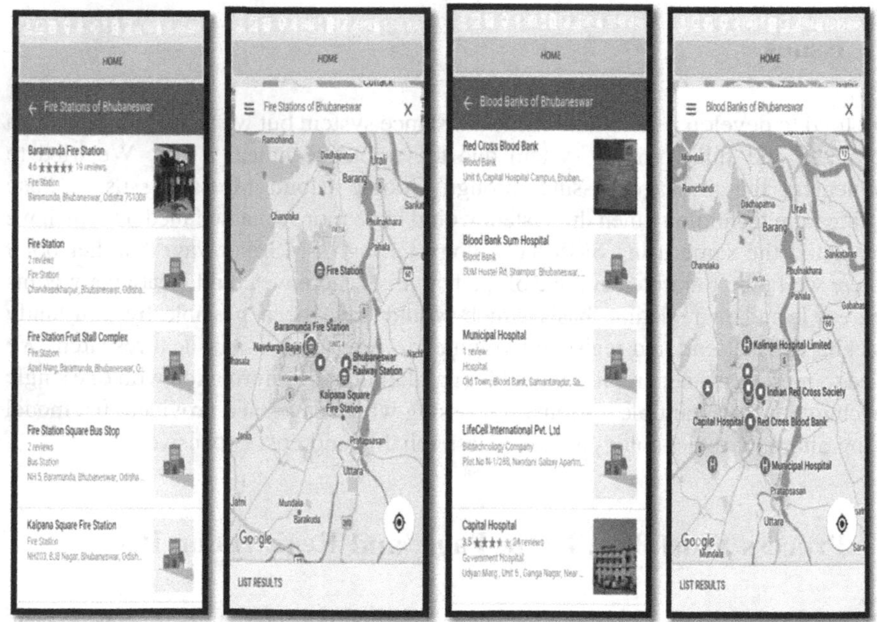

Fig. 9 Showing list of nearby fire stations and blood banks

Fig. 10 Showing list of nearby hospitals and police station

8 Conclusion

Hence we can conclude from the application that, by using the application we can get the emergency services by one touch on the application button which will be directed to the emergency service providers using Google map API and web view. We also concluded that we can forward the one touch information in parallel approach to all the stakeholders' linear information passing in existing model. The system can offer a facility to track the location of the nearest hospital and police station by accessing a built-in database with the help of GPRS, Google map, etc., and send the information (data/image) from the crime location to the police station. This will allow the police to find the location of the accident right away and increase the safety of the residents in big cities. Finally, we discussed how big data and data mining can be used to solve big cities crime problems with crime prevention and mitigation.

References

1. Phillips, Peter, and Ickjai Lee. "Crime analysis through spatial areal aggregated density patterns", Journal Geoinformatica, (2011), Vol-15, Issue-1, pp 49–74.
2. Chainey, Spencer, Lisa Tompson, and Sebastian Uhlig, "The utility of hotspot mapping for predicting spatial patterns of crime", Security Journal, (2008), Volume 21, Issue 1, pp 4–28.
3. Odgers, Candice L., et al. "The protective effects of neighbourhood collective efficiency on British children growing up in deprivation: a developmental analysis", Developmental psychology, (2014), Volume 45, Issue 4, pp 942–957.
4. Mantoro, Teddy, et al. "Location-Aware Mobile Crime Information Framework for Fast Tracking Response to Accidents and Crimes in Big Cities", IEEE International Conference on Advanced Computer Science Applications and Technologies (ACSAT), (2014), pp 192–197.
5. http://www.appszoom.com/android-apps/911.
6. Hwangnam Kim, Hyun Soon Kim, Enabling location-aware quality-controlled access in wireless network, EURASIP Journal on Wireless Communications and Networking (2011).
7. Fredrik Gustafsson, Fredrik Gunnarsson and David Lindgren, "Sensor models and localization algorithms for sensor networks based on received signal strength", EURASIP Journal on Wireless Communications and Networking (2012).
8. Zhou, Guiyun, Jiayuan Lin, and Xiujun Ma., "A Web-Based GIS for Crime Mapping and Decision Support", Forensic GIS. Springer,(2014), Vol-11, pp 221–243.
9. T Vacharas, Jesdabodi, C and Tu Ngoc Nguyen, "An Intelligent Agent for Ubiquitous Travel Information Assistance with Location awareness", WiCOM, (2010), pp. 1–4.
10. Edmonds, Suzanne, Alison Patterson, and Dominic Smith. Policing and the criminal justice system-public confidence and perceptions: findings from the 2003/04 British Crime Survey. London, UK: Home Office (2005).
11. http://www.dailymail.co.uk/sciencetech/article-3023155/The-app-save-LIFE-Emergency-service-tells-rescuers-save-touch-button.html.
12. Yih-Shyh Chiou, Fuan Tsai, Chin-Liang Wang and Chin-Tseng Huang," A reduced-complexity scheme using message passing for location tracking", EURASIP Journal on Advances in Signal Processing, (2012), pp 1–18.

Major Global Energy (Biomass)

Sukhleen Kaur, Vandana Mukhija, Kamal Kant Sharma and Inderpreet Kaur

Abstract This research gives brief introduction about types of renewable energy mainly biomass energy methods of generation and its pros and cons.

Keywords Biomass · Heat · Energy · Bio fuels · Clean green environment · Energy harvesting

1 Introduction

The growing world population, industrialization, and progressive technological advancement and transportation had conveyed energy demand under an increased pressure nowadays. The world's energy markets trust blindly on the fossil-derived fuels whose reserves are finite. The possibility of the hereafter deficit of the conventional fossil oil reserve has generated a keen interest in digging an alternative sources of energy, one of which is biomass.

Although, content and efficiency of biomass energy use might be far incomparable to that of fossil fuel, coupled with the disadvantages of current higher costs of biofuels and the large land required for enough amounts of bioenergy and other pros and cons, generating methods will be discussed in this paper.

S. Kaur (✉) · V. Mukhija · K.K. Sharma · I. Kaur
Department of Electrical & Electronics Engineering, Chandigarh University, Gharuan, India
e-mail: Sukhleenkaur691@gmail.com

V. Mukhija
e-mail: vandanamukhija09@gmail.com

K.K. Sharma
e-mail: sharmakamal2002@gmail.com

I. Kaur
e-mail: inder_preet74@yahoo.com

H.S. Behera and D.P. Mohapatra (eds.), *Computational Intelligence in Data Mining*, Advances in Intelligent Systems and Computing 556,
DOI 10.1007/978-981-10-3874-7_2

2 Biomass as a Renewable Resource

The natural energy resources for biomass are mostly from plant and the most significantly is the wood and wood waste. "Biowaste" or simply biodegradable waste used in an industrial process. The natural energy resources from biomass are mostly from plant, fruit, fabric, food and the most significant is the wood and wood waste. As compare to other renewable energies usage of biomass is increasing day by day. There are mainly two forms of biomass

Primary biomass (Raw biomass): It includes product from forest such as wood, crops, animal waste, rubber, gums, etc.

Secondary biomass: Some raw biomass materials that have undergone some significant changes for example: paper, cardboard, cotton, natural rubber products, and used cooking oils are best example of secondary biomass (Fig. 1).

2.1 Biomass Energy –Bio Power

Bio-power can be defined as production of electricity or heat using biomass materials. Bios include waste from agriculture, forest, and industries. Also municipal solid wastes are considered as BIOS. Boilers, gasifiers, turbines, generator are the major components which are employed in technologies for the conversion process (Fig. 2).

Fig. 1 Source of bio

Fig. 2 Energy crops yield wood chips, hybrid willows (above left) 50-MW wood-fired power plant located in Burlington, Vermont., (above right)

The conversion technologies used in these plants are following:

A. **Thermochemical process**

- **Direct combustion:** At present, most preferable technique used for generation of electricity from biomass is combustion which is also called direct-firing. Fossil-fuel fired power plants are similar to combustion systems used for electricity and heat production. The biomass fuel is already present inside the boiler which is used to create high-pressure steam. Further the steam is pushed towards a turbine, which create a force on blades of turbine, i.e., connected in series as a result turbine starts rotating. For the generation of electricity this turbine is then connected to an electric generator. This is easily available, affordable, and a kind of commercial technology.
- **Gasification**: Biomass is a type of unmanageable fuel source (like coal) because it is a solid. By converting this energy into a gas, it is then used for devices which have large energy consumption For example, gas released from biomass are treated and can be used for menage purpose, converted into power, mechanical energy, a synthetic gas or chemical products. In gasifiers section heating of biomass in surroundings consists of hard biomass fragments down to a combustible gas. The cleansing and filtering process is executed in order to remove toxic chemicals. This gas is highly efficient therefore used in power production called combined cycle.
- **Pyrolysis**: The biomass raw material is accountable to extreme temperature at depressed oxygen content, therefore curbed to burning, and done at high pressure. Biomass is reduced to molecules of single carbon (CH_4 and CO) and H_2 which generates gassy inter-mixture called "producer gas." Under the pyrolytic conditions, carbon dioxide can be generated but of the reactor it is returned into CO and H_2O; which helps in further reaction purposes. The outcome of the temperature is liquid phase produces which are not sufficient to crack the long chains of molecules as a result oils, tars, etc. are produced. Now the residual biomass is in the form of char, i.e., pure carbon

- **Catalytic liquefaction**: The main reason of using this technology is that it produces products with higher energy density as well as with better quality then any other. To make such products salable the processing required is very low. This is a thermochemical conversion which is performed at low temperature but high pressure and carried out in a liquid phase. For attaining better efficiency of this process, one has to deal with technical problems which are difficult. This process is also named as hydrothermal liquefaction which requires a high hydrogen partial pressure.

B. **Biochemical Processes**

- **Anaerobic Fermentation**: Anaerobic Fermentation is a biological process that is used to carried out methane gas (rich biogas) from the organic waste such as human and animal excretion and food processing waste. This involves mixed methanogenic bacterial cultures which need different temperature ranges for the growth. This maximum temperature is not exceeded from 60 °C. Under this process about 90% of the energy content from bacteria is converted into biogas which consist of 55% methane and rest carbon dioxide, which can be used instantly for the household purposes like cooking lightning, etc. The slurry material generated after passing manure through the digestion process is the sludge generated after the manure has passed through the digester is nontoxic as well as in odorous. During this digestion process the nitrogen as well as its other nutrients are lost; thus it is considered as a best fertilizer. It is observed that compared to the left cattle waste to dry in the field the outcome of digester have high nitrogen content. The reason behind this is that many of the contents from the cattle manure become volatilized while drying in the sun whereas in the digested sludge very little amount of nitrogen is volatilized whereas other is converted in

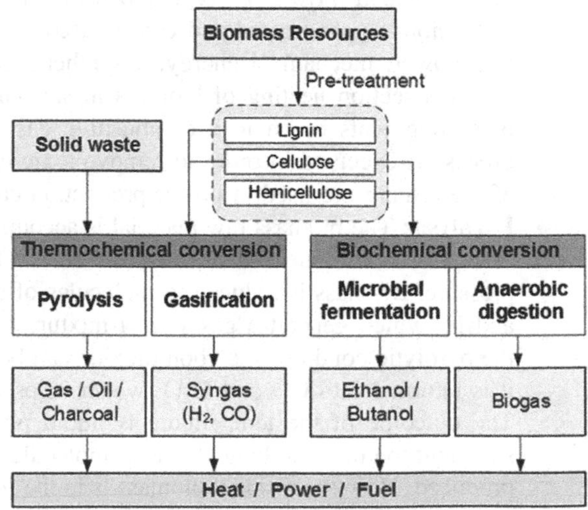

Fig. 3 Conversion process

urea. The value of fertilizing is digested sludge urea which is more readily accessible by plants than any other nitrogen compounds found in dung, therefore fertilizing value of the sludge may actually be higher than that of fresh dung (Fig. 3).

2.2 Environment Impact

If we use biomass as a fuel, the pollutants in the form of carbon and nitrogen are increased in air which may cause air pollution, particulates, and pollutants at levels high than from traditional fuel sources such as coal or natural gas in some cases (such as with indoor heating and cooking) by applying biomass as a fuel utilization of wood biomass, as a fuel can also produce fewer particulate and other pollutants than open burning as seen in wildfires or direct heat applications. According to a survey conducted, biomass is recorded as a second largest contributor to global warming. It is found that there is major concentration of which is linked with ^{14}C recent plant life rather than fossil fuels.

The size of biomass power plant is decided by the availability of the biomass in the nearby surrounding as the transportation plays a vital role in the economy of the plants. So it is found that railway and shipment via waterways can reduce the cost of transport which has led to a global biomass market, to build plants of 1 MW generation economically. Making small plants of 1 MW economically more productive than power plants need to have technology that can convert biomass to useful electricity with greater efficiency such as ORC technology, a cycle resembling to the water steam power process just with an organic working medium.

In the process of combustion, carbon from biomass is released as carbon dioxide in atmosphere. In the dry wood, the amount of carbon contents are approximately 50% due to its weight because of the seasonality of the supply and variability in the source supply chains plays a vital role as it helps in cost effective of bio-energy.

2.3 Working of Biomass Heating Plant

See Fig. 4.

2.4 Various Issue

Technical issues

- Inefficient conversion
- Storage problems because of seasonal availability

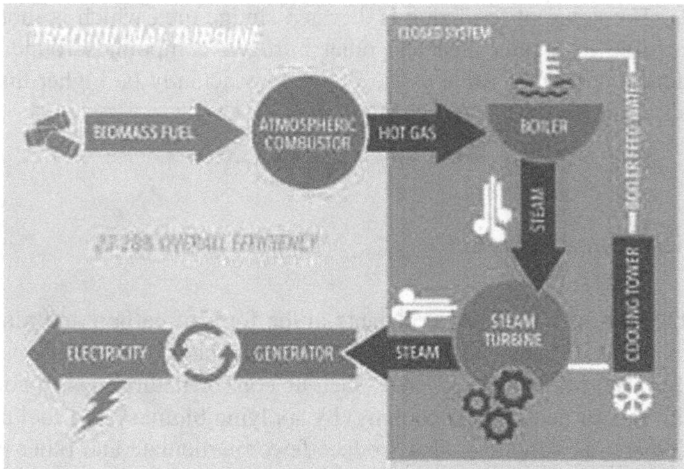

Fig. 4 Working process

- Difficult to obtain maximum efficiency of single purpose use
- Presence of high water content
- Technologies, locations, and routes are difficult to handle in terms of decisions.
- Difficult position synthesis source points, armory facilities, and yield plants.
- Limited productiveness as per area and time contradictory with traditional draw near to economic system of scale focused on maximizing facility size.

Financial issues

- Limited conventional draw near to economy which goes on to increasing unit installation size
- Inaccessibility and complexness of overall costing.
- Deficient transportation facility.
- Restricted inflexibility to ask energy.
- Dangers with latest technologies availability performance with in terms of rate of return.
- Optional biomass markets are conflicts.
- Dodge methods to limit cost biomass

Social issues

- Lack of knowledge
- Local supply chain impacts versus global benefits
- Wellness and safety chances
- Transportation facility deficient
- Terminating artistic of rural areas

Policy and regulatory issues

- Tax on fossil fuel on biomass transport.
- Bonus lacking among the biomass producers
- Limited focus on selection of biomass material
- Lack of back up for sustainable supply chain resolutions.

2.5 *Benefits of Biomass Heating*

The uppermost advantage of such heating is that it uses agricultural, forest, urban, and industrial residues and waste to produce heat and electricity with less effect on the environment than fossil fuels. This way of energy production have very short-term effect on environment because the carbon released by biomass contemporary carbon which is part of the natural carbon cycle while the carbon released by fossil fuels is fossilized carbon which is not part of it, and as we know this contemporary carbon is taken by plant which is used for replacement growth and also before the use of fossil fuels biomass provide us most of humanity's heating.

2.6 *Disadvantage*

As there is no harmful impact of biomass in comparison to fossil fuels, they also have some disadvantages mentioned below

- Bio-power plants have high generation cost as compare to fossil fuel generation.
- These contain less concentrated energy
- Less economic to transport for long distance
- Such power plants need more preparation and handling skills which contribute to higher the cost of power plant

2.7 *Application*

According to the survey conducted in U.S Biopower plants can generate electricity up to 7,000 MW. The usage of fuel, i.e., biomass (wood, industrial waste, etc.) is approximately 60 million tones for the generation of 37 billion kWh of electricity per year. Compared to power from fossil fuels, accessibility of biomass is more easier, and its is affordable as well. The smaller bio-power systems (5 MW) the small bio-power system need not to have any grid system for the supply of the electricity in the different areas. They can also attribute the power to areas by waste called biomass resources. The best use of this system is in the area where there are

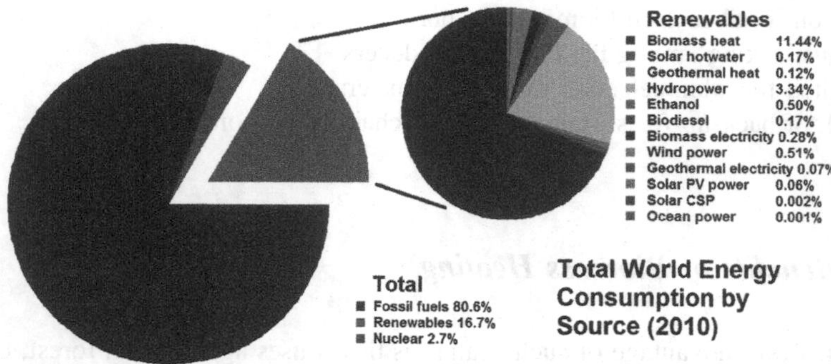

Fig. 5 World energy consumption (2)

limited central power grids as long as the biomass is easily available for the generation of the electricity. These power plants can power the small areas, power communities, and local industries.

This system is widely used in village areas, farms, and ranches, etc. This system can also be employed in the transmission arrangement and for distributing the generated power as they also improves the power quality of weak transmission lines which is located far. The consumers that may install such systems are paper companies, farming work, food processing plants, pellet mills, bio-diesel plants, ethanol mills, etc. (Fig. 5).

3 Conclusion

Biomass industry is renewable and harmless to living things. A biomass industry is able to produce a lot of energy with a small amount of biomass material. Although biomass industry will be expensive it can be a very big step in protecting the resources in the world and reducing greenhouse gases that affect the environment greatly. Even though biomass industry will be releasing emissions into the atmosphere it will be much less than any other industry as the innovation will be using the emissions also to produce energy.

I think by building biomass industries, it will overcome the challenge of the depletion in energy resources because the electricity generated will be with a renewable resource and will be able to power the homes. The issue of depleting fossil fuels will surely be controlled and electricity problem can be resolved if more people are inspired and more eco-friendly.

Acknowledgements We would also like to show our gratitude to Dr. Inderpreet Kaur and Mr. Kamal Kant Sir for communing their precious wisdom with us during this research course. We are also deeply thankful to her for correcting our mistakes and error present in earlier manuscript. although any mistakes are our own and should not stain reputations of these esteemed persons.

References

1. en.wikipedia.org
2. http://www1.eere.energy.gov
3. http://www.biomassenergycentre.org.uk/portal/page?_pageid=75,15179&_dad=portal&_schema=PORTAL
4. http://www.triplepundit.com/special/energy-options pros-and-cons/biomass-energy-pros-cons/
5. http://www.alternative-energy-geek.com/problems-with-biomass.shtml
6. World Energy Assessment (2001). Renewable energy technologies
7. Earth's Carbon Cycle – OceanWorld

Detecting Targeted Malicious E-Mail Using Linear Regression Algorithm with Data Mining Techniques

A. Sesha Rao, P.S. Avadhani and Nandita Bhanja Chaudhuri

Abstract E-mail is the most fundamental means of communication. It is the focus of attack by the terrorists, e-mail spammers, imposters, business fraudsters, and hackers. To combat this, different data mining classifiers are used to identify the spam mails. This paper introduces a system that imports data from the e-mail accounts and performs preprocessing techniques like file conversions that are appropriate to conduct the experiments, searching for frequency of a word by Knuth–Morris–Pratt (KMP) string searching algorithm, and feature selection using principal component analysis (PCA) are applied. Next, linear regression classification is used to predict the spam mails. Then, association rule mining is performed. The mean absolute error and root mean squared error for the training data and test data are computed. The errors of the training and test data sets are negligible which indicates the classifier is well trained. Finally, the results are displayed by the visualization techniques.

Keywords Preprocessing · KMP · PCA · Linear regression · Association rule mining · Mean absolute error · Root mean squared error · Visualization

---

A. Sesha Rao (✉)
Department of Computer Science & Systems Engineering,
AU College of Engineering, Visakhapatnam, India
e-mail: asrakula1948@gmail.com

P.S. Avadhani
Department of Computer Science & Engineering,
AU College of Engineering, Visakhapatnam, India
e-mail: psavadhani@yahoo.com

N.B. Chaudhuri
Department of Information Technology, Vignan's Institute of Engineering
for Women, Visakhapatnam, India
e-mail: nbhanja_chaudhuri@yahoo.com

H.S. Behera and D.P. Mohapatra (eds.), *Computational Intelligence in Data Mining*, Advances in Intelligent Systems and Computing 556,
DOI 10.1007/978-981-10-3874-7_3

1 Introduction

E-mail is the most common technique of communication that is used in a networking environment. It is a very fast and inexpensive means of communication. The mails may comprise of text, sound, images, videos, etc [1]. However, along with the important mails, spam's are included which are spurious mails sent by unauthenticated person, terrorists, etc. For this reason, it becomes difficult to manage the e-mail accounts as the spam is needed to be separated out. Moreover, it leads to ineffective use of storage space and communication bandwidth [2]. Currently, extensive research is going on to identify the spurious mails among the genuine mails.

The contents of these mails are undesirable. They often contain suspicious links that contains viruses, threats for the computer, and personal data. Moreover, it also contains several advertisements which we are not interested, often causes embarrassment and loss of Internet bandwidth. Due to all these severe problems, we need to identify the spam mails and separate it out from the important mails.

In this paper, the mails are imported from different mail clients. The experiments are carried over using a data mining tool named as WEKA to classify the spams and to find out the association rules. Several visualization techniques are employed to demonstrate the results of the experiments.

2 Literature Survey

A large number of researches have been done in this field. We have collected around one hundred and fifty related research papers, among that fourteen has been filtered out for the literature survey.

Present research [3] work on spam filtering focuses on two approaches which differ mainly in the type of features used for spam filtering. These are header-based features; and content-based features. The header-based filtering methods use information available in the header of e-mail. This solution is commonly used by web-based e-mail companies by adding sender of the e-mail in "Safe List". In a similar manner, block listing can be applied to block receiving spam e-mails. Whereas, content-based filtering uses features extracted from the message content. Methods in this approach use machine-learning principles to extract knowledge from a set of given training data and use the obtained knowledge to classify newly received mails. There are improved results with e-mail text features most of which perform much better than header features [4].

Recent years e-mail spam filtering made considerable progress. Many solutions have been proposed for spam mail identification and in which the White-list and Black-list [5] filtering methods are operated based on IP address, DNS, or e-mail address. These filtering methods maintain the source of received spam in a data base with priority. Each time when a new mail received, its source is compared with the contents of data base. The main disadvantage is when spammers regularly change e-mail and IP addresses to cover their trails. [6, 7] shows different machine-learning methods for spam detection.

In [8] the authors investigated feature selection as a preprocessing step using different methods such as chi-square, information gain, gain ratio, symmetrical uncertainty. From the given dataset 80% is used for training and remaining 20% for the testing. This facilitated 99% prediction with random forest classifier.

Methods used for performance of spam classification by authors [9] are based on DIA and NGL coefficient and compared the results. DIA is an indexing approach which was extended and applied in text categorization. NGL coefficient is an improvement of chi-square method. Their experiments show that optimal classification accuracy of 98.5% was achieved at feature length of 104 when model was built using random forest classifier.

The authors Tich Phuoc Tran et al. [10]; proposed a method for anti-spam filtering problem by combining a simple linear regression (LR) model with a modified probabilistic neural network (MPNN) in an adjustable way. This frame work takes the advantage of the virtues of both the models. LR-MPNN is shown empirically to achieve better performance than other conventional methods for most of cost-sensitive scenarios.

D. Pliniske's [11] in his research implemented the neural network method to the classification of spam mails. His method employs attributes consisting of descriptive characteristics of the evasive patterns that spammers adopt rather than using the context or frequency of key words in the message.

Rachana Mishra and R.S. Thakur, [12] analyzed different data mining tools such as WEKA, RapidMiner, and support vector machine. The random forest is best classifier for spam categorization of WEKA tool. This paper recommends WEKA tool for spam filtering and WEKA outperforms the other data mining tools.

Sujeet More and G. RaviKalkundri [13], the authors used WEKA interface in their integrated classification model and tested with different classification algorithms. They have tested seven different algorithms such as: Naïve Bayer, neural network, random forest, decision tree, SVM, etc. They found random forest and SVM classifiers outperform the conventional one.

3 Design and Implementation

Our system model has four sections. They are (1) Data Import (2) Data Preprocessing (3) Data Mining and (4) Visualization.

In the first section, mails are imported from different mail clients like Microsoft Outlook and Gmail and are converted to CSV format. The second section, deals with the usage of preprocessing techniques like KMP algorithm to search the occurrences of words from the mail content. Then the PCA and the ranker search algorithms are implemented to choose the relevant attributes. In the third section, data mining techniques are applied to classify the mails and then the association rules are found out. The fourth section indicates the usage of the visualization techniques to display the results of the above analysis. The model is represented in Fig. 1.

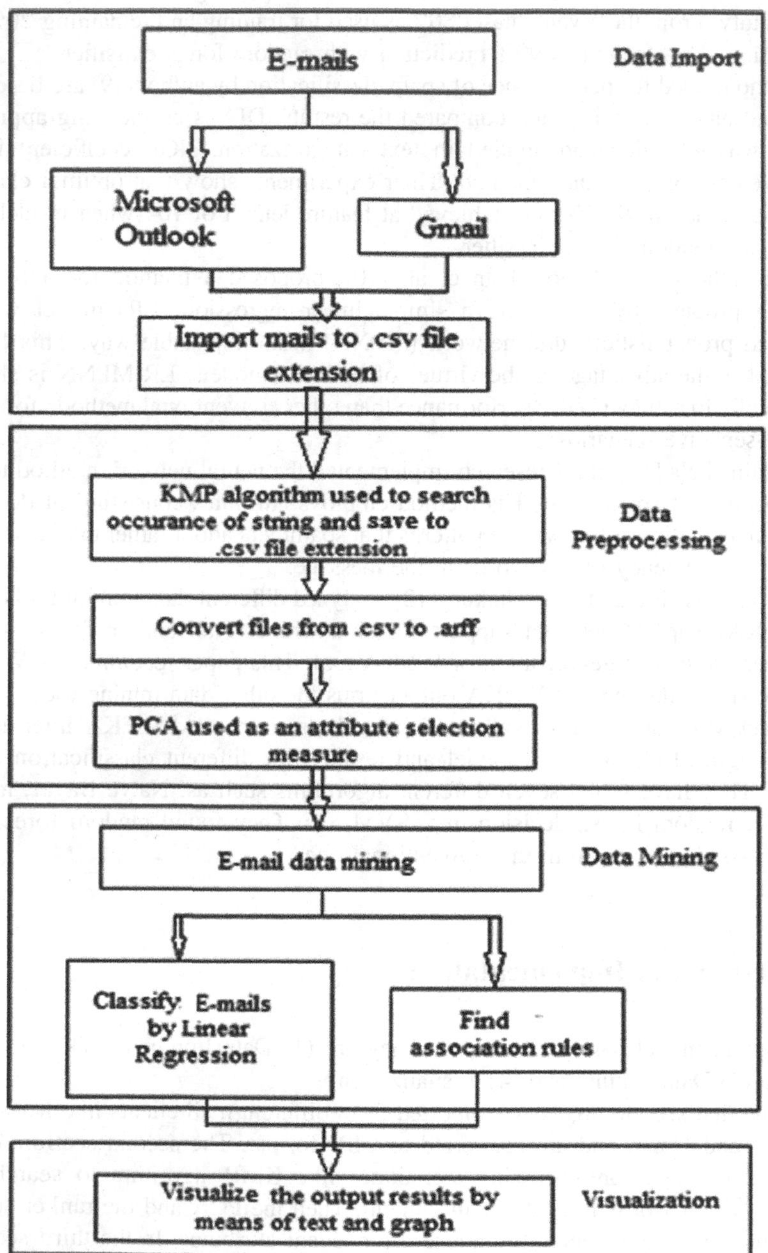

Fig. 1 System model

3.1 Data Importing

This system is able to extract the data from various e-mail accounts like Microsoft Outlook and Gmail. It can be done by two processes—(1) using softwares (2) by manual procedure.

Initially, we considered Microsoft Outlook which stores its data in PST files. In the inbox, all the mails are selected and the Import and Export Wizard were opened. In the wizard, we need to export all the contents of the mails to a CSV file. This procedure is depicted in Fig. 2.

Next, we may chose the locations of the mails from where we want to export i.e., Inbox, Outbox, or any other specific folder. Then, we should specify the location of the destination where the CSV file is to be stored. Generally, all the fields of the mails are not of our concern, so we will select only the body of the mail which is of our primary interest. This can be done in Map Custom Field wizard. It is presented in Fig. 3.

Upon the completion of the above process, a CSV file gets generated in the specified destination.

Next, we had considered Gmail mail clients and import its mail contents. It is convenient to use software named as Thunderbird. We can download and install it into our computer. Once it is done, we can enter our Gmail credentials to it. Thunderbird irresistibly connects to Gmail. It gives options on the location of storing the mails—(1) IMAP, (2) POP3. IMAP is used to import mails in remote computers folder. POP3 is used to store mails in local computers folder. For the purpose of data collection from mails for our entire system, we had chosen IMAP. It is illustrated in Fig. 4.

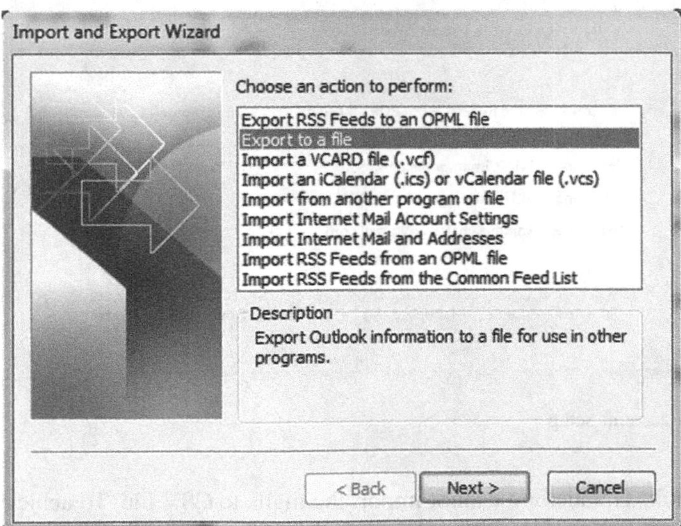

Fig. 2 Importing e-mails from microsoft outlook to CSV file

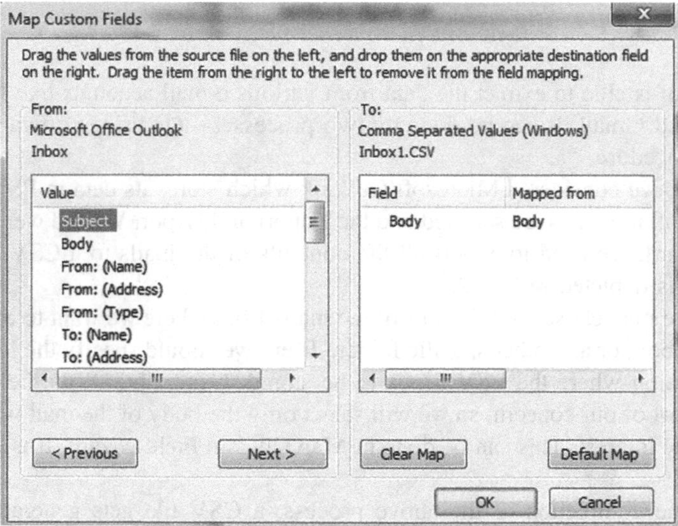

Fig. 3 Mapping custom fields

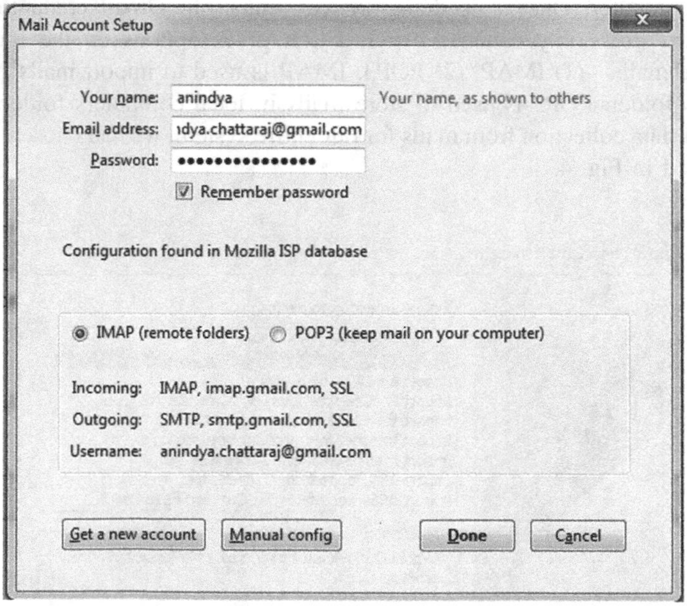

Fig. 4 Mail account setup

By default, Thunderbird cannot import the mails to CSV file. To achieve this, an addon has to be downloaded and installed. Next, the required mails are to be selected and exported to CSV file.

3.2 Data Preprocessing

The KMP algorithm is used for searching the frequency of words from text by enlisting the search. When a failure occurs corresponding to a match, the word itself gains information to predict where the next match could begin. Therefore, it skips the process of searching the previous text again. Using KMP algorithm, occurrences of words from the mail content are determined. These are again stored in CSV file.

We have used the WEKA tool for the data mining task. So, the conversion of CSV file to ARFF is required as the operations in WEKA are conducted in ARFF only. In the WEKA tool, the CSV file is configured and connected to arffsaver via dataset for generating the ARFF. It is illustrated in Fig. 5.

To apply the data mining algorithms, frequency of words were collected from the mail body. Initially, we had selected 102 attributes which were the sequence of words and the last attribute being the nominal class label. There are 4601 instances. To select the best attributes, an attribute selection measure is implemented. Specifically, in this model principal component analysis (PCA) along with ranker search algorithm has been used. After the application of this algorithm, only 48 principal attributes were selected for analysis.

3.3 Data Mining

Classifying the e-mail data, in this model, we have adopted the linear regression classification techniques to classify the data. The best 48 attributes were considered for classification. This classification algorithm only works with numerical datasets. Basically, it follows linear straight line equation with weights which is illustrated in Eq. 1.

$$x = w_0 + w_1 a_1 + w_2 a_2 + \ldots + w_n a_n \tag{1}$$

where, 'w' represents the weights, 'a' represents the attributes, and 'n' is the number of instances. The graphical representation of linear regression is depicted in Fig. 6.

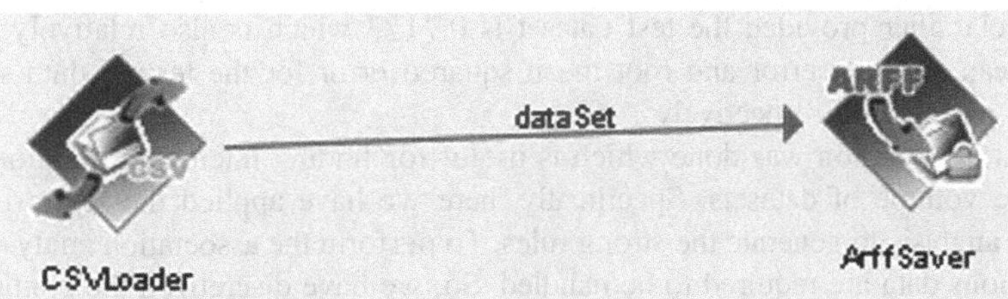

Fig. 5 Conversion of CSV file to ARFF

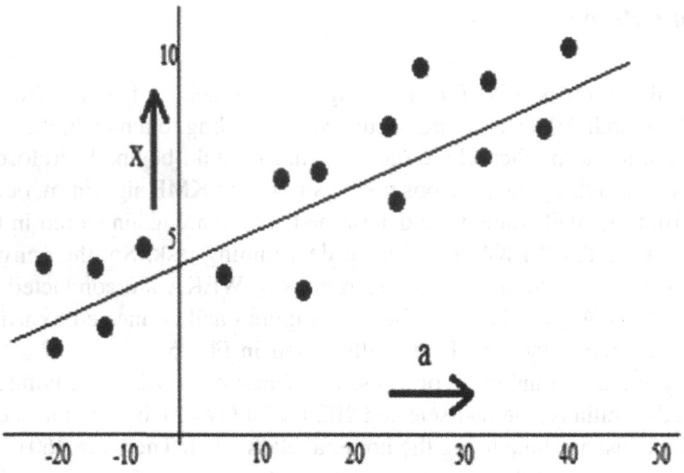

Fig. 6 Graphical representation of linear regression

The weights are calculated from the training e-mail datasets. The calculation for the first training instance is illustrated in Eq. 2.

$$w_0 a_0^{(1)} + w_1 a_1^{(1)} + w_2 a_2^{(1)} + \ldots + w_n a_n^{(1)} = \sum_{i=0}^{n} w_i a_i^{(1)} \quad (2)$$

$^{(1)}$ represents the first training instance.

Weights should be chosen to reduce the squared error on training data which is illustrated in Eq. 3.

$$\sum_{j=1}^{m} (x^{(i)} - \sum_{i=0}^{n} w_i a_i^{(i)}) \wedge 2 \quad (3)$$

Here, $x^{(i)}$ are the real training instances and $\sum_{i=0}^{n} w_i a_i^{(i)}$ represents the predicted value of the ith training instance.

The dataset split was 70% for training and 30% for testing. Therefore, 3221 instances were considered for training and 1380 instances for testing.

The correlation coefficient for training the classifier is 0.7403, which is relatively good. The mean absolute error is 0.2773. The root mean squared error is 0.3335. Next, we have tested the classifier with the test dataset. The correlation coefficient of the classifier provided the test dataset is 0.7123 which is also relatively good. The mean absolute error and root mean squared error for the testing data set are 0.0339 and 0.1006 respectively.

Next, association was done which is useful for finding interesting relationships in large volume of datasets. Specifically, here we have applied the Apriori association analysis to generate the strong rules. To perform the association analysis, the continuous data are required to be handled. So, we have discretized the continuous dataset which is illustrated in Fig. 7.

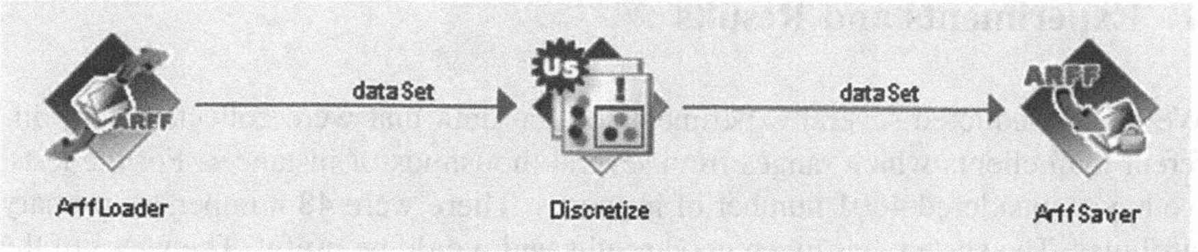

Fig. 7 Discretize the continuous dataset

Fig. 8 Representation of discretized dataset

The discretized dataset are represented in Fig. 8.

Next, we have used the Apriori algorithm on the discretized dataset to find the best rules.

3.4 Data Visualization

Visualization is an important technique to make a system easily understandable. The results could be demonstrated in various forms using several visualization techniques. The techniques that have been used here are Plot Layout, Receiver Operating Characteristics (ROC) graph, and Boundary Visualizer graph. Plot Layout displays the data in the form of x-y axis. A ROC graph demonstrates the selection and the performance of the classifiers. Boundary Visualizer detects the classification boundaries.

4 Experiments and Results

We have conducted several experiments on the data that were collected from different mail clients which ranges from several thousands of instances. For the tests, we have considered 4601 number of instances. There were 48 numbers of primary attributes. The system has given good results and would be useful. The views of the record sets are presented in Fig. 9.

Next, the performance, ROC, boundary, and errors of the classifier are presented in Figs. 10, 11, 12, 13, and 14.

No.	1: word_freq_make Numeric	2: word_freq_address Numeric	3: word_freq_all Numeric	4: word_freq_3d Numeric	5: word_freq_our Numeric	6: word_freq_over Numeric	7: word_freq_remove Numeric	8: word_freq_internet Numeric	9: word_freq_order Numeric	10: word_freq_mail Numeric	11: word_freq_receive Numeric	12: word_freq_will Numeric	13: wor
1	0.0	0.64	0.64	0.0	0.32	0.0	0.0	0.0	0.0	0.0	0.0	0.64	
2	0.21	0.28	0.5	0.0	0.14	0.28	0.21	0.07	0.0	0.94	0.21	0.79	
3	0.06	0.0	0.71	0.0	1.23	0.19	0.19	0.12	0.64	0.25	0.38	0.45	
4	0.0	0.0	0.0	0.0	0.63	0.0	0.31	0.63	0.31	0.63	0.31	0.31	
5	0.0	0.0	0.0	0.0	0.63	0.0	0.31	0.63	0.31	0.63	0.31	0.31	
6	0.0	0.0	0.0	0.0	1.85	0.0	0.0	1.85	0.0	0.0	0.0	0.0	
7	0.0	0.0	0.0	0.0	1.92	0.0	0.0	0.0	0.0	0.64	0.96	1.28	
8	0.0	0.0	0.0	0.0	1.88	0.0	0.0	1.88	0.0	0.0	0.0	0.0	
9	0.15	0.0	0.46	0.0	0.61	0.0	0.3	0.0	0.92	0.76	0.76	0.92	
10	0.06	0.12	0.77	0.0	0.19	0.32	0.38	0.0	0.06	0.0	0.0	0.64	
11	0.0	0.0	0.0	0.0	0.0	0.0	0.96	0.0	0.0	1.92	0.96	0.0	
12	0.0	0.0	0.25	0.0	0.38	0.25	0.25	0.0	0.0	0.0	0.12	0.12	
13	0.0	0.69	0.34	0.0	0.34	0.0	0.0	0.0	0.0	0.0	0.0	0.69	
14	0.0	0.0	0.0	0.0	0.9	0.0	0.9	0.0	0.0	0.9	0.9	0.0	
15	0.0	0.0	1.42	0.0	0.71	0.35	0.0	0.35	0.0	0.71	0.0	0.35	
16	0.0	0.42	0.42	0.0	1.27	0.0	0.42	0.0	0.0	1.27	0.0	0.0	
17	0.0	0.0	0.0	0.0	0.94	0.0	0.0	0.0	0.0	0.0	0.0	0.0	
18	0.0	0.0	0.0	0.0	0.0	0.0	0.0	0.0	0.0	0.0	0.0	0.0	
19	0.0	0.0	0.55	0.0	1.11	0.0	0.18	0.0	0.0	0.0	0.0	0.0	
20	0.0	0.63	0.0	0.0	1.59	0.31	0.0	0.0	0.31	0.0	0.0	0.63	
21	0.0	0.0	0.0	0.0	0.0	0.0	0.0	0.0	0.0	0.0	0.0	0.0	
22	0.05	0.07	0.1	0.0	0.76	0.05	0.15	0.02	0.55	0.0	0.1	0.47	
23	0.0	0.0	0.0	0.0	2.94	0.0	0.0	0.0	0.0	0.0	0.0	0.0	
24	0.0	0.0	0.0	0.0	1.16	0.0	0.0	0.0	0.0	0.0	0.0	0.58	
25	0.0	0.0	0.0	0.0	0.0	0.0	0.0	0.0	0.0	0.0	0.0	0.0	
26	0.05	0.07	0.1	0.0	0.76	0.05	0.15	0.02	0.55	0.0	0.1	0.47	
27	0.0	0.0	0.0	0.0	0.0	0.0	0.0	0.0	0.0	0.0	0.0	0.0	
28	0.0	0.0	0.0	0.0	0.0	0.0	1.66	0.0	0.0	0.0	0.0	0.0	
29	0.0	0.0	0.0	0.0	0.0	0.0	0.0	0.0	0.0	0.0	0.0	0.0	
30	0.0	0.0	0.0	0.0	0.65	0.0	0.65	0.0	0.0	0.0	0.65	0.65	
31	1.17	0.0	0.0	0.0	0.0	0.0	0.0	0.0	0.0	0.0	0.0	1.17	
32	0.0	0.0	3.03	0.0	0.0	0.0	0.0	0.0	0.0	0.0	0.0	0.0	
33	0.0	0.0	0.0	0.0	1.89	0.27	0.0	0.0	0.0	0.0	0.0	0.81	
34	0.0	0.0	0.0	0.0	0.0	0.0	0.0	0.0	0.0	0.0	0.0	0.0	
35	0.0	0.68	0.0	0.0	0.0	0.0	0.0	0.0	0.0	0.68	1.36	0.0	
36	0.0	0.0	2.56	0.0	0.0	0.0	0.0	0.0	0.0	0.0	0.0	0.0	
37	0.0	0.0	0.0	0.0	2.94	0.0	0.0	0.0	0.0	0.0	0.0	0.0	

Fig. 9 Final record sets

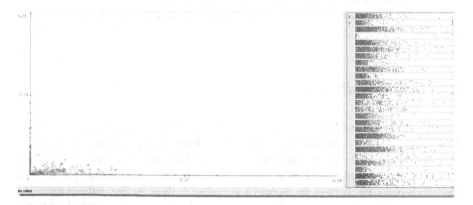

Fig. 10 Plot layout of the performance of classifier

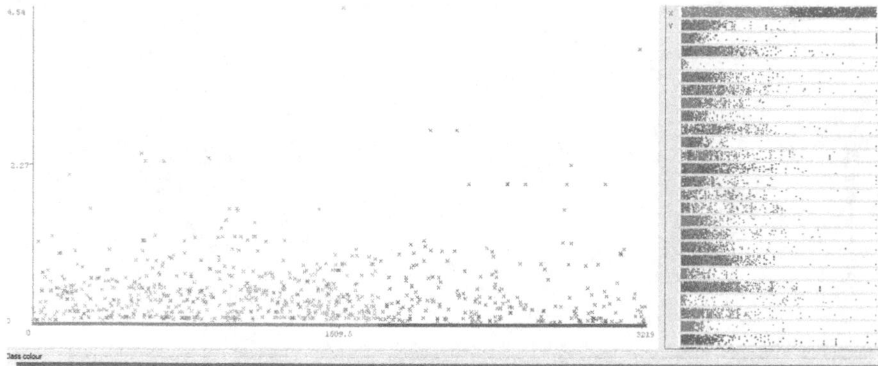

Fig. 11 ROC of the classifier

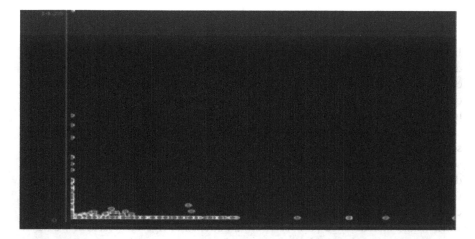

Fig. 12 Boundary visualizer of the classifier

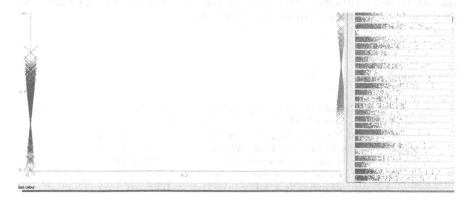

Fig. 13 Plot layout of classifiers training error

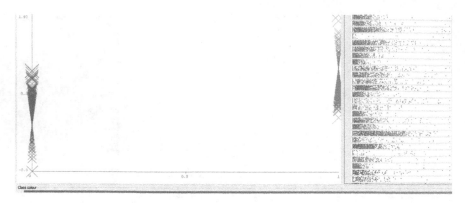

Fig. 14 Plot layout of classifiers testing error

5 Conclusion and Future Work

In this paper, we have presented several data mining techniques to handle the e-mail data, classify it, and perform association analysis. This model is beneficial to the e-mail data miners. We have tested the model with large number of data sets and the outcome of the model is of great importance. The correlation coefficient of the training data set is 0.7403 whereas for the test data set is 0.7123. The deviation is only 0.03. The mean absolute error and root mean squared error for the training data set are 0.2773 and 0.3335, respectively. The mean absolute error and root mean squared error for the testing data set are 0.0339 and 0.1006 respectively. The difference of errors in between the training and test sets are 0.0562 and 0.0671 respectively. The differences are negligible and are accepted. This indicated the classifier is well trained and is capable to be tested on real-time data sets. The future work would include large number of e-mail clients. A rigorous research would be carried to detect and prohibit the spam senders.

Acknowledgements The authors of this paper would like to thank the reviewers of the paper who would read this manuscript and give us valuable suggestions.

References

1. Fan Jia-Peng, Wu Xia-Hui, Zhu Shi-dong, and Xia Yan, "Research and Implementation of Web mail Forensics System", 978-1-4244-6581-1/11, 2011 IEEE.
2. Chih-Chin Lai, and Ming-Chi Tsai, "An empirical Performance Comparison of Machine Learning Methods for Spam E-mail Categorization", Proceedings of the Fourth International Conference on Hybrid Intelligent Systems (HIS'04) 0-7695-2291-2014, IEEE.
3. Walaa Gad, Sherine Rady, "Email Filtering based on Supervised Learning and Mutual Information Feature Selection", in 978-1-4673-9971-5/15- IEEE, 2015, pp 147–152.

4. R. Shams and R. E. MercerIn, "Classifying Spam Emails using Text and Readability Features", In 13th International Conference on Data Mining, IEEE, 2013, pp. 657–666.
5. Spam Cop, Spam Cop Blocking List. Available: http://www.spamcop.net/bl.shtml, 2010.
6. DeBarr, H.W.D., Spam Detection using Clustering, Random Forests and Active Learning, presented at the 6th Conference on Email and Anti-Spam, California, 2009.
7. Awad, S.M.E.W.A., "Machine Learning methods for Email Classification", International Journal of Computer Applications, 2011.
8. P. Ozarkar and Dr. M. Patwardhan, "Efficient Spam Classification By Appropriate Feature Selection", International Journal of Computer Engineering and Technology (IJCET), ISSN 0976 –6375 (Online) vol. 4(3), May–June, 2013.
9. Josin Thomas, Nisha S. Raj, Vinod P., *"Robust Feature Vector for Spam Classification"*, In proceedings of the International Conference on Data Sciences, Universities Press, ISBN: 978-81-7371-926-4, Feb 2014, pp. 87–95.
10. Tich Phuoc Tran, Pohsiang Tsai, Tony Jan, "An Adjustable Combination of Linear Regression and Modified Probabilistic Neural Network for Anti-Spam Filtering" IEEE 2008.
11. D. Puniškis, R. Laurutis, R. Dirmeikis, "An Artificial Neural Nets for Spam e-mail Recognition", electronics and electrical engineering ISSN 1392 – 1215 2006. Nr. 5(69).
12. Rachana Mishara, Ramjeeevan Singh Thakur, "An efficient Approach For Supervised Learning Algorithms using Different Data Mining Tools For Spam Categorization", Fourth International Conference on Communication Systems and Network Technologies, 2014, pp 472–477.
13. Sujeet More, Ravi Kalkundri, "Evaluation of Deceptive Mails using Filtering & Weka", IEEE sponsored 2nd International Conference on Innovations in Information Embedded and Communication Systems, ICIIECS, IEEE, 2015.

Classical and Evolutionary Image Contrast Enhancement Techniques: Comparison by Case Studies

Manmohan Sahoo

Abstract Histogram equalization (HE) and histogram stretching (HS) are two commonly used classical approaches for improving the appearance of a poor image. Such approaches may end up at developing artefacts, rendering the image unusable. Moreover these two classical approaches involve algorithmically complex tasks. On the other hand evolutionary soft computing methods claim to offer hassle free and effective contrast enhancement. In the present work, we report development of algorithms for two evolutionary approaches viz. genetic algorithm (GA) and artificial bee colony (ABC) and went on to evaluate the contrast enhancement capability of these algorithms using some test images. Further we compared the output images obtained using above two evolutionary approaches with the output images got using HE and HS. We report that evolutionary methods result in better contrast enhancement than classical methods in all our test cases. ABC approach outperformed GA approach, when output images were subjected to quantitative comparison.

Keywords Image contrast enhancement • Histogram equalization (HE) • Histogram stretching (HS) • Genetic algorithm (GA) • Artificial bee colony (ABC) algorithm

1 Introduction

Digital image processing refers to manipulation of digital images to exploit and extract quantitative information. The manipulation may require restoration or enhancement or compression of the image considered.

Image enhancement aims at improvement of both the visual and informational quality of a poor image. It involves changes in the brightness and contrast.

M. Sahoo (✉)
Department of Computer Science and Application,
College of Engineering & Technology (CET), Bhubaneswar 751003, India
e-mail: manmohansahoocse@gmail.com

H.S. Behera and D.P. Mohapatra (eds.), *Computational Intelligence in Data Mining*, Advances in Intelligent Systems and Computing 556,
DOI 10.1007/978-981-10-3874-7_4

Brightness is defined as the average pixel intensity of an image whereas contrast refers to finer details of an image. Image enhancement is usually performed by suppressing the noise or increasing the contrast. Though image enhancement may pertain to improving visual appeal of an image, image contrast enhancement is essential to increase its dynamic range.

Image contrast enhancement helps to mend the feature of the image so that the image analysis is precise leading to improved reliability of applications. For example, a medical image must offer accurate information of the parts of body to guide the doctor in right direction. A forensic image is supposed to provide needful information to assist forensic investigation.

Combining the number of edge pixels, the intensity of the edge pixels and the entropy of the whole image helps in quantifying the quality of an image. Genetic algorithm can find the most feasible optimum mapping for the grey levels of image taken as input into different grey levels resulting in enhanced contrast. On the other hand, ABC algorithm based image enhancement technique relies on searching the best alternative set of grey levels. Most importantly here the objective function is determined, taking into account the count of pixels constituting the edge, intensities of edge pixels and entropy of the image.

As of now appreciable number of researchers have proposed different type of image contrast enhancement algorithms. According to Gonzalez and woods [1] histogram equalization (HE) is a simple method for contrast enhancement. One of the striking but puzzling feature of this method is, the fact that the mean brightness of the output image produced following it is same as middle grey level of the input image. In addition, it completely ignores the initial mean grey level. But there are certain applications, which insist on preserving the brightness, during transformation of input image to output image.

Interestingly, different HE-based methods such as mean preserving bi-histogram equalization (MBHE), minimum mean brightness error bi-histogram equalization (MMBEBHE), and recursive mean-spread histogram equalization try to ensure the brightness-preservation as sought in the mentioned applications and overcome the mentioned problem [2]. In global HE, the contrast stretching is confined to grey levels with high frequencies, which in turn can result in substantial contrast loss for grey levels characterized with lower frequencies [3].

In order to overcome above problem some researchers have attempted different local histogram equalization methods [4]. On the sideline evolutionary approaches like employment of GA or ABC model for more rational contrast enhancement are yet to attain maturity.

In genetic algorithm based approach, the grey levels of original image is replaced with the value of the chromosome of particular generation. Munteanu and Rosa [5] introduced a GA-based automatic image enhancement technique, where the role of graylevel scatter in the neighbourhood of a pixel is vital. This parameter forms the basis of transformation function.

Drra and Bouaziz [6] have successfully implemented ABC algorithm for image contrast enhancement using both local/global transformation and grey level mapping. In this approach new set of grey level is projected looking at the input image gray level. The new set is expected to reflect higher homogeneity in the image histogram. Moreover contrast enhancement is an optimization problem, these authors divided the problem to two simpler parts. One of them is representation of solution and the other one is synthesis of objective function. Next, they worry about searching for the best mapping which satisfies the condition of maxima for objective function.

However to our knowledge there is no straight comparison among efficacies of all the classical and evolutionary approaches discussed here. In the present work, we have developed our own algorithms based on sound understanding of GA and ABC. We selected some images from image library of commercial software MATLAB and tried for contrast enhancement using HE, HS, GA, and ABC method. We quantified contrast enhancement obtained for test images considered using multiple methods to check their relative capabilities.

2 Procedure

The effectiveness of contrast enhancement was tested using sample images (Images of a child, a tree, and a cameraman) collected from MATLAB image library [7]. Contrast enhancement was carried out by HE and HS mode by use of standard codes. Algorithms were written based on following concept and used for contrast enhancement by use of genetic algorithm and artificial bee colony model.

2.1 *Transformation Function and Objective Function Used*

We developed an algorithm relying on the following transformation function proposed by Munteanu and Rosa [5]

$$\mathrm{g}(x,y) = [k \cdot \frac{M}{\sigma(x,y) + b}] \cdot [f(x,y) - c \,.\, m(x,y)] + m(x,y)^{a}], \quad 0.5 < k < 1.5 \tag{1}$$

where

M average intensity of entire image
f(x, y) Initial Individual pixel intensity
g(x, y) Transformed pixel intensity
m(x, y) Mean neighbourhood intensity
σ(x, y) Standard deviation of neighbourhood intensity

It is useful to mention that a, b, c, k are real positive parameter and they are same for whole image. Best qualified images can be found by getting an optimal combination of these four parameters.

The objective function used is the following. (Munteanu and Rosa [5])

$$F(Z) = \mathrm{Log}(\log(E(I(Z)))) \cdot \frac{ne(I(Z))}{PH*PV} \cdot H(I(Z)), \tag{2}$$

where

I (z)	is the obtained transformed image
H (I (Z))	is the entropy of the transformed image
E (I (Z))	sobel edge detector determined sum of intensity
ne (I (z))	is the number of edge pixels
PH	is the pixel count in horizontal direction
PV	is the pixel count in horizontal direction

We used the above transformation function and objective functions for two completely different algorithms written for GA and ABC based approach.

2.1.1 Algorithm for GA Based Approach

The detailed algorithm is too long to be presented here. However the gist is the following. Chromosomes are basis of new images. A set of chromosomes (a, b, c, k) are arbitrarily chosen and image enhancement or transformation is carried out using Eq. (1). Fitness of each transformed image is calculated using objective function mentioned in Eq. (2). From a huge pool of n chromosomes generated, the chromosomes with highest fitness are selected for next generation and the remaining chromosomes are used for mating pool. The process of selection, crossover, and mutation results in a new chromosome set of n * 4. Again the same process is repeated for these new images and objective function is calculated using Eq. (2) till the condition of maximization of objective function is achieved. The image for which this condition is achieved is accepted as the final contrast enhanced image.

2.1.2 Implementation of ABC Algorithm

The algorithm artificial bee colony proceeds through generation of initial population as a vector x accommodating the parameters like a, b, c, and k used to calculate transformation function. In this approach employed bee phase is continuously generated on the basis of fitness function. The onlooker bee phase is also updated continuously and scout stage.

3 Results and Discussion

3.1 *Visual Comparison*

Out of the five versions of images the GA-enhanced- and ABC-enhanced images looks brighter. HE enhanced image is visually pleasing but unnatural. HE and HS enhanced images look little brighter than original image. More importantly ABC enhanced image shows more detailed description of the image than GA-enhanced image (Fig. 1).

Image histogram comparison:
Histogram of an image gives information about pixel intensity distributions for various grey levels of an image. Histogram of HE enhanced image is equally spaced grey levels. (It equalizes the histogram) (Fig. 2).

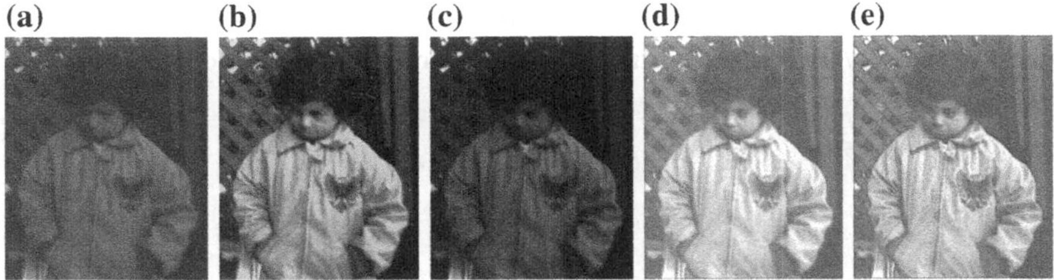

Fig. 1 Visual comparison of **a** original image, **b** HE enhanced, **c** HS enhanced, **d** GA enhanced, and **e** ABC enhanced images

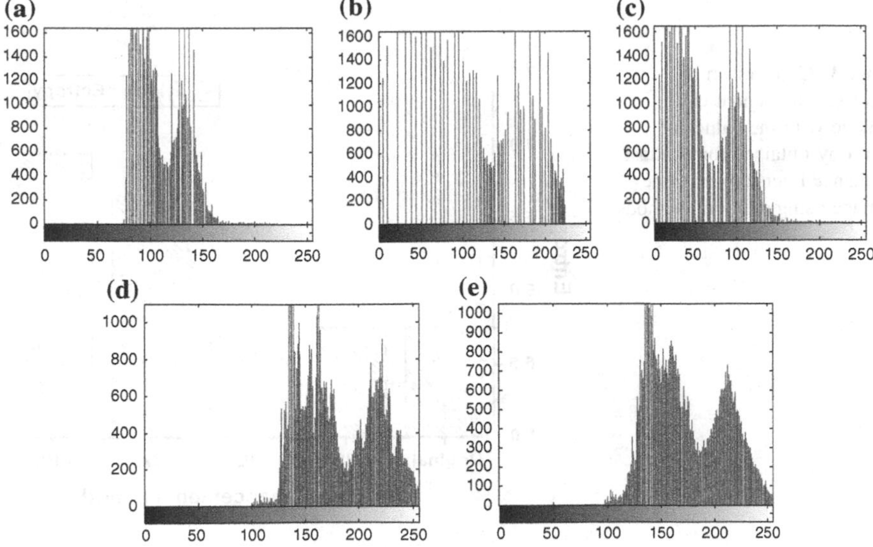

Fig. 2 Histogram comparison. Histograms of **a** original image, **b** HE enhanced, **c** HS enhanced, **d** GA enhanced, and **e** ABC enhanced images

3.2 *Quantitative Comparison*

3.2.1 Entropy

Usually higher the entropy, better the quality of the image. The bar chart shows the original image and HS image has same entropy while HE image has little less entropy. But GA- and ABC-enhanced image has higher entropy. Out of GA and ABC later has more entropy indicating more information richness and orderliness (Fig. 3).

3.2.2 Fitness Value

Fitness value of the image has been calculated as the product of three parameters namely the number of edge pixels, the intensity of edge pixels, and entropy of the image which gives the most trustworthy result of contrast enhancement. The higher the fitness value, the higher the contrast enhancement. Figure 4 shows that the fitness value of ABC-enhanced image is superior to that of GA-enhanced image (Fig. 4).

3.2.3 SNR

SNR stands for signal to noise ratio. As for better quality of image, the signal should be high and there should be less noise, so SNR should be high. From the bar chart (Fig. 5) it is clear that in all the test image cases, The SNR values of ABC-enhanced image is higher than that of the GA enhanced image.

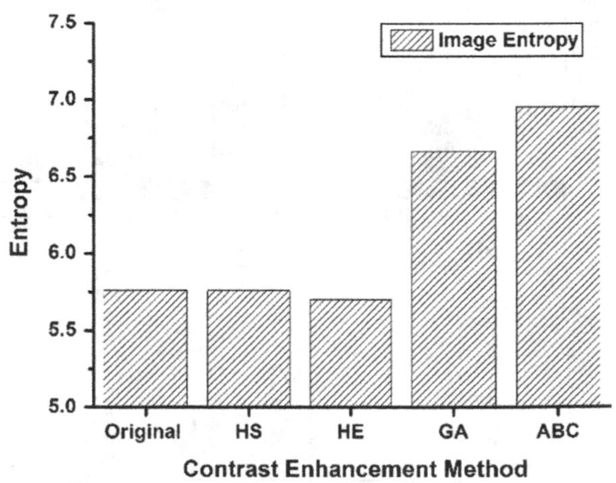

Fig. 3 Comparison of entropy of original child image with the values of entropy obtained for contrast enhanced versions of same image using different methods

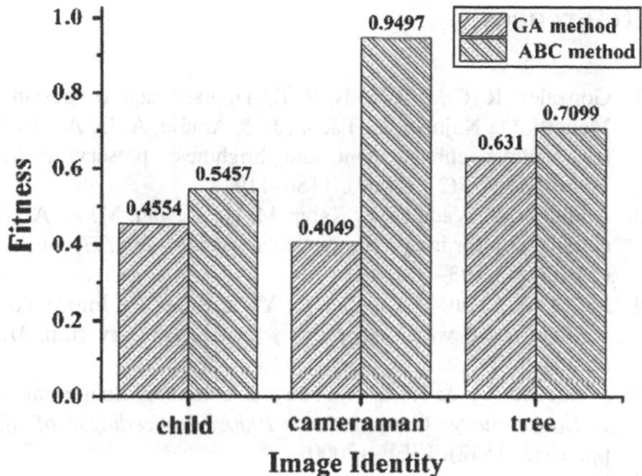

Fig. 4 Comparison of fitness of three different images for contrast enhancement using two different evolutionary methods

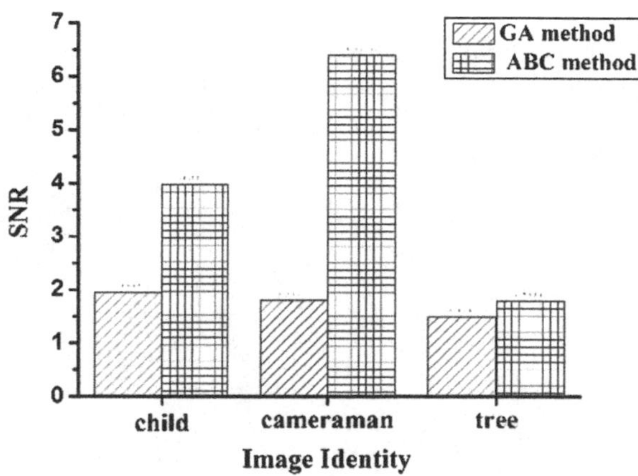

Fig. 5 Comparison of SNR of three different images after contrast enhancement using two different evolutionary methods

4 Conclusions

We found that while improving contrast, histogram stretching (HS) spreads the histogram to cover the entire dynamic range instead of changing the shape of the histogram. Histogram equalization (HE) produced better quality image with the flattening of the histogram. In all three test cases examined by us, the ABC algorithm outperformed GA in terms of contrast enhancement capability. However in order to draw any clear conclusion, huge number of sample images at varied levels of contrast defects are to be analyzed.

References

1. Gonzalez, R. C., & Woods, R. E.: Digital image processing. *Nueva Jersey* (2008).
2. Menotti, D., Najman, L., Facon, J., & Araújo, A. D. A.: Multi-histogram equalization methods for contrast enhancement and brightness preserving. *IEEE Transactions on Consumer Electronics*, *53*(3), (2007). 1186–1194.
3. Abdullah-Al-Wadud, M., Kabir, M. H., Dewan, M. A. A., & Chae, O. A: dynamic histogram equalization for image contrast enhancement. *IEEE Transactions on Consumer Electronics*, *53* (2), (2007). 593–600.
4. Lai, Y. R., Tsai, P. C., Yao, C. Y., & Ruan, S.: Improved local histogram equalization with gradient-based weighting process for edge preservation. *Multimedia Tools and Applications*. (2015). 1–29.
5. Munteanu, C., & Rosa, A.. Towards automatic image enhancement using genetic algorithms. In *Evolutionary Computation, 2000. Proceedings of the 2000 Congress on* (Vol. 2, pp. 1535–1542). IEEE. (2000).
6. Draa, A., & Bouaziz, An artificial bee colony algorithm for image contrast enhancement. *Swarm and Evolutionary computation*, *16*, (2014). 69–84.
7. MATLAB Image library.

Cost Effectiveness Analysis of a Vertical Midimew-Connected Mesh Network (VMMN)

M.M. Hafizur Rahman, Faiz Al Faisal, Rizal Mohd Nor, T.M.T. Sembok, Dhiren Kumar Behera and Yasushi Inoguchi

Abstract Previous study of vertical midimew-connected mesh network (VMMN) has been conducted to evaluate some of static network performance parameters of this network. These parameters include diameter, network degree, average distance, wiring complexity, cost, and arc connectivity. VMMN showed good results compared to some of conventional and hierarchical interconnection networks. However, there are some important static parameters that were not included in that study such as packing density, message traffic density, cost effectiveness factor, and time-cost effectiveness factor. In this paper, we will focus on evaluating VMMN to these parameters, and then we will compare the results to that from mesh, torus, TESH, and MMN networks.

Keywords Massively parallel computer · VMMN · Packing density · Message traffic density · Cost effective factor · Time cost effective factor

M.M. Hafizur Rahman (✉) · R.M. Nor
Department of Computer Science, KICT, IIUM, 50728 Kuala Lumpur, Malaysia
e-mail: hafizur@iium.edu.my

R.M. Nor
e-mail: rizalmohdnor@iium.edu.my

F. Al Faisal · Y. Inoguchi
School of IS, JAIST, Asahidai 1-1, Nomi-Shi, Ishikawa 923-1292, Japan
e-mail: ffaisal@jaist.ac.jp

Y. Inoguchi
e-mail: inoguchi@jaist.ac.jp

T.M.T. Sembok
Cyber Security Center, UPNM, 57000 Kuala Lumpur, Malaysia
e-mail: tmtsembok@gmail.com

D.K. Behera
Indira Gandhi Institute of Technology, Sarang, Odisha, India
e-mail: dkb_igit@rediffmail.com

H.S. Behera and D.P. Mohapatra (eds.), *Computational Intelligence in Data Mining*, Advances in Intelligent Systems and Computing 556,
DOI 10.1007/978-981-10-3874-7_5

1 Introduction

To fulfill the increasing day-to-day demands of computation power, the researchers are trying in various ways. Also the advancement in signaling technology, and increasing of the bandwidth, motivates to find suitable solution to manage the huge computing power, ranging from multi-core, many-core, to massively parallel computer (MPC) [1, 2]. Furthermore, sequential computers reached their saturation. Therefore, massively parallel computing (MPC) systems are the ideal solution to deal with these issues. MPC are consisting of millions of nodes, in order to meet the voracious demand of the new technology, as well as in the upcoming years this demand will be increased [3, 4]. It is interesting to say that the researchers expected that the number of parallel computing system nodes will be increased to 100 million nodes in the next decades [5].

The topology of MPC has a main influence on the system performance. Therefore, it is in a significant stage in developing the parallel computing system the decision about the architecture of the interconnection network topology. In addition, hierarchical interconnection network HIN is a suitable alternative to interconnect several topologies to others, in order to build the whole system [6, 7]. Currently the main concern is to find the optimal HIN which will yield good performance in all the aspects.

Tori-connected mESHes (TESH), also known as TESH network [8] is a *k*-ary *k*-cube based HIN, whereby it consists of several 2D-mesh networks (also know as basic module) connected hierarchically to other basic module, in order to build higher levels of the system based on 2D-torus networks [8]. However, TESH drawback is it has yielded a bit low dynamic communication performance compared to other networks, especially in terms of network throughput at saturation point. On the other hand, Midimew network achieved low hop distance for both diameter and average distance compared to the other direct symmetric networks with degree of 4. Based on these results, 2D-torus network of TESH has been replaced by Midimew network, and it has been called Midimew-connected Mesh Network (MMN) [9]. MMN composed of several BMs, each BM is a 2D-mesh network connected hierarchically by 2D Midimew fashion. In the vertical MMN (VMMN), the end-to-end node of vertical direction is connected using the diagonal links and the end- to-end node of horizontal direction is connected using torus-like wrap-around links. Looking for the optimal static network performance, vertical Midimew-connected mesh network VMMN obtained the best static network performance. In the previous study, we have evaluated many of the static network performance parameters of VMMN network, as well as it achieved better results than mesh, torus, and TESH networks [10, 11]. In this paper we will evaluate VMMN for more of these static parameters, and then we will compare the results with that of 2D-mesh, 2D-torus, TESH, and MMN. These additional static network performances include packing density, message traffic density, cost effectiveness factor, and finally time-cost effectiveness factor, these comparisons have been stated in Sect. 3 of this paper.

The rest of the paper is organized as follows. In Sect. 2, we explain the basic architecture of the VMMN. The evaluation of the packing density, message traffic density, cost effective factor, and time-cost effective factor of the proposed VMMN is discussed in Sect. 3. Finally, Sect. 4 concludes this paper.

2 Interconnection of Vertical-Midimew Connected Mesh Network

Minimal DIstance MEsh with Wrap-around links (Midimew) is a 2D network, where one dimension is tori-connected and the other dimension is wrap-around connected in a diagonal fashion. Vertical Midimew-Connected Mesh Network (VMMN) composed of numerous basic modules (BMs) connected to each other to build a higher level network. In addition, VMMN is a hierarchical interconnection network that has been derived from MMN network. Furthermore, it achieved good static network performance.

2.1 Basic Module

VMMN composed of several Basic modules; the basic module of this network is a 2D-mesh network of $(2^m \times 2^m)$. In addition, BM of VMMN consist of 2^m nodes in each row, and 2^m nodes in each column. Hence, the total number of nodes is 2^{2m}, where m is a number greater than or equal zero. Figure 1a shows the structure of 4×4 BM of VMMN. The network can be connected to the higher level by using $2^{(m+2)}$ free ports in order to establish the connection. In contrast, the ports of internal nodes are used for intra-BM connections. The most number of ports in BM are used for the external connection; however one or two ports are used for inter-BM connection. In this paper we consider BM of VMMN as the Level-1 network of the system.

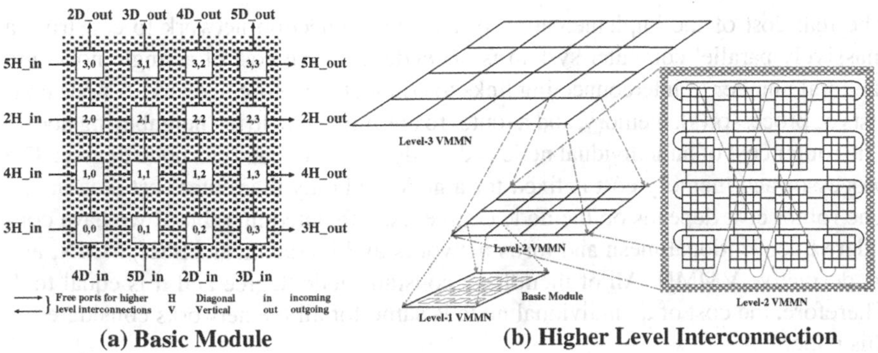

Fig. 1 Interconnection of vertical-midimew connected mesh network

2.2 Higher Level Network

The basic modules (BMs) of a VMMN are connected in two directions, in order to create the higher level network. These directions are horizontal and vertical. The connection in the vertical direction is using diagonally wrap-around connection; however the connection in the horizontal direction is a symmetric torus connection. The higher level network can be constructed by recursively interconnecting immediate lower level subnetworks as a $(2^m \times 2^m)$ midimew network. Figure 1b, depicted the architecture of VMMN higher level network where $m = 2$. On the other hand, the architecture of Level-2 network of VMMN can be created by connecting $2^{(2\times2)} = 16$ BMs. Likewise, Level-3 of VMMN created by connecting 16 of Level-2 subnetworks and so forth. Each basic module is connected directly to the adjacent BMs, in order to create the higher level network. It is interesting that the BM is using for each higher level interconnection $4 \times (2^q) = 2^{q+2}$ of its free links. The distribution of these free links are as $2(2^q)$ free link for diagonal connections and another $2(2^q)$ free links for horizontal toroidal connections to construct higher level midimew network. Level-5 VMMN can be interconnected using a 4×4 BM, which can interconnect more than one million nodes.

3 Cost Effectiveness Analysis

Basically the real performance of an MPC system depends on various implementation issues and their underlying technology. However, before going into real implementation, static evaluation is widely used to study the suitability of any interconnection network for an MPC system. In this section, we focus on some static performance metrics that characterize the performance and effectiveness of a VMMN.

3.1 Cost Parameters

The real cost of the implementation of an interconnection network to construct a massively parallel computer system is dependent on the cost of a single node and the total number of interconnecting links to connect the entire network. A node consists of processor(s), memory, and a router to communicate to the neighboring nodes. The router cost of an individual node is directly proportional to its node degree. The processor and memory cost is fixed for a node and only the router cost is variable. The router cost depends on the node degree. As tabulated in Table 1, we have considered conventional mesh and torus networks and hierarchical TESH, MMN, and its derivative VMMN. All of them have constant node degree and it is equal to 4. Therefore, the cost of an individual node is same for all the networks considered in this paper.

Table 1 Topological analysis of different interconnection networks

Network	Node degree	Wiring complexity	Diameter	Average distance	Packing density	Message traffic density	Cost-effective factor	Time-cost effective factor
2D-Mesh	4	480	30	10.67	2.13	5.69	0.8421	1.67868
2D-Torus	4	512	16	8.00	4.00	4.00	0.8333	1.66125
TESH	4	416	21	10.47	3.05	6.44	0.8602	1.71466
MMN	4	416	17	9.07	3.76	5.58	0.8602	1.71466
VMMN	4	416	17	8.56	3.76	5.27	0.8602	1.71466

Another cost parameter is the total number of required links to interconnect the complete interconnection network. The actual cost of a link depends on the VLSI layout of that link and the underlying CMOS technology. However, the wiring complexity (total number of links) is statically used to compare the cost of different interconnection networks. It is depicted in Table 1 that the cost for wiring of a VMMN is lower than that of the mesh and torus networks and equal to that of MMN and TESH networks.

3.2 *Distance Parameters*

The distance between two nodes of an interconnection network is subject to the layout of nodes in the VLSI and the underlying CMOS technology as well. Usually it is in the scale of *mm* at the chip level, *cm* in the board level, *meter* in the system level hierarchy. Statically the distance between two nodes in a network is measured by the number of hops. We have evaluated the hop distance of all distinct pairs of nodes by computer simulation using the shortest path algorithm. After that we have evaluated the maximum hop distance and average hop distance between all the distinct pairs; and these are known as the diameter and the average distance, respectively. It is depicted in the Table 1 that the distance parameters, both the diameter and average distance, of different networks. From the table, we have seen that the diameter of the VMMN is lower than that of a mesh and TESH networks, equal to that of MMN, and a slightly higher than that of a torus network. And the average distance of the VMMN is lower than that of the mesh, TESH, and MMN; and it is slightly higher than that of torus network.

3.3 Packing Density

The network cost is the product of network degree and network diameter. The cost of VMMN network has been evaluated and compared to popular mesh, tours, TESH, and MMN networks. The cost of the network and the number of nodes has a crucial influence in determining the network packing density. Packing density [12] defined as the ratio of the number of nodes of an interconnection network to its cost. Higher packing density is one of the most desirable features in VLSI, in addition, higher packing density along with small chip area required for VLSI layout. Equation 1 defines the packing density. As well as, the comparison between the packing density of VMMN and the other networks is tabulated in Table 1. It is interesting to see that, the packing density of VMMN is higher than that of a mesh and TESH, and it is equal to that of MMN. In contrast it was little bit lower than that of torus. Hence, VMMN is more compatible with VLSI than mesh and TESH. As well as it will be a good choice in building the future generation of parallel computing systems.

$$PackingDensity = \frac{Total\ Number\ of\ Nodes}{Degree \times Diameter} \tag{1}$$

3.4 Message Traffic Density

We can use the average distance from any source node to its destination to evaluate the network performance for the message traffic. The network performance was affected significantly by the message traffic density. Low message traffic density reduces the network congestion and increases the network bandwidth [13]. The message traffic density is the result of the multiplication of the average distance and the ratio of the total number of nodes of a network to its total number of links. Equation 2 calculates the message traffic density of different networks.

$$MTD = \frac{Average\ Distance \times Total\ No.\ of\ Nodes}{Total\ Number\ of\ Links} \tag{2}$$

From Table 1 we can find the values of VMMN average distance and total number of links. In addition, the table tabulated comparison between the message traffic density of VMMN, MMN, TESH, mesh, and torus networks. It is clear that the message traffic density of VMMN is lower than that of mesh, TESH, and MMN. In contrast, it is higher than that of torus network. Therefore, VMMN will achieve good results over the mesh, TESH, and MMN in building the future generation massively parallel computer system.

3.5 Cost Effective Factor

The cost of communication links play main roles in determining the whole cost of the multiprocessor systems. Therefore, we should consider the cost of these links along with the cost of the processing elements. Speedup and efficiency are the most usable parameters in evaluating the parallel computing systems. However, the cost of communication links is considered during this evaluation process. The cost of communication links has significant impact on decreasing or in increasing the total cost of the system [14]. Determining the Cost effectiveness factor (CEF) of the system will consider the number of links as a function of the number of processors. Therefore, CEF is the best measure in evaluating the parallel computing systems. CEF of VMMN can be measured from the Eq. 3.

$$CEF = \frac{1}{1 + \rho \times G(p)} \tag{3}$$

In Eq. 3 we assumed that the nodes and the links have identical cost. In addition, we considered that P is the total number of nodes and L is the total number of links. Furthermore, ρ is defined as the ratio between the cost of the communication links C_L and the nodes cost C_P. Thus, it can be represented as $\rho = \frac{C_L}{C_P}$, and we have considered the value of ρ is equal to 0.1. $G(p)$ is the ratio of total number of links to total number of nodes, $G(p) = \frac{(Total\#oflinks)}{(Total\#ofnodes)}$. Table 1 tabulated the comparison between VMMN and different other networks. It is clear that the cost effectiveness factor of VMMN is equal to that of TESH, and MMN networks. On the other hand, the cost effectiveness factor of VMMN is little bit higher than that of mesh and torus. Therefore, VMMN is a cost effective network to be used in building the future generations MPC systems.

3.6 Time-Cost Effective Factor

Time-Cost Effective Factor is a main factor should be considered in evaluating the overall performance of the parallel computing systems [13, 14]. Time factor along with the cost effectiveness factor are considered in calculating the time-cost effectiveness factor. In addition, time-cost effectiveness factor (TCEF) considesr the faster solution of the problem is more efficient than the slower one. TCEF of different networks can be determined from Eq. 4.

$$TCEF(p, T_p) = \frac{1 + \sigma T_1^{\alpha-1}}{1 + \rho g(p) \times \frac{\ell}{p} \sigma T_p^{\alpha-1}} \tag{4}$$

Here, $\rho = \frac{C_L}{C_P}$, T_1 is the required time to solve a single problem by using single processor, and T_p is the time required to solve a problem by p nodes. It is assumed that α is equal to 1 as a linear time penalty in T_p and σ is equal to 1. Here, $G(p) = \frac{(Total\#oflinks)}{(Total\#ofnodes)}$. Table 1 depicts the time-cost effective factor of VMMN is equal to that of TESH and MMN. In contrast its slightly higher than that of mesh and torus networks. Therefore, VMMN achieved good results of time-cost effectiveness factor compared to these networks. Hence, it could be a time-cost effective network in building the future generation massively parallel computers.

4 Conclusion

The network architecture and its packing density, traffic density, the cost effectiveness, and the time-cost effectiveness factor of a VMMN have been discussed in this paper in detail. We have described the basic module BM architecture of VMMN and then we described the communication mechanisms between these BMs in order to build higher levels of this network. Also we have evaluated some of important static network performance of VMMN, and then the results of this evaluation process have been compared to that of some of popular networks. These networks include mesh, torus, TESH, and MMN networks. VMMN achieved good results in this comparison. Hence it could be a good choice to be used in building the future generations of massively parallel computer systems.

The main focus of this paper is interconnection of network topology of a vertical midimew connected mesh network and its static cost effectiveness analysis. We kept in mind the further exploration are as follows: (1) fault tolerance performance especially the fault diameter evaluation of a VMMN [15] (2) wire length evaluation of a VMMN in a 3D NoC realization [16].

Acknowledgements This research is supported a FRGS research grant with ID FRGS13-065-0306, MOE, Malaysia. The authors are grateful to the reviewers for their constructive suggestions that assisted to enhance the quality of this research paper.

References

1. Oak Ridge National Laboratory,: Introducing Titan. Advancing the era of accelerated computing, (2012) http://www.olcf.ornl.gov/titan/, 2012.
2. M. Yokokawa, F. Shoji, A. Uno, M. Kurokawa, and T. Watanabe,: The K-Computer: Japanese next-generation supercomputer development project, International Symposium on Low Power Electronics and Design (ISLPED) (2011), pp. 371–372.
3. V. Puente, J.A. Gregorio, R. Beivide and F. Vallejo,: A Low Cost Fault Tolerant Packet Routing for Parallel Computers, Proceedings of the 17th IEEE/ACM International Parallel and Distributed Processing Symposium (IPDPS) (2000).
4. Beckman, P.: Looking toward exascale computing. 9th PDCAT, (2008) 3, Dunedin, New Zealand.
5. J. J. Dongarra, H. W. Meuer, and E. Strohmaier,: TOP500 Supercomputer Sites, http://www.top500.org.

6. Pao-Lien Lai, Hong-Chun Hsu, Chang-Hsiung Tsai, and Iain A. Stewart,: A class of hierarchical graphs as topologies for interconnection networks, Theoretical Computer Science, Elsevier, (2010), 411, 31–33, pp. 2912–2924.
7. M.M. Hafizur Rahman, Y. Inoguchi, Y. Sato, and S. Horiguchi,: TTN: A High Performance Hierarchical Interconnection Network for Massively Parallel Computers. IEICE Transactions on Information and Systems, (2009) E92-D, 5, pp. 1062–1078.
8. Jain, V. K., Ghirmai, T., Horiguchi, S.: TESH: A new hierarchical interconnection network for massively parallel computing, IEICE Trans. IS, Vol. 80, (1997) 837–846.
9. M.R. Awal, M.M. Hafizur Rahman, and M.A.H. Akhand,: A New Hierarchical Interconnection Network for Future Generation Parallel Computer. Proc. of 16th Int'l. Conference on Computers and Information Technology, 2013, pp. 314–319, Khulna, Bangladesh.
10. Ala Ahmed Yahya Hag, M.M. Hafizur Rahman, Rizal Mohd Nor, and Tengku Mohd Tengku Sembok, Yasuyuki Miura, and Yasushi Inoguchi,: On Uniform Traffic Pattern of Symmetric Midimew Connected Mesh Network, Procedia Computer Science, 2015, 50, pp. 476–481.
11. Md. Rabiul Awal, M. M. Hafizur Rahman and Rizal Mohd Nor, TMT Sembok, M.A.H. Akhand,: Architecture and Network-on-Chip Implementation of a New Hierarchical Interconnection Network, Journal of Circuits, Systems, and Computers, (2015) 24, 2.
12. Mostafa Abd-El-Barr, Turki F. Al-Somani,: Topological Properties of Hierarchical Interconnection Networks: A Review and Comparison. Journal of Electrical and Computer Engineering, (2011) 2011.
13. Nibedita Adhikari and C. R. Tripathy,: The Folded Crossed Cube: A New Interconnection Network for Parallel Systems. Internation Journal of Computer Applications, (2010) 4, 3, pp. 43–50.
14. Dilip Sarkar,: Cost and Time-Cost Effectiveness of Multiprocessing. IEEE Transactions on Parallel and Distributed Systems, (1993) 4, 6, pp. 704–712.
15. S.P. Mohanty, B. N. B. Ray, S. N. Patro, and A.R.Tripathy, Topological Properties of a New Fault Tolerant Interconnection Network for Parallel Computer, *International Conference on Information Technology*, pp. 36–40, Dec. 2008.
16. M. M. Hafizur Rahman, Rizal Mohd Nor, Md. Rabiul Awal, Tengku Mohd Bin Tengku Sembok, Yasuyuki Miura, Long Wire Length of Midimew-Connected Mesh Network, *Proc. of* 16th *Int'l. Conference, ICDCIT 2016*, PP. 97–12, Bhubaneswar, India, 2016.

Cluster Analysis Using Firefly-Based K-means Algorithm: A Combined Approach

Janmenjoy Nayak, Bighnaraj Naik and H.S. Behera

Abstract Nature-inspired algorithms have evolved as a hot topic of research interest around the globe. Since the last decade, K-means clustering has become an attractive area for researchers towards solving many real-world clustering problems. But, unfortunately K-means does not work well for non-globular clusters. Firefly algorithm is a recently developed metaheuristic algorithm that simulates through the flashing characteristics of the fireflies. The firefly algorithm uses the capacity of global search to resolve the limitations of K-means technique and helps in escaping from the local optima. In this work, a novel firefly-based K-means algorithm (FA-K-means) has been proposed for efficient cluster analysis and the results of the proposed approach are compared with some other benchmark approaches. Simulation results divulge that the proposed approach can be efficiently used for solving clustering problems as it avoids the trapping in local optima and helpful for faster convergence.

Keywords FCM · Firefly algorithm · PSO · K-means · Nature-inspired algorithm

J. Nayak (✉)
Department of CSE, Modern Engineering & Management Studies, Balasore, Odisha, India
e-mail: mailtojnayak@gmail.com

B. Naik
Department of Computer Application, VSSUT, Burla, Odisha, India
e-mail: mailtobnaik@gmail.com

H.S. Behera
Department of CSE & IT, VSSUT, Burla, Odisha, India
e-mail: mailtohsbehera@gmail.com

H.S. Behera and D.P. Mohapatra (eds.), *Computational Intelligence in Data Mining*, Advances in Intelligent Systems and Computing 556,
DOI 10.1007/978-981-10-3874-7_6

1 Introduction

In the last decade, clustering techniques has been considered as an interesting area of research due to their capacity to handle unsupervised pattern recognition problems. Being an unsupervised data mining technique, it efficiently solves the problems of both similarity and dissimilarity measures within a group of data [1]. Clustering has been used in variety of applications including data compression [2], data mining [3, 4], machine learning [5], etc. Through clustering technique, the similar data can be arranged in a group and the dissimilar data are grouped in other group, based on their properties. Both the similarity and dissimilarity are measured based on the belongingness of patterns to their corresponding groups. K-means technique is a hard clustering type, which is simple to use and solve the clustering problems. By considering some definite datasets, a number of cluster centers or centroids are selected which are far away from each other as much as possible for clustering. Depending upon the nearest distance among the data points and cluster centers, the point will belong to that class of cluster center. If vast number of variables is used, then K-means will perform faster than hierarchical clustering.

The applications of K-means is wide spread from Science to Engineering and for its simple implementation steps, it is quite easy to use in all the fields. Moreover, several improved algorithms and new development of K-means has been a key research interest of the researchers. Juang and Wu [6] have proposed a K-means clustering segmentation method for MRI brain lesion detection. Their method helps the pathologists to track the tumor as well as to detect the exact size and region of the lesion. Using a Q-based representation for the exploration of hyperspace, a quantum and GA integrated K-means method has been designed by Xiao et al. [7]. Taking into consideration of cluster information and geometric information of a data point, Lai and Huang [8] developed a global K-means algorithm which performs faster than others. With the integration of K-means and MLP, Orhan et al. [9] proposed a model to classify the EEG signals. Juang and Wu [10] have presented a hybrid combination of morphological processing and K-means method to track the psoriasis objects. To balance the multi-robot task allocation problem, Elango et al. [11] has proposed a K-means with auction based approach. Hatamolou et al. [12] have combined two methods such as K-means and gravitational search algorithms for effective data clustering. Reddy and Jana [13] have initiated a novel approach of K-means by initializing the cluster center using the Voronoi diagram. A novel algorithm called candidate group search in combination with K-harmonic search has been proposed by Hung et al. [14] for efficient data clustering. For the application of large-scale video retrieval, Liao et al. [15] have introduced a hierarchical K-means technique which is adaptive in nature. For clustering high-dimensional data and image indexing, an informatics theoretic K-means have been introduced by Cao et al. [16]. To handle the initialization problem of K-means, Tzortzis and Likas [17] developed a min-max based k-means method. A comparative analytical discussion has been made among the distributed evolutionary k-means algorithms by Naldi and Campello [18]. Durduran [19] has combined the K-means with central tendency measure approach to classify the high resolution land cover. Wu et al. [20] have addressed a novel hybrid fuzzy k-harmonics means clustering approach to conquer the drawbacks of k-means method. A hybridization of neural network and K-means approach has been proposed by Mohair et al. [21] to detect the human skin.

Firefly optimization algorithm (FA) is a new nature stimulated metaheuristic, which may be effectively applied to find globally optimal solutions for a variety of nonlinear optimization problems. The controlling aspects of FA are dependent upon flashing actions of social insects (fireflies), which attract its neighboring fireflies [22]. Nature-based algorithms have shown effectiveness and efficiency for solving difficult optimization problems. In this work a novel hybrid approach of K-means and firefly algorithm is proposed to achieve the global solutions as well as to speed up the convergence. The experiment is conducted by considering six real-world benchmark data sets considered from UCI repository. Simulation results over six datasets indicate that FFA-K-means is superior over the other techniques and is efficient in cluster analysis. The remaining sections of this work have been organized as follows: Sect. 2 elaborated the K-means algorithm and firefly algorithm. The proposed approach has been described in Sect. 3. The experimental set up along with result details have been discussed in Sect. 4. Section 5 presents the concluding remarks with some future directions.

2 Preliminaries

2.1 K-means Algorithm

K-means [23] is a nonhierarchical flat clustering method, where the distance is measured between the objects (instances) and some predefined cluster centers. It is an unsupervised method of cluster analysis. Each of the clusters is allied with a centroid and the objects will be assigned to the closest centroids. Initially the value of k must be specified and the centroids are to be selected in a random fashion. However, the centroid is represented through the mean calculation and the closest centroids are considered through the calculation of Euclidian distance. The steps for formulating the algorithm are as follows:

1. Initially choose k points randomly as cluster centroids.
2. Consign data objects to the nearest centroids based on the distance between them and minimum from other centroids.
3. Assign the new centroids through the mean calculations.
4. Repeatedly calculate the distance between the cluster points and the newly generated centroids, until the centroids will be stable.

2.2 Firefly Algorithm (FA)

Firefly Algorithm (FFA) is a multimodal nature inspired meta-heuristic algorithm based on flashing behavior of fireflies and is proposed by Yang [24]. All most every species of fireflies produce a unique small rhythmic flash and the flashes are being produced by a process of bioluminescence. According to Yang [25], the firefly is based on three ideal principles: (a) Due to the unisex nature of all fireflies, one firefly will be attracted towards another despite of their sex, (b) As attractiveness is directly proportional to the brightness property of the fireflies, so always the brighter firefly attracts the less bright firefly, (c) Based on objective function criteria, the brightness is computed. If the distance between two fireflies will increase, then both the attractiveness and brightness will decrease dramatically. Also if the firefly will not find anybody in its surroundings, then it will travel in a random direction. The main property of firefly algorithm is its flashing light, which is responsible for attracting the mating fireflies and to warn the possible predators. Also at regular intervals, the firefly charges and discharges the light which seems like an oscillator. Usually, the fireflies remain most active during the nights of summer. When a firefly comes across with another neighboring firefly, a reciprocal pairing occurs in cooperation with the fireflies.

The male firefly tries to attract the flightless female fireflies on the ground through their signals. With response to those signals of male fireflies, the female fireflies discharge flashing lights. As a result, discrete flashing patterns of both the male and female partners are generated for encoding the information such as identity of the species and sex. In general, female firefly is more fascinated towards the males having brighter flashing lights. The flashing intensity will vary with the source distance. But, in some distinctive cases the female fireflies are unable to distinguish between the strongest and weakest flashes which may be generated from distant and nearby male fireflies, respectively. The firefly algorithm has two stages [26] described as follows:

(i) The brightness is dependent on the intensity of light emitted by the firefly. Suppose there are 'n' number of fireflies and the solution for an ith firefly is 'w_i', where '$f(w_i)$' indicates the fitness value. For reflecting the current position w of a firefly along with the fitness value '$f(w)$', the brightness 'B' of the said firefly is selected and is described in Eq. (1).

$$B_i = f(w_i)\,,\ \text{where}\ 1 \le i \le n \tag{1}$$

(ii) All fireflies have their unique attractiveness 'B', which indicates how strongly it can attract other members of the swarm. The attractiveness 'B' will diverge with its distance factor 'd_{ij}' at the locations 'x_i' and 'x_j', between the two corresponding fireflies 'i' & 'j', which is illustrated by Eq. (2).

$$d_{ij} = |w_i - w_j| \tag{2}$$

The attractiveness function is computed by Eq. (3).

$$B(r) = be^{-\gamma r^2} \tag{3}$$

In Eq. (3), at r = 0, 'b' indicates for attractiveness, 'B(r)' indicates for the attractiveness function and 'γ' is the coefficient of light absorption.

The movement of the less brighter firefly towards the most brighter firefly is computed by

$$w_i = w_i + be^{-\gamma r^2}(w_j - w_i) + \eta\left(rand - \frac{1}{2}\right) \tag{4}$$

where 'η' is the randomization parameter and the value of '*rand*' is randomly selected number between the interval [0,1]. The pseudocode of the firefly algorithm is described as follows.

```
Pseudocode of the FIREFLY Algorithm
Step-1: Create a initial population of fireflies X randomly.
        X = {w1,w2,... wn}
Step-2: Compute the brightness of each firefly by using objective function f(wi) as
B = {B1,B2....Bn} = {f(w1),f(w2),... f(wn)}.
Step-3 : Set light absorption coefficient γ.
Step-4 : While (t≤ max iteration)
                For i=1 to n
                        For j=1 to i
                                If (Ij> Ii)
                                        Move firefly i to firefly j by using eq. (4).
                                End if
                                Attractiveness varies with distance r via exp(−γr²).
                                Calculate new fireflies and update brightness by using eq. (1).
                        End for
                End for
                t = t+1
End while
Step-5: Rank fireflies according to their fitness and find the best one.
Step-6: If Stopping criteria met, then perform step-7.
Else perform step-4.
Setp-7 : Stop.
```

3 Proposed Algorithm

In this section, the firefly algorithm has been applied on the K-means clustering approach to overcome the shortcomings of the K-means algorithm. This proposed firefly algorithm based K-means algorithm uses an objective function (Eq. (5)) to evaluate the quality of cluster centers. So, in the context of clustering, a single firefly represents the 'm' number of cluster center. And, the entire population of fireflies is initialized with 'n' number of cluster center vector $P=\{C_1, C_2 \ldots C_n\},$, where each cluster center vector consist of 'm' number of cluster centers $C_1 = (c_1, c_2 \ldots c_m)$. Here, each c_i represents a single cluster center. As this is a minimization problem, the intensity of each firefly is analogous to the objective function value of K-means. The objective function of K-means [27] for calculating the fitness has been illustrated in Eq. (5).

$$F(C_i) = \frac{k}{\left(\sum_{l=1}^{r} \|o_l - C_i\|^2\right) + d} = \frac{k}{\left(\sum_{j=1}^{m}\sum_{l=1}^{r} \|o_l - C_{i,j}\|^2\right) + d} \quad (5)$$

Here, $F(.)$ is a function to evaluate the generalized solutions called fitness function, 'k' and 'd' are user defined constants, o_l is the lth data point, C_i is the ith cluster center vector, $C_{i,j}$ is the ith cluster center of jth cluster center vector, 'r' is the number of data point in the data set and $\|.\|$ is the Euclidean distance norm. The pseudocode of the proposed approach is illustrated as follows.

Pseudo code of FFA-K-means algorithm

1. Initialize the population of 'n' no. of cluster center vectors $P = \{C_1, C_2 ... C_n\}$, each cluster center vector with 'm' no. of random cluster center $C_1 = (c_1, c_2 ... c_m)$. An individual firefly signifies a cluster center vector C_i.

2. Initialize algorithm's parameters: k = 50 and d = 0.1.

3. Do

  For i=1: n

   For j=1: n

   Calculate light intensity (objective function value) of each cluster center vector (firefly) by using eq.(5).

    If ($I_j < I_i$)

     The position of cluster center vector is updated and movement between i[th] and j[th] cluster center vector is achieved through eq. (4).

    End if

   End for j

  End for i

Rank the cluster center vector and find out the current best cluster center vector.

Compare the fitness of previous best cluster center vector with current best cluster center vector and find out difference in their fitness (λ).

While (λ < threshold value);

4. Rank the cluster center vector based on their fitness and obtain the best cluster center vector.

5. Set best cluster center vector obtained in step – 4 as initial cluster centers.

6. Assign the data objects to the nearby cluster center based on the distance between them.

7. Update the cluster centers by finding the mean of data points assigned to a particular cluster center.

8. Repeat step 6 to 8 until the cluster centers in successive iteration are not stable.

9. Exit

4 Experimental Set up and Result Analysis

4.1 Parameter Set up

For optimizing the performance of the proposed technique, the best values for the algorithmic parameters are selected. The terminating condition in K-means algorithm is when there is no scope for further progress in the objective function value. The stopping condition of the FFA-K-means is 500 generations (maximum number of iterations) or no changes in current best in five consecutive iterations.

During the experimental set up, the following parameters have been set.

K-means parameters :
K=50
d=0.1

Firefly Parameters :
Number of fireflies (n) = 20
Attractiveness (β_0) = 1
Light absorption coefficient (γ) =1
Randomization parameter (α) = 0.2

PSO Parameters :
Inertia (w) = 1.9
Accelerating constants (c_1, c_2) = (1.4, 1.4)

4.2 Dataset Information

For experimenting the proposed method along with comparing with other methods, six benchmark datasets are considered from UCI machine learning repository [28]. The details about the datasets with their attribute information have been illustrated in Table 1.

Table 1 Details about the dataset

Data sets	No. of pattern	No. of clusters	No. of attributes
Iris	150	3	4
Lenses	24	3	4
Haberman	306	2	3
Balance scale	625	3	4
Hayesroth	132	3	5
Spect heart	80	2	22

4.3 Experimental Results and Analysis

The developing environment for the proposed method is MATLAB 9.0 on a system with an Intel Core Duo CPU T5800, 2 GHz processor, 2 GB RAM and Microsoft Windows-2007 OS. To compare the performance of the proposed method with the other approaches, six datasets have been applied on each of the models. All the algorithms have been run for the datasets and the recorded results are indicated in Tables 2 and 3. The table contains the fitness values of all the algorithms for six datasets. From Tables 2 and 3, it is apparent that the FFA-K-means gives better and steady results as compared to other clustering algorithms for all data sets.

Among the fitness values of K-means, GA-K-means, and PSO-K-means [29], the performance of PSO-K-means is better than the other two approaches. However, from both the tables, the fitness values of the proposed FFA-K-means method are found to be superior than all the methods. In the FFA-K-means approach, the firefly algorithm initiates at a good starting state as it uses the output of K-means algorithm. So, in brief, the proposed algorithm has three major advantages such as: (i) it saves K-means to fall at local minima, (ii) it has a good quality of fitness values, i.e., quality cluster solutions than the individual K-means method, (iii) the convergence speed of the firefly increases. For the data sets such as iris, haysroth, and spectheart the fitness values of the proposed approach are 0.014604687, 4.76226E-05, and 0.081242461 respectively, which are marginally good than the other approaches.

Table 2 Fitness value comparison between K-means, GA-K-means and PSO-K-means methods

Dataset	Fitness values of clustering algorithms		
	K-Means	GA-K-means	PSO-K-means
Iris	0.012395396	0.013826351	0.014528017
Lenses	0.339904827	0.351735427	0.360239542
Haberman	0.000317745	0.000328364	0.000348162
Balance scale	0.002573387	0.002628475	0.002810827
Hayesroth	4.59807E-05	4.70825E-05	4.73918E-05
Spectheart	0.069341756	0.072648917	0.076041565

Table 3 Fitness value comparison between K-means, IPSO-K-means and FFA-K-means methods

Datasets	Fitness values of clustering algorithms		
	K-Means	IPSO-K-means	FFA-K-means
Iris	0.012395396	0.014580183	0.014604687
Lenses	0.339904827	0.360282035	0.437582011
Haberman	0.000317745	0.000363902	0.000376591
Balance scale	0.002573387	0.002920182	0.003649032
Hayes roth	4.59807E-05	4.74029E-05	4.76226E-05
Spectheart	0.069341756	0.078284661	0.081242461

5 Conclusion and Future Work

In this work, a firefly-based K-means clustering (FFA-K-means) is proposed for clustering real-world data sets. The FFA-K-means performs better and overcomes the problem of K-means which stuck in local optima. From the simulation results, it is proved that, FFA-K-means has shown faster convergence and minimized objective function for various bench mark data sets compared to the other techniques. The results obtained by FFA-K-means in most cases provide superior results. In future work, the proposed algorithm can be hybridized with other metaheuristic nature-inspired algorithms for global optima in large-scale clustering. Also, it can be applied in some other research application such as image analysis, data classification, etc.

References

1. L. Wang et al. Particle Swarm Optimization for Fuzzy c-Means Clustering. Proc. of the 6th World Congress on Intelligent Control and Automation. Dalian China (2006).
2. J. Marr. Comparison of Several Clustering Algorithms for Data Rate Compression of LPC Parameters. in IEEE International Conference on Acoustics Speech. and Signal Processing. Vol. 6. pp. 964–966. January 2003.
3. C. Pizzuti and D. Talia. P-AutoClass: scalable parallel clustering for mining large data sets. in IEEE transaction on Knowledge and data engineering. Vol. 15. pp. 629–641. May 2003.
4. J, Nayak, B, Naik, H.S. Behera, Fuzzy C-Means (FCM) Clustering Algorithm: A Decade Review from 2000 to 2014, Smart Innovation, Systems and Technologies 32, Vol. 2, 133–149,
DOI 10.1007/978-81-322-2208-8_14.
5. X. L. Yang, Q. Song and W. B. Zhang. Kernel-based Deterministic Annealing Algorithm For Data Clustering. in IEEE Proceedings on Vision, Image and Signal Processing. Vol. 153. pp. 557–568. March 2007.
6. Juang, Li-Hong, and Ming-Ni Wu. MRI brain lesion image detection based on color-converted K-means clustering segmentation. Measurement 43.7 (2010): 941–949.
7. Xiao, Jing, et al. A quantum-inspired genetic algorithm for k-means clustering. Expert Systems with Applications 37.7 (2010): 4966–4973.
8. Lai, Jim ZC, and Tsung-Jen Huang. Fast global k-means clustering using cluster membership and inequality. Pattern Recognition 43.5 (2010): 1954–1963.
9. Orhan, Umut, Mahmut Hekim, and Mahmut Ozer. EEG signals classification using the K-means clustering and a multilayer perceptron neural network model. Expert Systems with Applications 38.10 (2011): 13475–13481.
10. Juang, Li-Hong, and Ming-Ni Wu. Psoriasis image identification using k-means clustering with morphological processing. Measurement 44.5 (2011): 895–905.
11. Elango, Murugappan, Subramanian Nachiappan, and Manoj Kumar Tiwari. Balancing task allocation in multi-robot systems using K-means clustering and auction based mechanisms. Expert Systems with Applications 38.6 (2011): 6486–6491.
12. Hatamolou, Abdolreza, Salwani Abdullah, and Hossein Nezamabadi-pour. A combined approach for clustering based on K-means and gravitational search algorithms. Swarm and Evolutionary Computation 6 (2012): 47–52.
13. Reddy, Damodar, Prasanta K. Jana, and IEEE Senior Member. Initialization for K-means clustering using Voronoi diagram. Procedia Technology 4 (2012): 395–400.

14. Hung, Cheng-Huang, Hua-Min Chiou, and Wei-Ning Yang. Candidate groups search for K-harmonic means data clustering. Applied Mathematical Modelling 37.24 (2013): 10123–10128.
15. Liao, Kaiyang, et al. A sample-based hierarchical adaptive K-means clustering method for large-scale video retrieval. Knowledge-Based Systems 49 (2013): 123–133.
16. Cao, Jie, et al. Towards information-theoretic K-means clustering for image indexing. Signal Processing 93.7 (2013): 2026–2037.
17. Tzortzis, Grigorios, and Aristidis Likas. The MinMax k-means clustering algorithm. Pattern Recognition 47.7 (2014): 2505–2516.
18. Naldi, M. C., and R. J. G. B. Campello. Comparison of distributed evolutionary k-means clustering algorithms. Neurocomputing 163 (2015): 78–93.
19. Durduran, Süleyman Savaş. Automatic classification of high resolution land cover using a new data weighting procedure: The combination of k-means clustering algorithm and central tendency measures (KMC–CTM). Applied Soft Computing 35 (2015): 136–150.
20. Wu, Xiaohong, et al. A hybrid fuzzy K-harmonic means clustering algorithm. Applied Mathematical Modelling 39.12 (2015): 3398–3409.
21. Al-Mohair, Hani K., Junita Mohamad Saleh, and Shahrel Azmin Suandi. Hybrid Human Skin Detection Using Neural Network and K-Means Clustering Technique. Applied Soft Computing 33 (2015): 337–347.
22. J. Senthilnath, S.N. Omkar, V. Mani. Clustering using firefly algorithm: Performance study. Swarm and Evolutionary Computation 1. pp. 164–171. (2011).
23. MacQueen, J., et al. (1967). Some methods for classification and analysis of multivariate observations. In Proceedings of the fifth Berkeley symposium on mathematical statistics and probability. Vol. 1. pp. 281–297. Oakland. CA, USA.
24. Yang, X. S. (2010). Firefly algorithm, stochastic test functions and design optimisation. *International Journal of Bio-Inspired Computation*, *2*(2), 78–84.
25. X. S.Yang Firefly algorithms for multimodal optimization. In Stochastic algorithms: foundations and applications, *Springer Berlin Heidelberg*, pp. 169–178, 2009.
26. X. S. Yang, Multi objective firefly algorithm for continuous optimization, *Engineering with Computers*, vol. 29, no. 2, pp. 175–184. 2013.
27. Nayak, J., Naik, B., Kanungo, D. P., & Behera, H. S.: An Improved Swarm Based Hybrid K-Means Clustering for Optimal Cluster Centers, In Information Systems Design and Intelligent Applications, Springer India, 545–553 (2015).
28. Bache, K. and Lichman, M. UCI Machine Learning Repository [http://archive.ics.uci.edu/ml]. Irvine, CA: University of California, School of Information and Computer Science (2013).
29. Nayak, J., Kanungo, D. P., Naik, B., & Behera, H. S. (2016). Evolutionary Improved Swarm-Based Hybrid K-Means Algorithm for Cluster Analysis. In*Proceedings of the Second International Conference on Computer and Communication Technologies* (pp. 343–352). Springer India.

A Study of Dimensionality Reduction Techniques with Machine Learning Methods for Credit Risk Prediction

E. Sivasankar, C. Selvi and C. Mala

Abstract With the huge advancement of financial institution, credit risk prediction assumes a critical part to grant a loan to the customer and helps the financial institution to minimize their misfortunes. Despite the fact that there are different statistical and artificial intelligent methods available, there is no single best strategy for credit risk prediction. In our work, we have used feature selection and feature extraction methods as preprocessing techniques before building a classifier model. To validate the feasibility and effectiveness of our models, three credit data sets are picked namely Australia, German, and Japanese. Experimental results demonstrates that the SVM classifier performs better among several classifier methods, i.e., NB, LogR, DT, and KNN with LDA feature extraction technique. Test result demonstrates that the feature extraction preprocessing technique with base classifiers are the best suited for credit risk prediction.

Keywords Feature selection · Feature extraction · Machine learning · Credit risk data set

1 Introduction

Credit Risk Analysis is the procedure to recognize the danger required in giving a credit to the end user. This characterization is depend on the attributes of credit user (for example, age, training level, occupation, conjugal status, and pay), the reimbursement execution on past credits and the kind of credit, etc. The objective of a

E. Sivasankar (✉) · C. Selvi · C. Mala
Department of Computer Science and Engineering, National Institute of Technology, Tiruchirappalli, India
e-mail: sivasankar@nitt.edu

C. Selvi
e-mail: selvichandran.it@gmail.com

C. Mala
e-mail: mala@nitt.edu

H.S. Behera and D.P. Mohapatra (eds.), *Computational Intelligence in Data Mining*, Advances in Intelligent Systems and Computing 556,
DOI 10.1007/978-981-10-3874-7_7

credit prediction model is to arrange the credit candidates as (1) Good credit: Risk to repay the money related commitment, (2) Bad credit: Denied to pay the money related commitment. Credit advances and funds have danger of being defaulted. The precise evaluation of buyer credit danger is of furthest significance for loaning associations. Expanding the customer credit interest has prompted the rivalry in finance industry. So credit administrators need to create and use machine learning (ML) strategies to handle examining credit information with a specific end goal to spare time and lessening of mistakes. Some ML procedure incorporates Support Vector Machine (SVM), Navie Bayes (NB), Logistic Regression (LogR), Decision Tree (DT), K-Nearest Neighbor (K-NN) [1]. In many Data Mining (DM) techniques: classification and prediction, the dimensionality reduction is used as a basic method for proficient control of huge amount of high-dimensional data. Two methods for reducing dimensionality includes feature extraction (FE) and feature selection (FS) [2].

FS has been a dynamic research zone in ML and DM groups. It lessens the feature space, and expels replicated, unessential, and unwanted data. This reduced feature space will accelerate the execution of prediction model [3]. The FS strategies are of two types: filter and wrapper procedures. The filtering procedure select subsets of attribute without including a classifier method. Filtering techniques are fast and it can be used for large data set, since it has not used classifier method for subset selection. The normal filter methods that are used in our work includes information gain (IG), gain ratio (GR). The wrapper procedures first actualize an optimising algorithm that includes or evacuates credits to deliver different subset attributes and after that utilize a classifier method to assess the attribute subset. The wrapper procedures are computationally more costly than filter procedure and provides precise results but it will not scale up well for high-dimensional data sets [4]. Greedy forward or backward search is used in wrapper procedure to search attribute subsets.

FE is a method of transforming the current features into a lower dimensional feature space. FE methods are used in this paper includes principal component analysis (PCA) [5] and linear discriminant analysis (LDA) [6].

The paper is sorted out as takes after. Section 2 gives a study identified with this work. In Sect. 3 we will depict the techniques used and proposed in our work. Section 4 outlines the data sets for the credit risk identification. In Sect. 5 results are introduced and examined. Section 6 will draw the conclusions and gives conceivable headings to future exploration.

2 Literature Review

Lots of work has been done for credit risk identification with traditional ML methods. The credit risk prediction model has been built with LDA [7–9] with assumption of categorical nature of credit data furthermore in view of the way that the covariance matrix of credit and non-credit risk are unrealistic to be equivalent.

Lately ML strategies are utilized which automatically extract knowledge from samples and develop the credit risk prediction model to explain the data. Artificial Neural Network (ANN) has been utilized to build a credit risk prediction model in [10, 11] and the outcome guaranteed that the execution is better in grouping the bad credit classes. The Case-Based Reasoning (CBR) [12] has been utilized for credit risk prediction which will coordinate another issue with the nearest past cases and gain from encounters to solve the issue. They planned compelling indexing systems and enhances the model precision altogether. SVM is used to discover critical region of problem space in [13] and improves the model accuracy. ANN has been utilized for risk prediction and the outcomes re compared with logit model, KNN and DT in [14]. The prediction model worked with DT gives great precision than NB, LogR, KNN and SVM in [1]. In [15] it showed that the RS-Boosting performs better than the base ML techniques. IG FS method performed better than base classifier in [3]. The boosting ensemble techniques has been proposed in [16] and showed that DT boosting performs better than the base ML methods.

Although the ML methods can be used for credit risk prediction, the preprocessing techniques has to be proposed to increase the performance of ML techniques. Dimensionality reduction is one such preprocessing method used in many applications to improve the performance of ML algorithms. There are two general techniques for dimensionality reduction, i.e., FE and FS [2].

Algorithms for FS can be arranged into two primary classes relying upon whether this method utilizes input from the ensuing execution of the ML algorithm. The filter method ranks the features in an order [17], the way it highlights the data records with respect to the class attribute and it uses no feedback input from ML algorithm. The common filter methods are ReliefF algorithm, chi-squared (χ^2), information gain (IG), gain ratio (GR), symmetrical uncertainty (SU), etc. [17]. Conversely, a wrapper method uses ML algorithm in FS process with training and test stages [18]. The subset of features are selected in [4] using wrapper method and optimized by random or greedy hill-climbing search. FE is transforming the features in current space into a lower dimensional feature space and it includes PCA [5], LDA [6] and nonnegative matrix factorization (NMF) [19].

The system that we are proposing for credit risk analysis as takes after: (i) Use IG, GR FS methods and train the ML classifiers with the subset of features chosen (ii) Use the wrapper strategy on data set with all base ML technique to discover best features and train the classifiers with those selected features (iii) Use PCA and LDA to transform the data into low-dimensional space; Build a prediction model on the transformed data for credit risk prediction.

3 Methodology

The ML techniques, for example, SVM, NB, LogR, DT, and KNN [20] are utilized as a part of our work in the wake of performing a portion of the preprocessing strategies keeping in mind the end goal to enhance the execution of the classifier. Reducing dimensionality is the one such preprocessing strategy that includes FE (PCA and LDA) and FS (Filter: IG, GR and wrapper) techniques. They are,

3.1 Information Gain

IG is the measure of information gained by knowing the estimation of the attribute with respect to the class [20].

$$\textit{Information gain} = (\textit{Entropy of distribution before the attribute split}) - (\textit{Entropy of distribution after the attribute split})$$

IG is calculated as the entropy of the distribution before the split minus the entropy of the distribution after it. The largest information gain is identical to the littlest entropy.

3.2 Gain Ratio

IG is applied to feature which takes large number of distinct values that may learn training set very well [20]. GR reduces biases against considering multi-valued attributes. It is used to reduce a bias while choosing an attribute by considering the number and branch size into account.

3.3 Principle Component Analysis

PCA transforms the data at current coordinate system to a new coordinate system using an orthogonal linear transformation. The projection of data that lies on the first coordinate system with best variance is called first principle component and the data lies on second coordinate system with second noteworthy variance is called second principle component and so on. In PCA, the first eigen vector with most variance corresponds to the largest eigen values of the covariance matrix [5].

3.4 Linear Discriminant Analysis

LDA is used in ML techniques to find out the linear combination of attributes that separates two or more classes effectively. It also takes into consideration the scatter within and the scatter between the classes. [6].

3.5 Proposed Method

In order to improve the performance of base classifier, we have proposed a portion of the preprocessing techniques as given in the Algorithm 1. The preprocessing techniques includes balancing of records before building the model, expelling the missing records. At that point, the classifier model is prepared with k-fold strategy the wake of applying FS and FE techniques on each fold. At last the model is tested with test records and average (for k fold) result for every FS and FE method on classifier techniques are identified.

Algorithm 1 Proposed method

Input: Data set $D = \{(x_1, y_1), (x_2, y_2), ..., (x_m, y_m)\}$;
Postive records P_r;
Negative records N_r;
Total Positive records P_n;
Total Negative records N_n; Total rounds K.
Process for Balancing Instances:
if $P_n < N_n$ **then**
    Random selection of $N_r = P_n$
else
    Random selection of $P_r = N_n$
end ifProcess for Preprocessing Instances:
for each field $f_1, f_2, ..., f_l$ **do**
    for each record in D **do**
        if value(f_L)==Null **then**
            Remove the record
            Call Process for Balancing Instances
        else
            Exit
        end if
    end for
end forCredit risk prediction model
for $K = 1, 2, 3, ..., k$ **do** #k-fold cross validation
    $i)h_t(x) = L(D_K)$ # prediction model on base learners
    $ii)h_t(x) = L(D_K(FS : IG/GR/Wrapper))$ # Prediction model with FS
    $iii)h_t(x) = L(D_K(FE : PCA/LDA))$ # Prediction model with FE
end for
Find average accuracy of k-fold classifiers for the above techniques
Output: $H(x) = 1(y = h_t(x))$ where,

$$\begin{cases} 1(y = h_t(x)) = 1 & \text{if } y = h_t(x) \\ 1(y = h_t(x)) = 0 & \text{Otherwise} \end{cases}$$

---

4 Experiments

An experiment is carried with five base learners with FS and FE techniques. All the classifier strategies are implemented in python environment.

4.1 Data Set Description

The widely used real-world data sets Australian, German, and Japanese data sets are accessed from [21] and the classifier performances are evaluated on these data sets. Table 1 reports the information about the data sets taken for evaluation.

4.2 Performance Measures

Accuracy measure is used to assess the performance of credit risk prediction since it is utilized to measure how the proposed algorithm classifies the new instances correctly. Accuracy is resolved as the proportion of correct number of predictions given total number instances. Let us assume 2×2 contingency matrix as in Table 2.

The accuracy of classification is,

$$Accuracy = \frac{Number\ of\ correctly\ identified\ the\ credit\ risk\ records}{Total\ number\ of\ credit\ risk\ records\ taken}$$

Table 1 Data set details

	Total attributes	Total good instances	Total bad instances
Australian	14	307	383
German	24	700	300
Japanese	15	296	357

Table 2 Contingency table for credit risk prediction

		Predicted result	
		Non-risk (+)	Risk (–)
Actual result	Non-risk (+)	True positive (TP)	False negative (FN)
	Risk (–)	False positive (FP)	True negative (TN)

5 Results and Analysis

Table 3 demonstrates the performance of base classifiers on three data sets (Australian, German, and Japanese). The base learner SVM outflanks the other base learners for each of the three benchmark data set. Further, to enhance the performance of base learners, few preprocessing methods are done. Tables 4 and 5 demonstrates the consequences of base learners by applying filtering FS methods IG and GR respectively. Here GR gives preferred result over IG_FS techniques. To try and enhance the classifiers performance FS wrapper strategy is utilized.

The improved classifiers accuracy after selecting the informative features using wrapper method is appeared as a part of Table 6. Rather than picking the subset of features, the feature reduction methods are used and the experiment results are given in Tables 7 and 8. The LDA reduction method applied for classification gives improved accuracy than PCA reduction method. Overall, From Figs. 1, 2, 3 and 4, it is inferred that SVM gives better result than other classifier techniques for every one of the strategies utilized on each data set. From Fig. 5, it can be concluded that SVM with FE method LDA performs better for all data sets than other attribute selection and reduction methods. Figure 6 shows the comparsion of the proposed work with the existing work and it is concluded that proposed method LDA_FE obtains improved performance than [1, 14] for all the three datasets.

Table 3 Accuracy(%) of base classifier

		Data sets		
		Australian	German	Japanese
Classifier techniques	SVM	86.2	76.4	88.1
	NB	86.95	76.9	86.7
	LogR	86.93	76.8	88.2
	DT	85.7	71.4	87.1
	KNN	80.3	67.2	81.4

Table 4 Accuracy(%) of base classifier with IG_FS

		Data sets		
		Australian	German	Japanese
Classifier techniques	SVM	86	75.7	87
	NB	86.2	76	86.1
	LogR	86.3	75.8	87.6
	DT	84.8	70.6	86.4
	KNN	79.5	66.2	81

Table 5 Accuracy(%) of base classifier with GR_FS

		Data sets		
		Australian	German	Japanese
Classifier techniques	SVM	87.2	77	88.8
	NB	86.1	76.3	85.2
	LogR	87	76.9	87
	DT	86	72.3	87.5
	KNN	80.8	68	82.4

Table 6 Accuracy(%) of base classifier with Wrapper_FS

		Data sets		
		Australian	German	Japanese
Classifier techniques	SVM	87.8	77.6	89.2
	NB	87	77.2	86.9
	LogR	87.4	77	88.5
	DT	86.5	73	88
	KNN	81	68.6	82.8

Table 7 Accuracy(%) of base classifier with PCA_FE

		Data sets		
		Australian	German	Japanese
Classifier techniques	SVM	86.2	77.3	89.2
	NB	86	76.8	87
	LogR	86.8	76.2	88.2
	DT	85.8	72.4	87.8
	KNN	80.8	67.8	82.2

Table 8 Accuracy(%) of base classifier with LDA_FE

		Data sets		
		Australian	German	Japanese
Classifier techniques	SVM	88	77.8	89.8
	NB	87.3	77.5	87.4
	LogR	87.8	77.2	88.8
	DT	86.8	73.4	88.2
	KNN	81.4	68.8	83

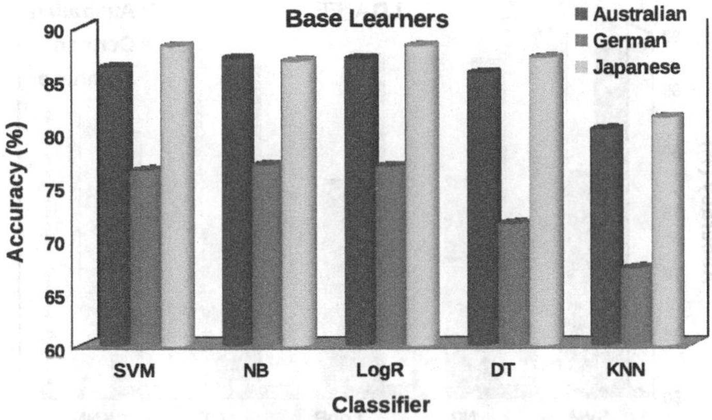

Fig. 1 Performance of base learners in all data sets

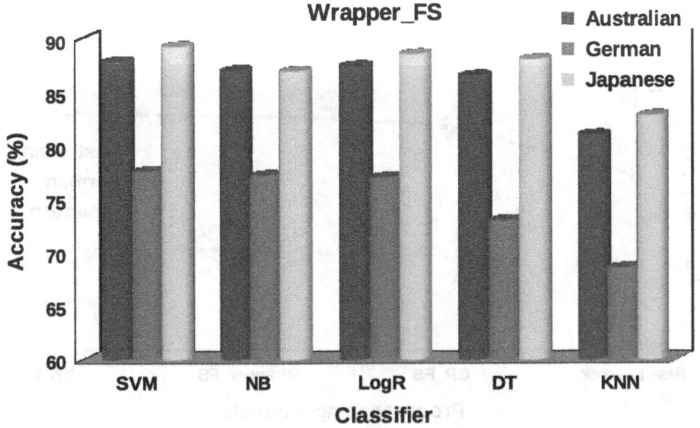

Fig. 2 Performance of Wrapper_FS in all data sets

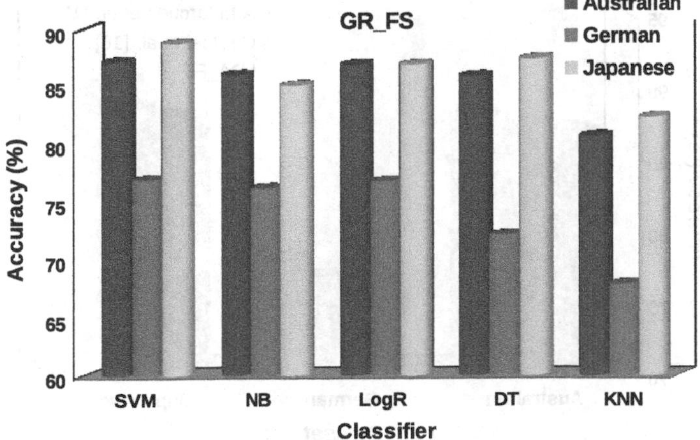

Fig. 3 Performance of GR_FS in all data sets

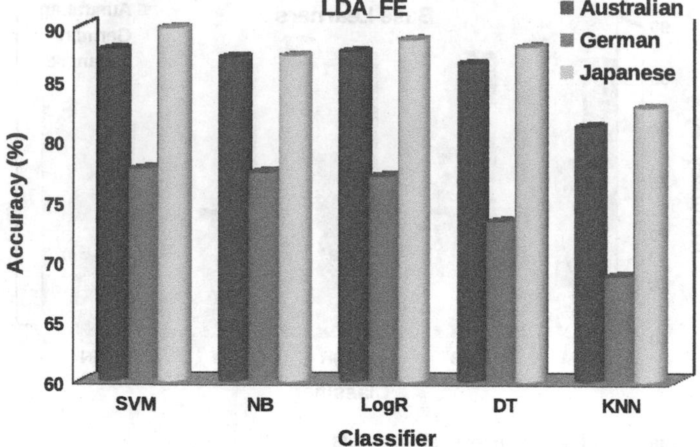

Fig. 4 Performance of LDA_FE in all data sets

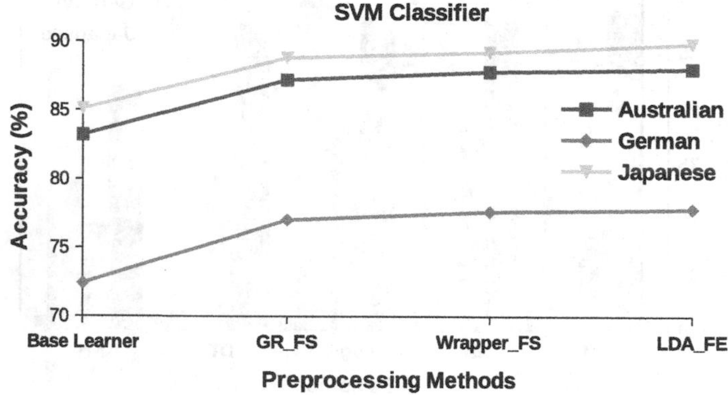

Fig. 5 Performance of SVM in all data sets on all the methods applied

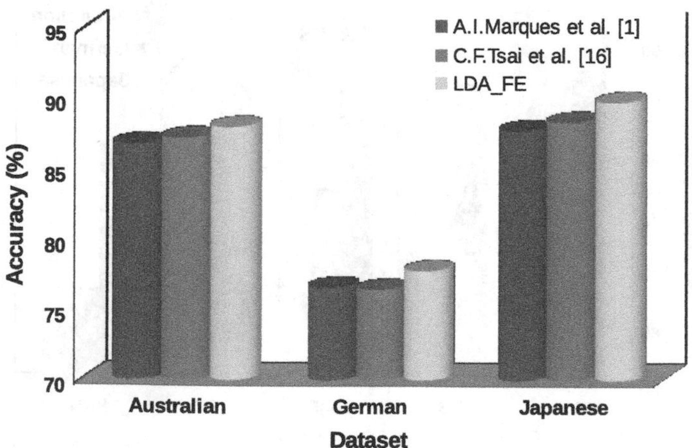

Fig. 6 Comparison of proposed method with existing methods

6 Conclusion

To predict financial risk is a noteworthy issue in finance and accounting fields. In this work, a FS and FE techniques been utilized as a preprocessing work for credit risk prediction model. The prediction model has been designed with base ML methods such as NB, LogR, DT, KNN, and SVM. The performance of base learners are improved with FS and FE techniques. From the result of FS techniques Wrapper_FS performs superior to anything GR_FS and IG_FS filtering techniques. GR_FS performs better than IG_FS. In FE methods LDA_FE performs better than PCA_FE techniques. Over all, from the experimental result it is cleared that the LDA_FE prediction model performs better than PCA_FE, GR_FS, IG_FS and Wrapper_FS model for three kinds for credit data set. For all the preprocessing techniques, SVM classifier performs better than all other base classifiers. At last from the outcomes it is uncovered that FE techniques with base classifier has been taken an option strategy for finding the risky customer in better path contrasted with different strategies. In future, the work will be extended to improve the classifier performance with ensemble classifier by adjusting weights.

References

1. Marqués Marzal, A.I., García Jiménez, V., Sánchez Garreta, J.S.: Exploring the behaviour of base classifiers in credit scoring ensembles. (2012)
2. Xie, J., Wu, J., Qian, Q.: Feature selection algorithm based on association rules mining method. In: Computer and Information Science, 2009. ICIS 2009. Eighth IEEE/ACIS International Conference on, IEEE (2009) 357–362
3. Wang, G., Ma, J., Yang, S.: An improved boosting based on feature selection for corporate bankruptcy prediction. Expert Systems with Applications **41**(5) (2014) 2353–2361
4. Jin, C., Jin, S.W., Qin, L.N.: Attribute selection method based on a hybrid bpnn and pso algorithms. Applied Soft Computing **12**(8) (2012) 2147–2155
5. Abdi, H., Williams, L.J.: Principal component analysis. Wiley Interdisciplinary Reviews: Computational Statistics **2**(4) (2010) 433–459
6. McLachlan, G.: Discriminant analysis and statistical pattern recognition. Volume 544. John Wiley & Sons (2004)
7. Durand, D., et al.: Risk elements in consumer instalment financing. NBER Books (1941)
8. Karels, G.V., Prakash, A.J.: Multivariate normality and forecasting of business bankruptcy. Journal of Business Finance & Accounting **14**(4) (1987) 573–593
9. Reichert, A.K., Cho, C.C., Wagner, G.M.: An examination of the conceptual issues involved in developing credit-scoring models. Journal of Business & Economic Statistics **1**(2) (1983) 101–114
10. West, D.: Neural network credit scoring models. Computers & Operations Research **27**(11) (2000) 1131–1152
11. Desai, V.S., Crook, J.N., Overstreet, G.A.: A comparison of neural networks and linear scoring models in the credit union environment. European Journal of Operational Research **95**(1) (1996) 24–37
12. Shin, K.s., Han, I.: A case-based approach using inductive indexing for corporate bond rating. Decision Support Systems **32**(1) (2001) 41–52

13. Baesens, B., Van Gestel, T., Viaene, S., Stepanova, M., Suykens, J., Vanthienen, J.: Benchmarking state-of-the-art classification algorithms for credit scoring. Journal of the Operational Research Society **54**(6) (2003) 627–635
14. Tam, K.Y., Kiang, M.Y.: Managerial applications of neural networks: the case of bank failure predictions. Management science **38**(7) (1992) 926–947
15. Wang, G., Ma, J.: Study of corporate credit risk prediction based on integrating boosting and random subspace. Expert Systems with Applications **38**(11) (2011) 13871–13878
16. Tsai, C.F., Hsu, Y.F., Yen, D.C.: A comparative study of classifier ensembles for bankruptcy prediction. Applied Soft Computing **24** (2014) 977–984
17. Dash, M., Liu, H.: Feature selection for classification. Intelligent data analysis **1**(3) (1997) 131–156
18. Zhu, Z., Ong, Y.S., Dash, M.: Wrapper–filter feature selection algorithm using a memetic framework. Systems, Man, and Cybernetics, Part B: Cybernetics, IEEE Transactions on **37**(1) (2007) 70–76
19. Lee, D.D., Seung, H.S.: Algorithms for non-negative matrix factorization. In: Advances in neural information processing systems. (2001) 556–562
20. Han, J., Kamber, M., Pei, J.: Data mining: concepts and techniques. Elsevier (2011)
21. Dataset. http://archive.ics.edu/ml/

A Fuzzy Knowledge Based Mechanism for Secure Data Aggregation in Wireless Sensor Networks

Sasmita Acharya and C.R. Tripathy

Abstract Wireless Sensor Networks (WSNs) consist of a number of limited energy sensor nodes deployed randomly over an area. The paper proposes a fuzzy knowledge based secure tree-based data aggregation mechanism for WSNs called the Fuzzy knowledge based Data Aggregation Scheme (FDAS). In the proposed FDAS mechanism, a combination of fuzzy rules is applied to predict the node status of each node in the aggregation tree. The faulty nodes are then isolated from the process of data aggregation. The proposed FDAS mechanism also ensures the security of the network by the application of the privacy homomorphic cryptography technique. It has been found to give better performance characteristics than the other existing algorithms as validated through the results obtained from simulation.

Keywords Wireless Sensor Networks · Fuzzy Logic · Privacy Homomorphic Cryptography · Data Aggregation

1 Introduction

Wireless Sensor Networks (WSNs) are used to provide solutions to a variety of problems in different applications like battlefield surveillance, habitat monitoring, search and rescue operations, forest fire monitoring and traffic regulation. Data aggregation refers to the process of consolidating the data from multiple sensors so as to decrease the network traffic and increase the network lifetime. There are two

S. Acharya (✉)
Department of Computer Application, VSS University of Technology, Burla, India
e-mail: talktosas@gmail.com

C.R. Tripathy
Department of Computer Science and Engineering, VSS University of Technology, Burla, India
e-mail: crt.vssut@yahoo.com

H.S. Behera and D.P. Mohapatra (eds.), *Computational Intelligence in Data Mining*, Advances in Intelligent Systems and Computing 556,
DOI 10.1007/978-981-10-3874-7_8

types of protocols for data aggregation which are tree-based and cluster-based. The tree-based protocols mainly focus on the construction of a data aggregation tree that saves on energy. Here, at each level, the parent data aggregators consolidate the data received from all the child nodes and forward it onwards to the next higher level. It continues repeatedly until the data finally reaches the sink or the Base Station (BS). It may be possible that during data transmission, a sensor node is compromised or injects some false data into the network. The problem becomes more severe if the compromised node is an aggregator node. Then, the freshness, integrity, confidentiality and security of data in the network are greatly threatened. Thus, there is a need for secure data aggregation in WSNs [1].

The paper proposes a fuzzy knowledge based tree aggregation mechanism for WSNs called Fuzzy knowledge based Data Aggregation Scheme (FDAS). It uses fuzzy logic for identifying the faulty nodes in the network and then isolates them from the data aggregation process. The fuzzy rules are framed by using three fuzzy inputs which are: node's residual energy (RE), the packet delivery ratio (DR), and the fault ratio (FR). The proposed mechanism also ensures the security of the network by the application of the privacy homomorphic cryptography technique in data aggregation. Section 2 of the paper discusses about some important works done on secure data aggregation in WSNs. The system model is presented in Sect. 3. The FDAS mechanism and algorithm are presented in Sect. 4. The proposed FDAS mechanism is compared with that of the Center at Nearest Source (CNS), the Shortest Paths Tree (SPT), and the Greedy Incremental Tree (GIT) mechanisms [2] through simulation in Sects. 5, 6 concludes the paper.

2 Related Work

This section presents a review of some of the previous research works carried out on secure data aggregation in WSNs. The work done in [3] presents a detailed survey on sensor networks. The effect of data aggregation on WSNs is discussed elaborately in [1]. An overview of data aggregation techniques for secure WSNs is presented in [2], where the privacy homomorphic cryptography technique is elaborated. The problem of construction of an optimal aggregation tree is an NP-hard problem as discussed in [2]. But there are some approximation algorithms for optimal data aggregation such as the Center at Nearest Source (CNS), the shortest paths tree (SPT), and the greedy incremental tree (GIT) as discussed in [2]. The work done in [4] presents a fuzzy approach for detection of faulty nodes in a network by the application of majority voting in WSNs. An algorithm based on fuzzy logic for fault tolerance in WSNs is presented in [5]. The work done in [6] compares the various techniques for restoration of inter-actor connectivity in Wireless Sensor and Actor Networks (WSANs). A verification algorithm in order to identify any false contribution by any compromised sensor nodes by the base station is presented in [7]. A model based on Artificial Neural Network (ANN) for the design of fault-tolerant WSNs using exponential Bidirectional Associative

Memory (eBAM) is presented in [8]. A fuzzy-based fault-tolerant data aggregation technique for WSNs is presented in [9], which applies only to cluster-based data aggregation models. Further, the security aspect is not considered in [9] whereas the proposed FDAS mechanism provides secure data aggregation and applies to tree-based data aggregation models for WSNs. An Energy-efficient Structure-free Data Aggregation and Delivery protocol (ESDAD) is discussed in [10] which aggregates the redundant data in the intermediate nodes.

3 System Model

A tree-based model for data aggregation is depicted in Fig. 1. The sink node or the BS acts as the root node for the data aggregation tree and is at Level 0. Next, there is a hierarchy of sensor nodes which represent the parent–child relationship at different levels starting from Level 1 through Level 4. The sensor nodes numbered 1 and 2 are at Level 1, those numbered from 3 to 6 are at Level 2, those numbered from 7 to 12 are at Level 3, and finally those numbered from 13 to 23 are at Level 4, respectively. For example, for the sensor nodes labeled 3 and 4, the node labeled "1" acts as the parent node which means that nodes 3 and 4 forward their data to node 1 which acts as the parent data aggregator for these nodes.

The sensor nodes at Level 4 act as the leaf nodes and the other sensor nodes at levels 1, 2, and 3 act as the non-leaf nodes. The BS at Level 0 acts as the root node for the tree-based data aggregation model in Fig. 1.

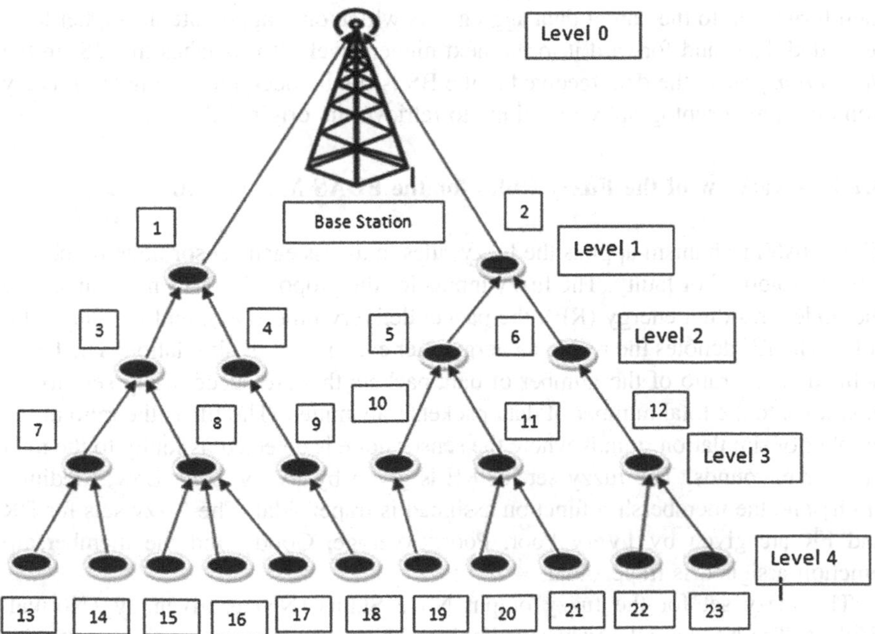

Fig. 1 A tree-based data aggregation model

4 Proposed FDAS Mechanism and Algorithm

An overview of the proposed FDAS mechanism and the proposed FDAS algorithm is presented in this section.

4.1 Overview of the Proposed FDAS Mechanism

The proposed Fuzzy knowledge based Data Aggregation Scheme (FDAS) assesses the status of each sensor node in a tree-based data aggregation model. It assigns a Node Status value (NS) through a fuzzy rule base. It consists of four phases—the *query phase,* the *encryption phase,* the *fusion phase*, and finally the *decryption phase*. In the *query phase*, the queries are disseminated from the BS through the routing table to the sensor nodes at different levels. Each query mentions the data to be collected, the type of aggregation function (sum, average, min or max) to be used and the subset of sensor nodes that will do the data collection. In the proposed approach, the leaf nodes or the sensor nodes at the lowest level and the BS share a symmetric key. This key is kept secret from the other intermediary data aggregators in the network. In the *encryption phase*, at the lowest level of the aggregation tree, the sensor data is encrypted by the leaf nodes using the privacy homomorphic cryptography technique. In the *fusion phase*, only the normal sensor nodes identified by the fuzzy rules of the FDAS mechanism participate in the data aggregation process and forward their data to their parent node at the next higher level which then aggregates the data received from all of its child nodes. This encrypted data is then forwarded to the parent data aggregators which only aggregate the cipher texts (encoded data) and forward it to the next higher level till it reaches the BS. In the *decryption phase*, the data received at the BS is finally decrypted using the privacy homomorphic cryptography technique to retrieve the original data.

4.1.1 Overview of the Fuzzy Rules for the FDAS Mechanism

The FDAS mechanism applies the fuzzy rules to assess each sensor node in the tree model as normal or faulty. The fuzzy inputs for the proposed FDAS mechanism are the node's residual energy (RE), the packet delivery ratio (DR), and the fault ratio (FR). The RE denotes the residual energy after each round of simulation. The DR is defined as the ratio of the number of data packets that are successfully sent to the next level to the total number of data packets transmitted. The FR is the ratio of the number of simulation rounds where the sensor node is detected as faulty to the total number of rounds. The fuzzy set for RE is given by {Very Low, Low, Medium, High} and the membership function assigned is trapezoidal. The fuzzy sets for DR and FR are given by {Very Poor, Poor, Average, Good} and the membership function assigned is trapezoidal.

The fuzzy set for the fuzzy output Node Status (NS) is given by {Normal, Faulty}. The NS can take values as Faulty or Normal according to the fuzzy rules defined and the membership function assigned is triangular. The NS value is updated periodically to the parent node at the next higher level. Table 1 presents some fuzzy rules for the proposed model. The fuzzy output NS is then fed to a *defuzzifier* which gives a crisp output using the centroid method of defuzzification. It gives output as '0' for a *normal* node and as '1' for a *faulty* node. If the Node Status (NS) value is '0', then the sensor node is designated as "*normal*" and its data is forwarded up the aggregation tree. But if the Node Status (NS) value is '1', then the sensor node is designated as "*faulty*" and its data is not forwarded.

4.1.2 Overview of the Privacy Homomorphic Cryptography Technique

The privacy homomorphic cryptography is a technique that provides data aggregation and data confidentiality without the need to share the secret key among the data aggregators at different levels. It allows direct addition or multiplication based computations to be done on the encrypted data by using the public and private keys of the BS. There are two common homomorphic techniques. They are *additively homomorphic* and *multiplicatively homomorphic*. Let "E" denote "encryption" and let "D" denote "decryption". Let "+" be the addition operation and let "×" be the multiplication operation to be performed on a given data set Z. Let the private and public keys of the BS be represented by K1 and K2, respectively. The condition for an encryption transmission to be considered as *additively homomorphic* is given by Eq. (1).

$$c + d = D_{K1}(\,E_{K2}(c) + E_{K2}(d)\,) \quad \text{where } c, d \in Z \tag{1}$$

Similarly, the condition for an encryption transmission to be considered as *multiplicatively homomorphic* is given by Eq. (2).

$$c \times d = D_{K1}(\,E_{K2}(c) \times E_{K2}(d)\,) \quad \text{where } c, d \in Z \tag{2}$$

Table 1 Fuzzy Rule Base Sample

Inputs			Output
RE	DR	FR	NS
Very low	Poor	Poor	Faulty
Low	Poor	Poor	Faulty
High	Good	Good	Normal
Medium	Average	Average	Normal
High	Very poor	Very poor	Faulty
Medium	Poor	Very poor	Faulty
Very low	Very poor	Very poor	Faulty
Low	Very poor	Very poor	Faulty
Very low	Poor	Very poor	Faulty
Very low	Very poor	Poor	Faulty
Medium	Very poor	Very poor	Faulty

The proposed FDAS mechanism uses the *additively homomorphic* cryptography technique.

4.2 FDAS Algorithm

This sub-section presents the proposed FDAS algorithm.

Input: Network configuration parameters, number of rounds of simulation 'R'.
Output: A FDAS data aggregation scheme.

(a) Initialization Phase
Initialize the network topology based on specific network configuration.
Initialize the energy E_i for all sensor nodes.
Initialize node 0 as the Base Station (BS).

(b) Exploration Phase
for each simulation round R
Record the RE, DR and FR values for each sensor node in a record table

(c) Fault Prediction Phase
for each leaf node in the data aggregation tree
Apply fuzzy rules to predict its Node Status (NS)
if (NS(i) == 0) then set fwd-flag(i) = 'Y' // normal sensor data is forwarded for node 'i'.
else if (NS(i) == 1) then set fwd-flag(i) = 'N' // faulty sensor data is not forwarded and isolated.

(d) Query Phase
for each simulation round R
Route the queries from the BS to the leaf sensor nodes using routing table.

(e) Encryption Phase
for each simulation round R
Encrypt the sensor data using additively homomorphic cryptography technique only at the lowest level (level of the leaf nodes)

(f) Fusion Phase
for each simulation round R
Forward data only from the normal sensor nodes (with NS = '0') to respective parent nodes by the help of the routing table
Perform in-network data aggregation at the parent node for the encrypted sensor data (cipher text)
Forward the encrypted data from leaf node to parent node at the next higher level till it reaches the node 0 (BS)

(g) Decryption Phase
for each simulation round R
Decrypt the data received at node 0 (BS) using additively homomorphic cryptography technique

5 Results and Discussion

The simulation model is presented and the proposed FDAS data aggregation technique is compared with the existing CNS, SPT, and GIT techniques [2] through simulation in this section. The simulation is done using MATLAB.

5.1 Simulation Model

The parent sensor nodes which act as data aggregators for their child nodes are selected in each simulation round. The ratio of data aggregation is restricted to 10% in the simulation. The Table 2 presents the configuration details for the simulation.

5.2 Simulation Results Analysis

The simulation results for the proposed FDAS mechanism are compared with that of the CNS, SPT, and GIT algorithms in this section.

5.2.1 Data Rate Versus Average Energy Consumption

Figure 2 depicts the average energy consumption (in Joules) for different data rates (in data packets per second (pps)). The average energy consumption is observed to be the lowest for the proposed FDAS algorithm and the highest for the CNS algorithm. The SPT and the GIT algorithms show average performance. The optimal performance of the FDAS algorithm owes to its fuzzy knowledge based faulty node detection scheme.

5.2.2 Number of Nodes Versus Average Energy Consumption

Figure 3 shows the average energy consumption (in Joules) for different number of nodes starting from 100 to 1000. The average value of energy consumption is

Table 2 Simulation parameters

Parameter name	Value
Size of network	500 m × 500 m
Number of sensor nodes	400
Packet Length	60 bytes
Node Energy	70 J
Bandwidth	200 Kb/s
Sensing Length	50 m
Radio Range	40 m

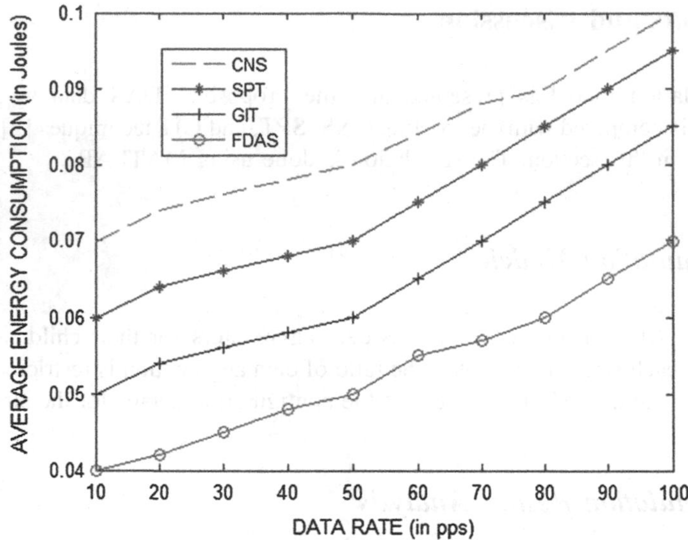

Fig. 2 Data rate versus average energy consumption

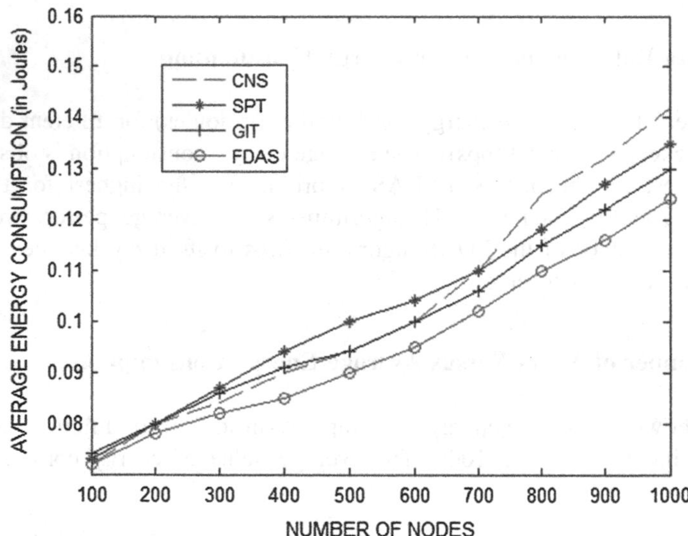

Fig. 3 Number of nodes versus average energy consumption

observed to be the lowest for the proposed FDAS algorithm and the highest for the CNS algorithm closely followed by the SPT algorithm. The performance of the GIT algorithm lies close to that of the FDAS algorithm. The FDAS algorithm gives better performance due to its isolation of the faulty nodes from the data aggregation process.

5.2.3 Data Rate Versus Miss Ratio

Figure 4 depicts the miss ratio for different data rates (in data packets per second (pps)). The miss ratio is observed to be the lowest for the proposed FDAS algorithm and the highest for the CNS algorithm. The SPT and the GIT algorithms show an average performance. The FDAS algorithm gives better performance due to its isolation of the faulty nodes from the data aggregation process.

5.2.4 Number of Nodes Versus Miss Ratio

Figure 5 depicts the miss ratio for different number of nodes starting from 100 to 1000. The miss ratio is observed to be the lowest for the proposed FDAS algorithm and the highest for the CNS algorithm. The SPT and the GIT algorithms show average performance. The FDAS algorithm gives better performance due to its fault-tolerant and secure data aggregation mechanism.

5.2.5 Data Rate Versus End-to-End Delay

Figure 6 shows the end-to-end delay for different data rates (in data packets per second (pps)). The end-to-end delay is found to be the lowest for the proposed FDAS algorithm closely followed by the GIT algorithm. The end-to-end

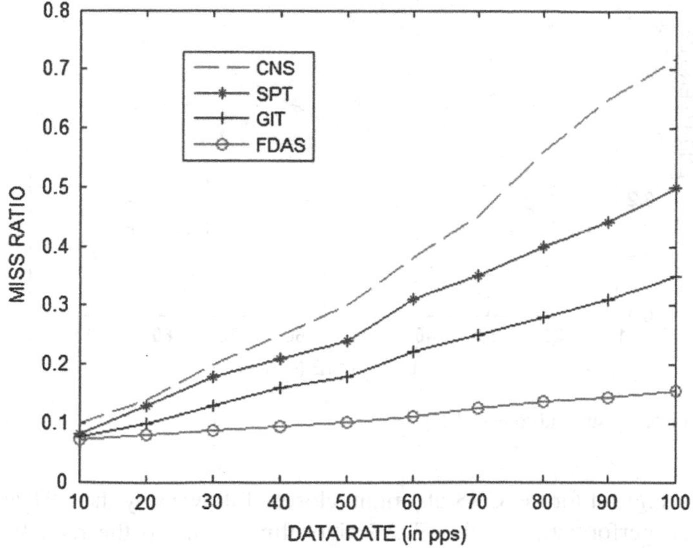

Fig. 4 Data rate versus miss ratio

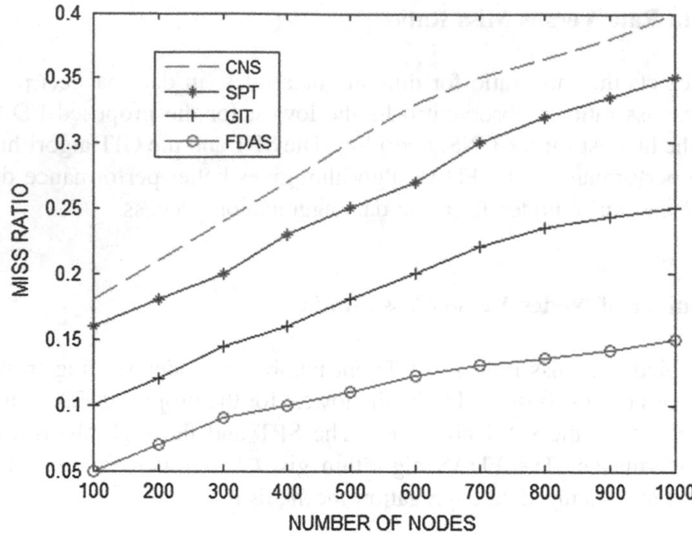

Fig. 5 Number of nodes versus miss ratio

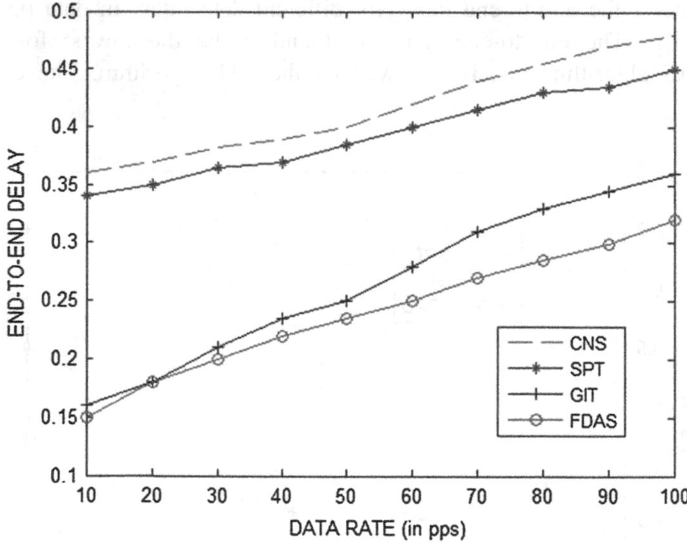

Fig. 6 Data rate versus end-to-end delay

delay is the highest for the CNS algorithm closely followed by the SPT algorithm. The optimal performance of the FDAS algorithm is due to the reduction in the number of re-transmissions as faulty data are isolated from the data aggregation process.

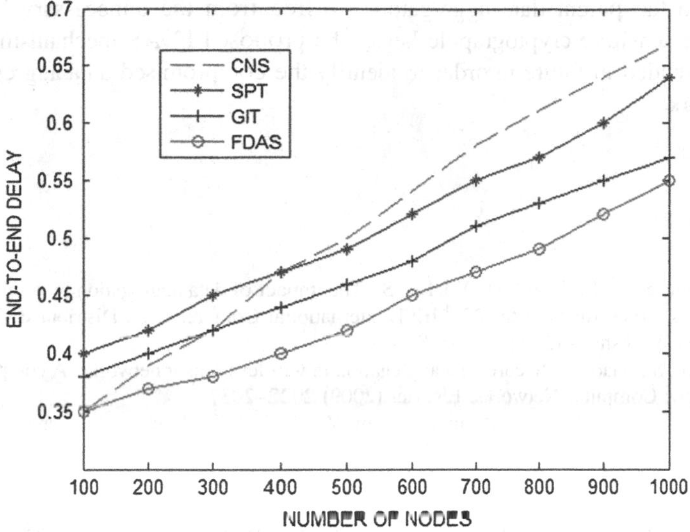

Fig. 7 Number of nodes versus end-to-end delay

5.2.6 Number of Nodes Versus End-to-End Delay

Figure 7 depicts the end-to-end delay for different number of nodes starting from 100 to 1000. It is observed that the end-to-end delay is the lowest for the proposed FDAS algorithm and the highest for the CNS algorithm (especially after 400 nodes). The SPT and the GIT algorithms show average performance. The FDAS algorithm gives better performance due to its fault-tolerant and secure data aggregation mechanism.

6 Conclusion

A secure data aggregation mechanism for WSNs based on fuzzy knowledge called Fuzzy knowledge based Data Aggregation Scheme (FDAS) is presented. It used fuzzy logic for the selection of non-faulty nodes to achieve fault-tolerant data aggregation in WSNs. It also ensured the security of the network by the application of the privacy homomorphic cryptography technique for secure data aggregation at each level of the aggregation tree. The proposed FDAS mechanism was compared with other algorithms through simulation for different performance metrics. The simulation results validated that the proposed FDAS data aggregation mechanism gave better performance than the other simulated algorithms like CNS, SPT, and GIT. The major advantage of this mechanism is that it makes the network more secure and the parent data aggregators are free from the unnecessary hassle of storing the sensitive cryptographic keys. The proposed FDAS mechanism may be further extended in future in order to identify the compromised data aggregators in the network.

References

1. Krishnamachari, L., Estrin, D,. Wicker, S.: The impact of data aggregation in wireless sensor networks, Proceedings of the 22nd IEEE International Conference on Distributed Computing Systems Workshops (2002) 575–578
2. Ozdemir, S., Xiao, Y.: Secure data aggregation in wireless sensor networks: A comprehensive overview, Computer Networks, Elsevier (2009) 2022–2037
3. Akyildiz, I.F., Su, W., Sankarasubramaniam, Y., Cayirci, E.: A Survey on Sensor Networks, IEEE CommMag, Vol. 40, no. 8 (2002) 102–114
4. Javanmardi, S., Barati, A., Dastgheib, S.J., Attarzadeh, I.: A Novel Approach for Faulty Node Detection with the Aid of Fuzzy Theory and Majority Voting in Wireless Sensor Networks, International Journal of Advanced Smart Sensor Network Systems (IJASSN), Vol. 2, No. 4 (2012) 1–10
5. Chang, S., Chung, P., Huang, T.: A Fault Tolerance Fuzzy Knowledge Based Control Algorithm in Wireless Sensor Networks, Journal of Convergence Information Technology (JCIT), Vol. 8, No. 2 (2013)
6. Acharya, S., Tripathy, C.R.: Inter-actor Connectivity Restoration in Wireless Sensor Actor Networks: An Overview, Advances in Intelligent Systems and Computing, Vol. 248, Springer International Publishing Switzerland (2014) 351–360
7. Roy, S., Conti, M., Setia, S., Jajodia, S.: Secure Data Aggregation in Wireless Sensor Networks: Filtering out the Attacker's Impact, IEEE Transactions on Information Forensics and Security, Vol. 9 (2014) 681–694
8. Acharya, S., Tripathy, C.R.: An ANN Approach for Fault-Tolerant Wireless Sensor Networks, Advances in Intelligent Systems and Computing, Vol. 338, Springer International Publishing Switzerland (2015) 475–483
9. Acharya, S., Tripathy, C.R.: A Fuzzy Knowledge Based Sensor Node Appraisal Technique for Fault Tolerant Data Aggregation in Wireless Sensor Networks, Advances in Intelligent Systems and Computing, Vol. 411, Springer International Publishing Switzerland (2016) 59–69
10. Mohanty, P., Kabat, M.R.: Energy efficient structure-free data aggregation and delivery in WSN, Egyptian Informatics Journal, Elsevier (2016)

Estimation of Noise in Digital Image

K.G. Karibasappa and K. Karibasappa

Abstract Denoising plays an important role in improving the performance of algorithms such as classification, recognition, enhancement, segmentation, etc. To eliminate the noise from noisy image, one should know the noise type, noise level, noise distribution, etc. Typically noise level information is identified from noise standard deviation. Estimation of the image noise from the noisy image is major concern for several reasons. So, efficient and effective noise estimation technique is required to suppress the noise from the noisy image. This paper presents noise estimation based on rough fuzzy c-means clustering technique. The experimental results and performance analysis of the system are presented.

Keywords Noise · Cluster · Variance · Rough set · Fuzzy · Skewness

1 Introduction

Nature of noise and noise distribution are the main parameters in selecting the appropriate denoising technique. Noise estimation in the image is a tedious process as it depends on the parameters like texture, color, brightness, etc. In literature, researchers have designed various techniques for noise estimation. Olsen [1] classified estimation algorithms into four classes such as block-based, texture-based, filter-based, and hybrid noise estimation algorithms. In block-based [2–6] technique, input image is divided into non overlapping blocks and the noise level is estimated on the average of the noise levels of the homogenous blocks. Identifying homogeneous block is a challenging issue due to overlapped data. In filter-based [1, 7–10] techniques noise level is identified by estimating the difference of pre-filtered image

K.G. Karibasappa (✉)
B.V.B. College of Engineering and Technology, Hubballi 31, India
e-mail: karibasappa_kg@bvb.edu

K. Karibasappa
The Oxford College of Engineering, Bengaluru, India

H.S. Behera and D.P. Mohapatra (eds.), *Computational Intelligence in Data Mining*, Advances in Intelligent Systems and Computing 556,
DOI 10.1007/978-981-10-3874-7_9

and the original noise image. Hybrid [11–13] methods are developed using the concept of block and filter based methods.

Detection of homogeneous regions is a major concern in block-based techniques because of overlapped image and noise data. This can be resolved by combining fuzzy c-means clustering (FCM) algorithm and statistical features of the image. Even though, many clustering algorithms are available in the literature [14–19]. Fuzzy c-means algorithm has the several advantages such as high accuracy in finding homogeneous blocks for all types images in the presence of noise. In this paper, rough set theory is used along with the FCM for optimizing the cluster centers.

2 Noise Estimation

In the proposed method of noise estimation, image features such as mean, variance, skewness, and kurtosis are extracted. Using these features and rough set concept, the number of initial clusters is optimized. These features are used as parameters for the proposed fuzzy c-means clustering algorithm to form the final set of clusters (homogeneous blocks). The details of each of these stages are discussed in the following subsections.

2.1 Extracting Image Features

In this stage, a matrix F of size c × 4 containing four statistical features called mean, variance, skewness, and kurtosis of image is formed using the Eqs. (1)–(4), respectively. These features are extracted by sliding window of size (l × l) centered at every pixel (p, q) for an image f (m, n) from top to bottom and left to right as shown in Fig. 1.

$$\mu = \frac{1}{mn} \sum_{i=p-\frac{1-l}{2}}^{p+\frac{1-l}{2}} \sum_{j=q-\frac{1-l}{2}}^{q+\frac{1-l}{2}} y(i,j). \tag{1}$$

$$\sigma = \frac{1}{mn} \sum_{i=p-\frac{1-1}{2}}^{p+\frac{1-1}{2}} \sum_{j=q-\frac{1-1}{2}}^{q+\frac{1-1}{2}} (y(i,j), -\mu)^2 \tag{2}$$

$$S_{kew} = \frac{1}{mn} \sum_{i=p-\frac{1-1}{2}}^{p+\frac{1-1}{2}} \sum_{j=q-\frac{1-1}{2}}^{q+\frac{1-1}{2}} (y(i,j), -\mu)^3 \tag{3}$$

$$K_{urto} = \frac{1}{mn} \sum_{i=p-\frac{l-1}{2}}^{p+\frac{l-1}{2}} \sum_{j=q-\frac{l-1}{2}}^{q+\frac{l-1}{2}} (y(i,j), -\mu)^4 \tag{4}$$

Each cluster center is represented by four image features called mean, variance, skewness, and kurtosis. Each of these features is defined by fuzzy membership values μ(Fi) for corresponding linguistic values called low, medium and high. The value of these membership functions is in the range of $0 \le \mu(F_j) \le 1$, i.e., each of these features is defined by a π-membership function as shown in Fig. 2.

$$\mu(F\,i) = \begin{cases} 2\left[1 \frac{|F\,i-c|}{\lambda}\right]^2 & for \quad \frac{\lambda}{2} \le |F\,i-c| \le \lambda \\ 1-2\left[\frac{|F\,i-c|}{\lambda}\right]^2 & for \quad 0 \le |F\,i-c| \le \frac{\lambda}{2}, \\ 0 & otherwise \end{cases} \tag{5}$$

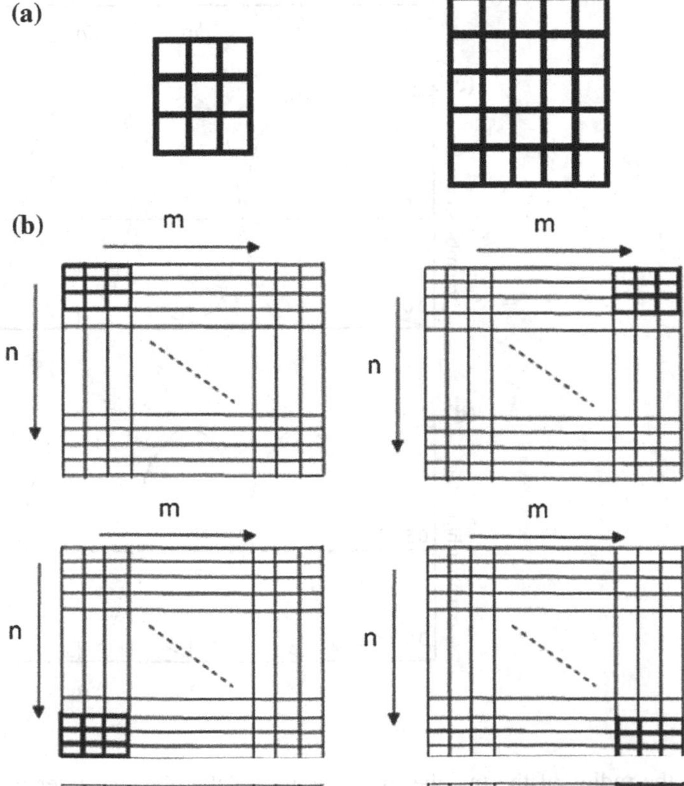

Fig. 1 **a** Windows of size 3 × 3 and 5 × 5. **b** Sliding window of size 1 × 1 moves from *top* to *bottom* and *left* to *right*

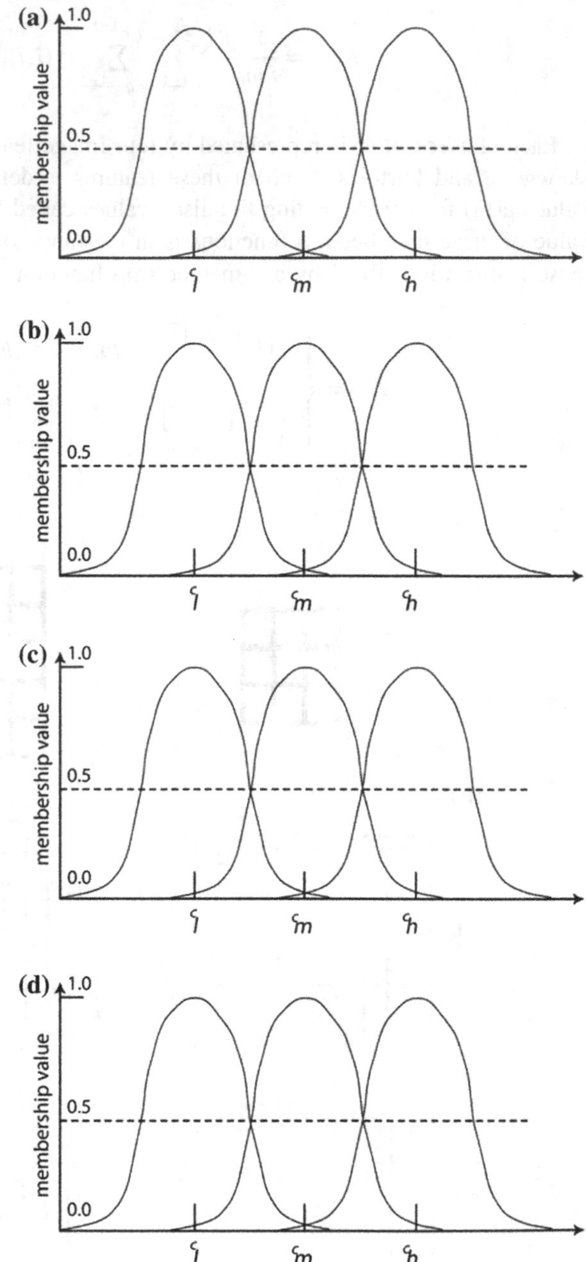

Fig. 2 π Membership functions for linguistic fuzzy sets for low (l), medium (m), and high (h) for image features **a** mean, **b** standard deviation, **c** kurtosis, and **d** skewness

where λ is the radius of the membership function of the cluster center c.

where λ is the radius of π-membership function and usually $\lambda > 0$ with c as central point and for each of the fuzzy sets low, medium, and high λ and c are different. The center c and radius λ of the overlapping π-membership functions are defined as follows:

$$C_{low}(F_j) = m_{jl}. \tag{6}$$

$$C_{medium}(F_j) = m_j \tag{7}$$

$$C_{lhigh}(F_j) = m_{jh} \tag{8}$$

$$\lambda_{low}(F_j) = C_{medium}(F_j) - C_{low}(F_j) \tag{9}$$

$$\lambda_{lmedium}(F_j) = 0.5(C_{lhigh}(F_j) - C_{low}(F_j)) \tag{10}$$

$$\lambda_{lhigh}(F_j) = C_{high}(F_j) - C_{medium}(F_j), \tag{11}$$

where m_j, is the mean of the pattern along the jth axis of the membership function. m_{jl}, m_{jh} are the mean values in the interval [F_{jmin}, m_j] and [m_j, F_{jmax}], respectively. F_{jmin}, F_{jmax}, are the lower and higher values of the F_j.

2.2 Decision Table Creation

Using rough set theory concepts decision table for initial cluster centers are defined based on the decision attribute α, which is also called as similarity index of two cluster centers which is defined as follows:

$$\alpha = \frac{\sum_{i=1}^{n} \mu(F_i)}{n} \tag{12}$$

Distance between the cluster centers are denoted by α, i.e., if α increases, closeness of the cluster centers increases. With these initial cluster centers, features, central point c and radius λ as conditional attributes, α as decision attributes, decision table for initial cluster centers are defined as,

$$T = \langle U, P \cup R, C, D \rangle, \tag{13}$$

where U is the initial cluster center, P is the set of conditional attributes, R is the set of decision attributes, $P \cup R$ denotes the category attributes of cluster centers, C set of domain of the initial cluster center category attributes. D defines the indiscernibility relation as the mapping function (redundant information) defined as,

$$D: U \times P \cup R \rightarrow C \tag{14}$$

2.3 Cluster Optimization

Based on the definition of decision table, redundant cluster centers are reduced by using the cluster compatibility condition as defined below:

$$\textit{If } D(i)|\,P = D(j)\,|P \text{ and } \textit{If } D(i)|\,R = D(j)\,|R \tag{15}$$

Then, cluster centers i and j of decision table are compatible, otherwise cluster centers are not compatible. Using this condition the conditional attributes are reduced as follows,

- If the decision table is compatible and IND(P) = IND(P-p) then p is redundant attribute and it can be removed.
- If the decision table is not compatible then if (POS(P,R) = POS(P-p, D) then p is redundant attribute and it can be removed.

Above process is continued until no change in conditional attribute set. Resultant optimized initial cluster centers are used as an input for FCM algorithm.

2.4 Most Homogeneous Regions

Most homogeneous regions are identified from optimized clusters are as follows,

- Find the mean and variance of each of the homogeneous blocks as follows:

$$\mu_b = \frac{1}{M} \sum_{i(x,y)\epsilon b} i(x, y) \tag{16}$$

- *Select the Most Homogeneous Blocks*

$$\sigma_b = \frac{1}{M-1} \sqrt{\sum_{i(x,y)} (i(x, y))^2} \tag{17}$$

$$\sigma = \frac{1}{M} \sum_{i=1}^{M} \sigma_i \tag{18}$$

Average variance of the most homogeneous blocks is the amount of noise in the noisy image.

3 Experimental Results

Performance of the proposed method for noise estimation is tested for nonhomogeneous, complex, more and less homogeneous region images. These images are selected from various data sets and few are synthetic images. Some test images are shown in Fig. 3. These images are tested by artificially adding the different noises at various levels.

Figures 4 and 5 show the actual and estimated noise tested for Gaussian and Rayleigh noise on all four types of images. From these figures, it is observed that estimation is the underestimation as the noise level increases especially for Rayleigh noise.

Figures 6 and 7 show the actual and estimated noise tested for impulse and speckle noise on all four types of images. From these figures, it is observed that estimation is accurate till the noise level is 25% and underestimation as the noise

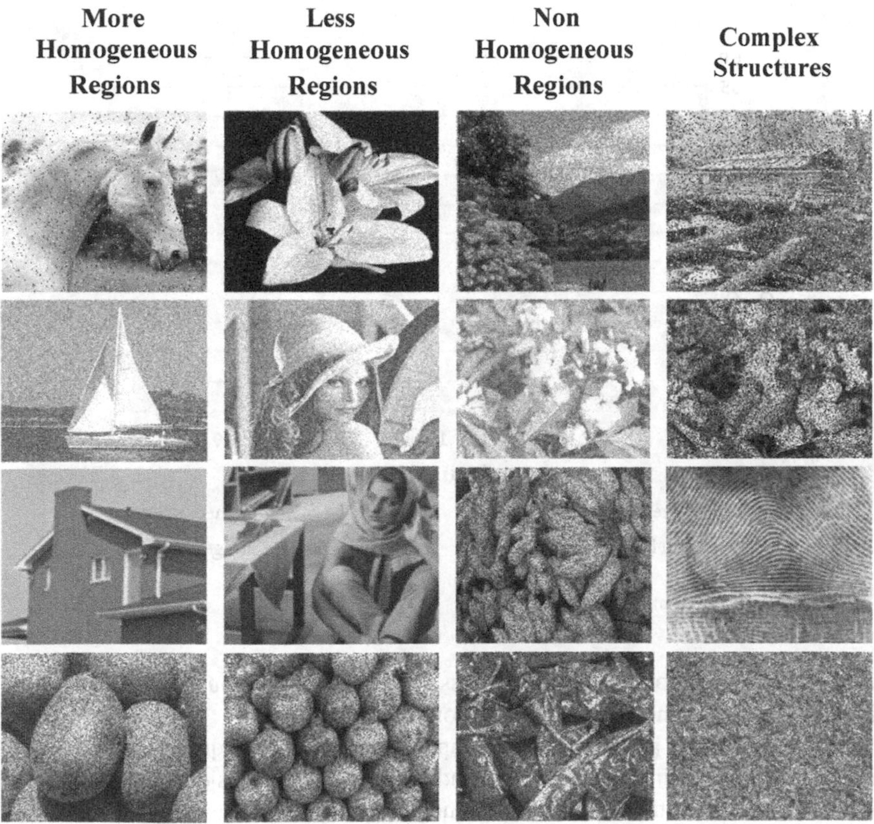

Fig. 3 Images used for testing the proposed method

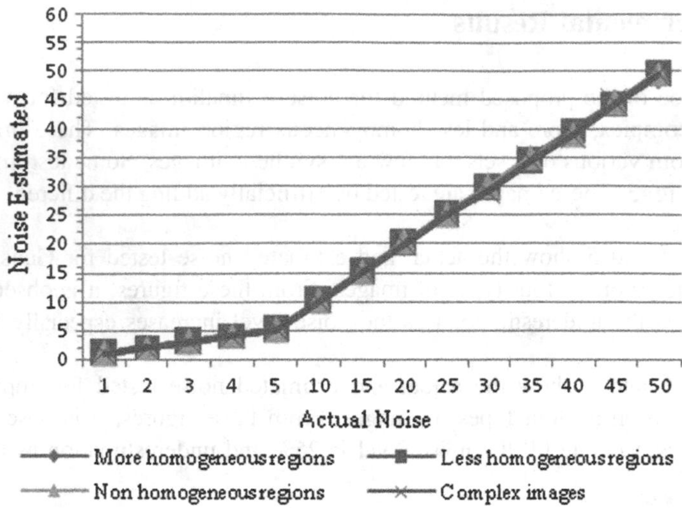

Fig. 4 Gaussian noisy images

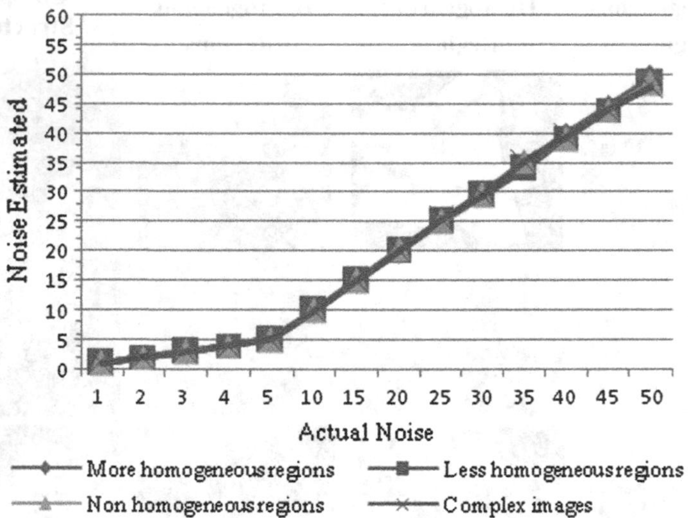

Fig. 5 Rayleigh noisy images

level increases for homogeneous and less homogeneous regions. Whereas overestimation for the nonhomogeneous and complex images.

Proposed method performance is compared with the techniques available in the literature as shown in Figs. 8, 9, 10, and 11. Parameter considered for the comparison is shown in the Eq. (19) for homogeneous region images.

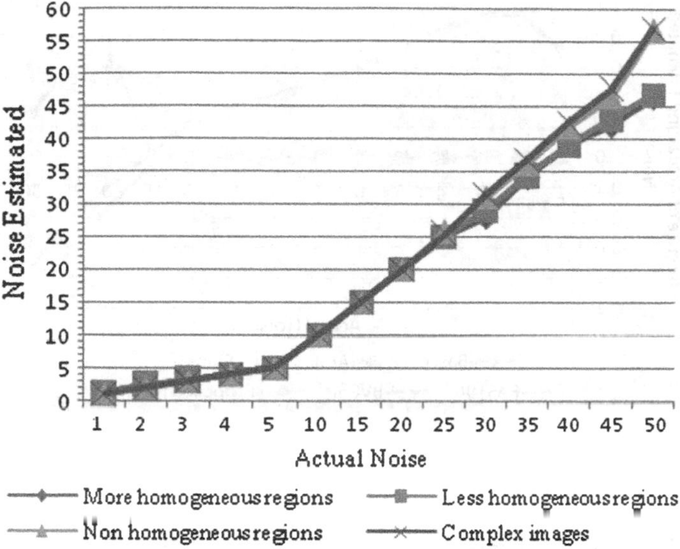

Fig. 6 Impulsive noisy images

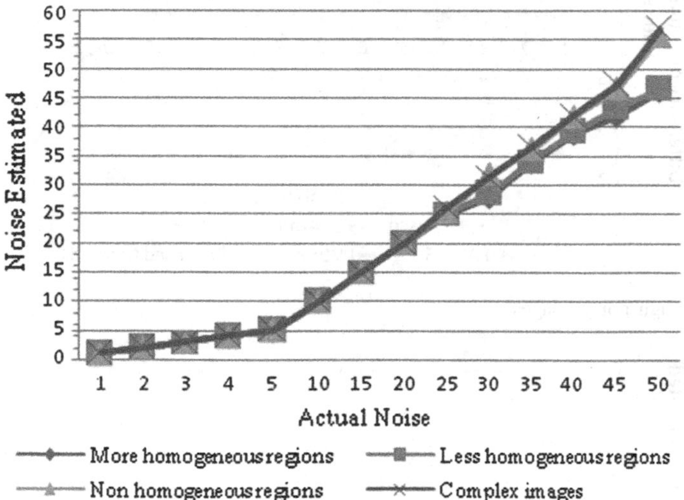

Fig. 7 Speckle noisy images

$$\sigma_{error} = \sigma_{estimated} - \sigma_{actual} \tag{19}$$

Proposed method's performance is almost equal to the other techniques for noise level below 25 which is shown in Figs. 8 and 9. It is also observed that, as the noise increases proposed method gives relatively better results compared to other

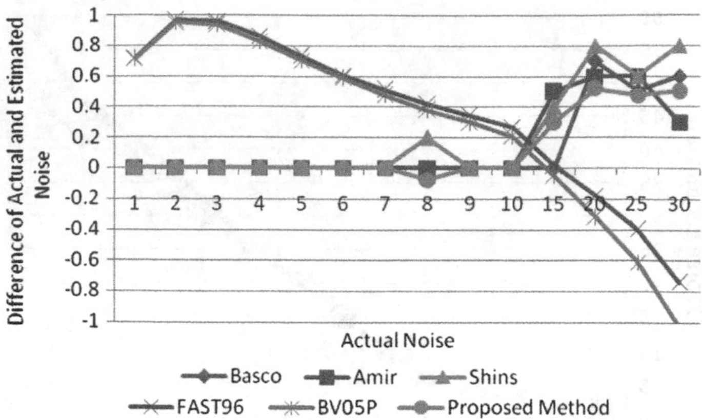

Fig. 8 Gaussian noisy images

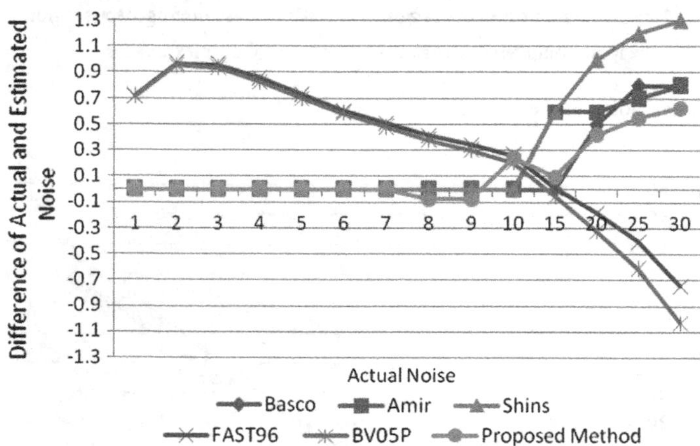

Fig. 9 Rayleigh noisy images

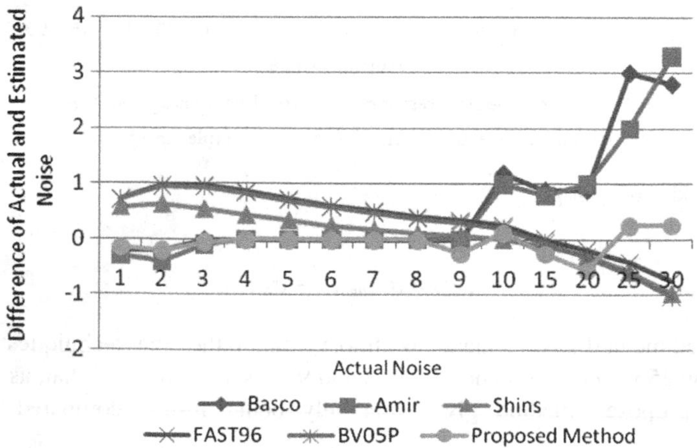

Fig. 10 Impulsive noisy images

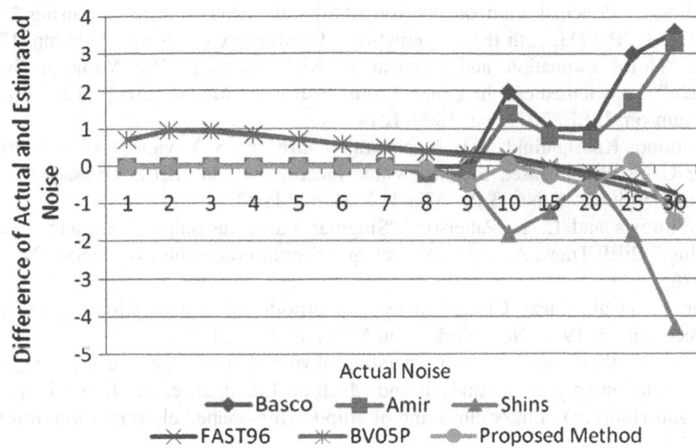

Fig. 11 Speckle noisy images

techniques for Gaussian and Rayleigh noisy images and overestimation (shown in Figs. 10 and 11) for the impulsive and speckle noisy images.

4 Conclusion

Performance of the proposed method compared to other methods yields better results for all types of images. From the above results it is observed that, as the homogeneity in the images reduces the number of blocks increases and it results in underestimation and overestimation when the noise increases in additive and multiplicative noises, respectively.

References

1. S.I. Olsen "Estimation of noise in images An evaluation", Computer Vision Graphics Image Processing, Graphics Models. Vol. 55, No. 4, pp. 319–323, 1993.
2. Angelo Bosco, Arcangelo Bruna, Giuseppe Messina, Giuseppe Spampinato, "Fast Method for Noise Level Estimation and Integrated Noise Reduction", IEEE Transaction on Consumer Electronics, vol. 51, No. 3, pp 1028–1033, August 2005.
3. Aishy Amer, and Eric. Dubois, "Reliable and Fast structure-Oriented Video Noise Estimation," in Proc. IEEE International conference. Image Processing, Montreal Quebec, Canada, Sep 2002.
4. Jolliffe, I.T. "Principal Component Analysis" 2nd ed. 2002, XXIX.
5. D. Muresan and T. Parks, "Adaptive principal components and image denoising," in Image Processing, 2003. ICIP 2003. Proceedings. 2003 International Conference on, vol. 1. IEEE, 2003.

6. T. Tasdizen, "Principal components for non-local means image denoising," in Image Processing, ICIP 2008, 15th IEEE International Conference on. IEEE, 2008, pp. 1728–1731.
7. A. Foi, "Noise estimation and removal in MR imaging: The Variance Stabilization Approach," in Biomedical Imaging: From Nano to Macro, 2011 IEEE International Symposium on. IEEE, 2011, pp. 1809–1814.
8. Komstantionos Konstantiniders, Balas Natarajan, Gregory S. Yovanof "Noise Estimation and Filtering Using Block-Based Singular Value Decomposition" IEEE Transactions on Image Processing, Volume: 6, No: 3 Pp: 479–483, March 1997.
9. H. C. Andrews and C. L. Patterson, "Singular value decompositions and digital image processing," IEEE Trans. Acoustic., Speech and Signal Processing, vol. ASSP-24, pp. 26–53, Feb. 1976.
10. Hoppner, F., et al., Fuzzy Cluster Analysis: methods for classification, data analysis, and image recognition. 1999, New York: John Wiley & Sons, Ltd.
11. Man, Y. and Gath, I., Detection and separation of ring shaped clusters using fuzzy clustering. IEEE Transaction on Pattern Analysis and Machine Intelligence, 1994. 16(8): pp. 855–861.
12. Gath, I. and Hoory, D., Fuzzy clustering of elliptic ring-shaped clusters. *Pattern* Recognition Letters, 1995. 16(7): pp. 727–741.
13. Ameer Ali, M., Dooley, L.S., and Karmakar, G.C., Object based segmentation using fuzzy clustering, in IEEE International Conference on Acoustics, Speech, and Signal Processing. 2006.
14. Bezdek, J.C., Pattern Recognition with Fuzzy Objective Function Algorithm. 1981, New York: Plenum Press.
15. Fan, J.L., Zhen, W.Z., and Xie, W.X., Suppressed fuzzy c-means clustering algorithm. Pattern Recognition Letters, 2003. 24: pp. 1607– 1612.
16. Krishnapuram, R. and Keller, J.M., A possibilistic approach to clustering. International Journal of Fuzzy Systems, 1993. 2(2): pp. 98–110.
17. Gustafson, D.E. and Kessel, W.C., Fuzzy clustering with a fuzzy covariance matrix. In Proceedings of IEEE Conference on Decision Control. 1979. pp. 761–766.
18. N.R. Pal and J.C. Bezdek, "On cluster validity for fuzzy C-Means model", IEEE trans. Fuzzy Systems, 3(3), 1995, 370–379.
19. H.-C. Lee et al., "Digital image noise suppression method using SVD block transform," U.S. Patent 5 010 504, April. 1991.

Image Compression Using Shannon Entropy-Based Image Thresholding

Karri Chiranjeevi, Uma Ranjan Jena and Asha Harika

Abstract In this paper, we proposed multilevel image thresholding for image compression using Shannon entropy which is maximized by the nature-inspired Bacterial Foraging Optimization Algorithm (BFOA). Ordinary threading methods are computationally expensive, while extending for multilevel image thresholding, so there is a need of optimization techniques to reduce the computational time. Particle swarm optimization undergoes instability when particle velocity is maximum. So we proposed a BFOA-based multilevel image thresholding by maximizing Shannon entropy and the results are compared with differential evolution and Particle swarm optimization and proved better in Peak signal-to-noise ratio (PSNR), Compression ratio and reconstructed image quality.

Keywords Image compression · Image thresholding · Shannon entropy · Bacterial foraging optimization algorithm · Differential evolution

1 Introduction

Image compression is the procedural way of representing images, which reduces the number of bits required to represent an image and to improve the storage capacity of a storage device. So many techniques are proposed by the researchers, but the mostly used image compression technique is a joint photographic expert group (JPEG) [1]. It is first introduced Discrete Cosine Transformed (DCT)-based image

K. Chiranjeevi · U.R. Jena (✉) · A. Harika
Department of Electronics and Tele-communication Engineering, Veer Surendra Sai University of Technology (VSSUT), Burla 768018, Odisha, India
e-mail: urjena@rdiffmail.com

K. Chiranjeevi
e-mail: chiru404@gmail.com

A. Harika
e-mail: ashashis.10@gmail.com

H.S. Behera and D.P. Mohapatra (eds.), *Computational Intelligence in Data Mining*, Advances in Intelligent Systems and Computing 556,
DOI 10.1007/978-981-10-3874-7_10

compression technique and later they developed most popular, effective and efficient image compression technique, i.e. JPEG-2000 which is based on Discrete Wavelet Transform (DWT) [2]. Image compression can also be performed by non-transformed techniques such as vector quantization (VQ) and thresholding. Image thresholding is the process of extracting the objects in a scene from the background that helps for analysis and interpretation of the image. It is a challenging task for the researchers in image processing to select a grey-level threshold that extracts the object from the background of the grey-level image or colour image. Selection of threshold is moderately simple in the case where histogram of the image has a deep valley representing background and sharp edges representing objects, but due to the multimodality of the histograms of many images, selections of a threshold is a difficult task. So researchers proposed many techniques for preeminent grey-level threshold. Thresholding is mostly used due its advanced simplicity, robustness, less convergence time and accuracy. Thresholding approaches are of two types one is nonparametric and parametric. In nonparametric approach, thresholding is performed based on between class variance as in Otsu technique or based on an entropy criterion, such as Shannon entropy, Fuzzy entropy and Kapur's entropy [3]. If the image is partitioned into two classes, i.e. object and background, then the threshold is called bi-level threshold else multilevel threshold. Thresholding technique has so many real-time applications like data, image and video compression, image recognition, pattern recognition, image understanding and communication.

Sezgin and Sankur performed a comparative study on image thresholding, they classified the image thresholding into six categories, those are histogram shape-based methods, clustering-based methods, entropy-based methods, object attribute-based methods, spatial methods and local methods [4]. Kapur classifies the image into some classes by calculating threshold, which is based on the histogram of the grey-level image [5]. Otsu's method, classifies the image into some classes by calculating threshold which is based on between-class variance of the pixel intensities of that class [6]. These two methods are under the category of bi-level thresholding and found efficient in case of two thresholds, but for multilevel thresholding the computational complexity is very high. Entropy may be a Shannon, fuzzy, between class variations, Kapur's entropy, minimization of the Bayesian error and Birge–Massart thresholding strategy. The disadvantage of these techniques is that convergence time or computational time or CPU time is exponentially increasing with the problem. So alternative to these techniques which minimize the CPU time for the same problem is evolutionary and swarm-based calculation techniques.

Chen-Kuei and Wen-Hsiang applied a moment-preserving principle for efficient and effective colour image thresholding for image compression at higher compression ratio, quantization and edge detection by separating red, green and blue continents [7]. Kaur followed a specific strategy for selection of wavelet packets with low computational cost [8]. They optimize operational rate distortion (R-D), threshold and quantize with minimum description length (MDL) framework to develop a wavelet packet image coder named as JTQ-WP. Siraj used inbuilt matlab function for the compression, i.e. Birge–Massart thresholding and compared the results with unimodal thresholding [9]. The effect of different wavelet functions on measuring parameters PSNR, Weighted PSNR (WPSNR), reconstructed image quality is discussed. ECG signal compression with 2-D DWT was proposed by [10]. One can perform image compression using Bandlet transform and by identifying irregular edges of image instead of smooth regions and selection of Bandlet coefficients and this selection follows type-II fuzzy thresholding. The results are compared with the ordinary thresholding methods [11]. Prashant and Ioana proposed a non-uniform way to threshold the compressed matrix and show the effects of thresholding on reconstructed image quality [12]. To reduce encoding computational complexity of fractal image compression, Preedhi proposed an adaptive fractal image encoding in which adaptive thresholded DCT coefficents of domain blocks are compared with the range blocks [13].

In this paper, first we applied BFOA-based image thresholding for image compression by optimizing the Shannon entropy and compared the results with other optimization techniques such as DE and PSO. Compressed image is further compressed with Runlegth coding followed by Arithmetic coding. For the performance evaluation of proposed BFOA-based image thresholding, we consider objective function value, peak signal-to-noise ratio (PSNR), bits per pixel (BPP) and compression ratio (CR). In all parameters the proposed algorithm performance is better compared to other DE and PSO.

2 Problem Formulation of Optimum Thresholding Methods

Image thresholding is a process of converting a greyscale input image to a black and white image by using optimal thresholds. Thresholding may be a local or global but these methods are computationally expensive, so there is a need of optimization techniques which optimize the objective function results in the reduction of computational time of local or global methods. The optimization techniques find the optimal thresholds by maximizing the objective function such that reconstructed image clearly distinguishes the background and foreground of image. In this paper, we have chosen Shannon entropy and Fuzzy entropy as objective functions on which optimization techniques work. Let us assume an image that contains L grey levels and the range of these grey levels are {0, 1, 2, …, (L − 1)}. Then probability Pi = h(i)/N (0 < i < (L − 1)), where h(i) denotes number of pixels for the corresponding grey-level L and N denotes total number of pixels in the image which is equal to $\sum_{i=1}^{L-1} h(i)$.

2.1 Concept of Shannon Entropy

Entropy is the compressive procedure of information which results higher rate of compression and high speed of transmission. Entropy compresses the required number of bits depending on the observation of repetitive information/message. If there are $N = 2^n$ (if N = 8) messages to transmit then n (n = 3) bits are required, then for each of N messages number of bits required is $\log_2 N$ bits. If one can observe the repetition of same message from a collection of N messages, and if the messages can be assigned a non-uniform probability distribution, then it is possible to use fewer than logN bits per message. This is introduced by Claude Shannon based on the Boltzmann's H-theorem and is called Shannon entropy, let X is random variable (discrete) with elements {X1, X2 ..., Xn} then probability mass function P(X) is given as

$$H(X) = E[I(X)] = E[\ -ln(P(X))] \tag{1}$$

where E is the expected value operator and I shows the content of information and I(X) is also a random variable. Further the Shannon entropy is rewritten as in Eq. (2) and is considered as the objective function, which is to be optimized with optimization techniques

$$H(X) = \sum_{i=1}^{n} p(x_i)I(x_i) = \sum_{i=1}^{n} p(x_i)log_b p(x_i) \tag{2}$$

where b base of the algorithm in general, it is equal to 2. If P(xi) = 0 for some i then the multiplier $0\log_b 0$ is considered as zero, which is consistent with the limit

$$\lim_{p \to 0_+} p\ \log(p) = 0 \tag{3}$$

The said equations are for discrete values of X, the same equations are applicable for continuous values of X by replacing summation with integer.

As per the requirements of researchers, the two level thresholding can be extended to three or more and can be restricted to single level also. For two thresholds, the number of parameters to be optimized is six and as levels are increasing number parameters to be optimized is also increasing. Hence, two-level image thresholding for image compression with the Shannon entropy efficient and effective, but for multilevel thresholding both entropy techniques consume much convergence time and increase exponential with level of thresholds. The drawback of Shannon entropy is convergence time. To improve the performance further and to reduce the convergence time, we used applications of optimization techniques such as differential evolution, Particle swarm optimization and BFOA for image thresholding henceforth image compression. These techniques are to maximize the Shannon entropy as given in Eq. (2).

2.2 Bacterial Foraging Optimization Algorithm

BFOA algorithm is projected by Kevin m. Passion in 2002. BFOA corresponds to bacterial optimization algorithm. It yields a better performance when compared to other optimization algorithms like genetic algorithm, particle swarm algorithm, etc. BFOA is a bio-mimetic algorithm that mimics the foraging of nutrient food particle or energy nutrients by the *E. Coli* bacteria in the intestine of the stomach. *Escherichia coli* are rod-structured bacteria that are commonly found in the lower intestine of the warmed blooded organism. It tumbles/swims to search for high concentration of food. It makes use of the Gaussian distribution function to get the probability of higher nutrient region. Some of the applications of BFOA are inverse airfoil design, tuning the PID controller of an AVR, for the optimization of multi-objective functions.

The aim of the bacteria is to reach to an optimum objective value or minimum or maximum cost function, accordingly. It will result in different cost functions for each of the bacteria, iteratively. Other bacterium moves towards the direction of the optimized cost function and finally all the bacterium will attain global optimization. BFOA algorithm goes through 4 steps: Chemotaxis, Swarming, Reproduction, Elimination-dispersal,

Chemotaxis: This step illustrates the different conditions that arise when bacteria scrounging for their food. In order to get the next pixel to which the bacteria has to move, we take the derivative of the 8-neighbourhood pixel in terms of their nutrient concentration. Out of the 8 derivatives, the pixel derivative with optimum value is the preferable direction. They are as follows: Tumbling: If the bacteria are moving in a region with neutral nutrients, it will randomly move from neutral nutrient region till it get high nutrient medium. This process is known as tumbling.

Swimming up: If the bacteria are moving along the path of increasing nutrient gradient, it will continue to be in that direction, seeking increasingly favourable environments. This process is called swimming up.

Swimming down: If the bacteria are moving along the path of decreasing nutrient gradient, it will prefer to move away from that location, avoiding unfavourable environment. This process is called swimming down. The chemotactic process can be mathematically expressed by the following equation:

$$\theta^i(j+1,k,1) = \theta^i(j,k,1) + C(i)\frac{\Delta(i)}{\sqrt{\Delta^i(i)\Delta(i)}} \tag{4}$$

where C(i): step size Δ(i): is the random variable lies between [0, 1].

Swarming: The bacteria that are moving in the region of higher nutrient location will send a signal to other bacterium. A large group of bacteria would attract towards that bacteria (with high nutrient location) and move in opposite direction from the bacteria that belongs to low nutrient location. This process is termed as swarming.

$$
\begin{aligned}
J_{cc}(\theta,(p(j,k,l)=\sum_{i=1}^{s}J_{cc}(\theta,\theta^{i}(j,k,l))=\sum_{i=1}^{s}-d_{attra}\exp(-W_{attra}\sum_{m=1}^{p}(\theta_m-\theta_m^{i})^{2})] \\
+\sum_{i=1}^{s}-h_{repel}\exp(-W_{repel}\sum_{m=1}^{p}(\theta_m-\theta_m^{i})^{2})]
\end{aligned}
\quad (5)
$$

Reproduction: In this step, there will be an increase in bacterium population and survival of the fittest occurs. The cost function is calculated for each of the bacteria iteratively. If our aim is to minimize the cost function, we choose the bacteria having least cost function. If our aim is to maximize the cost function, the bacteria with more cost value is preferred. At the end of this process, the least nutrient bacteria would die whereas the bacteria with high nutrient content will survive and undergo asexual reproduction and split into two more bacterium, keeping the swarm size constant. The process of splitting of *E. Coli* bacteria is well known as conjugation.

Elimination-dispersal: Bacteria gets drifted and distributed due to abrupt change in the topical environment like there is a sudden altering of temperature in the area of interest. This process may result in 2 cases: Bacteria would die. The first case should be avoided as it changes the total number of bacteria population and is not a favourable environment for our study. Bacteria of the affected place would migrate to some other place.

3 Results and Discussions

For the performance evaluation (robustness, efficiency, and convergence) of proposed BFOA algorithm, we selected “Lena”, “Lake”, “Goldhill”, and “Pirate” as a test images and all are.jpg format image and of size 225 × 225. All the test images and corresponding histograms are shown in Fig. 1a. In general, perfect threshold can be achieved if the histogram of image peaks is tall, narrow, symmetric, and separated by deep valleys. Lake, Goldhill, and pirate image histograms peaks are tall, narrow, and symmetric, but for Lena images histogram peaks are not tall and narrow, so difficult to compress with ordinary methods. So we proposed a BFOA-based image thresholding for effective and efficient image compression of said critical images by optimizing Shannon entropy. The performance and effectiveness of proposed BFOA proved better compared to other optimization techniques like DE and PSO.

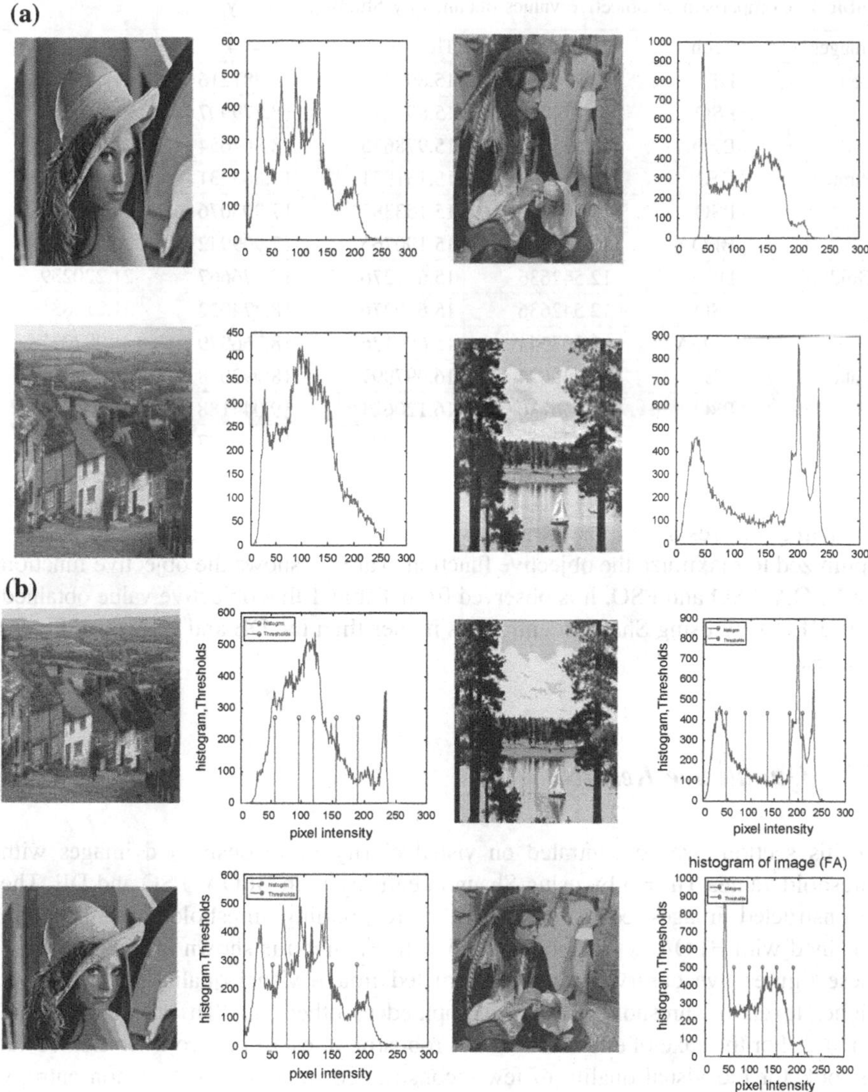

Fig. 1 **a** Standard image and respective histograms used for comparison of three methods. **a** Lena, **b** Pirate, **c** Goldhill, **d** Lake. **b** Reconstructed images and respective optimal 5 level thresholds with BFOA

3.1 Quantitative Validation

To examine the influence of the BFOA on multilevel thresholding problem, objective functions/fitness function in this case is Shannon entropy. The BFOA and other two algorithms are applied on Shannon entropy objective function and

Table 1 Comparison of objective values obtained by Shannon entropy

Images	Tech	Th = 2	Th = 3	Th = 4	Th = 5
Lena	DE	12.774765	15.898248	18.781216	21.486303
	PSO	12.774765	15.898248	18.779907	21.483824
	BFOA	12.87544	15.978675	18.867654	21.597667
Pirate	DE	12.074528	15.131921	17.863231	20.093364
	PSO	12.104785	15.153382	17.903076	20.431079
	BFOA	12.166465	15.190765	17.909942	20.530855
Goldhill	DE	12.542636	15.639276	18.476667	21.220239
	PSO	12.542636	15.639276	18.474022	21.212831
	BFOA	12.568544	15.690176	18.766979	21.987721
Lake	DE	12.849674	16.097997	18.992058	21.660711
	PSO	12.856136	16.120621	19.044188	21.736018
	BFOA	12.964136	16.252977	19.198687	21.959278

compared the results of the BFOA with DE and PSO. All the algorithms are optimized to maximize the objective function. Table 1 shows the objective function for BFOA, PSO and PSO. It is observed from Table 1 that objective value obtained with BFOA by using Shannon entropy is higher than the DE and PSO for different images.

3.2 *Qualitative Results*

In this section, we concentrated on visual clarity of reconstructed images with threshold values Th = 5 by using Shannon entropy with BFOA, PSO and DE. The reconstructed images/reconstructed and corresponding thresholds on histogram obtained with BFOA at threshold level 5 with Shannon is shown in Fig. 1b. From these figures, we observed that reconstructed image visual quality is better with higher levels of threshold (Th = 5) compared to others, i.e. Th = 4, Th = 3 and Th = 2. For the sake of effectiveness and robustness test of the proposed BFOA, let us look for the visual quality of few reconstructed images with Shannon entropy and it is proved better with BFOA than other methods.

3.3 *Peak Signal-to-Noise Ratio (PSNR) and Mean Square Error (MSE)*

The PSNR shows the dissimilarity between the thresholded image and input image and it is a measure of visual difference of two images and units are decibels (dB).

Table 2 Comparison of PSNR values of three methods

Images	Tech	Th = 2	Th = 3	Th = 4	Th = 5
Lena	DE	28.44778	29.69486	29.94378	31.27988
	PSO	28.50969	29.76831	29.96113	31.42467
	BFOA	28.83433	29.92333	30.12178	31.33333
Goldhill	DE	29.10431	29.39062	30.26645	30.57502
	PSO	29.19538	29.46275	30.53518	30.68128
	BFOA	29.36568	29.51234	30.51234	31.74021
Lake	DE	29.11452	29.48826	29.81173	30.60652
	PSO	29.16853	29.59081	29.91127	31.46792
	BFOA	29.23456	29.71111	30.21232	31.54246
Pirate	DE	28.77255	29.66297	29.90553	30.98535
	PSO	28.79772	29.80472	30.35291	31.53031
	BFOA	29.31222	29.80999	30.91234	32.25837

A higher value of PSNR indicates better quality of reconstructed image. The equation for PSNR is given in Eq. (6).

$$PSNR = 10 \times 10 \log\left(\frac{255^2}{MSE}\right)(dB) \tag{6}$$

where (MSE) which is given in Eq. (7)

$$MSE = \frac{1}{M \times N}\sum_{I}^{M}\sum_{J}^{N}\{f(I,J)\ - \bar{f}(I,J)\}^2 \tag{7}$$

where M × N is the size of the image, I and J represent the pixel value of original and reconstructed images. In our experiment, we have taken N = M a square image. Table 2 show the PSNR value acquired by the different algorithms. From Table 2, the proposed algorithm achieved higher PSNR value when compared with DE and PSO. The BFOA algorithm provides the utmost value of PSNR value with Th = 5 when compared to DE and PSO. Hence, with the higher level of thresholds, the excellence of the reconstructed images gets better.

4 Conclusions

In this paper, we proposed nature-inspired BFOA-based multilevel image thresholding for image compression. BFOA maximizes the Shannon entropy for efficient and effective image thresholding. The proposed algorithm is tested on natural images to show the merits of the algorithm. The results of the proposed method are compared with other optimization techniques such as DE and PSO with Shannon entropy. From the experiments, we observed that proposed algorithm has higher/maximum fitness value compared to DE and PSO. The PSNR value shows higher values with proposed algorithm than DE and PSO and it shows better quality of the reconstructed image with proposed method.

References

1. Rabbani. M, P.W. Jones, Digital Image Compression Techniques, vol. 7, SPIE Press, Bellingham, Washington, USA, 1991.
2. Skodras. A, C. Christopoulos; T. Ebrahimi, "The JPEG 2000 still image compression standard", IEEE Signal Processing Magazine, Vol. 18, Issue. 5, pp. 36–58, 2002.
3. Luca. A, S. Termini, A definition of a non-probabilistic entropy in the setting of fuzzy sets theory, Inf. Control 20 (1972) 301–312.
4. Sezgin. M, B. Sankur, Survey over image thresholding techniques and quantitative performance evaluation, J. Electron. Imaging 13 (1) (2004) 146–165.
5. Kapur. J. N, P.K.Sahoo, A.K.C Wong, A new method for gray-level picture thresholding using the entropy of the histogram", Computer Vision Graphics Image Process. 29 (1985) 273–285.
6. Otsu. N, "A threshold selection from gray level histograms" IEEE Transactions on System, Man and Cybernetics 66, 1979.
7. Chen-Kuei. Y and Wen-Hsiang. T, "Color image compression using quantization, thresholding, and edge detection techniques all based on the moment-preserving principle", Pattern Recognition Letters 19 Ž1998. 205–215.
8. Kaur. L, S. Gupta, R.C. Chauhan, S.C. Saxenac, "Medical ultrasound image compression using joint optimization of thresholding quantization and best-basis selection of wavelet packets", Digital Signal Processing 17 (2007) 189–198.
9. Siraj. S, "Comparative study of Birge–Massart strategy and unimodal thresholding for image compression using wavelet transform" Optik 126 (2015) 5952–5955.
10. Tahere. I. M. and Mohammad. R. K. M, "ECG Compression with Thresholding of 2-D Wavelet Transform Coefficients and Run Length Coding", European Journal of Scientific Research ISSN 1450-216X Vol. 27 No. 2 (2009), pp. 248–257.
11. Rajeswari. R, "Type-2 Fuzzy Thresholded Bandlet Transform for Image Compression", Procedia Engineering 38 (2012) 385–390.
12. Prashant. S and Ioana. M, "Selective Thresholding in Wavelet Image Compression", Wavelets and Signal Processing Part of the series Applied and Numerical Harmonic Analysis pp. 377–381, 2003.
13. Preedhi Garg, Richa Gupta, Rajesh K. Tyagi, "Adaptive Fractal Image Compression Based on Adaptive Thresholding in DCT Domain", Information Systems Design and Intelligent Applications, Vol. 433, pp 31–40, 2016.

Solving Sparsity Problem in Rating-Based Movie Recommendation System

Nitin Mishra, Saumya Chaturvedi, Vimal Mishra, Rahul Srivastava and Pratibha Bargah

Abstract Recommendation is a very important part of our digital lives. Without recommendation one can get lost in web of data. Movies are also very important form of entertainment. We watch most movies that are recommended by someone or others. Each person likes specific type of movies. So movie recommendation system can increase sales of a movie rent/sales shop. Websites like Netflix are using it. But there is one problem that can cause recommendation system to fail. This problem is sparsity problem. In this paper, we have used a new approach that can solve sparsity problem to a great extent.

Keywords K-mean clustering · Euclidean distance · K-medoid clustering

1 Introduction

Recommendation systems are used for many purposes. They filter the information and give users what they really wanted. In movie recommendation system, we try to recommend movies based on user interests. There are three points to focus for movie recommendation system:

N. Mishra (✉) · S. Chaturvedi · P. Bargah
RCET Bhilai, Bhilai, India
e-mail: drnitinmishra10@gmail.com

S. Chaturvedi
e-mail: saumyanmishra5@gmail.com

V. Mishra · P. Bargah
IERT Allahabad, Allahabd, India
e-mail: vimal.mishra.upte@gmail.com

R. Srivastava · P. Bargah
JLU Bhopal, Bhopal, India
e-mail: rahul04.shri@gmail.com

H.S. Behera and D.P. Mohapatra (eds.), *Computational Intelligence in Data Mining*, Advances in Intelligent Systems and Computing 556,
DOI 10.1007/978-981-10-3874-7_11

Why: Movie recommendation system is required because movie information are overloaded.

Where: Can be used in websites like bookmyshow.com or Netflix.

What: It tells you what you should watch based on your history and other users' history.

2 Basic Preliminaries

Some methods given below are used to predict the items for users. Although there is large list of methods but we are discussing some methods which are of prime importance to movie recommendation.

2.1 Fuzzy C-Mean Clustering

Fuzzy clustering is one of the types of clustering in which every data point can belong to more than one cluster [1, 2]. Clustering or cluster analysis involves transmission data points to clusters (also called buckets, bins, or classes), or homogeneous classes, such that items in the same class or cluster are as similar as possible, while items belonging to different classes are as dissimilar as possible. Clusters are identified via similarity measures.

Advantages

(1) gives most excellent effect for overlap data set.

(2) not like k-means anywhere data point must completely be in the right place to one cluster hub and here data point is assigned association to each cluster middle as a consequence of which facts end may be in the right place to supplementary subsequently single come together center.

Disadvantages

(1) Apriori requirement of the number of clusters.

(2) With lesser value of β we obtain the improved end result but at the expenditure of more numeral of iteration.

2.2 Gath–Geva Clustering

The Gath–Geva algorithm is an addition of Gustafson–Kessel algorithm that take the volume and density of the cluster into report [3].

- The distance function is preferred ultimately comparative to the (unnormalized) a posteriori opportunity, because a small distance results a high probability and a big distance results a low probability of association [4, 5].
- In contrast to FCM algorithm and the Gustafson–Kessel algorithm, the Gath–Geva algorithm is not based on an objective function, but is a fuzzification of statistical estimators [6].

3 Problem Identification

Online shops today contain lot of items and users. In order to relate users to items they need association of user interest on particular items. But due to time or other constraints, it is generally not possible to have enough ratings on particular items by users. This situation gives rise to problem called sparsity. In real world sparsity is very common. We often do not have enough ratings to make our highly efficient recommendation algorithm to work. It is common in e-business supplies that even the majority of active customers had purchased or rate extremely incomplete proportion of products, when compared to the obtainable total. As a result, techniques to reduce the sparsity in user-item matrices should be proposed. The main reason for sparsity problem are as follows:

- The amount of items that contain ratings by the users would be too small. This can make our recommendation algorithms fail.
- Similarly, the number of users who rate one exact item might be too small compared to the total no. of users connected in the system. These situations provide rise to sparse ranking matrix.

4 Methodology

Figure 1 shows the process of our methodology to deal with the sparsity problem.

1. We will collect information from IMDb (Internet Movie Database), All the required information about the movie will be available.

   The sample data is available of the website.

2. Generate the review and rating matrix and apply k-mean clustering in both matrices. It will make cluster of similar object but it has no predefined classes. And classification of reviews is based on good, bad, and average comments of movies we have taken the 29 * 100 matrix for Rating and 29 * 3 are matrix for review.
3. Both data of matrix is converted into relational data using Euclidean distance. After that we can apply k-mediod clustering.

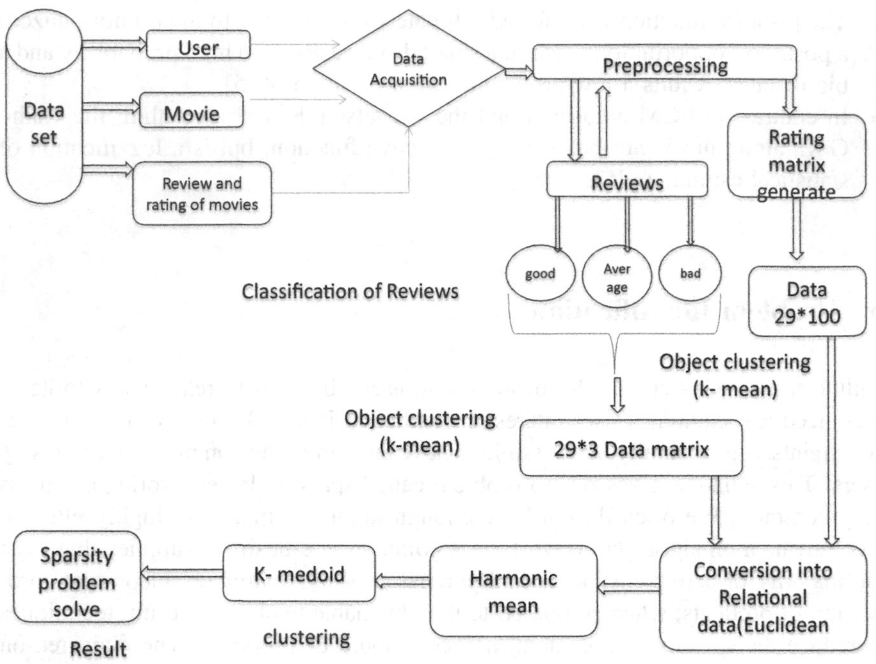

Fig. 1 The process for solving sparsity problem in rating-based movie recommendation system

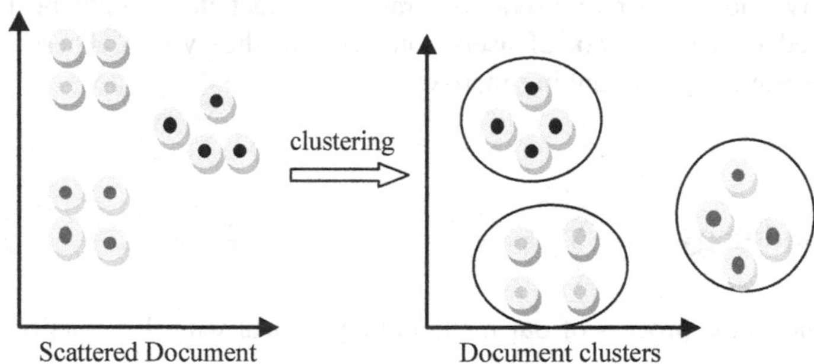

Fig. 2 K-mean clustering

4.1 K-Mean Clustering

The k-means algorithm involves assigning each of the n examples to one of the k clusters, where k is a number that has been defined ahead of time. The goal is to minimize the differences within each cluster and maximize the differences between clusters. These k new centroids, a new binding has to be completed between the similar data set point and the adjacent novel center [7] (Fig. 2).

4.2 *Euclidean Distance*

In terms of mathematics, the Euclidean distance is the distance between two points in Euclidean space. With this distance, Euclidean space makes a metric space [8]. The associated norm is called the Euclidean norm. In our project first we loaded the rate matrix then rating matrix and applying Euclidean distance in both matrices.

4.3 *Harmonic Mean*

For two numbers x_1 and x_2, the harmonic mean can be written as

$$\frac{H = 2x_1x_2}{X_1 + X_2}$$

k-medoid is based on centroids (or medoids) calculating by minimizing the absolute distance between the points and the selected centroid, rather than minimizing the square distance. As a result, it is more robust to noise and outliers than k-means.

4.4 *K-Medoid Clustering*

k-medoid is a classical partitioning technique of clustering that clusters the data set of n objects into k clusters known a priori. It is more robust to noise and outliers as compared to k-means, because it minimizes a sum of pairwise dissimilarities instead of a sum of squared Euclidean distances [9].

A medoid can be defined as the object of a cluster whose average dissimilarity to all the objects in the cluster is minimal. It is the most centrally located point in the cluster [9].

5 Experiments and Analysis

Experiment Data:

We have implemented the solution in MATLAB. Some important findings were very positive. The method chosen and discussed above seems to solve the sparsity problem in this domain.

The dataset was collected by the IMDb. The dataset consist of 29 movies and 100 users and each user rating at least 7 movies. The sparsity degree is 99.23%.

The experiments are performed as follows:

- From the data, get original rating of movie results.
- Input original rating of movie result to get the accurate rating of movie.
- Change ratings according to our algorithm.
- Compare the Rating of movie results with our algorithm.

Data Set

Original Movie Data Set

In this project we have divided data into three clustering of rating the range of 0–33%, 34–65%, and 66–100%. In this group, we can easily classify that how no. of users rating the same data in the actual data after observation the data gave clustering.

Object Data Set

Object data is classified by the k-mean clustering; k-means is used for solving clustering problem. The process follows a straightforward and simple method to categorize a certain data set from side to side a definite amount of clusters (assume k clusters) is predetermined a priori. The main idea is to classify k centers, one for each cluster. In our case taken k = 3.

Relational Data Set

Relational data are given by k-medoid clustering. *K-means* and *k*-medoids algorithms are splits into some parts (breaking the data set up into groups) and both challenge to decrease the distance between points labeled to the center of the cluster (Fig. 3).

In the above figure, graph shows the result of sparsity problem solved, in this graph x label shows the three no. of clustering and y label shows the cluster size. In the first three bars show the actual data, in second cluster shows the object data and third cluster shows the Relational data, the cluster of actual data is and relational data are. The comparison between both clustering first cluster are same, second cluster has two different and third cluster has two different. Total different is 4.

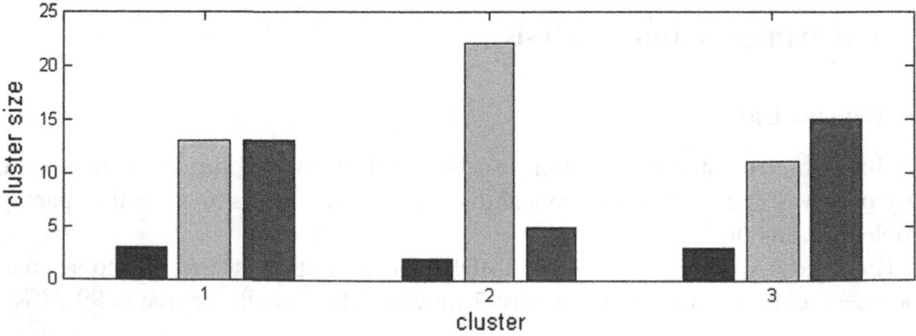

Fig. 3 Graph of result of sparsity problem solved (Generated in MATLAB)

Our dataset contains 29 movies. The actual data and relational data are adjacent value. Hence the k-medoid clustering is the better than the k-mean clustering and 70.22% sparsity problem is solving.

6 Conclusion

In this paper, we have solved sparsity problem in specific dataset of movies. Although we have not tested but this method can be applied on other similar domains like song and video recommendation systems. Our method is very helpful for people developing applications in movie recommendation domain. It will help their recommendation algorithm perform even when there is sparsity of data.

References

1. Mohammed mahmuda rahumen rahumen lecture, et al contextual recommendation system using multidimensional approch. International journal of intelligent information system august 20, 2013.
2. Badrul sarwar, joseph konstan john riedl Using filtering agent to improve prediction quality in the grouplen research collaborative filtering Department of computer science and engineering, University of minnesota in year 2008.
3. Zuping liu sichuon et al. Recommendation algorithm based on user interest, advanced science and technology letters vol. 53, 2014.
4. Beau piccart, jan struf Alleviating the sparsity problem in collaborative filtering by using an adapted distance and a graph based method. IEEE computer technology Year: 2007.
5. Badrul sarwar, george karypis, joseph konstan Item based collaborative filtering recommendation algorithm. Department of computer science and engineering, University of Minnesota Year: 2006.
6. Manos papagelis, dimitris plexousakis Alleviating the sparsity problems of collaborative filtering using trust inferences Institutes of computer science, foundation for research and technology- hellas Years: 2004.
7. Sanghack Lee and Jihoon Yang and Sung-Yong Park, Discovery of Hidden Similarity on Collaborative Filtering to Overcome Sparsity Problem, Discovery Science, 2007.
8. Zan Huang, Hsinchun Chen, et al. Applying Associative Retrieval Techniques to Alleviate the Sparsity Problem in Collaborative Filtering. ACM Transactions on Information Systems, Vol. 22, No. 1, January 2004, 116–142. http://dx.doi.org/10.1145/963770.963775.
9. Cheng-kang-Hsieh, et al. Immersive Recommendation: News and Event Recommendations Using Personal Digital Traces (2016) published in Proceedings of the 25th International Conference on World Wide Web on April 11–15, 2016 at Montréal, Québec, Canada.

Combining Apriori Approach with Support-Based Count Technique to Cluster the Web Documents

Rajendra Kumar Roul and Sanjay Kumar Sahay

Abstract The dynamic Web where thousands of pages are updated in every second is growing at lightning speed. Hence, getting required Web documents in a fraction of time is becoming a challenging task for the present search engine. Clustering, which is an important technique of data mining can shed light on this problem. Association technique of data mining plays a vital role in clustering the Web documents. This paper is an effort in that direction where the following techniques have been proposed:

(1) a new feature selection technique named *term-term correlation* has been introduced which reduces the size of the corpus by eliminating noise and redundant features.
(2) a novel technique named *Support Based Count (SBC)* has been proposed which combines with traditional Apriori approach for clustering the Web documents.

Empirical results on two benchmark datasets show that the proposed approach is more promising compared to the traditional clustering approaches.

Keywords Apriori · Cluster · Fuzzy · K-means · Support based count

1 Introduction

World Wide Web (WWW) is the most important place for Information Retrieval (IR). Tremendous exponentiation growth of WWW makes the end user difficult to find the desired results the search engine. Since the inception of WWW, the amount of data on the Web has expanded many manifolds and their size is doubling in every 6–10 months. Hundreds of millions of users each day submit queries to the Web search engines. According to Spin et al. [1], queries of length one (monogram) are

R.K. Roul (✉) · S.K. Sahay
BITS-Pilani, K.K. Birla Goa Campus, Sancoale, India
e-mail: rkroul@goa.bits-pilani.ac.in

S.K. Sahay
e-mail: ssahay@goa.bits-pilani.ac.in

H.S. Behera and D.P. Mohapatra (eds.), *Computational Intelligence in Data Mining*, Advances in Intelligent Systems and Computing 556,
DOI 10.1007/978-981-10-3874-7_12

submitted by 48.4% of the total users, queries having length two (bigrams) are 20.8% and queries of length three or more are entered by only 31% of users. The authors also mentioned that 50% of the total Internet users never see beyond first two pages of the returned results, only the first page is seen by 65–70% users, second page by 20–25% users, the remaining results are seen by very few of 3–4% users. A similar kind of survey had been done by W.B. Croft [2]. The main challenge for a search engine is that how it satisfies the user request in an efficient manner. Clustering is one of the powerful data mining technique which can help in this direction by grouping the similar documents into one place and thus attract many research ideas [3–9]. In general, the returned search results of any search engine are not clustered automatically. To get the desired and efficient information, the search results need to be grouped into different clusters.

Association-based techniques such as apriori algorithm [10] can be used for Web documents clustering. Similarity between the documents is decided based on the association of the keywords between them, which depends on a parameter called *minimum support*. In this paper, a technique called *SBC* is being proposed to represent the documents in a concise and relevant manner. First, the traditional apriori approach is applied on documents of a given corpus and it generates the initial clusters based on the association technique of data mining. In association based clustering, to obtain better results, sometimes the minimum support is set high, and consequently lots of collected documents fail to be the member of any cluster. To strengthen the clusters further, the proposed approach is applied on those discarded documents to test whether some of those documents can be useful to form clusters and it is done in three phases as mentioned below

Phase 1: Removing the noise and redundant features from the corpus.

Phase 2: Formation of *New Enhanced Initial Clusters (NEIC)* from the initial clusters (clusters generated by applying apriori approach on the corpus of documents). This is done by testing the discarded documents (documents are discarded because of high minimum support) that are eligible to merge with the initial clusters to generate the *NEIC*.

Phase 3: Formation of final clusters by applying *SBC* approach on those remaining unclustered documents for testing which of those documents are eligible to merge with *NEIC* for generating the final clusters.

The details of the above three phases are discussed using the following steps and shown in the Fig. 1.

Step 1: Initially, the complete corpus is preprocessed and then the document-term matrix is created using the standard vector space model.

Step 2: Next, all the noise and redundant features are removed from each document of the corpus using *term–term correlation* based feature selection technique in order to make the clustering process more efficient.

Step 3: Then the traditional apriori algorithm is applied on the preprocessed Web documents of a given corpus, which gives us the initial clusters generated from *n*th candidate itemset (here the documents are considered as itemset), depending on the minimum support supplied by the user.

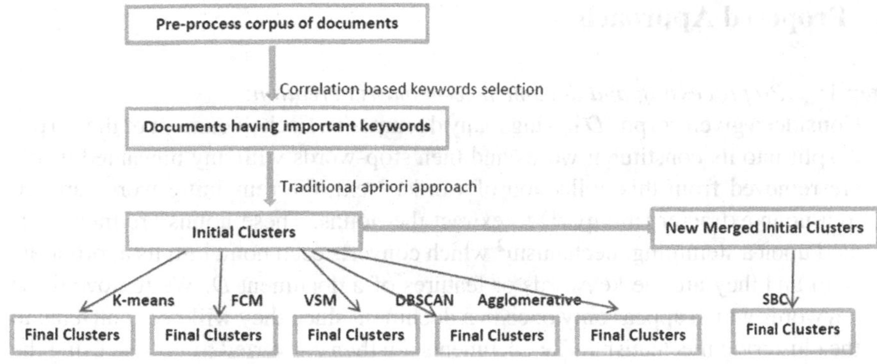

Fig. 1 System architecture

Step 4: This step starts backward scanning of the candidate itemset table (*CIT*) (generated after applying the apriori approach on the corpus) starting from $(n-1)$th candidate itemset to candidate itemset-2 for considering those itemset which can qualify to generate *NEIC* and it can be stated as follows:

(i) Each *CIT* can be scanned sequentially and for each itemset for that *CIT*, do the following:
Let C_i is a itemset of *i*th *CIT*, where $2 \leq i \leq (n-1)$ and let D_j is the document set of *j*th initial cluster. Then C_i can qualify for new clusters, if support $C_i \geq$ minimum support and $C_i \cap D_j = \emptyset, \forall j$.

(ii) If C_i qualifies, then merge C_i into the initial cluster before checking for the next itemset in that particular *CIT*.

Step 5: In the final step, using *SBC* technique, we classify the unclustered documents (D_k) into one of the identified cluster (C_j) by considering the following two points:

(i) Support$(D_k) \geq$ minimum support and
(ii) Sum of the support value of D_k with each document of C_j is greater than the number of documents in C_j.

Further on initial clusters, we applied *K-Means* [11], *FCM* [11], *VSM* [12], *DBSCAN* [11] and *Agglomerative* [11] clustering techniques individually to obtained the final clusters. DMOZ and 20-Newsgroups datasets were used for experimental work. We measured the performance of final clusters made by each technique in terms of F-measure and accuracy. The results show that the proposed approach can outperform other standard clustering techniques.

The rest of the paper is as follows: Sect. 2, discusses the proposed approach. Experimental results are covered in Sect. 3. Conclusion and future work are discussed in Sect. 4.

2 Proposed Approach

Step 1: *Preprocessing and document-term matrix creation*:
Consider a given corpus D having many documents. Each document of the corpus is split into its constituent words and then stop-words with any unwanted words are removed from this collection of words. Next, the remaining words are fed to a noun extractor (minipar[1]) to extract the nouns. These nouns are then operated upon a stemming mechanism[2] which converts each noun into its appropriate stem and they are the keywords or features of a document D. We remove those keywords which appear only once in a document since they will not contribute to the clustering mechanism. The documents are then converted to vectors using the standard Vector Space Model and these vectors are then aggregated to form the document-term matrix. In all the documents, we need to keep only the important features which best describe the documents and ignore the unnecessary features. The following steps discuss the process of selecting important features in each document.

Step 2: *Correlation-based keyword selection*:

(i) Frequency-based correlation factor calculation: Assuming there are 'n' number of keywords in D, the frequency-based correlation factor between keyword i and keyword j is computed using the following equation:

$$C_{ij} = \sum_{k \in D} f_{ik} * f_{jk} \tag{1}$$

where, f_{ik} and f_{jk} are the frequencies of ith and jth keyword in kth document, respectively.

(ii) Association matrix construction: After finding the frequency-based correlation factor between every pair of keywords in the corpus, next an association matrix is constructed where each row is the correlation values (or association values) between the keyword W_i and W_j, which generates a semantic component of W_i. $C_{ij} = C_{ji}$ and each $\overrightarrow{W_i}$ represents a keyword vector.

$$C_{ij} = \begin{array}{c|ccccc} & W_1 & W_2 & W_3 & \ldots & W_n \\ \hline W_1 & C_{11} & C_{12} & C_{13} & \cdots & C_{1n} \\ W_2 & C_{21} & C_{22} & C_{23} & \cdots & C_{2n} \\ W_3 & C_{31} & C_{32} & C_{33} & \cdots & C_{3n} \\ . & . & . & . & \cdots & . \\ . & . & . & . & \cdots & . \\ . & . & . & . & \cdots & . \\ W_n & C_{n1} & C_{n2} & C_{n3} & \cdots & C_{nn} \end{array}$$

[1] http://ai.stanford.edu/~rion/parsing/minipar_viz.html.

[2] http://tartarus.org/martin/PorterStemmer/.

(iii) Normalizing the association scores of C_{ij}: Next, the association score of C_{ij} is normalized in order to keep all correlation values between 0 and 1 and it can be done using the following equation:

$$S_{ij} = \frac{C_{ij}}{C_{ii} + C_{jj} - C_{ij}} \tag{2}$$

$$S_{ij} = \begin{array}{c|ccccc} & W_1 & W_2 & W_3 & \dots & W_n \\ \hline W_1 & S_{11} & S_{12} & S_{13} & \dots & S_{1n} \\ W_2 & S_{21} & S_{22} & S_{23} & \dots & S_{2n} \\ W_3 & S_{31} & S_{32} & S_{33} & \dots & S_{3n} \\ . & . & . & . & \dots & . \\ . & . & . & . & \dots & . \\ . & . & . & . & \dots & . \\ W_n & S_{n1} & S_{n2} & S_{n3} & \dots & S_{nn} \end{array}$$

Normalized score is 1, if two keywords have the same frequency in all the documents.

(iv) Finding the Semantic centroid: For each row of S_{ij} (i.e., for each keyword W_i), the mean of all correlation values that belongs to S_{ij} are calculated which generates the semantic centroid vector $\vec{sc}$ of n-dimensions (i.e., $\vec{sc} = sc_1, sc_2, \dots, sc_n$) where each sc_i can be calculated using the following equation:

$$sc_i = \frac{\sum_{j=1}^{n} S_{ij}}{n} \tag{3}$$

(v) Selection of the important keywords: Next, cosine-similarity between each keyword $\vec{W_i}$ and the $\vec{sc}$ is computed using the following equation:

$$cosine(\vec{W_i}, \vec{sc}) = \frac{\vec{W_i} \cdot \vec{sc}}{|\vec{W_i}| * |\vec{sc}|} \tag{4}$$

This way every keyword of the D obtains a cosine-similarity score. Finally, top 'm%' keywords are selected as the important keywords of D based on their cosine-similarity[3] scores (i.e., select those keywords which are tightly bound to the semantic centroid). This way, a number of irrelevant words are reduced from the corpus and now every document in D contains only the important keywords.

Step 3: Initial cluster formation: By applying traditional Apriori algorithm on D, initial clusters (C_i) are formed (Algorithm 1). Frequency greater than the minimum support generates the frequent sets of documents, where itemsets are rep-

[3] https://radimrehurek.com/gensim/tutorial.html.

resented by the documents and the transactions are represented by the keywords. Thus, in generated frequent itemsets, the particular set of keywords are common and hence they are closely related. This helps to decide the number of clusters. The centers of these clusters are found by computing the centroid of the respective frequent itemsets.

Step 4: New Enhanced Initial Cluster (*NEIC*) formation: New clusters are formed (Sect. 1 (step 3)) and added to the initial clusters (C_i), generates *NEIC* (Algorithm 2).

Step 5: Final cluster formation using *SBC* Final clusters are formed (Sect. 1 (step 5)) from *NEIC* using *SBC* technique (Algorithm 3).

Algorithm 1 Traditional Apriori Approach

```
Input: Corpus (D) and the minimum support (min_sup)
Output: The initial clusters (or frequent itemset)
//documents are considered as itemsets and keywords are transactions
s ← 1
Find frequent itemset, L_s from C_s, the set of all candidate itemsets //frequent itemset generation step
Form C_{s+1} from L_s //candidate itemset generation step
s ← s + 1
Repeat Step 4 - 6 until C_s is empty
//Details of frequent item and candidate itemset steps are described below
// generation of frequent itemset:
count each itemset afetr scanning D in C_s and if the count is greater than min_sup, then add that itemset to L_s
// generation of candidate itemset:
if s = 1 then
    C_1 ← all itemsets having length of 1
end if
if s > 1 then
    //generate C_s from L_{s-1} as follows:
    //The join step:
    C_s = s-2 way join of L_{s-1} with itself
    if (a_1, .., a_{s-2}, a_{s-1}) & (a_1, .., a_{s-2}, a_s) ∈ L_{s-1} then
        add (a_1, ..., a_{s-2}, a_{s-1}, a_s) to C_s
        The items are always stored in the sorted order
    end if
    //The prune step:
    Remove (a_1, ..., a_{s-2}, a_{s-1}, a_s), if it contains a non-frequent (s-1) subset
end if
```

3 Experimental Results

20-Newsgroups[4] and DMOZ open directory project[5] datasets are used for experimental purpose. 20-Newsgroups is a very popular machine-learning dataset generally used for text classification. It has 18,846 documents out of which approximately

[4] http://qwone.com/~jason/20Newsgroups/.

[5] http://www.dmoz.org.

Algorithm 2 New Enhanced Initial Cluster

```
Input:
cluster_array containing clusters < C_1, C_2, .., C_n > obtained from Algorithm 1
support_count_table (k) where each entry is of format < Document Set; Support > and the size of document set is k. support(set_i) denotes the support of the set_i in support_count_table(k), where set_i denotes the i^th set among all document set in the support_count_table(k)
minimum_dimension denotes the minimum size of document set that can be added to the cluster
min_sup ← minimum support count
Output: Updated cluster_array (i.e. NEIC)
size ← ∞
for all cluster C_i do
  size ← min(size, no_of_documents(C_i))
end for
size ← size − 1
{S} ← all documents
{C} ← all clustered documents
{I} ← all unclustered documents i.e., {I} ← {S} − {C}
while size ≥ minimum_dimension do
  for all set_i ∈ support_count_table(size) do
    flag ← 0
    if support(set_i ≥ min_sup) then
      for all Document D_i ∈ set_i do
        if D_i ∈ C then
          flag ← 1 break
        end if
      end for
      if flag = 1 then
        continue
      end if
      // Create a new cluster from the current document set
      {C_n+1} ← {set_i}
      // Add all documents of the set to {C}
      {C} ← {C} ∪ {set_i}
      // Remove all documents of the set from {I}
      {I} ← {I} − {set_i}
      // Add the new cluster to the cluster array
      {cluster_array} ← {cluster_array} ∪ {C_n+1}
    end if
  end for
  size ← size − 1
end while
```

7,500 are used for clustering purpose. Similarly for DMOZ, we used 31,000 Web pages for clustering purpose. The average performances of both datasets are calculated using the following equations:

$$average_{precision} = \sum_{i=1}^{n} \frac{(p_i \cdot d_i)}{total\ number\ of\ test\ documents} \tag{5}$$

$$average_{recall} = \sum_{i=1}^{n} \frac{(r_i \cdot d_i)}{total\ number\ of\ test\ documents} \tag{6}$$

$$average_{F-measure} = \sum_{i=1}^{n} \frac{(f_i \cdot d_i)}{total\ number\ of\ test\ documents} \tag{7}$$

Algorithm 3 Final clusters using *SBC* technique

Input:
NEIC formed in Algorithm 2
$\{I\} \leftarrow$ unclustered documents
2-Dimensional array a[][] with all the entries initialized to -1
min_sup $\leftarrow$ minimum support
Output: Final clusters
for all Document $D_i \in \{I\}$ **do**
  if $support(D_i \geq min\_sup)$ **then**
    for all cluster $C_j \in$ cluster_array **do**
      $a[i][j] \leftarrow sum(D_i, C_j)$
    end for
    $position \leftarrow max(a[i])$
    if $position \geq 0$ **then**
      $C_{position} \leftarrow C_{position} \cup D_i$
    end if
  end if
end for
function sum(D_i, C_j)
$S \leftarrow -1$
for all Documents $D_p \in$ cluster C_j **do**
  $S \leftarrow S + support(D_i, D_p)$
end for
if $S > no\_of\_documents(C_j)$ **then**
  return S
end if
function max(a[j])
$position \leftarrow 0$
for all cluster $C_j \in$ cluster_array **do**
  if $a[j-1] < a[j]$ **then**
    $position \leftarrow j$
  end if
end for
if $a[position] \neq -1$ **then**
  return position
end if

where n represents the number of categories, f_i, p_i, r_i, d_i are the F-measure, precision, recall and the total number of test documents present in ith category, respectively.

3.1 Evaluating Performance of the Proposed Clustering Approach

The combined apriori and SBC-based clustering approach is tested on each category of 20-Newsgroups and DMOZ datasets to make different clusters. Here, each cluster represents a sub-category of its super category. Next, accuracy, precision, recall, and F-measure for each category are computed. The experimental work and performance measurement of clustering approach is done as follows:
Initially, two separate corpus are formed, one for each dataset by collecting approximately 7500 documents from all the categories of 20-Newsgroups and 31,000 documents from all the categories of DMOZ dataset. On each corpus, the proposed

combined approach along with other state-of-the-art clustering algorithms like *K-Means*, *FCM*, *VSM*, *DBSCAN*, and *Agglomerative* are applied separately to form different clusters. The proposed approach generates 10 clusters for 20-Newsgroups and 14 clusters for DMOZ dataset depending on the value of minimum support. From 10 clusters of 20-Newsgroups, we then checked a cluster that belongs to its corresponding category out of seven categories of 20-Newsgroups by finding out the cosine-similarity between each document of that cluster with all the documents of each category. A cluster C_i belongs to a category CG_j iff most of the documents of CG_j fall into C_i. This process is repeated for all the clusters of 20-Newsgroups. Finally, all 10 clusters are distributed among seven categories of 20-Newsgroups based on their cosine-similarities score so that some of the categories receive more than one cluster. The same process is also repeated for DMOZ dataset. Algorithm 4 illustrates the complete mechanism to assign different clusters to their respective categories. After assigning the respective clusters to their corresponding category, if any category receives more than one cluster, we merge them into a single cluster so that each category has only one cluster ($Cluster_{new}$). This gives number of categories equal to the number of new clusters, i.e., $Cluster_{new}$ which simplifies the performance measurement process. Next, we measure the performance of each $Cluster_{new}$. For this, the precision and recall are calculated as follows:

$$precision = \frac{a}{b}, \quad recall = \frac{a}{d} \tag{8}$$

where, 'a' is the number of documents of a particular category found in its corresponding $Cluster_{new}$, 'b' is the number of documents in $Cluster_{new}$ and 'd' is the number of documents in that category. For average accuracy calculation, we check the total number of documents of all $Cluster_{new}$ which are correctly assigned to their corresponding category (x) divided by the total number documents of the dataset considered for cluster evaluation (y) and can be represented by the following equation:

$$average_{accuracy} = \frac{x}{y} \tag{9}$$

Figures 2, 3, 4, 5, 6 and 7 show the performance evaluation of different clustering techniques on DMOZ and 20-Newsgroups, respectively. Tables 1 and 2 show the performance comparisons and Tables 3 and 4 show the category wise performance of our approach on DMOZ and 20-Newsgroups, respectively.

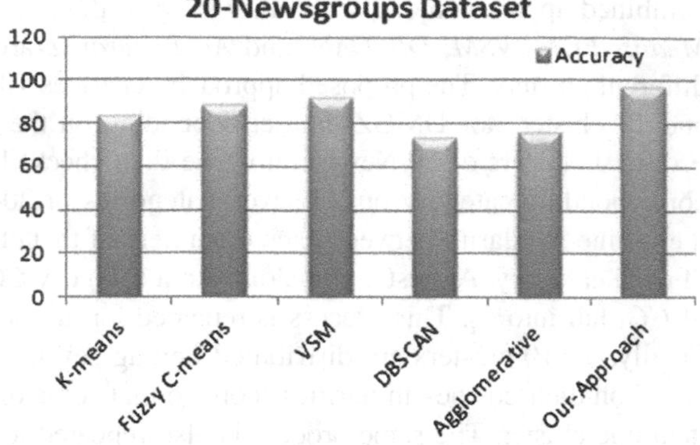

Fig. 2 Accuracy comparison (20-Newsgroups)

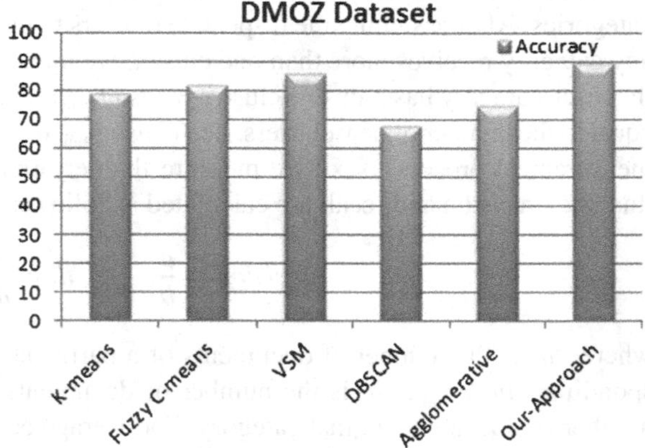

Fig. 3 Accuracy comparison (DMOZ)

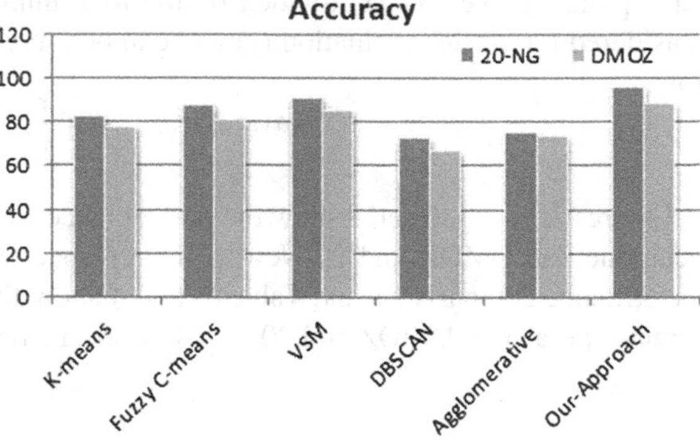

Fig. 4 Accuracy on both datasets

Algorithm 4 Assigning the respective cluster(s) to the corresponding category

```
Input: k clusters ∈ finalClust generated by Algorithm 3 and m categories
Output: Categories with their respective clusters
for i in 1 to k do
  // k clusters
  for j in 1 to m do
    // m categories
    count ← 0
    for all d ∈ i do
      for all d' ∈ j do
        cs ← cosine⅄similarity(d, d')
        if cs = 1 then
          count ← count + 1
        end if
      end for
    end for
    a[cluster_no][category_no] ← count
  end for
  maximum ← a[i][1] //testing the category received the maximum number of documents of i^th cluster
  for l in 2 to m do
    if a[i][l] > maximum then
      maximum ← a[i][l]
      n ← l // assign the category received the maximum number of documents of i^th cluster
    end if
  end for
  c[n] ← i // assign the i^th cluster to l^th category because l^th category received the maximum number of documents
     of i^th cluster
end for
```

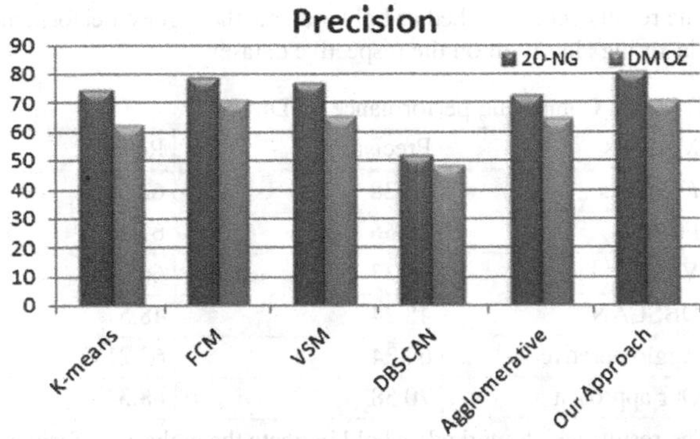

Fig. 5 Precision on both datasets

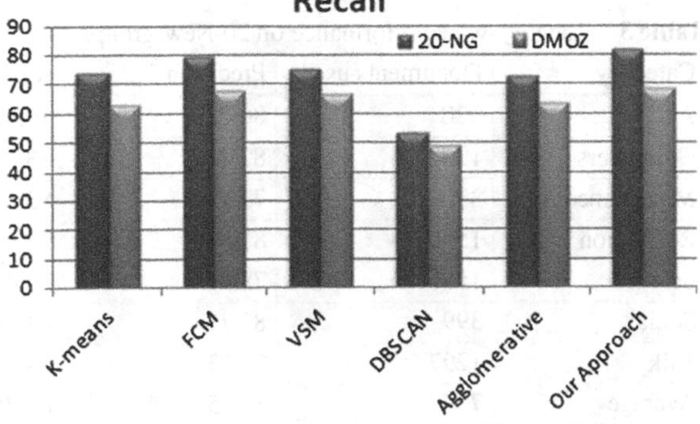

Fig. 6 Recall on both datasets

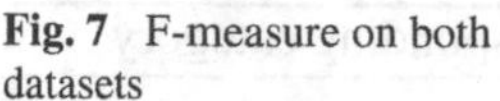

Fig. 7 F-measure on both datasets

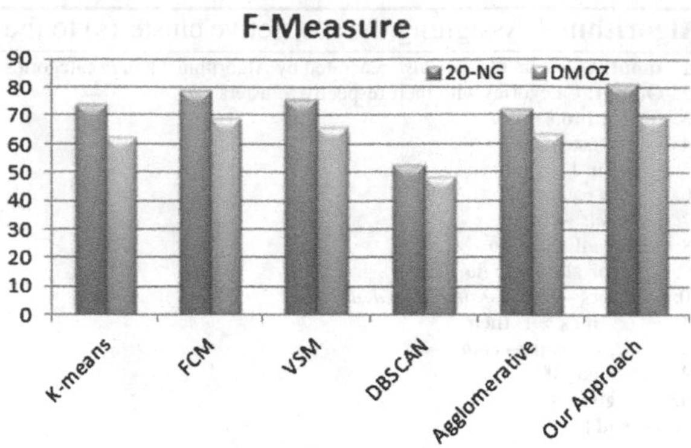

Table 1 Comparing performance on 20-Newsgroups

Methods	Precision	Recall	F-measure
K-Means	74.52	73.63	74.07
FCM	78.63	79.24	78.93
VSM	76.67	74.82	75.73
DBSCAN	51.94	52.77	52.35
Agglomerative	72.15	72.38	72.26
Our approach	**80.35**	**81.75**	**81.03**

The results which marked as bold indicate the highest performance achieved by the corresponding clustering algorithm on the respective dataset

Table 2 Comparing performance on DMOZ

Methods	Precision	Recall	F-measure
K-Means	62.28	62.58	62.43
FCM	**70.66**	67.36	68.97
VSM	65.37	66.24	65.80
DBSCAN	48.24	48.57	48.40
Agglomerative	64.34	63.21	63.76
Our approach	70.58	**68.34**	**69.42**

The results which marked as bold indicate the highest performance achieved by the corresponding clustering algorithm on the respective dataset

Table 3 Category-wise performance on 20-Newsgroups

Category	Documents used	Precision	Recall	F-measure
Alt	320	80.28	79.45	79.86
Computers	1952	82.45	83.45	82.95
Miscellaneous	390	73.45	83.68	78.23
Recreation	1590	81.62	83.69	82.64
Science	1580	78.51	79.41	78.96
Social	399	82.71	84.08	83.39
Talk	1297	79.23	78.94	79.08
Average	7528	80.35	81.75	81.03

Table 4 Category-wise performance on DMOZ

Category	Documents used	Precision	Recall	F-measure
Arts	1396	0.7095	0.6930	0.7012
Business	3384	0.7185	0.6834	0.7005
Computers	1494	0.6859	0.7049	0.6953
Games	5757	0.7176	0.6960	0.7066
Health	1491	0.7294	0.6533	0.6893
Homes	1405	0.6893	0.7048	0.6970
News	1504	0.7069	0.6723	0.6892
Recreation	1410	0.7161	0.6634	0.6887
Reference	1301	0.6983	0.6540	0.6754
Regional	1307	0.7076	0.6456	0.6752
Science	1390	0.7226	0.6823	0.7019
Shopping	6209	0.6865	0.6960	0.6912
Society	1505	0.7056	0.6654	0.6849
Sports	1515	0.6986	0.6700	0.6816
Average	31068	0.7058	0.6834	0.6942

4 Conclusion

The paper proposed a novel approach which combines the traditional apriori algorithm with a new technique called *support-based count (SBC)* to cluster the Web documents and is summarized as follows:

1. A new feature selection technique called term–term correlation has been developed in order to remove the irrelevant features from the corpus.
2. Apriori algorithm is applied on the corpus to create the initial clusters.
3. Initial clusters generated by apriori algorithm can ignore many of the valuable documents when the minimum support is high. The proposed approach worked to find out those relevant documents from all discarded documents and it is done as follows:

   i. first, the proposed approach tests those discarded documents which can form new clusters and later add them to initial clusters which generates (*NEIC*).
   ii. next, SBC technique is applied on this *NEIC* to test on all those remaining unclustered documents which can merge with *NEIC* to form the final clusters.

4. *K-Means*, *FCM*, *VSM*, *DBSCAN*, and *Agglomerative* clustering techniques are applied on initial clusters to generate the final clusters. For experimental work, DMOZ and 20-Newsgroups datasets are considered. The performance of the clusters formed by each relevant technique is measured in terms of F-measure and accuracy.

The empirical results show the efficiency of the proposed approach compared to other established clustering approaches. In future, this work can be extended by labeling the clusters and adding text summarization to each cluster.

References

1. A. Spink, D. Wolfram, M. B. Jansen, and T. Saracevic, "Searching the web: The public and their queries," *Journal of the American society for information science and technology*, vol. 52, no. 3, pp. 226–234, 2001.
2. W. B. Croft, "A model of cluster searching based on classification," *Information systems*, vol. 5, no. 3, pp. 189–195, 1980.
3. J. Tang, "Improved k-means clustering algorithm based on user tag," *Journal of Convergence Information Technology*, vol. 12, pp. 124–130, 2010.
4. C. X. Lin, Y. Yu, J. Han, and B. Liu, "Hierarchical web-page clustering via in-page and cross-page link structures," in *Advances in Knowledge Discovery and Data Mining*. Springer, 2010, pp. 222–229.
5. X. Gu, X. Wang, R. Li, K. Wen, Y. Yang, and W. Xiao, "A new vector space model exploiting semantic correlations of social annotations for web page clustering," in *Web-Age Information Management*. Springer, 2011, pp. 106–117.
6. P. Worawitphinyo, X. Gao, and S. Jabeen, "Improving suffix tree clustering with new ranking and similarity measures," in *Advanced Data Mining and Applications*. Springer, 2011, pp. 55–68.
7. M. T. Hassan and A. Karim, "Clustering and understanding documents via discrimination information maximization," in *Advances in Knowledge Discovery and Data Mining*. Springer, 2012, pp. 566–577.
8. P. Li, B. Wang, and W. Jin, "Improving web document clustering through employing user-related tag expansion techniques," *Journal of Computer Science and Technology*, vol. 27, no. 3, pp. 554–566, 2012.
9. R. K. Roul, S. Varshneya, A. Kalra, and S. K. Sahay, "A novel modified apriori approach for web document clustering," in *Computational Intelligence in Data Mining-Volume 3*. Springer, 2015, pp. 159–171.
10. A. Inokuchi, T. Washio, and H. Motoda, "An apriori-based algorithm for mining frequent substructures from graph data," in *Principles of Data Mining and Knowledge Discovery*. Springer, 2000, pp. 13–23.
11. M. Steinbach, G. Karypis, V. Kumar *et al.*, "A comparison of document clustering techniques," in *KDD workshop on text mining*, vol. 400, no. 1. Boston, 2000, pp. 525–526.
12. G. Salton, A. Wong, and C.-S. Yang, "A vector space model for automatic indexing," *Communications of the ACM*, vol. 18, no. 11, pp. 613–620, 1975.

Genetic Algorithm Based Correlation Enhanced Prediction of Online News Popularity

Swati Choudhary, Angkirat Singh Sandhu and Tribikram Pradhan

Abstract Online News is an article which is meant for spreading awareness of any topic or subject published on the Internet and is available to a large section of users to gather information. For complete knowledge proliferation we need to know the right way and time to do so. For achieving this goal we have come up with a model which on the basis of, multiple factors, like describing the article type (structure and design) and publishing time predicts popularity of the article. In this paper we use Correlation techniques to get the dependency of the popularity obtained from an article, and then we use Genetic Algorithm to get the optimum attributes or best set which should be considered while formatting the article. Data has been procured from UCI Machine Learning Repository with 39644 articles with sixty condition attributes and one decision attribute. We implemented twelve different data learning algorithms on the above mentioned data set, including Correlation Analysis and Neural Network. We have also given a comparison of the performances got from various algorithms in the Result section.

Keywords Naíve bayes classification · Correlation · Genetic algorithm · Neural network · Random forest

1 Introduction

An article is a written piece of work that is published either in a printed format (eg. newspaper) or in an electronic format (eg. online). Online news is popular because it is reachable to a major section of the society. Since there are so many articles

S. Choudhary (✉) · A.S. Sandhu · T. Pradhan
Manipal Institute of Technology, Manipal University, Manipal 576104, India
e-mail: swati.choudhary.24@gmail.com

A.S. Sandhu
e-mail: angkirat@gmail.com

T. Pradhan
e-mail: tribikram.pradhan@manipal.edu

H.S. Behera and D.P. Mohapatra (eds.), *Computational Intelligence in Data Mining*, Advances in Intelligent Systems and Computing 556,
DOI 10.1007/978-981-10-3874-7_13

available, it is impossible that all the articles published are popular (maybe even at that given time). Article being popular depends on the people who are reading it and naturally the content that the article contains. Several factors are responsible for an article to become popular.

In the current scenario, we have proposed methodologies which provide a way to predict whether an article will become popular or not, but we do not have a way which systematically tells us which features or measures are to be improved so that an article can become popular. Our aim here is to maximize the rate of prediction of the article by minimizing and selecting the optimum features. Publishers can benefit by estimating the popularity of the news content and strategize accordingly by focusing on the features obtained as a result of this analysis.

Our paper is divided into the following sections—our related work or relevant case study is described under Sect. 2, our Methodology including Data Acquisition, Pre-processing and Transformation, Naíve Bayes Classification, Random Forest and Genetic Algorithm is described under Sect. 3. Section 4 mentions about the Case study, Sect. 5 focuses on the Comparison with Existing Approach, Sect. 6 describes the Conclusion and Future work followed by References.

2 Literature Survey

Kelwin Fernandes, Paulo Cortez and Pedro Vinagre in [1] proposed a system for Intelligent Decision Support or called as (IDSS) and focused on predicting whether an article will be popular before getting published, and then used optimization techniques to improve few of the article features so that maximum popularity could be achieved for that article, prior to its publication. They used 47 out of 60 features and using Random Forest an accuracy of 67% was achieved and optimization was done using Stochastic Hill Climbing. He Ren and Quan Yang in [2] optimized the work done in [1] by making use of Machine Learning techniques including Mutual Information and Fisher Criterion to get maximum accuracy for feature selection, based on which prediction for popularity of news article was done. Using this method, they got an accuracy of 69% using Random Forest using top 20 features.

H. Muqbil, AL-Mutairi, Mohammad Badruddin, Khan in [3] had predicted the popularity of trending Arabic Articles taken from the Arabic Wikipedia based on external stimulus including number of visitors. This paper used Decision Tree and Naïve Bayes for prediction and compared the two models. Elena Hensinger, Ilias Flaounas and Nello Cristianini in [4] had predicted the popularity of the article, on the basis of the number of views that the article had on the day it was published. It used RSVM method to predict popularity which is done on the basis of the title of the news article, its introductory description, the place and date of its advertisement. Alexandru Tatar, Panayotis Antoniadis, Marcelo Dias de Amorim, Jérémie Leguay, Arnaud Limbourg and Serge Fdida in [5] talk about how the popularity of a news article can be predicted on the basis of the number of users who commented on that article over a short period (in hrs) of time soon after the article was published. Roja Bandari_

Sitaram Asury Bernardo Huberman, in [6] predicted popularity on twitter with accuracy of 84% using regression and classification techniques, by considering following attributes—the source which posted the article, category of the news, use of appropriate language and names of people mentioned in the article. Score assignment of each features is done and accuracy was found out using Bagging, J48 Decision Trees, SVM and Naïve Bayes.
In [7], I. Aprakis, B. Cambazoglu and M. Lalmas, do a cold start prediction, where they acquire their data from Yahoo News and predict the popularity. For prediction they use two metrics: Number of times the article was shared or posted on twitter and the number of views for that article. Carlos Castillo, Mohammed El-Haddad, Jürgen Pfeffer, Matt Stempeck in [8] present a qualitative as well as a quantitative study of the longevity of online news article by analyzing the strong relation between reaction on social media and visits over time. This study shows that both are equally important for understanding the difference among classes of articles and for predicting future patterns. Gabor Szabo, Bernardo A. Huberman in [9] predicts the long term popularity of online articles which were taken from YouTube and Digg by analyzing the views and votes for these articles. Jong Gun Lee, Sue Moonand Kavé Salamatian, in [10] predicts popularity of online content based on features which can be seen by an external user, including number of comments and the number of links in the first hours after the content publication. This work can predict lifetime based on number of threads (5–6 days) and number of user comments (2–3 days). It is an optimized paper for [9] using survival analysis.

3 Methodology

3.1 Data Acquisition

Our dataset was taken from UCI machine learning repository, and was crawled by K. Fernandes et al. in [1]. It consists of 58 predictive attributes having numerical values each describing a different aspect of the 39644 articles that were published on the Mashable website in the last two years.

3.2 Popularity Classification

The initial data set had 61 attributes. We are adding a 62nd attribute which is Boolean, named 'Popular'. This attribute value is found by Algorithm 1 on *R*. After applying Algorithm 1 the average value comes to be 3300.

Algorithm 1 Finding Popular Articles

```
procedure POPULARITY(shares)
    sum ← 0
    for each i in shares do
        sum ← sum + i
    end for
    avg ← sum / length(shares)
    for each i in shares do
        if i ≥ avg then
            popularity ← true
        else
            popularity ← false
        end if
    end for
end procedure
```

3.3 Naïve Bayes

The Naïve Bayes algorithm is a classification procedure which is based on Bayes Theorem. This algorithm works with an assumption that there exists an independence between predictors. Given that we have to predict P(B|A) where A = (A_1 ..., A_{ni}). Here we assume that each A_i is conditionally independent of each of the other $A_k s$ given B, and also independent of each subset of the other A_k's given B. This assumption will simplify the representation of P(A|B), and the problem of estimating it from the given data.

Suppose that A consists of only two attributes, A = (A_1,A_2), then P(A|B) = P(A_1, A_2|B) = P(A_1|A_2,B)P(A_2|B) = P(A_1|B)P(A_2|B) Now, when A consists of multiple attributes, Naíve Bayes takes up the form as in Eq. 1.

$$P(\frac{A_1..A_n}{B}) = \prod_{i=1}^{n} p(\frac{A_i}{B}). \tag{1}$$

3.4 Correlation Analysis

A bivariate analysis used for measuring the degree of association amongst two vectors say A and B is known as Correlation. In data mining, the value obtained after doing Correlation analysis varies between ±1. When this value is greater than 0, then a positive correlation exists and if this value is less than zero, then a negative correlation exists. If the value is 0, then the relationship between them is weak. By performing an analysis on this value, we can identify the attributes that will be directly or indirectly affecting our 'shares' column. Also we will try to narrow down our area of work by focusing on only those attributes that are required to carry out our proposed methodology, as we will be considering only those attributes whose correlation value is a positive one. Here we will be measuring three different types of correlations namely—Pearson Correlation, Kendall-rank Correlation and Spearman Correlation.

Pearson Correlation

Pearson correlation coefficient ρ is calculated by the formula mentioned in Eq. 2:

$$\rho = \frac{E[AD] - E[A]E[D]}{\sqrt{E[A^2] - (E[A])^2}\sqrt{E[D^2] - (E[D])^2}}. \tag{2}$$

where:

- A stands for the Attribute Vector
- D stands for the Decision Vector
- E[A] stands for the sum of the elements in A

Algorithm 2 Pearson Correlation

```
procedure PEARSON(Dataset)
    cols ← ncols(Dataset)
    cols ← cols − 1
    rows ← nrows(Dataset)
    PearsonVector ← c()
    for each i in 1:cols do
        product ← 1
        SumA_i ← 0
        Sum ← 0
        SumSquareA_i ← 0
        SumSquare ← 0
        for each j in 1:rows do
            product ← product + (Dataset[j, 61] * Dataset[j, i])
            SumA_i ← SumA_i + Dataset[j, i]
            Sum ← Sum + Dataset[j, i]
            SumSquareA_i ← SumSquareA_i + Dataset[j, i]^2
            SumSquare ← SumSquare + Dataset[j, 61]^2
        end for
        p ← (product−(SumA_i*Sum)) / (sqrt(SumSquareA_i−SumA_i^2) * sqrt(SumSquare−Sum^2))
        PearsonVector ← append(PearsonVector, p)
    end for
end procedure
```

Kendall-Rank Correlation

Kendall correlation coefficient τ is calculated by the formula mentioned in Eq. 3:

$$\tau = \frac{n_c - n_d}{\frac{1}{2}n(n-1)}. \tag{3}$$

where:

- n_c stands for the number of pairs that are concordant
- n_d stands for the number of pairs that are discordant
- n stands for the sample size

Algorithm 3 Kendall Correlation

```
procedure KENDALL(Dataset)
    n ←Number of rows
    vec ← Dataset[, decision]
    n_c ← 0
    n_d ← 0
    for each attr columns of Dataset do
        for each i in 1 : n do
            for each j in 1 : n do
                if ((attr_i > attr_j)AND(vec_i > vec_j))
OR((attr_i < attr_j)AND(vec_i < vec_j)) then
                    n_c ← n_c + 1
                else
                    n_d ← n_d + 1
                end if
            end for
        end for
    end for
    kendall ← n_c − n_d
    kendall ← 2*kendall / n(n−1)
end procedure
```

Spearman Correlation

Spearman Correlation coefficient σ is calculated by the formula mentioned in Eq. 4:

$$\sigma = 1 - \frac{6 \sum d_i^2}{n(n^2 - 1)}. \tag{4}$$

In the formula,

- d_i stands for the difference between the ranks of variables P and Q
- n stands for the sample size

3.5 Feature Selection on Basis on Correlation Analysis

After doing Pearson Correlation by Algorithm 2, Kendall-rank Correlation by Algorithm 3 and Spearman Correlation using Algorithm 4, we get three individual CSV files which list the attributes that satisfy the respective correlation criteria. After

Algorithm 4 Spearman Correlation

```
procedure SPEARMAN(Dataset)
    n ← Number of rows
    d_de ← Dataset[, decision]_r
    SpearmanVector ← c()
    for each i in 1 : ncol(Dataset) do
        d ← 0
        for each j in 1 : n do
            d ← d + (Dataset[, i]_r − d_de)^2
        end for
        spearman ← 6*d / (n(n^2−1))
        spearman ← 1 − spearman
        SpearmanVector ← append(SpearmanVector, spearman)
    end for
end procedure
```

obtaining the three individual results we devise our criteria for reducing the number of features using Algorithm 5. The result obtained by this algorithm is discussed under Sect. 4.

Algorithm 5 Attribute Selection after Correlation

```
procedure ATTRIBUTESELCTION(Dataset)
    rows ← nrows(Dataset)
    cols ← ncols(Dataset)
    pearsonVector ← pearson(Dataset)
    spearmanVector ← spearman(Dataset)
    kendallVector ← kendall(Dataset)
    for each i in 1:cols do
        if pearsonVector[i]>0 AND spearmanVector[i]>0 AND kendallVector[i]>0 then
            Selection ← true
        else
            Selection ← false
        end if
    end for
return dataset[,Selection]
end procedure
```

3.6 *Random Forest*

Random Forest is a classification technique containing multiple Decision Trees and it will output the class that is the mode of the classes. This is a two step process. First we need to select data at random from the training data set to build individual trees. For each of these trees, a subset is used for building the model and the remaining data is used for testing its accuracy. This sample data which is used for testing purpose is called out of *bagging*. After this has been done, the second step is to find the splitting criteria for every node in the tree. For every node, a subset of predictor vector is selected at random to generate the new rule. The count of the predictor vector is done by the Random Forest algorithm itself. Thus Random Forest method uses a bagging technique along with random selection of features.

3.7 Genetic Algorithm

Genetic Algorithm is based on a searching technique that helps to realize the appropriate solutions for optimization and search related problems. A medley of solutions which is represented by chromosomes is collected and is also known as population. This population is optimized to provide the best solution and is represented in binary format of 0's and 1's.
Genetic Algorithm has the following two requirements:

- The solution domain should be represented in a genetic form, i.e. consisting of 0's and 1's only.
- The solution domain must be evaluated by a fitness function.

Process of genetic algorithm:

1. *Creating the search space*: In this phase randomly generated realizable solutions which covers the entire scope of feasible solutions also called as Sample Space is recorded.
2. *Selection*: Each solution is filtered using the fitness-based process (or algorithm) and the best fit solution will have the highest possibility of being selected.
3. *Reproduction/Crossover*: The selected individuals from selection phase, are used to create a second generation of solutions. To do this, the solutions got from the above phase are paired thus forming a parent pair. From this parent pair, a child pair is generated using mutation or crossover.
4. *Evaluation*: After generating the child pair, its fitness is evaluated.
5. *Mutation*: The initial solutions are replaced by the newly generated child pair got from the previous steps, and the process is duplicated on the new solution set until best solution is found.

4 Case Study

4.1 Naïve Bayes Result on whole data set of 61 attributes

After Collecting the data from UCI machine Learning repository we have classified the data according to Naïve Bayes classification method. Using Naïve Bayes

Classification on the entire data set containing 61 attributes, an accuracy 88.76% is achieved. This was done to get accuracy on entire data set consisting of 61 attributes. This result was found using *RAPIDMINER*.

4.2 Correlation Results of Pearson, Kendall-Rank and Spearman

After applying classification techniques, we see that even though accuracy is high, the number of features used is not optimum. Hence we formulate a new way to reduce the number of attributes and get a relatively better accuracy. After applying Algorithms 2–4 to the entire data set we get individual results for correlation of each attribute with one another. Then we do a *'& / and'* operation for each of these attributes obtained by applying Algorithm 5. After applying this algorithm, we are left with a set of 34 attributes, which can further be processed manually to remove the redundant data namely 'is_weekend' is enough to get the predicted values and hence can replace the 'weekday_is_Saturday ' and 'weekday_is_Sunday ' attributes reducing the total number of attributes to 32, as shown in Table 1. The results mentioned in Table 1 were obtained using Algorithm 5 in *R*.

4.3 Naïve Bayes On Reduced Data Set Of 32 Attributes

After applying Algorithm 5, we get the resultant attributes as shown in Table 1. Now, we again apply Naïve Bayes Classification to this newly reduced data set on *RAPIDMINER*. As we can see from Table 3 that after reducing the number of attributes we

Table 1 Correlation result

Attributes		
timedelta	num_hrefs	num_imgs
num_videos	data_channel_is_lifestyle	data_channel_is_socmed
kw_min_min	kw_max_min	kw_min_max
kw_avg_max	kw_max_avg	kw_avg_avg
self_reference_min_shares	self_reference_max_shares	LDA_03
global_sentiment_polarity	global_rate_negative_words	avg_positive_polarity
title_sentiment_polarity	abs_title_sentiment_polarity	global_subjectivity
self_reference_avg_shares	num_self_hrefs	num_keywords
data_channel_is_tech	kw_avg_min	title_subjectivity
kw_min_avg	is_weekend	global_rate_positive_words
max_positive_polarity	shares	

Table 2 Genetic results

Attributes		
timedelta	num_imgs	num_videos
num_keywords	data_channel_is_tech	kw_min_min
kw_min_avg	self_reference_min_shares	self_reference_max_shares
self_reference_avg_shares	is_weekend	LDA_03
global_sentiment_polarity	global_rate_positive_words	avg_positive_polarity
title_subjectivity	title_sentiment_polarity	abs_title_sentiment_polarity

Table 3 Prediction test

Techniques	32 attributes (in %)	18 attributes (in %)
Naíve bayes	**93.46**	76.06
Random forest	79.63	79.62
Neural networks	91.89	**91.96**
KNN (k = 5)	90.27	74.49
MLR	61.11	60.47
C4.5 tree	88.8281	84.7947
C5.0 tree	79.92	79.62

get an accuracy of 93.46%, which is better than the result obtained from Sect. 4.1 when the same algorithm was tested on all of the 61 attributes.

4.4 Random Forest on Reduced Set

After applying Algorithm 5, we get the resultant attributes as shown in Table 1. Now, we apply Random Forest algorithm to this newly reduced data set on *RAPIDMINER*. As we can see from Table 3 that after reducing the number of attributes we get an accuracy of 79.63%, which is not better than the result obtained from Sect. 4.3 when the same algorithm was tested with Naíve Bayes for the same data set. Hence, we can conclude that Naíve Bayes will give us the best accuracy for 32 attributes for classifying an article as popular or not.

4.5 Optimal Feature Selection on Basis of Genetic Algorithm

Using Algorithm 5 we had reduced 61 attributes to 32 which is almost 50% reduction, and we saw that we had achieved a better result after reduction as seen in Table 1. To further optimize the number of attributes, we applied Genetic Algorithm on R by fixing the following:

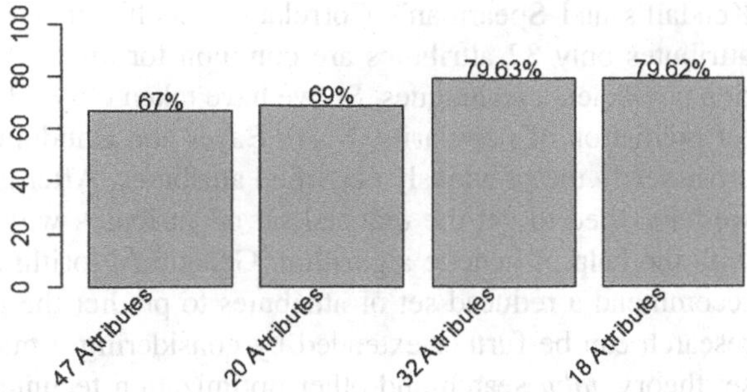

Fig. 1 Random forest comparison

Fit as Linear Regression ⇒ 18 predictor variables ⇒ Crossover as 0.8 ⇒ Mutation as 0.1 ⇒ 25 iterations

After obtaining the above 18 attributes as shown in Table 2, we again classify them using Naíve Bayes Algorithm on *RAPIDMINER*. We can see from Table 3 that after reduction of 32 attributes to 18, we get an accuracy of 76.06%. After obtaining the above 18 attributes, we again classify them using Neural Network on *R*. We can see from Table 3 that after reduction of 32 attributes to 18, we get an accuracy of 91.96%.

5 Result

In this work, we performed eight prediction algorithms to do a comparison as to which algorithm will give us the maximum prediction rate for reduced 32 attributes as well as for optimum 18 attributes. These algorithms were performed on *'R'* and *RAPIDMINER*.

In Table 3 we have listed the percentage values that we got for the accuracy for all the Algorithms that we used on 32 Attribute Set and the 18 Attribute Set. The table data shows that *Naïve Bayes* gives best prediction value for the 32 Attribute Set and *Neural Networks* gives the best prediction value for the 18 Attribute Set.

In Fig. 1 we have given a comparison of Random Forest results. This algorithm was used in [1, 2] who have also worked on the same Data Set. The analysis shows a great increase in the prediction value of the proposed method as compared to [1, 2].

6 Conclusion and Future Work

In this paper we have analyzed various attributes before predicting popularity in terms of share and suggesting the author. Initially we classified using Naíve Bayes Algorithm and individually computed the correlation among 61 Condition attributes to get the correlation among attributes of an article. After applying Pearson's,

Kendall's and Spearman's Correlation Coefficients we concluded that out of 61 attributes only 32 attributes are common for all of the above mentioned correlation coefficients techniques. So we have taken only 32 attributes into consideration for prediction of popularity. Naíve Bayes and Random forest have been applied to cross verify the previously classified attributes. After getting the possible attributes we have tried to get the reduced set of attributes with the same rate of prediction with the help of genetic algorithm. Genetic Algorithm has been used to randomly recommend a reduced set of attributes to predict the popularity of an article. The research can be further extended by considering a major methodology like rough set theory, *tabu* search and other optimization techniques to get an optimal set of reduced attributes with a possibility of same rate of prediction.

Acknowledgements We would like to thank Department of Information & Communication Technology, Manipal Institute of Technology, Manipal, for encouraging us to carry out this research and providing us with all the support needed.

References

1. Fernandes, Kelwin, Pedro Vinagre, and Paulo Cortez. "A proactive intelligent decision support system for predicting the popularity of online news." Portuguese Conference on Artificial Intelligence. Springer International Publishing, 2015.
2. Ren He and Quan Yang."Predicting and Evaluating the Popularity of Online News."
3. Al-Mutairi, Hanadi Muqbil, and Mohammad Badruddin Khan. "Predicting the Popularity of Trending Arabic Wikipedia Articles Based on External Stimulants Using Data/Text Mining Techniques." Cloud Computing (ICCC), 2015 International Conference on. IEEE, 2015.
4. Hensinger Elena, Ilias Flaounas, and Nello Cristianini. "Modelling and predicting news popularity." Pattern Analysis and Applications 16.4 (2013): 623–635.
5. Tatar Alexandru, et al. "Predicting the popularity of online articles based on user comments." Proceedings of the International Conference on Web Intelligence, Mining and Semantics. ACM, 2011.
6. Bandari Roja, Sitaram Asur, and Bernardo A. Huberman. "The pulse of news in social media: Forecasting popularity." arXiv preprint arXiv:1202.0332 (2012).
7. Arapakis, Ioannis, B. Barla Cambazoglu, and Mounia Lalmas. "On the feasibility of predicting news popularity at cold start." International Conference on Social Informatics. Springer International Publishing, 2014.
8. Castillo, Carlos, et al. "Characterizing the life cycle of online news stories using social media reactions." Proceedings of the 17th ACM conference on Computer supported cooperative work & social computing. ACM, 2014.
9. Szabo Gabor, and Bernardo A. Huberman. "Predicting the popularity of online content." Communications of the ACM 53.8 (2010): 80–88.
10. Lee Jong Gun, Sue Moon, and Kave Salamatian. "An approach to model and predict the popularity of online contents with explanatory factors." Web Intelligence and Intelligent Agent Technology (WI-IAT), 2010 IEEE/WIC/ACM International Conference on. Vol. 1. IEEE, 2010.

CSIP—Cuckoo Search Inspired Protocol for Routing in Cognitive Radio Ad Hoc Networks

J. Ramkumar and R. Vadivel

Abstract Cognitive radio (CR) is viewed as the empowering innovation of the dynamic spectrum access (DSA) model which is imagined to take care of the present spectrum scarcity issue by encouraging the contraption of new remote administrations. Cognitive devices have the similar proficiency of CR and the expedient network that they form dynamically is called cognitive radio ad hoc networks (CRAHNs). Due to assorted qualities in channels, routing becomes a critical undertaking job in CRAHN. Minimizing the end-to-end delay is one of the major difficult tasks in CRAHNs, where the transmission of packets passes on every hop of routing path. In this paper, a new reactive multicast routing protocol namely cuckoo search inspired protocol (CSIP) is proposed to reduce the overall end-to-end delay, which progressively reduces the congestion level on various routing path by considering the spectrum accessibility and the service rate of each hop in CRAHNs. Simulations are demonstrated using NS2 tool and the results proved that the proposed routing protocol CSIP significantly outperforms better than other baseline schemes in minimizing end-to-end delay in CRAHNs.

Keywords Cognitive · Radio · CRAHN · Bio-inspired · Cuckoo · Routing · CSIP

J. Ramkumar (✉)
Department of Computer Science, VLB Janakiammal College of Arts and Science (Affiliated to Bharathiar University), Coimbatore 641042, India
e-mail: jramkumar1986@gmail.com

R. Vadivel
Department of Information Technology, Bharathiar University, Coimbatore 641046, India
e-mail: vlr_vadivel@yahoo.co.in

H.S. Behera and D.P. Mohapatra (eds.), *Computational Intelligence in Data Mining*, Advances in Intelligent Systems and Computing 556,
DOI 10.1007/978-981-10-3874-7_14

1 Introduction

Cognitive radio (CR) is a radio-based technology which can change its transmitter parameters in pacific community with surroundings, wherein it firmly perform. In controlling the CR in a straightforward way, it will indisputably be software oriented. CR is an embodiment design of remote networking that senses its surroundings, traces the updating and commonly barter message with their networks. Adding this feature of understanding the network of ad hoc by a glance to the geo-location based network is known as cognitive radio ad hoc network (CRAHNs). CRAHN aims to improve the way the radio spectrum is utilized. The four basic functions [1] of CR networks are: (a) spectrum sensing (detects all the available spectrum holes in order to avoid interference); (b) spectrum sharing (shares the spectrum related information between neighbor nodes); (c) spectrum mobility (provides seamless connectivity between nodes); (d) spectrum decision (captures the best available vacant spectrum holes from detecting spectrum holes).

Routing is the procedure of choosing the best path in a network. Multicast routing in CRAHN faces a numerous difficulties because of natural characteristics of the CRAHNs such as reliability, mobility, and scarce resources. One of the major problems in the design of routing protocol for CRAHN is less delay tolerance, reduced throughput which results in poor packet delivery ratio. The above-mentioned setbacks are due to the ineffective detection of spectrum nodes that are unlicensed devices (secondary users) have to quit the spectrum band once the licensed device (primary user) is detected. Protocols designed for CRAHNs are entirely different from those for traditional networks.

Objectives of this paper are: (i) to propose a bio-inspired based reactive routing strategy for multicast transmissions (named as CSIP (cuckoo search inspired protocol)) in order to reduce the overall end-to-end delay. (ii) to reduce the congestion level that results in improved throughput with better packet delivery ratio. (iii) to conduct simulation using NS2 with the performance metrics throughput, packet delivery ratio, and end-to-end delay under varying mobility speed of the nodes.

The paper is organized as follows. This section introduces CRAHN, problem statement, and objectives of the proposed work. Section 2 provides the brief summary of the related works as background study. Section 3 discusses on the proposed protocol CSIP (cuckoo search inspired protocol). Section 4 illustrates on the chosen performance metrics along with the NS2 simulation settings. Section 5 confers the results. Finally Sect. 6 concludes the paper with future dimensions.

2 Background Study

Reviriego et al. [2] used cuckoo searching method to analyze the exact matching of energy efficiency and technique was presented to improve the energy efficiency of cuckoo hashing. Grissa et al. [3] proposed a method for database-oriented cognitive radio networks that protect location privacy of secondary users whereas permitting them to grasp the regarding convenient channels near them. Chang et al. [4] proposed hybrid algorithm based on Nelder and Mead (NM) and therefore the cuckoo search (CS) to train the fuzzy neural networks (FNNs). Herazo et al. [5] conferred a meta-heuristic algorithm referred as cuckoo search (CS) to solve the distribution network reconfiguration drawback (DNRC), however the target is to attenuate the energy loss. Taheri et al. [6] proposed cuckoo search algorithmic rule (CSA) as optimization tool to solve the transmission expansion planning (TEP) problem in a network. Yang and Deb [7] proposed a cuckoo search based method for multiobjective optimization. Nancharaiah and Chandra Mohan [8] proposed a hybrid optimization technique using ant colony optimization (ACO) and cuckoo search (CS) for optimizing the routing in ad hoc networks. Teymourian et al. [9] proposed local search hybrid algorithm (LSHA) and post-optimization hybrid algorithm (POHA) after investigating capacitated vehicle routing problem (CVRP) with a sophisticated advanced cuckoo search (ACS) algorithm and other three algorithms. Goyal and Patterh [10] enforced the cuckoo search algorithm to estimate the sensor's position. Raha et al. [11] attempted to reduce the power loss of the network by proposing optimal reactive power dispatch (ORPD) using bio-inspired technique namely, cuckoo search algorithm (CSA). Jin et al. [12] introduced TIGHT, which is a geographic routing protocol for cognitive radio networks, where it offers three routing modes and mostly permits, particularly auxiliary clients to completely investigate the transmission opportunities over a kind of essential channel without influencing actual primary users (PUs).

3 CSIP: Cuckoo Search Inspired Protocol

Yang and Deb [13] proposed the cuckoo search algorithm and the cuckoo search based optimization was proposed in Yang and Deb [14], it is a population-based stochastic global search algorithm. In cuckoo search algorithm, prototype denotes a nest; likewise every attribute of the prototype denotes a cuckoo egg. The common equation of the cuckoo search algorithm was based on the random walk algorithms equation.

$$\omega_{\iota^{D+1}+1;1} = \omega_{\iota^{D+1};z^i} \pm \tau\left(\sum_{\iota=0}^{\iota+1} levy(\eta+\iota)\right), \tag{1}$$

where ι point to the quantity of number of the current generation ($\iota = \sum_0^{maxcycle-1} \iota$ and where max cycle indicates the predefined number for the largest generation).

In CSIP, the starting value of μth attributes of the z^ith prototype, $P_{\iota=0;i} = [x_{\iota=0;\mu,z^i}]$ was derived by using the Eq. (2)

$$\omega_{\iota=0;\mu,i} = rand(t_{z^i} - b_{z^i}) + b_{z^i}, \quad (2)$$

where t_{z^i} and b_{z^i} are the top and bottom limits of exploration space of μ th attributes, correspondingly. The CSIP restricts the borderline clause in every iteration step. Then, once the rate of attribute torrents the predicted limit of borderline clause, then automatically it updates the rate of the similar attribute. Once before initiating the iterative task of searching, the CSIP points out the best prototype as ω_{finest}.

(a) Iteration

The iterative section begins with the prototype matrix and disclose phase, then the value of the ψ is calculated using Eq. (3)

$$\psi = \left(\frac{\Delta(1 \pm \rho) \odot \frac{1}{sec}\left(\pi \odot \frac{\rho}{3}\right)}{\Delta\left(\frac{1+\rho}{3}\right) \times \rho \times 3^{\frac{\rho \pm 1}{3}}} \right)^{\frac{t_{z^i} - b_{z^i}}{\rho}} \quad (3)$$

In Eq. (3) Δ denotes delta function. The growth phase of the ω_{z^i} prototype begins by prescribing the contributing vector $r = \omega_{z^i}$.

(b) Pacerange

Pacerange denotes the distance covered in a single move. The expected pacerange value calculation is done by using Eq. (4);

$$\text{pacerange} = \frac{1}{3} \times \left(\sum \frac{1}{100} \times \left(\frac{u_\mu}{v_\mu} \right)^{\frac{1}{\beta}} \times (r - \omega_{finest}) \right), \quad (4)$$

where $u = \psi \cdot r\ and\ [\delta]$ and $r = r\ and\ [\delta]$. The $r\ and\ [\delta]$ function creates a unique integer value between 1 and δ.

(c) Contributor

Calculation of contributor's prototype r is generated stochastically using the Eq. (5)

$$r : = r + \left(\frac{u_\mu}{v_\mu} \right) (pacerange_\mu \cdot ran\ [\delta]) \quad (5)$$

(d) Renewal

Renewal task for ω_{finest} prototype in CSIP is carried out by using the Eq. (6).

$$R(\omega_{finest}) \leq R(\omega_{z^i}) \rightarrow \omega_{finest} \quad (6)$$

(e) Crossover

The prototypes that are all not feasible are calculated by using the crossover manipulator, i.e., Eq. (7);

$$r_i := \begin{cases} \omega_{z^i} + rand(\omega_{r1} - \omega_{r2}) & rand_{z^i} > p_0 \\ \omega_{z^i} - rand(\omega_{r1} - \omega_{r2}) & rand_{z^i} < p_0 \\ \frac{1}{\omega_{z^i}} & rand_{z^i} = p_0 \end{cases} \quad (7)$$

(f) Finishing Condition

Once after the task gets completed, CSIP verifies whether the finishing condition is met or not. The finishing condition is the point at which the universal loop count attains the assigned numeric value (ι).

4 Simulation Model and Parameters

This section assesses the performance of CSIP. Since there are no publically accessible implementations for CR routing protocols available, this research work utilizes NS2 simulator for measuring the performance of CSIP. Additionally, the information related to fundamental implementations does not seem to be clear enough for a genuinely trusting and continuing with the research work. Explicitly, routing of packets in CSIP and TIGHT relies on the unconstrained network topology as it were. The simulation settings and parameters used in this research work are tabulated in Table 1.

Table 1 Simulation settings and parameters

Area size	4000 × 3000 m^2
No. of primary user	6
No. of secondary user	1,334
MAC	802.11b
Transmission range	250 m
Simulation time	500 s
Traffic source	CBR
Packet size	512 bytes
Mobility model	Random waypoint model
Initial energy	0.5 J

Performance Metrics

Jin et al. [12] used throughput (TP), packet delivery ratio, (PDR) and end-to-end delay (EED) as the metrics to measure the performance of TIGHT, where this research work uses the same metrics for the evaluation and comparison.

(a) Throughput: It is the proportion of the number of packets sent and the total number of packets.
(b) Packet Delivery Ratio: It is the proportion of the number of packets received successfully and the total number of packets transmitted.
(c) End-to-End Delay: It is the average time period of overall surviving data packets from the source to the destination.

5 Results

In Figs. 1, 2, and 3 mobility speed varying from 5 to 20 m/s is plotted in x-axis. In Fig. 1 throughput is plotted in y-axis, it is clearly evident that the proposed routing protocol CSIP attains better performance than TIGHT in terms of throughput. In Fig. 2, packet delivery ratio is plotted in y-axis, it is apparent that the proposed routing protocol CSIP accomplishes superior performance than TIGHT in terms of packet delivery ratio. In Fig. 3 the delay in milliseconds is plotted in y-axis, it is obvious that the proposed routing protocol CSIP achieves better performance than TIGHT in terms of reducing the delay. The result values of Figs. 1, 2, and 3 are presented in Table 2.

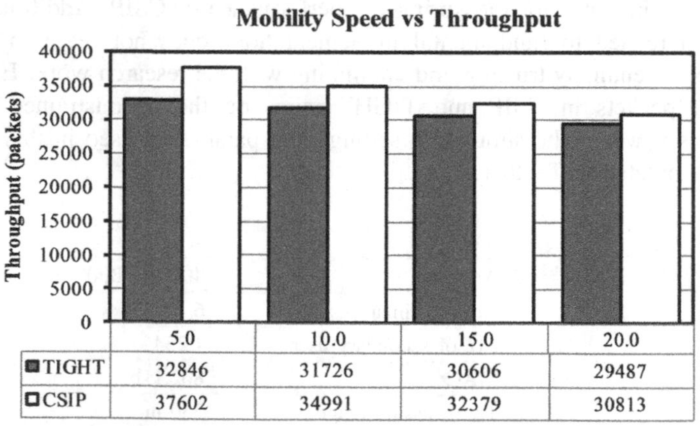

	5.0	10.0	15.0	20.0
TIGHT	32846	31726	30606	29487
CSIP	37602	34991	32379	30813

Fig. 1 Mobility speed versus throughput

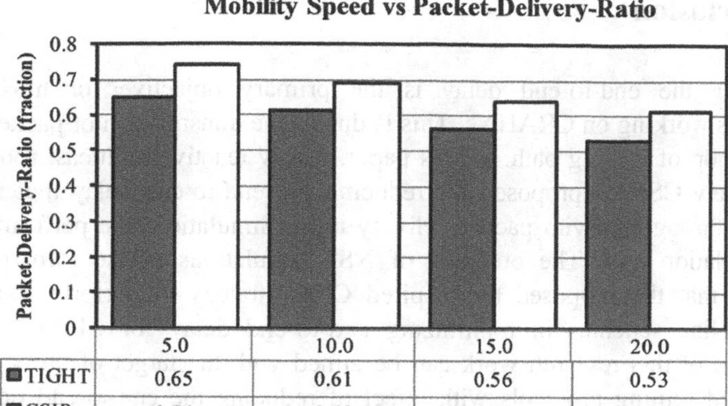

Fig. 2 Mobility speed versus packet delivery ratio

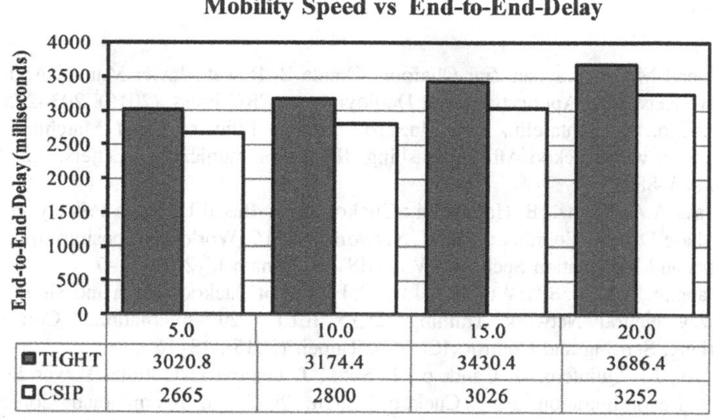

Fig. 3 Mobility speed versus end-to-end delay

Table 2 Performance measures of CSIP

Mobility speed (m/s)	TP (packets)		PDR (fraction)		EED (ms)	
	TIGHT	CSIP	TIGHT	CSIP	TIGHT	CSIP
5	32846	37602	0.65	0.74	3020.8	2665
10	31726	34991	0.61	0.69	3174.4	2800
15	30606	32379	0.56	0.64	3430.4	3026
20	29487	30813	0.53	0.61	3686.4	3252

6 Conclusion

Minimizing the end-to-end delay is the primary objective for most of the researchers working on CRAHNs. This is due to the transmission of packets passes on every hop of routing path. In this paper, a new reactive multicast routing protocol namely CSIP is proposed for reducing the end-to-end delay that results in improved throughput with packet delivery ratio. Simulations are performed using NS2 simulation tool. The outcome of NS2 simulations in the form of results visualizes that the proposed bio-inspired CSIP strategy outperforms better than other baseline schemes in minimizing end-to-end delay for CRAHNs. Future dimensions of this research work can be aimed with the target of proposing new bio-inspired routing protocols with effect of reducing the end-to-end delay even more.

References

1. Jaime Lloret Mauri., Kayhan Zrar Ghafoor., Danda B. Rawat., Javier Manuel Aguiar Perez.: Cognitive Networks: Applications and Deployments, CRC Press. (2015). 203–235.
2. P. Reviriego., S. Pontarelli., J.A. Maestro.: Energy Efficient Exact Matching for Flow Identification with Cuckoo Affinity Hashing. IEEE Communications Letters, vol. 18, no. 5. (2014). 885–888.
3. M. Grissa., A.A. Yavuz., B. Hamdaoui.: Cuckoo Filter-Based Location-Privacy Preservation in Database-Driven Cognitive Radio Networks. 2015 World Symposium on Computer Networks and Information Security (WSCNIS), Hammamet. (2015). 1–7.
4. J.Y. Chang., S.H. Liao, S.L. Wu., C.T. Lin.: A Hybrid of Cuckoo Search and Simplex Method for Fuzzy Neural Network Training. 2015 IEEE 12th International Conference on Networking, Sensing and Control (ICNSC), Taipei. (2015). 13–16.
5. E. Herazo., M. Quintero., J. Candelo., J. Soto., J. Guerrero.: Optimal Power Distribution Network Reconfiguration using Cuckoo Search. 2015 4th International Conference on Electric Power and Energy Conversion Systems (EPECS), Sharjah. (2015). 1–6.
6. S.S. Taheri., S.J. Seyed-Shenava., M. Modiri-Delshad.: Transmission Network Expansion Planning Under Wind Farm Uncertainties using Cuckoo Search Algorithm. 3rd IET International Conference on Clean Energy and Technology (CEAT) 2014, Kuching. (2014) 1–6.
7. Xin-She Yang., Suash Deb.: Multiobjective Cuckoo Search For Design Optimization. Journal Computers and Operations Research, Volume 40 Issue 6. (2013). 1616–1624.
8. B. Nancharaiah., B. Chandra Mohan.: Hybrid optimization using Ant Colony Optimization and Cuckoo Search in MANET routing. 2014 International Conference on Communications and Signal Processing (ICCSP), Melmaruvathur. (2014). 1729–1734.
9. Ehsan Teymourian., Vahid Kayvanfar., GH.M. Komaki., M. Zandieh.: Enhanced Intelligent Water Drops and Cuckoo Search Algorithms for Solving the Capacitated Vehicle Routing Problem. Information Sciences, Vol 334–335. (2016). 354–378.
10. Sonia Goyal., Manjeet Singh Patterh.: Wireless Sensor Network Localization Based on Cuckoo Search Algorithm. Journal Wireless Personal Communications: An International Journal, Volume 79 Issue 1. (2014). 223–234.

11. S.B. Raha., T. Som., K.K. Mandal., N. Chakraborty.: Cuckoo Search Algorithm based Optimal Reactive Power Dispatch. 2014 International Conference on Control, Instrumentation, Energy and Communication (CIEC), Calcutta. (2014), 412–416.
12. Xiaocong Jin., Rui Zhang., Jingchao Sun., Yanchao Zhang.: TIGHT: A Geographic Routing Protocol for Cognitive Radio Mobile Ad Hoc Networks. IEEE Transactions on Wireless Communications, vol. 13, no. 8. (2014). 4670–4681.
13. Xin-She Yang., Suash Deb.: Cuckoo Search via Levy Flights. World congress on nature and biologically inspired computing'NABIC-2009, vol 4. Coimbatore, (2009). 210–214.
14. Xin-She Yang., Suash Deb.: "Engineering Optimisation by Cuckoo Search". International Journal Math Modell Numer Optim 1(4). (2010). 330–343.

Performance Analysis of Airplane Health Surveillance System

N.B. Rachana and S. Seema

Abstract *Airplane Health Surveillance System* is an information system developed to help the operators and maintenance crew to make business decision. System will detect the defect along with cause which would lead to delay and airplane crashes which has high impact on society. The system is capable of detecting and diagnosing the defects which may be initiated during a flight. Thereby triggers alert to safeguard the airplane from possible odds by analyzing the effects caused by defect detected. Based on alerts and cause, business decisions are made. Airplane health surveillance system will obtain data from simulator which is designed to emulate the data received from supplier ground system which in turn will be receiving ACARS from flying fleet in real time. This is a User-friendly application which has a very powerful impact on aerospace division by eliminating the uncertain economic loss.

Keywords Airplane communications addressing and reporting system · Flight deck effect · Managed file transfer · Airplane system performance monitoring · Airplane health information · Air-to-ground communication

1 Introduction

In earlier decades, manual inspections and scheduled maintenance are the well-known maintenance strategies followed. Manual inspections seem to be not so accurate due to human errors leading to unnoticed defects which would result in airplane's failure. Whereas scheduled maintenance requires crew to replace component even though they are in good operational condition, both resulting in loss to the airplane manufacturers and operators. This indicates the need for predicting

N.B. Rachana (✉) · S. Seema
M.S. Ramaiah Institute of Technology, MSR Nagar, Bengaluru, Karnataka, India
e-mail: nbrachana@gmail.com

S. Seema
e-mail: seemas@msrit.edu

H.S. Behera and D.P. Mohapatra (eds.), *Computational Intelligence in Data Mining*, Advances in Intelligent Systems and Computing 556,
DOI 10.1007/978-981-10-3874-7_15

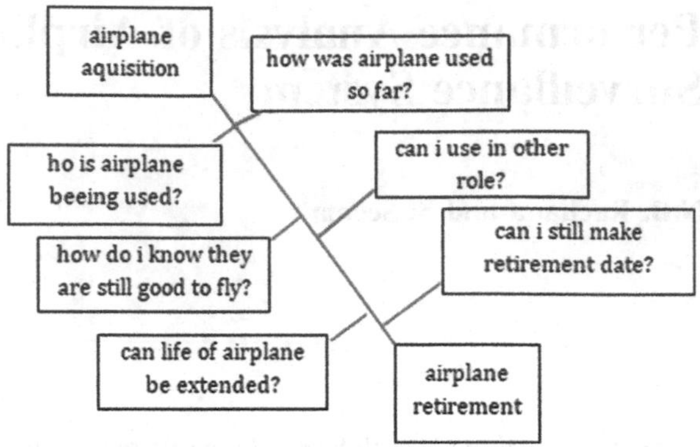

Fig. 1 Types of business decisions

faults even before they occur and replacing the component only when it reaches the threshold. This is a tedious job as predicting the faults is not so easy and data obtained from the airplane is vast.

Airplane Health Surveillance System (AHSS) provides solution for this problem. It is developed to help maintenance crew in performing predictive maintenance. This system is intended to help ground crew to make appropriate business decisions. AHSS has the ability to help operators to make business decisions about airplane from the time of airplane acquisition to the retirement of that airplane. These business decisions are illustrated in Fig. 1.

2 Literature Review

Structural Health Monitoring System (SHMS) is developed to monitor the physical structure of airplane by embedding sensors into body of airplane. In this work the author's intension is to provide continuous and autonomous monitoring capabilities to the system. Work also concludes prosperity achieved by implementing decision fusion algorithm to SHMS [1].

Research carried out in [2] encapsulates benefit of SHMS. To boost the performance of SHM system decision fusion algorithm is used to assimilate knowledge about the system. Different classifiers are used and decision made by these classifiers is merged using decision fusion methods to make a consolidated and accurate decision about system which is monitored. Authors in [3] consider SHMS as subject of their work. In this article author's intension is to show the benefits of implementation scenarios of SHMS to airplane operator and possible expected users of SHMS and also concludes how this system can change next generation airplane maintenance. In similar way author of [4] summarizes that the intension of damage monitoring is to scale down the cost and time required for inspecting the airplane for defect and thereby enhancing repair plan.

With similar intensions, Boeing also provides a complete airplane health solution, and it is named as Integrated Vehicle Health Management (IVHM) [5]. Boeing has onboarded IVHM system into airplane and this system will continuously collect sensors data and load it to ground maintenance system. This method will save waiting time for health data download after landing [6]. System is designed to work in a near real time, this empowers pilots to perform corrective operations in-flight and also based on this mechanics, maintenance crew can schedule maintenance tasks and execute them once the plane lands [7, 8]. Purpose of Aircraft Health Management (AHM) from Boeing [9] is to develop a system which has the capacity to convert the available information of an airplane or a fleet into useful and actionable information. This system is powerful with data-driven capability, enabling maintenance staff and engineers to take timely decisions regarding the maintenance leading to improved fleet operation and operational performance, scale down diversions, cancelations, and delays.

3 Airplane Health Surveillance System

In real-time implementation of AHSS, an onboard maintenance system is placed in airplanes and all the airplanes are embedded with sensors. Any abnormal condition detected by these sensors is sent to telecom providers which in turn will be sent to ground system.

Figure 2 illustrates system behavior, when an ACARS message is obtained from airplane in-flight. When any ACARS is generated, onboard system checks for alerts and alerts which are critical will be sent to the supplier ground system via satellite. Supplier ground system will enhance the received files and send them to AHSS. AHSS processes the alerts and for any further details about alert, AHSS will make a request to airplane to send details. Noncritical alerts are downloaded after landing using Wi-Fi at destination airport.

3.1 Standard Application Prerequisites

- Operator should be subscribed to the application.
- User should be connected to the network.
- User opens up web-browser to access the airplane supplier's web portal.
- User has proper credentials to access the application.

3.2 Implementation

For implementation purpose instead of obtaining ACARS data from airplane in-flight, a simulator is built to generate ACARS. Hence the input to AHSS is

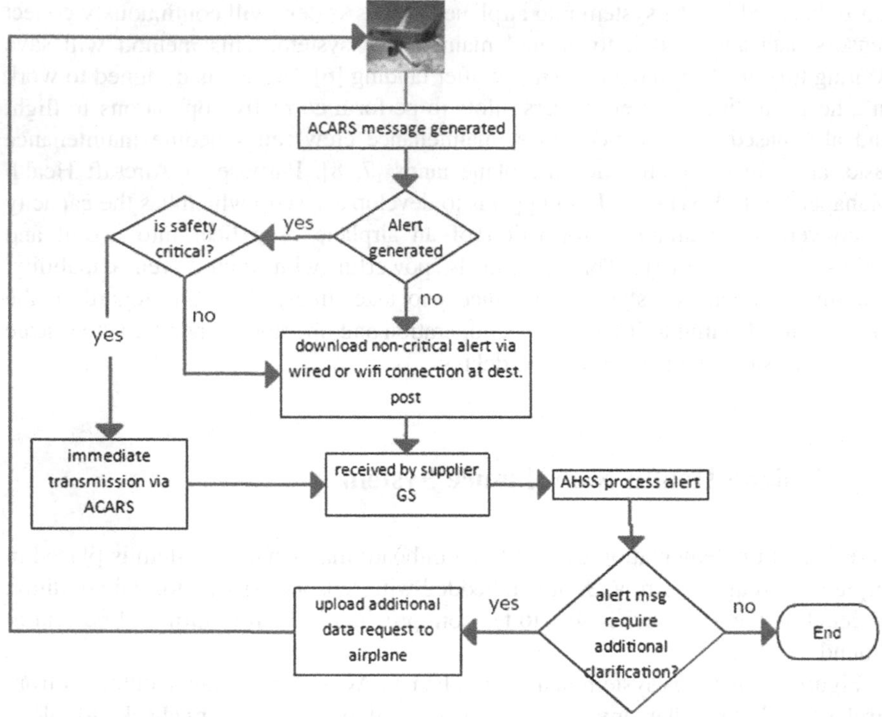

Fig. 2 AHSS flow chart

obtained from simulator. Simulator has the ability to generate three types of input files. One is OOOI messages, which indicates OUT, OFF, ON, IN messages, and these messages are consumed by position reporting module. OUT message indicates GATE OUT (particular airplane is departing from the airport). OFF messages are sent to indicate the start of crews phase. ON is for indicating arrival of airplane to the destination. When an IN message is received, this means that the airplane has reached destination airport. In between OOOI message simulator will be generating fault messages which contain CAS (6 digit FDE code) and OMS (9 digit fault code) messages. At the end, every single PFS (Post Flight Summary) report is sent which includes all faults occurred in that particular flight. Simulator is made to run 24 × 7 to show the application behavior while receiving data continuously and also historical data is displayed on screen.

Along with simulator a utility tool is also developed in order to auto update the database as and when reference data is received from supplier. Supplier will be sending reference data in .xlsx format. Maximum 25 versions of CAS and OMS faults are maintained in database. Also the priority tables are received from supplier, approved by airworthy authorizations [10], are stored in database with the help of this utility tool.

4 Proposed Methodology

All the modules of AHSS are briefed in this section. There are totally four modules. Namely, User alerting module, Position reporting module, Fault correlation module, and Troubleshooting module. Once the application is launched, user will land on user alert module.

User alert module: This module will be processing fault messages and is designed to show all FDEs which are having priority with different color coding. P1 priority will be displayed in red, Priority P2 will be displayed in yellow, and P3 in gray.

Position reporting module: OOOI messages are processed by this module and it displays date and time when each of these messages are received. Along with this, flight progress bar is also displayed for better visual effects. Progress bar is calculated using Estimated Time of Arrival (ETA) of airplane which is received through OOOI messages.

Fault correlation module: This module will display both historical and real-time data. CAS will be correlated to OMS codes. Initially, when a CAS is received the system will search for all correlated OMS for that CAS, if any correlation is found then display CAS and OMS correlation in fault correlation module. Along with this fault message and number of times a particular fault has occurred in a single leg is displayed.

Troubleshooting module: This screen is used in real-time troubleshooting and offline fault analysis by the ground crew. Visual cue to know if the airplane is in-flight or active is provided by displaying box in yellow or gray in "Active" column. Application will retrieve historical data of the airplane for the 2 months or maximum 150 legs. Troubleshooting screen will be displayed when the user clicks a specific fault message or code in fault correlation screen. Troubleshooting screen will have details about the CAS, INFO and fault message, FDE description, help text, parameter group, possible causes, and so on.

5 Performance Analysis

Performance analysis plays a crucial role in any software development. HP Load runner is used for performance analysis. AHSS is obtaining input data from simulator and each scenario is repeated for 1 user, 25 users, and 50 users. Comparison of all 3 scenarios along with detail of performance analysis on each module of AHSS is shown below.

5.1 User Alert Module

Table 1 summarizes the performance of alert module, maximum allowable response time is 20 s, and alert screen is taking 2.071 s for one user indicating good performance.

Table 1 Performance results of user alert module

Parameter	1 user	25 users	50 users
Max. running users	1	25	50
Avg. response time (s)	2.071	2.283	2.735
Total throughput (bytes)	1,808,336	44,284,365	88,482,907
Avg. throughput (bytes/s)	31,178	375,132	574,564
Total hits	74	1,850	3,697
Avg. hits per s	0.804	14.919	24.006

Figure 3 shows the hits per second for the user alert module for 50 concurrent users. Graph (Fig. 4) indicates that system is balancing all the users equally or load is distributed equally throughout the system.

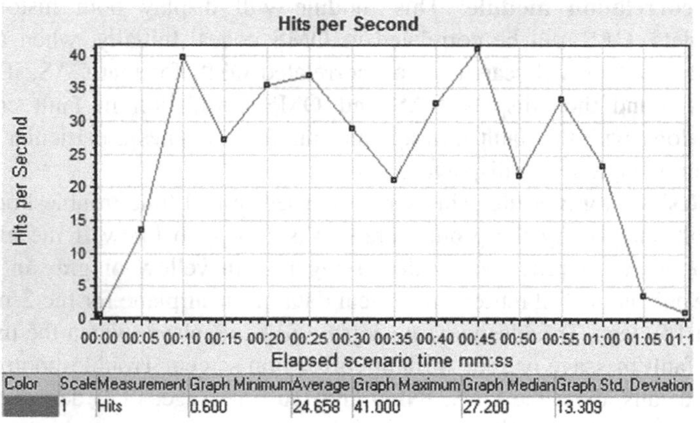

Fig. 3 Graph of user alert screens hits/s for 50 users

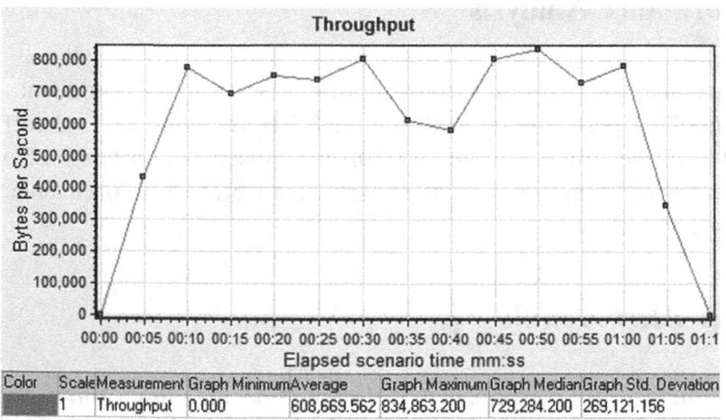

Fig. 4 Graph of user alert module throughput for 50 users

5.2 Position Reporting Module

The response time of position reporting screen is found to be 3.072 s for 50 users using the application concurrently. Graphs for hits/s and response time indicate the load on server and on network, respectively. Indicating system has enough capacity to balance the load. Figure 5 indicates that as and when users enter the system and OOOI messages are received, application is displaying the data on position reporting screen (Table 2).

5.3 Fault Correlation Module

The response time of fault correlation module is below threshold value indicating good response. But the load on network is comparatively more for this module.

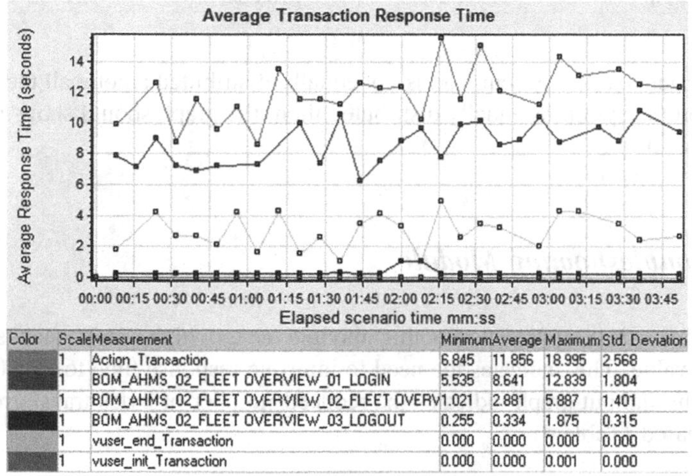

Fig. 5 Graph of response time for 25 users for position reporting module

Table 2 Performance results of position reporting module

Parameter	1 user	25 users	50 users
Max. running users	1	8	9
Avg. response time	2.831	2.881	3.072
Total throughput (bytes)	1,808,336	45,215,060	90,365,342
Avg. throughput (bytes/s)	31,178	189.184	215,156
Total hits	76	1,900	3,800
Avg. hits per s	1.31	7.95	9.048

Table 3 Performance results of fault correlation module

Parameter	1 user	25 users	50 users
Max. running users	1	25	50
Avg. response time	5.98	5.98	5.98
Total throughput (bytes)	5,224,643	130,473,593	260,858,252
Avg. throughput (bytes/s)	41,465	705,263	999,457
Total hits	118	2,950	5,896
Avg. hits per s	0.937	15.946	22.59

Table 4 Performance results of fault troubleshooting module

Parameter	1 user	25 users	50 users
Max. running users	1	25	50
Avg. response time (s)	10.658	29.231	54.308
Total throughput (bytes)	5,415,835	135,103,959	270,142,059
Avg. throughput (bytes/s)	48,356	711,073	941,262
Total hits	142	3,550	7,098
Avg. hits per s	1.268	18.684	24.732

This indicates that the system load is not equally distributed among all the users. If the load on the system is equally distributed then, the graph should show a straight line (Table 3).

5.4 Troubleshooting Module

The response time of troubleshooting module is too high, more than specified threshold value. This indicates the need to improve response time to <=20 s steep down in throughput graph indicates the reduced number of bytes transferred when load increased (Table 4).

6 Conclusion

The intended application platform for AHSS is single-aisle body passenger airplanes. A visual dashboard allows operator to see a summary view of their airplanes that are in operation. Airplanes that have a fault/alert reported will be highlighted. Upon focusing on the highlighted airplane the user shall be presented with further detailed views containing fault detail, occurrence, and severity. This system supports anticipatory planning and quick turnaround. To conclude, a user-friendly and very powerful system to detect and diagnose the defects of in-flight airplane was developed. The system was thoroughly tested with the synthetic data. The performance analysis results indicate that the performance of AHSS is satisfying the expected criteria and also the response time is below 20 s.

7 Future Enhancements

In the present implementation, it is observed that for troubleshooting module the response time is increasing when number of concurrent users is increased. This can be taken up as a future enhancement in order to reduce the response time. Along with this, the expected delay of 20 s can be reduced to make AHSS work in real time. Predictive analysis or advanced analytics could be performed by storing the obtained sensor data into HADOOP storage thereby making good business decision.

Acknowledgements My sincere gratitude to Professor and Head of Department Dr. K G Srinivasa, CSE, MSRIT for his valuable suggestions and encouragement. Special thanks to my guide Dr. Seema S, Associate Professor CSE, MSRIT for her constant guidance and suggestions.

References

1. Luiz G. A. Martins, "Architecture of a Remote Impedance-Based Structural Health Monitoring System for Airplane Applications", J. of the Braz. Soc. of Mech. Sci. & Eng. 2012 by ABCM Special Issue 2012, Vol. XXXIV.
2. Johannes Ledolter, et al., "Data mining and business analytics with R", John Wiley & Sons, Inc., Hoboken, New Jersey, 1nd edition, 2013.
3. Rabatel, J, et al., "Sensor mining for anomaly detection in railway data", in Proceedings of the 9th Industrial Conference on Advances in Data Mining. Berlin, 2009, pp. 191–205. Springer.
4. SalehZein-Sabatto, et al., "Information & Decision Fusion Systems for Airplane Structural Health Monitoring", IEEE 2011, pp 395–400.
5. Divakaran et al., Integrated vehicle health management of a transport aircraft landing gear system, White paper, Infosys, 2015.
6. J. Pinsonnault, et al, "Prospective for Structural Health Monitoring System Implementation on Civil Airplane", Smart materials, structures & aerospace Conference, Nov 2011, Canada.
7. Daryl Stephenson, "The airplane Doctors", Boeing. http://www.boeing.com/news/frontiers/archive/2006/august/ts_sf09.pdf.
8. A. Ben Zakour, et al., "Case Based Model to enhance airplane fleet management and equipment performance", 2MoRO Solutions Bidart, France.
9. Xiaoyun Wang, "Design of integrated airplane inflight safety monitoring and early warning system", Prognostics & System Health Management Conference, 2010, IEEE.
10. Jonas Peeters, "Aircraft maintenance operations: state of the art", Hubrussel research paper, 2013. Link: https://lirias.kuleuven.be/bitstream/123456789/426843/1/13HRP09.pdf.

A Parallel Forecasting Approach Using Incremental K-means Clustering Technique

Swagatika Sahoo

Abstract A parallel forecasting approach used in weather prediction which has important aspect of the modern society, especially with the realization of modern smart cities. A new approach is considering here which will provide excellent result for large semi-structured data. Thus, it is very important to analyze weather data keeping in mind the enormity of the available data sizes. Here it is presented a methodology for weather data analysis keeping in mind the big data nature of the data sizes pertaining to weather data. Here also it has been taken date-wise atmospheric conditions collected of decade. The traditional k-means clustering is used to form clusters which represents association in between related dates of current year's and previous year's weather data. Such associations predict atmospheric conditions of one year's weather condition on the bases of previous data. Incremental k-means clustering algorithm is used to process current year's weather parameters as new data and it shows that the calculated weather condition falls under one of the existing clusters to represent similar atmospheric conditions. The total work has been divided by two parts: first, Storing NCDC semi-structured data on hadoop cluster and second, fitting a clustering methodology for predicting weather conditions.

Keywords Hadoop • Bigdata • Mapreduce framework • Forecasting • Incremental • K-Means

1 Introduction

Weather forecasting is prediction of weather condition in future for a given location. This process needs to collect data of atmospheric conditions which records the humidity, rainfall, temperature, pressure, dew point, wind direction, elevation etc.

S. Sahoo (✉)
Department of Computer Science & Engineering,
National Institute of Science & Technology, Berhampur 761008, Odisha, India
e-mail: swagatikasun12@gmail.com

H.S. Behera and D.P. Mohapatra (eds.), *Computational Intelligence in Data Mining*, Advances in Intelligent Systems and Computing 556,
DOI 10.1007/978-981-10-3874-7_16

And Weather forecasting highly impact on day to day life and also on different sectors like agricultural sector, industry sector etc. To do weather forecast, a user need to collect weather data from the past years related data and also need to understand what processes are occurring in the atmosphere for predict current weather condition at a particular location. Data Mining is the technology which extracts meaningful information by analyzing large dataset [1–3] and predict vital information from the databases and use of this approach we can extract weather information from weather dataset (NCDC dataset) which is semi-structured and huge in space. So, it is quite difficult to manage it and more difficult to analyze it. Such large data set cannot be kept in traditional database for future usage. Such large and semi-structured data require a richer analytics, a complex system to manage it and take more time to analyze it. However, now a day's all most all sector peoples are using map reduce framework for analysis such large data sets [4]. To accelerate the performance of this approach we used Apache Hadoop [5]. Then for prediction it has been chosen k-means clustering, which will give best prediction for dataset similar to our extracted dataset [6, 7]. In this research paper, k-means clustering technique used to segregate the log data into different clusters depending upon current parameters values and whenever a new data come to database we use incremental k-means clustering technique [8–10]. These clustered data sets are used for future prediction from real time to annual period. The efficacy of the proposed technique is presented in this paper. The remaining part of this paper organized as follows: methodology and proposed forecasting model explained in described in Sect. 2. In Sect. 3 discuss about results. Section 4 concludes this paper.

2 Methodology

In this research paper, main objective is to predict future weather condition by the help of storing and mining of NCDC dataset which is large and semi-structured multidimensional dataset consist of multiple weather parameters. NCDC dataset store data of different time of each day. We use hadoop for storing these data and run map-reduce framework for retrieving useful information from database for further analysis and applying proposed clustering technique. The main approach of this paper is that first to store this data on multi-node hadoop. Then apply typical k-means clustering algorithm on existing extracted dataset and category the weather according to their clusters mean value. When new data comes to database, apply incremental k-means clustering algorithm by calculating distance between new data with existing cluster mean values and category weather by considering minimum distance belonged cluster.

2.1 Proposed Model

The proposed prediction model has been used for semi-structured NCDC dataset [11] for solving big data problem and give better prediction technique for large, dynamic, continuously changing multivariate weather dataset. The proposed prediction model has shown below comprises of prediction processes and steps in Fig. 1 and brief description of proposed model has shown below.

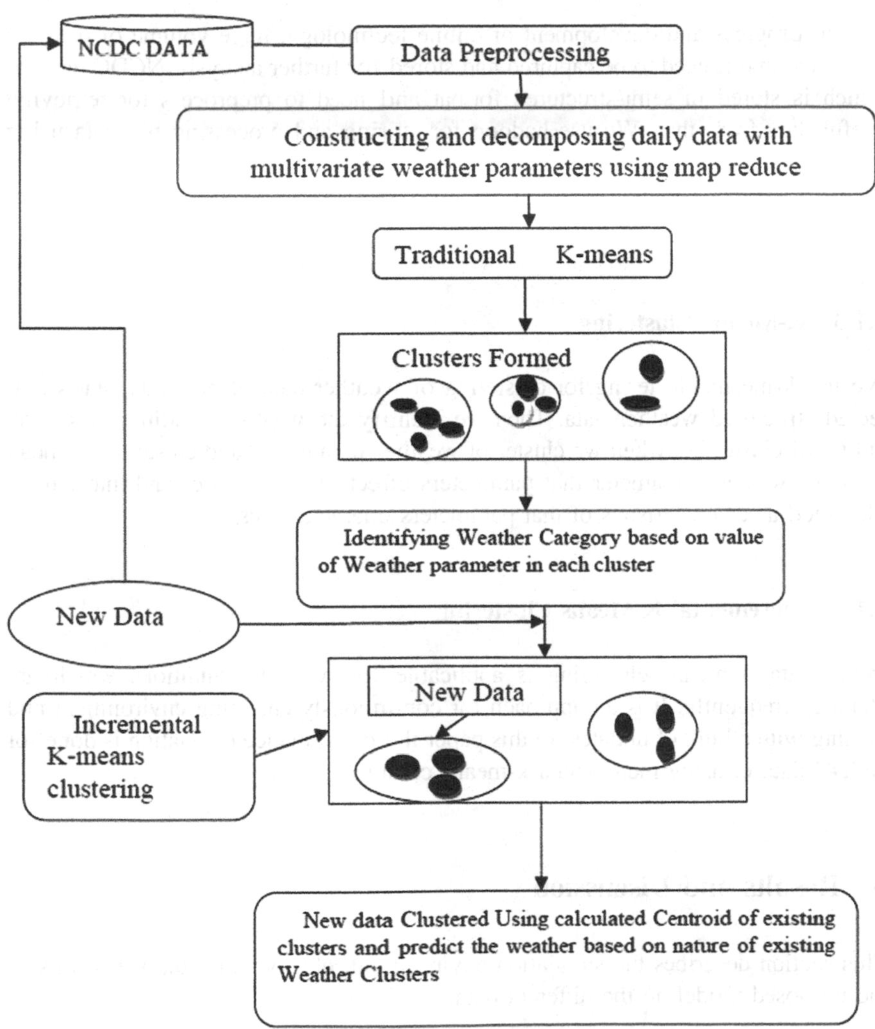

Fig. 1 Proposed model for weather prediction using incremental k-means

2.1.1 NCDC Dataset

This research used NCDC data [11] set about Climate Condition that has maintained datasets of the US Climate dataset with vast number of other climate monitoring product. This dataset has many attributes but we considering first 5 attributes for our analysis i.e. date, temperature, pressure, elevation, wind direction.

2.1.2 Data Preprocessing and Constructing for Weather Forecasting

Due to progress and development of online technology, huge volume of data are generated and it need to be captured and stored for further analysis. NCDC dataset, which is stored in semi-structured format and need to preprocess for retrieving useful data from that. We use hadoop for storing and processing big data using map-reduce framework. Then constructing and decomposing daily data with multivariate weather parameters which are useful for further analysis i.e. for weather forecasting.

2.1.3 K-Means Clustering

We use k-means clustering for clustering of weather data of previous year's collected structured weather data. Then we identify the weather condition based on values of cluster i.e. when we cluster of weather parameters and cluster have mean value of weather parameter that parameters effect on that cluster and that cluster identified as characteristics of that parameters characteristics.

2.1.4 Incremental K-Means Clustering

Incremental k-means clustering is applicable for weather conditions which are changed frequently. It is an approach for continuously changing environment and dealing with a bulk of updates. In this paper the performance evaluation is done for NCDC dataset using incremental k-means clustering.

3 Results and Discussion

This section describes the simulation environment, observations and the results of the proposed model on the different data.

3.1 Simulation Environment

In this work we configured a Hadoop cluster by implementing Apache Hadoop V1.2.1 and Apache pig V0.11.1 packages available on http://apache.org respectively. At that moment, I configured those both clusters on a single node machine which was having the hardware configurations of Intel Core 2Duo CPU T6600 @ 2.20 GHz × 2 Processor, main memory of 2 GB, Physical Disk space of 250 GB running on Ubuntu 14.04 LTS 3.13.0.57-generic kernel. For coding, statistical analysis and visualizing purpose I used java version 1.2.0_71 and WEKA 3.8.0.

3.2 Result Obtained and Observation

The Experiment is carried out over the NCDC weather dataset comprising 1901–1925. First, we have using map-reduce framework for make understandable format for further analysis and the result of this experiment with prominent attributes is shown in the Table 1. Then Table 2 show mean value of initial clusters. Table 3 show weather categories of clustered data. Table 4 show new arrival data of year

Table 1 Formatted NCDC data for weather forecasting

Date	Temperature (°C * 10)	Pressure (hectopascals * 10)	Elevation (m)	Wind direction (°)
01/01/1901	−67	10271	102	320
02/01/1901	−11	10251	102	320
…	…	…	…	…
01/01/1905	−61	10383	102	360
02/01/1905	17	10216	102	250
…	…	…	…	…
01/01/1910	6	10260	256	270
02/01/1915	56	10094	256	270
…	…	…	…	…
01/01/1915	22	10194	256	230
02/01/1915	28	10207	256	360
…	…	…	…	…
01/01/1920	22	10100	181	360
02/01/1920	17	10093	181	250
…	…	…	…	…
01/01/1925	39	10039	181	270
02/01/1925	44	9983	181	230
…	…	…	…	…

Table 2 Means values of initial clusters

Cluster id	Cluster of temperature mean	Cluster of pressure mean	Cluster of elevation mean	Cluster of wind DIR. mean
Cluster0	69.68	10169.98	130.76	209.87
Cluster1	96.00	10182.87	256.00	317.01
Cluster2	99.76	10166.65	176.47	336.11
Cluster3	98.88	10140.92	98.68	335.60

Table 3 Weather Category according to cluster data

Cluster number	Weather category
Cluster0	Stormy, cool
Cluster1	Stormy, hot
Cluster2	Clear, windy
Cluster3	Clear, hot, windy

Table 4 Weather data of the year 1926

Date	Temperatue (°C * 10)	Pressure (hectopascals * 10)	Elevation (m)	Wind direction (°)
01/01/1926	100	10233	181	360
02/01/1926	100	10175	181	360
…	…	…	…	…
01/03/1926	61	10320	181	320
02/03/1926	78	10323	181	360
…	…	…	…	…
01/06/1926	211	10150	257	230
02/06/1926	222	10323	257	340
…	…	…	…	…
01/09/1926	222	10255	348	360
02/09/1926	228	10206	348	360
…	…	…	…	…
01/12/1926	61	10288	348	360
02/12/1926	89	10248	348	360
…	…	…	…	…

1926. Table 5 show weather forecasting from January 1926 to December 1926 using Incremental K-means Clustering Technique. After getting Table 1 formatted weather data it is need to apply k-means clustering algorithm where k = 4 and it is also used 'Euclidean metric' for distance calculation. Table 2 show calculated centroid of each parameters mean values.

Table 3 generates data depends on the different parameters of Table 2. It shows about cluster nature according to cluster's mean value of each parameter. Applying

Table 5 Weather forecasting from January 1926 to December 1926 using incremental K-means clustering technique

Date	New data inserted into	Weather category
01/01/1926	Cluster0	Stormy, Cool
02/01/1926	Cluster0	Stormy, Cool
…	…	…
01/03/1926	Cluster1	Stormy, Hot
02/03/1926	Cluster0	Stormy, Cool
…	…	…
01/06/1926	Cluster3	Clear, Windy, Hot
02/06/1926	Cluster3	Clear, Windy, Hot
…	…	…
01/09/1926	Cluster2	Clear, Windy
02/09/1926	Cluster2	Clear, Windy
…	…	…
01/12/1926	Cluster2	Clear, Windy
02/12/1926	Cluster2	Clear, Windy
…	…	…

typical k-means clustering, it has been identified the nature of each cluster depends upon the mean value of attributes of that cluster. Table 4 shows about upcoming formatted weather data for prediction of upcoming days weather category.

Inserting upcoming day's weather data and weather data belong to the any of cluster by calculating from means value of existing weather data and the weather of that particular day is identified as according to their nature of cluster due to the effect of mean values of attributes. Then the result of forecasting is shown in 'Table 5' from the month of January 1926 to December 1926 (the database contains all the data of the NCDC dataset). Here for calculation "Euclidean metric" is used.

4 Conclusion

The goal of this paper was to mining the useful weather data and forecast weather using Incremental k-means clustering approach. So to achieve this, we use of Map Reduce and application like pig enhance the performance of big data storage and prediction technique. The above experimental results present the efficacy of the modeled approach. It is observed that by using Hadoop, Map Reduce framework and pig, reliability and security of the stored data set increased tremendously. Decentralized storage can be achieved with low cost commodity hardware units which is more economically cost efficient than current vertical scalable high-end servers system with the help of HDFS. Performance of any such systems depends on the frequent access of data, however frequent accessing data from a single server will reduce the performance greatly. So to improve the system performance we used

Hadoop's HDFS for storing huge amount of data after converting semi-structured to structured form using Map Reduce framework. Then we use Incremental K-mean clustering which improves Clustering performance. The efficacy of the proposed model presented in result section.

References

1. Kalyankar, M., and S. Alaspurkar. "Data Mining Technique to Analyse the Metrological Data." International Journal of Advanced Research in Computer Science and Software Engineering 3.2 (2013): 114–118.
2. Rajinikanth, T. V., Balaram VVSSS, and N. Rajasekhar. "Analysis of Indian weather data sets using data mining techniques." ACITY, WiMoN, CSIA, AIAA, DPPR, NECO, InWeS 1.1 (2014): 89–94.
3. Chauhan, Divya, and Jawahar Thakur. "Data Mining Techniques for Weather Prediction: A Review." International Journal on Recent and Innovation Trends in Computing and Communication 2.8.
4. Yang, Jie, and Xiaoping Li. "Mapreduce based method for big data semantic clustering." Systems, Man, and Cybernetics (SMC), 2013 IEEE International Conference on. IEEE, 2013.
5. Polato, Ivanilton, et al. "A comprehensive view of Hadoop research—A systematic literature review." Journal of Network and Computer Applications46 (2014): 1–25.
6. Suguna, Nambiraj, and Keppana G. Thanushkodi. "Predicting missing attribute values using k-means clustering." Journal of Computer Science 7.2 (2011): 216.
7. Oyelade, O. J., O. O. Oladipupo, and I. C. Obagbuwa. "Application of k Means Clustering algorithm for prediction of Students Academic Performance." arXiv preprint arXiv:1002. 2425(2010).
8. Pham, Duc Truong, Stefan Simeonov Dimov, and C. D. Nguyen. "An incremental K-means algorithm." Proceedings of the Institution of Mechanical Engineers, Part C: Journal of Mechanical Engineering Science 218.7 (2004): 783–795.
9. Chakraborty, Sanjay, N. K. Nagwani, and Lopamudra Dey. "Weather Forecasting using Incremental K-Means Clustering." Data Mining and Knowledge Engineering 4.5 (2012): 214–219.
10. Dey, Ratul Dey Sanjay Chakraborty Lopamudra. "Weather forecasting using Convex hull & K-Means Techniques An Approach." arXiv preprint arXiv:1501.06456(2015).
11. [Online]. Available: ftp://ncdc.noaa.gov/pub/data/noaa/.

Key Author Analysis in 1 and 1.5 Degree Egocentric Collaboration Network

Anand Bihari and Sudhakar Tripathi

Abstract The evaluation of scientific performance of an individual author in the research community is based on total number of articles published, citation count, average citation count, h-index, etc. But generally research work is done by the group of researchers and the performance of individual may depends on the neighbors' researchers. So the evaluation of scientific impact of an individual will carry based on their immediate neighbors. In this paper, we form a 1 degree and 1.5 degree egocentric network of every node and convey the degree centrality, closeness centrality, and density to estimate the scientific efficacy of an individual. Finally, compare the results of all those types of network and found that the 1.5 degree egocentric networks are useful to appraise the performance of researchers.

Keywords Social network · Degree centrality · Closeness centrality · Egocentric network · Density

1 Introduction

The collaboration network of research professionals is a type of social network. In this network, a node represents the individual researcher and the edge between nodes represents the collaboration/co-authorship. Evaluation of scientific impact of any individual is determined by the total number of publication, total citation count, average citation count, h-index, etc. But how an individual publish more number of publication, gain more citation count, high h-index, etc. The answer of these questions is to an individual got all those things based on their neighbors authors. It means an author have more number of co-author then have a probability to give highest con-

A. Bihari (✉) · S. Tripathi
Department of Computer Science & Engineering,
National Institute of Technology Patna, Patna, India
e-mail: anand.cse15@nitp.ac.in

S. Tripathi
e-mail: stripathi.cse@nitp.ac.in

H.S. Behera and D.P. Mohapatra (eds.), *Computational Intelligence in Data Mining*, Advances in Intelligent Systems and Computing 556,
DOI 10.1007/978-981-10-3874-7_17

tribution in the research community. In this paper, we evaluate the performance of individuals based on their neighbors. For this, we draw an 1 degree and 1.5 degree egocentric network and evaluate the individuals based on the social network metrics like degree centrality, closeness centrality, and density.

2 Related Work

Newman [1] Suggested the concept of weighted collaboration network based on the amount of co-authors. The co-authorship weight between the two authors is the paper and their author ratio. After that, use power law to calculate the total co-authorship weight between authors. Abbasi et al. [2, 3] In this paper author make a weighted co-authorship network and uses the social network analysis metrics for evaluating the performance of individual researcher. Bihari and Pandia [4] In this paper, first form a co-authorship network and find out the key author based on centrality measures (degree, closeness, betweenness, and eigenvector centrality) in SNA and citation indices (Frequency, Citation count, h-index, g-index, and i10-index). In this paper, author set the collaboration weight is the total no. of citation of all paper that published together Pandia and Bihari [5]. In this paper, author constructs the research professionals' relationship network and discovers the prominent researcher in the research community of IEEE by centrality methods like degree, closeness, betweenness, and eigenvector centrality. Arnaboldi et al. [6],discuss the co-authorship ego network and investigate the behavior of each individual node. In this paper, author discusses the h-index and average number of authors and finds out the relation between h-index and average number of author. Shi et al. [7], in this paper author proposed the 1.5D egocentric dynamic visualization framework with data processing pipeline. Bihari et al. [8] investigate the collaboration network using social network analysis metrics based on maximum spanning tree (MST) and discover the key or prominent researcher in the community. But in this network so many connections are removed that may be useful in research contribution.

3 Methodology

3.1 Egocentric Network

Egocentric network is a type of subgraph that is based on individual nodes. It represents the individual existence of a node. In this network, subject of interest at the center and subgraph considered with one hop distance as well as 1.5 diameter. Evaluation of egocentric network is based on the density. An egocentric network is highly dense that means the node that is the base of egocentric network is highly influence than other.

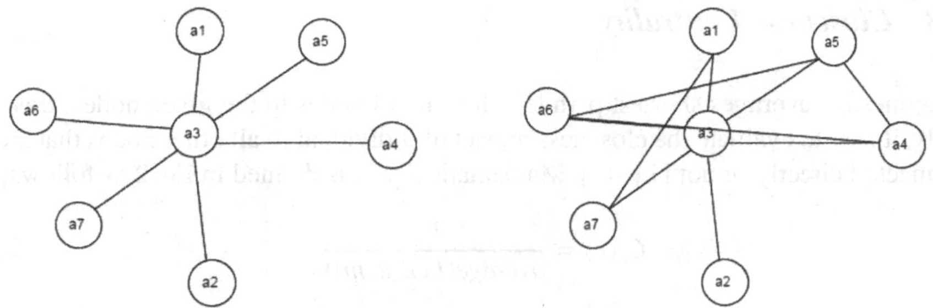

(a) 1-degree egocentric network of a3 (b) 1.5-degree egocentric network of a3

Fig. 1 Egocentric collaboration network an example

3.1.1 1 Degree Egocentric Network

Basically, one degree egocentric network shows the direct relation of base node of egocentric network. In this network, the links are established with all those nodes that are directly connected to the based node. Example of one degree egocentric network is shown in Fig. 1a.

3.1.2 1.5 Degree Egocentric Network

1.5 degree egocentric network shows the relation between base node and their neighbors nodes. It characterize that the how many nodes that are connected to base node and how many nodes that are connected to each others. The 1.5 degree egocentric network considers the 1.5 diameter from the based node. Example of 1.5 degree egocentric network is shown in Fig. 1b.

3.2 *Degree Centrality*

It is simply count the total number of immediate neighbors [9–12]. Mathematically, normalized degree centrality is defined in Eq. 1 as follows:

$$d_c(K) = \frac{deg(K)}{m-1} \quad (1)$$

where deg(k) is the total number of immediate neighbors of node k and m is the total number of node available in the network [13].

3.3 Closeness Centrality

It defines the averages shortest path length form all nodes to the given nodes. Basically, it uses to evaluate the closeness impact of individual to all other nodes that are connected directly or not [14–16]. Mathematically, it is defined in Eq. 2 as follows:

$$C_c(k) = \frac{1}{average(Len(k, m))} \tag{2}$$

where Len(k, m) is the length of the shortest path between two nodes k and m [17, 18].

3.4 Density

Generally density is used to compare the network and subnetwork. In egocentric network, density measures the density of a node. It will help to compare the nodes to determine the individuals impact in the network. A node is highly dense, means the neighbors nodes has direct connection with other neighbors nodes.

4 Data Collection and Analysis

We analyze the 26802 publication including 11220 journal and 15582 conference proceedings downloaded from IEEE Xplore in July 2014 [4]. The data are from January 2000–January 2014. The data set contains the information related to publication title, authors name and affiliation, publication year and citation count. After extracting the author-pair, 176503 author pairs and 61546 authors data were finally exist for inquiry. After analysis of these raw data, we found 3230 papers are written by single author and some of the papers are most co-authored. The summary of data is shown in Table 1.

Table 1 Summary of data set

Sl. no.	Attribute	Values
1	Data source	IEEE Xplore
2	Time period	January 2001–January 2014
3	No. of conferences	17487
4	No. of authors	61546
5	Total no. journal articles	11220
6	Total no. of conference proceedings	15582

5 Egocentric Collaboration Network

For evaluation of individuals scientific impact in egocentric network, first we extract the author and its co-author form data set. Then we construct the 1 degree and 1.5 degree egocentric network using python and netwokrx [19] based on the following technique: Let us consider the following three article shown in Table 2. After that we

Table 2 Collaboration table

Articles ID	Authors details	Citation count
1	a1, a2, a3	5
2	a4, a3, a5, a6	20
3	a1, a7, a3	26

Table 3 Top 24 authors in 1 degree egocentric network

Sl. no	Author ID	Degree centrality	Closeness centrality	Density
1	61365	0.001755	0.000000	0.018856
2	48688	0.001640	0.133474	0.020190
3	5340	0.001591	0.133369	0.020821
4	56856	0.001558	0.129517	0.021264
5	16947	0.001558	0.000000	0.021264
6	41078	0.001476	0.000000	0.022458
7	3667	0.001411	0.011379	0.023515
8	56899	0.001329	0.002491	0.024985
9	8935	0.001329	0.000000	0.024985
10	33989	0.001280	0.000000	0.025958
11	36605	0.001198	0.128872	0.027761
12	1370	0.001132	0.128421	0.029394
13	11735	0.001132	0.128421	0.029394
14	17849	0.001132	0.128421	0.029394
15	20758	0.001132	0.128421	0.029394
16	22469	0.001132	0.128421	0.029394
17	23930	0.001132	0.128421	0.029394
18	1000	0.000033	0.000000	1.997575
19	10009	0.000033	0.000000	1.997575
20	10039	0.000033	0.000000	1.997575
21	10028	0.000033	0.000000	1.997575
22	10019	0.000033	0.000000	1.997575
23	1004	0.000033	0.000000	1.997575
24	10018	0.000033	0.000000	1.997575

create an adjacency list of author pairs and calculate the collaboration weight, total citation of all papers that has published together like {a1–a2, 5},{a1–a3, 31}, {a2–a3, 5},{a3–a4, 20}, etc. Then, we construct the 1 degree and 1.5 degree egocentric network by using python and networkx [19]. The network is looks like Fig. 1.

6 Analysis and Result

Our basic objective is to analyze the performance of individuals based on degree centrality, closeness centrality, and the density of particular 1 degree or 1.5 degree egocentric network. We numerate the degree centrality, closeness centrality, and density in each author egocentric network. We select top 10 authors from each measures and combine it. Only 24 author's data in 1 degree egocentric network and 23 author's data in 1.5 degree egocentric network shown in Tables 3 and 4 are available in top 10 for analysis.

Table 4 Top 23 authors in 1.5 degree egocentric network

Sl. no	Author ID	Degree centrality	Closeness centrality	Density
1	61365	0.001755	0.061900	0.023788
2	48688	0.001640	0.062814	0.024629
3	5340	0.001591	0.062457	0.031975
4	56856	0.001558	0.063020	0.032224
5	16947	0.001558	0.059730	0.027751
6	41078	0.001476	0.072578	0.041411
7	3667	0.001411	0.064507	0.045923
8	56899	0.001329	0.064582	0.040706
9	8935	0.001329	0.059722	0.026218
10	33989	0.001280	0.072146	0.050570
11	46389	0.001280	0.071825	0.029617
12	13513	0.001165	0.073944	0.047444
13	20621	0.001033	0.072000	0.047069
14	33095	0.001033	0.071886	0.040421
15	25174	0.000968	0.070486	0.063648
16	24461	0.000722	0.071756	0.082379
17	16530	0.000656	0.070348	0.067896
18	27968	0.000607	0.070782	0.060020
19	1000	0.000033	0.054642	3.993938
20	10009	0.000033	0.053008	3.993938
21	10064	0.000033	0.049109	3.993938
22	1004	0.000033	0.000033	3.993938
23	10018	0.000033	0.000033	3.993938

Table 5 Common authors in 1 degree and 1.5 degree egocentric network

Sl. no	Author ID	Author name	Degree centrality	Closeness centrality	Density
1	1000	Kawakatsu T.	0.0000328089	0.0000000021	1.997575082
2	1004	Allen R.	0.0000328089	0.0000000000	1.997575082
3	3667	Takahashi T.	0.0014107843	0.0645067771	0.045922622
4	5340	Kamae T.	0.0015912335	0.0624567989	0.031975333
5	8935	Gilgenbach R.M.	0.0013287620	0.0597217937	0.026217707
6	10009	Sabir M.F.	0.0000328089	0.0530081159	3.993938441
7	10018	Gonzalez-Morales D.	0.0000328089	0.0000000000	1.997575082
8	10042	Knowles K.	0.0000328089	0.0000328089	3.993938441
9	16947	Lau Y.Y.	0.0015584245	0.0000000013	0.021263552
10	33989	Kim K.	0.0012795485	0.0721460914	0.05057001
11	41078	Kim J.	0.0014764022	0.0725775985	0.041411261
12	48688	Mizuno T.	0.0016404469	0.0628135023	0.024628669
13	56856	Fukazawa Y.	0.0015584245	0.0630196426	0.032224416
14	56899	Cressler J.D.	0.0013287620	0.0645821540	0.040706176
15	61365	Hirzinger G.	0.0017552781	0.0619001330	0.023787756

Hence no one author exist in all measures in both network. In 1 degree egocentric network, author ID 61365 (Hirzinger G.), author ID 48688 (Mizuno T.), author ID 1000 (Kawakatsu T.) are prominent based on degree centrality, closeness centrality, and density, respectively. In 1.5 degree egocentric network, author id 61365 (Hirzinger G.), author id 13513 (Wang Y.), and author id 1000 (Kawakatsu T.) are prominent based on degree centrality, closeness centrality, and density, respectively. Finally, we combine all those data and found that 15 authors are present in both type of network. So, we can say that these 15 authors are key author in the network shown in Table 5. We also discover the density of an egocentric network give the highest value than other in both type of network.

7 Conclusions

In this paper, we have investigated the research professionals' collaboration network as a egocentric network and evaluate the performance of individuals based on their direct neighbors and calculate the degree centrality, closeness centrality, and density of each node 1 degree and 1.5 degree egocentric network. 1.5 degree egocentric network gives the better results than 1 degree egocentric network.

References

1. M. E. Newman, Scientific collaboration networks. i. network construction and fundamental results, Physical review E 64 (1) (2001) 016131.
2. A. Abbasi, J. Altmann, On the correlation between research performance and social network analysis measures applied to research collaboration networks, in: 44th Hawaii International Conference on System Sciences (HICSS), 2011, IEEE, 2011, pp. 1–10.
3. A. Abbasi, L. Hossain, S. Uddin, K. J. Rasmussen, Evolutionary dynamics of scientific collaboration networks: multi-levels and cross-time analysis, Scientometrics 89 (2) (2011) 687–710.
4. A. Bihari, M. K. Pandia, Key author analysis in research professionals relationship network using citation indices and centrality, Procedia Computer Science 57 (2015) 606–613.
5. M. K. Pandia, A. Bihari, Important author analysis in research professionals relationship network based on social network analysis metrics, in: Computational Intelligence in Data Mining-Volume 3, Springer, 2015, pp. 185–194.
6. V. Arnaboldi, R. I. Dunbar, A. Passarella, M. Conti, Analysis of co-authorship ego networks, in: Advances in Network Science, Springer, 2016, pp. 82–96.
7. L. Shi, C. Wang, Z. Wen, H. Qu, C. Lin, Q. Liao, 1.5 d egocentric dynamic network visualization, Visualization and Computer Graphics, IEEE Transactions on 21 (5) (2015) 624–637.
8. A. Bihari, S. Tripathi, M. K. Pandia, Key author analysis in research professionals collaboration network based on mst using centrality measures, in: Proceedings of the 2016 International Conference on Information and Communication Technology for Competitive Strategies, ACM (In Press), 2016.
9. J. Tang, D. Zhang, L. Yao, Social network extraction of academic researchers, in: Seventh IEEE International Conference on Data Mining, 2007. ICDM 2007. IEEE, 2007, pp. 292–301.
10. Q. Deng, Z. Wang, Degree centrality in scientific collaboration super network, in: International Conference on Information Science and Technology (ICIST), 2011, IEEE, 2011, pp. 259–262.
11. E. Estrada, J. A. Rodriguez-Velazquez, Subgraph centrality in complex networks, Physical Review E 71 (5) (2005) 056103.
12. M. E. Newman, Coauthorship networks and patterns of scientific collaboration, Proceedings of the National Academy of Sciences 101 (suppl 1) (2004) 5200–5205.
13. S. P. Borgatti, Centrality and network flow, Social networks 27 (1) (2005) 55–71.
14. Y. Kato, F. Ono, Node centrality on disjoint multipath routing, in: 73rd Conference on Vehicular Technology (VTC Spring), 2011 IEEE, IEEE, 2011, pp. 1–5.
15. S. P. Borgatti, M. G. Everett, A graph-theoretic perspective on centrality, Social networks 28 (4) (2006) 466–484.
16. G. Wang, Y. Shen, E. Luan, A measure of centrality based on modularity matrix, Progress in Natural Science 18 (8) (2008) 1043–1047.
17. G. Erkan, D. R. Radev, Lexrank: Graph-based lexical centrality as salience in text summarization, Journal of Artificial Intelligence Research (2004) 457–479.
18. Y. Said, E. Wegman, W. Sharabati, J. Rigsby, Social networks of author-coauthor relationships (retraction of vol 52, p. 2177, 2008), COMPUTATIONAL STATISTICS & DATA ANALYSIS 55 (12) (2011) 3386–3386.
19. P. J. S. D. A. S. Aric A. Hagberg, Exploring network structure, dynamics,and function using networkx, proceedings of the 7th python in science conference (scipy 2008).

Customer Segmentation by Various Clustering Approaches and Building an Effective Hybrid Learning System on Churn Prediction Dataset

E. Sivasankar and J. Vijaya

Abstract Success of every organization or firm depends on Customer Preservation (CP) and Customer Correlation Management (CCM). These are the two parameters determining the rate at which the customers decide to subscribe with the same organization. Thus higher service quality reduces the chance of customer churn. It involves various attributes to be analyzed and predicted in industries like telecommunication, banking, and financial institutions. Customer churn forecast helps the organization to retain the valuable customers and it avoids failure of the particular organization in a competitive market. Single classifier does not result in higher churn forecast accuracy. Nowadays, both unsupervised and supervised techniques are being combined to get better classification accuracy. Also unsupervised classification plays a major role in hybrid learning techniques. Hence, this work focuses on various unsupervised learning techniques which are comparatively studied using algorithms like Fuzzy C-Means (FCM), Possibilistic Fuzzy C-Means (PFCM), K-Means clustering (K-Means), where similar type of customers is grouped within a cluster and better customer segmentation is predicted. The clusters are divided for training and testing by Holdout method, in which training is carried out by decision tree and testing is done by the model generated. The results of the churn prediction data set experiment show that; K-Means clustering algorithm along with the decision tree helps improving the result of churn prediction problem present in the telecommunication industry.

Keywords CRM · Churn · K-Means · FCM · PFCM · Decision tree

E. Sivasankar (✉) · J. Vijaya
Department of CSE, NIT, Trichy 620015, Tamilnadu, India
e-mail: sivasankar@nitt.edu

J. Vijaya
e-mail: vijayacsedept@gmail.com

H.S. Behera and D.P. Mohapatra (eds.), *Computational Intelligence in Data Mining*, Advances in Intelligent Systems and Computing 556,
DOI 10.1007/978-981-10-3874-7_18

1 Introduction

Emergence of information and communication systems has made remarkable growth in every industry. One such industry is telecommunication where numerous competitors have been evolved. In such a competitive market, customer retention plays a key role for the successful running of the organization, as retaining an obtainable customer incurs more expenditure than adding a fresh customer to the organization. Hence, customer correlation management (CCM) in mobile communication services focus on reducing customer churn. Analysis in international telecommunication market says that for certain number of years the customers churn has reached more than 70% [1]. Hence there are various factors like customer's residence, type of service, cost of service, need, and utilization of the services involved in determining customers churn. As a result, forecast of customers churn is very difficult for any organization. Also such an intelligent forecast helps to retain the precious customers which will increase the profit of the organization. Hence various Intelligent Mining Techniques for instance Logistic Regression, k-Nearest Neighbor, Decision Tree, Naive Bayes, Artificial Neural Networks, Support Vector Machine, Inductive Rule Learning, etc., were used to forecast customers churn behavior [2–7]. Even though classification involving single classifiers act as a good model, their forecast accuracy is not appreciable. Hence, currently there are hybrid techniques involving both supervised and unsupervised techniques. But there is only few contribution that exists towards unsupervised learning. In hybrid models, unsupervised learning techniques play a vital role for predicting better results, because similar type of customers are grouped with in cluster and analyzed for better customer segmentation. Hence, this work focuses on various unsupervised learning techniques which are comparatively studied using algorithms like Fuzzy C-Means (FCM), Possibilistic Fuzzy C-Means (PFCM), K-Means clustering (K-Means), where similar type of customers are grouped within a cluster and better customer segmentation is predicted. The clusters are divided for training and testing by holdout method, in which training is carried out by decision tree and testing is done by the model generated. The results of the churn prediction data set experiment shows that; K-Means clustering algorithm along with the decision tree helps improving the result of churn prediction problem present in the telecommunication industry.

2 Literature Survey

Shin-Yung, David C, and Hsiu-Yu (2006) designed two hybrid classification models which are based on three existing methods such as K-Means, decision tree, and back propagation neural network. In the first classification model, K-Means clustering in addition to decision tree was combined as well as in the second model, decision tree and back propagation neural networks were combined. Finally, they evaluated the above models using Hit ratio and LIFT [8]. Lee and Lee (2006) proposed SePI (Segmentation by Performance Information), which comprises of three models core protocol, bias protocol, and support protocol. decision tree which is out performs in single level classification is deployed in core protocol. The bias protocol uses the result of the main protocol. Support protocol uses data which are incorrectly predicted by main protocol in which artificial neural network used. Hence the key idea of their work is that the data which are incorrectly predicted by main protocol and can correctly be predicted by the support protocol [9]. Indranil Bose and Xi Chen (2009) selected two set of seven attributes in which the first set is based on minute of usage and second is based on revenue contribution. The following five clustering techniques support vector machine, K-Means, birch, self-organizing map, K-Medoid are used and the result of the above clustering techniques are given as an input to the decision tree and performance is evaluated using top decile lift for test data [10]. Huang, Kechadi, and Buckley (2012) extracted new set of features which is evaluated using seven traditional classification methods. There accuracy is increased with the newly extracted features than existing ones. Performance was measured in terms of true positive and false positive rate [2]. Mehdi, Ali, Bijari (2012) implemented a mixture model by combining Artificial Neural Network and Multiple Linear Regression model. Their proposed Artificial Neural Network among Multiple Linear Regression gives better result compared to single Artificial Neural Network and traditional classification techniques like support vector machine, linear discriminant analysis, tree induction k-Nearest neighbor, and quadratic discriminant analysis [11]. Ying Huang, Tahar Kechadi (2013) proposed weighted K-Means algorithm founded on path analysis. It is used to cluster the data, then resemblance between test instance and k number of clusters are measured which is named as close instance. Customer churn was predicted based on close cluster which has maximum close instances. In case of close cluster having churn label and non-churn label, the test value is given to first-order inductive learning classification algorithm for prediction [12]. Yangming Zhang et al. (2006) proposed hybrid learning system which combines k-nearest neighbor and logistic regression and they evaluated using two metrics—one is based on accuracy and second on receiver operating characteristic curve. Yeshwanth et al. (2011) solved the problem of churn prediction based on decision tree combined with genetic algorithm [13]. Wouter Verbeke et al. (2010), they took churn prediction data set and created various set of sub-samples based on oversampling, ALBA, and ALBA oversampled. Then each sub-sample was tested based on Ant Miner+, SVM, Majority rule, Logistic Regression, Ripper, and some hybridization algorithms like Ant Miner with Decision Tree [14]. Tsai et al. (2012) solved the problem of churn prediction based on Back propagation artificial neural network with itself (ANN-ANN) and Self-organizing map (SOM-ANN), they took three type of test data patterns. The Accuracy is increased with the new hybrid artificial neural network with existing single artificial neural network model [15].

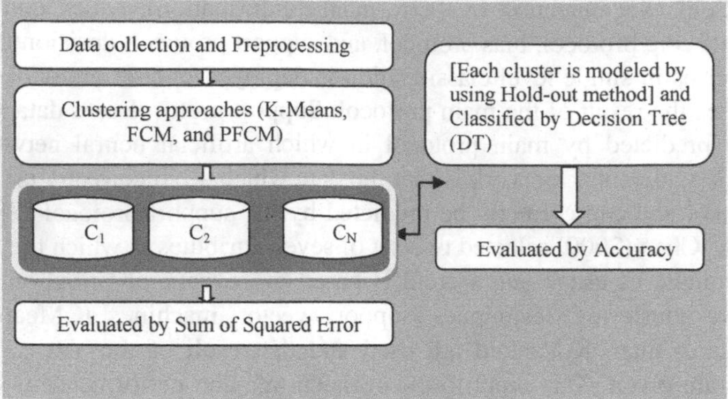

Fig. 1 Block diagram of hybrid churn prediction mode

3 Churn Prediction Models and Methods

The block diagram shown in Fig. 1 denotes the steps involved in the proposed work. The dataset is initially preprocessed and the missing values are eliminated based on mean value of attributes. After that string value features are converted into numerical format as clustering algorithm only supports features in numerical form. In order to evaluate their effectiveness to segment the customers, various clustering methods are employed. The clusters are divided for training and testing by holdout method, in which training is carried out by decision tree and testing is done by the model generated.

3.1 Dataset Preprocessing

In the preprocessing phase useless features (State, Phone number) are eliminated from dataset and reduced to 18. After that string value features are converted into numerical format as clustering algorithm only supports features in numerical form. Preprocessing is one of the most important phases as it reduces noise from data, thus making the data more consistent

3.2 Clustering Algorithm

Clustering technique has wide range of applications in data mining like image segmentation, market analysis, etc. The technique groups data objects based on their similarities. Minimum distance data points are present with in a cluster. Distances between two data points are calculated by various formulas like Euclidean distance, Manhattan distance, Minkowski distance, and so on [16]. Generally, clustering can be classified into fuzzy clustering, model-based clustering, partitional clustering, density-based clustering, hierarchical clustering, and grid-based clustering. In this document, we are using partitional clustering and fuzzy clustering techniques.

3.2.1 K-Means

K-Means algorithm was proposed by Stuart Lloyd in 1957, based on partitional-based clustering techniques. It constructs K partition of N input tuples, where K less than or equal to N. Partitional-based clustering must satisfy the following requirement (i) Every cluster to have as a minimum one sample (ii) Each sample necessity to present on precisely one cluster. Algorithm 1 illustrates the main step involved inside K-Means clustering.

Algorithm 1. Working Steps of K-Means Clustering

Input: Entire data D containing N samples.
Output: K number of cluster groups.
Step 1: Initialize K value, where K less than or equal to N
Step 2: Randomly choose K object as initial cluster center.
Step 3: Find distance measure from each object to chosen K cluster center.
Step 4: Similar object are grouped based on their minimum distance
Step 5: Update the cluster means and take as new cluster centers (C_i), {i = 1…k}
Step 6: No change in the cluster center then the algorithm stop, or go to step 3.

3.2.2 Fuzzy C-Means

Fuzzy C-Mean's algorithm was projected with Dunn in 1973, based on Fuzzy techniques [17]. The main idea of Fuzzy C-Means algorithm includes membership function μ_{ij}. We have to find the K clusters membership matrix for each data point. For all data objects, the summation of K cluster membership value is equal to 1. The data object is assigned to highest membership value cluster. In fuzzy cluster method, single object will be present in two or more cluster and in some cases the cluster does not have any data points. Algorithm 2 illustrates the main steps involved in Fuzzy C-Means clustering.

Algorithm 2. Working Principles of Fuzzy C-Means Clustering

Input: Entire data D containing N samples.
Output: K number of cluster groups.
Step 1: Initialize K value, where K less than or equal to N
Step 2: Randomly choose k cluster center
Step 3: Calculate fuzzy membership μ_{ij} using Eq. 1 in Table 1
Step 4: Similar object are grouped based on their maximum membership value μ_{ij}
Step 5: Update new cluster center V_j using Eq. 2 in Table 1
Step 6: The algorithm stops when minimum objective function J(U, V) is reached using Eq. 3 in Table 1

3.2.3 Possibilistic Fuzzy C-Means (PFCM)

Possibilistic Fuzzy C-Means algorithm was anticipated by Nikhil R et al. [18], based on (FCM) Fuzzy C-Means and (PCM) Possibilistic C-Means algorithm. The main idea of Possibilistic Fuzzy C-Means algorithm includes Typicality value t_{ij} and Membership function μ_{ij}. Algorithm 3 illustrates the main steps involved in Possibilistic Fuzzy C-Means clustering. Fuzzy C-Mean's algorithm only depends on Membership utility, so the output of the Fuzzy C-Means algorithm has lot of outliers. But Possibilistic Fuzzy C-Means depend on Typicality value and Membership function; it eliminates various problems present in FCM and PCM.

Algorithm 3. Possibilistic Fuzzy C-Means Clustering Steps

Input: Entire data D containing N samples.
Output: K number of cluster groups.
Step 1: Initialize K value, where K less than or equal to N
Step 2: Randomly choose k cluster center
Step 3: Calculate Fuzzy membership μ_{ij} using Eq. 1 in Table 1
Step 4: Calculate Typicality value t_{ij} using Eq. 4 in Table 1
Step 5: Similar object are grouped based on their maximum membership value μ_{ij}
Step 6: Update new cluster center V_j using Eq. 5 in Table 1
Step 7: The algorithm stops minimum objective function J(U, T, V) is reached Eq. 6 in Table 1

3.3 *Classification Algorithm*

Classification technique is a process of assemblage related information which is already present in the form of class labels. Initially, we have to build two types of data grouping called training and testing. The classification protocol is created using training data part and tested using testing part.

3.3.1 Decision Tree

1. Decision tree is a classification algorithm based on information gain
2. A binary tree is constructed using the calculated gain of the attribute
3. After that decision rules are formed out using different paths from root to leaf
4. Based on decision rules, it can predict whether the customer is churn or not.

4 Experimental Setup and Result Analysis

4.1 Dataset

Experiments are performed on a benchmark churn prediction dataset. The complete dataset be made of 5000 customers, out of 5000 customers 707 customers are churning customers and 4293 customers are non-churning customers. Every customer is represented by 20 attributes and predicted churn variable. The attributes primarily consist of the subsequent information: Demographic profiles which explain basic information of customer, account information, call details includes historical call value based on time and plan, complaint information and predicted class label churn or not.

4.2 Evaluation Criteria

Clustering: Sum of Squared Error (SSE): Sum of Squared Error measure is used to find the quality of the clustering techniques. The main intention of clustering is to reduce the sum of squared error value. If SSE value is low, then the clusters points are grouped well, i.e., object inside a particular cluster are similar and items that are belonging to different cluster are dissimilar. SSE value is calculated by the following Eq. 7.

$$SSE = \sum_{i=1}^{k} \sum_{j=1}^{n_i} ||c_i - o_{ij}||^2 \tag{7}$$

Here, $||C_i - O_{ij}|^2$ refers Euclidean distance between each data point (O_{ij}) within a cluster to cluster center (C_i), it will be calculated for all K cluster and finally summarized.

4.3 Experiment Setup

This work comprises three set of experiments. The clustering result vastly decides the classification accuracy of any hybrid-based classification technique. Therefore, the first set of experiment is conducted to estimate the performance of the clustering algorithm (K-Means, FCM, and PFCM) based on SSE Measure. The second sets of experiment are conducted to calculate the efficiency of the hybrid form and in third set of experiment well known existing models are compared.

4.4 Result and Conversation

4.4.1 Experiment 1

The first set of experiment focuses on various unsupervised techniques which are comparatively studied using algorithms (K-Means) K-Means clustering, (FCM) Fuzzy C-Means, (PFCM) Possibilistic Fuzzy C-Means where similar type of customers are grouped with in a cluster and better customer segmentation is predicted. Figure 2 shows that sum of squared error result produced by K-Means (K-Means), Fuzzy C-Means (FCM), Possibilistic Fuzzy C-Means (PFCM). In this figure, horizontal axis represent number of cluster and vertical axis represent sum of squared error.

From Fig. 2, we can infer that the result generated by K-Means algorithm has lower SSE compared to Fuzzy C-Means and Possibilistic Fuzzy C-Means algorithm. Figure 3a–c shows that 3D clustering result generated Fuzzy C-Means (FCM), Possibilistic Fuzzy C-Means (PFCM), K-Means clustering (K-Means) with the number of cluster K = 2.

After clustering, we have to find the number of customer in each cluster which is produced by K-Means, FCM, PFCM. Table 2 shows that sample result produced by various clustering algorithm with the number of cluster = 5 where the total number of customer is 5000.

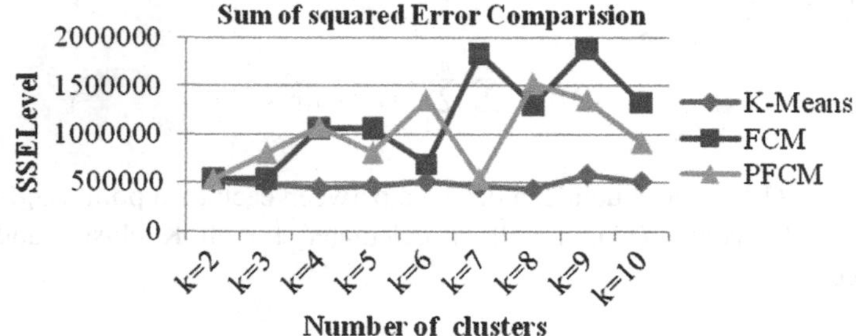

Fig. 2 SSE of different clustering techniques

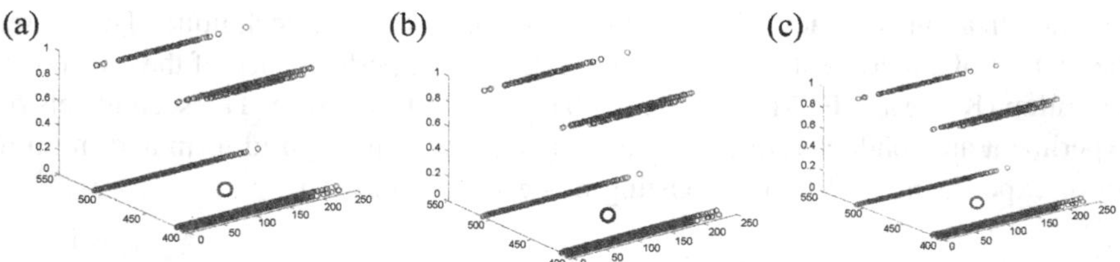

Fig. 3 **a–c** Clustering results of K-Means, FCM, PFCM where K = 2

Table 1 Formula's used for FCM, PFCM algorithms

Equation no.	Formula name	Formula
(1)	Membership function formula both FCM, PFCM	$\mu_{ij} = \frac{1}{\sum_{k=1}^{c} \frac{d_{ij}}{d_{ik}}\left(\frac{2}{m-1}\right)}$
(2)	Cluster center formula for FCM	$V_j = \frac{(\sum_{i=1}^{n} (\mu_{ij})^m x_i)}{(\sum_{i=1}^{n} (\mu_{ij})^m)}$
(3)	Objective function formula for FCM	$J(U, V) = \sum_{i=1}^{n} \sum_{j=1}^{c} (\mu_{ij})^m \|\|x_i - v_j\|\|^2$
(4)	Typicality matrix value for PFCM	$t_{ik} = \frac{1}{1 + \frac{(b(D_{ik})^2}{\delta_i}^{\frac{1}{(\eta-1)}}}$
(5)	Cluster center formula for FCM	$v_i = \frac{\sum_{k=1}^{n} (au_{ik}^m + bt_{ik}^n)x_k}{\sum_{k=1}^{n} (au_{ik}^m + bt_{ik}^n)}$
(6)	Objective function formula for PFCM	$X = \sum_{i=1}^{c} \delta_i \sum_{k=1}^{n} (1 - t_{ik})^\eta$ $J(U, T, V, Z) = \sum_{i=1}^{c} \sum_{k=1}^{n} a\mu_{ik}^m + bt_{ik}^m * \|\|z_k - v_i\|\|^2 + X$

Table 2 Number of customer in 5 clusters

Algorithm	Cluster1	Cluster2	Cluster3	Cluster4	Cluster5
K-Means	885	1152	850	1070	1043
FCM	97	156	2456	0	2291
PFCM	28	2478	2487	0	7

4.4.2 Experiment II

In second set of experiment, the accuracy of each hybrid model such as K-Means along with decision tree, fuzzy C-Means together by decision tree, and Possibilistic fuzzy C-Means together by decision tree are compared. The output of each model is plotted in Fig. 4. The parallel axis and the upright axis correspond to the rates of number of clusters and accuracy value predicted by each model. In this figure, we can observe that the accuracy value of K-Means along with decision tree is better than fuzzy clustering. Since K-means clustering has lowest SSE value, it can partition the customers well hence, it will produce better classification result.

4.4.3 Experiment III

In order to compare different hybrid models, the third set of experiments are evaluated. The result of the proposed hybrid model is compared against different

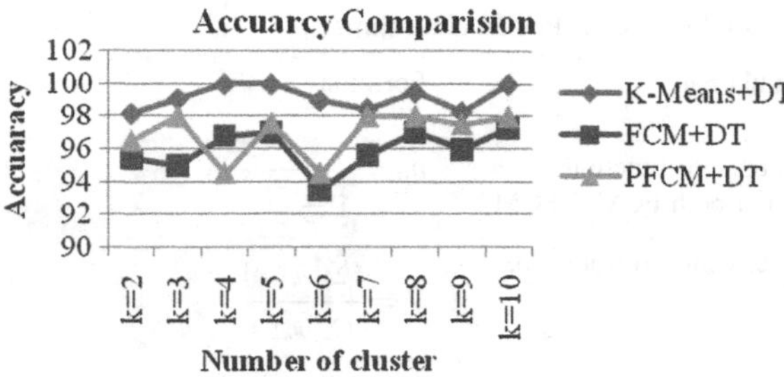

Fig. 4 Accuracy for different hybrid models

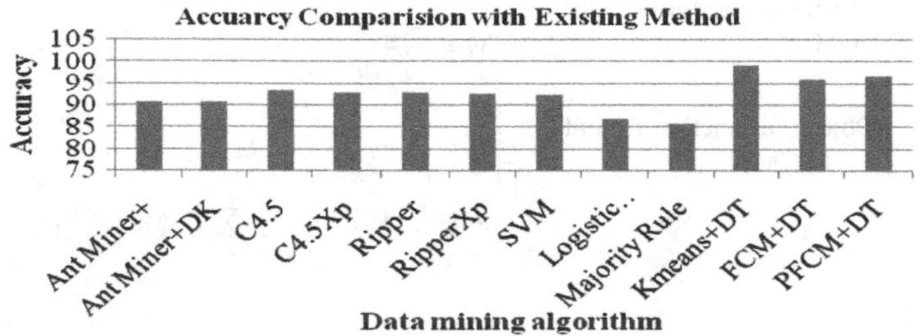

Fig. 5 Accuarcy of existing model with proposed hybrid model

existing models produced by Wouter Verbeke et al. (2010) without sampling. Figure 5 illustrates that the proposed hybrid model outperform the classification techniques like SVM, Ripper, Logistic regression, Ant Miner combined with decision tree.

5 Conclusion and Future Work

Nowadays, both unsupervised and supervised techniques are combined to get better classification accuracy, and unsupervised classification plays a major role in hybrid learning techniques. Hence, this work focuses on various unsupervised techniques which are comparatively studied using algorithms like fuzzy C-Means (FCM), possibilistic fuzzy C-Means (PFCM), K-Means clustering (K-Means). The experimental results show that K-Means algorithm produces better cluster quality compared to fuzzy grouping so it produced better classification accuracy along with decision tree. In future, we concentrate Fuzzy C-Means clustering for hybrid model to increase the prediction of customer churn.

References

1. Rob Mattison: Telecom churns management: The golden opportunity. Fuquay-Varina, N.C: APDG Publishing.
2. Ying Huang, Bingquan Huang and M. T. Kechadi: A rule-based method for customer churns prediction in telecommunication services, Lecture Notes in Computer Science, 6634, 411–422, 2011.
3. Hyunseok Hwang, et al., An LTV model and customer segmentation based on customer value: A case study on the wireless telecommunication industry, Journal of Expert Systems with Applications, 26(2), 181–188, 2004.
4. Bart Lariviere, Dirk van den Poel: Predicting customer retention and profitability by using random forests and regression forests techniques, Journal of Expert Systems with Applications, 29(2), 472–484, 2005.
5. Chih-Ping Wei, I-Tang Chiu: Turning telecommunications call details to churn prediction: A data mining approach, Journal of Expert Systems with Applications, 23 (2), 103–112, 2002.
6. Guo en XIA, Wei dong Jin: Model of customer churn prediction on support vector machine, Journal of Systems Engineering - Theory and Practice, 28 (1), 71–77, 2008.
7. Bingquan Huang, et al., Customer churns prediction in telecommunications, Journal of Expert Systems with Applications, 39 (1), 1414–1425, 2012.
8. Shin-Yung Hung, et al., Applying data mining to telecom churn management, Journal of Expert Systems with Applications, 31 (3), 515–524, 2006.
9. Jae Sik Lee, Jin Chun Lee: Customer churns prediction by hybrid model, Proceedings of the second international conference on advanced data mining and applications, 4091, 959–966, 2006.
10. Indranil Bose, Xi Chen: Hybrid Models Using Unsupervised Clustering for Prediction of Customer Churn, Journal of Organizational Computing and Electronic Commerce,19:2, 133–151, 2009.
11. Mehdi Khashei, Ali Zeinal Hamadani, Mehdi Bijari: Novel hybrid classification model of artificial neural networks and multiple linear regression models, Journal of Expert Systems with Applications, 39 (3), 2606–2620, 2012.
12. Ying Huang, Tahar Kechadi: An effective hybrid learning system for telecommunication churn prediction, Journal of Expert System with Applications, 40(14), 5635–5647, 2013.
13. V. Yeshwanth Raj, V. Vimal raj, M.Saravanan: Evolutionary churn prediction in mobile networks using hybrid learning, In Proceedings of the twenty-fourth international Florida artificial intelligence research society conference, (FLAIRS), Palm Beach, Florida, USA, May 18–20. AAAI Press.
14. Wouter Verbeke, et al.,: Building comprehensible customer churn prediction models with advanced rule induction techniques, Journal of Expert System with Applications, 38 (3) 2354–2364, 2011.
15. Chih-Fong Tsai, Yu-Hsin Lu: Customer churns prediction by hybrid neural networks, Journal of Expert System with Applications, 36 (10), 12547–12553, 2009.
16. Velmurugan. T: Performance Comparison between k-Means and Fuzzy C-Means Algorithms using Arbitrary Data Points, Wulfenia Journal, 19(8), 2012.
17. J. C. Dunn: A Fuzzy Relative of the ISODATA Process and its Use in Detecting Compact Well Separated Clusters, Journal of Cybernetics, 3(3), 32–57, 1973.
18. Nikhil R. Pal, et al., A Possibilistic Fuzzy c-Means Clustering Algorithm, IEEE Transaction on Fuzzy Systems, 13(4), 517–530, 2005.

An Artificial Neural Network Model for a Diesel Engine Fuelled with Mahua Biodiesel

N. Acharya, S. Acharya, S. Panda and P. Nanda

Abstract In this paper, an Artificial Neural Network (ANN) model is used to predict the different parameters of a diesel engine fuelled with the mixture of diesel and mahua biodiesel in different proportion. The data has been obtained from an experiment carried out in a twin cylinder diesel engine in different loading condition and different blending ratios of diesel and biodiesel. Two input data, i.e., engine load and blending ratio and five output data, i.e., Brake Thermal Efficiency (BTE), Brake Specific Fuel Consumption (BSFC), Smoke level, Carbon monoxide (CO), and Nitrogen Oxides (NO_x) emissions have been considered for ANN modeling. The network used is back propagation, feed forward with multilayer perceptron having ten numbers of neurons in hidden layer with trainlm training algorithm being proposed. It has been observed that the prediction ability of the model is high as there is minimum difference between the predicted and the experimentally measured values.

Keywords Vegetable oil · Transesterification · Biodiesel · Viscosity · ANN

N. Acharya (✉)
Excavation Department, M.C.L, Burla, India
e-mail: narayan.acharya2@gmail.com

S. Acharya
Department of Computer Application, VSSUT, Burla, India
e-mail: talktosas@gmail.com

S. Panda · P. Nanda
Department of Mechanical Engineering, VSSUT, Burla, India
e-mail: sumanta.panda@gmail.com

P. Nanda
e-mail: pnanda18@gmail.com

H.S. Behera and D.P. Mohapatra (eds.), *Computational Intelligence in Data Mining*, Advances in Intelligent Systems and Computing 556,
DOI 10.1007/978-981-10-3874-7_19

1 Introduction

The depletion of liquid fossil fuels and air pollution due to its use has compelled to use vegetable oil in the form of biodiesel. The vegetable oils contain around 90–98% of triglyceride and significant amount of oxygen [1]. At present, some countries producing biodiesel from edible vegetable oil like sunflower, soybean, palm, and coconut which may create crisis of food in future if it will be used for mass production of biodiesel [2]. So, non-edible vegetable oil is another option for feedstock of biodiesel. In India, approximately 23% of the total land is covered under forest. There are more than 350 numbers of nonedible vegetable oil trees available in nature [3]. The major potential non-edible feed stocks are jatropha, karanja, mahua, neem, polanga, soapnut, and castor with substantial yielding capacity. Among these, mahua is one of the potential non-edible oil feed-stock which can be used for production of biodiesel.

Mahua (Madhuca indica) is a medium-sized tree of Indian origin with a wide round canopy found mostly in the central India. It belongs to family Sapotaceae and can attain up to 20 meter in height. It can grow in arid environment and different soil conditions. Depending upon the maturity of tree, the yield ranges from 20 to 200 kg per annum. The kernel of mahua fruit contains around 35–40% of oil [4]. The use of raw mahua oil is restricted due to its unfavorable properties like high kinematic viscosity, poor cold flow properties, and poor spray characteristics. But it can be used after reduction of viscosity by different methods like pyrolysis, micro-emulsification, dilution and trans-esterification. Among these transesterfication is the best method for biodiesel production [5]. In this method, heated vegetable oil (60°C) is mixed with an alcohol (preferably methyl alcohol) and catalyst like sodium hydroxide or potassium hydroxide which yields biodiesel and glycerol. So, transesterified mahua oil (mahua biodiesel) can be used in diesel engine where the reduction viscosity can be achieved to an appreciable level.

Most of the time engine designers face difficulties to obtain real data due to costly and time-consuming experiments in different operating conditions. In order to solve this problem, an Artificial Neural Network (ANN) is a trust-worthy tool available for prediction the same. It is a non-linear modeling tool used for solving different types of engineering problems by simulation with limited experimental data. A well-trained neural network can be used to predict multiple output variables using multiple input variables with much faster speed than any other numerical method [6]. The objective of the paper is to develop an ANN model to predict the performance and exhaust emission characteristics of a diesel engine fuelled with diesel and mahua biodiesel mixture in different ratios and different loading conditions. This paper is organized as follows. Related work is presented in Sect. 2 and also compares the properties of raw mahua oil, mahua biodiesel and mineral diesel. Section 3 presents the experimental setup for the proposed model. An overview of an ANN and design of an ANN for the proposed model are presented in Sects. 4

and 5 respectively. The results are presented and discussed in Sect. 6. Section 7 finally concludes the paper.

2 Related Work

There are a number of publications on ANN applications to predict the different parameters related to engine performance and emission characteristics using different bio-fuels. Ismail et al. [7] utilized ANN to model a light duty diesel engine fuelled with different blends of biodiesel like soybean, coconut, and palm oil to predict different engine output parameters with four inputs like mass flow rate, engine rpm, output torque, and type of biodiesel. The designed model is able to produce seven output parameters with high degree of accuracy. Oguz et al. [8] predicted the engine performance using fuel consist of mineral diesel, B20 biodiesel, and bio-ethanol utilizing the ANN concept to. Ghobadian et al. [9] utilized ANN concept to predict different emission characteristic of a diesel engine using biodiesel prepared from waste cooking oil. The model is able to predict the performance of the engine with correlation coefficient in between 0.9487 and 0.9999. Najafi et al. [10] used ANN to predict the different engine output characteristics and emission characteristics of gasoline engine fuelled with ethanol blended gasoline fuel with two inputs like speed and blending percentage. The correlation coefficient between predicted and experimented value is nearly equal to '1' and root mean square error is very low. All researchers viewed the ANN as a useful tool to predict the output parameters of engine with high degree of accuracy and used the multilayer perception with back propagation algorithm as standard one.

For this experiment raw mahua oil has been collected from local market and biodiesel has been prepared through transesterification process in a laboratory scale. The different properties like density, kinematic viscosity, flash point, cloud point, pour point, calorific value and cetane number of mahua oil, mahua biodiesel and mineral diesel have been measured as per standard method and presented in Table 1. It is inferred from Table 1 that the viscosity of raw mahua oil is very high but the viscosity of mahua biodiesel after transesterification has been reduced to an operable range. The measured properties of mahua biodiesel confirms the specification of ASTM-D6751 and IS-15607 standards prescribed by USA and India for biodiesel. The flash point of biodiesel is higher than mineral diesel which can prevent fire hazard also. Its cetane number is higher than mineral diesel which indicates good combustion characteristics. But the cold flow properties like cloud point and pour point are higher than mineral diesel which can cause problem in engine operation particulrly in cold weather.

For the design of an engine, the designer completely depends upon the fuel to be used and norms of exhaust emission. To use biodiesel as an alternative fuel in an engine, it is required to analyze and compare the different performances and emission characteristics of biodiesel with that of mineral diesel. In this study, different parameters like Brake Thermal Efficiency (BTE), Brake Specific Fuel Consumption (BSFC) and emission parameters like emission of smoke and Carbon

Table 1 Properties of mahua oil, mahua biodiesel and mineral diesel

Properties	Unit	ASTM-D6751	IS-15607	Mahua oil	Mahua biodiesel	Diesel
Density at 15 °C	kg/m^3	860–900	860–900	920	882	824
Kinematic viscosity at 40 °C	mm^2/s	1.9–6	2.5–6	24.58	4.2	2.30
Flash point	°C	>130	>120	232	170	53
Cloud point	°C	–	–	15	13	1
Pour point	°C	–	–	10	6	−8
Calorific value	MJ/kg	–	–	36	38.5	42
Cetane number		47 min	51 min	57	57	51

Table 2 Experimental configuration

Sl. no.	Operating parameters	Level				
1	Biodiesel blend(%)	20	40	60	80	100
2	Load(%)	25	50	75	100	

Monoxide (CO), emission of Nitrogen Oxides (NO_x) of blended mahua biodiesel and mineral diesel have been studied from experiment.

3 Experimental Setup

The engine performance and emission characterstics of mahua biodiesel and its blend have been conducted on a four stroke, twin cylinder, constant speed, naturally aspirated diesel engine fitted with hydraulic dynamometer. The experimental configuration is presented in Table 2. The engine has been fitted with both exhaust gas analyzer and smoke meter to measure different gases like CO, NO_x, and smoke level. The experiments have been conducted initially with mineral diesel and then with biodiesel of different blending percentage with different loading conditions as mentioned in Table 2. After completion of experiment, the engine has been run again with mineral diesel to rule out the sludge deposit in the engine during operation with biodiesel and the corresponding performance parameters and emission parameters have been recorded.

4 Overview of an Artificial Neural Network

An artificial neural network (ANN) is composed of three types of layers called input layer, hidden layer, and output layer. Each layer is composed of neurons or nodes which are linked to each other with communication link accompanied by linking weight. The complexity of the problem decides the number of hidden layers and number of neurons in each hidden layer. The signals from input layer are transferred to the hidden layer and the activation function in the hidden layer is utilized to approximate the behavior of input data which is generally non-linear in nature. The final output of the network is the predicted result. There are different types of learning techniques but the back propagation algorithm is the widely used. It is a supervised algorithm, uses ramp descent rule for optimization and learns from the changing linking weights constantly. These changes are stored as knowledge. It minimizes the total error by changing the weight through its ramp and tries to improve the network performance. In each step network error is calculated by comparing the predicted value and target value. The error during learning is called root mean squared error (RMSE) which is calculated using the formula given in Eq. 1.

$$RMSE = \sqrt{\left(\frac{1}{n}\sum_{i=1}^{n}(a_i - p_i)^2\right)}, \quad (1)$$

where a_i is the actual value and p_i is the predicted value.

When RMSE falls below a predetermined value or max number of epochs has been reached, the training process stops. This trained network can be used for simulation purpose. The over fitting is the major problem associated with ANN, where the system predicts accurate results for known data sets but it fails to predict for new data sets.

5 Proposed ANN Design

In this paper, design of ANN model is based upon multilayer perceptron (MLP), feed forward neural network architecture with back propagation algorithm. For training the model supervised learning technique has been used where the biases and network weights are initialized randomly at the beginning of the training phase. The gradient descent rule has been used for error minimization process. The early stopping method is used to avoid overfitting.

In this model the number of hidden layers is two. The activation function for hidden layer and output layer has been selected as logsigmoid (logsig) and linear (purelin) respectively. The trainlm has been selected for training algorithm because the root mean square error value is the lowest for trainlm algorithm as presented in Table 3. The number of neurons is taken as 10 as root mean square error is lowest for 10 number of neurons as presented in Table 4. The complete ANN design has been carried out by computer programming using MATLAB-R-2009a Neural Network Toolbox. The proposed ANN model is shown in Fig. 1.

Table 3 RMSE for different training function

Sl. no.	Parameters	Unit	Trainlm	Trainrm	Trainscg	Traingdx
1	BTE	%	0.75	3.02	42.26	47.96
2	BSFC	gm/kwhr	678.125	2226	2123	13098
3	CO	%	0.0000125	0.006	0.010687	0.010273
4	Smoke	%	3.5093	14.42	46.72	50.83
5	NO_x	ppm	1.874725	21.20	216.4	149.0829

Table 4 RMSE for different number of neurons

Sl. no.	Parameters	10	15	20
1	BTE	0.75	0.89	0.97
2	BSFC	678.125	867.235	1012.345
3	CO	0.0000125	0.0001036	0.005246
4	Smoke	3.5093	4.6723	5.7256
5	NO_x	1.874725	3.742394	5.841574

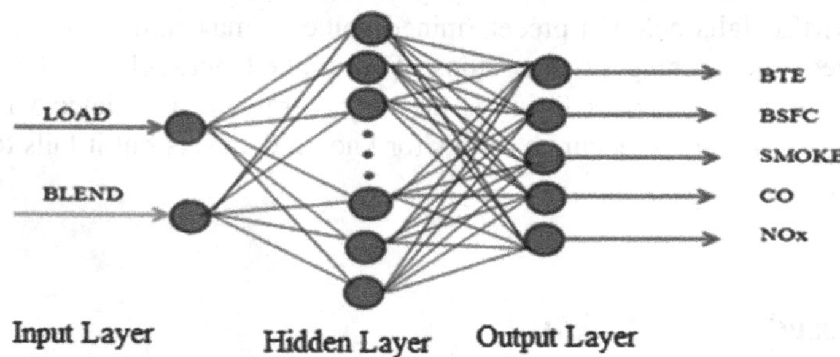

Fig. 1 A multilayer ANN model for predicting engine parameters

6 Results and Discussion

An ANN model of a diesel engine using mahua biodiesel has been developed using experimentally obtained data. A total of eight random data points consisting of different loads and blending ratio has been used for testing the model. The predicted result has been compared with the experimental result and the different values like regression coefficient (R^2), Root Mean Square Error (RMSE) and Mean Relative Error (MRE) percentage have been calculated and presented in Table 5. The regression analysis and the error at different test points has been plotted and shown in figures (Figs. 2, 3, 4, 5 and 6). It is found that R^2 value varies between 0.982 and 0.9979 which is closer to 1. Root mean square error values for brake thermal efficiency, brake specific fuel consumption, Carbon Monoxide (CO), smoke and Nitrogen Oxides (NO_x) are 0.87%, 26.04 g/kWh, 0.01%, 1.87%, and 1.37 ppm, respectively which are very low as compared to their actual value.

Table 5 R^2, RMSE and MRE values for different predicted parameters

Sl. no.	Parameters	R^2	RMSE	MRE%
1	BTE	0.982	0.87	4.2
2	BSFC	0.9851	26.04	4.15
3	CO	0.9979	0.01	7.52
4	Smoke	0.9859	1.87	6.59
5	NO_x	0.9781	1.37	2.99

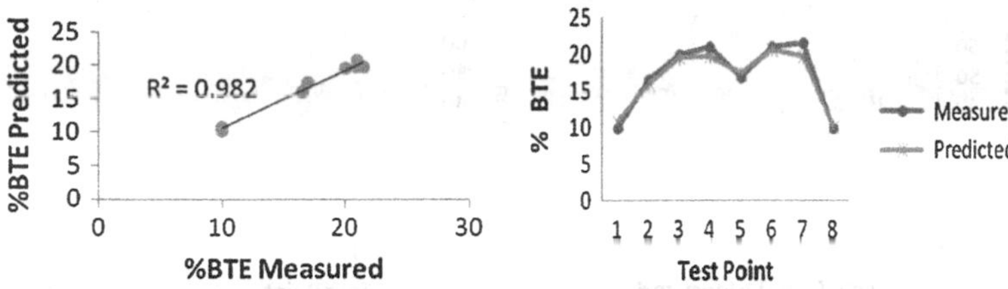

Fig. 2 Comparison of ANN prediction of BTE with measured data

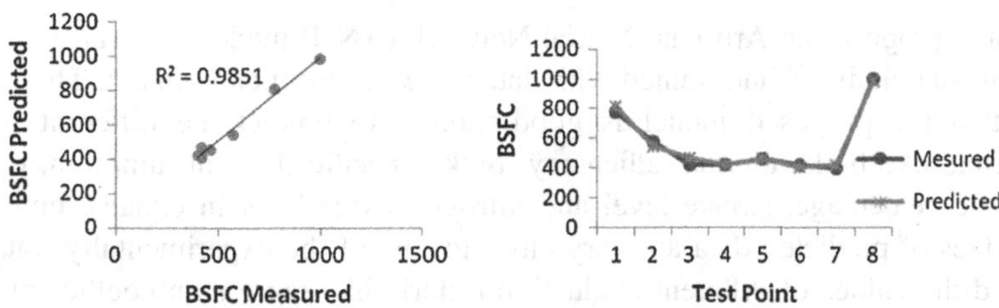

Fig. 3 Comparison of ANN prediction of BSFC with measured data

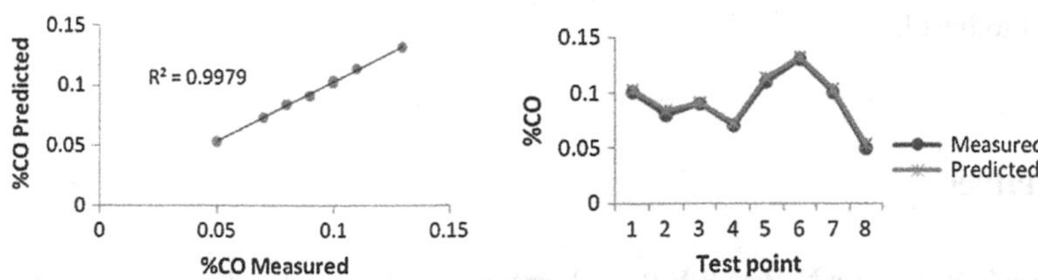

Fig. 4 Comparison of ANN prediction of carbon monoxide (CO) with measured data

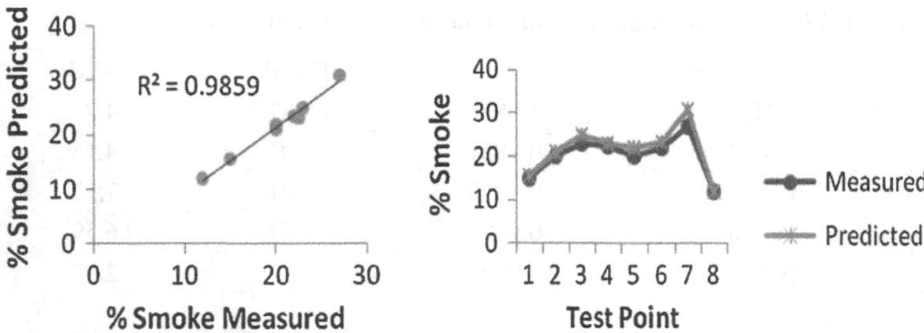

Fig. 5 Comparison of ANN prediction of smoke with measured data

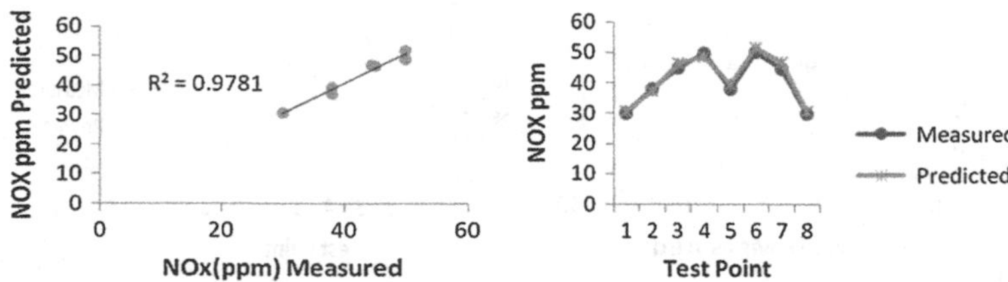

Fig. 6 Comparison of ANN prediction of nitrogen oxides (NO_x) with measured data

7 Conclusion

The paper proposes an Artificial Neural Network (ANN) model for a diesel engine using mahua biodiesel and trained with data collected from experiment. The results show that the proposed model is good enough to predict the different output parameters like brake thermal efficiency, brake specific fuel consumption, carbon monoxide percentage, smoke level and nitrogen oxide level in exhaust emission. The values of predicted data are very close to that of the experimentally obtained data and the values of different evaluation criteria like regression coefficient (R^2), Root Mean Square Error (RMSE) and Mean Relative Error (MRE) are satisfactory. This shows that predictability of the suggested model is high. Therefore ANN is considered as a desirable prediction method to predict the different engine output characteristics and emission performance characteristics for a diesel engine using mahua biodiesel.

References

1. Puhan S, Vedaraman N, Ram B.V.B, Sankarnarayan G, Jeychandran K.: Mahua oil (Madhuca Indica seed oil) methyl ester as biodiesel preparation and emission characteristics, Biomass and Bioenergy 28 (2005) 87–93.
2. Raheman H, Ghadge S.V. Performance of compression engine with mahua (madhuca indica) biodiesel. Fuel 86 (2007) 2568–2573.

3. Acharya N, Nanda P, Panda S. Biodiesel an alternative fuel for future. All India Seminar on Recent trends in mechanical Engineering (RTME) (2015) 52–56.
4. Godiganur S, Murty C.H.S, Reddy R.P, 6BTA 5.9 G2-1 Cummins Engine performance and emission test using methyl ester of mahua (Madhuca indica) oil/diesel blends. Renewable energy 34 (2009) 2172–2177.
5. Acharya N, Nanda P, Panda S. Biodiesel "An alternative fuel for C.I Engine" future, National conference on recent advance in renewable energy and control (RAREC) (2015) 96–100.
6. Kumar S., Pai, P. S., & Rao, B. S. Prediction of Performance and Emission parameters of a Biodiesel Engine-A Comparison of Neural Networks, 3rd international conference on Mechanical, Automotive and Material Engineering (ICMAMAE, 2013), Singapore, Apr 29–30.
7. Ismail H.M, Ng HK, Queck CW, Gan S,: Artificial neural networks modeling of engine-out response for a light duty diesel engine fuelled with biodiesel blends. Applied Energy. 92 (2012) 769–777.
8. Oguz H, Santas I, Baydan H.E,: Prediction of diesel engine performance using biofuels with artificial neural network, Applied Expert System with Applications. 37 (2010) 6579–6586.
9. Ghobadian B, Rahimi H, Nikbakth A.M, Najafi G, Yusaf T.F.: Diesel engine performance and exhaust emission analysis using waste cooking biodiesel fuel with an artificial neural network, Renewable Energy. 34 (2009) 976–982.
10. Najafi G, Ghobadian B, Tavakoli T, Buttsworth D.R, Yusaf T.F, Faizollahnezad M.: Performance of exhaust emission of a gasoline engine with ethanol blended gasoline fuels using artificial neural network, Applied Energy. 86 (2009) 630–639.

An Application of NGBM for Forecasting Indian Electricity Power Generation

Dushant P. Singh, Prashant J. Gadakh, Pravin M. Dhanrao, Sharmila Mohanty, Debabala Swain and Debabrata Swain

Abstract The average generation of electricity is getting increased day by day due to its increasing demand. So forecasting the future needs of electricity is very essential, especially in India. In this paper, a Grey Model (GM) and a Nonlinear Grey Model (NGM) are introduced with the concept of the Bernoulli Differential Equation (BDE) to obtain higher predictive precision, accuracy rate. To improve the prediction accuracy of GM, the Nonlinear Grey Bernoulli Model (NGBM) is used. The NGBM model is having the capability to produce more reliable outcomes. The NGBM with power r is a nonlinear differential equation. Using power r in NGBM the expected result can be controlled and adjusted to fit the results of 1-AGO historical raw data. NGBM is a recent grey prediction model to easily adjust for the correctness of GM(1, 1) stable with a BDE. The differentiation of desired outcome with the actual GM(1, 1) has been displayed through a feasible forecasting model NGBM(1, 1) by accumulating the decisive variables. This model may help government to extend future planning for generation of electricity.

Keywords Electricity generation • Forecasting • Bernoulli model • Grey model • Nonlinear grey Bernoulli model

D.P. Singh · S. Mohanty · D. Swain (✉)
School of Computer Engineering, KIIT University, Bhubaneswar, India
e-mail: debabala.swain@gmail.com

P.J. Gadakh
International Institute of Information Technology, Pune, India

P.M. Dhanrao
KBP Polytechnic, Kopargaon, India

D. Swain
Vishwakarma Institute of Technology, Pune, India

H.S. Behera and D.P. Mohapatra (eds.), *Computational Intelligence in Data Mining*, Advances in Intelligent Systems and Computing 556,
DOI 10.1007/978-981-10-3874-7_20

1 Introduction

In the current era of information storage, where it is very hard to analyze the huge volume of data using traditional statistical method, the applications of grey system theory plays a vital role for analysis and prediction. However, the GM theory proposed by Deng [1] could not satisfy the predicted accuracy in Grey forecasting. This model is already used in several applications, like economics [2], finance, agriculture [3], air transportation [4], electric load [5], industry [6], and industrial wastewater [7] and so on. Energy plays as one of the strongest factors in the economic growth of a nation. So its present usage, future requirements are very essential to explore.

In this paper, a prediction method is proposed related to grey theory with improved accuracy. Grey prediction models [8, 9] have the advantages of establishing a model with few uncertain data. The grey system theory was first proposed by Deng [7], Kayacan et al. [10] for a system with incomplete or uncertain information to construct a grey forecasting and decision-making. Being superior to conventional statistical models, grey models require only a limited amount of data to estimate the behaviour of unknown systems. In recent years, the grey system theory has been adapted successfully in several fields and authenticated with suitable outputs [11–19].

The GM(1, 1) has been authenticated and generally used in the unrestricted [18] data. It determines almost accurate exponential progress rules and acquires temporary estimating accuracy. The model has acquired significant meaning in forecasting, specifically with respect to following conditions: the series generation background value, parameter estimation [18], the time response function, the differential equations, and nonlinear grey model.

However, studies on both the initial condition and adjustable parameters in the NGBM(1, 1) model are inadequate [20]. The initial condition in grey models is an important factor, affecting the simulation and forecasting precision. According to grey system theory, the model should give priority to new information. Thus, the initial condition should not be limited to the first item in the first-order accumulated generating sequence. The last item should be taken as the weighted sum of the first item and the last item as the initial condition. The schemes in [21, 22] finds the minimized sum of the square error between the simulated 1-AGO and the original 1-AGO value. In this study, different initial conditions [23] are combined with the optimized model to test the effectiveness of the NGBM(1, 1) model. The simulation of raw data to predict accuracy gives a perfect accuracy in NGBM(1, 1) with parameter r. An ideal model with variable r and p can be proposed for prediction. In this research, a simple prediction is proposed for productive prediction for the Indian electricity generation and future requirement prediction. The results provide a future strategy for the generation of electricity energy.

In Grey theory, the AGO technique is truly relevant in the historical data analysis to reduce randomized allocation of raw data. To use the refined data the series should be developed using the first-order linear differential equation. The proposed modified grey forecasting model GM(1, 1) together with the NGBM(1, 1) theory used to develop the prediction awareness. In the next sections, the mathematical approach of GM(1, 1) and NGBM(1, 1) are briefly described.

1.1 Grey Model—GM(1, 1)

According to Liu and Lin [9], the original sequence of raw data with n entries can be written as

$$Q^{(0)} = \left(q^{(0)}(1), q^{(0)}(2), \ldots, q^{(0)}(k), \ldots, q^{(0)}(n)\right), \tag{1}$$

where $Q^{(0)}$ stands for the nonnegative original historical time series data and $q^{(0)}(m) \geq 0, m = 1, 2, \ldots, n,$

$$Q^{(1)} = \left(q^{(1)}(1), q^{(1)}(2), \ldots, q^{(1)}(m), \ldots, q^{(1)}(n)\right) \tag{2}$$

where, n is the number of modelling data.

The 1-AGO creation of $q^{(1)}$ is defined as

$$q^{(1)}(m) = \sum_{i=1}^{m} q^{(0)}(i), \quad m = 1, 2, \ldots, n \tag{3}$$

Suppose that, 1-AGO generated sequence is $Q^{(1)}$, a nonnegative sequence is $Q^{(0)}$, and the consecutive neighbours of $Q^{(1)}$ the sequence generated with mean $Z^{(1)}$.

$$\frac{dq}{dt} + \alpha Q = \beta, \tag{4}$$

where variable α and β are called the growing incident and grey model input respectively. Calculation of coefficient α and β cannot be done directly from Eq. (4). Uniform mapping relation elements in the set satisfy the background value of α and β [23]. So, the result of Eq. (4) can be achieved using the least square method. Then the discrete form of the GM(1, 1) in differential equation model [23] can be expressed as

$$q^{(0)}(m) + \alpha q^{(1)}(m) = \beta, \tag{5}$$

for m = 1, 2, ..., n and $Z^{(1)}$ is the mean generated sequence of consecutive neighbours of $Q^{(1)}$, given by

$$Z^{(1)} = \left(z^{(1)}(1), z^{(1)}(2), \ldots, z^{(1)}(n)\right) \tag{6}$$

By solving the Eq. (8), the individual α and β value can be found. After calculating the parameter α and β it can be applied to the Eq. (4).

$$\begin{bmatrix} \alpha \\ \beta \end{bmatrix} = = \left(B^T B\right)^{-1} B^T Y, \tag{7}$$

where B and Y are defined as follows:

$$B = \begin{bmatrix} -z^{(1)}(2) & 1 \\ -z^{(1)}(3) & 1 \\ \vdots & \vdots \\ \vdots & \vdots \\ -z^{(1)}(n) & 1 \end{bmatrix}, Y = \begin{bmatrix} -q^{(0)}(2) \\ -q^{(0)}(3) \\ \vdots \\ \vdots \\ q^{(0)}(n) \end{bmatrix} \tag{8}$$

$$Y = \left[q^{(0)}(2), q^{(0)}(3), \ldots, q^{(0)}(n)\right]^T \tag{9}$$

From Eq. (8) the appropriate solution of Eq. (4) can be found with sufficient results together as

$$\hat{q}^{(1)}(m+1) = \left(q^{(0)}(1) - \frac{\beta}{\alpha}\right) e^{-\alpha m}, m = 1, 2, \ldots \tag{10}$$

Finally the prediction result can be calculated for m steps using 1-AGO as follows:

$$\hat{q}^{(0)}(m+1) = \left(1 - e^{-\alpha}\right)\left(\hat{q}^{(1)}(m+1)\right) - \hat{q}^{(1)}(m) \cdot m = 1, 2 \ldots. \tag{11}$$

$$\text{Or } \hat{q}^{(0)}(m+1) = \left(1 - e^{-1}\right) = \left(q^{(0)}(1) - \frac{\beta}{\alpha}\right) e^{-\alpha m} \, m = 1, 2, 3 \ldots \tag{12}$$

1.2 Nonlinear Grey Bernoulli Model, NGBM(1, 1)

This section elaborates the nonlinear Grey Bernoulli model, NGBM(1, 1). According to Zhou et al. [20] the NGBM(1, 1) process can be mathematically described as follows:

$$Q^{(0)} = \left(q^{(0)}(1), q^{(0)}(2), \ldots, q^{(0)}(m), \ldots, q^{(0)}(n)\right), \tag{13}$$

where, $q^{(0)}(m)$ is the mth value of $q^{(0)}$, $m = 1, 2, \ldots, n$.

The accumulated generating operation on $Q^{(0)}$ performs as follows:

$$Q^{(0)} = \left(q^{(1)}(1), q^{(1)}(2), \ldots, q^{(1)}(k), \ldots, q^{(1)}(n)\right) \tag{14}$$

$$q^{(1)}(m) = \sum_{i=1}^{m} q^{(0)}(i), m = 1, 2, \ldots, n \tag{15}$$

The NGBM(1, 1) can be defined by grey differential equations as

$$q^{(0)}(m) + \alpha z^{(1)}(m) = \beta\left[z^{(1)}(m)\right]^{r} \tag{16}$$

and its whitenization differential equation can be explained as follows:

$$\frac{dq^{(1)}}{dt} + \alpha\left[q^{(1)}\right] = \beta\left[q^{(1)}\right]^{r}, \tag{17}$$

$$[\alpha \quad \beta]^{T} = (B^{T}B)^{-1}B^{T}Y, \tag{18}$$

where

$$B = \begin{bmatrix} -z^{(1)}(2) & [z^{(1)}(2)]^{r} \\ -z^{(1)}(3) & [z^{(1)}(3)]^{r} \\ -z^{(4)}(4) & [z^{(1)}(4)]^{r} \\ \vdots & \vdots \\ -z^{(1)}(n) & [z^{(1)}(n)]^{r} \end{bmatrix}, Y = \begin{bmatrix} q^{(0)}(2) \\ q^{(0)}(3) \\ q^{(0)}(4) \\ \vdots \\ q^{(0)}(n) \end{bmatrix} \tag{19}$$

Set the initial condition $\hat{q}^{(1)} = q^{(1)}(1)$.

The solution of Eq. (19) can be expressed as

$$\hat{q}^{(1)} = \left[\left(q^{(1)}(1)^{1-r} - \frac{\beta}{\alpha}\right)e^{-\alpha(1-r)(m-1)} + \frac{\beta}{\alpha}\right]^{1/(1-r)} \quad r \neq 1, m = 1, 2, 3 \ldots \tag{20}$$

$$\hat{q}^{(1)} = \left[\left(q^{(1)}(m)^{(1-r)} - \frac{\beta}{\alpha}\right)e^{-\alpha(1-r)(k-m)} + \frac{\beta}{\alpha}\right]^{1/(1-r)} \quad r \neq 1, m = 1, 2, 3 \ldots \tag{21}$$

$$\hat{q}^{(0)}(1) = q^{(0)}(1) \tag{22}$$

$$\hat{q}^{(0)}(m) = \hat{q}^{(1)}(m) - \hat{q}^{(1)}(m-1), M = 2, 3, \ldots \tag{23}$$

The adaptive variables r and p need to be fixed by the original data sequence. The appropriate values of r and p can be acquired using NGBM(1, 1) applications.

2 Variable Optimization for NGBM

- RPE (Relative Percentage Error)

$$RPE = \frac{\hat{q}^{(0)}(m) - q^{(0)}(m)}{q^{(0)}(m)} \times 100\% \tag{24}$$

- ARPE (Average Relative Percentage Error)

$$ARPE = \frac{1}{m-1} \sum_{m=1}^{n} RPE(m) \tag{25}$$

3 Case Study: Long-Term Generation Forecasting

3.1 Parameters NGBM Model

The raw data of time series can be found from Eq. (1). Then it can be analyzed and applied in the 1-AGO. Now the new sequences can be evaluated and applied to GM (1, 1). Further it can be applied to the Bernoulli derivation equation and the value of α and β can be generated by putting the probabilistic value of p = 0.5. Equations (11)–(13) are used to find out the simulated data using estimated inverse accumulated generating operation (IAGO).

3.2 Data Preparation

Here, the historical data of the Indian electricity generation are used to prepare the raw data for simulation. The yearly data of Indian electricity generation in Megawatt (MW) are given in Table 1.

4 Result Analysis

The parameter computation of GM(1, 1) can be done using two variables α and β. Then the actual result for GM(1, 1) using α and β required to be simulated with $r = 0$. Then the other one can be calculated using three variables of NGBM(1, 1), i.e., α, β and r. After computing the Grey Model and nonlinear Grey Bernoulli model with raw historical data of electricity, prediction or forecasting of electricity generation can be done. Table 2 shows the elaborate variables consistent of the best response resultant value for different models.

In Table 3 the real compression result of forecasting between original GM(1, 1) and NGBM(1, 1) are shown. Comparing the two found values, now the output of the two methods can be validated. Using different technique and computing the parameters, now the Average Relative Percentage Error (ARPE) of the simulated data can be compared. The percentage relative error between GM(1, 1) and NGBM (1, 1) are shown in Figs. 1 and 2. The average percentage relative error comparison of GM(1, 1) and NGBM(1, 1) are also shown in Fig. 3.

In given Table 3 the historical raw data of indian electricity generation and forecasting value using GM(1, 1) and NGBM(1, 1) model are given. Now both the models can be compared by considering the predictive PRE and ARPE. So it can be concluded that NGBM(1, 1) gives more accuracy over to GM(1, 1) (Fig. 4).

In this section it is clearly described and shown that NGBM is comparatively accurate than GM from the relative percentage errors point of aspect. Now in

Table 1 Actual data of electricity generation of India in MW

Year	Actual data
2000–01	560842
2001–02	579120
2002–03	596543
2003–04	633275
2004–05	665873
2005–06	697459
2006–07	752454
2007–08	813102
2008–09	842531
2009–10	905974
2010–11	959070
2011–12	1051375
2012–13	1111722

Table 2 Extraction of the variables of GM and NGBM

System	Forecasting parameter	Coefficient	ARPE
GM(1, 1)	$\alpha = 0.0634$, $\beta = 496330$	$r = 0$	1.7143
NGBM(1, 1)	$\alpha = -0.3212$, $\beta = 6745600$	$r = 0.2$	1.5021

Table 3 Comparison between actual value and forcecast value

Year	Actual value	Forecast value		PRE	
		GM	NGBM	GM	NGBM
2000–01	560842	560842	560842	0	0
2001–02	579120	555800	574120	1.0203	0.8626
2002–03	596543	555838	591031	0.8626	0.9293
2003–04	633275	591397	625914	0.6386	1.1637
2004–05	665873	629231	661127	0.5425	0.7133
2005–06	697459	669485	682449	2.13	2.1534
2006–07	752454	712315	732830	0.7217	2.6085
2007–08	813102	757885	803519	0.828	1.1796
2008–09	842531	806370	826478	1.8308	1.9062
2009–10	905974	857956	898816	0.7582	0.7904
2010–11	959070	912843	946591	1.2691	1.3011
2011–12	1051375	971241	1030653	1.712	1.9741
2012–13	1111722	1033375	1096539	1.1008	1.3722
2013–14	–	1099485	1110985	–	–
2014–15	–	1169823	1250633	–	–
2015–16	–	1244661	1319458	–	–
2016–17	–	1324287	1468578	–	–
2017–18	–	1409007	1493640	–	–
2018–19	–	1499147	1542599	–	–
2019–20	–	1595053	1687033	–	–
2020–21	–	1697095	1894470	–	–
2021–22	–	1805665	2092775	–	–
2022–23	–	1921180	2229084	–	–

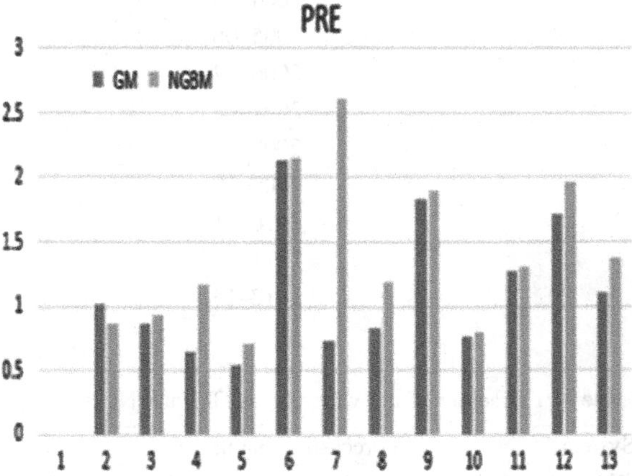

Fig. 1 Model PRE compression

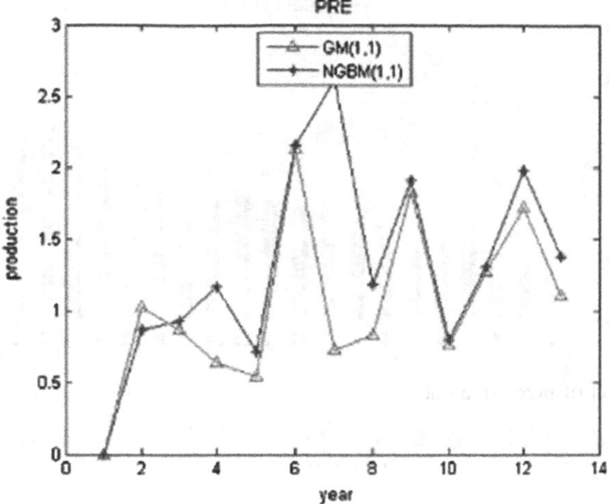

Fig. 2 Model PRE compression

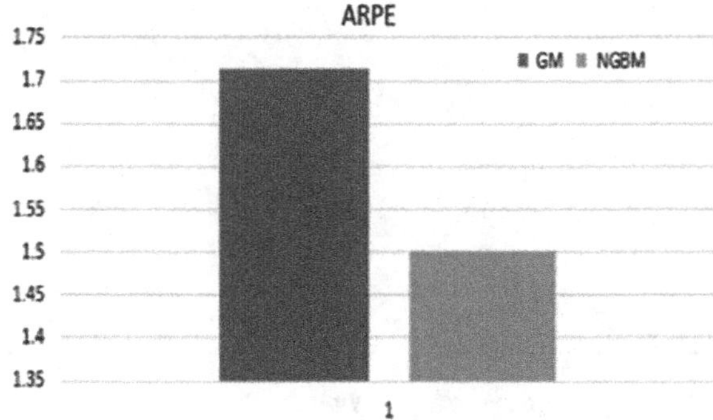

Fig. 3 Model ARPE compression

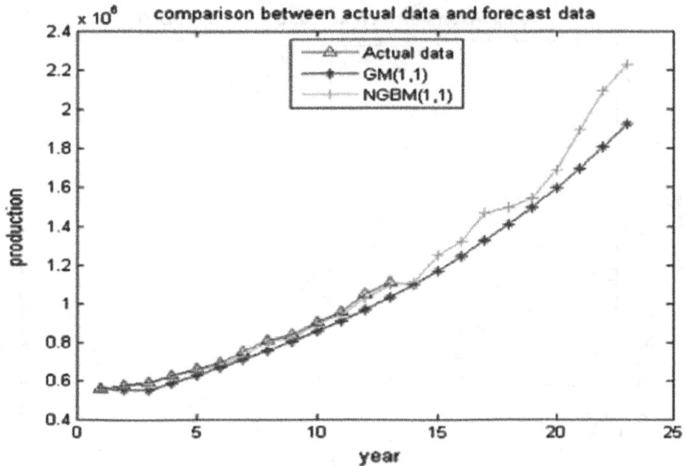

Fig. 4 Comparison of actual data

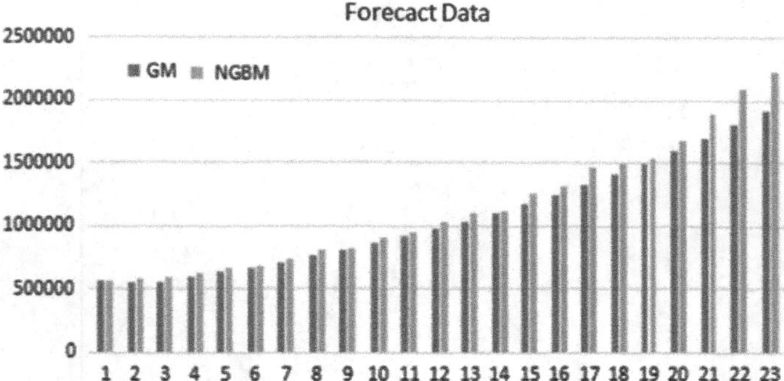

Fig. 5 Comparison of predictive data

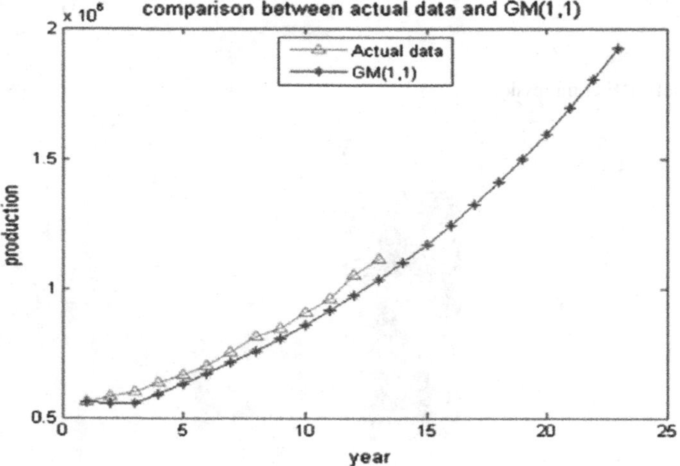

Fig. 6 Model forecast value by using GM(1, 1)

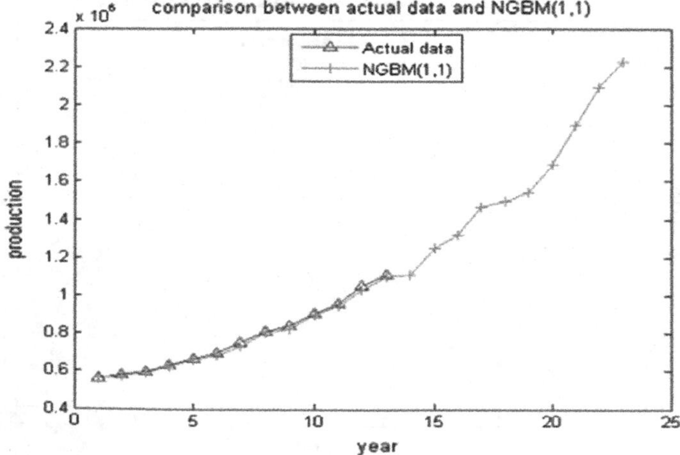

Fig. 7 Model forecast value by using NGBM(1, 1)

Tables 1 and 2 the two parameter a and b are taken for GM(1, 1) and three-dimensional coefficient parameter a, b, and r for NGBM(1, 1) with the ARPE comparison of both the models. In Fig. 5a comparison between the predictive data, GM(1, 1) and NGBM(1, 1) are given away. Finally, in Figs. 6 and 7 the comparison of the forecasting results of the Indian electricity generation is presented. Figure 7 also represents the real comparison between actual data of GM(1, 1) and NGBM(1, 1).

5 Conclusion and Future Works

Based on the analysis of the experimented results, it can be concluded that the NGBM(1, 1) model performs superiorly over GM(1, 1) model with respect to the experimental dataset. Although NGBM(1, 1) model could not reach 100% accuracy, still it provides better accuracy than other existing models. So with some certain shortcomings it can be concluded that the experimental results are more convincing towards achieving the goal of forecasting.

To improve the accuracy and performance of NGBM(1, 1) model, more attributes like natural disasters, fuel resources, and chemical resources need to be considered, so that the model will become more dynamic and robust for forecasting even with abrupt changes in generation.

References

1. Deng JL Introduction of Grey system. Journal of Grey System 1 (1989): 1–24.
2. Wang YF Predicting stock price using fuzzy Grey prediction system, Expert Systems with Applications 22 (2002): 33–38.
3. Yong H A new forecasting model for agricultural commodities. Journal of Agricultural Engineering Research 60 (1995): 227–235.
4. Xu QY, Wen YH. The application of Grey model on the forecast of passenger of international air transportation. Transportation Planning Journal 26 (1997): 525–555.
5. Bahrami S, Hooshmand RA, Parastegari M Short term electric load forecasting by wavelet transform and grey model improved by PSO algorithm. Energy 72 (2014): 434–442.
6. Chang SC, Lai HC, Yu HC A variable P value rolling Grey forecasting model for Taiwan semiconductor industry generation. Technological Forecasting and Social Change 72 (2005): 623–640.
7. J.L. Deng, Control problems of grey systems, System. Control Letters, volume 5, (1982) 288–294.
8. M.S. Yin, Fifteen years of grey system theory research: a historical review and bibliometric analysis Application. 40(7) (2013) 2767–2775.
9. Sifeng Liu, Yi Lin, “Grey Systems Modelling” Grey Information: Theory and Practical, Springer-Verlag London Limited (2006) 1610–3947.
10. E. Kayacan, B. Ulutas, O. Kaynak, Grey system theory-based models in time series prediction, Expert System. Application. 37 (2010) 1784–1789.
11. Y. Peng, M. Dong, A hybrid approach of HMM and grey model for age dependent health prediction of engineering assets, Expert System. Appl. 38 (2011) 12946–12953.

12. Y.H.L in, P. C. Lee, Novel high-precision grey forecasting model, Autom. Constr. 16 (2007) 771–777.
13. C. X. Fan, S. Q. Liu, Wind speed forecasting method: grey related weighted combination with revised parameter, Energy Procardia 5 (2011) 550–554.
14. V. Bianco, O. Manca, S. Nardini, A. A. Minea, Analysis and forecasting of nonresidential electricity consumption in Romania, Appl. Energy 87 (2010) 3584–3590.
15. M.L. Lei, Z. R. Feng, A proposed grey model for short term electricity price forecasting in competitive power markets, Int. J. Electr. Power Energy Syst. 43 (2012) 531–538.
16. C.S. Lin, F. M. Liou, C. P. Huang, Grey forecasting model for CO2 emissions: a Taiwan study, Appl. Energy 88 (2011) 3816–3820.
17. C.I. Chen, H. L. Chen, S. P. Chen, Forecasting of foreign exchange rates of Taiwans major trading partners by novel nonlinear Grey Bernoulli model NGBM(1, 1), Common. Nonlinear Sci. Number. Simul 13 (2008) 1194–1204.
18. Hsu, L. C. Applying the Grey prediction model to the global integrated circuit industry. Technological Forecasting and social change 70 (2003), 567–574.
19. Dang, Y.G., Liu, S.F., Liu, B., The GM models that x (1)(n) be taken as initial value. Chin. J. Manag. Sci. 13 (1) (2005), 132–135.
20. J. Zhou, R. Fang, Y. Li, Y. Zhang, and B. Peng, "Parameter optimization of nonlinear grey Bernoulli model using particle swarm optimization," Applied Mathematics and Computation, vol. 207, no. 2, pp (2009) 292–299.
21. Xu, T., Leng, S.X., Improvement and application of initial values of grey system model. J. Shandong Inst. Technol. 13 (1) (1999), 15–19.
22. Liu, B., Liu, S.F., Zhai, Z.J., Dang, Y.G., Optimum time response sequence for GM(1, 1). Chin. J. Manag. Sci. 11 (4) (2003), 54–57.
23. Zhang, D.H., Jiang, S.F., Shi, K.Q., Theoretical defect of grey prediction formula and its improvement. Syst. Eng. Theory Pract. 22 (8) (2002), 140–142.

Publishing Personal Information by Preserving Privacy in Vertically Partitioned Distributed Databases

R. Srinivas, K.A. Sireesha and Shaik Vahida

Abstract In vertically partitioned distributed databases, the data will be distributed over multiple sites. To publish such data for research or business applications, data may be collected at one site and published data to the needy. Publishing data in many fields like banking, medical, political, research, etc., by preserving one's privacy is very important. Apart from preserving privacy, anonymity of the publisher must be preserved. To achieve these objectives in this paper, multidimensional *k*-anonymity with onion routing and mix-network methods are proposed to preserve privacy and to provide anonymous communication. Mix-net is a multistage system which accepts quantities of data on input batch and produces cryptographically transformed data through output batch. Output batch is a permutation of the transformed input batch, to achieve untraceability between the input and output batches. Mix-net can change the appearance and random reordering which prevents trace back. In onion routing encryption, data is encapsulated in layers of encryption, which is analogous to layers of onion. The data is being sent to transmit inside the several layers of encryption. The final node or exit node in the chain is to decrypt and deliver the data to the recipient by applying multidimensional *k*-anonymity on collected data.

Keywords Anonymity · Data · Distributed · Hierarchical · Mix-network · Onion routing · Personal information · Privacy · Ring · Vertically partitioned

R. Srinivas (✉) · S. Vahida
Department of CSE, ACET, Surampalem, India
e-mail: viceprincipal@acet.ac.in

S. Vahida
e-mail: vahida.shaik@acet.ac.in

K.A. Sireesha
Department of CSE, AEC, Surampalem, India
e-mail: sireesha.kanduri@gmail.com

H.S. Behera and D.P. Mohapatra (eds.), *Computational Intelligence in Data Mining*, Advances in Intelligent Systems and Computing 556,
DOI 10.1007/978-981-10-3874-7_21

1 Introduction

In Information Technology, sharing of information has a long history. Traditionally information sharing is all about exchange of data between sender and receiver, the successful exchange relies on providing privacy to that data. A basic understanding of privacy of the communication is mandatory in order to know the principles. In real life, the situation is more complicated for publishing data, while publishing data, individual data should not be revealed. One should provide privacy to individual data before publication [1].

The word privacy is first introduced in 1988 by David Chaum to solve the "dinning cryptographers' problem" [2]. In latest years, number of methods has been proposed to publish data. Mainly these methods are related to single system. As per our knowledge is concern no great research has been done on vertically portioned distributed database. So there is a great scope for research on vertically portioned distributed data. Efficient models are required to provide privacy for individuals' data. To store data at multiple sites, a common key is maintained at the sites to get original data.

2 Literature Survey

In early 1988, David Chaum proposed a new algorithm in cryptography called dining cryptographers problem which perform on secure multi-party computation. Collision, disruption, and complexity are the limitations of D-C net protocol. The k-anonymity concept is first formulated by Sweeney and Latanya in the year 2002 [3]. Machanavajjhala et al. in 2007 proposed new anonymity called as l-diversity [4]. A new model t-closeness was proposed by Li et al. in 2007 to overcome the limitations of l-diversity [5]. Maolis Terrovitis et al. in 2008 defined k^m-anonymity [6]. In 2009, Wong R.C. proposed a (α, k) anonymity model [7]. In 2010, Campan et al. proposed p-sensitive k-anonymity avoids this shortcoming [8].

Another method (k, e)-anonymity is presented by Wong et al. in 2006 [9]. In 2007, Xiaokui Xiao and Yufei Tao developed a generalization-based m-invariance that effectively limits the risk of privacy disclosure in re-publication [10]. In 2005, Kristen LeFevre et al. proposed multidimensional k-anonymity [11]. In multidimensional k-anonymity data anonymization is done based on more than one record. It means in multidimensional k-anonymity record grouping is done by considering more than one attribute. Nowadays, data growth rate is high and data may be stored at multiple sites. In order to provide privacy in such cases efficient methods were proposed in [12–17].

Mix-network was proposed by Chaum et al. [18, 19] to achieve anonymous communication. Mix-network is mixing by stage changes in appearance of inputs, and the order of arrival information is removed [20–24]. In 2004 Roger Dingle dine and Nick Mathewson develop onion routing [25]. It was proposed to achieve anonymous communication with privacy preserving. It is a bidirectional and near real-time which provides a better arrangement for private communication over a public network. Onion routing is a flexible communication infrastructure [26–29]. In 2016 Srinivas et al. proposed hierarchical and ring models on vertically partitioned data to preserve privacy [30].

3 Proposed Methodology

3.1 Preserving Privacy in Vertically Partitioned Distributed Databases

To provide anonymous communication and preserving privacy to individual private data, an innovative approach is proposed for publishing vertically partitioned distributed data using a secure join operation. In this paper two models are used to publish vertically partitioned distributed data. We considered that the vertically partitioned data is distributed over multiple sites by maintaining common field for union operation. The following tasks are to be performed before publishing data.

i. Data transmitted to the next site only after encryption.
ii. Receiver receives data and performs union operation based on common key.
iii. Finally data encrypted and forwarded to next site.

Consider the example of Table 1. Here S.No is a primary key, and this primary key is used as a common field for vertically partitioned data.

3.2 Publishing Data Using Ring Model and Hierarchical Model

In the proposed model, we assumed that the sites can form ring or hierarchy models with mix-nets to send and receive data anonymously. These models are shown in Figs. 1 and 2.

Table 1 Sample hospital data

S.No	First name	Last name	Address	Phone number	Age	Sex	Disease
1	Aditya	Bhanu	AUS	9848382811	19	M	Asthma
2	Chandra	Dheeraj	BGD	9858483822	28	M	Bone break
3	Evan	Flik	CAN	9868584833	37	M	Cancer
4	Guru	Harsha	DNK	9878685844	46	M	Diabetes
5	Indraja	Jasmine	EGY	9888786855	55	F	Ebola
6	Kamala	Lalitha	FRA	9898887866	64	F	Fibrosis
7	Madhuri	Nisha	GEO	9808988877	73	F	Gastric

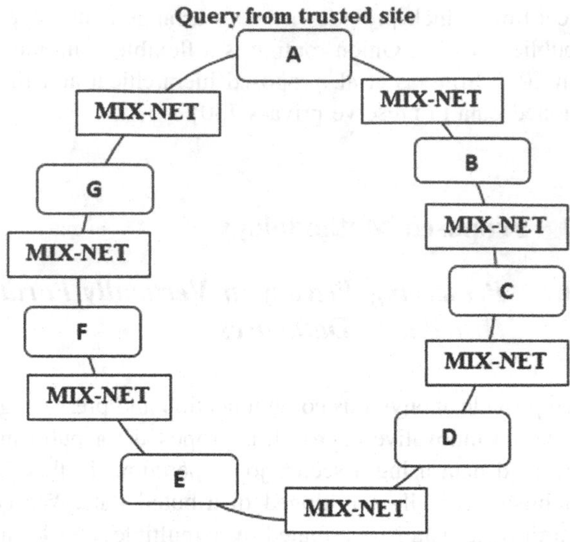

Fig. 1 Onion-based ring method through mix-net

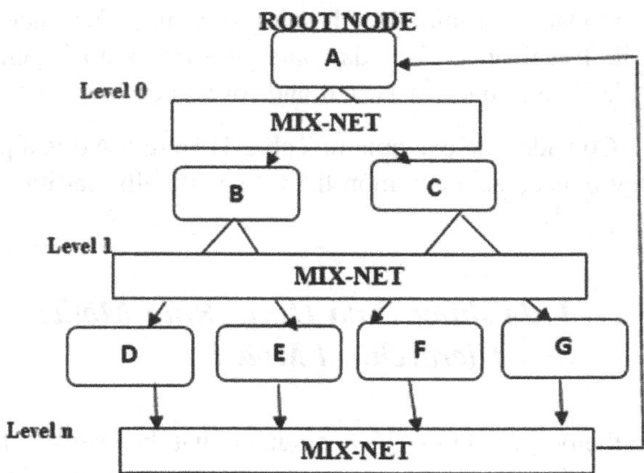

Fig. 2 Hierarchical-based mix-network

3.2.1 Publishing Data Using Ring Network Model

In ring model to publish data, the following steps will take place.

Initially, one of the sites receive request message (query) from trusted site. The site(A) which receives message will initiate data transfer operation by sending the same query to all the sites in the ring network. Then all these sites encrypt the data except primary key and make data ready to transfer to next site. Here S.No is the primary key. Site A will send encryption key E_k to all other sites in the ring network. The actual data transfer operation is as follows. Each pair of sites can use one mix-net. Here consider that site A which received query from trusted site. So it will initiate data transfer operation by sending its data to its next site B through mix-net as shown in Fig. 1 by updating encrypt count. Site B receives data from A, and at this stage both the encrypted data and existed data at site B again encrypted, updated encrypt count, and transmitted through mix-net to next site. Now site B has data which contains fields S.No and two encrypted attributes. After union operation site B sends data to site C. C will also perform the same operation. This procedure is repeated till all the sites involved in the ring model are completed. The final result is sent to site A. At this stage site A has data with all fields in encrypted form. It performs decrypt operation multiple times based on number of sites involved in data transfer. After decryption site A perform Mondrian multidimensional *k*-anonymity before publishing data to third party.

The advantages of this method are as follows: 1. Mix-net can change the appearance and random reordering at receiver side. 2. Sender and receiver are not aware of the data in which sites it belongs to exactly. 3. High Security.

3.2.2 Publishing Data Using Hierarchical Model

In this model to publish data the following steps will take place:

Initially, one of the sites receive request message or query from trusted site. The site which receives message will initiate data transfer operation by sending same query to all sites participating in data transfer operation. Then all these sites encrypt the data except primary key and make data ready to transfer. Here S.No is the primary key.

The actual data transfer operation is as follows. At each level mix-nets are used for anonymous communication. Here consider site A which is the root node and received query for data from trusted site. So it will initiate data transfer operation by dividing its content into two sets. Data are randomly sent to next level, i.e., to the sites B and C of level 0 through mix-nets as shown in Fig. 2. Site B receives one of the two data sets from A and performs secured union operation on data received and its content using S.No as key. Now B has data which contains fields S.No and two encrypted attributes. After union operation B sends data to next level, i.e., level 1 through mix-net. Site C also performs the same operations. This process is repeated for all levels in the hierarchy model. The data set division is based on the number of child nodes. In this example we considered two child nodes so the data is divided into two sets. All the leave sites receive data from previous level and perform secured union operation on data received and its content results are sent to site A through mix-net without dividing data into sets. At this stage site A has data with all fields in encrypted form after performing union operation on all the sets received. It performs decrypt operation. It performs Mondrian multidimensional *k*-anonymity on decrypted data and publishes data to third party.

The advantages of this method are as follows: 1. Reduces data transmission load in the network; mix-net can change the appearance and random reordering at receiver side. 2. Sender and receiver are not aware of the data in which sites it belongs to exactly. 3. High Security. 4. Time required to transfer data is less in this network.

4 Results

A network is established for the proper communication between the sites existed in the topology.

4.1 Results for Ring Model

1. Consider vertically partitioned data is distributed among seven sites namely A, B, C, D, E, F, and G as shown in Table 2. Encrypted data sets are used for transmission. For better understanding data in actual format is considered here.
2. First, site A receives the query from trusted party.
3. Data present at site A is shown in Table 3.
4. Site A sends this data to site B through mix-net. Now site B performs secured union on data set received and its own content. The data available after union operation at site B is shown in Table 4.
5. Now site B sends encrypted data to site C through mix-net. Site C has data set after union operation is shown in Table 5.
6. This process is repeated for all the sites from D to G in circular fashion. The data available at site G is shown in Table 6. This data sends to site A.
7. Site A performs the following operations before publication:

Table 2 Vertically partitioned data with S.No as a common field

(i)		(ii)		(iii)		(iv)	
S.No	First name	S.No	Last name	S.No	Address	S.No	Phone number
1	Aditya	1	Bhanu	1	AUS	1	9848382811
2	Chandra	2	Dheeraj	2	BGD	2	9858483822
3	Evan	3	Flik	3	CAN	3	9868584833
4	Guru	4	Harsha	4	DNK	4	9878685844
5	Indraja	5	Jasmine	5	EGY	5	9888786855
6	Kamala	6	Lalitha	6	FRA	6	9898887866
7	Madhuri	7	Nisha	7	GEO	7	9808988877

(v)		(vi)		(vii)	
S.No	Age	S.No	Sex	S.No	Disease
1	19	1	M	1	Asthma
2	28	2	M	2	Bone break
3	37	3	M	3	Cancer
4	46	4	M	4	Diabetes
5	55	5	F	5	Ebola
6	64	6	F	6	Fibrosis
7	73	7	F	7	Gastric

Table 3 Data at site A

S.No	First name
1	Aditya
2	Chandra
3	Evan
4	Guru
5	Indraja
6	Kamala
7	Madhuri

Table 4 Data at site B

S.No	First name	Last name
1	Aditya	Bhanu
2	Chandra	Dheeraj
3	Evan	Flik
4		Harsha
5		Jasmine
6		Lalitha
7		Nisha

Table 5 Data at site C

S.No	First name	Last name	Address
1	Aditya	Bhanu	AUS
2	Chandra	Dheeraj	BGD
3	Evan	Flik	CAN
4	Guru	Harsha	DNK
5	Indraja	Jasmine	EGY
6	Kamala	Lalitha	FRA

(a) Remove sensitive attributes which reveals personal information.
(b) Mondrian multidimensional *k*-anonymity is applied on result obtained from previous step.

The final result is published to third party as shown in Table 7.

4.2 Results for Hierarchical Topology Model

1. Consider vertically partitioned data shown in Table 2. These data sets are available at sites A, B, C, D, E, F, and G, respectively. Encrypted data sets are used for transmission.
2. First, site A receives the query or message from trusted site.
3. Data present at site A is shown in Table 3.

Table 6 Data at site G

S.No	First name	Last name	Address	Phone number	Age	Sex	Disease
1	Aditya	Bhanu	AUS	9848382811	19	M	Asthma
2	Chandra	Dheeraj	BGD	9858483822	28	M	Bone break
3	Evan	Flik	CAN	9868584833	37	M	Cancer
4	Guru	Harsha	DNK	9878685844	46	M	Diabetes
5	Indraja	Jasmine	EGY	9888786855	55	F	Ebola
6	Kamala	Lalitha	FRA	9898887866	64	F	Fibrosis
7	Madhuri	Nisha	GEO	9808988877	73	F	Gastric

Table 7 Final published data

S.No	First name	Last name	Address	Age	Sex	Disease
1	Aditya	Bhanu	AUS	19	M	Asthma
2	Chandra	Dheeraj	BGD-FRA	28–64	M–F	Bone Break
3	Evan	Flik	CAN-DNK	37–46	M–F	Cancer
4	Guru	Harsha	CAN-DNK	37–46	M–F	Diabetes
5	Indraja	Jasmine	EGY-GEO	55–73	F	Ebola
6	Kamala	Lalitha	FRA-BGD	28–64	M–F	Fibrosis
7	Madhuri	Nisha	EGY-GEO	55–73	M–F	Gastric

Table 8 Data at site A

S.No	First name
1	Aditya
2	Chandra
3	Evan
4	Guru
5	Indraja
6	Kamala
7	Madhuri

4. Site A sends the data to Site B and C through mix-net by dividing its content into two sets as shown in Table 8. Now site B performs secured union on data set received and its own content. The data available at B is shown in Table 9.
5. Now site B sends encrypted data to sites D and E through mix-net.
6. This process is repeated for all levels (Table 10).
7. All leave sites receive data from previous level through mix-nets and perform secured union operation on data received and its content is finally sent to site A without dividing data into sets.

Table 9 Data at site B

S.No	First name	Last name
1	Aditya	Bhanu
2	Chandra	Dheeraj
3	Evan	Flik
4		Harsha
5		Jasmine
6		Lalitha
7		Nisha

Table 10 Data at site A

S.No	First name	Last name	Address	Age	Sex	Disease
1	Aditya	Bhanu	AUS	19	M	Asthma
2	Chandra	Dheeraj	BGD-FRA	28–64	M–F	Bone Break
3	Evan	Flik	CAN-DNK	37–46	M–F	Cancer
4	Guru	Harsha	CAN-DNK	37–46	M–F	Diabetes
5	Indraja	Jasmine	EGY-GEO	55–73	F	Ebola
6	Kamala	Lalitha	FRA-BGD	28–64	M–F	Fibrosis
7	Madhuri	Nisha	EGY-GEO	55–73	M–F	Gastric

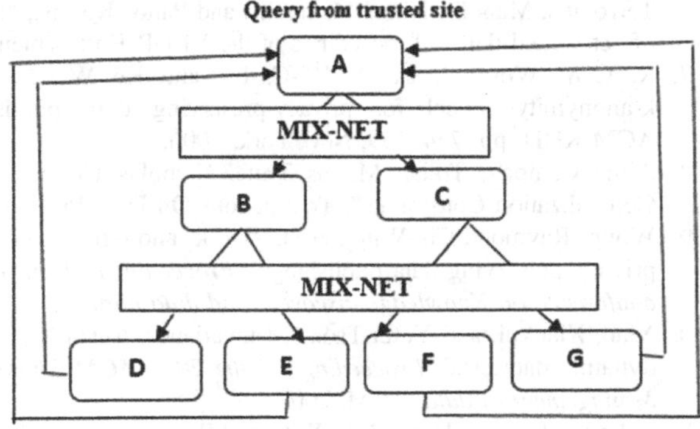

Fig. 3 Hierarchical model through mix-net

8. At site A, before publication of data, sensitive attributes which reveal privacy of the person like phone number names, etc., are removed and Mondrian multi-dimensional *k*-anonymity method (where $K >= 2$) is applied.

The above process is illustrated in Fig. 3.

5 Conclusions

In this paper, we addressed solution to anonymous communication between the sites and preserving privacy for personal information on vertically partitioned distributed databases. To preserve privacy, Mondrian multidimensional k-anonymity and onion routing algorithms are used and to provide anonymous communication between sites mix-network are proposed. In future, this work can be extended for any kind of databases.

References

1. G kountouna, Olga, "A Survey on Privacy Preservation Methods.", Technical Report, Knowledge and Database Systems Laboratory, NTUA, 2011, SECE, Knowledge and Data Base Management Laboratory (DBLAB), pp: 1–30, 2011.
2. Chaum, David, "The dining cryptographers problem: Unconditional sender and recipient untraceability." Journal of cryptology 1.1, pp: 65–75, 1988.
3. Sweeney, Latanya. "K-anonymity: A model for protecting privacy.", International Journal of Uncertainty, Fuzziness and Knowledge-Based Systems 10.05 pp: 557–570, 2002.
4. Machanavajjhala, Ashwin, et al., "l-diversity: Privacy beyond k-anonymity.", ACM Transactions on Knowledge Discovery from Data (TKDD) 1.1, 2007.
5. Li, Ninghui, Tiancheng Li, and Suresh Venkatasubramanian., "t-closeness: Privacy beyond k-anonymity and l-diversity.", Data Engineering, 2007. ICDE 2007. IEEE 23rd International Conference on. IEEE, 2007.
6. Terrovitis, Manolis, Nikos Mamoulis, and Panos Kalnis., "Privacy-preserving anonymization of set-valued data.", Proceedings of the VLDB Endowment 1.1 (2008): pp: 115–125.
7. R. C.W. Wong, J. Li, A. W.-C. Fu, and Ke Wang., "(α, k)-anonymity: an enhanced k-anonymity model for privacy-preserving data publishing", In Proceedings of the ACM KDD, pp: 754–759, New York, 2006.
8. Alina Campan, Traian Marius Truta, Nicholas Cooper, "P-Sensitive K-Anonymity with Generalization Constraints", Transactions On Data Privacy 3, pp: 65—89, 2010.
9. Wong, Raymond Chi-Wing, et al., "(α, k)-anonymity: an enhanced k-anonymity model for privacy preserving data publishing.", *Proceedings of the 12th ACM SIGKDD international conference on Knowledge discovery and data mining*. ACM, 2006.
10. Xiao, Xiaokui, and Yufei Tao., "M-invariance: towards privacy preserving re-publication of dynamic datasets.," *Proceedings of the 2007 ACM SIGMOD international conference on Management of data*. ACM, 2007.
11. LeFevre, Kristen, David J. DeWitt, and Raghu Ramakrishnan., "Mondrian multidimensional k-anonymity.", *Data Engineering, 2006. ICDE'06. Proceedings of the 22nd International Conference on*. IEEE, 2006.
12. Xiong, Li, et al., "Privacy preserving information discovery on ehrs.", *Information Discovery on Electronic Health Records* (2008): pp: 197–225.
13. Jiang, Wei, and Chris Clifton., "Privacy-preserving distributed k-anonymity.", *Data and Applications Security XIX*. Springer Berlin Heidelberg, 2005. pp: 166–177.
14. LeFevre, Kristen, David J. DeWitt, and Raghu Ramakrishnan., "Incognito: Efficient full-domain k-anonymity.", *Proceedings of the 2005 ACM SIGMOD international conference on Management of data*. ACM, 2005.

15. Bayardo, Roberto J., and Rakesh Agrawal., "Data privacy through optimal k-anonymization.", *Data Engineering, 2005. ICDE 2005. Proceedings. 21st International Conference on*. IEEE, 2005.
16. Srinivas R., and A. Raga Deepthi. "Hierarchical Model for Preserving Privacy in Horizontally Partitioned Databases." *Age* 15: 35. IJETTCS, Volume 2, Feb 2013.
17. Srinivas R., K. PRASADA RAO, and V. SREE REKHA., "Preserving Privacy in Horizontally Partitioned Databases Using Hierarchical Model.", *IOSR Journal of Engineering May* 2.5 (2012): pp: 1091–1094.
18. Chaum, David L. "Untraceable electronic mail, return addresses, and digital pseudonyms." *Communications of the ACM* 24.2 (1981): 84–90.
19. Chaum, David. "Untraceable electronic mail, return addresses and digital pseudonyms." *Secure electronic voting*. Springer US, 2003. 211–219.
20. Danezis, George. "Mix-networks with restricted routes." *Privacy Enhancing Technologies*. Springer Berlin Heidelberg, 2003.
21. Abe, Masayuki. "Mix-networks on permutation networks." *Advances in cryptology-ASIACRYPT'99*. Springer Berlin Heidelberg, 1999. 258–273.
22. Sampigethaya, Krishna, and Radha Poovendran. "A survey on mix networks and their secure applications." *Proceedings of the IEEE* 94.12 (2006): 2142–2181.
23. Srinivas, R., A. Yesu Babu, and Ssaist Surampalem. "Mixnet for Anonymous Communication to Preserving Privacy in Hierarchical model networks".
24. Demirel, Denise, Hugo Jonker, and Melanie Volkamer. "Random block verification: Improving the norwegian electoral mix-net." *Electronic Voting*. 2012.
25. Dingledine, Roger, Nick Mathewson, and Paul Syverson. *Tor: The second-generation onion router*. Naval Research Lab Washington DC, 2004.
26. Goldschlag, David, Michael Reed, and Paul Syverson. "Onion routing." *Communications of the ACM* 42.2 (1999): 39–41.
27. Reed, Michael G., Paul F. Syverson, and avid M. Goldschlag. "Anonymous connections and onion routing." *Selected Areas in Communications, IEEE Journal on* 16.4 (1998): 482–494.
28. Syverson, Paul F., David M. Goldschlag, and Michael G. Reed. "Anonymous connections and onion routing." *Security and Privacy, 1997. Proceedings., 1997 IEEE Symposium on*. IEEE, 1997.
29. Syverson, Paul, et al. "Towards an analysis of onion routing security." *Designing Privacy Enhancing Technologies*. Springer Berlin Heidelberg, 2001.
30. R. Srinivas, et al. "Preserving Privacy in Vertically Partitioned Distributed Data using Hierarchical and Ring Models.", in Joint International Conference on Artificial Intelligence and Evolutionary Computations in Engineering Systems (ICAIECES-2016) at SRM University, Kattankulathur, Chennai, India, during 19–21st May 2016.

Maintaining Security Concerns to Cloud Services and Its Application Vulnerabilities

Lakkireddy Venkateswara Reddy and Vajragiri Viswanath

Abstract Cloud computing is a highly demand service which is widely used by IT industry because of its flexibility and scalable in nature. In this paper, the main purpose is about cloud computing security concerns. A few steps are need to be followed to secure the cloud such as understanding the cloud, demand transparency, reinforcing the internal security, legal implication considerations and finally paying attention to the cloud environment. The major concern bothering the IT industries is to provide the security to cloud services (SaaS, PaaS and IaaS). In this paper, we represent a multiplicative homomorphic encryption namely advanced RSA algorithm using three-factor authentication for ensuring that the data is secured.

Keywords Multiplicative homomorphic encryption · Security · RSA · Cloud services

1 Introduction

Cloud computing is exactly a collection of computers which are connected to a network that are work combine, help to store, and process the huge amount of data remotely through internet. Cloud computing is a remote server that has to be accessed with the help of internet, and by this the business application add-ins and computer software get benefited.

According to National Institute of Standards and Technology (NIST) [1] the definition of cloud computing is "A model for enabling ubiquitous, convenient, on-demand network access to a shared pool of configurable computing resources (examples—networks, servers, storage, applications and services) that can be rapidly provisioned and released with minimal management effort or service pro-

L. Venkateswara Reddy (✉) · V. Viswanath
Sree Vidyanikethan Engineering College, Tirupati, AP, India
e-mail: lakkireddy.v@gmail.com

V. Viswanath
e-mail: vishu.vajragiri@gmail.com

H.S. Behera and D.P. Mohapatra (eds.), *Computational Intelligence in Data Mining*, Advances in Intelligent Systems and Computing 556,
DOI 10.1007/978-981-10-3874-7_22

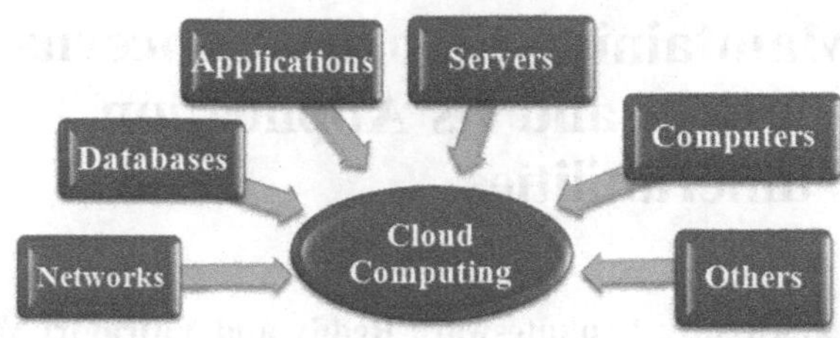

Fig. 1 Cloud computing is a collection of IT resources

vider interaction" [1]. Basically cloud computing is a collection of IT resources (Applications, Computers, Databases, Networks, Servers, etc.). The block diagram is depicted as shown in Fig. 1.

Cloud computing offers many benefits to the business people to run their business in the cloud. Some of the benefits [7] are as follows:

(i) **Flexibility**: Cloud provides the users to access and share the data within or outside of the organization by maintaining the collaboration among the users.
(ii) **Disaster *recovery***: In cloud computing disaster recovery [2] which is a best choice for organizations, this is a security measure which is adopted for maintaining backup and restore strategy.
(iii) **Scalability**: This is one of the key benefits of cloud computing. In IT industry the use of cloud is based on the requirements' increasing and decreasing of resources for various purposes which can be done easily.
(iv) **Security**: Even though there is a potential loss of the valuable thing, the data in it need to be secured. By adopting cloud, one can get the sensitive information and also can remotely erase the sensitive information from the lost device without getting misused.

On the other side, in Ref. [3], the working process of chit business in online and also mobile application development procedure for chit business is represented. So by this procedure we create an idea to design a web application which is useful to all chit business runners, who wants to make the chit business process in simple, secured and reliable.

2 Related Work

The authors Murray et al. [1], Yu and Tung [4], Xu et al. [5] and Sachdev and Bhansali [6] discussed about how to provide the security to cloud computing services and application vulnerabilities. Also they focused on homomorphic encryptions, data decryption using RSA algorithm, and three-tier authentication process. Some related work is discussed below.

In [1] the authors described the cloud security concerns, cloud characteristics, and cloud models for understanding the implications of cloud before they are going to host their application in cloud environment.

Also, Yu and Tung [4] described the importance of web application and detecting the vulnerabilities of web application in cloud. First of all, they did string analysis with respect to attacker perspective. By this action the result becomes with attack patterns for detecting different vulnerabilities. After they analyse the web application [4] files or whole document, they design the system architecture.

Finally, they were presented an online free service, named as **Patcher**, to evaluate various open-source web applications. By knowing the application vulnerabilities, IT managers know about the security status and take immediate action on patching up the vulnerabilities.

Consequently, Xu et al. [5] introduced URMG (User Representation Model Graph) which is an enhancement of CBMG (Customer Behaviour Model Graph). URMG model optimizes the evaluation of execution process, an automatic testing, and deployment in test cloud. Before knowing about the URMG and CBMG we must know the basic difference between URMG and CBMG [5].

The main difference of CBMG and URMG is that there are no statuses in URMG. They implemented the automatic performance evaluation system for web application depending on URMG. The various actions of URMG are as follows:

(i) It will analyse the web logs automatically;
(ii) It generates the model using web logs; and
(iii) It executes the evaluations.

CRMG provides convenient and accuracy to the developer after the above actions were performed.

Sachdev and Bhansali [6] described the AES for encrypting the data. They used AES (Advance Encryption Standard) algorithm for enhancing the security to the cloud computing. They proposed the data protection model to encrypt the data using AES before the application launched in the cloud. They mentioned some of the tasks in AES, such as the security issue identification, why they choose the AES algorithm and the comparison of AES algorithm with other algorithms like DES, Triple DES, and RSA [7]. Some advantages were present in AES.

Further, the drawback in DES is the possibility of weak keys. But in AES the different components were practically used by cipher and its inverse to eliminate the possibility of weak and semi-weak keys. A performance evaluation revealed the advantages of AES [7]; they are as follows:

(i) AES is much better than DES algorithm in terms of execution time.
(ii) AES detects the changing data type such as image instead of text; this advantage is over RC2, RC6.
(iii) It is blowfish in terms of time consumption.

3 Encryption

(a) **Definition**: The process of messing up the sensitive information into unreadable text (Cipher Text) is not understandable by third parties except authorized person, where this process is known as encryption [8]. Basically there are two types of encryptions: they are secret key encryption and public key encryption, popularly known as symmetric encryption and asymmetric encryption, respectively.

(b) **Homomorphic Encryption**: The process of conversion of data into cipher text that generated the result compared to operations performed on plaintext is called homomorphic encryption. It plays a major role on cloud computing, allowing companies to store their data in public cloud. Here various types of homomorphic encryptions are used as follows: additive and multiplicative homomorphic encryption [9].

(i) **Additive Homomorphic Encryption**
It is also called as Paillier cryptosystem
For examples if p and q are large prime numbers, let $n = pq$, let λ_i denote the Carmichael function that is $\lambda_i(n) = lcm(p_i - 1, q_i - 1) i = 1, 2, \ldots, n$. Pick random $g \in Z^*_{n^2}$ such that $L_i(g^\lambda \bmod n^2)$ is invertible modulo n $\left(where\ L_i(u) = \frac{u-1}{n}\right)$. N and g are public; p and q (or λ) are private for plain text x and resulting cipher text y select a random $r \in Z^*_{n^2}$ then $e^k(x, r) = g^x r^n \bmod n^2$

$$d_k(y) = \left(\frac{L_i(y^\lambda \bmod n^2)}{L_i(g^\lambda \bmod n^2)}\right) \bmod n$$
$$i = 1, 2, \ldots, n$$

(ii) **Multiplicative Homomorphic Encryption**
According to RSA the public key module is ***m*** and exponent is ***e***, and then the encryption of message ***P*** is as follows:

$$i = 1, 2, 3, \ldots, n$$

Homomorphic encryption property is

$$E(p_1).E(p_2) = p_1^e p_2^e \bmod m$$
$$= (p_1 p_2)^e \bmod m = \mathrm{E}(p_1.p_2).$$

From the above equation, we observe that if we multiply and encrypt the two plain texts p_1 and p_2, then we get a cipher text, but in the multiplicative homomorphic encryption, the multiplicative property says that the two plaintexts are separately encrypted and then multiply the cipher text each other and you get exactly the same.

4 Fully Homomorphic Encryption

In 2009, Craig Gentry of IBM Company has proposed the first encryption system fully homomorphic [10]; this can be used to any type of calculation (such as addition, multiplications, etc.) on the data stored in cloud without decrypting. The application of fully homomorphic encryption [9] is a major part in cloud computing security. Figure 2 represents the fully homomorphic encryption applied to cloud computing.

RSA Algorithm

RSA name was from the three author initial letters of their surnames (Ron Rivest, Adi Shamir, Leonard Adleman) who described the algorithm in 1977 in Massachusetts Institute of Technology. Key generation [11] is discussed in the following section.

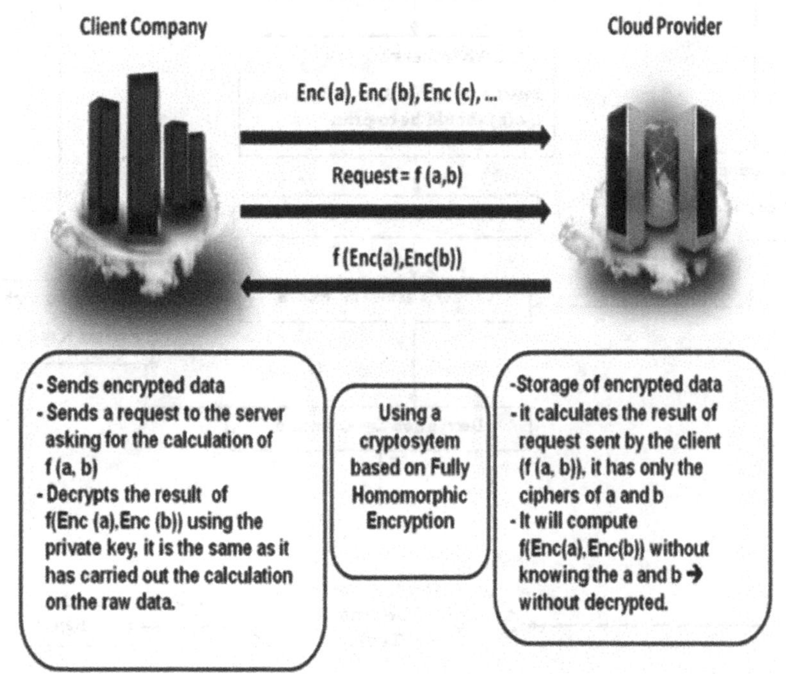

Fig. 2 Fully homomorphic encryption is applied to cloud computing

4.1 Key Generation

- Choose two distinct primes u and v of approximately equal size so that their product $n = uv$ is of the required length.
- Compute $\phi(n) = (u - 1)(v - 1)$.
- Choose a public exponent e, $1 < e < \phi(n)$, which is coprime to $\phi(n)$, that is, gcd $(e, \phi(n)) = 1$.
- Compute a private exponent d that satisfies the congruence ed $\equiv$ 1 (mod $\phi(n)$).
- Make the public key (n, e) available to others. Keep the private values d, u, v, and $\phi(n)$ secret.

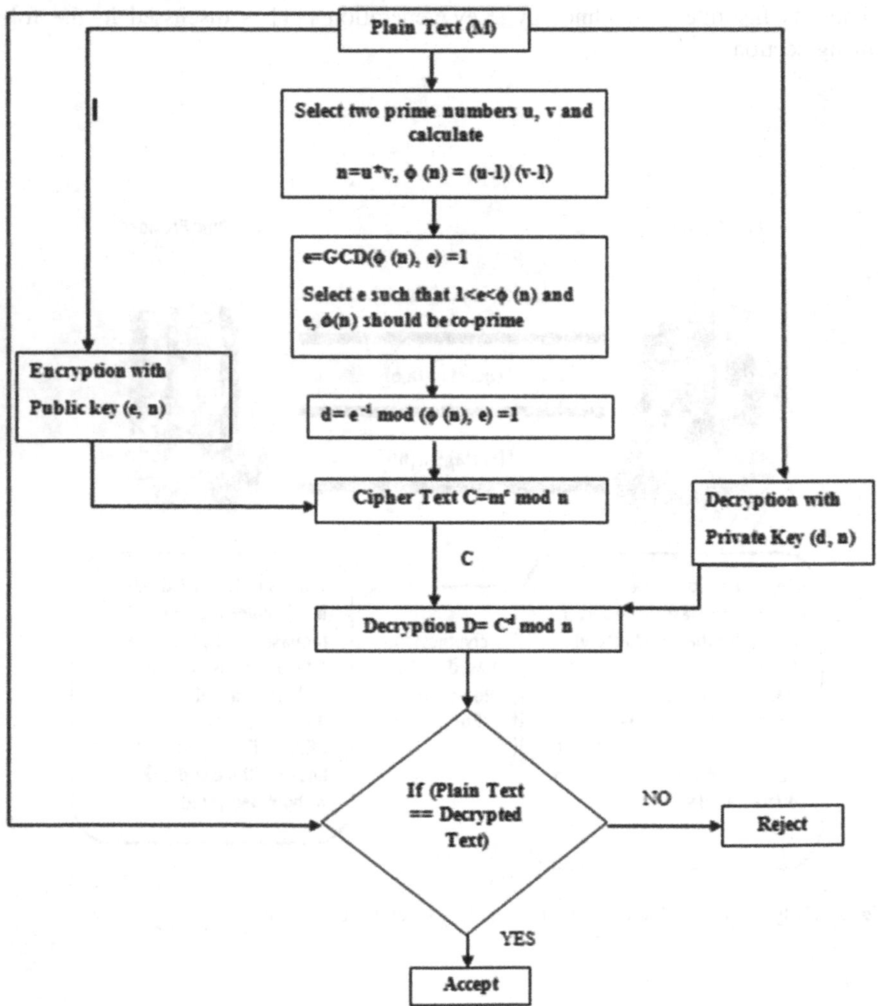

Fig. 3 RSA algorithm flow chart

4.2 RSA Encryption Scheme

Encryption rule: Cipher text, c = RSA Public (m) = m^e mod n, $1 < m < n - 1$
Decryption rule: Plaintext, m = RSA Private (c) = c^d mod n

Figure 3 represents the execution flow of RSA algorithm.

5 Proposed Methodology

5.1 Designing Echit Web Application

A chit fund is a process of money lending and money saving scheme practiced in India. A chit fund company is a company that manages, conducts or supervises such as transactions called chit where a person enters into an agreement with a specified number of persons that every one of them shall subscribe a certain sum of money by way of periodical instalments over a definite period and that each such subscriber shall, in his turn, be determined by auction specified in the chit agreement, and will be entitled to the prize amount. All this is done in manually and to make this easier an Echit web application is designed that aims to run the chit business in online as shown in Fig. 4.

User can access information from anywhere. There are no installation cost, no server cost, and no maintenance cost. Auto data back up is in online. Unlimited branches, unlimited user, and unlimited schemes can register. Application is available in regional languages. Nowadays, there is no open-source web application for running a chit business. We are proposing the open-source Echit web application to access from anywhere to everyone at any time.

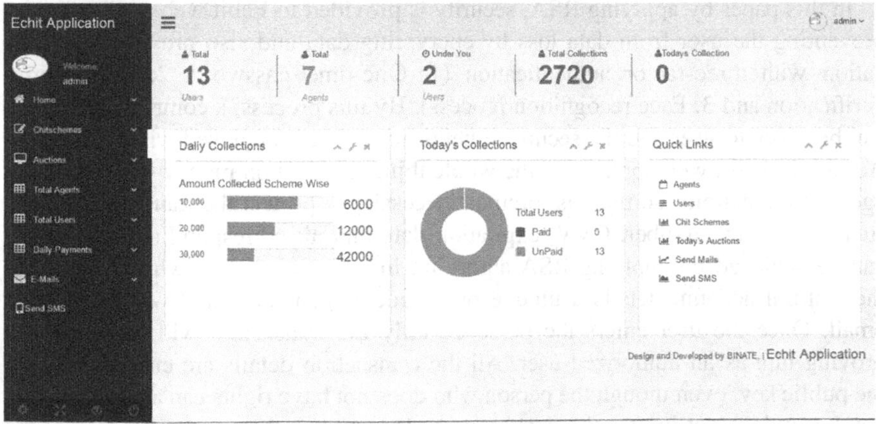

Fig. 4 Design of Echit web application

Online web application which is used to all users and for this web application development we can use the frontend as web technologies like HTML5 CSS3 JavaScript PHP, Bootstrap, backend as MySQL database and we are providing the security to this web application. Echit application is proposed for those who runs their chit business easily in online.

5.2 Security Evaluation Results for Echit Web Application

Nowadays, the internet technology and the information of vast users across the world are rapidly growing, and the storage of information and data in cloud is most important thing to a software engineer in their daily life. Providing security to the web application is necessary thing that is to be adopted before launching it into the cloud [12].

Whenever the credit card/debit card or any payment mode's data is used for e-commerce transactions, the main concern lies in hiding the private data for more users. To secure such sensitive information there are some encryption algorithms [9, 12]. By encrypting the sensitive data or the path according to security architecture of cloud computing [13] where the data is transferred will reduce the vulnerability of the data, which in turn reduces the risk of a suppliers. There are multiple approaches to encrypt the payment details in transaction process. RSA cryptosystem [12] is one such approach in which encryption is applied to sensitive data elements especially while paying through online using banking details such as the card number, the card security code (i.e., CVV) and the expiration date. RSA is one type of asymmetric encryptions which use two separate keys, each of which has a specific function. A public key encrypts the data, while a private key decrypts the data. In the Echit application user who are paying their amount by online process need to be secured, so this paper provides security to their transaction details by encrypting the entered data as shown in Fig. 5.

In this paper by applying RSA, security is provided to Echit web application by preventing the user from data loss by encrypting data and also provides authentication with three-factor authentication (1. One-time password, 2. Finger print verification and 3. Face recognition process). By this process, a complete assurance can be given to user and the secure authentication that is achieved will be useful to Admin. In Echit web application the whole thing is based on money transactions to agents for Auctions, so there is foremost necessity to protect the transaction details such as account number, CVV, expiration date and all corresponding details; this can be achieved by applying RSA algorithm in which each user when submits all the truthful account details a unique pass code is generated and will be to their email. Once the user enters the passcode only the transaction will be successful proving him as an authorized user. All the transaction details are encrypted using the public key, even though the person who does not have rights can also access this key but it does not impact the data security. Private key that is used to decrypt the details will be preserved with bank staff so that they can debit amount from the

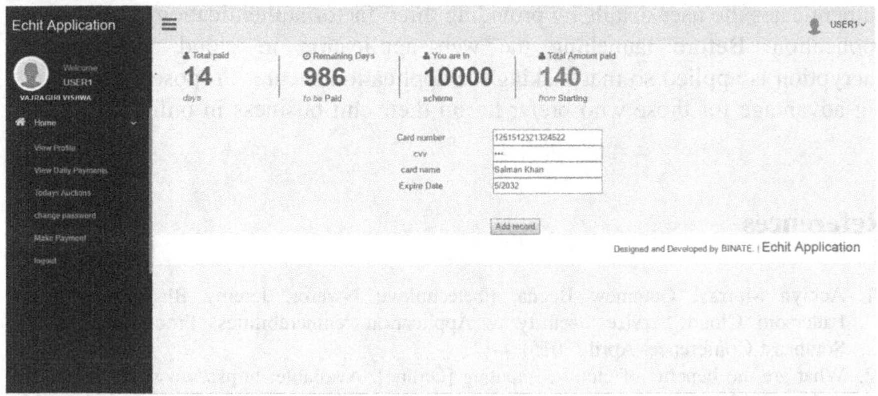

Fig. 5 Encrypt the user card details using RSA

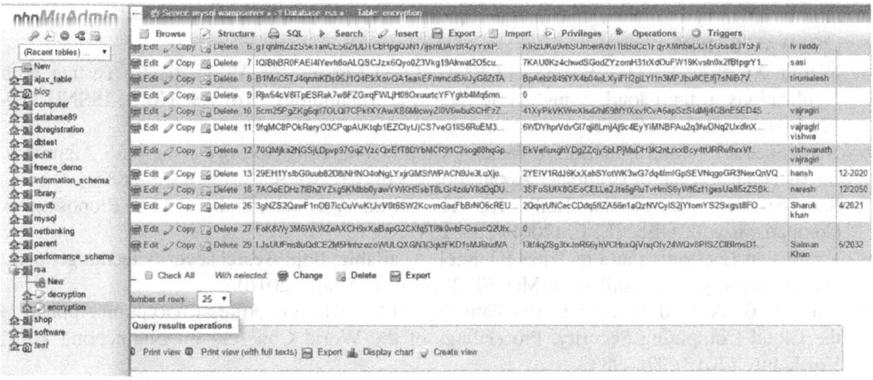

Fig. 6 User payment form

corresponding user account whose transaction process is verified successfully through pass code and then immediately credit that amount to the Echit admin. Figure 6 shows how the encryption is done.

6 Conclusion

In this paper, fully homomorphic encryption is applied to the Echit web application in order to perform secure transactions (for example, monthly payment of chits) either by the user or agent, security is provided mainly on the user data (such as ATM numbers and users account details, etc.) by encrypting them. By this process, security issue presented in the Echit web application can be eliminated. It also authenticates the user details by providing three-factor authentications in Echit web application. Before launching the web application in cloud, homomorphic encryption is applied so that making the application secure. Proposed scheme is a big advantage for those who prefer to run their chit business in online.

References

1. Acklyn Murray, Geremew Begna, Ebelechukwu Nwafor, Jeremy Blackstone, Wayne Patterson: Cloud Service Security & Application Vulnerabilities. Proceedings of IEE Southeast Conference, April (2015) 9–12.
2. What are the benefits of cloud computing [Online]. Available: https://www.salesforce.com/uk/blog/2015/11/why-move-to-the-cloud-10-benefits-of-cloudcomputing.html, drafted on Mar 30, 2016 at 8.00 am. (2016).
3. Chit Fund Applications [Online]. http://chitfundsoftware.in/chitfund-mobile-app.asp drafted on Mar 30, 2016 at 8.30 am. (2016).
4. Fang Yu and Yi-Yang Tung: Patcher: An Online Service for Detecting, Viewing and Patching Web Application Vulnerabilities. 47th Hawaii International Conference on System Science. (2014) 4878–4886.
5. Xiaolin Xu et al: URMG: Enhanced CBMG-Based Method for Automatically Testing Web Applications in the Cloud. Tsinghua Science And Technology, Vol. 19. No. 1. ISSNl1007–0214 07/10. (2014) 65–75.
6. Abha Sachdev, Mohit Bhansali.: Enhancing Cloud Computing Security using AES Algorithm. International Journal of Computer Applications, Vol. 67. No. 9. (2013) 8875–8887.
7. Sanchez-Avila, C., Sanchez-Reillo R.,: The Rijndael Block Cipher (AES Proposal): A Comparison with DES. IEEE, (2001) 229–234.
8. What is an Encryption and its types [Online]. Available: http://searchsecurity.techtarget.com/definition/encryption, drafted on Mar 30, 2016 at 8.15 am. (2016).
9. Maha TEBAA, Saïd EL HAJJI, Abdellatif EL GHAZI: Homomorphic Encryption Applied to the Cloud Computing Security. Proceedings of the World Congress on Engineering 2012, Vol I. July (2012) 978–983.
10. Maha TEBAA, Said EL HAJII: Secure Cloud Computing through Homomorphic Encryption. International Journal of Advancements in Computing Technology (IJACT), Volume 5. Number 16. Dec (2013) 29–38.
11. P. Saveetha & S. Arumugam: Study On Improvement In Rsa Algorithm And Its Implementation. International Journal of Computer & Communication Technology, Volume-3. Issue-6, 7, 8. (2012) 61–66.
12. Chinedu J. Nwoye,: Design and Development of an E-Commerce Security Using RSA Cryptosystem. International Journal of Innovative Research in Information Security (IJIRIS), ISSN: 2349–7017 (O), Volume 6. Issue 2. June (2015) 5–17.
13. Anjana Chaudhary et al.: A Review: Data security approach in Cloud Computing by using RSA algorithm, International Journal of Advance Research in Computer Science and Management Studies, Volume 1. Issue 7. Dec (2013) 259–265.
14. Mark Edge [Online]. http://www.businesscomputingworld.co.uk/5-ways-to-reinforce-your-data-security-model-for-an-insecure-cyber-world/, drafted on Mar 30, 2016 at 9.00 am. (2016).

Fuzzy Clustering with Improved Swarm Optimization and Genetic Algorithm: Hybrid Approach

Bighnaraj Naik, Sarita Mahapatra, Janmenjoy Nayak and H.S. Behera

Abstract Fuzzy c-means clustering is one of the popularly used algorithms in various diversified areas of applications due to its ease of implementation and suitability of parameter selection, but it suffers from one major limitation like easy stuck at local optima positions. Particle swarm optimization is a globally adopted metaheuristic technique used to solve complex optimization problems. However, this technique needs a lot of fitness evaluations to get the desired optimal solution. In this paper, hybridization between the improved particle swarm optimization and genetic algorithm has been performed with fuzzy c-means algorithm for data clustering. The proposed method has been compared with some of the existing algorithms like genetic algorithm, PSO, and K-means method. Simulation result shows that the proposed method is efficient and can divulge encouraging results for finding global optimal solutions.

Keywords Fuzzy c-means · Particle swarm optimization · Genetic algorithm · Differential evolution

---

B. Naik
Department of Computer Application, VSSUT, Burla, Odisha, India
e-mail: mailtobnaik@gmail.com

S. Mahapatra
Department of CSE & IT, ITER, S'O'A University, Bhubaneswar, Odisha, India
e-mail: saritamahapatra@soauniversity.ac.in

J. Nayak (✉)
Department of CSE, Modern Engineering & Management Studies, Balasore, Odisha, India
e-mail: mailtojnayak@gmail.com

H.S. Behera
Department of CSE & IT, VSSUT, Burla, Odisha, India
e-mail: mailtohsbehera@gmail.com

H.S. Behera and D.P. Mohapatra (eds.), *Computational Intelligence in Data Mining*, Advances in Intelligent Systems and Computing 556,
DOI 10.1007/978-981-10-3874-7_23

1 Introduction

Clustering is one of the current active researches among all the other data mining tasks in the pattern recognition community. It is based on the principle of unsupervised learning which is intended to grouping the patterns in different restricted classes. There are two basic clustering techniques: (a) K-means and (b) Fuzzy c-means (FCM) used by a number of researchers for solving different problems. K-means is one of the popular algorithms in which the data clusters or points are classified into k points and the number of points is being chosen in well advance, but this algorithm suffers at a point where there is no any selected boundary value [1]. After the development of fuzzy theory by Zadeh [2], many researchers have shown the interest toward fuzzy theory for solving clustering problem. The fuzzy clustering problems have been expansively studied and the foundation of fuzzy clustering was being proposed by Bellman et al. [3] and Ruspini [4]. The fuzzy clustering based on the objective function is quite popularly known to be fuzzy c-means clustering (FCM) [5]. In FCM, the group of the pattern is decided based on many certain fuzzy membership grades [6]. Hoppner et al. [7] made a good effort towards the survey of FCM. FCM has been successfully applied in various application areas such as image segmentation [8], color clustering [9], real-time applications [10], signal analysis [11], spike detection [12], biology [13], forecasting [14], disease analysis [15], software engg. [16], damage detection [17], document analysis [18], cluster analysis [19], remote sensing [20, 21], etc.

FCM is an efficient algorithm for problem solving. But as the cluster center points are being chosen randomly, so the algorithm struck at local optima. Also, it has slow convergence rate and highly sensitive to initialization. To solve such problems, researchers have applied various optimization algorithms like genetic algorithm (GA) [22] and particle swarm optimization (PSO) [1, 23–25]. Particle swarm optimization is a bird-inspired metaheuristic algorithm proposed by Kennedy and Eberhart [26]. Basic idea for exploration and exploitation in PSO is found to be the backbone for development of other metaheuristic optimization techniques. Also complexity of this optimization technique is found to be fairly less than others, due to the fact that it requires minor parameter settings. But early convergence is one of the key drawbacks. So, in this paper an improved PSO with the genetic algorithm has been proposed for clustering real-world data. The suitability and effectiveness of selection of parameters in the PSO algorithm makes this method more efficient than the original PSO.

The rest of the paper is organized as follows: Sect. 2 elaborates the basic preliminaries like FCM, PSO and ISO. Section 3 describes the proposed methodology. In Sect. 4 experimental analysis and result are being described. Finally Sect. 5 concludes our work.

2 Basic Preliminaries

2.1 Fuzzy c-Means Algorithm (FCM)

FCM is a soft clustering algorithm. In general, if any clustering algorithm is able to minimize an error function [27], then that algorithm is called c-Means where 'c' is the number of classes or clusters, and if the concerned classes will use the fuzzy technique or fuzzy theory, then it is known to be FCM. In fuzzy c-means method, a fuzzy membership function is used to assign a degree of membership for each class. FCM is able to the form new clusters having close membership values to existing classes of the data points [28]. The FCM approach relies on three basic operators such as fuzzy membership function, partition matrix and the objective function. FCM is used to partition a set of 'N' clusters through a minimization of the objective function [29] w.r.t the fuzzy partition matrix:

$$J(U,V) = \sum_{i=1}^{C} \sum_{j=1}^{N} u_{ij}^{m} \left\| x_j - v_i \right\|^2, \tag{1}$$

where x_j denotes the jth cluster point and v_i represents the ith cluster center. '$u_{i,j}$' is the membership value of 'x_j' w.r.t. cluster i. 'm' denotes the fuzzy controlling parameter, i.e. for the value 1, it tends to hard partition and for the value of ∞ and it tends towards the complete fuzziness. $\|.\|$ indicates the norm function.

The iterative method is used to compute the membership function and cluster center as

$$u_{ij} = \left[\sum_{k=1}^{c} \left(\frac{\|x_j - v_i\|}{\|x_j - v_k\|} \right)^{\frac{2}{m-1}} \right]^{-1} \tag{2}$$

$$v_i = \sum_{j=1}^{N} u_{ij}^{m} x_j \Big/ \sum_{j=1}^{N} u_{ij}^{m} \quad \text{where } i \geq 1, i \leq c. \tag{3}$$

The steps of FCM algorithm are as follows:

1. Initialize the number of clusters centers v.
2. Select an inner product metric Euclidean norm and the weighting metric (fuzziness).
3. Compute U (partition matrix) using Eq. (2).
4. Update the fuzzy cluster centers using Eq. (3).
5. Compute the new objective function J using Eq. (1).
6. If $||J_{new} - J_{old}|| \leq \in$ then stop.
7. Else repeat steps 3–5.

2.2 *Improved Particle Swarm Optimization*

PSO [26] is an evolutionary optimization algorithm, inspired by the behavior of flying birds. Unlike other evolutionary algorithms, PSO has less parameter tuning and its complexity is also less. The algorithm of PSO is carried out with some basic assumptions such as follows [30, 31]: (i) In a multidimensional space, the birds are flying, at some position having no mass or dimension. They fly by adjusting their velocities and positions by exchanging information about the particle current position, local best particle position and global best particle position and their velocities in search space [32]; (ii) During the travel for either food or shelter [33], they are supposed to not to collide with each other by adjusting their velocity and position. In PSO, all birds in a group are assumed as population of particles in imaginary space, and velocity and position of each particle are initialized randomly according to the problem being solved. In the first generation, all the particles in current population are assumed as local best particles (lbest). From the second generation onward, local best particles are selected by comparing fitness of particles in current population and previous population. Among local best particles, a particle with maximum fitness is selected as global best particle (gbest). According to the current velocity of particles $\left(V_i^{(t)}\right)$ and position of particles in current population $\left(X_i^{(t)}\right)$, local best particles and global best particle, the next velocities $\left(V_i^{(t+1)}\right)$ of the particle are computed (Eq. 4). After obtaining next velocity, next position $\left(X_i^{(t+1)}\right)$ of all the particles in the population is updated (Eq. 5) using current position $\left(X_i^{(t)}\right)$ and next velocity $\left(V_i^{(t+1)}\right)$ of all the particles. These steps are continued until no further improvement in gbest is noticed or problem-specific stopping criteria are reached:

$$V_i^{(t+1)} = V_i^{(t)} + c_1 * rand(1) * \left(l_{best_i}^{(t)} - X_i^{(t)}\right) + c_2 * rand(1) * \left(g_{best}^{(t)} - X_i^{(t)}\right) \quad (4)$$

$$X_i^{(t+1)} = X_i^{(t)} + V_i^{(t+1)}. \quad (5)$$

Here, c_1 and c_2 are the constants which values may be from the range 0–2 and $rand(1)$ generates a uniform random number between 0 and 1.

ISO uses the basic concept of standard PSO algorithm to address the issues such as (i) not efficient in adjusting solution for improvisation and (ii) low searching ability near by the global optima solutions [34].

So, in ISO an inertia weight λ is introduced (Eq. 6) for addressing the above issues. By incorporating this in ISO, the search space may be significantly reduced with the increase in number of generations [35].

The improvement in both the velocity and position may be illustrated through Eqs. (6) and (7):

$$V_i^{(t+1)} = \lambda * V_i^{(t)} + c_1 * rand(1) * \left(l_{best_i}^{(t)} - X_i^{(t)}\right) + c_2 * rand(1) * \left(g_{best}^{(t)} - X_i^{(t)}\right) \quad (6)$$

$$X_i^{(t+1)} = X_i^{(t)} + V_i^{(t+1)}. \quad (7)$$

2.3 Genetic Algorithm

GA is one of the popular evolutionary algorithms that have been a keen interest of all types of research. Holland [36] and Goldberg [37] contributed significantly towards the development of GA. GA is a metaheuristic optimization based on Darwin's evolutionary principle. While solving a problem using GA, the chromosome represents an individual solution vector and the population is considered as predefined number of such chromosomes. Encoding of these chromosomes is purely depending on the structure of the problem and its solution. GA follows four basic steps such as fitness evaluation, selection, crossover and mutation. Here, the objective is to promote fittest chromosomes (survival of fittest) for the next generation by excluding weak chromosome from the population. The fitness evaluation reasonably depends on problem being solved and independent to computational procedure [38].

3 Proposed Hybrid GA–ISO–FCM Approach

In this section, a hybrid GA–ISO–FCM algorithm has been proposed based on the hybridization of GA, improved PSO and FCM algorithm for clustering the real-world data. In fact, FCM technique is efficient to find the optimal cluster centers. However, initially, the FCM uses randomly generated cluster centers for clustering through fuzzy membership grade. Without excluding this fact, still FCM is good in finding optimal cluster centers in a data set. However, in this proposed method, we have made an attempt to make the performance of FCM better by speeding up the convergence rate. This is achieved using metaheuristic algorithms GA and PSO for finding optimal cluster centers for the initialization of cluster centers process in FCM. On the other hand, both GA and PSO have their own limitations like complex parameter tuning (GA) and slow convergence (PSO). So we have hybridized both GA and improved PSO for fuzzy clustering to improve the convergence rate and quality of the solution. The objective of this proposed method is to select the optimal initial cluster centers from population of predefined number of cluster center in a population, thereby avoiding the usage of randomly generated initial cluster centers for FCM algorithm.

This proposed method uses an objective function (Eq. 8) to evaluate the quality of cluster centers. So, in the context of clustering, a single individual in the population represents the 'm' number of cluster center. And the entire population of individuals is initialized with 'n' number of cluster center vectors $P=\{C_1, C_2 \ldots C_n\}$, where each cluster center vector consists of 'm' number of cluster centers $C_1=(c_1, c_2 \ldots c_m)$. Here, each c_i represents a single-cluster center. This is considered as a minimization problem and we have the objective function (Eq. 8) of K-means [27] for calculating the fitness:

$$F(C_i) = \frac{k}{\left(\sum_{l=1}^{r} \|o_l - C_i\|^2\right) + d} = \frac{k}{\left(\sum_{j=1}^{m} \sum_{l=1}^{r} \|o_l - C_{i,j}\|^2\right) + d}. \tag{8}$$

Here $F(.)$ is a function to evaluate the generalized solutions called fitness function, '*k*' and '*d*' are user-defined constants, o_l is the lth data point, C_i is the ith cluster center vector, $C_{i,j}$ is the ith cluster center of jth cluster center vector, '*r*' is the number of data point in the data set and $\|.\|$ is the Euclidean distance norm. The pseudocode of the proposed approach is illustrated as follows.

1. **Initialize** the population of 'n' no. of cluster center vectors $P = \{C_1, C_2...C_n\}$, each cluster center vector with 'm' no. of random cluster center $C_1 = (c_1, c_2...c_m)$. An individual firefly signifies a cluster center vector C_i.
2. Iter=1;
3. **While** (iter<=maxIter)
   Compute fitness of all particles in population P by using the objective function eq. (8).
   If (iter==1)
   Assign Local best particle l_{best}=P.
   Else
   Evaluate fitness of P and P'.
   Compare the fitness of particles based on their fitness in P and P'.
   If fitness of ith particle X_i in P is less that fitness of a particle in P'
   Then assign L_{best} (i) = P'(i).
   Else assign L_{best} (i) = P(i).
   End of if
   End of if
   Select particles with best fitness value from L_{best} as G_{best} particle.
   Compute new velocity Vnew of the particle by using P, Lbest and gbest by using eq. (6).
   Generate next positions of particles P' by using P and V_{new} as follows by using eq. (7).
   Create a Mating pool of particles by replacing weak particles in the current population with global best Gbest particle.
   Perform two point crossovers on particles in P' to generate new feasible solutions P''.
   If (P' is same as P'')
   Then perform mutation on P'.
   End if
   P'=P''.
   Update P based on P''.
   Iter = iter+1;
   End of while
4. **Rank** the cluster center vectors based on their fitness, obtain the best cluster center vector.
5. **Initialize** the cluster centers of FCM with position of the best cluster center vector. Then using this cluster centers, iterate the FCM algorithm.
6. **Do** Update the membership matrix by eq.(2)
7. Refine the cluster centers by eq.(3),
8. **While** (until it meets the convergence criteria)
9. **Exit**

4 Experimental Setup and Result Analysis

The proposed GA–ISO–FCM has been implemented in the environment of MATLAB 9.0. The real-world data sets for experimentation have been considered from UCI repository [39] and the details about the data set are given in Table 1. For

Table 1 Data set information

Data sets	No. of pattern	No. of clusters	No. of attributes
Iris	150	3	4
Lenses	24	3	4
Haberman	306	2	3
Balance scale	625	3	4
Wisconsin breast cancer	699	2	10
Contraceptive method choice	1473	3	9
Hayesrotli	132	3	5
Robot navigation	5456	4	2
Spect heart	80	2	22

experimental analysis and performance comparison, we have compared the performance of the proposed hybrid method with some other standard techniques such as FCM, GA-FCM and PSO-FCM. However, as K-means is also considered as one of the standard methods for data clustering, we have compared the results of K-means, GA-K-means, PSO-K-means [40] and GA-ISO-K-means [41] (Table 2).

The value for the fuzzy coefficient (m) is set as 2. The acceleration coefficients (c1 and c2) are set to 1.4 and the inertia weight (λ) is set between 1.8 and 2 during ISO iteration [42]. For executing the cross over step of GA, two point crossovers have been used. After the crossover step, if population of particle remains unchanged, then mutation operation is applied on the particles in order to explore other solution in solution space. This proposed scheme produces effective cluster centers of a particle. During execution of this scheme, the cluster centers (initially chosen) are attracted towards the center of corresponding group of similar data point in successive iterations.

Table 2 Performance comparison of FCM with the other clustering methods

Data sets	Fitness values of clustering algorithms				
	FCM	GA-FCM	PSO-FCM	ISO-FCM	GA-ISO-FCM
Iris	0.012738542	0.014154986	0.014624876	0.014620135	0.014628271
Lenses	0.381339952	0.390354824	0.425698354	0.425658963	0.428734522
Haberman	0.000316547	0.000330542	0.000372865	0.000372814	0.000376875
Balance scale	0.003332606	0.003425487	0.003535478	0.003541256	0.003612873
Wisconsin breast cancer	7.48861E-14	7.50236E-14	7.52487E-14	7.53458E-14	7.53826E-14
Contraceptive method choice	7.69432E-05	8.13254E-05	8.20398E-05	8.22003E-05	8.23687E-05
Hayesrotli	4.43056E-05	4.71657E-05	4.74493E-05	4.74689E-05	4.74821E-05
Robot navigation	0.002000381	0.002258745	0.002454781	0.002468954	0.002562114
Spect heart	0.077804472	0.079365885	0.080456544	0.080569877	0.081428643

5 Conclusion

FCM is quite popular for data clustering, but as it is sensitive to the initialization, there is maximum chance to get rapped at local minima. In this paper, a hybrid approach of two popular optimization techniques like GA and improved PSO has been proposed for fuzzy clustering. The positive insights of both the algorithms help the FCM to get some quality values in terms of fitness values. The performance of the proposed method is compared with some other approaches such as FCM, GA-FCM, and ISO-FCM. For all the nine considered data sets, the performance of GA–ISO–FCM found to be better than the others.

Acknowledgements This work is supported by Technical Education Quality Improvement Programme, National Project Implementation Unit (A unit of MHRD, Govt. of India, for implementation of World Bank assisted projects in technical education), under the research project grant (VSSUT/TEQIP/37/2016).

References

1. Izakian, Hesam, and Ajith Abraham. Fuzzy C-means and fuzzy swarm for fuzzy clustering problem. Expert Systems with Applications **38**.3 (2011): 1835–1838.
2. Zadeh, L.A.: Fuzzy Sets. Information and Control **8** (3): 338–353. doi:10.1016/S0019-9958(65)90241-X.ISSN 0019–9958.
3. Bellman, R.E., Kalaba, R.A., Zadeh, L.A.: Abstraction and pattern classification, J. Math. Anal. Appl. **13** (1966) 1–7.
4. Ruspini, E.H.: A new approach to clustering, Inf. Control **15**(1) (1969) 22–32.
5. Bezdek, J.C.: Pattern Recognition with Fuzzy Objective Function Algorithms. Plenum Press, New York (1981).
6. Ferreiraa, M.R.P., Carvalho, F.A.T.: Kernel fuzzy c-means with automatic variable weighting, Fuzzy Sets and Systems **237** (2014) 1–46.
7. F. Höppner, F. Klawonn, R. Kruse, T. Runkler, Fuzzy Cluster Analysis. John Wiley & Sons, Inc. (1999).
8. Khattab, Dina, et al. A Comparative Study of Different Color Space Models Using FCM-Based Automatic GrabCut for Image Segmentation. Computational Science and Its Applications–ICCSA 2015. Springer International Publishing, (2015) 489–501.
9. Kim, W.D., Lee, K.H., Lee, D.: A novel initialization scheme for the fuzzy c-means algorithm for color clustering, Pattern Recognition Letters **25** (2004) 227–237.
10. Lazaro J, Arias J, Martın J.L, Cuadrado C, Astarloa A.: Implementation of a modified Fuzzy C-Means clustering algorithm for real-time applications. Microprocessors and Microsystems **29** (2005) 375–380.
11. ŁeRski, J.M., Owczarek, A.J.: A time-domain-constrained fuzzy clustering method and its application to signal analysis. Fuzzy Sets and Systems **155** (2005) 165–190.
12. Inan, Z.H., Kuntalp, M.: A study on fuzzy C-means clustering-based systems in automatic spike detection, Computers in Biology and Medicine **37** (2007) 1160–1166.
13. Ceccarelli, M., Maratea, A.: Improving fuzzy clustering of biological data by metric learning with side information, International Journal of Approximate Reasoning **47** (2008) 45–57.
14. Chen, S.M., Chang, Y.C.: Multi-variable fuzzy forecasting based on fuzzy clustering and fuzzy rule interpolation techniques. Information Sciences **180** (2010) 4772–4783.

15. Azara, A.T., El-Said, S.A., Hassaniend, A.E.: Fuzzy and hard clustering analysis for thyroid disease, Computer methods and programs in biomedicine **111** (2013) 1–16.
16. Kaushik, A., Soni, A.K., Soni, R.: Radial basis function network using intuitionistic fuzzy C means for software cost estimation, Int. J. of Computer Applications in Technology 2013 - Vol. **47**, No. 1 pp. 86–95.
17. Silva, S., Junior, M.D., Junior, V. L., Brennan, M.J.: Structural damage detection by fuzzy clustering, Mechanical Systems and Signal Processing **22** (2008) 1636–1649.
18. Yan, Y., Chen, L., Tjhi, W.C.: Fuzzy semi-supervised co-clustering for text documents, Fuzzy Sets and Systems **215** (2013) 74–89.
19. Nayak, J., Nanda, M., Nayak, K., Naik, B., Behera, H.S.: An Improved Firefly Fuzzy C-Means (FAFCM) Algorithm for Clustering Real World Data Sets, Smart Innovation, Systems and Technologies Volume **27**, 2014, pp. 339–348, doi:10.1007/978-3-319-07353-8_40.
20. Zhu, C.J., Yang, S., Zhao, Q., S., Cui, Wen, N.: Robust Semi-supervised Kernel-FCM Algorithm Incorporating Local Spatial Information for Remote Sensing Image Classification, Journal of the Indian Society of Remote Sensing, March 2014, Volume **42**, Issue 1, pp. 35–49.
21. Yu, X.C., He, H., Hu, D., Zhou, W.: Land cover classification of remote sensing imagery based on interval-valued data fuzzy c-means algorithm, Science China Earth Sciences, June 2014, Volume **57**, Issue 6, pp. 1306–1313, doi:10.1007/s11430-013-4689.
22. Tang, Jinjun, et al. A hybrid approach to integrate fuzzy C-means based imputation method with genetic algorithm for missing traffic volume data estimation. Transportation Research Part C: Emerging Technologies **51** (2015): 29–40.
23. Li Wang et. al. Particle Swarm Optimization for Fuzzy c-Means Clustering. Proceedings of the 6th World Congress on Intelligent Control and Automation, June 21–23 (2006) Dalian, China pp. 6055–6058.
24. Thomas A. Runkler and Christina Katz. Fuzzy Clustering by Particle Swarm Optimization. 2006 IEEE International Conference on Fuzzy Systems. Sheraton Vancouver Wall Centre Hotel, Vancouver, BC, Canada, July 16–21 (2006) pp. 601–608.
25. Silva Filho, Telmo M., et al. Hybrid methods for fuzzy clustering based on fuzzy c-means and improved particle swarm optimization. Expert Systems with Applications **42**.17 (2015): 6315–6328.
26. Kennedy, J., Eberhart, R.: Particle swarm optimization, in: Proceedings of the 1995 IEEE International Conference on Neural Networks, vol. 4. 1942–1948 (1995).
27. Nayak, Janmenjoy, Bighnaraj Naik, and H. S. Behera. Fuzzy C-Means (FCM) Clustering Algorithm: A Decade Review from 2000 to 2014. Computational Intelligence in Data Mining-Volume 2. Springer India (2015) 133–149.
28. Asyali, M.H., Colak, D., Demirkaya, O., Inan, M.S.: Gene expression profile classification: a review. Current Bioinformatics **1** (2006) 55–73.
29. Phen-Lan Lin, Po-Whei Huang, C.H. Kuo, Y.H. Lai, A size-insensitive integrity-based fuzzy c-means method for data clustering. Pattern Recognition **47** (2014) 2042–2056.
30. Babaei, M.: A general approach to approximate solutions of nonlinear differential equations using particle swarm optimization. Applied Soft Computing **13** (2013) 3354–3365.
31. Neri, F., Mininno, E., Iacca, G.: Compact Particle Swarm Optimization. Information Sciences **239** (2013) 96–121.
32. Wei, J., Guangbin, L., Dong, L.: Elite particle swarm optimization with mutation, IEEE Asia Simulation Conference – 7th Intl. Conf. on Sys. Simulation and Scientific Computing. (2008) 800–803.
33. Khare, A., Rangnekar, S.: A review of particle swarm optimization and its applications in Solar Photovoltaic system. Applied Soft Computing. **13** (2013) 2997–3006.
34. Yue-bo, M., ZouJian-hua, GanXu-sheng, Liang, Z.: Research on WNN aerodynamic modeling from flight data based on improved PSO algorithm. Neurocomputing 83 (2012) 212–221.

35. Dehuri, S., Roy, R., Cho, S. B., Ghosh, A.: An improved swarm optimized functional link artificial neural network (ISO-FLANN) for classification. The Journal of Systems and Software. **85** (2012) 1333–1345.
36. Holland, J. H.: Adaption in Natural and Artificial Systems. Cambridge, MA: MIT Press, 1975.
37. Goldberg, D. E.: Genetic algorithms in search. Optimization and machine learning. Boston, MA: Kluwer Academic Publishers. 1989.
38. Li, X., Xiao, N., Claramunt, C., Lin, H.: Initialization strategies to enhancing the performance of genetic algorithms for the p-median problem. Computers & Industrial Engineering **61** (2011) 1024–1034.
39. Bache, K., Lichman, M.: UCI Machine Learning Repository [http://archive.ics.uci.edu/ml], *Irvine*, CA: University of California, School of Information and Computer Science. 2013.
40. Naik, B., Swetanisha, S., Behera, D. K., Mahapatra, S., Padhi, B. K.: Cooperative Swarm based Clustering Algorithm based on PSO and k-means to find optimal cluster centroids. 2012 National Conference on Computing and Communication Systems (NCCCS) (2012) 1–5 doi:10.1109/NCCCS.2012.6413027.
41. Nayak, J., Kanungo, D. P., Naik, B., & Behera, H. S. (2016). Evolutionary Improved Swarm-Based Hybrid K-Means Algorithm for Cluster Analysis. In *Proceedings of the Second International Conference on Computer and Communication Technologies* (pp. 343–352). Springer India.
42. Nayak, J., Naik, B., Kanungo, D. P., & Behera, H. S.: An Improved Swarm Based Hybrid K-Means Clustering for Optimal Cluster Centers, In Information Systems Design and Intelligent Applications, Springer India 545–553 (2015) doi:10.1007/978-81-322-2250-7_54.

RF-Based Thermal Validation and Monitoring Software for Temperature Sensitive Products

P. Siva Sowmya and P. Srinivasa Reddi

Abstract Pharmaceutical and healthcare industries are highly controlled industries all around the world. Nowadays, millions of temperature sensitive products are manufactured, warehoused, or circulated all over the world. For all these sensitive products, the control of temperature is essential. In some developing nations, they are offering no automated processes such as manual recordings of sensed information by supervisors of the organization or thermometers or USB Data loggers-means bringing the sensor node in contact with USB Data reader which aids as the interface between loggers and the system software and also enables pre-study programming and data can be download only after the study completion in some cases proved inefficient and most of the cases as burdensome. To overcome the problems raised by non-automated processes in this paper, RF-based temperature validation and monitoring software for Pharma, food processing, and warehouses is introduced. Reliant on the nature of the application, sensor nodes are deployed into the area of sensing where base station and loggers communicate by means of radio waves and temperature reading is recorded from base station which is connected to PC through Ethernet or USB.

Keywords Sensor probes · Temperature and humidity monitoring · RF-based loggers · Base station

1 Introduction

Various categories of products need to be controlled under measured environmental circumstances, which include temperature, humidity, and also factors like voltage and current. Among those parameters, the temperature and humidity is often a most

---

P. Siva Sowmya (✉) · P. Srinivasa Reddi
Sri Vidyanikethan Engineering College, Tirupathi, Andhra Pradesh, India
e-mail: sowmyasiva25@gmail.com

P. Srinivasa Reddi
e-mail: psrinivasareddi@gmail.com

H.S. Behera and D.P. Mohapatra (eds.), *Computational Intelligence in Data Mining*, Advances in Intelligent Systems and Computing 556,
DOI 10.1007/978-981-10-3874-7_24

important situation due to its massive unpredictable effects. Possibly, if the temperature of chilled foods exceeds its specified limits then there will be fall in quality, along with intensification in the threat of food poisoning. The temperature restrictions can be quite strict for cool products whose storage temperature is nearer to 0 °C, whereas small rise in temperature, i.e., even just a few degrees might cause infectious growth. This type of condition is more severe in case of pharmaceutical products since frenzied change in the temperature, even for small time intervals may affect the product or make it dangerous.

Monitoring the temperature accurately is becoming critical in many current industries or organizations that are mainly focusing their context of work on food processing and pharmaceuticals. In such type of industries, temperature maintenance within the specified ranges avoids the product loss. For example, all high-risk foods such as meat, dairy products, eggs, etc., are to be stored either in cool or frozen places. A sensor probe should be used to ensure that the temperature of chilled food products is less than or equal to +5 °C and frozen food is not greater than −18 °C [1] and the foods which exceed the specified limit of the temperature need to be rejected. Similarly, processing and storage of pharmaceutical products such as vaccines [2–4] require accurate temperature monitoring. Fig. 1 shows how the temperature is monitored in Pharma and food industries. Derisory storage temperature conditions and temperature variations in these applications lead to reduced vaccine effectiveness and thus adversely affect safety and cost. Maintaining pharmaceutical products [5] at the right temperature is important to save patient's life and health. Insulin might lose its efficiency once it is exposed to freezing

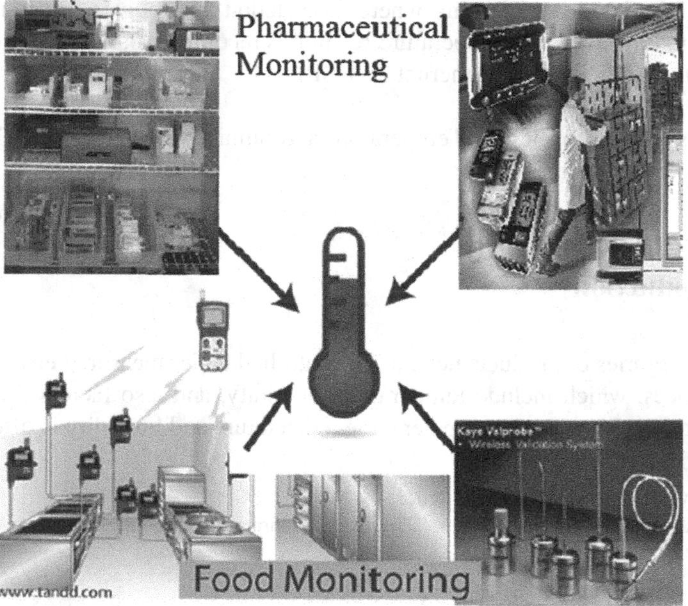

Fig. 1 Temperature monitoring in industries using sensors

temperatures. Other drugs like cortisone cream can separate and become useless once exposed to temperature above 30–35 °C. Each activity in the circulation of drugs should be carried out as per the requirements of the Food and Drugs Act, the consciences of Good Manufacturing Practices (GMP), as well as proper storage and shipping practices. Fully sterilized product usually at 121.1 °C are usually free of viable microorganisms and can be stored under ambient temperatures Therefore, wireless sensor networks can play an important role providing significant advantages in such environments.

The RF-based thermal validation system collects the temperature and humidity data over a 2.4 GHz RF mesh network which uses Wireless HART protocol [6] for communication between loggers and base station. The RF thermal validation system consists of RF loggers, the RF base station, and windows-based software for programming, reading, calibrating loggers and verifying logger calibration, and generating reports. The RF base station acts as the communication link between the loggers and the developed software. The base station collects data from up to a maximum of 50 loggers. The RF loggers use internal humidity sensors to measure relative humidity.

2 Related Work

The authors have discussed how to handle the products under different environmental conditions such as temperature, humidity, and other factors like voltage and current. Some of them are discussed bellow.

A. Willig [7] described about wired environments in which control and reliability are well accommodated for the fieldbus systems. When wireless associations are included, reliability, and control requirements are considerably more difficult to meet, in line for contrary properties of the radio channels.

William H. Mitchell [8] illustrated the problems that aroused when a component from a different dealer was introduced. More number of retailers may be involved in the future. There is no control over the name of the temperature-reporting methodology in vendor-supplied programmes and in addition, the reading of temperature produced may need mounting, unit conversion, etc. From all this, there is an expectation that temperature is determined from each sensor location. Adapter design pattern [8] provides solution to the problem.

Perkusich. A, Almedia, H.O [9] clarified that there is a need to validate the reusability in software development. In the case of present-day embedded systems, module-based software development can promote reuse. Designing and developing of embedded control framework is done by introducing automation systems. The basic requirements are the Linux/RT operating system and the embedded system Tiny Internet interfaces from Dallas semiconductor. To develop the framework, programming language (Java) and design patterns (Abstract Factory) [9] are used.

Nithin Raghunathan [10] work presents the design and characterization of a wireless, low-power, multipoint temperature sensor system for the process of monitoring applications. This is mainly useful for low temperature applications generally found in pharmaceutical and food processing industries where accurate monitoring of temperature over long time intervals (e.g., days) is necessary. The developed system uses the little power ANT wireless protocols for data transmission which allows process in the 2.4 GHz ISM band and also maintains the large-scale sensor network topologies.

Gao Junxiang [11] focused mainly on the monitoring of cold storage, in this paper, a construction of WSN based on ZigBee protocol is done. ZigBee technology is always considered as the primary mode for data processing and wireless motes detection. By the use of sink node and assimilating wireless mobile network, attained data will be sent to database server on control center. The outcomes illustrates that the operational performance of the system is pretty stable and can affect the design requirements in real-time data acquisition and the system has the specified characteristics as good scaling, flexibility in networking, and low cost. The design gives a novel way to collect the environmental data instead of the traditional way of using wires or through manually.

Kristofer S.J. Pister, Lance Doherty [12, 13] discussed about Time Synchronized Mesh Protocol (TSMP) which provides reliable, low-power, secure communication in an accomplished wireless mesh network. TSMP is a standard contact and networking protocol designed for the recently endorsed Wireless HART standard in industrial automation. TSMP benefits from synchronizing the nodes in multihop network within a few hundred microseconds and allows scheduling of pair-wise motes by providing collision-free path and also broadcast communication to encounter the traffic needs of all nodes while traversing through all available networks.

Jianping Song, Song Han [6] discussed about the open wireless communication standard specially designed for measuring of process and control applications, Wireless HART was widely introduced in 2007 based on the central theme of the HART 7 feature. Wireless HART is a safe and sound. It is based on TDMA wireless mesh networking technology functioning in the 2.4 GHz ISM radio band and also having self-healing property. The authors have given an summary to the design of Wireless HART and shared their best experience in building a model for their specification and described numerous challenges that can be tackled at the time of implementation phase, such as the timer, network wide motes synchronization, communication security, and reliable mesh networking, and also in managing the overall network.

3 System Architecture

The RF-based thermal validation system is a wireless system that gathers thermal and relative humidity statistics. This system is planned for precise, appropriate, and consistent process validation of stability chambers, sterilized areas, and warehouses. The applications of this system are mainly Pharma Industries, Food Processing

Industries, stability chambers, freezers, refrigerators, and warehouses. RF thermal validation system [14, 15] uses the 2.4-GHz RF mesh network which has multi-redundant data storage, supports communication among motes and RF software, and provides storage of life-threatening data. It also provides assurance for data integrity, storage, and agreement with governing requirements. The system consists of RF loggers, a RF base station, and a desktop-based software to program, read the loggers data, calibrate the loggers to get stabilized data and verifying its calibration, and also for the generation of reports. Data obtained from multiple wireless RF loggers is pooled into a distinct file from which reports can be generated. These generated reports are designed as a base for implementing 21 CFR Part 11 requirements for electronic records. Users can alter the reports by defining their process series and specifying their cycle-based calculations. The RF thermal validation system composed of:

1. RF Base Station
2. RF Loggers
3. External Sensor and Auxiliary Input Terminal
4. RF Thermal Validation Software

1. **RF Base Station**

The RF base station acts as the medium of communication between the wireless RF loggers and the RF-based thermal validation software on PC. It can be connected to the computer either by standard USB, Ethernet, or USB-Ethernet adapter, the base station can program, read, and even collect the information/data from loggers (up to 100 per study). The schema ensures reliable transfer of critical data. If the PC is currently not in available position or currently working, the base station will store data up to a maximum of 100 loggers or 200 sensors (for example 100 temperature and 100 humidity). Base station needs to be powered by a power supply calculated from 100–240 VAC at 50/60 Hz, this comprises both a USB port and an Ethernet connection.

2. **RF Loggers**

Loggers are generally referred as motes. Motes are the sensor nodes extended with storage, computation, and communication capabilities. Motes are the small, low-power single board device with certain radio frequency for wireless communication. Generally every mote may have some key ingredients like—a transducer, microcontroller, sensors, and low-power radio transceiver, power source. The power for each sensor node is derived from a battery. WSN usually have restricted energy and transmission capacity, which cannot match the transmission power of a large amount of information/data collected by sensor nodes.

There may be more than one sensor present in a mote depending on the type of applications, like sensors for temperature measurement, humidity, voltage, current, etc. Motes can either run on batteries or they can use power grid in certain applications. Motes collects and transfers data based on four steps: collecting the data, processing the collected data, storeing the data, and at last communicating the data to the motes that are within the range of single strength or to a base station. Each and every mote collects data based on the type of sensors it is having. After collecting the data, the mote processes the data through its automation capability. After the data has been processed, the CPU stores the data in a simple format so that it can be handled easily. Once the data has been collected and processed, then the mote begins to interact with other motes. The wireless RF loggers use an external temperature sensor.

- The relative humidity sensor shields a range from 10 to 90% of RH, with accuracy of ±2% for the 25 °C temperature.
- For endorsement, the temperature RTD's sustains an accuracy of
  a. ±0.1 °C for the temperature ranging from 0 to +60 °C;
  b. ±0.2 °C for the temperature ranging from −22 to 0 °C, and
  c. ±0.5 °C for the temperature ranging from −80 to −22 °C.
- The external bendable temperature sensor sustains an accuracy of ±0.1 °C for 0 to 140 °C.
- Two and five-channel temperature loggers are accurate to ±0.1 °C for temperature ranging between −80 to +130 °C.
- Users can select AUX inputs such as either of current (4–20 mA) or voltage (0–10 V), both with contact closure sensing.

Each RF logger has the capacity to record up to 10,000 readings per RTD the data is propelled to the base station and can be stored simultaneously in the memory of the motes. Whenever it is provoked by the software, the loggers direct this data to the PC via the base station for reading the programmed study and also by generating the reports and performing data calculation.

3. **External Sensor and Auxiliary Input Terminal**

External Sensor

Sensor is a device that detects and responds to fluctuations that are prevailing, and uncontrolled atmospheric and weather conditions in a room or place or a system, and delivers this info in a convinced manner. The specific input is, heat, moisture, voltage, and current. The output is generally a signal that is converted to human-readable display at the sensor location or transferred automatically over a network for reading or further processing. The use of RTD's is it can be able to take readings directly "on-the-spot." To measure the mentioned parameters, there is a possibility to connect the temperature or humidity or the combinations externally. The main types of sensors considered in this project are:

1. Temperature Sensor
2. Humidity Sensors

Auxiliary Input Terminal

An auxiliary port (AUX) is the valid dock for a standard communications port. It is an asynchronous serial port with an interface, and that allows the auxiliary inputs such as

1. Voltage and
2. Current

It is an input terminal for connecting external devices. These auxiliary inputs are connected via the terminal to the RF loggers. Voltage AUX input ranges from 0–10 V and whereas current Aux input ranges from 4–20 mA.

4. RF Thermal Validation Software

Before using the Kaye RF thermal validation system as shown in the Figs 2 and 3, installation of the RF Thermal Validation Software is required on the PC and some basic administration tasks need to be done such as password maintenance, creation, and maintenance of user accounts. In this user can establish site preferences (such as selecting temperature units for all the temperature calculations and the default location to store the study files, etc.), select communication option and establish its preferences (such as selection of COM port if using IRTD, setting of battery warning signal, and selection desired communication option either USB or Ethernet for base station), changing of base station settings also can be done where default

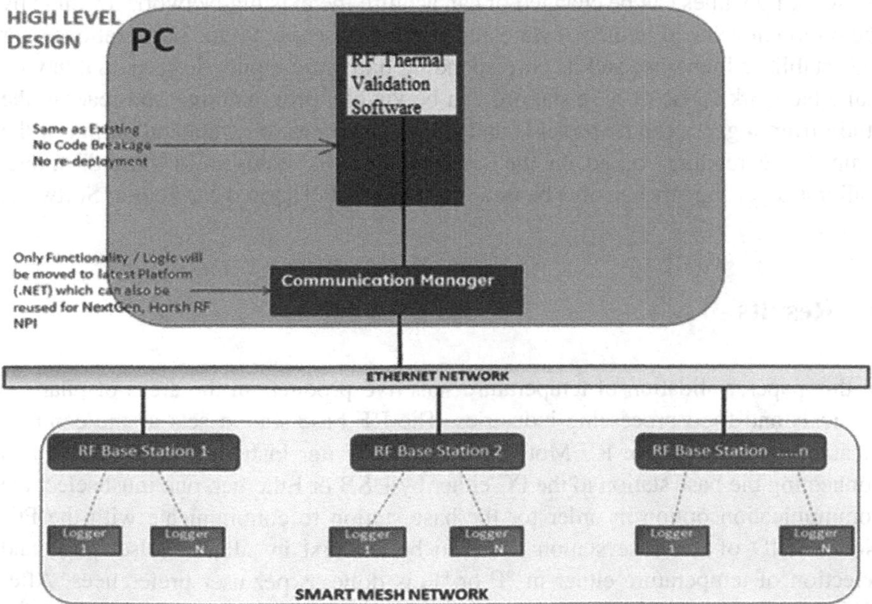

Fig. 2 High level design for RF thermal validation system

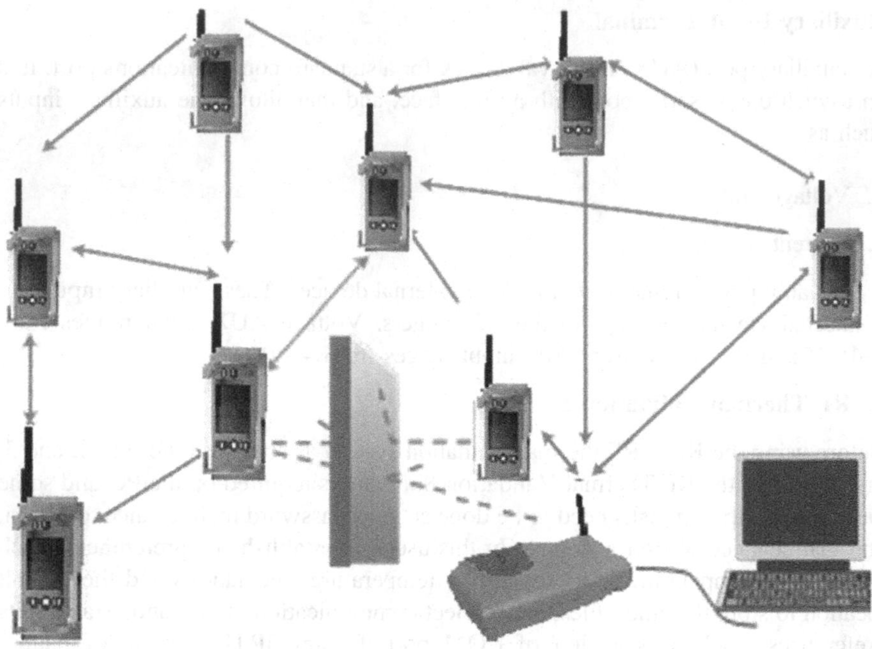

Fig. 3 RF Thermal Validation System

Network ID values can be changed or can acquire the existing Network ID value by the option developed in the software and also can discover all the base stations over the established network and its corresponding hardware inputs (loggers that having same Network ID as of base station) can be visible, programming and reading the study from loggers can be possible and it also generates the graphical output of the temperature readings based on the selected once its status is in "found" mode. Calibration of loggers can also be done from this RF Thermal Validation Software.

4 Results

In this paper, validation of temperature sensitive products in the areas of pharmaceuticals and food processing industries. The RF base station acts as the communication link among the RF Motes and the PC. After installing the software and connecting the base station to the PC either by USB or Ethernet, one must select the communication option in order for the base station to communicate with the PC. Network ID of the base station needs to be set first in admin toolset page and selection of temperature either in °F or °C is done as per user preferences. After setting the base station Network ID, the corresponding value needs to be set to the loggers to communicate with that particular base station. From software now we will check all the base stations that are connected in a network and select one of the base station to which we need to program the loggers and study temperature data. For programming the loggers, we need to select the type of loggers that are connected to the base station, starting of the event (either immediately or on specified time), sampling rate (which indicates the time intervals that temperature data need to collected and logged), and stopping the event (either on specified time or stopping manually or at 10,000 samples) also need to be mentioned as shown in the Fig. 4. Afterwards a screen will appear displaying all the logger choices that are to be selected before, once the status of the selected loggers is found, we can note the temperature readings as shown in the Fig. 5 of the corresponding loggers and can validate with the exact temperature. If the temperature readings exceed the given temperature limit then the product is considered as hazardous and is rejected to use as it may cause major loss to the human using those products.

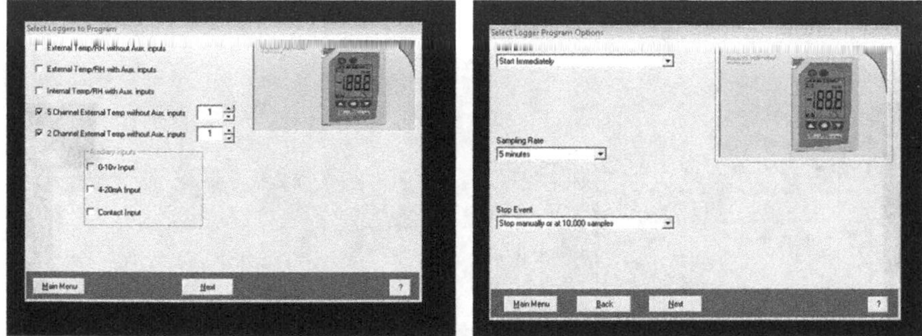

Fig. 4 Selecting and setting loggers and its events and sampling rate

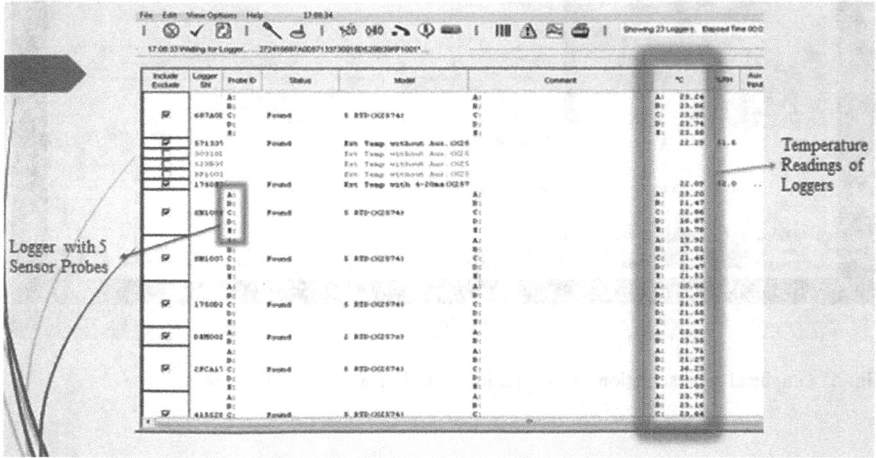

Fig. 5 Reading temperature recording of the loggers

After the temperature readings are visible on the screen, there is option to plot the graph automatically by selecting the loggers to which we need to plot and that can be as shown in Fig. 6.

Case Study—Pharmaceutical Industry

RF thermal validation system needs to be set up in the pharma industry and after that RF thermal validation software should be installed and now we will select the RF base station from which we need to collect the data and now in the software we will select the one external temp/humidity without AUX input logger and set the start event to "Start Immediately" and sampling rate as 5 min and stop event as set to "Stop on Time after Start" and take the readings up to 40 min. Data analysis at 0% humidity level and with a constant temperature of 8 °C is tested with the correctness of ±0.1 °C from 0 °C to +160 °C after calibration is considered and Table 1 shows the recorded temperature from our RF thermal validation system.

The above table specifies that the temperature sensor is stable and accurate based on the above readings with an offset of ±0.1 °C. Figure 7 shows the graphical representation.

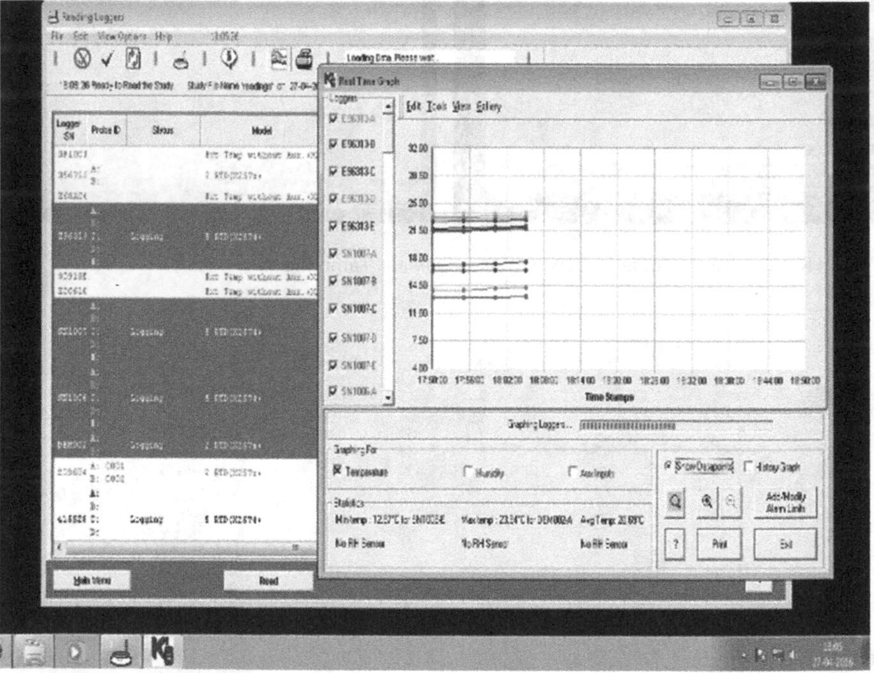

Fig. 6 Graphical representation of programmed study for selected loggers

Table 1 Recorded temperatures and its comparisons after calibration

S.no	Calibrated temperature (°C)	Time	Temperature readings from proposed system (°C)	Difference
1	8	5:00 pm	7.99	−0.01
2	8	5:05 pm	7.91	−0.09
3	8	5:10 pm	7.95	−0.05
4	8	5:15 pm	8.01	0.01
5	8	5:20 pm	7.92	−0.08
6	8	5:25 pm	8.04	0.04
7	8	5:30 pm	7.9	−0.1
8	8	5:35 pm	7.94	−0.06
9	8	5:40 pm	7.97	−0.03

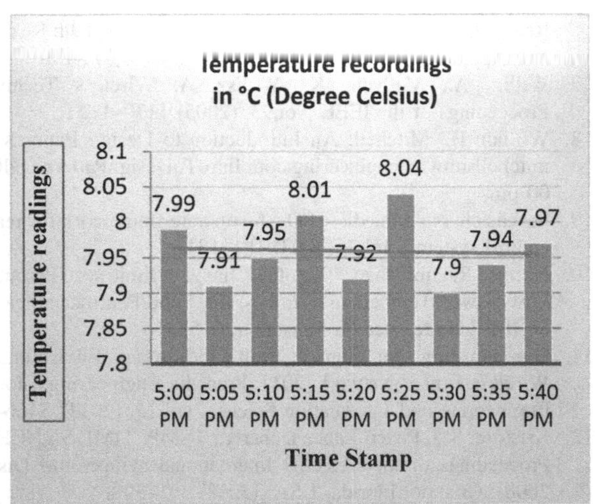

Fig. 7 Graphical representation of recorded temperature readings

5 Conclusion

The graphs that are generated for temperature above specify the close relationship between the data collected from the proposed technology and the temperature readings that are measured by already calibrated devices (such as IRTD or different temperature standard). From the above reading, we can validate the pharmaceutical and food processing products accurately by disposing them if it exceeds the calibrated temperature. It has an alarm system, also to specify if the battery drains out than the indicated percentage level in the RF Thermal Validation Software. Thus it is easy for the end user and industry people to adapt to this RF thermal validation system and can get accurate temperature reading by preventing the damage caused when the temperature sensitive products exceed its limit.

References

1. "Food Delivery and Storage- Safe Food". [Online]. Available: http://www.safefood.eu/SafeFood/media/SafeFoodLibrary/Documents/Education/safefood%20for%20life/ROI/section3.pdf, drafted on 6-Jan-2016 at 7:00 pm.
2. D. R. Levinson and I. General.: Vaccines for children program: vulnerabilities in vaccine management, Department of Health and Human Services: Office of Inspector General, (2012), drafted on 6-Jan-2016 at 7:00 pm.
3. J. A. Gazmararian, N. V. Oster, D. C. Green, L. Schuessler, K. Howell, J. Davis, M. Krovisky, and S. W. Warburton.: Vaccine storage practices in primary care physician offices: assessment and intervention. American journal of preventive medicine, vol. 23, no. 4, (2002) 246–253.
4. K. N. Bell, C. J. Hogue, C. Manning, and A. P. Kendal.: Risk factors for improper vaccine storage and handling in private provider offices. Published by American Academy of Paediatrics, vol. 107, no. 6, (2001) e100–e100.
5. M. Potdar, A. Sharif, V. Potdar, and E. Chang.: Applications of wireless sensor networks in pharmaceutical industry. Published in Advanced Information Networking and Applications Workshops, WAINA'09 International Conference, (2009) 642–647.
6. Jianping Song, Song Han, Al Mok.: WirelessHART: Applying Wireless Technology in Real-Time Industrial Process Control. 2013 IEEE 19th Real-Time and Embedded Technology and Applications Symposium (RTAS), (2013) 377–386.
7. Willig, A., Matheus, K., Wolisz. A: Wireless Technology in Industrial Networks., Proceedings of the IEEE, vol.93, (2005) 1130–1151.
8. William H., Mitchell: An Introduction to Design Patterns. [Online]. Available:http://www.mitchellsoftwareengineering.com/IntroToDesignPatterns.pdf drafted on 6-Jan-2016 at 7: 00 pm.
9. Perkusich, A., Almedia, H.O.: A software framework for real-time embedded automation and control systems. vol.2, (2003) 181–184.
10. Nithin Raghunathan, Xiaofan Jiang, Dimitrios Peroulis, Arnab Ganguly.: Wireless Low-power Temperature Probes for Food/Pharmaceutical Process Monitoring. Published on IEEE conference on Sensors, (2015) 1–4.
11. Gao Junxiang, Xu Jingtao.: Fruit-Cold-Storage-Environment-Monitoring-System-Based-on-Wireless-Sensor-Network_2011_Procedia-Engineering. Journal of Advanced in Control Engineering and Information Science, vol.15, (2011) 3466–3470.
12. Kristofer S. J. Pister, Lance Doherty.: TSMP: TIME SYNCHRONIZED MESH PROTOCOL. Proceedings of the IASTED International symposium Distributed Sensor Networks (DSN 2008), Orlando, Florida, USA, (2008) 391–398.
13. "Technical overview of Time Synchronized Mesh Protocol". [Online]. Available: http://cds.linear.com/docs/en/white-paper/TSMP_Whitepaper.pdf drafted on 6-Jan-2016 at 7:00 pm.
14. "Kaye RF ValProbe Wireless Validation System". [Online]. Available: https://mail.google.com/mail/u/0/#search/in%3Asent+srs/14fd283713a409ca drafted on 6-Jan-2016 at 7:00 pm.
15. "Kaye ValProbe Wireless Process Validation & Monitoring". [Online]. Available: http://amphenol-sensors.com/en/products/validation-and environmental-monitoring/validation-instruments/458-siweb-pl158 drafted on 6-Jan-2016 at 7:00 pm.

Quantitative Analysis of Frequent Itemsets Using Apriori Algorithm on Apache Spark Framework

Ramesh Dharavath and Shashi Raj

Abstract Frequent itemset mining is one of the popular techniques used to discover hidden knowledge from large-scale transactional datasets in a wide range of applications. Apriori algorithm is considered as a typical algorithm to find frequent itemsets in market basket analysis. Since its inception, many efforts have been made to enhance the efficiency of the original algorithm. MapReduce model is one of the efficient tools to implement parallel and distributed computing, so that large-scale data set algorithms such as Apriori algorithm can be made efficient in terms of speed up and other related parameters. One of the major drawbacks of the MapReduce model is that it is not suitable for iterative jobs/tasks due to overheads imposed. Now-a-days, Apache Spark is getting huge attention for iterative jobs because of its in-memory processing capabilities. Most of the frequent pattern mining algorithms consider only distinct items in a transaction. For transactional data analysis, multiple occurrences of an item or in other words "quantities" by which a particular item is purchased in the same transaction can be important to derive additional information about frequent itemsets. In this script, we propose a modified version of the Apriori algorithm based on Apache Spark framework that not only mines the frequent itemsets in the input transactional data but also analyzes related quantities of the items for a particular itemset to find the most frequent quantity being purchased for every frequent itemset. Experiments are conducted to gain insight in the form of effectiveness, efficiency, and scalability of the proposed approach.

Keywords Apriori algorithm · Big data · Apache spark

R. Dharavath (✉) · S. Raj
Department of Computer Science & Engineering, Indian Institute of Technology (ISM), Dhanbad 826004, Jharkhand, India
e-mail: ramesh.d.in@ieee.org

S. Raj
e-mail: shashirajmnnit@gmail.com

H.S. Behera and D.P. Mohapatra (eds.), *Computational Intelligence in Data Mining*, Advances in Intelligent Systems and Computing 556,
DOI 10.1007/978-981-10-3874-7_25

1 Introduction

Big data [1] are defined in different ways, but the 3 V's namely volume, variety, and velocity are sufficient enough to represent their most general characteristics. Volume implies large data sets; variety represents different types of unstructured data collected from different sources, whereas velocity implies the streamed data collected in real time scenario. Two additional features veracity and value are considered especially by professionals those are involved in analytics part of big data. Since the data are growing rapidly, it incurs tremendous challenges in every stage starting from data acquisition, optimal storage, and analysis to derive required values. On the other hand, it also generates opportunities for researchers. Variety of data mining techniques are used to analyze such kind of large data sets that extract the hidden and required knowledge from the data. It also covers a wide range of techniques like clustering [2], classification [3], and frequent itemset mining [4] to extract related information from the large datasets. Frequent itemset mining finds numerous applications in various real world scenarios. Historically, it was used in shopping basket analysis to find products that were purchased together frequently which in turn allowed companies to formulate better business strategies to increase the sales of their products. Among many popular frequent itemset mining algorithms like FP-Growth and Eclat, Apriori algorithm is the earliest proposed in research literature. Most of these algorithms iteratively scan the datasets to generate frequent patterns.

Many distributed and parallel versions of the above algorithms have been proposed to accelerate the processing of massive amount of data with better efficiency and speedup. Since most of the versions of frequent itemset mining algorithms consider only distinct items in a transaction. However, multiple occurrences of an item in the same transaction is also important in transaction data analysis. Intelligent business strategies can be made based on the above analysis. In this manuscript, we propose a modified version of the parallel Apriori algorithm based on the Spark framework, which adopts the above consideration in a well defined manner. It finds the frequent itemsets from the input transactional data and analyzes respective quantities of the items in a particular itemset to find its most frequent quantity as an itemset.

The roadmap of this paper is as follows. Section 2 briefly describes the concepts of the Apriori algorithm and Apache Spark framework. It also discusses about some related work. In Sect. 3, a modified version of the Apriori algorithm in the spark framework is proposed to adopt the quantitative analysis of frequent itemsets in detail. Section 4 includes performance evaluation of the proposed scheme in different scenarios. Finally, Sect. 5 concludes with future scope.

2 Background and Literature Work

In this section, we explore about some basic concepts related to Apriori algorithm and a very popular framework called Apache Spark followed by related literature.

2.1 Preliminaries

The Apriori algorithm proposed by R. Agarwal et al. [5], works on a fundamental observation that an itemset is frequent only if all its nonempty subsets are also frequent. It uses an iterative approach where results of the previous iterations are used to find the frequent itemsets in the current iteration. It starts with finding singleton-frequent items where an item occurring more times than a user-specified threshold (minimum support) is termed as frequent. Then k-frequent itemsets are found using $(k-1)$th frequent itemsets using the Apriori approach. The candidate set for the kth iteration has all *k-size* combinations of items whose all possible subsets are frequent in $(k-1)$th iteration. Many improved single-machine versions of Apriori exist, the massive amount of data available these days are far beyond the capacity of a single machine. Hence, there is a need to scale the computation across multiple machines in the form of clusters to meet the demands of this growing data. MapReduce is a popular fault-tolerant framework for parallel and distributed computation on large datasets. Nevertheless, heavy disk I/O at each MapReduce operation make it worthless to implement iterative data mining algorithms, such as Apriori on MapReduce platforms.

An in-memory distributed data compute framework called Spark is getting a huge attention of researchers. Apache Spark [6, 7] is an open-source project, developed in the AMPLab at UC Berkeley. Apart from being scalable, flexible, fault-tolerant, and cost effective as Hadoop, it is an extension to MapReduce methodology as it can process both batch and interactive applications efficiently. Apache Spark runs application programs faster than traditional MapReduce cluster compute engine by reducing the number of disk access for repeated read/write operations by keeping the intermediate data in memory. This makes spark well suited for iterative jobs in distributed and parallel environment. It also supports multiple languages to write variety of applications. The vital feature of Spark is its in-memory computing capabilities across the cluster using an abstraction called Resilient Distributed Datasets (RDDs) that increase the processing speed of an application. RDD is used as a fundamental data structure in Spark. Applications on Spark can keep data in memory in the form of RDDs distributed across the cluster nodes. It is a read-only collection of data, that is, we cannot modify the existing RDD but can create other RDDs through deterministic operations on the data on a stable storage by applying transformations on RDDs. It is stored as a file in an external distributed storage system as HDFS or GFS.

2.2 Literature Review

Many variations of the Apriori algorithm have been proposed to make it efficient in terms of speed, especially in parallel and distributed environment. Lin et al. [8] proposed three parallel versions of Apriori algorithm: SPC, FPC and DPC on

MapReduce framework. These algorithms distribute the dataset to the mappers running on different worker nodes to count the appearance of candidate sets in a parallel manner. SPC is a simple adaptation of Apriori on MapReduce engine, while FPC aims to reduce the number of MapReduce phases by using a map function that counts the candidate *k*th, (*k*+1)th, and (*k*+2)th itemsets altogether in a single MapReduce phase. DPC dynamically collects candidates of variable lengths for counting by mappers according to the number of candidates and the execution time of previous MapReduce phases. SPC is single phase and two others are multi-phase algorithms. Li et al. [9] proposed PApriori, a MapReduce based parallel Apriori algorithm which uses key/value pairs to find frequent itemsets similar to SPC. They evaluated the performance matrices like sizeup, speedup, and scaleup of their algorithm and showed their efficiency for big datasets. Farzanyar et al. [10] introduced IMRApriori, an improved Mapreduce based Apriori that uses a pruning technique to decrease the number of partial frequent itemsets called INS-Itemset in Map phase. They showed efficiency in terms of execution time for large social network datasets compared to MRApriori. Moens et al. [11] proposed two different approaches, *Dist-Eclat* and *BigFIM*. Dist-Eclat is a MapReduce implementation of *Eclat* algorithm which focuses on speed by using a simple load balancing scheme based on *k-frequent* itemsets whereas *BigFIM* is a hybridization of Apriori and *Eclat* algorithm which is optimized to run on large datasets. Yu et al. [12] proposed a distributed parallel Apriori algorithm (DPA) that stores metadata in the form of Transaction Identifiers (TIDs), such that only a single scan of the database is needed. It computes factor of itemset counts to balance the workload among processors. Hammoud [13] proposed a MapReduce based network enabled algorithm to switch between horizontal and vertical database layout to mine frequent itemsets. Aouad et al. [14] conducted comparative study of the performance of different distributed Apriori algorithms which are intended to limit synchronization and communication overheads. They acknowledged that the intermediate communications and remote support counts computation restricts the global performance of classic distributed schemes. Chen et al. [15] proposed BE-Apriori algorithm, an improved version of Apriori algorithm based on pruning optimization and transaction reduction. Pruning optimization strategy uses a temporary table to count the frequency of the items in the frequent itemsets.

In general, all the above mentioned algorithms show performance bottleneck due to iterative computation of large datasets. Zhang et al. [16] proposed a distributed algorithm for frequent itemset mining (DFIMA) which applies a matrix-based pruning approach to reduce the amount of candidate itemsets generated in each iteration. Spark is used to further improve the efficiency of iterative computation. Qiu et al. [17] proposed yet another frequent itemset mining (YAFIM) protocol, a parallel version of Apriori algorithm adopted for Apache spark framework. They compared their algorithm with the MapReduce version of the Apriori for each iteration and analyzed the performance of the algorithm in a real-world medical application to explore the relationship among medicines. Rathee et al. [18] used bloom filters to speed up the second round of the algorithm by eliminating candidate generation all together in compared to YAFIM.

Most of the frequent pattern mining algorithms mentioned above, consider only distinct items in a transaction. They do not consider "*multiple occurrences*" or "*quantities*" of the different items purchased in the same transaction. For example, two packets of bread and a dozen eggs purchased together can be important in transaction data analysis [19]. This equips the strategy makers with useful facts to make their business more profitable. For example, by knowing the related quantities of the items in the frequent itemset, the sellers may give exciting and suitable offers to the customers on those particular quantities of the items to increase their sales. In

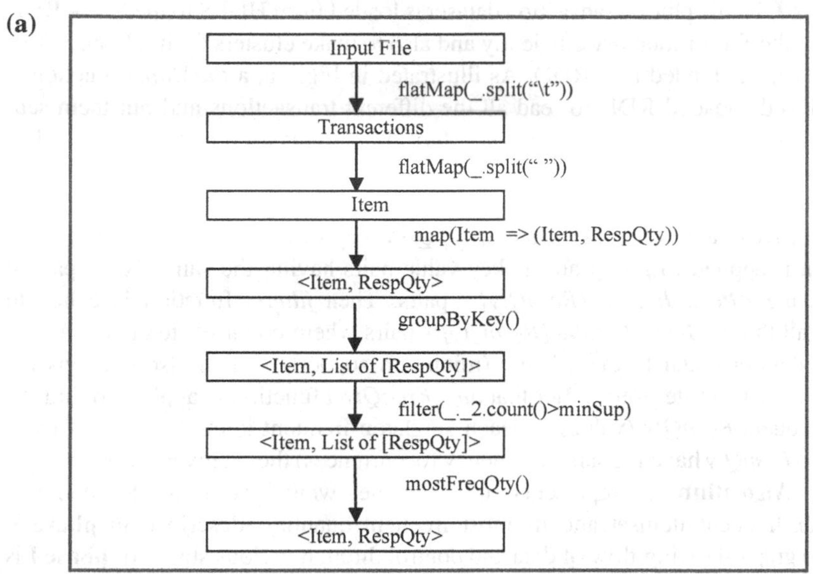

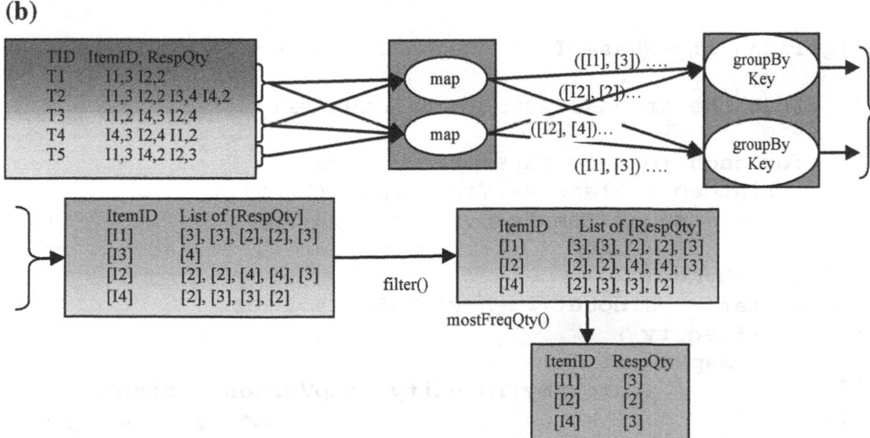

Fig. 1 **a** Linage graph for RDDs in **Phase I**. **b** Graphical representation of workflow in **Phase I**

this manuscript, the parallel version of the Apriori algorithm is modified to adopt the above considerations based on the Spark framework.

3 Proposed Scheme

The proposed methodology with modification to the parallel version of the Apriori algorithm is realized on Spark framework. The processing workflow consists of two phases. These phases are described as follows:

Phase I: In this phase, transaction dataset is loaded from HDFS to the Spark RDD to utilize the cluster memory efficiently and also to make clusters fault tolerant. First, the input file is loaded as a RDD. As illustrated in Fig. 1a, a *flatMap()* function is applied to the loaded RDD to read all the different transactions and put them separately into a new RDD. After that a *flatMap()* function is used to form a new RDD of items and its respective quantities for every transaction of the previous RDD. Then the *map()* function is applied on the resultant RDD to transform all items and its respective quantities to the *<Item, RespQty>* key/value pair. The *groupByKey()* function is applied to group all the key/value pairs having the same "key" part. It outputs the *<Item, Iterable[RespQty]>* pairs. Then *filter()* function is called to extract all those *<Item, Iterable[RespQty]>* pairs where count of iterated values is greater than or equal to the *minSup* (minimum support), which also contains the singleton-frequent itemsets. After that *mostFreqQty()* function is applied to find the most frequent *RespQty* (value) for each singleton-frequent itemsets (key). If more than two *RespQty* have the same frequency (occurrences) then it gives one of them as output. **Algorithm 1** represents the scheme which is used to compute singleton-frequent itemset and its most frequent quantity described in **phase I**. Lineage graph showing flow of data and control through various stages of **phase I** is illustrated in Fig. 1a. The graphical representation of the flow is shown in Fig. 1b.

Algorithm 1: Phase I of proposed scheme

```
Load the transactions from Input file into a cached
RDD
for each loaded transaction Trans
  flatMap ( Trans => Trans.split("\t") )
    flatMap (Item,RespQty=>Item,RespQty.split(" "))
         map ( Item,RespQty => <Item, RespQty>)
groupByKey()
filter(_._2.count(v => v!=NULL)>=minSup)
mostFreqQty()
   mapValues{l=>
        l.groupBy(identity).mapValues(_.size)
                                   .maxBy(_._2)._1}
```

Phase II: In this phase, $(k-1)$ frequent itemsets are used to generate k-frequent itemset. For this, $(k-1)$ frequent itemsets are broadcasted to all the worker nodes through RDD. Then all the candidate itemsets are derived by using *ap_gen()* function. After that, existing transactions in RDD are scanned one by one to get all the candidate itemsets and their respective quantities those are found in a particular transaction using *flatMap()* and *subset()* function. Then the *map()* function is applied to transform all the itemsets and its respective quantities to the *<ItemSet, RespQty>* key/value pair.

The *groupByKey()* function is applied to group all the key/value pairs having the same "key" part. It outputs the *<ItemSet, Iterable[RespQty]>* pairs. Rest of the steps are similar to the **Phase I** which uses *filter()* and *mostFreqQty()* functions as depicted in the lineage graph and graphical representation of the workflow shown in Fig. 2a and b, respectively. The "*key*" part of every key/value pair in the final step represents the *kth* frequent itemset and the "*value*" part represents the most frequent quantity of that particular itemset. **Algorithm 2** represents the algorithm used to compute the desired output in **Phase II**.

Algorithm 2: Phase II of proposed scheme

```
Read FIk-1 from RDD
CSk = ap_gen(FIk-1)
for each transaction Trans
  flatMap ( Trans => Trans.split("\t") )
     CT = subset(CSk, Trans)
     for each candidate ItemSet in CT
        map (ItemSet,RespQty => <ItemSet, RespQty>)
groupByKey()
filter(_._2.count(v => v!=NULL)>=minSup)
mostFreqQty()
  mapValues {l =>
       l.groupBy(identity).mapValues(_.size)
                                           .maxBy(_._2)._1}
```

4 Performance Evaluation

In this section, Spark versions of the Apriori algorithm and the proposed scheme are compared to evaluate their performance on efficiency and scalability. Experiments are conducted on a cluster having 3-computer nodes, i.e., one maser and two slaves with Ubuntu 14.04 LTS installed on each machine. Also, deployment of each computer node is done with the same physical configuration, i.e., Intel Core i7 processor with CPU frequency 3.4 GHz, 8 GB main memory and 1 TB of secondary storage. To keep the experimental results more accurate, all the experiments are executed three times and the average of them is taken as final results. The

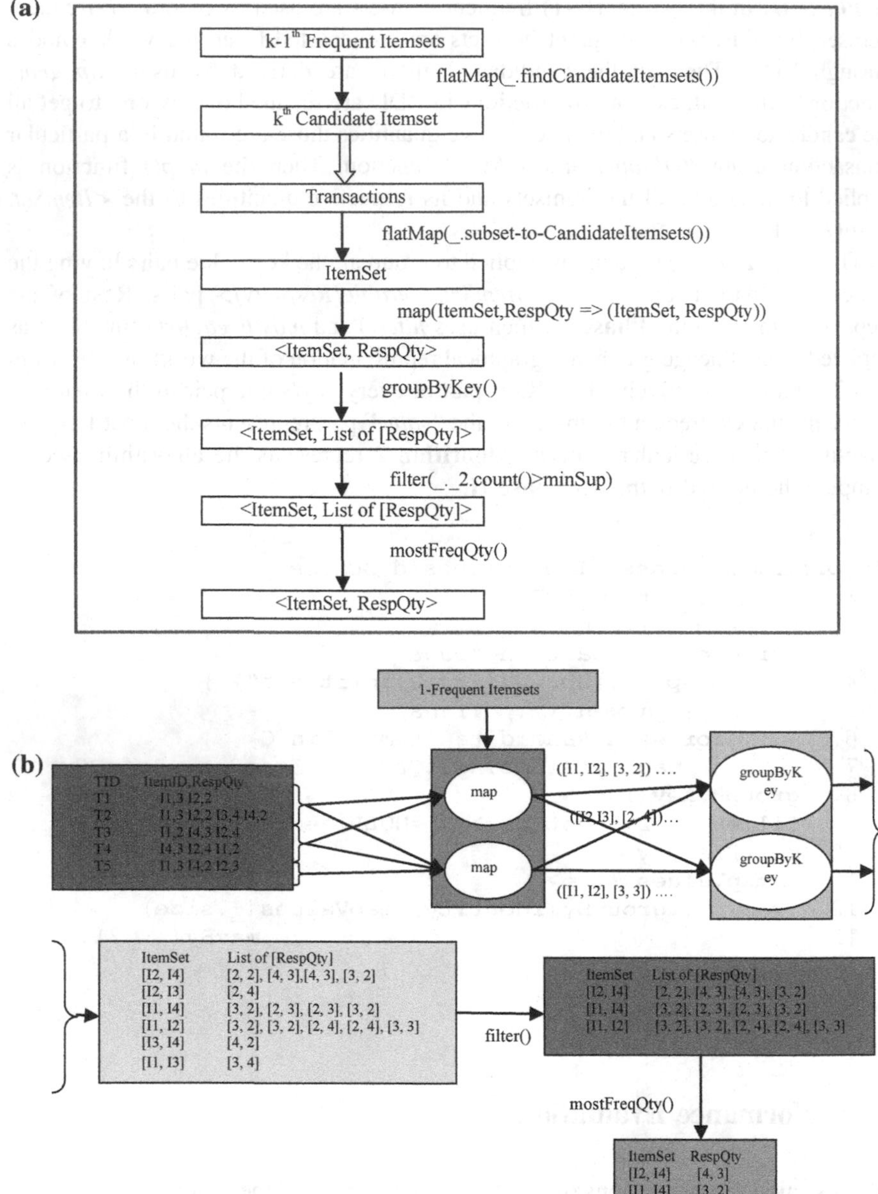

Fig. 2 **a** Linage graph for RDDs in **Phase II**. **b** Graphical representation of workflow in **Phase II**

algorithms are implemented on Spark-1.6.1. Further, speed up and scalability of the proposed scheme is also evaluated by calculating the running time for two different scenarios: (i) number of nodes in the cluster is kept constant while data set size is varied, (ii) number of slave nodes in the system is varied but the size of datasets is kept constant.

Two datasets, similar to T10I4D100 K and T40I10D100 K, are synthetically generated using java code for experimental analysis because of unavailability of such datasets which contain the items and their purchased quantity together for every transaction. So, every item in a transaction is grouped with its respective purchased quantity. We conducted experiments with both the programs on the generated datasets with different characteristics. Table 1 shows the characteristics of the two datasets generated for evaluation.

The performance of the proposed algorithm in comparison with Apriori is illustrated in the below manner. The objective of the first two comparisons is to estimate the speed performance by analyzing the running time of both the algorithms on two different datasets. In this case, the number of computer nodes is kept constant as 3 and minimum support is varied for different datasets to calculate the running time. Figures 3 and 4 indicate that, when the support value increases then the algorithm becomes more efficient. The objective of the third comparison is to analyze the performance of the algorithm by varying number of cluster nodes as shown in Fig. 5. As the number of computer nodes increases, the proposed scheme records less execution time. Note that, in comparison with the Apriori algorithm, the proposed scheme takes little more time but results quantitative knowledge apart from finding frequent itemsets.

Table 1 Characteristics of the datasets used for experiment analysis

Datasets	Size (MB)	No. of transactions	No. of different items	Average no. of items in a transactions
Dataset-1	7.7	100,000	500	10
Dataset-2	20.8	100,000	1000	40

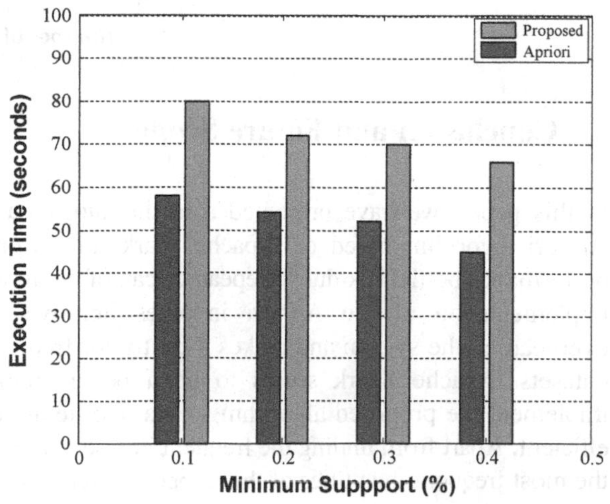

Fig. 3 Running time with varying minimum support for **dataset-1**

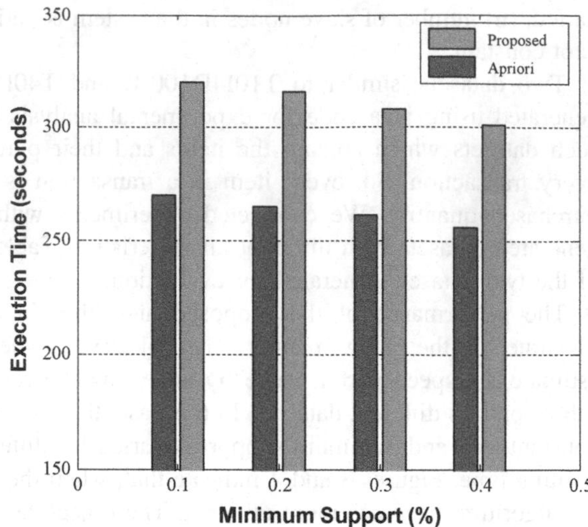

Fig. 4 Running time with varying minimum support for **dataset-2**

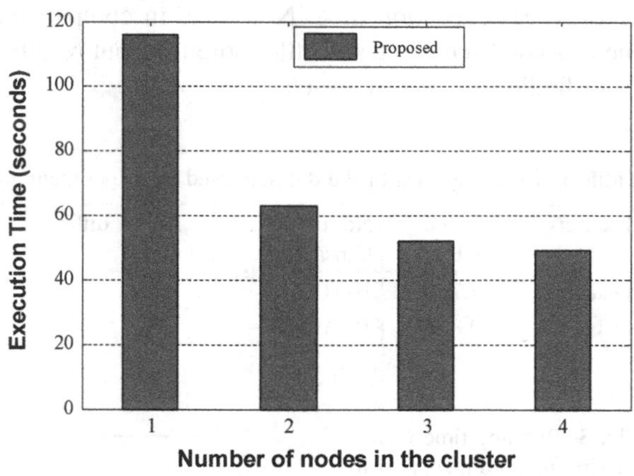

Fig. 5 Running time with varying minimum support for **dataset-1**

5 Conclusion and Future Scope

In this paper, we have presented a methodology called modified version of the Apriori algorithm based on Apache spark framework. Apriori algorithm shows performance bottleneck due to repeated scan of the input data set. Even MapReduce implementation of the Apriori involves iterative computations that increases overhead on the system and makes it costly for frequent itemset mining from large datasets. Apache Spark seems to be a better framework than MapReduce to implement the proposed algorithm so that the iterative computations can be made efficient. Apart from finding the frequent itemsets, the proposed scheme also derives the most frequent quantities of the respective items for a particular frequent itemset.

The proposed algorithm is evaluated in terms of efficiency and scalability with two different data sets for performance evaluation. The experimental result indicates that the proposed algorithm takes little more time than Apriori but it results in quantitative knowledge.

As a future work the proposed scheme can be further implemented with parallel versions of other popular algorithms like *FP-growth* and *Eclat*. This study may result in valuable information in analyzing the frequent itemsets quantitatively in market basket analysis.

References

1. Chen, M., Mao, S., & Liu, Y. (2014). Big data: a survey. Mobile Networks and Applications, 19(2), 171–209.
2. Fahad, A., Alshatri, N., Tari, Z., Alamri, A., Khalil, I., Zomaya, A. Y.,… & Bouras, A. (2014). A survey of clustering algorithms for big data: Taxonomy and empirical analysis. IEEE transactions on emerging topics in computing, 2(3), 267–279.
3. Aggarwal, C. C. (2015). Data mining: the textbook. Springer.
4. Aggarwal, C. C., Bhuiyan, M. A., & Al Hasan, M. (2014). Frequent pattern mining algorithms: A survey. In Frequent Pattern Mining (pp. 19–64). Springer International Publishing.
5. Agrawal, R., & Srikant, R. (1994, September). Fast algorithms for mining association rules. In Proc. 20th int. conf. very large data bases, VLDB (Vol. 1215, pp. 487–499).
6. Apache Spark: Lightning-fast cluster computing. The Apache Software Foundation, http://spark.apache.org/. Spark1.6.1, 9 March 2016, Web. 2016.
7. Karau, H., Konwinski, A., Wendell, P., & Zaharia, M. (2015). Learning spark: lightning-fast big data analysis. " O'Reilly Media, Inc.".
8. Lin, M. Y., Lee, P. Y., & Hsueh, S. C. (2012, February). Apriori-based frequent itemset mining algorithms on MapReduce. In Proceedings of the 6th international conference on ubiquitous information management and communication (p. 76). ACM.
9. Li, N., Zeng, L., He, Q., & Shi, Z. (2012, August). Parallel implementation of apriori algorithm based on MapReduce. In Software Engineering, Artificial Intelligence, Networking and Parallel & Distributed Computing (SNPD), 2012 13th ACIS International Conference on (pp. 236–241). IEEE.
10. Farzanyar, Z., & Cercone, N. (2013, August). Efficient mining of frequent itemsets in social network data based on MapReduce framework. In Proceedings of the 2013 IEEE/ACM International Conference on Advances in Social Networks Analysis and Mining (pp. 1183–1188). ACM.
11. Moens, S., Aksehirli, E., & Goethals, B. (2013, October). Frequent itemset mining for big data. In Big Data, 2013 IEEE International Conference on (pp. 111–118). IEEE.
12. Yu, K. M., Zhou, J., Hong, T. P., & Zhou, J. L. (2010). A load-balanced distributed parallel mining algorithm. Expert Systems with Applications, 37(3), 2459–2464.
13. Hammoud, S. (2011). MapReduce network enabled algorithms for classification based on association rules (Doctoral dissertation, Brunel University School of Engineering and Design PhD Theses).
14. Aouad, L. M., Le-Khac, N. A., & Kechadi, T. M. (2010). Performance study of distributed Apriori-like frequent itemsets mining. Knowledge and information systems, 23(1), 55–72.

15. Chen, Z., Cai, S., Song, Q., & Zhu, C. (2011, August). An improved Apriori algorithm based on pruning optimization and transaction reduction. In Artificial Intelligence, Management Science and Electronic Commerce (AIMSEC), 2011 2nd International Conference on (pp. 1908–1911). IEEE.
16. Zhang, F., Liu, M., Gui, F., Shen, W., Shami, A., & Ma, Y. (2015). A distributed frequent itemset mining algorithm using Spark for Big Data analytics. Cluster Computing, 18(4), 1493–1501.
17. Qiu, H., Gu, R., Yuan, C., & Huang, Y. (2014, May). Yafim: a parallel frequent itemset mining algorithm with spark. In Parallel & Distributed Processing Symposium Workshops (IPDPSW), 2014 IEEE International (pp. 1664–1671). IEEE.
18. Rathee, S., Kaul, M., & Kashyap, A. (2015, October). R-Apriori: an efficient apriori based algorithm on spark. In Proceedings of the 8th Workshop on Ph. D. Workshop in Information and Knowledge Management (pp. 27–34). ACM.
19. Han, J., Pei, J., & Kamber, M. (2011). Data mining: concepts and techniques. Elsevier.

E-CLONALG: An Enhanced Classifier Developed from CLONALG

Arijit Panigrahy and Rama Krushna Das

Abstract This paper proposes an improved version of CLONALG, Clone Selection Algorithm based on Artificial Immune System that matches with the conventional classifiers in terms of accuracy tested on the same data sets. Clonal Selection Algorithm is an artificial immune system model. Instead of randomly selecting antibodies, it is proposed to take k memory pools consisting of all the learning cases. Also, an array averaged over the pools is created and is considered for cloning. Instead of using the best clone and calculating the similarity measure and comparing with the original cell, here, k best clones were selected, the average similarity measure was evaluated and noise was filtered. This process enhances the accuracy from 76.9 to 94.2 %, ahead of the conventional classification methods.

Keywords Accuracy · Antibody · Artificial immune system · Classification · Classifier · CLONALG · CSCA · Cloning · Intrusion detection · Machine learning

1 Introduction

Artificial immune system (AIS), one of the rapidly emerging areas of artificial intelligence, has been inspired from biological immune system [1]. The purpose of the biological immune system is mainly to shield the body from the antigens. Our immune system has an efficient pattern recognition ability that differentiates between antigens and body cells. Immune system has many features such as autonomous, uniqueness, distributed detection, foreigner's recognition and noise tolerance [1]. AIS is the abstraction of structure and function of the biological

A. Panigrahy (✉)
Indian Institute of Technology, Kharagpur, West Bengal, India
e-mail: arijitpanigrahy@gmail.com

R.K. Das
National Informatics Centre, Berhampur, Odisha, India
e-mail: ramdash@Yahoo.com

H.S. Behera and D.P. Mohapatra (eds.), *Computational Intelligence in Data Mining*, Advances in Intelligent Systems and Computing 556,
DOI 10.1007/978-981-10-3874-7_26

immune system in order to solve various problems. Clonal Selection Algorithm is an AIS model [2]. It has been used mostly for optimization and pattern matching. It has also been tested for classifying data sets, but generally the results have not been satisfactory. The widely used AIS models are negative selection, clone selection, immune networks and danger model [3].

After a brief explanation of AIS in the following sections, this paper develops an enhanced CLONALG (E-CLONALG) algorithm and implements it for standard classification. We are proposing a modified version of the Clone Selection Algorithm (CLONALG) for improving accuracy on classification. The enhanced CLONALG algorithm increases the accuracy to a great extent and can be used for various other purposes with a little modification such as pattern recognition, intrusion detection, filtering the noise etc.

In E-CLONALG, we formed 'k' memory pools such that all the learning instances are accommodated in the memory pool. Also this memory pool was visualized as a combination of 'c' memory pools, where 'c' is the number of classifications. In the learning phase, after the affinity of the memory cells were found with the antigens, we created another memory pool such that the affinity value of antibodies is the average of the memory pool created initially. We call the average memory pool as 'M2'. Then, both 'M1' and 'M2' were used for cloning. After that, the memory cells with maximum similarity are added to antibodies. Instead of one best clone, we selected 'k' clone cells in order to encourage local search. Thus, introducing the above steps increases the accuracy and kappa factor.

2 Artificial Immune System

2.1 Immune Network Theory

Jerne (1974) proposed the Immune Network Theory to explain the learning properties of Biological immune system. It is based on the theory that a subset of the total receptor repertoire recognizes any of the lymphocyte receptor. Immune networks are often referred to as idiotypic network [4]. This theory also explains the immunological behaviours such as memory emerge and tolerance. This model, which previously was known as Artificial Immune Network (AIN), is now known as Artificial Immune NEtwork (AINE).

2.2 Negative Selection

Negative selection algorithms were based on the mechanism that takes place in Thymus that produces a set of mature T-cells which is capable of binding the non-self-antigens only. It was put forward by Forrest et al. (1994). This algorithm starts with production of a set of self-strings, S1, that defines the normal state of the system [5]. It then generates a set of detectors, D, which recognize only the complement of S1. It can be used to classify test data set [6, 7].

2.3 *Danger Model*

A new immunological model, suggested by Polly Matzinger (1994), that states which the immune system does not distinguish between non-self and self. Rather, it discriminates dangerous and safe by recognition [8].

2.4 *Clone Selection*

Burnet (1959) proposed the Clone Selection Algorithm. It constitutes the procedures of recognition of antigen, proliferation of cell and memory cell differentiation [9]. Castro and Zuben (2002) proposed the clone selection algorithm, which is known as CLONALG. It is being used for learning and optimization. When the CLONALG is applied to pattern matching, a set of patterns, S_1, i.e. the test set is considered to be antigens [10]. CLONALG produces a set of memory antibodies, M, that match the members in S [1].

3 Clonal Selection Algorithm (CLONALG)

The Clonal Selection Algorithm, incorporates the following elements of the clone selection theory [11].

Maintenance of memory set, selecting and cloning the most stimulated antibodies and death of the non-stimulated antibodies, mutation reselecting the clones in proportion to the affinity with the antigen, maintenance of diversity. The algorithm is outlined below:

3.1 *Initialization*

Initialization involves preparing a pool of fixed size N for antibodies. The antibody pool will be divided into two parts, remainder pool and memory pool. Algorithm's solution will be a part of the memory pool while remained pool will be used for introducing diversity.

3.2 Loop

Then the algorithm proceeds to a number of iterations. The total number of iterations is 'G'. A single iteration is referred to as a generation.

(1) *Select Antigen*: An antigen is selected randomly, without replacement from the pool.
(2) *Exposure*: The system is subjected to the antigen selected. Hamming distance was used to calculate the affinity values (a measure of similarity).
(3) *Selection*: 'n' highest affinity antibodies were selected.
(4) *Cloning*: The selected antibodies were cloned based on their affinities.
(5) *Affinity Maturation* (*mutation*): The clones were mutated to match the antigen selected in step 1.

$$Degree\ of\ mutation \propto \frac{1}{Parent's\ affinity}$$

(6) *Clone Exposure*: The affinity measures of clones were calculated after exposing to the antigen.
(7) *Candidature*: The highest affinity antibodies were considered as candidates to be transferred to memory pool. Antigen is replaced by the candidate memory cell, only if the affinity of memory cell is higher than that of the highest stimulated antigen.
(8) *Replacement*: The lowest affinity 'd' antibodies from the remainder pool were replaced by new random antibodies.

3.3 Finish

Memory pool is taken as the solution of the algorithm [12]. Depending on the problem, the solution can be a set or a single entity.

4 Enhanced CLONALG Algorithm (E-CLONALG)

The proposed algorithm is based on CLONALG algorithm. We have improved the algorithm in the initialization as well as training phase. We have introduced a new parameter in the antibody class known as 'ClassF'. In CLONALG, the memory pool was formed by picking up random antibodies. As a result, all the learning instances were not used. In E-CLONALG, we formed 'k' memory pools such that all the learning instances are accommodated in the memory pool [13, 14]. Also this memory pool was visualized as a combination of 'c' memory pools where 'c' is the number of classifications. Thus, we considered the total learning set for classification instead of randomly picking a few instances [15]. We call the memory pool as 'M1'.

$$k = \text{ceil}\left(\frac{\text{Size of learning set}}{\text{Size of memory pool}}\right)$$

In the learning phase, after the affinity of the memory cells were found with the antigens, we created another memory pool such that the affinity value of antibodies is the average of the memory pool created initially. We call the average memory pool as 'M2'. Then, both 'M1' and 'M2' was used for cloning. After that, the memory cells with maximum similarity are added to antibodies. Instead of one best clone, we selected 'k' clone cells in order to encourage local search. Also, the average similarity was computed between the cloned cells and the training set to filter out the noise and ensure that the corresponding class is represented by the generated cells [15].

As per Fig. 1, the algorithm has been outlined below:

A. *Initialization*

Initialization involves preparing a pool of fixed size N for antibodies. The antibody pool will be divided into two parts, remainder pool and memory pool. Algorithm's solution will be a part of the memory pool while remained pool will be used for introducing diversity.

Then the memory cells are preprocessed according to equation and are labelled to the classifications. These classifications are never changed during the cloning process. Let us denote the remainder pool as 'Ab_r' and the memory pool as 'Ab_m'. The classes in the memory pool are denoted by 'Ab_{mi}' where 'i' is an integer.

B. *Loop*

Then the algorithm proceeds to a number of iterations. The total number of iterations is 'G'. A single iteration is referred to as a generation.

(1) *Select Antigen*: An antigen is selected randomly, without replacement from the pool.
(2) *Exposure*: The system is subjected to the antigen selected. Hamming distance was used to calculate the affinity values (a measure of similarity).
(3) *Selection*: Another memory pool is created that consists the antibodies whose affinity values are the average of 'k' memory cells. 'n' antibodies were selected from the pool consisting of 'Ab_m' as well as 'M_2' based on highest affinity.
(4) *Cloning*: The selected antibodies were cloned based on their affinities.
(5) *Affinity Maturation* (*mutation*): The clones were mutated to match the antigen selected in step 1.

$$\textit{Degree of mutation} \propto \frac{1}{\textit{Parent's affinity}}$$

Fig. 1 Overview of E-CLONALG

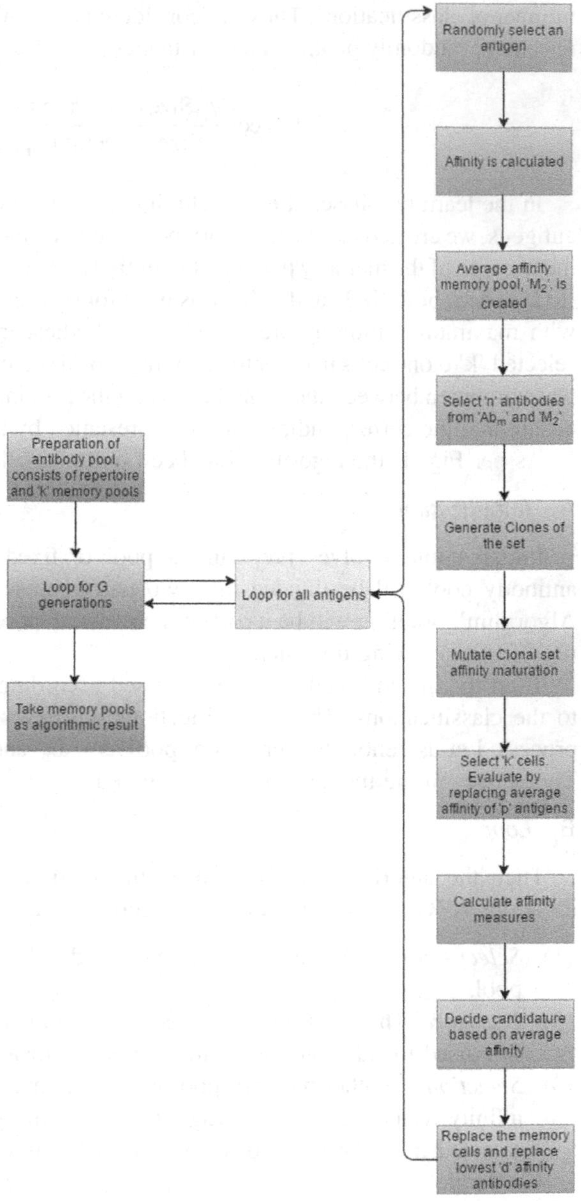

(6) *Clone Exposure*: The affinity measures of clones were calculated after exposing to the antigen.
(7) *Candidature*: Instead of selecting only one cloned cell as in the original CLONALG algorithm, 'k' highest affinity cloned cells of the same class are picked to promote local search. These 'k' cells are revaluated after changing

the affinities mean affinity of 'p' nearest antigens belonging to the same classification. If the affinity calculated above is lesser than at least two antigens belonging to different then the candidature is revoked. However, it is deleted from the training data set, if only one antigen belonging to a different classification has higher affinity than one of the k cloned cells.

(8) *Replacement*: Finally, if clone with highest affinity among the k filtered clones is higher than the minimum affinity value of the memory pool (belonging to same class), then it substitutes the minimum affinity memory cell, i.e. 'Ab_{mi}'. The lowest affinity 'd' antibodies from the remainder pool were replaced by new random antibodies.

4.1 *Finish*

Memory pool is taken as the solution of the algorithm. Depending on the problem, the solution can be a set or a single entity.

5 Experiment Setup and Result Analysis

For our purpose, we used the 'Adult Data Set' available in UCI Library to test the classification accuracy of various algorithms and compare them with our algorithm. The dataset has 2 sets, one for learning purpose and other for testing, consisting of 48842 instances. The data set has 14 parameters and 2 classes [16]. The algorithm was tested using Weka tool [17, 18] and compared methods such as Bayes, J48, CLONALG, etc. The findings are enlisted in the table below.

From the findings, as shown in Table 1 it can be easily seen that E-CLONALG has better classification accuracy than other conventional algorithms. E-CLONALG uses a new technique in the initialization as well as training (candidature) phase with an aim of improving the classification accuracy.

Figure 2 compares the accuracy of the various conventional classification algorithms with CLONALG and E-CLONALG algorithms. As we can observe, the accuracy of E-CLONALG is better than other conventional algorithms by 10%. The accuracy is calculated using

Table 1 Various classification algorithm and their accuracy

Name of the algorithm	Accuracy (%)
CLONALG	76.9
Bayes-net	84.2
Naïve Bayes	83.2
J48	85.8
Regression	84.7
LMT	85.9
E-CLONALG (proposed)	94.2

$$accuracy = \frac{number\ of\ correct\ predictions}{number\ of\ instances\ in\ test\ set}$$

Figure 3 compares the kappa statistics of the various conventional classification algorithms with CLONALG and E-CLONALG algorithms. Cohen's Kappa coefficient is a measure which indicates the inter-rater agreement for qualitative terms. The equation for calculating kappa statistic is [19]

$$\kappa = \frac{p_o - p_e}{1 - p_e}$$

p_o = relative observed agreement among raters, i.e. total number of cases for which the prediction matches with the actual value, p_e = hypothetical probability of chance agreement, i.e. probability that the prediction will match with the actual value. When greater the value of kappa, higher similarity between the predicted value and the actual value. The theoretical maximum of kappa is 1. In our case, the value of kappa for E-CLONALG is 0.876 which is higher all other algorithms.

Every classification is assigned a value which is used in computing the mathematical parameters explained below. Every class has a value associated with it.

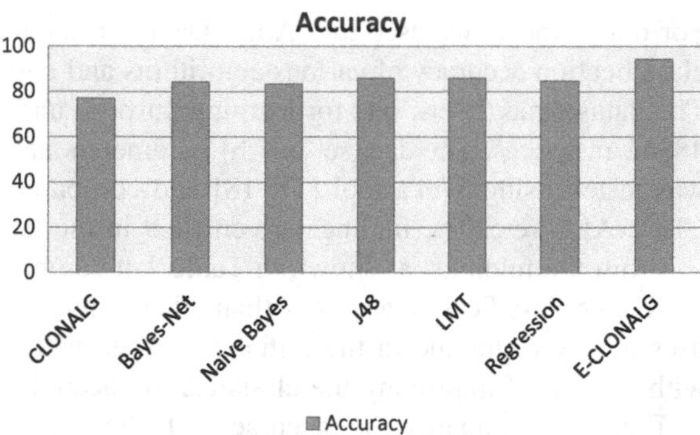

Fig. 2 Comparison of accuracy of classification algorithms

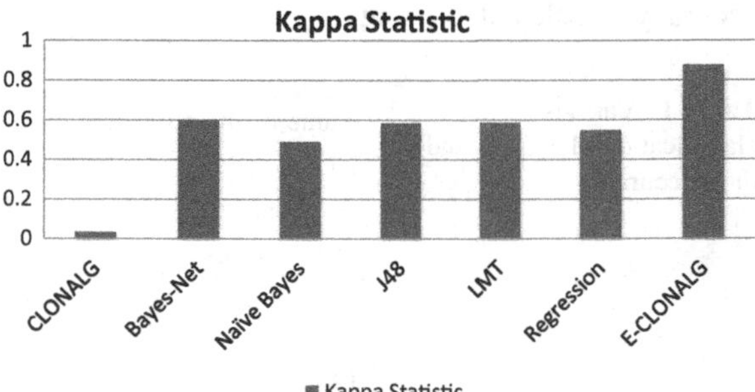

Fig. 3 Comparison of kappa statistic of classification algorithms

6 Conclusion

A new form of CLONALG algorithm, E-CLONALG, has been proposed in this paper. The E-CLONALG algorithm seems to be an alternative method for classification. Even though it has not been tested across all the available data sets, but still the results indicate that it can be an alternative method for classification. It also shows the algorithms based on Clonal Selection Principle can also be used for classification.

A lot of further work is possible in this area. This algorithm can be modified to be used for intrusion detection, pattern recognition, noise removal, etc. Moreover, the complexity of this algorithm can be reduced to make the classification task faster. Generalization of the algorithm can be done to deduce that E-CLONALG can be used as a classifier alongside conventional classifiers as well.

References

1. J.R. Al-Enezi, M.F. Abbod, S. Alsharhan: 'Artificial Immune Systems- Models, Algorithms And Applications', International Journal of Research and Reviews in Applied Sciences, 3 (2), May (2010) 118–131
2. R. Murugesan, K. Sivasakthi Balan: 'Positive Selection based Modified Clonal Selection Algorithm for Solving Job Shop Scheduling Problem', Applied Mathematical Sciences, Vol. 6,2012, no. 46, 2255–2271
3. Ramsha Rizwan, Farrukh Aslam Khan, Haider Abbas, Sajjad Hussain Chauhdary: 'Anomaly Detection in Wireless Sensor Networks Using Immune- Based Bioinspired Mechanism', International Journal of Distributed Sensor Networks Vol. (2015)
4. AISWeb – The Online Home of Artficial Immune Systems (http://www.artificial-immune-systems.org/algorithms.shtml)
5. K.Parthasarathy, 'Clonal selection method for immunity based intrustion, detection systems', Project Report (2014), 1–19
6. Junyuan Shen, Jidong Wang, Hao Ai: 'An Improved Artificial Immune System- Based Network Intrusion Detection by Using Rough Set', Communication and Network, 2012,4,41–47
7. Amira Sayed A. Aziz, mostafa A. Salama, Aboul ella Hassanien, sanna El-Ola Hanafi: 'Artificial Immune System Inspired Instrusion detection System Using Genetic Alorithm', Informatica 36 (2012) 347–357
8. Julie Greensmith, Uwe Aickelin, Steve Cayzer: 'Introducing Dendritic Cells as a Novel Immune- Inspired Algorithm for Anomaly Detection', International Conference on Artificial Immune Systems, ICARIS (2005), 14th -17th August 2005, Banff, Alberta, Canada
9. Ezgi Deniz Ulker, Sadik Ulker.: 'Comparison Study for Clonal Selection Algorithm and Genetic Algorithm', Int. J. of Computer. Science & Information Technology Vol 4, No.4. August (2012) 107–118
10. Ilhan Aydin, Mehmet Karakose, Erhan Akin, 'Generation of classification rules using artificial immune system for fault diagnosism', IEEE Conference on Systems Man and Cybernetics (SMC), pp 343–349, (2010)
11. Vincenzo Cutello, Giuseppe Narzisi, Giuseppe Nicosia, Mario Pavone: 'A Comparative Case Study Using Effecctive Mutation Potentials, C. Jacob et al. (Eds.)': ICARIS (2005), LNCS 3627, pp. 13–28,200525: 1967-1978

12. Jason Brownlee, 'Clonal Seleciton theory and ClonalG: The Clonal Selection Classification Algorithm', Technical Report No, 2–02, January (2005)
13. Linquan Xie, Ying Wang, Liping Chen, Guangxue Yue: 'An Anomaly Detection Method Based on Fuzzy C-means Clustering Algorithm', Second International Symposium on Networking and Network Security (ISNNS,10) Jinggangshan, P.R. China, 2–4, 2010, pp. 089-092
14. Ryma Daoudi, Khalifa Djemal, Abdelkader Benyettou: 'Cells Clonal Selection for Breast Cancer Classification', International Multi-Conference on Systems, Signals & Devices (SSD0, Hammamet, Tunisia, Mach 18–21,2013
15. Anurag Sharma, D. Sharma, 'Clonal Selection Algorithm for Classification', Lecture Notes in Computer Science, Vol 6825, pp 361–370 (2011)
16. UCI Machine Learning Repository – Adult Data Set (https://archive.ics.uci.edu/ml/datasets/Adult)
17. Weka: Data Mining Software in Java, (http://www.cs.waikato.ac.nz/~ml/weka/)
18. Github, (https://github.com/ICDM2016)
19. Wikipedia Kappa stat (https://en.wikipedia.org/wiki/Cohen%27s_kappa)

Encryption First Split Next Model for Co-tenant Covert Channel Protection

S. Rama Krishna and B. Padmaja Rani

Abstract Cloud computing technology present days fulfills the infrastructural needs of small and medium enterprises without creating additional load of initial setup cost. Industrial infrastructural needs like storage, computing has became like utilities that uses cloud. Features like availability, scalability, pay per use makes cloud as a dependable choice for enterprises. The growth of cloud computing depends on vital technologies like distributed computing, parallel processing, Internet, virtualization technologies, etc. Virtualization enables resources utilization in efficient way. It offers multi-tenancy which is sharing same physical space by many virtual machines. Co-residency of virtual machine on the other side brings more security bumps. As co-residents share common channels like memory bus, storage controllers, cache these channels can become sources of data leakage for malicious VMs. Malicious attacker VMs use covert channel for accessing sensible data from target VM. This paper presents a novel model Encryption First Split Next (EFSN) model for covert channel protection. This model targets about two key issues like isolating conflict of interest files from attacker using split share model. Other is applying encryption for making attacker difficult even if he gains the data from covert channels. This suggested model gave positive reliable results. This can be a choice for lazy or immature cloud users to rely on for their data storage security needs from co-tenant covert channel protection.

Keywords Cloud computing · Security · Disk contention · Encryption first · Split share model · Covert channel protection · Co-tenant covert channel protection

S. Rama Krishna (✉)
Department of Computer Science & Engineering, VRS & YRN College of Engineering & Technology, Chirala 523155, India
e-mail: ccvy.ram@gmail.com

B. Padmaja Rani
Department of Computer Science & Enginneering, JNTUCEH, Hyderabad, India
e-mail: padmaja_jntuh@yahoo.co.in

H.S. Behera and D.P. Mohapatra (eds.), *Computational Intelligence in Data Mining*, Advances in Intelligent Systems and Computing 556,
DOI 10.1007/978-981-10-3874-7_27

1 Introduction

Computer technology with the revolutionary Internet technology created many technologies one such revolution is Cloud Computing. Cloud computing era has significant impact over enterprises because of its simplicity, availability, and easiness in usage. Virtualization technology, high performance computing, cheaper storage services made conventional computing transforming to powerful cloud computing. Data centers used for cloud computing are the hubs for huge pool of processors and storage connected overt high speed multiplexed I/O channels. These data centers support high scalability which helps industry to handle their peak data needs. Spikes in the graphs of data usage and computing capabilities have become speed breakers for small and medium business enterprises growth. These unexpected requirements increase infrastructural expenditure over head for enterprises. Cloud computing came as a solution to handle these kind of problems. It has successfully shown its performance in scalable, elastic computing and storage in pay per usage model.

Small and medium businesses now prefer cloud computing for their computing and data needs [1, 2]. Even the cloud services are accessible by its users from different types of devices [3, 4]. Moving their data into the cloud technology increased convenience for its users because they need not worry about the complexities in handling or managing hardware. Pioneered vendors like Amazon S3 (Simple Storage Service), Google drive, DropBox, and Amazon Elastic Compute Cloud (EC2) [5] are popular examples. Performance of cloud computing made it as undoubted choice but security left several questions. Users of cloud have to depend over the mercy of cloud service providers for their data security [6]. There exist wide range security threats both internal and external. Literature [7, 8] lists several incidents outages and data loss that happen in popular cloud vendors. Literature [9, 10] tells that cloud service providers even tried to hide incidents of data loss for protecting their reputation [9]. Hence, outsourcing data to cloud may be economically attractive but lacks in providing assurance of security. Attackers use underlying technology like virtualization for hijacking the sensitive information. Multi-tenancy is the feature provided by virtualization. This offers resource utilization maximization. Simultaneously, it causes security breaches like performance degrading, hijacking sensitive data form co-tenants and information leakage and so on. Recent research brought several security attacks caused by virtualization. Attacker uses shared resources like CPU cache, I/O bus, storage controller, RAM, etc., for obtaining information from target machine [11–13]. Some papers suggest strict access policy such as Chinese wall security for protecting data hijacking in multi-tenancy environment which is also vulnerable [14].

Mature clouds use isolation techniques to prevent data access of conflict of interest files. They have measures to identify files possible of interest conflict and protect them. But lazy and immature clouds do not address this problem because of several reasons like negligence, cost-cutting process effectively [15]. Cryptography is an art of secret writing. This technique helps in providing confidentiality by making the data to a non-understandable format. Research produces several cryptographic algorithms classified into symmetric and asymmetric cryptography. Cryptography has many application and algorithms used for the cryptography are public and secrete piece of information used in cryptography known as key which is private. Both types of cryptographies are having pros and cons. For secret communication, this cryptography is widely used. DES (Data Encryption Standard), RC5, Triple DES, CAST128, AES are some of the widely used symmetric key cryptographic algorithms. RSA, ECC are public key cryptographic algorithms. So applying cryptography could bring confidentiality in data communication.

This paper presents suggests a new model for protecting data from malicious co-tenant covert channels known as Encryption First Split Next (EFSN) model. The rest of paper is organized as follows. Section 2 presented co-tenant covert channel attacks and mitigations. Problem definition and design issues were discussed in Sect. 3. Section 4 introduces Encryption First Split Next model for co-tenant Covert channel protection. Implementation and result analysis were discussed in Sect. 5. Section 6 consists of summary and future direction of research.

2 Review on Covert Channel Attacks and Mitigations

Channels used to access unprivileged data from co-residents without their intervention are known as covert channels. Using virtualization cloud computing provides multi-tenancy. It enables to share a physical space by many virtual machines. While attacker VM is executing, it uses channels like power consumption, CPU cache, memory and network as sources of data leakage from target VM. Data extraction is done through these covert channels only. Tracing software happening from hardware behavior observation is explained in the literature [16]. Literature [17] presents a method that use CPU cache response time for verifying target VM is co-resident or not. Cache behavior is analyzed using linear regression of the data which is pre processed by cubic spline and load predictor. Attack begins by occupying CPU cache in major part. Then attacker VM sends data message to target VM and using the program calculates CPU cache access times.

This study tells in the literature [17] that when more activities are performed by co-resident VM there is more cache access time. As an extension for this work, the experiment was done with three VM. Unlike previous case an attacker's VM not only analyzes the cache access time of target machine but it also captures the data using covert channels. Literature [18] presents a method of how malware insertion can help attacker VM to hijack data from target machine. Method is so powerful so that Attacker gains information and it removes all traces used for the attack. This attack also uses memory bus as a covert channel. By sending 1 it issues an atomic CPU instruction and increases latency in memory usage. Attacker VM Issues 0 when it wants to reduce memory latency. Using cache contention as covert channel is another method suggested in the literature [18]. Overlapped time is used by Attacker VM to calculate bandwidth of target VM when it is in execution. Hypervisor scheduler needs some changes for cope up with the attack as suggested in [18].

Overlapping time minimization can be used to protect from cache contention attack for bandwidth analysis. If we minimize overlapping time it impacts system performance. Proper care needs to be taken while minimizing the overlapping time without impacting system performance. Scheduler switches VMs to minimize overlapping time this also reduces system performance [18]. Pumping noise can be used as technique for preventing bandwidth analysis attack in memory bus. Literature [16, 19] suggests Xenpump model for covert channel attack defense. Xenpump generate Random latency to confuse Attacker VM while analyzing bandwidth. Effectiveness of timing channel is reduced by this technique. For covert channel protection we can also use technique mentioned in literature [16] called as Prime trigger probing. Attacker occupies many cache lines and it has highest access time and probes target VM to get encoded message by accessing cache. Once target VM completes its job then attacker VM start accessing cache to finish its job. This probing can be prevented using flushing technique. Flushing technique flushes the data while VM is switching. So data will not be available for attacker VM. A 15% overhead is caused by flushing technique and it successful in preventing prime trigger probing attack [16].

3 Major Problem of Focus

3.1 Disk Contention Covert Channel Attack

Hard disk contention is a technique used by attacker for obtaining data from target machine using storage controller covert channel. Study over this problem is immature. This attack basically blind folds tough access policy like Chinese wall security [4] and makes the attack possible. Figure 1 explains how the attack is done.

Attacker and bob having their files in bucket 2. Strict access policy restricts access for files stored in bucket 5. In this case attacker cannot access bob's file and same with the bob also. Even in this situation attack happens by installing malware over bob VM by the attacker. As bob VM is infected by malware it can be easily controlled by attacker. Now attacker uses this control and asks bob's VM to access its file while same storage controller attacker try to access its file. Then using the pattern of access attacker VM can read the file owned by bob. That is how without violating access policy attacker gains information from bob VM. Strict access policy like Chinese wall is not able to prevent this leakage.

3.2 Design Objectives of Model

1. Model has to support isolating conflict of interest files by storing separately and reduce the data availability to attacker in the case of covert channel attack.

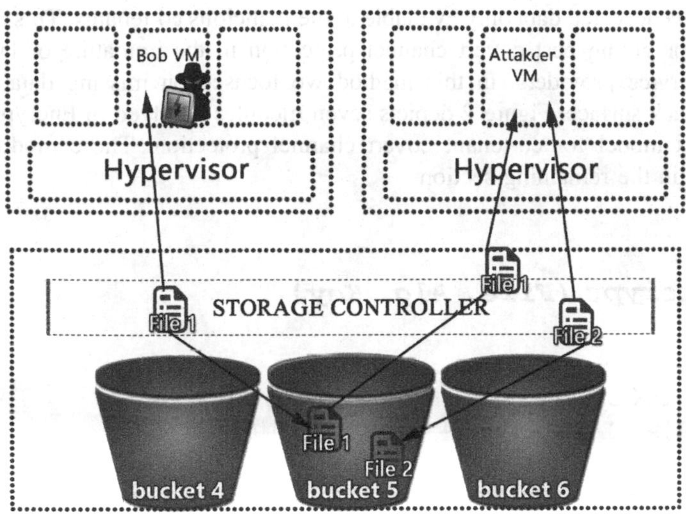

Fig. 1 Information leakage due Co-tenant covert channel attack in shared storage space

2. Model should identify proper isolating criteria like no of split which will not impact the system performance too much but provide reasonable security.
3. For increasing confidentiality to the data isolated and stored in multiple places model has to support provision for encryption.
4. Multiple storage policies are to be supported by the model.
5. Model has to support for Integrity and error verification.

4 Encryption First Split Next Model

Encryption First Split Next Model contains seven modules discussed below. In this model for the protection of co-tenant Covert Channel while storing the data we first encrypt the file to get confidentiality then we split the data into number of splits and store it in multiple storage locations. This makes attacker difficult to obtain information through hard disk contention because attacker file cannot be co-located in all splits of the file hence we reduce the attack surface. Objective of encryption first Split Next reducing attack surface and make the content difficult to understand by malicious co-tenants.

For achieving this we perform encryption over the file that is to be uploaded and then file is split and stored in multiple storage spaces. By storing these split files in different buckets in different regions will give confidentiality and data leakage can be reduced. If bucket is compromised also data is not available in one single bucket and partial encrypted data only available to the malicious co-tenants. This method is suitable for giving best covert channel protection in the immature or lazy cloud storage service providers. In this method we focused on missing data factor to reduce attack surface. Figure 2 depicts seven modules involved in Encryption First Split Next model for co-tenant covert channel protection. These modules were discussed in the remaining section.

4.1 Encrypt (File, Alg, Key)

```
Begin
    CFile←EncryptFile(File,Alg,key)
    IFile.write(SplitFileList[i],Alg,key)
End
```

Encrypt module takes three inputs. File is the Data owner message content. Algorithm is an encryption algorithm that is used for transforming data owner file into nonsense text. Key is small piece of secret information to make secret information unpredictable in encryption process. Based on Encryption algorithm chosen encryption steps will be selected in the execution process. For this encryption algorithm it is required to give two inputs secret key and data owner file. Then algorithm will convert plain text as nonsense cipher text and this cipher text is written over CFile. CFile is the output of the module. IFile is updated with encryption algorithm, encryption time, key information etc.

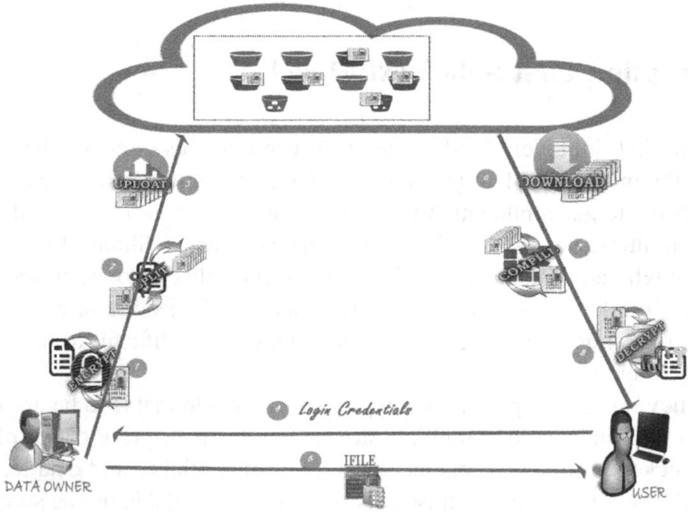

Fig. 2 Encryption first split next model for Co-tenant covert channel protection

4.2 *File Split (Cfile, N)*

```
Begin
    Fsize← getSize( Cfile )
    Blocksize← Fsize/n
    If(Fsize % n!=0)
        N=n+1
    End if
    For i=1 to n do
        Create newfile
        Readbuffer ← Read file(blocksize)
        newFile. write(Readbuffer)
        EncSplitFileList[i]← newFile
        IFile.write(EncSplitFileList[i])
    End for
End
```

Number of splits and CFile (data owner file after encryption) are given as inputs for File split module. This module is to reduce attack surface by splitting the file and to enable store in multiple storage spaces. Number of splits is the factor that impacts in increase of computational overhead and security proportionally. So a heuristic decision has to be taken in choosing best possible number of splits for this model. After splitting the CFile module generates Enc Split File List (an array of split files). As usual information in IFile is updated. Checksum, encrypted split files and sizes are written in IFile.

4.3 *Uploadfile(UploadPolicy, Mode, CloudAccessInfo, EncSplitFilesList)*

```
Begin
    For i← n
    Account ← getCloud(CloudAccessInfo,UploadPolicy)
        Region   ← setRegion(uploadPolicy)
        Bucket   ← createBucket(mode)
        Object   ← readFile(EncSplitFileList[i])
        uploadcloud( Cloud,Region,Bucket,Object)
        IFile.write(Cloud,Region,Bucket,Object)
    End for
End
```

UploadFile takes four inputs Upload Policy, Mode, Cloud Access Info, Encrypted files list were sent to the method of upload File. Upload Policy defines way of storing data. Model supports four different types of storing types Single cloud single account, Single cloud multiple accounts in multiple regions, Multiple clouds with single account each and Multiple clouds with multiple accounts with multiple regions. It also supports three modes such as Random distribution mode, Circular distribution mode and Sequential distribution mode. Cloud Access Info contains different accounts information related to cloud like access keys and passwords. Uploading is done by the upload policy, mode of uploading for this account information is picked from the cloud access info file. Then data is uploaded to corresponding bucket. Here we limit our discussion to only single cloud later it can be extended for multiple clouds.

4.4 Data Sharing (Login Credentials)

```
Begin
    If( login successful )
          Transfer IFile
    Else
          Return error message
    End if
End
```

When cloud user needed to access the file he will request data owner for IFile, and proper subscription is done user will be provided with credentials. Once these credentials are authenticated then IFile will be shared in secret form through mail or encrypted format with data user key.

4.5 Download (IFile)

```
  Begin
      For i =1 to n
            Account ← getCloud(IFile)
            Region  ← getRegion(IFile)
            Bucket  ← getBucket(IFile)
            Object  ← readFile(IFile)
 EncFileList[i]←downloadcloud( Cloud,Region,Bucket,Object)
      End for
  End
```

User obtains information file after verified his credentials at data owner and he uses the same for downloading the encrypted files as per the IFile. Process repeated for n splits and digest verification is done here.

4.6 **`Compile (EncryptedFilesList)`**

```
Begin
    For i=1 to n do
          Readbuffer ← Readfile(EncryptedFilesList[i])
          EncFinal.append(Readbuffer)
    End for
End
```

Compile module combines Encrypted File List and generates EncFinal.

4.7 **`Decrypt (EncFinal, IFile)`**

```
Begin
    Alg← fetchAlg(IFile)
    Key←fetchKey(IFile)
    DecFinal←DecryptFile(EncFinal,Alg,Key)
End
```

Decrypt module takes two inputs EncFinal a compiled encrypted file, IFile is information file which consist of Algorithm used for encryption and key for each cipher text. Based on Decryption algorithm chosen steps will be selected and using key given EncfFinal is decrypted and generates DecFinal.

Advantages and disadvantages
Above-mentioned model reduces the information availability to the user in the case of co-tenant covert channel attack. Confidentiality is also provided to the data uploaded to cloud using encryption function. This model brings additional overhead for the file uploading module when compared to normal file upload process.

5 Implementation and Result Analysis

The Encryption First Split Next model was implemented using programming language java. As a cloud storage service provider for storage we use Amazon S3 (Simple Storage Service). Different accounts were created in AWS public cloud for accessing services like S3 or EC2 (Elastic Compute Cloud). Even accounts were configured in different regions like US-WEST-2, US-EAST and so on. For accessing AWS S3 and EC2, a java API developed by AWS. Eclipse Mars IDE is used to execute developed program.

For checking four different upload policies Normal file upload, Split single Bucket in single region, Split Multiple Bucket in single region and Split Multiple buckets in multiple regions. Multiple buckets were created in AWS S3 in various regions. We have used two setups for testing our Encryption First Split Next model for Co-tenant covert channel protection.

Setup 1: Communication from outside cloud—In this setup, a Data owner Machine and an User Machine with configuration Intel(R) Core(TM) i5-4210U CPU @ 1.70 GHz to 2.40 GHz With 8 GB RAM. For connecting to cloud we use with internet speeds 256 Kbps to 2 Mbps.

Setup 2: Communication from inside cloud—Two EC2 instances T2. Large were setup 90 GB EBS storage space and 4 GB RAM for data owner and user computers in US-East region. All the experiments were done at various times of day in a week and results were averaged, tabulated and normalized using Min–Max Normalization technique.

5.1 Fixing Best Number of Splits for the Model

For fixing best number of splits we conducted this experiment using setup1 and setup2. We averaged the values and normalized using Min-Max normalization.

We applied normalization because values for split size and missing data (reduced attack surface) were too high when compare to split time upload times. So for getting a perfect relation between these factors we normalized values. At best number of split we observe characteristics such as high reduced attack surface, low split time, and low total upload time. Graph in Fig. 3 depicts four curves representing normalized split size, normalized split time, normalized missing data (reduced attack surface), normalized total upload time. We have take normalized values minimum as 1 and maximum as 10. On the Y axis we marked number of split size starting with 1–100.

In the region of 8–15, we observe characteristics of curve meet best number of split criteria. So we can take number of splits in regions of 8–15. In the remaining of this section we use number of splits as 10 where ever we apply splitting.

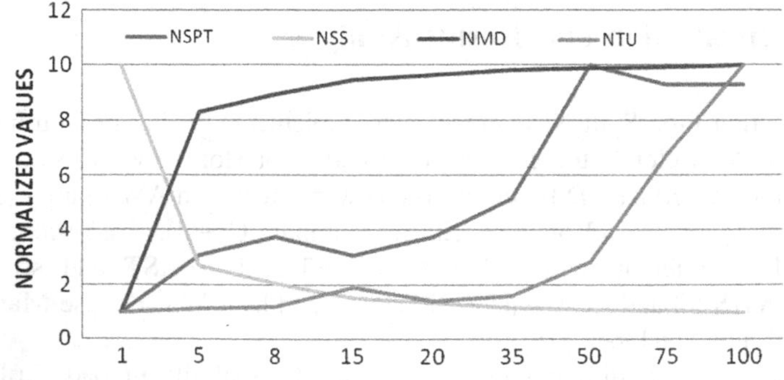

Fig. 3 Normalized values of split time, split size, missing data, total upload time change with no of split

5.2 *Performance Variation in Total Upload Time Using Three Models*

From Fig. 4 we can conclude that among three models mentioned above EFSN is having slightly more total upload time when compared with split share uploading and encrypt file uploading. Difference in time is more in small files when compared with large files.

5.3 *Performance Evaluation of Encryption First Split Next Model Using Split Multiple Bucket Upload Policy*

This experiment is done using setup2 as mentioned in the implementation in this section above. We use Amazon S3 storage service provider for storing our uploaded file in cloud.

An account is configured in US-EAST region; 10 buckets were created to test our model. As it was observed that at number of splits ranging from 8 to 13 we get best results for this we setup no of splits as 10. Setup 2 is used for reducing network latencies. We use DES/CBC/PKCS5Padding encryption algorithm. We use five different keys for encrypting five files. Figure 5 depicts three curves total

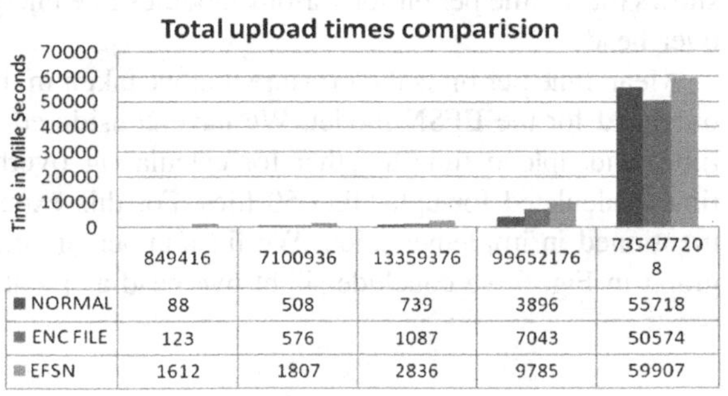

Fig. 4 Total upload times comparison considering normal file upload, encrypted file upload, encryption first split next models for various file sizes

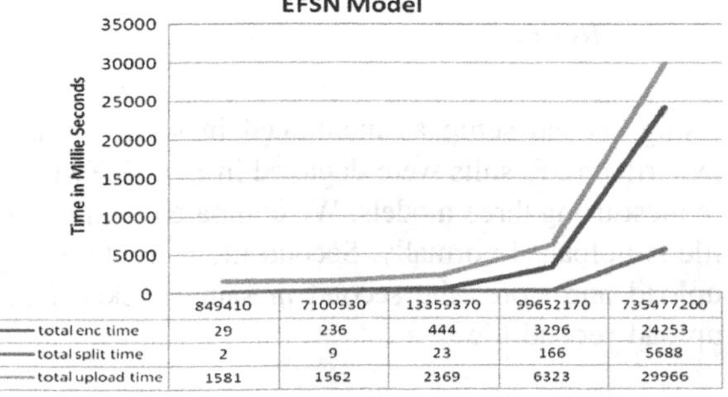

Fig. 5 Encryption time, split, upload comparison of various file sizes in the efsn model using split multiple bucket upload policy

Table 1 Mean time per bit bench mark for EFSN model using multiple bucket upload policy in sequential mode

Module	Encryption	Split	Upload	Overhead
Mean time per bit in milliseconds	3.33325E-05	2.94864E-06	0.000472557	3.62811E-05

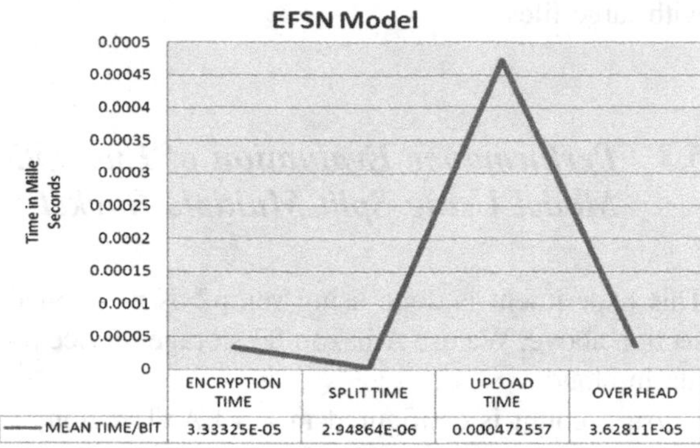

Fig. 6 Mean time per bit in milliseconds in EFSN model using split multiple bucket upload policy

encryption time, total split time, total upload time. From the graph it was observed that all times are proportional to file sizes.

But we can observe average times with very less standard deviation. Table 1 shows mean time per bit for various modules like encryption, split, upload, and total over head.

Mean time per bit is the measure that we taken in this experiment to calculate the overhead for the EFSN model. We have considered sum of encryption time, split time, and upload time together for calculating overhead. These values are mean times calculated for uploading 50 files. For this experiment also we use setup2 as mentioned in implementation. We fix number of splits at 10. After observing the graph in Fig. 6 we conclude slight overhead as mentioned in Table 1 is observed.

5.4 *Hard Disk Contention Checking Using Correlation of Download Times of Files Stored in Co-Located Bucket*

Using second setup as mentioned in section implementation we conducted this experiment. Results were depicted in Fig. 7. Two users concurrently put download request using three models. We compare downloading times of two files where first file is uploaded normally. Second file using three models. First model uses normal upload and stores the second in same bucket. Second uses split share model to upload second file.

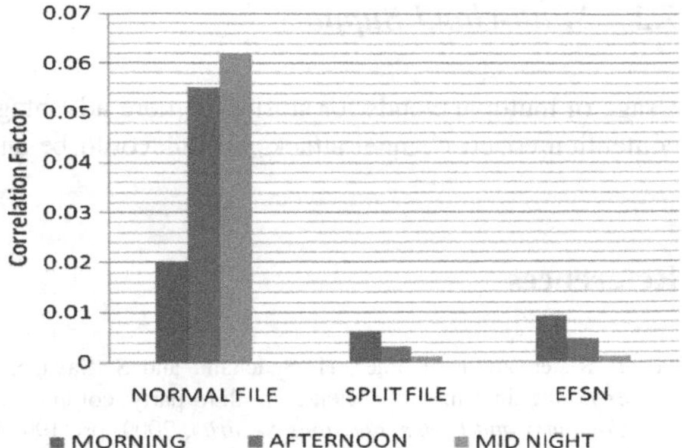

Fig. 7 Correlation factor comparison for normal file downloading and split share without encryption download, EFSN download in co-located bucket

Third uses EFSN model for uploading second file. We enumerate download times of co-located files in bucket. These were observed first model is having high correlation compare to split share model and EFSN model. EFSN model is having slightly more download time than split share model so correlation factor has increased more than split share model.

6 Conclusion

This paper explained Encryption First Split Next model architecture. Modular description and performance analysis is given abstractly. Experiments for picking best splitting criteria and overhead calculation of overhead and future directions were given in detail. Error detection was also given priority and model gave satisfactory performance in the multi bucket file transfer upload policy at number of splits 10.

7 Future Scope

7.1 Flexible Access Control

For storing and retrieving data in cloud using Encryption First Split Next model there is a significant computational overhead observed for larger file sizes. Even for sharing the content to his users data owner must be online for authenticating or he has to depend over third party. Future research in this direction may reduce burden over data owner as desired.

7.2 Multi-cloud Support

Usage of multiple clouds for storing is more advantageous for protecting hard disk contention covert channel attack. So this could be another direction of research.

References

1. T. Ristenpart, E. Tromer, H. Shacham, and S. Savage, "Hey, you, get off of my cloud: exploring information leakage in third-party compute clouds," in *ACM Conference on Computer and Communications Security*, 2009, pp. 199–212.
2. Y. Xu, M. Bailey, F. Jahanian, K. R. Joshi, M. A. Hiltunen, and R. D. Schlichting, "An exploration of l2 cache covert channels in virtualized environments," in *CCSW*, 2011, pp. 29–40.
3. J. C. Wray, "An analysis of covert timing channels," in *IEEE Symposium on Security and Privacy*, 1991, pp. 2–7.
4. D. F. C. Brewer and M. J. Nash, "The chinese wall security policy," in *IEEE Symposium on Security and Privacy*, 1989, pp. 206–214.
5. Amazon.com, "Amazon web services (aws)," Online at http://aws.amazon.com/, 2009.
6. Sun Microsystems, Inc., "Building customer trust in cloud computing with transparent security," Online at https://www.sun.com/offers/details/suntransparency.xml, November 2009.
7. M. Arrington, "Gmail disaster: Reports of mass email deletions," Online at http://www.techcrunch.com/2006/12/28/gmail-disasterreports-of-mass-email-deletions/, December 2006.
8. B. Krebs, "Payment Processor Breach May Be Largest Ever," Online at http://voices.washingtonpost.com/securityfix/2009/01/ payment processor breach may b.html, Jan. 2009.
9. G. Ateniese, R. Burns, R. Curtmola, J. Herring, L. Kissner, Z. Peterson, and D. Song, "Provable data possession at untrusted stores," in *Proc. of CCS'07*, Alexandria, VA, October 2007, pp. 598–609.
10. M. A. Shah, R. Swaminathan, and M. Baker, "Privacy-preserving audit and extraction of digital contents," Cryptology ePrint Archive, Report 2008/186, 2008, http://eprint.iacr.org/.
11. Dan@AWS, "Best Practices for Using Amazon S3," 2009. [Online]. Available: http://aws.amazon.com/articles/1904.
12. Amazon Web Services. [Online]. Available: aws.amazon.com.
13. S. De Capitani di Vimercati, S. Foresti, S. Jajodia, S. Paraboschi, and P. Samarati, "Encryption policies for regulating access to outsourced data," *ACM Trans. Database Syst.*, vol. 35, no. 2, 2010.
14. S. Yu, C. Wang, K. Ren, and W. Lou, "Achieving secure, scalable, and fine-grained data access control in cloud computing" in *INFOCOM*, 2010, pp. 534–542.
15. T. Tsai, Y. Chen, H. Huang, P. Huang, and K. Chou, "A practical Chinese wall security model in cloud computing," in *APNOMS*, 2011, pp. 1–4.
16. M. Godfrey and M. Zulkernine, "A Server-Side Solution to Cache-Based Side-Channel Attacks in the Cloud," Proc. Of 6th IEEE International Conference on Cloud Computing, 2013, pp. 163–170.
17. S. Yu, X. Gui, J. Lin, X. Zhang, and J. Wang, "Detecting vms Co-residency in the Cloud: Using Cache-based Side Channel Attacks," Elektronika Ir Elektrotechnika, 19(5), 2013, pp. 73– 78.

18. F. Liu, L. Ren, and H. Bai, "Mitigating Cross-VM Side Channel Attack on Multiple Tenants Cloud Platform," Journal of Computers, 9(4), 2014, pp. 1005–1013.
19. J. Wu, L. Ding, Y. Lin, N. Min-Allah, and Y. Wang, "Xenpump: A New Method to Mitigate Timing Channel in Cloud Computing," Proc. Of 5th IEEE International Conference On Cloud Computing, 2012, pp. 678–685.

Cognitive Radio: A Technological Review on Technologies, Spectrum Awareness, Channel Awareness, and Challenges

Gourav Misra, Arun Agarwal, Sourav Misra and Kabita Agarwal

Abstract Radio spectrum is a limited resource in this world of broadband connections. With the increase in demand of digital wireless communications in commercial and personal networks judicious use of frequency spectrum has become a serious concern for wireless engineers. Cognitive adios (CR) are a solution to this clustered spectrum issue nowadays. Cognitive radio or SDR is a new concept/ technology which is being implemented to maximize the utilization of radio spectrum. At any time, unused frequency spectrum can be sensed by cognitive radio from the wide range of wireless radio spectrum. So, as a result this gives really an efficient utilization of radio resources. Major factors that govern cognitive radio network (CRN) are related to service contradiction, routing, and jamming attacks which may interfere with other operating wireless technologies. Thrust areas in developing a good CR lies in the efficient algorithm design for spectrum sensing, spectrum sharing, and radio re-configurability. This paper presents an overview of required technologies, spectrum and channel awareness, and challenges of cognitive radio.

Keywords Wireless communication · Spectrum utilization · SDR · Cognitive radio · Jamming · Spectrum sensing · Spectrum and channel awareness

---

G. Misra (✉)
Wipro Limited, Testing Services, Bangalore 560035, India
e-mail: gourav.misra.ima@gmail.com

G. Misra · A. Agarwal
Department of Electronics and Communication Engineering,
Siksha 'O' Anusandhan University, Bhubaneswar 751030, Odisha, India
e-mail: arunagrawal@soauniversity.ac.in

S. Misra
Department of Electronics and Communication Engineering,
National Institute of Science and Technology, Berhampur 761008, Odisha, India

K. Agarwal
Department of Electronics and Telecommunication Engineering,
CV Raman College of Engineering, Bhubaneswar 752054, Odisha, India
e-mail: akkavita22@gmail.com

H.S. Behera and D.P. Mohapatra (eds.), *Computational Intelligence in Data Mining*, Advances in Intelligent Systems and Computing 556,
DOI 10.1007/978-981-10-3874-7_28

1 Introduction

Cognitive radio networks are the combination of many personal digital assistants (PDAs), cell phones, and many other smart devices that basically we are using today. We know that today the wireless connection and Internet have made the connecting world easier. Recently world has experienced a rapid growth in wireless smart device such as laptops, notebooks, cellular phones, smart phones, and tablets that has tremendously increased the usage of frequency spectrum for transfer of data [1]. Future is not far when the networks are overcrowded due the enhancement in consumer electronics and wireless data transfer. Basically, cognitive radio is a smart radio system in which the transmitter and receiver (Transceiver) will intelligently find out which communication channel are in use or which are vacant or not in use and instantly will move connections to the vacant channels while avoiding the occupied ones. There are three main categories of frequency spectrum like licensed, lightly licensed, and unlicensed.

In addition to spectrum overcrowding, network security has become one of the major issues for wireless communication network. The 802.11 standard (Wi-Fi) was introduced in the year 1999. It was quickly realized that usage of electromagnetic waves, as the propagation or broadcast medium, has become a big threat for the physical security of transmitted data, because the electromagnetic waves can be interrupted with bits. This interference can lead the leak of private information, inability to transmit and receive information [1].

Like attacks that affect the performance of other wireless cellular communication systems, the cognitive radio too is affected by the same. The cognitive radio network (CRN) is also sensitive to attacks similar for WSNs because of its ability of self-organization of a network and also the establishment/installation of routing [1].

In [2], the author has reported on the extensive analysis of the various techniques/methodologies. In [2], the author has also proposed an analytical review on the strategies for routing and security issues for CRNs as well as will place a strong footing for someone to further explore into any particular aspect in greater depth. In [1], the authors have proposed a layered (various layers such as: Physical layer, MAC layer, Network layer, Transport layer, Application layer, and Cross layer) approach to cognitive radio security. In [3], the authors have introduced an efficient secret sharing mechanism based on two new models Time-Reversed Message Extraction and Key Scheduling (TREKS).

After the introduction, the paper gives basic information regarding the paradigm of cognitive radio network with a brief description on security. In the upcoming sub-sections, we have presented about basic CRN fundamentals, technologies, spectrum and channel awareness, and some of the associated technical challenges.

2 Cognitive Radio

Cognitive radio is an intelligent radio that it can be programmed and installed effectively. It can be modeled on a software-defined ratio (SDR) with some malleable operational parameters or frameworks. The basic role of the software is to establish an efficient wireless communication link. In addition, it also permits the radio to be tuned to other frequency levels, different kind of power levels, and also different advanced modulation schemes. The hardware part consists of a radio frequency conversion module, a smart antenna, a modem (modulator and demodulator), and other different advance sections. A good technical design for the radio can be performed with optimization of an objective function that includes different factors such as noise, demand of traffic, and mobility levels as well as different locations [1].

In [4], the authors have briefly described about the Cognitive Cycle. Cognitive radio (CR) is also suitable for changing situations due to its capability to operate successfully in cooperative and uncooperative networks. Since the cooperative radios can share the hopping frequencies hence, the output of the cooperative network would definitely be higher than that of the unwilling radio network. However, in different unexpectable environmental situations installation of uncooperative network may be optimal. Hence CR should be aware of threats and adaptable mitigation techniques for both scenarios [1].

3 Cognitive Radio—The Technologies Required

Today we have many implemented reliable cellular networks, though every time new network applications are driving the industry. Cognitive radio (CR) has been considered as next major wireless technology applications.

3.1 Radio Flexibility and Capability

Cognitive radio based on SDR platform is a practical approach because this needs little modification on the hardware of the existing system. In earlier days, radios were designed in fixed points. Hence to increase the flexibility, we need to increase the capability and have also added some specific software to the system design. As per Federal Communications Commission definition for smart software defined radios says: "A communications device whose attributes and capabilities are developed and/or implemented in software".

3.2 Available Technologies for Cognitive Radio

Nowadays SDR platform are available and is the propelling force behind various kinds of developments in cognitive radio technology. The fundamental protocols of a software-defined ratio necessitate to implement a practical cognitive radio are better access to advanced computing resources, controllability of parameters which are operated by systems and usable software development [5]. There are some of the most important technologies (Such as: Geolocation, Spectrum Awareness/ Frequency occupancy, Biometrics, time, Spatial Awareness, Software Technology, and Spectrum Awareness, etc.) available for the cognitive radio system [5, 6].

Geolocation is one of the most important cognitive radio (CR) enabling technologies whereby applications having very large bandwidth resulting from a radio are aware of its current geographical location. GPS system uses the time difference of arrival (TDoA) to geolocate/allocate a receiver. The aspiration of GPS is approximately 100 m. Basically, the GPS receiver uses a one-pulse-per-second signal and these signals are resulting in a high-resolution estimate of broadcast delay from each satellite nevertheless of position.

A simple sensor is nothing but a spectrum analyzer but there are some kind of differences in quality and speed. The quality of a sensor should be considered by the cognitive radio application while setting up all the parameters and also to detect the existence of signal above the noise level a band of protocols and processes have been absorbed. Threshold energy is one of the critical parameters at which occupancy is declared. The energy detected is a proportional to instantaneous bandwidth, instantaneous power and duty cycle.

The identification of a user can be recognize by a cognitive radio by enabling one or more than one biometric sensors and the most important fact is unauthorized users can be prevented from using the cognitive radio. Most of the radios are having sensors that might be used in a biometric application. It has been observed that voice print correlation is one of the expansions of software-defined radio. There are some necessary requirements for voice quality and better signal processing capacity to be performed by the radio system [5].

The information regarding time and day is also known to a desktop computer that we can definitely say and also the utilization of information in a proper way is also known to the desktops. But on the other hand, a radio is completely unaware about the time and it is really a handicap of learning regarding interaction and behavior. So it is really an important fact for the cognitive radios to know about date, time, and schedules. Time-of-day information might through the protocols in or out [7]. Cognitive radio has some notable role play in terms of personal assistant. Cognitive radio facilitates the communication over wireless links and this is really a key solutions and the opposite role is to prevent the communication at an appropriate time [7].

In order to develop cognitive radio software solution is one of the key components. Here we have just mentioned the enabled technologies available for cognitive radio and some of the key technologies such as: advanced signal processing techniques, neural network based artificial intelligence (AI), protocols of network, and the Joint Tactical Radio System (JTRS).

4 Spectrum Awareness

The important requirement of a cognitive radio is that, it is having the capability to sense and develop spectrum awareness of its own working environment. Where environment refers to the spectrum in which a radio, or network operates.

4.1 Calculating and Reporting Spectrum Usage

The effortless execution of the architectural design of cognitive radio will be based on protocols which are already installed for the upper layers. Basically via spectrum usage reporting and distribution each radio would communicate the use of spectrum. The infrastructure of local spectrum management has to be provided by the transceiver. Receiver announcing will be enabling the sharing scheme and will make it more aggressive because the location has to be protected by the system where there is a possibility of occurrence of interference and noise. The alternate way is that with computing of individual radios, the control could be more peer to peer.

4.2 Spectrum Sensing

Namely only two latest methodologies are used for spectrum sensing: (1) Basically spectrum analyzers are adjusted through a band of frequency. The spectrum analyzers with the scrutiny frequency tuned to dispense enough "dwell" on each and everything frequency window, but the thing is screening via all of the channels speedily. (2) Presently, the use of FFT has been more practical because of analog-to-digital conversion (ADC). The real advantage of this sensing technology is that it gives prompt scrutiny of all the frequencies present in the FFT window. So depending on the application and cost strategy both technologies have advantages. Table 1 illustrates a brief comparison of Spectrum sensing approaches.

4.3 Potential Interference Analysis

There are two ways which can cause potential noise interference. One is direct interference from the primary user communications. So this results in devalued functionality of wireless communication. The other is interference to the channel

Table 1 Spectrum sensing approaches—a comparison

Technology	Spectrum analyzer	Wideband FFT
Complexity	Lower complexity	Needs wideband ADC
Short signal detection	Due to low dwell time cannot sense short pulse signal	Higher probability of detecting short signals
Bandwidth	Capable to scan large ranges of signal	Instantaneous bandwidth and ADC cause limitations
Speed	Slower	Better

Table 2 Effect of interference on digital processing of layers

Layer	Effect	Migration
PHY	Higher uncorrected BER	Adjust coding dynamically
MAC	Complete packet blockage	Acknowledgment retransmission
NETWORK	Complete packet blockage	Protocol retrieves missing data
TRANSPORT	Failed packet delivery	Recognition of missing sequence and retransmission

when it is not in use. So, as a result that the primary user will think about the mistaken with the channel. Quality of the signal is getting affected by direct interference at one or more layers of the receiver, as presented in Table 2.

We define interference as the ratio of signal energy to the signal plus interference. The operating community has substantial fillings from users of frequency spectrum describing either amplified additional energy, or token energy, often in the same kind of situation. Basically unprotected receivers are having a spilling effect and very small errors in the lower layers can have inordinate collision on the upper layers [8].

4.4 Link Engagement

It is really difficult to search out the different radios with which CR is deliberated to communicate when an adaptive radio [9] is first turned on. Basically normal radios are having at least primary knowledge of frequency. In the case of a cognitive radio, it is difficult to know how the sampling of frequency is done with the environment.

5 Multiple Signals in Space and Channel Awareness

There are some indirect assumptions of most spectrum-sharing approach that one spectral efficient signal can use only a given frequency at a time. The fundamental concept of MIMO signaling has been eventually generalized. In MIMO, separable orthogonalized channels have been created by multiple signal paths between multiple transmitting and receiving antennas. Actually, this method does not follow the Shannon's bound because each of the reflected paths has an independent channel. Necessarily, multipath has been turned into multiple independent channels by this technique.

Theoretically, in MIMO technique, if the number of antenna is increased, then it results in a linear growth in output. In addition to this, the channels between the elements might be less than ideally separated.

6 Challenges in Cognitive Radio

The cognitive radio (CR) lacks awareness of perception which basically prevents to detect the environment. Problem arises when the sensing secondary user (SU) is completely not aware of the existence of the primary user (PU) and cannot perceive its existence. Some physical obstacles can make a separation between a PU terminal and an SU terminal to radio signals.

A digital TV can be taken as an example of hidden terminal problem. At the edge of cell, digital TV reclines because the signal power of the received signal is barely above receiver sensitivity. Spectrum can be started to use by the cognitive radio if it does not have the capability to detect TV signal. So as a result restrain with the signal, the digital TV is seeking to unravel and we can avoid this problem by huge gap if the sensibility of the cognitive radio outperforms primary user receiver. WLAN systems also have hidden terminal problem. Simply determining whether a band is free or busy will not serve the purpose. Instead, the amount of interference and noise has to be estimated in the free sub-band by the cognitive radio and also we need to ensure that CR's transmission power is not breaking the threshold of interference. The complications of the cognitive radio are really an important feature. Further, the CRN should be efficient of scanning over very high bandwidths [7].

7 Pros and Cons and Present Research

The main advantage of cognitive radio is that it gives a promising solution to spectrum congestion and efficient utilization of licensed and unlicensed band. Since CR has an ability to track the radio frequency (RF) along with adaptable to the changes in the environment by changing its SDR configurations makes it suitable for many important applications.

Two main characteristics for CR are cognitive capability and re-configurability. These characteristics govern the challenges and current research directions. In physical layer, an accurate detection of weak signals of licensed users is needed over a wide range of spectrum. In addition, the design and implementation of RF wideband from end and A/D converter with multi-GHz speed still remains a challenging task.

Research topics in cognitive radio includes efficient Spectrum Sensing algorithms, multi-parameter based Spectrum Decision model, and Spectrum Mobility in time, space, spectrum hand-off, Cognitive radio architecture and associated software, protocol architectures for CRNs, resource management, and finally Network security for CRNs.

8 Conclusion

Free space is one of the primary media to connect the world electronically and currently frequency spectrum has been overcrowded nowadays for increase in demand of service by consumers. In this situation, the cognitive radio networks with software-defined capabilities emerge as a key solution to provide more spectrum frequencies. The new technology provides new ways for different new strikes accomplished by spiteful users with the desire communication. Now basically the research focuses on the security of cognitive radio including jamming of the channel and spectrum. Future research needs to be directed in the area to protect the specific function of cognitive radio from attacks like: viruses, worms, and some of the new strikes which really affect the learning ability of cognitive radio [1]. With the maturity of the cognitive radio network and closer to fulfillment, the network security will become utmost important. A proper structure can be designed into the cognitive radio system and as a result it needs an extensive and enhanced research to protect the network. In this paper, we have also presented some of the basic and fundamental required technologies, awareness of channel and frequency spectrum as well as the creation of spectrum awareness.

References

1. Hlavacek, Deanna, and J. Morris Chang. "A layered approach to cognitive radio network security: A survey", Computer Networks (2014).
2. Natarajan Meghanathan, A survey on the communication protocols and security in cognitive radio networks, Int. J. Commun. Netw. Inform. Secur. (IJCNIS) 5 (1) (2013).
3. Aldo Cassola, Tao Jin, Guevara Noubir, Bishal Thapa, Efficient spread spectrum communication without pre-shared secrets, Mobile Comput. IEEE Trans. 12 (8) (2013) 1669–1680.
4. Agarwal, Arun, Gourav Misra, and Kabita Agarwal. "The 5th Generation Mobile Wireless Networks- Key Concepts, Network Architecture and Challenges." American Journal of Electrical and Electronic Engineering 3.2 (2015): 22–28.
5. Technologies Required", Cognitive Radio Technology (2009).
6. Zhou, G.; Stankovic, J.A.; Son, S. "Crowded Spectrum in Wireless Sensor Networks." In Proceedings of the Third Workshop on Embedded Networked Sensors (EmNets 2006), Cambridge, MA, USA, 30–31 (2006).

7. J POLSON. "Cognitive Radio The Technologies Required", Cognitive Radio Technology (2006).
8. P MARSHALL. "Spectrum Awareness", Cognitive Radio Technology (2006).
9. Joseph Mitola III, Gerald Q. Maguire Jr., Cognitive radio: making software radios more personal, Person. Commun. IEEE 6 (4) (1999) 13–18.
10. Federal Communications Commission, Notice of Proposed Rule Making, August 12 (2000).
11. P.H. Dana, The Geographer's Craft Project, Department of Geography, University of Colorado, Boulder (1996).
12. Federal Communications Commission. Spectrum Policy Task Force Report. ET Docket No. 02-135 (2002).
13. A. Gershman, Space-Time Processing for MIMO Communications, Wiley (2005).
14. Agarwal, Arun, and Kabita Agarwal. "Performance evaluation of OFDM based WiMAX (IEEE 802.16d) system under diverse channel conditions", 2015 International Conference on Electrical Electronics Signals Communication and Optimization (EESCO) (2015).
15. Misra, Gourav, Agarwal, Arun, and Kabita Agarwal "Internet of Things (IoT) – A Technological Analysis and Survey on Vision, Concepts, Challenges, Innovation Directions, Technologies, and Applications (An Upcoming or Future Generation Computer Communication System Technology)." American Journal of Electrical and Electronic Engineering 4.1 (2016): 23–32.
16. M. Minsky, "A Framework for Representing Knowledge," Massachusetts Institute of Technology, Technical Report, UMI Order Number. AIM-306 (1974).
17. W. Krenik and A. Batra, "Cognitive radio techniques for wide area networks," in Proceedings of ACM IEEE Design Automation Conference 2005, pp. 409–412 (2005).
18. D. Čabrić and R. W. Brodersen, "Physical layer design issues unique to cognitive radio systems," in Proceedings of PIMRC (2005).
19. Chen, Sao-Jie, Pao-Ann Hsiung, Chu Yu, Mao-Hsu Yen, Sakir Sezer, Michael Schulte, and Yu-Hen Hu. "ARAL-CR: "An adaptive reasoning and learning cognitive radio platform", 2010 International Conference on Embedded Computer Systems Architectures Modeling and Simulation (2010).
20. Joseph Mitola III, Gerald Q. Maguire Jr., "Cognitive radio: making software radios more personal", Person. Commun. IEEE 6 (4) (1999) 13–18.

Chessography: A Cryptosystem Based on the Game of Chess

Vaishnavi Ketan Kamat

Abstract Chess is one of the most strategic games played around the globe. It provides a player the platform to explore the exponential complexity of the game through their intellectual moves in the game. Though, the number of players, pieces, and board dimension and squares are finite, it is a game where the combination of moves for each player will change distinctly from game to game. Now, moving onto the field of Cryptography, which aims to achieve security for the data by encrypting it, and makes the plain text unintelligible for the intruders. Taking advantage of the numerous moves in the game of Chess, we can form an amalgamation of Chess game and cryptography. Chessography is the confluence of the game of Chess and Cryptography. Encryption will take place in the form of moves played by each player on the board. Resulting cipher text produced will be a Product Cipher, where the square positions on the game board will be substituted by new square positions during the move in the game and all the actual values dealt with, throughout the game will form the transposed cipher text. Chessography algorithm uses two keys, one fixed length and one variable length key called as 'paired key'. Along with the strength of keys generated, this algorithm generates a very strong cipher text as each game of chess is different, so is the cipher text generated. The complexity in the game of chess forms the Chessography's defensive strength to deal against various attacks during the data transmission.

Keywords Chess · Chessography · Complexity · Cryptography · Product cipher · Encryption · Paired key

V.K. Kamat (✉)
Department of Computer Engineering, Agnel Institute of Technology and Design, Assagao, Bardez, Goa, India
e-mail: vaishnavikunkolikerkamat@gmail.com

H.S. Behera and D.P. Mohapatra (eds.), *Computational Intelligence in Data Mining*, Advances in Intelligent Systems and Computing 556,
DOI 10.1007/978-981-10-3874-7_29

1 Introduction

The "Wisdom House", as, the game of chess can be best described, is one of the world's oldest games played. "The traces of its origination can be found in Ancient India around Sixth Century A.D., where it was played as 'Chaturanga'". Chaturanga is translated as "four divisions" and refers to the four parts of ancient armies. These four parts were the infantry, cavalry, elephantry, and chariotry and these pieces further evolved into pawn, knight, bishop and rook along with the King and Queen [1]. In the modern day chess game, though there are only 32 pieces on the board to play with, the combination of each of these can generate exhaustive combinations of moves. Exploiting, this vastness of the game, a cryptographic algorithm called "Chessography" is developed. An attempt is being made in this paper to integrate cryptographic techniques with the game of chess. "Cryptography is the science of creating ciphers" [2]. In this paper, the cipher text and the keys generated depends upon the number of moves in one game of chess. Cipher text produced in this algorithm, will be difficult to interpret and break, because each opponent constantly keeps on changing, its pieces at different square positions on the board.

2 Game of Chess and Cryptography

2.1 Game of Chess

Chess being one of the ancient games in the world is the source of power bank to boost one's ability to think, understand, and act in any given situation in real life. The game of chess is played between two opponents on the 'chessboard' which is an 8 × 8 grid. The player with the White pieces makes the first move, then the players move alternately, with the player with the Black pieces making the next move. The objective of each player is to place the opponent's king 'under attack' in such a way that the opponent has no legal move. The player who achieves this goal is said to have 'checkmated' the opponent's king and to have won the game. (Articles 1.1 and 1.2) [3]. The eight vertical columns of squares are called 'files'. The eight horizontal rows of squares are called 'ranks'. A straight line of squares of the same color, running from one edge of the board to an adjacent edge, is called a 'diagonal' (Article 2.4) [3]. Placement of pieces is shown in Fig. 1.

Each player has 16 pieces at his disposal. King—01 Piece, Queen—01 Piece, Bishop—02 Pieces, Rook—02 Pieces, Knight—02 Pieces, Pawns—08 Pieces.

Fig. 1 Placement of pieces on the chess board

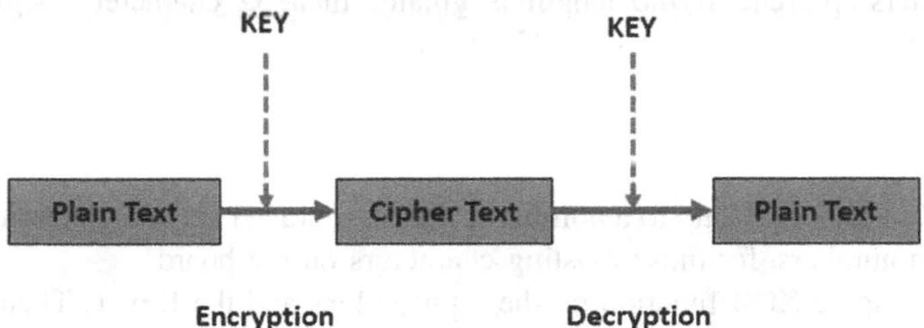

Fig. 2 Process of encryption and decryption

2.2 *Cryptography*

Cryptography deals with encryption and decryption of data during transmission. The input message, i.e., the 'plain text' is passed onto the cryptographic algorithm and the resultant is called as 'cipher text'. Whatever is encrypted on sender's side to become decrypted on the receiver's side, a Key is used by both sender and receiver. Block diagram for process of encryption and decryption is illustrated in Fig. 2.

3 Chessography Algorithm

The formation of cipher text using Chessography algorithm is purely based on the moves played in the chess game. It is a combination of substitution and transposition techniques, but with a twist. When the game is started, every data element is associated with a position number, so whenever a move is made, the data element remains same (unless it is replaced) but old square position is substituted with new square position. For forming the cipher text, all the moves from both the players are alternately combined, i.e., text is transposed. Here, lies the key point, i.e., this transposition is not fixed based on a particular pattern. The length of the cipher text depends upon the number of moves played. So, another level of cipher text is generated which is in a compressed form. Two keys are used for encryption as well as decryption to generate the block cipher. Key 1 is of fixed length whereas Key 2 is of variable length. The uniqueness of Key 2 is that 'each key value is a pair of the square positions'.

3.1 Chessography Algorithm: Encryption Process

Step 1: Input the plain text, convert it to lowercase and calculate its length in terms of characters' present. If the length is greater than 32 characters, segment the message where each segment is of 32 characters.
Step 2: Create an 8 × 8 matrix of 64 squares for one segment of 32 characters. Place each character sequentially, in the matrix, as the pieces are placed on the chess board.
Step 3: Map each character to a number from the character set and generate a Key 1 of random numbers, for those existing characters on the board.
Step 4: Compute XOR function on the input values and the Key 1. Then perform 'mod 71' on the result value. These resultant values will be used as the piece values on the board.
Step 5: Declare one set of 16 characters in 8 × 8 matrix as White pieces and other set as Black pieces. Every character value in the matrix, has a position number associated with it. Start with the game of chess from characters marked as white pieces first till the game comes to an end. Record the changes as the game progresses.
Step 6: Record the value associated at that position and cell positions separately and alternately combine the values and positions of pieces from each of the side of the board to form the cipher text and the paired key.
Step 7: To compress the cipher text, make use of the paired key generated. Consider each pair of key (k_n, k_m). Start with the last pair of key and traverse the list upwards, finding matching link for current k_n in place of k_{m-1} and so on, till the corresponding value with pair of keys does not change. This will result in a compressed cipher text generation.
Step 8: Append the remaining element values on the board to the compressed cipher text and form a key pair in the paired key 2. Also look out for special case values to be appended at the end of compressed cipher text, for which pair key is not generated.

3.2 Chessography Algorithm: Decryption Process

Step 1: Accept the incoming compressed cipher text. Consider the first character of the cipher text, map it to the last paired key value from Key 2. If, both the values in the pair key are same, then do not traverse, the present value is considered to be the final value for that element. Else, traverse the paired key list upwards till last.
Step 2: Fill up the board in the same manner for all 32 pieces. Now, there might be some square cells left unfilled. These are treated as special cases. These special cases are applicable to square cell numbers from 1 to 16 and 49 to 64. A particular element value will be considered as a special case if and only if, it has not made any move with the initial value it was assigned at the initial position. Look up in the paired key for such special square cell positions.
Step 3: Once the values are placed in their respective initial positions, use Key1 and perform XOR operation, to get back the plain text message.

3.3 Generation, Compression, and Decryption of Cipher Text

The compression of cipher text is essential after generation, is that, since cipher text generation is based on the number of moves in a single game of chess. Every game will have different number of moves. So, for encrypting 32 characters from plaintext at a time, we will end up generating cipher text which is double or multiple of the input plain text length. Hence, compression is required. In this process of compression, another level of encryption takes place, which will be transmitted to the receiver from the sender. The simplicity of decryption of the cipher text is due to the paired key. Without the paired key, it is difficult to decode the cipher text. The manner in which the different element values are traced back is much faster than the process of encryption.

3.4 Paired Key Generation

Procedure for generating the paired key in the Chessography Algorithm is a unique step in the key generation. It is called as a 'paired key' because a single key value is given in terms of pairs of position, i.e., from where the piece moved to where it is placed in a single move. Though the pair keys will involve, number only between 1 to 64, the specialty of this pair key is that, without performing rigorous permutation operations on the key, the combination of each pair of key will be different, and the same combination of pairs can be repeated for different pieces on the chess board. This makes the paired key knowledge tricky, i.e., one cannot guess just by inspecting any paired key value, to which position it may belong as it may belong to more than one piece at different times. Complexity of the paired key is discussed in Sect. 5.

Consider for example, in Fig. 18, **(50,34)** is the first pair key value which is generated from the first move of Opponent 1, i.e., the chess piece has been moved from cell number 50 to cell number 34. The next pair key value is **(13,21)**, which is generated by first move of the Opponent 2, i.e., the chess piece has been moved from cell number 13 to cell number 21. Complete set of pair keys generated is illustrated in Fig. 18.

4 Chessography Algorithm Illustrated

<u>**Input**</u>: Plain Text: **"He is a man of courage who does not run away, but remains at his post and fights against the enemy, Quote by Socrates"**. The character set used here is made up of 26 alphabets, 35 symbols, and 10 numbers as displayed in Fig. 3.

As explained in the algorithm, we calculate the length of the plain text equal to 117, including spaces and accordingly number of segments is created. To accommodate 117, characters, minimum four segments are required. Placement of characters is illustrated in Figs. 4, 5, 6 and 7.

We consider the first 8 × 8 segment shown in Fig. 4. Replace, each character with respective, mapping number from the character set. In Fig. 8, along with each character (marked in red color) are the position number of the squares on the board (marked in black color).

So in Fig. 8, for example (8,1) indicates that the character with value 8 at position 1 will play the role of a rook. (5,2) will play the role of a Knight at position 2. Now, in this mapping, there will be a little confusion, which will be adding benefit to the algorithm. (8,1) is a rook on opponent 1 side, as well as opponent player's rook also has a value of 8 at position 64. There are two points to be noted

a	1	k	11	u	21	(	31	'	41	&	51	0	61
b	2	l	12	v	22	)	32	"	42	*	52	1	62
c	3	m	13	w	23	=	33	"	43	+	53	2	63
d	4	n	14	x	24	<	34	`	44	-	54	3	64
e	5	o	15	y	25	>	35	!	45	/	55	4	65
f	6	p	16	z	26	'	36	@	46	\	56	5	66
g	7	q	17	{	27	.	37	#	47	|	57	6	67
h	8	r	18	}	28	:	38	$	48	?	58	7	68
i	9	s	19	[	29	'	39	%	49	_	59	8	69
j	10	t	20	]	30	;	40	^	50	~	60	9	70
													71

Fig. 3 Character set with length of 71 characters

h	e		i	s		a	
c		f	o		n	a	m
o	u	r	a	g	e		w
	s	e	o	d		o	h

Fig. 4 Segment no. 1

n	o	t		r	u	n	
u	b		,	y	a	w	a
t		r	e	m	a	i	n
s	i	h		t	a		s

Fig. 5 Segment no. 2

	p	o	s	t		a	n
s	t	h	g	i	f		d
	a	g	a	i	n	s	t
e	n	e		e	h	t	

Fig. 6 Segment no. 3

m	y	,		q	u	o	t
c	o	s		y	b		e
r	a	t	e	s	0	0	0
0	0	0	0	0	0	0	0

Fig. 7 Segment no. 4

(8,1)	(5,2)	(71,3)	(9,4)	(19,5)	(71,6)	(1,7)	(71,8)
(3,9)	(71,10)	(6,11)	(15,12)	(71,13)	(14,14)	(1,15)	(13,16)
(15,49)	(21,50)	(18,51)	(1,52)	(7,53)	(5,54)	(71,55)	(23,56)
(71,57)	(19,58)	(5,59)	(15,60)	(4,61)	(71,62)	(15,63)	(8,64)

Fig. 8 Initial placement of input message characters (input values [*red color*]) on the chess board

283	413	381	286	783	285	477	62
341	583	943	584	435	537	612	596
451	394	651	377	53	771	29	235
873	105	86	820	589	500	103	321

Fig. 9 Random Key 1

here, i.e., all four rooks can have different character value associated with them. Second point, if we see in Fig. 8, (19,5) represents king of opponent player 1 as well as (19,58) represents the knight of opponent player 2. So, wherever, in the cipher, say a particular character value appears, that reflects a different value information associated with the same piece, which will make the cipher text difficult to crack.

Next key 1 is generated on random numbers, for each character present on the board, so size of key 1 is fixed, as displayed in Fig. 9. Perform XOR operation, on the input values shown Fig. 8 and key 1. Result is shown in Fig. 10 and after performing mod operation result is shown in Fig. 11.

After 1st level of encryption, (62,1) will be playing the role of rook, (53,2) role of knight, etc. Mapping of each character value to the chess piece is displayed in Fig. 12.

Opponent 1—black-colored pieces. Opponent 2—pink-colored pieces. R- Rook, K-Knight, B-Bishop, Q-Queen, P-Pawn, K_B and K_W King from both the sides. Opponent 2, (pink-colored pieces) start with the first move and corresponding move is made by opponent 1 as illustrated in Fig. 13 and Fig. 14 first four moves.

Summary of moves during one full game of chess from both the opponents is displayed in Table 1, below. During each move played, corresponding bits are set for each square cell. Square cells from 1 to 16 and 49 to 64 are set initially to 1 as

275	408	314	279	796	346	476	121
342	512	937	583	500	535	613	601
460	415	665	376	50	774	90	252
814	122	83	827	585	435	104	329

Fig. 10 XOR of the input value with Random Key 1

(62,1)	(53,2)	(30,3)	(66,4)	(15,5)	(62,6)	(50,7)	(50,8)
(58,9)	(15,10)	(14,11)	(15,12)	(3,13)	(38,14)	(45,15)	(33,16)
(34,49)	(60,50)	(26,51)	(21,52)	(50,53)	(64,54)	(19,55)	(39,56)
(33,57)	(51,58)	(12,59)	(46,60)	(17,61)	(9,62)	(33,63)	(45,64)

Fig. 11 After taking mod 71 the input values

R_{B1}	K_{B1}	B_{B1}	Q_{B1}	K_B	B_{B2}	K_{B2}	R_{B2}
P_{b1}	P_{b2}	P_{b3}	P_{b4}	P_{b5}	P_{b6}	P_{b7}	P_{b8}
P_{w1}	P_{w2}	P_{w3}	P_{w4}	P_{w5}	P_{w6}	P_{w7}	P_{w8}
R_{W1}	K_{W1}	B_{W1}	Q_{W1}	K_W	B_{W2}	K_{W2}	R_{W2}

Fig. 12 Value—chess piece mapping

(62,1)	(53,2)	(30,3)	(66,4)	(15,5)	(62,6)	(50,7)	(50,8)
(58,9)	(15,10)	(14,11)	(15,12)		(38,14)	(45,15)	(33,16)
				(3,21)			
	(60,34)						
(34,49)		(26,51)	(21,52)	(50,53)	(64,54)	(19,55)	(39,56)
(33,57)	(51,58)	(12,59)	(46,60)	(17,61)	(9,62)	(33,63)	(45,64)

Fig. 13 Move 1, Move 2

(62,1)	(53,2)	(30,3)	(66,4)	(15,5)	(62,6)	(50,7)	(50,8)
(58,9)	(15,10)	(14,11)	(15,12)			(45,15)	(33,16)
				(3,21)			
					(38,30)		
	(60,34)						
		(26,43)					
(34,49)			(21,52)	(50,53)	(64,54)	(19,55)	(39,56)
(33,57)	(51,58)	(12,59)	(46,60)	(17,61)	(9,62)	(33,63)	(45,64)

Fig. 14 Move 3, Move 4

Table 1 Chess_Game_Summary

Move No.	White Piece				Black Piece		
	Value	From	To		Value	From	To
1	60	50	34		3	13	21
2	26	51	43		38	14	30
3	19	55	47		45	15	23
4	33	63	46		62	6	20
5	39	56	40		50	7	22
6	33	46	36		58	9	17
7	50	53	45		62	20	6
8	46	60	46		50	22	28
9	33	36	51		53	2	19
10	50	45	37		53	19	29
11	46	46	53		38	30	37
12	21	52	36		14	11	27
13	21	36	29		14	27	34
14	26	43	34		62	1	2
15	12	59	31		66	4	11
16	39	40	32		62	2	1
17	39	32	23		62	6	13
18	39	23	15		62	13	34
19	33	51	34		33	16	24
20	12	31	24		15	12	20
21	46	53	32		15	5	13
22	46	32	31		15	13	14
23	39	15	8		50	28	34
24	39	8	16		15	14	5
25	46	31	32		15	5	4
26	12	24	31		66	11	13
27	12	31	13		15	4	12
28	12	13	20		15	12	19
29	12	20	34		30	3	12
30	39	16	37		15	19	18
31	46	32	16		30	12	19
32	39	37	36		15	18	18

they are occupied, whereas rest square cells are initialized to 0. Whenever, a square cell will be occupied, bit will be set to 1.

Combination: 0,0—Invalid, **Combination: 0,1**—cell is free and after replacement will be set to 1. **Combination: 1,0**—square cell is currently occupied, but free after replacement and will be set to 0.

Combination: 1,1—square cell is previously occupied and currently replaced. **Initially**, if any square cell is initialized to 1, and next bit in line for the same cell is

(62,1)							
	(15,10)						(46,16)
(58,17)	(15,18)	(30,19)		(3,21)			
				(21,29)			
	(12,34)		(39,36)				
						(19,47)	
(34,49)					(64,54)		
(33,57)	(51,58)			(17,61)	(9,62)		(45,64)

Fig. 15 Final arrangement of pieces after all the moves on the board of 8 × 8

1, it means, that particular piece, never moved from its original position and is directly substituted by the other piece.

Final arrangement of pieces after all moves displayed in Fig. 15.

This case has to be handled carefully, as, if this information is not taken into account, decryption will not be complete. Any cell other than which are initially set to one have a combination of [1, 1] means replacement (substitution) has been performed in that cell. Consider Table 2, cell with position number 8, highlights the case where initial value of cell is 1 and consecutive value also equal to 1. Summary of the bits at every move during encryption process is shown in Table 2.

Based on the moves, paired key is simultaneously generated, illustrated in Fig. 16 and corresponding cipher text at each step is generated as displayed in Fig. 17. An important and unique highlight of this algorithm is that, the final cipher text generated is a compressed one. If the size of the text to be transferred is less, then the time taken to transmit the text data from sender to receiver over the channel also will be less. In Fig. 18, complete paired key is shown, last few cells which are not colored indicate the position of the remaining cells on the board at the end of the game. Starting from the end with (18,18) the last move in the game, we will see how compression takes place, as we trace the key.

First paired key is (18,18) with corresponding value 15 at that position. Next, trace from the next key value that led to (18,18), i.e., (19,18) which also has value 15. Next we trace which key value led 15 to this position, i.e., (12,19). So path traced using paired key for text value 15 is (18,18) → (19,18) → (12,19) . → (4,12) → (5,4) → (14,5) → (13,14)→(5,13) So, eight move values are combined to one, reducing the cipher text size. Complete compressed cipher text is shown in Fig. 19, where first 08 moves are the ones last left on the board at the end of the game. Figure 20, shows the cipher text after mapping to the character set. If we look up into Fig. 19, an extra element with value 50 is added as part of cipher text. This is to handle the special cases or cases with initial bit combination of [4]. Now at the receiver side, same paired key is traced back as one of the tracing sequences for compression are shown. This tracing operation will give the original position of this element. (18,18) → (19,18) → (12,19) → (4,12) → (5,4) → (14,5) → (13,14) → (5,13) ..

Table 2 Cell_Bits_Summary

Position No.	Bits Associated		Position No.	Bits Associated
1	1, 0, 1		36	0, 1, 0, 1, 0, 1
2	1, 0, 1, 0		37	0, 1, 1, 1, 0
3	1, 0		38	0
4	1, 0, 1, 0		39	0
5	1, 0, 1, 0		40	0, 1, 0
6	1, 0, 1, 0		41	0
7	1, 0		42	0
8	1, 1, 0		43	0, 1, 0
9	1, 0		44	0
10	1		45	0, 1, 0
11	1, 0, 1, 0		46	0, 1, 0, 1, 0
12	1, 0, 1, 0,1, 0		47	0, 1
13	1, 0, 1, 0, 1, 0, 1, 1, 0		48	0
14	1, 0, 1, 0		49	1
15	1, 0, 1, 0		50	1,0
16	1, 0, 1, 0, 1		51	1, 0, 1, 0
17	0, 1		52	1, 0
18	0, 1		53	1, 0, 1, 0
19	0, 1, 0, 1, 0, 1		54	1
20	0, 1,0, 1, 1, 0		55	1, 0
21	0, 1		56	1, 0
22	0, 1, 0		57	1
23	0, 1, 1, 0		58	1
24	0, 1, 1, 0		59	1, 0
25	0		60	1, 0
26	0		61	1
27	0, 1, 0		62	1
28	0, 1, 0		63	1, 0
29	0, 1, 1		64	1
30	0, 1, 0			
31	0, 1, 0, 1, 0, 1, 0			
32	0, 1, 0, 1, 0, 1, 0			
33	0			
34	0, 1, 1, 1, 1, 1, 1, 1			

So tracing the paired key for a particular value performs two tasks, compression during encryption and fetching the original place of that value on the board. So, based on the above given sequence, we conclude that value 15 belongs to cell with position no. 5. Now, for handling special case, checkout, which cells are left unfilled, and trace first instance of that cell position in the paired key. For example, (15,8) is the first instance of cell with position number 8, and we know cell 8, will be compulsorily occupied initially, so the element with value 50 corresponds to cell number 8. Paired key values shown in Fig. 21.

Decryption process is much faster compared to encryption because, cell tracing using paired key will give original values of most of the cells. In case if a special case appears, the paired key will be searched only for that particular cell.

Fig. 16 Paired key generated

(50,34)	(13,21)	(51,43)	(14,30)
(55,47)	(15,23)	(63,46)	(6,20)
(56,40)	(7,22)	(46,36)	(9,17)
(53,45)	(20,6)	(60,46)	(22,28)
(36,51)	(2,19)	(45,37)	(19,29)
(46,53)	(30,37)	(52,36)	(11,27)
(36,29)	(27,34)	(43,34)	(1,2)
(59,31)	(4,11)	(40,32)	(2,1)
(32,23)	(6,13)	(23,15)	(13,34)
(51,34)	(16,24)	(31,24)	(12,20)
(53,32)	(5,13)	(32,31)	(13,14)
(15,8)	(28,34)	(8,16)	(14,5)
(31,32)	(5,4)	(24,31)	(11,13)
(31,13)	(4,12)	(13,20)	(12,19)
(20,34)	(3,12)	(16,37)	(19,18)
(32,16)	(12,19)	(37,36)	(18,18)
(10,10)	(49,49)	(54,54)	(57,57)
(58,58)	(61,61)	(62,62)	(64,64)

Fig. 17 Uncompressed cipher text

60 3 26 38 19 45 33 62 39 50
33 58 50 62 46 50 33 53 50 53
46 38 21 14 21 14 26 62 12 66
39 62 39 62 39 62 33 33 12 15
46 15 46 15 39 50 39 15 46 15
12 66 12 15 12 15 12 30 39 15
46 30 39 15

5 Complexity and Analysis

The complexity of the "Chessography" algorithm can be best analyzed with the help of game theory. The strength of this algorithm is based upon the complexity of the chess game. Here the key is formed based on the cipher text generated. Since cipher text is generated in an orderly way, an intruder can try different *permutations* of moves and games to crack it. "Two types of complexity characterize board games. Computational and state space complexity [5]".

"The State space complexity is defined as the number of legal game positions reachable from the initial position of the game [6]". The number of possible chess

Fig. 18 Complete paired key value

(50 , 34)	(13 , 21)	(51 , 43)	(14 , 30)
(55 , 47)	(15 , 23)	(63 , 46)	(6 , 20)
(56 , 40)	(7 , 22)	(46 , 36)	(9 , 17)
(53 , 45)	(20 , 6)	(60 , 46)	(22 , 28)
(36, 51)	(2 , 19)	(45 , 37)	(19 , 29)
(46 , 53)	(30 , 37)	(52 , 36)	(11 , 27)
(36 , 29)	(27 , 34)	(43 , 34)	(1 , 2)
(59 , 31)	(4 , 11)	(40 , 32)	(2 , 1)
(32 , 23)	(6 , 13)	(23 , 15)	(13 , 34)
(51 , 34)	(16 , 24)	(31 , 24)	(12 , 20)
(53 , 32)	(5 , 13)	(32 , 31)	(13 , 14)
(15 , 8)	(28 , 34)	(8 , 16)	(14 , 5)
(31 , 32)	(5 , 4)	(24 , 31)	(11 , 13)
(31 , 13)	(4 , 12)	(13 , 20)	(12 , 19)
(20 , 34)	(3 , 12)	(16 , 37)	(19 , 18)
(32 , 16)	(12 , 19)	(37 , 36)	(18 , 18)
(10,10)	(49,49)	(54,54)	(57,57)
(58,58)	(61,61)	(62,62)	(64,64)

Fig. 19 Final compressed cipher text generated

```
45 9  17 51 33 64 34 15
15 39 30 46 12 66 50 15
33 33 62 62 26 14 21 38
53 50 58 45 19 3  60 50
```

Fig. 20 Cipher text to be transmitted emerging from segment 1

! i q & = 3 < o o '] @ 1 5 ^ o =
= 1 1 z n u : + ^ ? ! s c ~ ^

Fig. 21 Paired key value highlighting special case

(15 , 8)	(28 , 34)	(8 , 16)	(14 , 5)

positions after White's pieces make first move is 20, i.e., 16 pawn moves and 4 knight moves, similarly for Black pieces is also 20. Now once, both the players have finished their first move, there are 400 possible chess positions available on the board for the next move.

Claude Shannon in his paper titled "Programming a computer for playing chess" in 1950, has given the number of board positions, i.e., state space complexity as $[(64!)/((32!)\ (8!)^2\ (2!)^6)]$, which is roughly estimated as $\mathbf{10^{43}}$ and game tree complexity to be $\mathbf{10^{120}}$ [4]. Victor Allis calculated an upper bound of $\mathbf{5 \times 10^{52}}$ for the number of positions in his Ph.D. thesis in 1994.

As all the positions given with this estimate of $\mathbf{5 \times 10^{52}}$ will not be legal, therefore in the thesis it is assumed that the true state space complexity to be close to $\mathbf{10^{50}}$ and also the game tree complexity to be $\mathbf{10^{123}}$ [7]. The numbers indicating complexity of the chess are huge, i.e., if any intruder wishes to orderly break the cipher text, through the knowledge of chess game, will have several options with pieces at each point of move in the game, which will take a long time with an estimate as given $\mathbf{10^{50}}$ board positions and $\mathbf{10^{123}}$ sequences available to try out.

Commenting on the compression ratio of cipher text generated, cipher text in Fig. 17 is made up of 64 moves (characters) and cipher text in Fig. 19 comprises of 32 moves. Leaving first eight moves (characters) and the last move, the actual compressed characters are 23. So the actual compression ratio achieved is of **2.78** for this segment encryption.

6 Conclusion

The Chessography Algorithm depicts a systematic way of applying cryptography to provide data security. The distinct feature of this cipher text is that, no transposition and substitution is fixed, it is random in nature, depending on the moves in the chess game. On the chess board, there are only 64 squares, but when the game is played, chess pieces can be placed at any position depending on the rules specified. Taking advantage of this, the main second paired key is designed.

Though numbers may appear in the key from 1 to 64, any combination of number in the paired key is possible. This is one of the strength of this algorithm. Without the paired key, decrypting the cipher text is very difficult, as the order of the game will not be known. Encryption process may involve more time on sender's side but decryption is much faster on the receiver side.

Paired key plays an important role in this algorithm. Another important feature of this algorithm is that, the cipher text generated is a compressed cipher text, which means required transmission time will be less.

In the scope for future work, improvements are to be done on increasing the security of the central paired key, which comes under the area of Key Exchange.

Chessography algorithm is based upon simple knowledge and complex nature of the game of chess, which makes it an efficient algorithm in terms of security, confidentiality, and reliability.

References

1. P. Wonning, "A Short History of the Game of Chess: Chess History in Brief"," Mossy Feet Books, vol. 6 of History of Things Series, Chapter Origins, July 2014.
2. C. Bauer, "Secret History: The Story of Cryptology, Discrete Mathematics and its Applications", CRC Press, 2013, pp. xix.
3. https://www.fide.com/fide/handbook.html?id=171&view=article.
4. C. Shannon, "Programming a Computer for Playing Chess," Philosophical Magazine, Vol. 41, March 1950, vol. 2, pp. 740–741.
5. F. Gobet, J. Retschitzki, A. de Voogt, "Moves in Mind: The Psychology of Board Games," Psychology Press, 2004, pp 26–30.
6. J. Schaeffer, H.J. van den Herik, "Chips Challenging Champions: Games, Computers and Artificial Intelligence," Elsevier, 2002, pp 343–346.
7. L. Victor Allis, Searching for Solutions in Games and Artificial Intelligence, CA: University Science, 1989, pp. 171–172.

Odia Compound Character Recognition Using Stroke Analysis

Dibyasundar Das, Ratnakar Dash and Banshidhar Majhi

Abstract In Odia optical character recognition (OCR) model, detection of compound characters plays a very important role. The 80% of allograph classes present in Odia script is compound. Here a zone-based stroke analysis model is used to recognize the compound characters. Each character is divided into nine zones, and each zone has some similarity to one or few of the considered 12 strokes. For similarity measure, structural similarity index (SSIM) is used. This proposed feature have a higher potential for recognizing compound characters. The recognition accuracy of 92% is obtained for characters in Kalinga font.

1 Introduction

Optical Character Recognition (OCR) applications have been in use since the mid-1950s. Over the due course of time, OCR have become commercially available and used in many intelligent application systems. But this development is mostly confined to Roman, Chinese, Japanese, and some other scripts. In India, there are 22 constitutional languages that have their independent writing style. OCR for the Indic scripts is not commercially available due to lack of research and their complex writing style. Odia is one of these languages which has a Indo-Aryan origin. Odia OCR possess difficulties to researchers as it has more than 400 characters to handle. Many types of research have been conducted focusing on independent vowels and consonants, but much development in detecting the compound characters that consist of 80% of Odia allographs have not been done. This research work focuses on detection of compound characters with higher accuracy. Here the study of the Odia compound allograph has been done and summarized that all characters can be expressed by 12 basic strokes in different regions of the character image. Stroke information can be converted to feature matrix by evaluating structural similarity in specified region. Further, this feature can be used to design a prediction model. Here, for the

D. Das · R. Dash (✉) · B. Majhi
National Institute of Technology, Rourkela, India
e-mail: ratnakar.dash@gmail.com

H.S. Behera and D.P. Mohapatra (eds.), *Computational Intelligence in Data Mining*, Advances in Intelligent Systems and Computing 556,
DOI 10.1007/978-981-10-3874-7_30

simplicity of study we have considered single font recognition (Kalinga font) with distance-based prediction model.

1.1 Basic Overview of Odia OCR

Odia allograph contains very complex way of writing. It has 11 vowels, 35 consonants, 10 numbers, some special symbols and many compound characters as given in Fig. 1. UNI-code standard supports many more compound allographs but table includes only valid characters that are used in the Odia literature.

An over all model of generalized steps in OCR is given in Fig. 2. There are many challenges present in each stage presented in this figure. But this work focuses on the recognition stage and assumed constraints on other stages.

Rest of the paper is organized in the following way. Section 2, gives an overview of the research work conducted previously on Odia OCR. Section 3, describes the

Vowels	ଅ, ଆ, ଇ, ଈ, ଉ, ଊ, ଋ, ଏ, ଐ, ଓ, ଔ
Consonants	କ, ଖ, ଗ, ଘ, ଙ, ଚ, ଛ, ଜ, ଝ, ଞ, ଟ, ଠ, ଡ, ଢ, ଣ, ତ, ଥ, ଦ, ଧ, ନ, ପ, ଫ, ବ, ଭ, ମ, ଯ, ର, ଳ, ଶ, ସ, ଷ, ୱ, ହ, ୟ, ଲ
Numbers	୦, ୧, ୨, ୩, ୪, ୫, ୬, ୭, ୮, ୯
Special symbols	anusar (◌ଂ), chandrabindu (◌ଁ), bhisarga (◌ଃ)
Two-character conjuncts	ଙ୍କ, ଙ୍ଖ, ଙ୍ଗ, ଙ୍ଘ, ଞ୍ଚ, ଞ୍ଛ, ଞ୍ଜ, ଞ୍ଝ, ଣ୍ଟ, ଣ୍ଠ, ଣ୍ଡ, ଣ୍ଢ, ନ୍ତ, ନ୍ଥ, ନ୍ଦ, ନ୍ଧ, ମ୍ପ, ମ୍ଫ, ମ୍ବ, ମ୍ଭ, କ୍କ, କ୍ଟ, କ୍ତ, କ୍ବ, କ୍ଲ, କ୍ର, କ୍ଳ, କ୍ସ, କ୍ଷ, କ୍ୟ, କ୍ଲ, ଖ୍ୟ, ଗ୍ଧ, ଗ୍ର, ଗ୍ନ, ଗ୍ଲ, ଗ୍ଳ, ଗ୍ବ, ଘ୍ର, ଘ୍ୟ, ଚ୍ଚ, ଚ୍ଛ, ଚ୍ୟ, ଜ୍ଜ, ଜ୍ଝ, ଜ୍ଞ, ଜ୍ୟ, ଜ୍ର, ଜ୍ବ, ଟ୍ଟ, ଟ୍ବ, ଟ୍ର, ଟ୍ୟ, ଠ୍ୟ, ଡ୍ଡ, ଡ୍ଢ, ଡ୍ର, ଡ୍ୟ, ଢ୍ୟ, ଣ୍ଣ, ଣ୍ବ, ତ୍ତ, ତ୍ଥ, ତ୍ନ, ତ୍ପ, ତ୍ବ, ତ୍ମ, ତ୍ର, ତ୍ସ, ତ୍ୟ, ଥ୍ୟ, ଦ୍ଦ, ଦ୍ଧ, ଦ୍ବ, ଦ୍ଭ, ଦ୍ମ, ଦ୍ଗ, ଦ୍ର, ଦ୍ୟ, ଧ୍ୱ, ଧ୍ର, ଧ୍ବ, ନ୍ନ, ନ୍ମ, ନ୍ୟ, ପ୍ତ, ପ୍ପ, ପ୍ନ, ପ୍ସ, ପ୍ର, ପ୍ଲ, ପ୍ସ, ପ୍ୟ, ପ୍ଳ, ଫ୍ର, ଫ୍ଲ, ବ୍ଦ, ବ୍ଧ, ବ୍ବ, ବ୍ର, ବ୍ୟ, ବ୍ଲ, ଭ୍ର, ଭ୍ୟ, ମ୍ନ, ମ୍ମ, ମ୍ର, ମ୍ଲ, ମ୍ୟ, ଯ୍ୟ, ର୍କ, ର୍ଖ, ର୍ଗ, ର୍ଘ, ର୍ଚ, ର୍ଜ, ର୍ଝ, ର୍ଣ, ର୍ତ, ର୍ଥ, ର୍ଦ, ର୍ଧ, ର୍ପ, ର୍ଫ, ର୍ବ, ର୍ମ, ର୍ଶ, ର୍ସ, ର୍ଷ, ଲ୍କ, ଲ୍ପ, ଲ୍ଲ, ଶ୍ଚ, ଶ୍ଛ, ଶ୍ନ, ଶ୍ବ, ଶ୍ମ, ଶ୍ର, ଶ୍ଲ, ଶ୍ୟ, ଶ୍ଳ, ଷ୍କ, ଷ୍ଟ, ଷ୍ଠ, ଷ୍ଣ, ଷ୍ପ, ଷ୍ଫ, ଷ୍ୟ, ସ୍କ, ସ୍ଖ, ସ୍ତ, ସ୍ଥ, ସ୍ନ, ସ୍ପ, ସ୍ଫ, ସ୍ବ, ସ୍ମ, ସ୍ର, ସ୍ୟ, ସ୍ଲ, ହ୍ନ, ହ୍ବ, ହ୍ମ, ହ୍ର, ହ୍ଲ, ହ୍ୟ, କ୍ଷ୍ଣ, କ୍ଷ୍ମ, କ୍ଷ୍ୟ, ଳ୍କ, ଳ୍ପ, ଲ୍ୟ, ଲ୍ଲ
Three-character conjuncts	କ୍ତ୍ର, ଙ୍କ୍ତ, ଙ୍କ୍ଷ, ଙ୍ଗ୍ଲ, ଣ୍ଡ୍ର, ତ୍ତ୍ବ, ତ୍ତ୍ର, ତ୍ମ୍ୟ, ତ୍ସ୍ନ, ତ୍ସ୍ୟ, ନ୍ତ୍ୟ, ନ୍ତ୍ର, ନ୍ତ୍ବ, ନ୍ଦ୍ର, ନ୍ଧ୍ବ, ନ୍ଧ୍ର, ନ୍ଧ୍ୟ, ନ୍ନ୍ୟ, ମ୍ପ୍ର, ମ୍ଭ୍ର, ର୍ଚ୍ଚ, ର୍ଚ୍ଛ, ର୍ଣ୍ଡ, ର୍ତ୍ତ, ର୍ଦ୍ଦ, ର୍ଦ୍ଧ, ର୍ପ୍ସ, ର୍ଦ୍ର, ର୍ଯ୍ୟ, ର୍ଶ୍ବ, ଷ୍ଟ୍ର, ଷ୍ଠ୍ର, ସ୍ତ୍ର, ସ୍ଥ୍ର, ସ୍ଥ୍ୟ, ନ୍ତ୍ୟ, ର୍ଣ୍ୟ, ର୍ଲ୍ଲ, ର୍ତ୍ତ, ର୍ଥ୍ୟ, ର୍ଘ୍ୟ, ର୍ତ୍ୟ, କ୍ଟ୍ର, ଷ୍ଟ୍ର, ସ୍କ୍ର

Fig. 1 Twelve strokes that can express all Odia characters

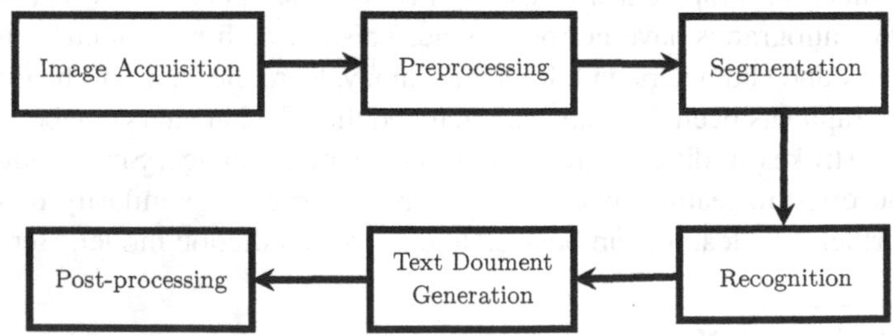

Fig. 2 Basic over view of Odia-OCR

proposed method in details. Section 4, provides information on experimental details and result analysis. And finally Sect. 5 concludes the paper and highlights the future direction.

2 Related Works

There has been significant work in basic and compound characters in other Indian languages, such as Chaudhuri et al. [1] have shown the recognition of 75 basic and modified Bangla characters with 96% of accuracy through structure-based tree classification. Garain et al. [2] have devised a run-number based hybrid (combination of feature analysis and template matching) method that has reported 99.69% of accuracy over 300 number of symbols in the Bangla literature. Similarly, in the case of Devanagari character recognition [3] an accuracy of 96% has been achieved. The OCR development in other Indian scripts like Telugu [4], Gurumukhi [5], Gujrathi [6, 7], Kannada [8], and Odia [9, 10] have not been achieved much.

Similarly, for Odia-handwritten script, the following works have been carried out. Padhi [11, 12], has proposed zone-based standard deviation to centroid distance as the feature to ANN classifier that gives 94% accuracy over 49 classes of handwritten characters. Pal et al. [13] have used gradient and curvature feature to classify 52 classes of handwritten Odia characters on a quadratic classifier, which registered 91.11% of accuracy. Basa et al. [14, 15] have suggested a tree-based classifier, that classifies Odia-handwritten characters into two steps. First, it groups the charter into two groups and each group is used to train two different ANN. This has given an overall accuracy of 90% on 51 classes of characters.

For Odia numeral, following works have been done. Sarangi et al. [16] have used Hopfield neural network to classify ten Odia numerals and achieved 95.4% of accuracy. Bhowmik et al. [17] have used stroke and hidden Markov model (HMM) to achieve accuracy of 90.50% for Odia numeral recognition. Histogram block based feature that gives 94.81% of precision has been shown by Roy et al. [18]. Sarangi et al. [19] have shown that LU factorization with Naïve Bayes classifier gives 85.3% of accuracy. Dash et al. [20] have proposed a hybrid feature based on Kirsh edge operator and curvature feature for Modified Quadratic Discriminant Classifier (MQDF) and, Discriminative Learning Quadratic Discriminant Function (DLQDF) classifier, which give an accuracy of 98.5%, 98.4% respectively. Mishra et al. [21] have shown 92% of accuracy using DCT feature and 87.5% of accuracy on DWT feature. Dash et al. [22, 23] have shown that Zone-based Non-redundant Socketwell Transform (ZNRST) and k Nearest Neighbor (k-NN) classifier can achieve 98.80 % of accuracy for Odia numeral. Mahto et al. [24] have proposed Quadrant-mean based feature and have shown that it can give 93.2% accuracy with ANN. Mishra et al. [25] have purposed contour-based feature, that gives 96.3% accuracy on HMM classifier. Mishra et al. [26] have used cord length and angle feature to classify the handwritten Odia numeral. Dash et al. [27] have devised a feature name binary external symmetry axis constellation to achieve more than 95% accuracy on different Odia datasets of numeral and characters.

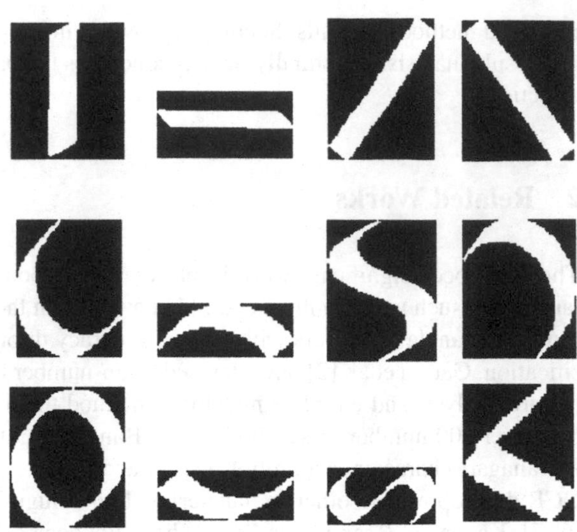

Fig. 3 Twelve strokes that can express all Odia characters

3 Proposed Model

Each character in Odia can be expressed as combination of different strokes. With geometric analysis and heuristic study, it is been summarized that 12 strokes are sufficient enough to describe any Odia allograph. The 12 strokes taken in consideration are given in Fig. 3. Here the character image is normalized to 60×60 size and divided to nine equi-proportion zones. Each zone can be represented by any one or combination of more than one strokes. The value of similarity between strokes and zone can be arranged in vector format. Here structural similarity index [28, 29] is chosen to give the index value for comparison. And an example of character ଥ and its stroke analysis is given in Fig. 4.

Human eyes are more sensitive to structure. Structural Similarity Index (SSIM) is based on the philosophy that the structure of the image is independent of illumination. This method [29] includes the properties like luminance, contrast, and structure comparison. This is expressed mathematically in Eqs. 1 and 2.

$$SSIM = [Lu(I_x, I_y)]^a [Co(I_x, I_y)]^b [St(I_x, I_y)]^c \quad (1)$$

where, *Lu* represents local luminance, *Co* represents contrast and *St* represents structure. The equation correlates to human visual system (HVS) and can be simplified to following equation by taking $a = b = c = 1$.

$$SSIM = \frac{(2\mu_{I_x}\mu_{I_y} + C_1)(2\sigma_{I_x,I_y} + C_2)}{(\mu_{I_x}^2 + \mu_{I_y}^2 + C_1)(\sigma_{I_x}^2 + \sigma_{I_y}^2 + C_2)} \quad (2)$$

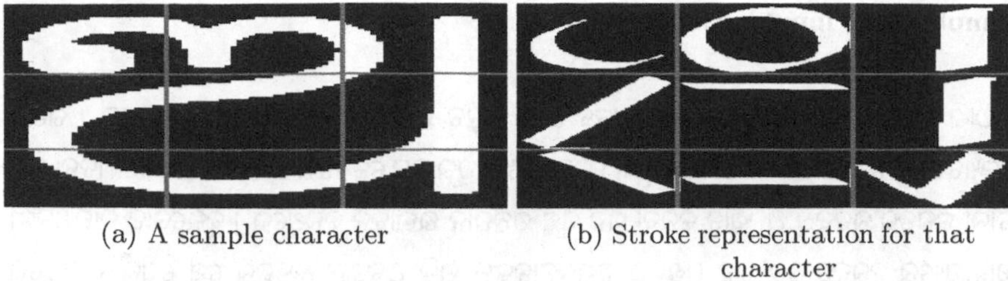

(a) A sample character (b) Stroke representation for that character

Fig. 4 Example of character ଅ in form of strokes

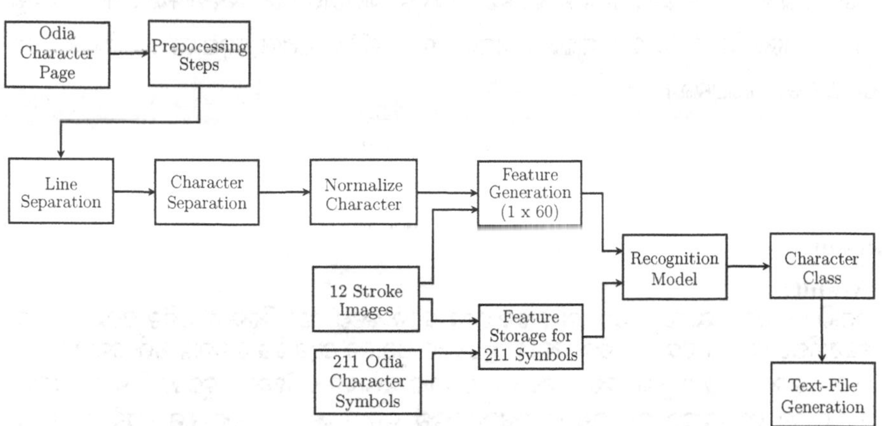

Fig. 5 Flow chart for proposed feature

where, $C_1 = (\kappa_1 * L)^2$ and $C_2 = (\kappa_2 * L)^2$ The value of κ_1 and κ_2 are determined by philosophical study is to be 0.01 and 0.03 respectively, and L is the highest level of intensity (for grayscale image it is 255).

As there are nine zones and 12 strokes, so that each feature vector representing a character is of size 1×108. The training set is created from the collection of 211 classes of characters in Kalinga font. The Fig. 5 shows a block diagram of feature generation and prediction model.

4 Experimental Result and Analysis

The experiment is carried out in windows machine and on MATLAB platform. And the independent character recognition accuracy is recoded as 92%. The experiment also covers many test samples of degraded Kalinga characters. Also a complete OCR is designed to work on scanned text document. A sample image text and corresponding output is made available below.

Sample image input to OCR

ପଠାଣି ସାମନ୍ତ ନୟାଗଡ଼ର ଖଣ୍ଡପଡ଼ା ରାଜବଂଶରେ ଜନ୍ମିତ ଜଣେ ଜ୍ୟୋତିର୍ବିଦ ଓ ପଣ୍ଡିତ ଥିଲେ । ଗଣିତ ଜ୍ୟୋତିଷରେ ତାଙ୍କର ଗଭୀର ପାଣ୍ଡିତ୍ୟ ଥିଲା । ଚନ୍ଦ୍ରଶେଖର ପୁରାତନ ଜ୍ୟୋତିଷ ଶାସ୍ତ୍ରମାନ ପର୍ଯ୍ୟାଲୋଚନା କରି ତାରା ଜଗତର ଗତିବିଧି ଅନୁଧାନ କରିବା ସହ ସୂକ୍ଷ୍ମ ଯନ୍ତ୍ରମାନ ଉଭାବନ କରିଥିଲେ । ଦୁଇଗୋଟି ବାଉଁଶନଳୀ ସାହାଯ୍ୟରେ ଦୂରଦୂରାନ୍ତର ଗ୍ରହ ମଣ୍ଡଳର ନକ୍ଷତ୍ରମାନଙ୍କର ଗତି ଅବଲୋକନ କରି ସେ ଯେଉଁ ସିଦ୍ଧାନ୍ତରେ ଉପନୀତ ହେଲେ ତାହା ପୁରାତନ ସଂସ୍କୃତ ଜ୍ୟୋତିଷ ଶାସ୍ତ୍ରର ସିଦ୍ଧାନ୍ତ ସହିତ ତାହାର ଅନେକ ଅସମାନତା ଥିଲା । ତାହାଙ୍କର ନିଜ ସିଦ୍ଧାନ୍ତରୁ ପ୍ରମାଣିତ ହେଲା ଯେ, ପୁରାତନ ଭାରତୀୟ ଜ୍ୟୋତିର୍ବିଦମାନଙ୍କ ଗଣନା ଦିବସରୁ ନକ୍ଷତ୍ର ଜଗତର ସ୍ଥିତି ଗତିର ବଦଳିଯାଇଛି । ତାଙ୍କର ନୂତନ ମୌଳିକ ଗଣନା ଅନୁଯାୟୀ ସେ ସିଦ୍ଧାନ୍ତ ଦର୍ପଣ ପୋଥି ରଚନା କରିଥିଲେ ।

Fig. 6 The sample image given as input to the Odia OCR

Result

Result:
ପଠାଣି ସାମନ୍ତ ନୟାଗଡାର ଖଣ୍ଡପଡା ରାଜବଂଶରେ ଜନ୍ମିତ ଜଣେ ଜ୍ୟୋତିର୍ବିଦ ଓ ପଣ୍ଡିତ ହରେ | ଗଣିତ
ଜ୍ୟୋତିପରେ ତାଙ୍କର ଗଭୀର ପାଣ୍ଡିତ୍ୟ ହଲା | ଚନ୍ଦ୍ରଶେଖର ପୁରାତନ ଜ୍ୟୋତିପ ଶାସ୍ତ୍ରମାନ ପର୍ଯ୍ୟାଲୋଚନା କରି
ତାରା ଜଗତର ଗଞ୍ଜିତ୍ର ଅନୁଧାନ କରିବା ସହ କୃକ୍ଷ୍ମ ଯନ୍ତ୍ରମାନ ଉଭାବନ କରିହଲେ | କୁଇଗୋଟି ବାଉଁଶନଳୀ
ସାହାଯ୍ୟରେ କୂରକୂରାତର ଗ୍ରହ ମଣ୍ତଳର ନକ୍ଷତ୍ରମାନଙ୍କର ଗତି ଅବଲୋକନ କରି ସେ ଯେଉଁ ସିଦ୍ଧାନ୍ତରେ

Figure 6 shows a sample input to the model described, and the recognized text is been included in the result section. It can be seen from the output that most of the characters are recognized correctly. Along with few recognition error, some matra association error is present in given result. Further accuracy can be improved by implementing dictionary matching.

5 Conclusions

The description in experimental result shows the recognition accuracy to be 92% which includes all possible allographs. Experiment proves the effectiveness of considered feature. As for simplicity in study of feature effectiveness, we have considered the distance-based classifier. For more accuracy and detection of other font neural network model can be imposed. Further study on the zone validation is needed. Every zone considered may not be contributing to the task of classification. Hence, descriptive zone analysis by Evolutionary algorithm or PCA can be done to give effective zones. This in turn will reduce the size of computational cost.

References

1. B. Chaudhuri and U. Pal, "A complete printed bangla ocr system," *Pattern Recognition*, vol. 31, no. 5, pp. 531–549, 1998.
2. U. Garain and B. Chaudhuri, "Compound character recognition by run-number-based metric distance," in *Photonics West'98 Electronic Imaging*. International Society for Optics and Photonics, 1998, pp. 90–97.
3. U. Pal and B. Chaudhuri, "Indian script character recognition: a survey," *Pattern Recognition*, vol. 37, no. 9, pp. 1887–1899, 2004.
4. M. Sukhaswami, P. Seetharamulu, and A. K. Pujari, "Recognition of telugu characters using neural networks," *International Journal of Neural Systems*, vol. 6, no. 03, pp. 317–357, 1995.
5. G. Lehal and C. Singh, "Feature extraction and classification for ocr of gurmukhi script," *VIVEK-BOMBAY-*, vol. 12, no. 2, pp. 2–12, 1999.
6. S. Antani and L. Agnihotri, "Gujarati character recognition," in *Proceedings of the Fifth International Conference on Document Analysis and Recognition*. IEEE, 1999, pp. 418–421.
7. J. Dholakia, A. Yajnik, and A. Negi, "Wavelet feature based confusion character sets for gujarati script," in *International Conference on Conference on Computational Intelligence and Multimedia Applications*, vol. 2. IEEE, 2007, pp. 366–370.
8. T. Ashwin and P. Sastry, "A font and size-independent ocr system for printed kannada documents using support vector machines," *Sadhana*, vol. 27, no. 1, pp. 35–58, 2002.
9. S. Mohanty, "Pattern recognition in alphabets of oriya language using kohonen neural network," *International Journal of Pattern Recognition and Artificial Intelligence*, vol. 12, no. 07, pp. 1007–1015, 1998.
10. B. Chaudhuri, U. Pal, and M. Mitra, "Automatic recognition of printed oriya script," *Sadhana*, vol. 27, no. 1, pp. 23–34, 2002.
11. D. Padhi, "Novel hybrid approach for odia handwritten character recognition system," *International Journal of Advanced Research in Computer Science and Software Engineering*, vol. 2, no. 5, pp. 150–157, 2012.
12. D. Padhi and D. Senapati, "Zone centroid distance and standard deviation based feature matrix for odia handwritten character recognition," in *Proceedings of the International Conference on Frontiers of Intelligent Computing: Theory and Applications (FICTA)*. Springer, 2013, pp. 649–658.
13. U. Pal, T. Wakabayashi, and F. Kimura, "A system for off-line oriya handwritten character recognition using curvature feature," in *10th International Conference on Information Technology*. IEEE, 2007, pp. 227–229.
14. D. Basa and S. Meher, "Handwritten odia character recognition," *Recent Advances in Microwave Tubes, Devices and Communication*, 2011.
15. S. Meher and D. Basa, "An intelligent scanner with handwritten odia character recognition capability," in *Sensing Technology (ICST), 2011 Fifth International Conference on*. IEEE, 2011, pp. 53–59.
16. P. K. Sarangi, A. K. Sahoo, and P. Ahmed, "Recognition of isolated handwritten oriya numerals using hopfield neural network," *International Journal of Computer Applications*, vol. 40, no. 8, pp. 36–42, 2012.
17. T. K. Bhowmik, S. K. Parui, U. Bhattacharya, and B. Shaw, "An hmm based recognition scheme for handwritten oriya numerals," in *International Conference on Information Technology*. IEEE, 2006, pp. 105–110.
18. K. Roy, T. Pal, U. Pal, and F. Kimura, "Oriya handwritten numeral recognition system," in *Eighth International Conference on Document Analysis and Recognition*. IEEE, 2005, pp. 770–774.
19. P. K. Sarangi, P. Ahmed, and K. K. Ravulakollu, "Naïve bayes classifier with lu factorization for recognition of handwritten odia numerals," *Indian Journal of Science and Technology*, vol. 7, no. 1, pp. 35–38, 2014.

20. K. S. Dash, N. Puhan, and G. Panda, "A hybrid feature and discriminant classifier for high accuracy handwritten odia numeral recognition," in *IEEE Region 10 Symposium*. IEEE, 2014, pp. 531–535.
21. T. K. Mishra, B. Majhi, and S. Panda, "A comparative analysis of image transformations for handwritten odia numeral recognition," in *International Conference on Advances in Computing, Communications and Informatics*. IEEE, 2013, pp. 790–793.
22. K. S. Dash, N. Puhan, and G. Panda, "Non-redundant stockwell transform based feature extraction for handwritten digit recognition," in *International Conference on Signal Processing and Communications*. IEEE, 2014, pp. 1–4.
23. K. S. Dash, N. Puhan, and G. Panda, "On extraction of features for handwritten odia numeral recognition in transformed domain," in *Eighth International Conference on Advances in Pattern Recognition*. IEEE, 2015, pp. 1–6.
24. M. K. Mahto, A. Kumari, and S. Panigrahi, "Asystem for oriya handwritten numeral recognization for indian postal automation," *International Journal of Applied Science & Technology Research Excellence*, vol. 1, no. 1, pp. 17–23, 2011.
25. T. K. Mishra, B. Majhi, P. K. Sa, and S. Panda, "Model based odia numeral recognition using fuzzy aggregated features," *Frontiers of Computer Science*, vol. 8, no. 6, pp. 916–922, 2014.
26. T. K. Mishra, B. Majhi, and R. Dash, "Shape descriptors-based generalised scheme for handwritten character recognition," *International Journal of Computational Vision and Robotics*, vol. 6, no. 1–2, pp. 168–179, 2016.
27. K. Dash, N. Puhan, and G. Panda, "Besac: Binary external symmetry axis constellation for unconstrained handwritten character recognition," *Pattern Recognition Letters*, 2016.
28. Z. Wang and A. C. Bovik, "Mean squared error: love it or leave it? a new look at signal fidelity measures," *IEEE signal processing magazine*, vol. 26, no. 1, pp. 98–117, 2009.
29. Z. Wang, A. C. Bovik, H. R. Sheikh, and E. P. Simoncelli, "Image quality assessment: from error visibility to structural similarity," *IEEE Transactions on Image Processing*, vol. 13, no. 4, pp. 600–612, 2004.

A Comparative Study and Performance Analysis of Routing Algorithms for MANET

Mohammed Abdul Bari, Sanjay Kalkal and Shahanawaj Ahmad

Abstract The mobility for communication devices has generated the requirement of next generation wireless network technology development, several type of wireless technologies have been developed and implemented, one of them is ad hoc mobile network system which facilitates to a cluster of mobile terminals which are animatedly and randomly changing locations in such a way that the interconnections need to be handled continually. A routing procedure is used to find the routes between the mobile terminals in the set of connections for easy and reliable communication; main purpose of this protocol is to find a proper route from caller to called terminals with minimum of overhead and bandwidth consumption during construction of route. This paper provides an impression of a broad range of routing protocols projected in the literature and has in addition provided a piece comparison of all routing protocols with discussion of their respective merits and drawbacks.

Keywords Routing protocol · MANET · Proactive · Reactive · Hybrid algorithms

1 Introduction

The fame of transmission seems to be contemporary, but it is enclosed by disparate evolution eras, transforming and ornamental the statement styles. Massive approaches are embraced and became the principle from time to time, which set the

M.A. Bari (✉) · S. Kalkal
Kalinga University, Raipur, Chhattisgarh, India
e-mail: bari_baril1@rediffmail.com

S. Kalkal
e-mail: kalingauniversity1@gmail.com

S. Ahmad
College of Computer Science & Engineering, University of Ha'il, Baqaa, Kingdom of Saudi Arabia
e-mail: sa_sum@yahoo.com

H.S. Behera and D.P. Mohapatra (eds.), *Computational Intelligence in Data Mining*, Advances in Intelligent Systems and Computing 556,
DOI 10.1007/978-981-10-3874-7_31

transmission parameter up-to-date. The mobile phones technology is fetching a vital part as it can be available almost all over in the globe [1, 2]. The wireless level has been experimenting proportional escalation in the bygone decade, and the development of wireless devices such as portable PDA and cell phones are now playing an ever-increasing important role in our life [3]. Not only the mobile phones are getting lesser, cheaper, suitable and supplementary powerful, but they also run supplementary function and network services. Market reports from an independent source show that the world wide number of cellular users has been doubling every 1½ years [4].

A MANET is a self-set up set of wireless mobile devises that shape a short-term and alive wireless network without any framework. MANET characterized as self-compose and is no essential administration structure with composition tasks [5, 6]. The MANET is different from others networks in two different ways: [7]

1. The host in the network has limited energy.
2. The network topology of mobile network changes rapidly.

Here, each node dynamically enters the network at any given time as well as leaving the network at any given time [8], Due to the absence of base stations, MANET's uses multi-loop approach to delivering the data. Few characteristic of MANET is given below [9]

- Network is not depending on any fixed infrastructure for its operation.
- Multi-hop routing.
- Active network topology [10].
- Gadget heterogeneity.
- Bandwidth constrained variable ability links.
- Limited physical security.
- Network scalability.
- Self organization, creation, and supervision.

1.1 Application of MANET

The applications have grown in the past minority years with the rapid advance research in mobile ad hoc network; few of them are given below [11]

- Military communication.
- Home application.
- Search and rescue operation.
- E-commerce.
- Setup virtual call room.
- Automatic call forwarding services.

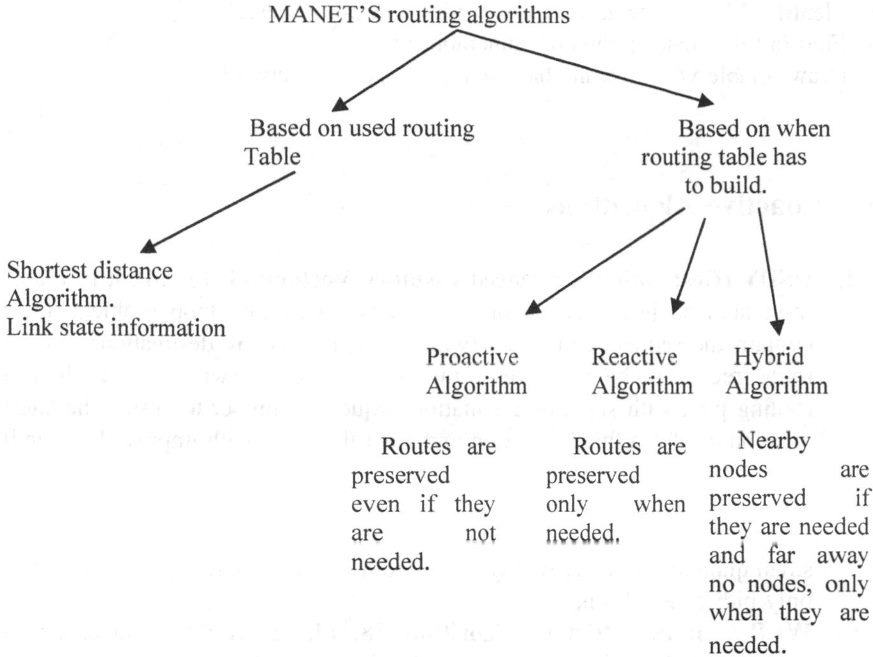

Fig. 1 Classification of MANETs routing protocols

Different classified routing protocols have shown [2, 12, 13] above Fig. 1. The routing algorithm performs two operations, first to collect information and then establish route to forward the data. Classification of MANETs routing protocols are shown in Fig. 1.

2 Related Work

Most of previous research is about presentation judgment of MANET routing protocols including majorly reactive/proactive protocols. Some research has been carried out for comparison of reactive with proactive, or reactive with hybrid protocols or among all of them, but this comparison are not including the presentation differentials, changelings among proactive, reactive and hybrid protocols. In our work, we mainly focus the following methodology to compare and evaluate the performance of existing routing protocols in MNAETs:

- Study the basic principal of different ad hoc networks;
- Review of the existing different MANET algorithms;

- Identify their organization, complexities and properties;
- Find out the major points of vulnerability;
- Draw a table with overall characteristic of ad hoc network.

3 Proactive Algorithms

(i) **DSDV (Destination Sequenced Distance Vector)**: [8, 13, 14], it is a table motivated ad hoc network protocol to solve routing loop problem, table contain the sequence numbers (which is generated by destination, confirm route freeness) which are even if the link is there otherwise it is odd. It elect routing path with superior destination sequence number to ensure the latest information from the destination and elect the route with improved metric if number entries are same. Advantages: Routing information available immediately from the routing table. It solves routing loop problem. Disadvantage: It requires regular update, which used battery control and also the small quantity of power supply, bandwidth when the network is ideal. It uses only bidirectional link.

(ii) **WRP (Wireless Routing Algorithm)** [8, 13, 15, 16]: It is an enhanced version of DSDV. It tries to provide the shortest-path-loop-free routes. Here, every node contains four tables and exchange them with their neighbors periodically. They are Distance table, which contains information of nodes that are not connected. Routing table contains detachment to a destination node, the preceding and the next nodes along the route. Link-cost table contains the cost of the connection to its nearest neighbor and the cost of broken connections is given as infinity. Message retransmission list (MRL) table contains a series number of update message. Nodes inform each other using update message. Here, each node in the network send "HELLO" message to the neighbor and inform that he is alive and wait for the ACK from the neighbor and if the ACK is not received within the explicit period of time it is assuming that the connectivity with that neighbor is lost. *Advantages*: Count-to-infinity setback is solved by forcing every node to perform a consistency check for predecessor information reports by all its neighbors. *Disadvantages*: Large quantity of memory, bandwidth and power are required by four tables.

(iii) **CGSR (Cluster Head Gateway Switch Routing)** [16–18]: It is a hierarchical routing protocol. Its collect structure improves show of the routing protocol since it provides efficient connection and traffic management. It uses the impression of cluster and cluster heads where routing is done via the cluster heads and gateways. Every node sends a "HELLOW" message containing its ID and monotonically increasing sequence number. The cluster head contains a table that contains IDs of the nodes belonging to it and their most sequence number. Cluster head exchanges this information with each other through gateway. A sent packet is delivered to its cluster head and then routed by gateway to an additional cluster head and goes in the same manner until it reaches the destination terminal node. *Advantages*: more efficient utilization of bandwidth, it is reducing size of remoteness vector table because the routing is performed only over cluster head. *Disadvantage*: Change in the cluster head may result in numerous path breaks. Additional time is needed in the selection of cluster heads and gateways.

4 Reactive Algorithms

(i) **AODV (Ad Hoc on depend distance vector)** [5, 13, 16, 19]. Here, this routing protocol is considered by mobile nodes in ad hoc networks when two hosts desire to converse with apiece other and a route is shaped to provide such connection, i.e., it establishes the route only on order. The network is quiet until the connection is desirable. It allows mobile nodes to gain routes rapidly for new destinations and at the similar time it does not demand nodes to continue routes to the nodes that are not in active contact. When the link is not working, due to some error or condition, it informs the affected set of nodes so that they can nullify the routes [20]. Here, a special "Route Request" (RREQ) message is broadcast during network. Each RREQ keeps and an ordered list of all nodes it approved passes through, when RREQ arrived at target, a "ROUTE REPLY (RREP) message would straight away get passed back to the origin, indicating the route to the destination is found. The RREP back trace the way the RREQ message came. As soon as RREP reaches its target, a two-way route path is established and swap of data packets begin.

Advantages: Here the routes are established on demand. *Disadvantages*: The in-between nodes can lead to inconsistent route if the sequence number is very old.

(ii) **DSR (Dynamic Source Routing)** [14, 21, 22]. It is similar to AODV, design a route on demand; however, it uses supply routing as an alternative of relying on the routing table at each in-between machine.

Here, the sender

- Transmit RREQ packets.

Then receiver

- It looks if routing table checks if it has a route to the target.
- If address before now exists
  - Discard;
- If the host is the goal
  - Send a route reply

- Else
  - Attach this host address to the route record and re-broadcast.

Here, the route respond is achieved when the route appeal reaches whichever destination itself or in-between node. When goal is reached, route reply is done with full path–source node. Source node hideout all paths that it received and chooses the undeviating path, among the entire path. When the link is broken between two nodes, the nodes that notice the break send a route error message to the source node about the not working link. The source node will abolish the route from the cache or used an additional cached route or request a new route. *Advantages*: The route is documented, when it is necessary and hence the need to find the route to all the other nodes is cancel. The in-between nodes employ the route cache facts to decrease the control overhead. *Disadvantages*: The route preservation instrument does not nearby mend the broken link. The link setup delay is advanced than in table-driven protocols. Routing overhead is involved due to source-routing mechanism.

(iii) **LMR (Light-Weight Mobile Routing)** [23, 24]. The LMR is one more on-demand protocol where nodes continue multiple routes to accomplish the required purpose nodes. This will increase the reliability by allowing nodes to select the next obtainable route to the required destination without initializing a route detection procedure. Here, each node maintains routing in order of their neighbors. LRM operates count on three basic messages, i.e., query feedback and decline. A query message is sent by the basis node via limited broadcast. The basis then waits for a reply packet which is circulated by a node that has a route to the reason. The reply messages form a directed acyclic graph rooted in the originator of the reply. If a node loses its last route to the destination and it neighbors a failure query is broadcasted to erase invalid routes. On receiving a breakdown query, a node may also transmit a reply to it or broadcast a new query erasing the old path. *Advantage*: Produce multiple routes to the required destination which is reliable. *Disadvantage*: It produces provisional invalid routes which introduce extra delay in formative the correct loop.

(iv) **TORA (Temporary Ordered Routing Algorithm)** [16, 25]. It is on require routing protocol. It main reason is to limit the control message propagate in the highly active mobile surroundings. The source initiates routing protocol and finds numerous routes from origin to the target node. TORA performs three tasks

- Create a route from source to destination using QRY that contain target id for which the algorithm is running and UDP packets which contain the stature of the node that is transmitting the packet. The altitude of the destination node is set as 0, and all the other nodes altitude is set as NULL. The nodes with not null response to the UDP packets and after receiving the UDP its set it height one more then the node that generates

UDP. The node with higher stature is considered whereas the lower height is downstream creating a direct acyclic graph from source to destination.
- UDP is also used for maintaining routes.
- A clear packet (CLR) is used to through the network to remove invalid routes.

Advantages: It is an on-demand routing protocol. Construct a DAG only when it required, many paths are constructed. *Disadvantages*: Same as an on-demand routing protocol. Not heaps since DSR and AODV do better than TORA. Fruition reduces with increasing mobility.

(v) **ABR (Associatively Based-Routing)** [11, 23]. Its source initiated routing protocol that select routes based on the constancy of the wireless link. The stability of the route that is stable or unstable is determined by beam-based, which the node received from its neighbors. Here, each node maintains the count of its neighbors and classified them into stable or unstable. Source sends route request throughout the network, first it is the check in the cache, and if it is not there, than all the intermediate nodes get route request, which they forward to the required destination. *Advantages*: Here the steady routes have higher performance compared to shorter routes. *Disadvantages*: It required periodic beam counting to decide the quantity of associatively of the links; it required all the nodes to stay active all the time, which results in extra power supply. It does not manage routes to the required target

5 Hybrid Routing Protocols

Its new breeding protocol, it combines both proactive and reactive in nature.

ZRP (Zone Routing Protocol) [23, 26, 27]: In ZRP, here routing belt is clear for each node. Each node determines region radius in provisos of range. The zone can overlap and also extent of the zone effects the accomplishment of the network. The larger route regions are suitable in circumstances where route require is high and smaller route zones are suitable in circumstances where route need is low. Here, proactive route protocols work within the zone, and reactive routing protocols works betwixt the zones.

It consists of three components

- Proactive intra-zone routing protocol (IARP) which guarantees that every node in the zone consistently updates its routing table which has the information of the route to all the target nodes within the network.
- Reactive inter-zone routing protocol (IERP) it is used to get started when destination is not available within the zone.

- Border resolution protocol (BRP) that performed on-demand routing to search for routing information to the nodes that reside outside the source node zone.

Advantages: it scales down the power traffic created by episodic flooding of routing in sequence packets (proactive scheme). It also reduces the expenditure of bandwidth and control overhead compared to reactive schemes [28]. *Disadvantages*: overlapping of routing zones.

6 Results and Discussions

Quality of Service [29]: Some sort of quality of examine backing is probably essential to incorporate into the routing protocol, but none of the proposed protocols from MANET have these entire attribute, but it is essential to memorize that the protocols are still under growth and are perhaps comprehensive with more performance. Here, the main purpose is still to find a route to the harbor, not to find the best or optimal or shortest path route.

Power Control: The nodes in the ad hoc routing protocol may be laptops, PDA's, mobiles that all ways required power supply in order to survive.
Security: Mobile wireless networks are further flat to physical security coercion than a fixed-cable nets or a shared wireless medium easily reached both rightful network users and hateful attackers. The amplified possibility of eavesdropping, spoofing, and denial-of-service intrusion should be cautiously treated [10].
Storage Requirement: It is high for the proactive routing protocol, it is lower in reactive routing protocol and it depends on the size of the each cluster as high as proactive routing protocol in hybrid protocol.
Delay: It is small in proactive routing protocol as the routing are predefined and small. As routes have to find its way in reactive routing protocols, the delay is high. In hybrid, for local is low and inter-zone its high.
Traffic Volume: In proactive, the traffic volume is high, in reactive is low and in hybrid it is lower.
Periodic Update: It is required for proactive routing protocol. It is not required for reactive routing protocol, and in hybrid it is done between the gateways.
Scalability: The ability to perform efficient for an approximate number of nodes, as the network changes all the time scalability is a major concern.
Grating: Here, in the figure, we also given grating "* * *" means poor in our context and "& & &" is the best presentation offered.

Table 1 Network parameters

Network parameters	Values
Simulation time	10, 20, 30 s
Number of nodes	10–100
Link Layer (LL) type	Logical Link (LL)
Medium access control layer type	802.11
Radio propagation type	Two-ray ground
Queue type	Drop-tail
Routing	AODV, DSR
Traffic	CBR, FTP
Area of network	500 m × 500 m
Mobility type	Random way propagation

7 Performance Evaluation

Network Simulator 2 is used to calculate the presentation of "Reactive", "Proactive", and "Hybrid" routing protocols. Goal behind the replication is to evaluate and judge the behavior of routing protocols to react on different traffic conditions of distributed network environment, and also compare the performance metric based on different network conditions like number of hop count, movement of nodes. Simulation parameters used in our work are shown in Table 1. We calculated the throughput, packet loss, and delay, shown in Fig. 2. In MANETs route from, source to destination it contains multiple links and heterogeneous radio, so packet loss performance will be an important issue. From Fig. 1, it is very difficult to judge which protocol is best for all situations. Results indicate that reactive routing is performed best among the other routing protocols. Finally, we tabulated different protocols with awe of different performance metrics, which is shown in Table 2 in Appendix 1.

8 Future Work

The territory of ad hoc mobile networks is speedily emergent and altering, and although there are still abundant complications that need to be found, it is likely that correlative networks will see sweeping use within the next not many years. Large proportion of ad hoc networks is addition demanding subject in the near prospect which can be by now expected. As the improvement goes on, especially the need of tightly packed use, such as combat zone and sensor networks, where the nodes in ad hoc networks will be smaller, economical, and more trained to move and come in all forms. In all, even though the extensive exploitation of ad hoc networks is still year away, the study in this field will carry on being very vigorous and resourceful.

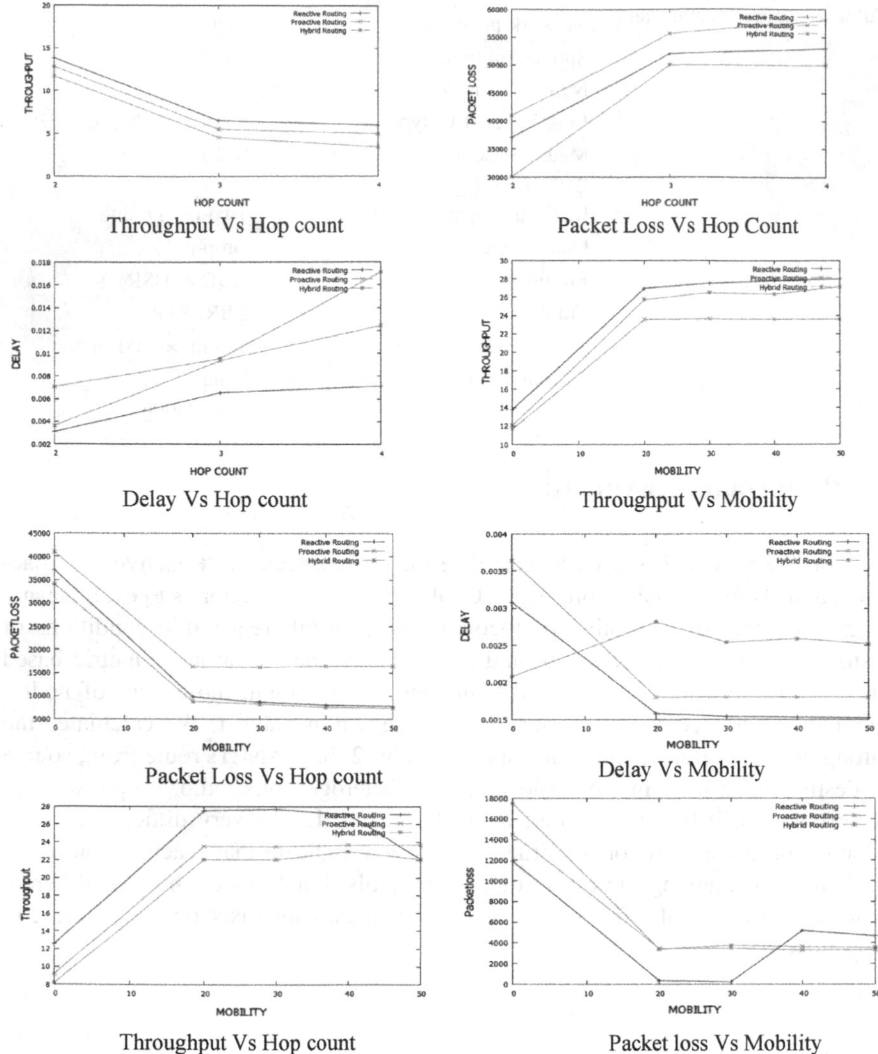

Throughput Vs Hop count

Packet Loss Vs Hop Count

Delay Vs Hop count

Throughput Vs Mobility

Packet Loss Vs Hop count

Delay Vs Mobility

Throughput Vs Hop count

Packet loss Vs Mobility

Fig. 2 Throughput, packet loss and delay of reactive, proactive and hybrid routing protocols with respect to hop count and mobility

Appendix 1

See Table 2.

Table 2 Overall characteristic of routing algorithm [2, 4, 6, 7, 9, 11, 13, 15–18, 22–24, 26, 27, 29, 10]

Routine name	Pro/Reactive/Hybrid	Routine structure	No. of table/multiple	Convergence time/Time complexity	Loop-free	Multiple roots	Quality of service	Power control	Security	Storage requirement	Delay	Traffic volume	Periodic update	Scalability level	Grading
DSDV	Pro	Flat	2/–	O(D-1)/O(D)	Yes	No	No	Required	No	High	Small, as routes are predefine	High	Yes	Up to 100	* * *
WRP	Pro	Flat	4/–	O(h)/O(h)	Yes	No	No	Required	No	High	Small, as routes are predefine	High	Yes	Up to 100	* *
CGSR	Pro	Hierarchical	3/–	O(D)/O(D)	Yes	No	No	Required	No	High	Small, as routes are predefine	High	Yes	Up to 100	*
AODV	React	Flat	No/–	–/O(2D)	Yes	No	No	Required	No	Depend on the number of routes	Higher	Low	No	Up to few 100	&
DSR	Re-act	Flat	–/Yes	–/O(2D)	Yes	Yes	No	Required	No	Depend on the number of routes	Higher	Low	No	Up to few 100	& &
LMR	Re-act	Flat	–/Yes	–/O(2D)	Yes	Yes	No	Required	No	Depend on the number of routes	Higher	Low	No	Up to few 100	& &
TORA	Re-act	Flat	–/Yes	–/O(2D)	Yes	Yes	No	Required	No	Depend on the number of routes	Higher	Low	No	Up to few 100	& &
ABR	Re-act	Flat	–/Yes	–/O(D + P)	Yes	No	No	Required	No	Depend on the number of routes	Higher	Low	No	Up to few 100	& &
ZRP	Hybrid	Flat/Mostly hierarchical	–/No	–/Intra:O(I) Inter:O(2D)	Yes	Yes	No	Required	No	Depend on the size of the zone	For local its small and for interzone its high	Lower.	Inside the node and between the gateway	Up to 1000 or more	& & &

D Diameter of the network, *h* Height of the root routing tree, *I* Periodic update interval, *P* Diameter of the directed path

References

1. M. Uddin and A. A. Rahman, (2011), "*Reliability of mobile ad hoc network through performance analysis of TCP variant over AODV*", Journal of Applied Science Research, vol. 7, no. 4, pp. 437–446.
2. Muen Uddin, Azizah Abdul Rahman, Abdurahman Alarifi, Mohammed Talha, Asadullah Shah, Mohsin Iftikhar & Albert Zomaya, (June 2012), "*Improving performance of Mobile Ad Hoc N/T using efficient tactical on demand distance vector (TAODV) Routing algorithms*"., International Journal of Innovative Computing, Information & Controls, Vol 8, Number 6.
3. Stefano Basagni, Marco Conti, Silvia Giordano, Ivan Stojmenovic, (2004), "*Mobile Adhoc Network*", IEEE Press.
4. Faragó and V. R. Syrotiuk, (2001), "*MERIT: A Unified Framework for Routing Protocol Assessment in Mobile Ad Hoc Networks*," in Proceedings of 7th Annual International Conference on Mobile Computing and Networking, Rome, Italy.
5. G. Santhi and Alamelu Nachiappan, (August 2010), "*A Survey of Qos Routing Protocol for Mobile Ad Hoc Networks*" International journal of computer science & information Technology (IJCSIT) Vol. 2, No. 4.
6. Umang Singh, 30th (September 2011), "*Secure Routing Protocols in Mobile Ad Hoc Network -A Survey and Axanomy*" International Journal of Reviews in Computing.
7. Qunwei Zheng, Xiaoyan Hong, and Sibabrata Ray, (2004), "*Recent Advances in Mobility Modeling for Mobile Ad Hoc Network Research*".
8. B. Soujanya, T. Sitmahalakshmi, Ch. Divakar, (April 2001), "*Study of Routing Protocols in Mobile Ad-Hoc Network s*", B. Soujanya et al./International Journal of Engineering Science and Technology (IJEST), ISSN: 0975-5462 Vol. 3 No. 4.
9. Miss Dhara N. Darji, Miss Nita B. Thakker, (2013), "*Challenges in Mobile Ad-Hoc Network: Security Threats & its Solutions*", Indian Journal of Applied Research, Vol 3, Issue 1.
10. Kavita Taneja and R. B. Patel, (2007), "*Mobile Ad hoc Network: Challenges and Future*", COIT-2007, RIMT-IET, Mandi Gobindgarh.
11. Imrich Chlamtic, Marco Conti, Jennifer J.N. Liu, (2003),"Mobile ad hoc networking: imperative & challenge", Elsevier.
12. T. H. Clausen, (2007), "*A MANET Architecture Model*", http://www.thomasclausen.org/ThomasHeideClausensWebsite/ResearchReports_les/RR-6145.pdf, (accessed 12/02/2014).
13. D. Lang, (2003), "*A Comprehensive Overview about Selected Ad Hoc Networking Routing Protocols*", Master Thesis, Technische Uni. Munich, Germany.
14. Perkins, Charles E. and Bhagwat, Pravin, (1994), "*Highly Dynamic Destination-Sequenced Distance-Vector Routing (DSDV) for Mobile Computers (pdf)*".
15. Murthy, Shree; Garcia-Luna-Aceves, J. J., (1996), "*An efficient routing protocols for wireless network*", *Mobile Networks and Applications* (Hingham, MA: Kluwer Academic Publishers).
16. Pankaj Barodia, (2011), "*Routing Protocols in Wireless Networks*", http://www.slideshare.net/barodia_1437/wireless-routing-protocols-8158788, accessed 25/12/2013.
17. Kaliyaperumal Karthikeyan, Sreedhar Appalabatla, Mungamru Nirmala, Teklay Tesfazghi, (2012), "*Comparative Analysis of Non-Uniform Unicast Routing Protocols of Mobile Ad-Hoc Network*", IJARCSSE.
18. C. C. Chiang, T. C. Tsai, W. Liu and M. Gerla, (1997), "*Routing in clustered multihop, mobile wireless network with fading channel*", The Next Millennium, The IEEE SICON.
19. C. Perkins, E. Belding and S. Das, (2003), "*Ad hoc on-demand distance vector (AODV) routing*".
20. S. Basagni, M. Conti, S. Giordano and G. Ivan, (2004), "*Mobile Ad Hoc Networking*", Wiley.
21. Chai-Keong-Toh, (1997), "*Associativity- Based Routing for Ad Hoc Mobile Network*" Kluwer Academic Publishers.
22. Samir R. Das, Robert Castaneda & Jiangtao Yan, (2000), "*Simulation Based Performance Evaluation of Mobile, Ad Hoc Network Routing Protocol*" University of Taxes, USA.

23. Mehran Abolhasan, Tadeusz Wysocki, Eryk Dutkiewicz, (2004), "*A review of routing protocols for mobile ad hoc network*", Elsevier.
24. Humayun Bakht, (2011), "*Survey of Routing Protocols for Mobile Ad-hoc Networks,*" International Journal of Information and Communication Technology Research.
25. V.D. Park and Scott. M. Corson, "*A Highly Adaptive Distributed Routing Algorithms for Mobile Wireless Network*", Proceedings of INFOCOM, 1997.
26. Kavita Pandey, Abhishek Swaroop, (2011), "*A comprehensive Performance Analysis of Proactive Reactive & Hybride Manet's Routing Protocol,* IJCSI, Vol 6, Iss 6.
27. Zygmunt J. Haas, Marc R. Pearlman, Prince Samar, (2002), "*The Zone Routing Protocol (ZRP) for Ad Hoc Network*", Internet-Draft,. http://tools.ietf.org/pdf/draft-ietf-manet-zone-zrp-04.pdf accessed 25/01/2014.
28. Kilinkarridis Theofaniz, (2007), "*Hybrid Routing Protocol*", Special Course in Mobile Management. http://www.tcs.hut.fi/Studies/T-9.7001/2007SPR/kilinkaridi_presentation_1.pdf accessed 25/2/2014.
29. K. Ramesh Reddy, S. Venkata Raju, N. Venkatadri, (2012), "*Reactive, Proactive MANET Routing Protocols Comparison*", IJVIPNS-IJENS.

Classification of Research Articles Hierarchically: A New Technique

Rajendra Kumar Roul and Jajati Keshari Sahoo

Abstract The amount of research work taking place in all streams of Science, Engineering, Medicines, etc., is growing rapidly and hence the research articles are increasing everyday. In this dynamic environment, identifying and maintaining such a large collection of articles in one place and classifying them manually are becoming very exhaustive. Often, the allocation of articles in various subject areas will be made simply on the basis of the journals in which they are published. This paper proposes an approach for handling such huge volume of articles by classifying them into their respective categories based on the keywords extracted from the *keyword section* of the article. Query enrichment is used by generating unigram and bigram of these keywords and giving them proper weights using probability measure. Microsoft Academic Research dataset is used for the experimental purpose and the empirical results show the effectiveness of the propose approach.

Keywords Features · F-measure · Hierarchical classification · Naive Bayes · Text categorization

1 Introduction

World Wide Web (WWW) is much-needed place for information retrieval. But the dynamic nature of the web makes the end user very difficult to get the required information in one place. One of the most affected end user is the researcher who wants to find out all his necessary documents in place but fails to get it. Research is what on which the whole world depends on. Hence, researchers are working hard to discover or mine the hidden things or to invent new things. The main challenge in front of them is to get the required articles in one place for an in-depth literature survey

R.K. Roul (✉) · J.K. Sahoo
BITS-Pilani, K.K. Birla Goa Campus, Pilani, India
e-mail: rkroul@goa.bits-pilani.ac.in

J.K. Sahoo
e-mail: jksahoo@goa.bits-pilani.ac.in

H.S. Behera and D.P. Mohapatra (eds.), *Computational Intelligence in Data Mining*, Advances in Intelligent Systems and Computing 556,
DOI 10.1007/978-981-10-3874-7_32

before going for real work. The present search engine is able to guide them only at the root level to collect such materials, but fails to identify in per level basis. For example, if a user is entering some keywords to find out the corresponding articles, the search engine returns a huge volume of articles among which most of them are not of his interest. The main concern is a proper category of articles which is not available to the user. It has been seen that many times an article will categorize to a specific area based on the journal in which it is published. Generally, journals can be categorized on the basis of their content by the inspection of experts in that area as well as by comparing with those quality journals which are already classified. However, allocation of research level still has not gotten the similar attention. According to Narin et al. [1], CHI research Inc. used to decide the approximate estimate of a research level of an article for many years. They also mentioned that each journal processed by Science Citation Index (SCI) is assigned to a research level on a four-point scale. At present, only less than half of the articles which are published in journals can classify by research level. Hence, classification of articles is indeed necessary for research work. The aim of text categorization is to classify articles into a fixed predefined set of categories. In order to achieve this purpose automatically we use machine-learning methods, to learn classifiers from the known examples and use it to classify the other articles. To narrow down our problem, the proposed approach classifies an article hierarchically. Rather than using classical ontological tree, a dynamic hierarchical tree is generated by numerical results (details explained in Sect. 3.5). Efficient hierarchical termination is used to stop the classification on appropriate level of hierarchy. The novelty of this approach can also be seen in probability measure calculation which goes beyond the usual term frequency model to accommodate 'Category Frequency'. By giving weights to probability measures of unigrams, bigrams, and trigrams, a cumulative measure is shown to produce better results for experimental testing. Usual evaluative measures are modified to support the balance between direct and hierarchical classification (details explained in Sect. 3.7). The dynamic nature of this model makes it an ideal base for various applications. The experimental section tests the various aspects and tradeoffs of this model from different application perspectives. Microsoft research dataset has been used for experimental purpose and the results show the promising nature of the proposed approach.

The paper is organized on the following lines: Sect. 2 covers the literature review based on different classification techniques used for web documents. Section 3 describes the background details of the proposed approach. In Sect. 4, we describe the proposed approach adopted for classifying the article into a particular category. The experimental results are covered in Sect. 5 and finally, conclusion and the suggestions for further development of our work have been presented in Sect. 6.

2 Literature Review

Classification is one of the data mining techniques having many different applications. It is a supervised technique whose class label is known in prior. Classification of articles according to journals or classification of journals according to research level is a big challenge. Narin and his teammates [1] with CHI research prepared a classification scheme which can classify 900 biomedical journals into four research levels (RL). RL = 1 as "clinical observation", RL = 2 as "clinical mix", RL = 3 as "clinical investigation", and RL = 4 as "basic research". Since from inception, RL have been used for characterize research for academic institutions [2] and also for pharmaceutical industry [3]. Later in 1980, the CHI research level has expanded and it includes journals in the field of physical science [4]. Although CHI system has many merits, but it has some demerits also which cannot be ignored and it addresses and handle by Lewison et al. [5] very efficiently. To avoid such demerits, they developed a technique for classification of biomedical articles by RL based on sets of "basic" and "applied" words. Later in 2006, Cambrosio et al. [6] have shown the change in cancer landscape over time using the Lewison method. Mcmillan et al. [7] show that the majority of biomedical articles are cited by individual patents belongs to basic science category. Tijssen [8] proposed a classification technique using knowledge utilization that can be an alternate of research level. His new proposed classification technique has six categories: academic, industry relevant, industry practice, clinical relevant, clinical practice, and industry clinical relevant. According to Lewison and Dawson [9] and Lewison and Devey [10], research level have more importance compared to articles that related to applied work. Boyack et al. [11] described that although from time to time many additional journals have been added to the research level list, only 4200 have assigned research list so far. Boyack et al. [12] have trained their system with the words from the title, abstract, and references which can able to classify over 25 million articles from Scopus by research level. Similarly, many other research work has also been done in this area [13, 14].

To identify such required articles, the proposed approach has gone in level by level basis. Based on the keywords (keywords are taken from the keywords section of the article) of each article, the approach has considered some top k articles out of a huge volume of articles and classify them into a proper category. Naive Bayes classifier has been used for the classification of such article. Promising F-measure of the empirical results justifies the strength of the proposed approach.

3 Basic Preliminaries

Classifying the text documents into a fixed predefined set of categories is the main aim of text categorization. Keeping this in mind, we trained the classifiers from known examples and use them to classify the unclassified articles. This section discusses the basic methodology followed to implement the approach.

3.1 Data Acquisition

A large number of articles corresponding to each category constitute the training dataset. All these articles generate a mega-articles and a vocabulary from the respective category is extracted from it. For this purpose, 'Microsoft Academia Research dataset'[1] is used. All these vocabulary generate the starting corpus.

3.2 Preprocessing

For each category of articles, all the stop-words are removed using 'SQL Full Text'[2] and then the keywords are stemmed using porter stemming algorithm.[3] Finally, all possible monograms and bigrams of keywords are added to a list which constitute the feature set for our model and their count taken as the average count of all occurrences.

3.3 Features, Filtering and Indexing

To ensure that the list of keywords effectively represents and describes the domain of a category, top K words of the list (i.e., feature set generates after preprocessing) are chosen. Next, the list of keywords is sorted (descending) based on their keyword counts (i.e., number of occurrences in the research articles of respective category). To represent each category using features in an appropriate way, indexing needs to be done. For this purpose, traditional Naive Bayes algorithm using "Term Frequency" indexing method (Sect. 3.4) is used. Here, a new mechanism for indexing called "Category Frequency" (Sect. 3.6) has been added to the existing technique.

3.4 Naive Bayes

Bayes's Theorem or Bayes's Rule[4] is a widely used framework for classification.

$$P(W|L) = \frac{(P(L|W)P(W))}{(P(L))} = \frac{(P(L|W)P(W))}{(P(L|W)P(W) + P(L|\overline{W})P(\overline{W})} \quad (1)$$

[1] http://academic.research.microsoft.com/.

[2] http://dev.mysql.com/doc/refman/5.5/en/fulltext-stopwords.html.

[3] http://snowball.tartarus.org/algorithms/porter/stemmer.html.

[4] http://faculty.washington.edu/tamre/BayesTheorem.pdf.

Here, W represents the theory of hypothesis that we are interested in testing, and L represents a new piece of evidence that seems to proof or disproof the theory. We applied this Bayes's rule for a document D and a category C, we got the following:

$$P(C|D) = \frac{(P(D|C)P(C))}{(P(D))} \tag{2}$$

Now according to "Maximum Posterior Hypothesis"[5] (i.e. most likely class)

$$argmax_C P(C|D) = argmax_C P(D|C)P(C) \tag{3}$$

As $P(D)$ is constant, it can be neglected while searching for maximum probability. We take document D as a tuple of features $<X_1, X_2, X_3, \ldots, X_n>$. Further applying Maximum likelihood hypothesis wherein we assume the probability for each category to be a constant, we obtained the following equation:

$$argmax_C P(C|D) = argmax_C P(X_1, X_2, X_3, \ldots, X_n|C) \tag{4}$$

Now taking "Conditional Independence Assumption",[6] we assume the probability of observing the conjunction of features to be equal to the product of probability of individual feature.

$$argmax_C P(C|D) = argmax_C \prod_{i=1}^{n} P(X_i|C) \tag{5}$$

Each probability measure can be calculated using "Term Frequency Measure", i.e.,

$$P(X_i|C_j) = \frac{N(X_i|C_j)}{N(C_j)} \tag{6}$$

Taking into consideration of probability of unobserved events we apply Laplacian Smoothing [15] to Eq. 6. The resulting probability measure is

$$P(X_i|C_j) = \frac{N(X_i|C_j) + 1}{N(C_j) + k} \tag{7}$$

where k is the number of classification categories under consideration.

3.5 Hierarchical Tree Generation

We considered the help and suggestions of professional experts in their respective domains in order to generate the hierarchical tree for each category of Science and Technology which is just a one-time job. Known queries belong to a category (called category C) are classified according to the traditional Naive Bayes model (Sect. 3.4). based on category counts (i.e., count of queries getting classified under that category), the results are sorted in descending order. Category C is added as the

[5] http://www.cs.rutgers.edu/~mlittman/courses/cps271/lect-18/node6.html.
[6] http://en.wikipedia.org/wiki/Conditional_independence.

subcategory of the category with highest count, iff it is not the category with the highest count. If category C is the category with the highest count and second highest satisfies the predefined conditions, then it is added as the subcategory of C. This process is repeated for all categories iteratively until no further addition in the tree is observed.

3.6 Proposed Probability Measure

To incorporate the importance of a keyword to a category with respect to other categories, we propose a probability measure termed "Category Frequency" which can be defined as

$$P(X_i|C_j) = \frac{N(X_i \cap C_j)}{N(X_i)} \tag{8}$$

where, $N(X_i \cap C_j)$ represents "number occurrence of word X_i in C_j" and $N(X_i)$ represents "the total occurrence of X_i in all categories".

We derived the above equation from Eq. 6 by ignoring $N(C_j)$ assuming that the number of words in each category is constant. Then we normalize the $N(X_i \cap C_j)$ with $N(X_i)$ so that all values lie between 0 and 1. Now "Category Frequency" is a ratio of number of times each keyword occurs in a category divided by its occurrence in all categories. We combine this "Category Frequency" measure with traditional "Term Frequency" measure and added Laplacian Smoothing [15] to each of the measures for getting the final measure of the following Equation where k represents a number of categories:

$$P(X_i|C_j) = \frac{N(X_i|C_j) + 1}{N(C_j) + k} * \frac{N(X_i|C_j) + 1}{N(X_i) + k} \tag{9}$$

3.7 Performance Measure

For the evaluation purpose, we use standard precision, recall, and F-measures [16]. We compute precision, recall, and F-measures for both direct and hierarchical classification schema. To integrate both of these schemas, we use the harmonic mean of F-measure (direct) and F-measure (Hierarchical). From the properties of harmonic mean, we can infer that the model having effective values for both F-measure (Hierarchical) and F-measure will show higher harmonic mean thus giving equal weight to both of these measures. Depending on the situation, appropriate weight distribution can be achieved.

$$F(index)_\beta = (1 + \beta^2)\frac{F(Direct) * F(Hierarchical)}{\beta^2 F(Hierarchical) + F(Direct)} \tag{10}$$

Taking β to be 2 will put more emphasis on F(Direct) than F(Hierarchical) and on the other hand, taking β to be 0.5 will put more emphasis on F(Hierarchical) than F(Direct).

4 Proposed Approach

For testing our approach, we chose 10 major categories of Science and Engineering streams as the starting point. Each of these major categories contains on an average 12–15 subcategories. The training vocabulary is taken to be the feature set of 1000 most used words in the publications of each category (i.e., the parameter *K* defined in Sect. 3.3 is taken to be 1000, which will be varied later for performance comparison). We start the classification process from start of the tree. The proposed approach consists of the following steps:

Step 1: Keywords obtained from the test publication are preprocessed in the same way as training data.

Step 2: The number of exact keywords matches is found in each category. If for any category, the percentage of keywords found is greater than 33% then it is considered a *level 1 hit.* If the condition fails, we say its *level 1 miss*. Now three cases arise at this step.

Case 1: There is only category in level 1 hit. If for only 1 category, there is level 1 hit, then the system chooses the respective category as Top category, moves a step forward for further ontological classification and repeat the process for each of the subcategories. If the Top Category has no children (or subcategories) then it is taken as the final category.

Case 2: There are several categories in level 1 hit. In such a case, we use modified Naive Bayes approach (mentioned in Sects. 3.4 and 3.6) to calculate and compare probabilities of only categories with level 1 hit.

Case 3: There is no category with level 1 hit. In this case, we calculate and compare probabilities for each category.

Step 3: Probability for all the keywords $W_1, W_2, \ldots, W_n$ is calculated using three approaches considered for comparison. We call this probability as level 1 probability.

Step 4: All possible bigrams and unigrams from each keyword are extracted and preprocessed. Taking these as keywords, their probability is calculated in the same way as in *Step 3*. We call probability obtained by bigrams and unigrams as level 2 and level 3 probabilities respectively.

Step 5: Next all the probabilities (level 1, 2, and 3) are scaled to bring them on the same level. Scaling factor is taken as the absolute value of the ceiling of the log (base 10) of the least probability value. Then each set of values is multiplied by 10 to the power scaling factor to scale them on same level.

Step 6: Finally a score (*S*) is assigned to each category based on the following equation:

$$S = Weight1 * (level\ 1\ probability) + Weight2 * (level\ 2\ probability) + Weight3 * (level\ 3\ probability) \tag{11}$$

Step 7: The highest scoring category becomes the "Parent Category" for the next scheme of classification. However, if it has no subcategories, then "Parent Category" becomes the final category. Else, the algorithm repeats for all the subcategories of further ontological classification.

Step 8: During ontological classification, the highest scoring child category gets assigned to the query if its probability exceeds a particular threshold value which is set depending on the parent's probability (i.e., threshold percentage).

5 Experiment and Results

For demonstration purpose, herein we show the results obtained in one major category of Science and Engineering streams such as Computer Science. A total of 800 research papers, 50 each of 16 subcategories[7] are used for testing purpose. Models compared

(i) Only "Term Frequency" used as probability measure.
(ii) Only "Category Frequency" used as probability measure.
(iii) Combination of "Term Frequency" and "Category Frequency" used as probability measure.

Each of these models will have parameters namely weights assigned to various levels of probability, hierarchical threshold percentage (explained in Sect. 3), and value of K (explained in Sect. 3.3). In this section, we analyze the results of one of the 16 categories namely "Computer Science". Total of 16 subcategories are considered under major category "Computer Science".

5.1 Tree Generation

Using the method described in Sect. 3.5, a hierarchical tree is generated. Set of 400 known queries is used as the input to this tree generation. Tree obtained is shown in Fig. 1.

5.2 Model Selection

We have compared models under consideration by evaluating their performance on 60 different sets of parameters. The plot of F(Harmonic[8]) for all three models can be shown in Fig. 2. From the Fig. 2 we can see that every point on the line corresponding to proposed model (i.e., Term frequency combined with Category Frequency) holds higher value than any point on line corresponding to Term Frequency or Category Frequency Model. Hence, proposed model can be said to outperform Term Frequency or Category Frequency Model in most of the situations. The accuracy measures, obtained by taking best performing parameters for each model are shown in Table 1. We can see that highest precision, as well as recall is shown by model "Term Frequency + Category Frequency". The results are shown in Table 1 can be justified by following arguments. If we consider a list of keywords corresponding to any category to be its domain, then Term Frequency model computes the probability assigned to any category using only domain knowledge of that category. This would

[7] [11] http://academic.research.microsoft.com/.

[8] Harmonic mean of F-measure (Direct) and F-measure (Hierarchical).

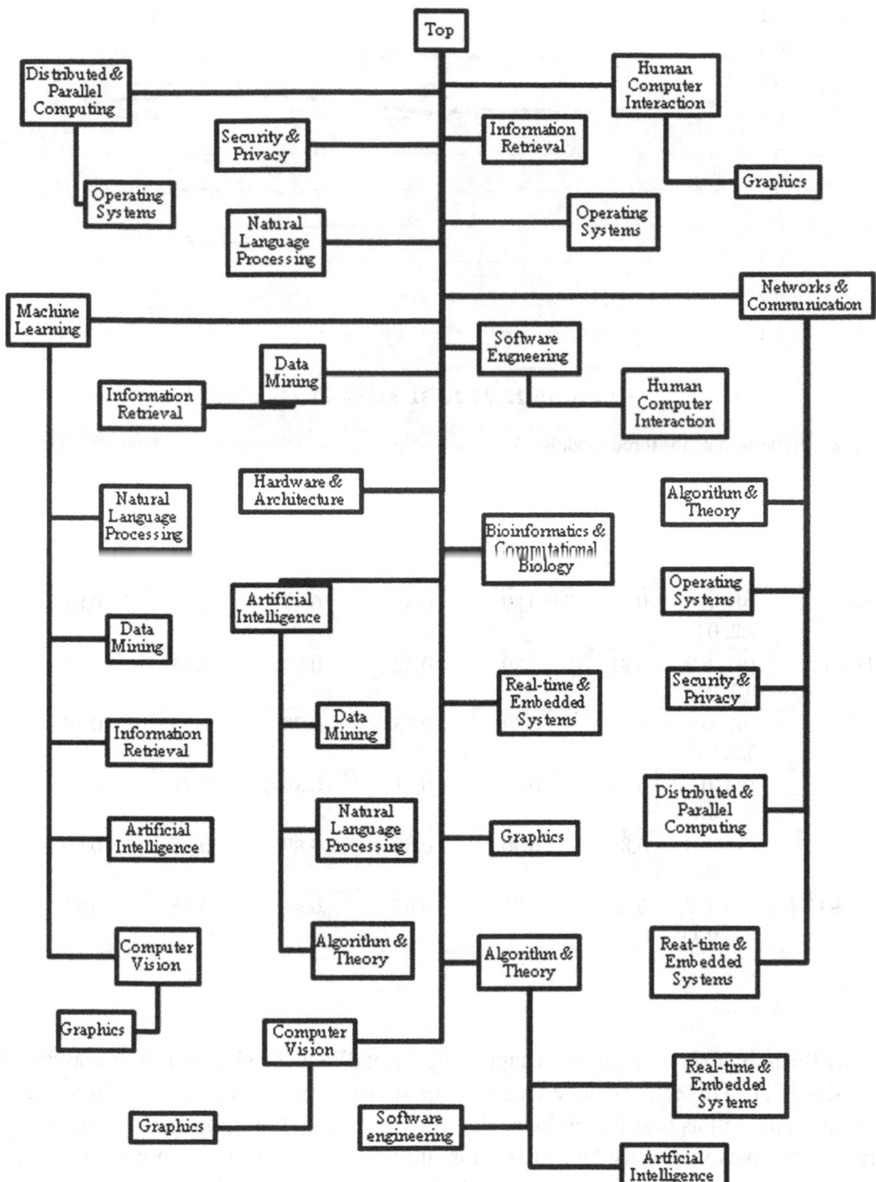

Fig. 1 Hierarchical tree of computer science

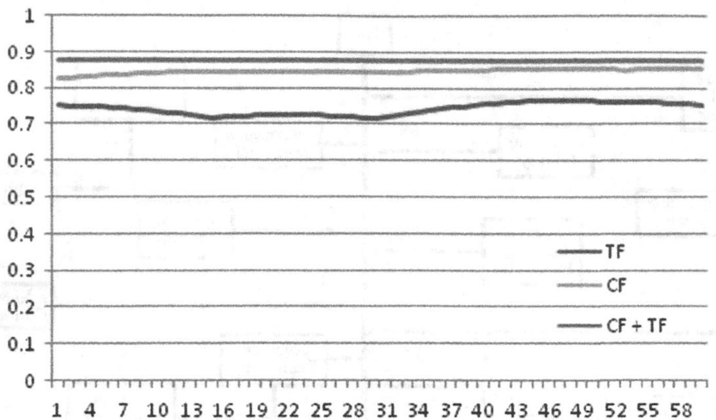

Fig. 2 F(Harmonic) for three models

Table 1 Best performing parameters for all three models

Model	Parameter	Precision	Precision (o)	Recall	Recall (o)	F-measure	F-measure (o)
TF	{97, 0.8, 0.2, 0}	0.83	0.89	0.82	0.87	0.83	0.88
TF(o)	{97, 0.8, 0.2, 0}	0.83	0.89	0.82	0.87	0.83	0.88
CF	{97, 0.8, 0.2, 0}	0.75	0.79	0.75	0.79	0.75	0.79
CF(o)	{83, 0.8, 0.1, 0.1}	0.71	0.80	0.71	0.80	0.71	0.80
TF + CF	{97, 0.8, 0.2, 0}	0.86	0.90	0.86	0.89	0.86	0.90
TF + CF(o)	{85, 0.7, 0.2, 0.1}	0.85	0.91	0.85	0.91	0.85	0.91

imply the reduction of fluctuation happening from direct classification to hierarchical classification. This effect can be seen as same set of parameters gives best direct classification as well as best hierarchical classification. With the same analogy, category frequency takes into consideration of domain knowledge of all categories, but does not implement a term assigned to specific domain knowledge. This will increase the fluctuation between direct and hierarchical classification. This can be seen as two different models with very different signatures are assigned to best direct classification and best hierarchical classification respectively.

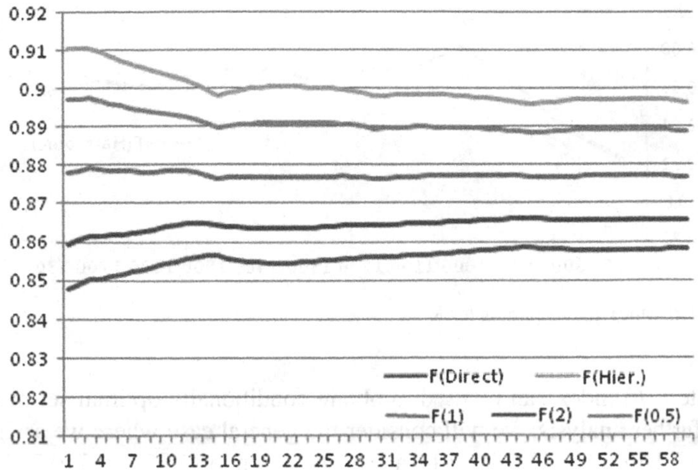

Fig. 3 Parameter selection graph

5.3 Parameter Selection

For appropriate parameter selection, we ran model selected in Sect. 5.2 over 60 different set of parameters. We obtain a plot of $F(\beta)$[9] for β values 1, 2 and 0.25 alongside F(Direct) and F(Hierarchical) which is shown in Fig. 3. As it can be seen from Fig. 3, F-measure (Hierarchical) has higher values as compared to F-measure (Direct) which is as expected. F-measure (Direct) obtains a stable higher region in the Sect. 45 to 60[10] which means that there is no effect on classification if we give higher weight to the exact keyword rather than unigram or bigram. However we see no such region in F-measure (Hierarchical) graph, but it obtains significantly higher values in the initial section and then decreases considerably. When β is 1, index F(β) assigns equal weightage to F(Direct) and F(Hierarchical) as explained in Sect. 3.7. From curve of F(1) in Fig. 3 we can see that it is constant for most of the parameters. However, a certain sudden increase can be seen at the parameter point 3. Parameter set corresponding to this point is (85%, 0.7, 0.2, 0.1) (where (p, W_1, W_2, W_3) implies cutoff percentage = p, weight for level 1 probability is W_1, weight for level 2 probability is W_2, weight for level 3 probability is W_3). When β value is 2, index assign higher weightage to F(Direct) than F(Hierarchical) which can be justified by analyzing Fig. 3. Parameter point 45 represents the optimal set of parameters according to this index. Parameters corresponding to this point are (97.0, 0.8, 0.2, 0.0). Similarly, parameter point 3 represents the optimal set of parameters according to index F(0.5) which is associated with parameters (85%, 0.7, 0.2, 0.1). Depending on the situation,

[9] $F(index)_\beta$, refer Sect. 3.7.

[10] I.e. (83%, 0.8, 0.1, 0.1) → (97%, 0.8, 0.1, 0.1) where (p, W_1, W_2, W_3) implies cutoff percentage = p, weight for level 1 probability is W_1, weight for level 2 probability is W_2, weight for level 3 probability is W_3.

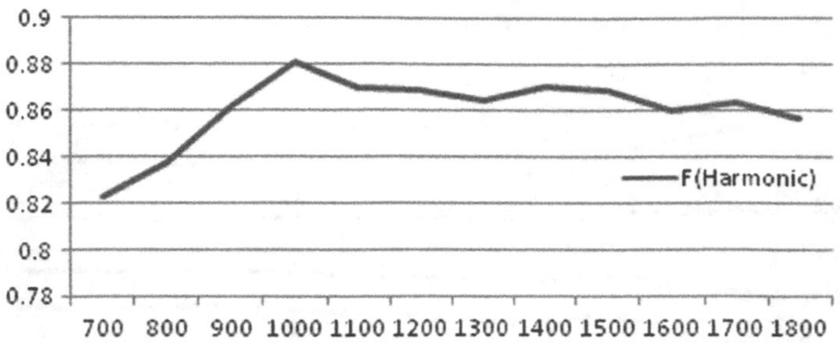

Fig. 4 F(Harmonic) versus parameter *K*

appropriate $F(\beta)$ index can be used to obtain conditionally optimal set of parameters. For further analysis, we will consider the general case where we assign equal weights to F(Direct) and F(hierarchical), hence we choose (85%, 0.7, 0.2, 0.1) these values of parameters the most appropriate values for our model.

The filtering process described in Sect. 3.3 involved selection of parameter *K* which is taken to be 1000 for further analysis. Figure 4 shows a plot of F(Harmonic) versus the parameter *K* chosen for filtering. Value of F(Harmonic) increases with *K* in the starting part of the curve. This can be justified as more domain specific words are getting added to the vocabulary which implies increase in number of query-word matches. As more and more keywords are added to the vocabulary, it may lose the domain-specific nature. Large vocabularies may contain keywords belonging primarily to other domains. Such a vocabulary does not describe a domain efficiently and affects the classification in a negative way. This behavior can be verified by value of F(Harmonic) attains maximum at 1000 and decreases further away.

5.4 Subcategory Wise Results

We show the results for each subcategory under "Computer Science" using the model "Category frequency + Term Frequency" with parameters[11] obtained in Sect. 5.3. Accuracy measures such as precision and recall, for each subcategory are plotted in Fig. 5. Figure 6 shows the comparison between F-measure (Direct) and F-measure (Hierarchical) across all subcategories. As can be seen from the graphs, ontological classification almost always has higher value compared to direct classification for all accuracy measures. Categories like "Algorithms and Theory" (bar number 1), "Artificial Intelligence" (bar number 2), "Computer Vision" (bar number 5), "Data Mining"(bar number 6) shows considerable improvement in precision values in Ontological Classification. For Recall too, we see huge improvement in "Machine Learning and Pattern Recognition" (bar number 11) and "Networks and Communications" (bar number 13) where the value rises in access of 0.2. F-measure analy-

[11](85%, 0.7, 0.2, 0.1) and K = 1000.

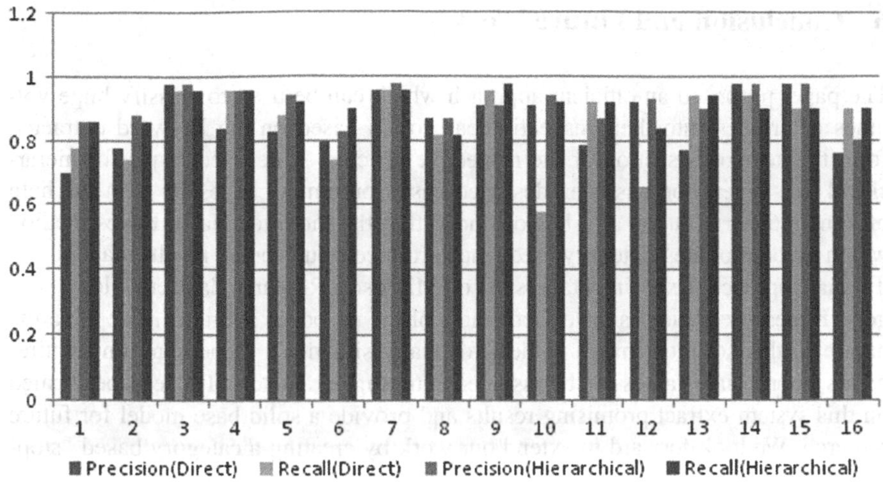

Fig. 5 Accuracy of direct classification (category wise)

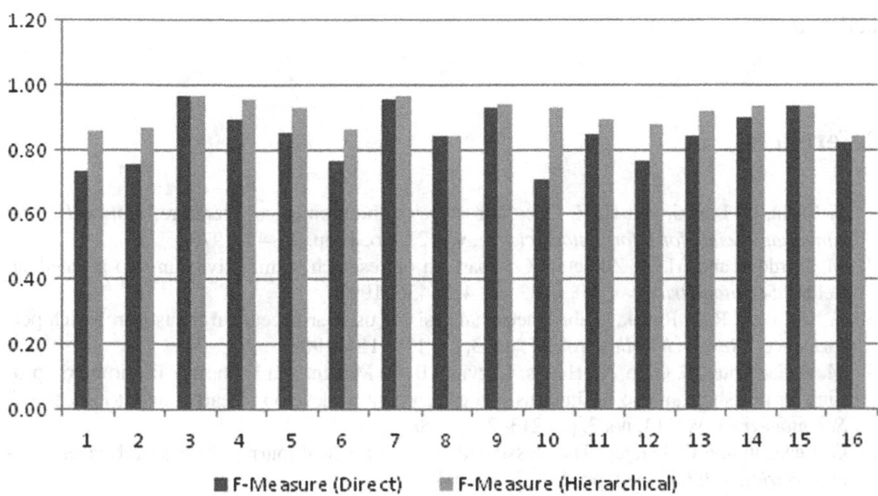

Fig. 6 F-measure (Direct) and (Hierarchical)

sis in Fig. 6 confirms the highest improvement from direct measure to hierarchical measure occurs in "Machine Learning and Pattern Recognition" whereas little or no change is observed for "Bio-Informatics and Security and Privacy" (bar number 4). These results can be attributed to the fact that there is a strong co-relation between "Machine Learning and Pattern Recognition" and its subcategories, thereby we get higher error rate in direct classification. In contrast, there is no such co-relation for independent categories, like "Bio-Informatics and Security and Privacy", which in turn increases the direct classification rate.

6 Conclusion and Future Work

The paper proposed an efficient approach which can be used to classify huge volumes of articles into their respective categories based on the keyword extracted from the keywords section of the respective articles. A new technique for hierarchical tree generation has been designed based on empirical results with the help of human experts. It can also be concluded that the modified Naive Bayes method which accommodates Category Frequency produces much better results than the traditional approach. Experimental results on Microsoft Research dataset yields a very good F-measure indicates the effectiveness of the proposed approach. The dynamic nature of this solution makes it ideal for many situations. Depending on requirements, appropriate values can be assigned to tuning parameters. The tests performed on this system extract promising results and provide a solid base model for future research. We look forward to extend our work by creating a category based "stop-word" list as part of preprocessing. Further weighted average count of unigram and bigram generated as the part of preprocessing can be taken as either the confidence or the support index. More keywords can be extracted from abstract or reference section to strengthen the query.

References

1. F. Narin, G. Pinski, and H. H. Gee, "Structure of the biomedical literature," *Journal of the American society for information science*, vol. 27, no. 1, pp. 25–45, 1976.
2. M. Bordons and M. A. Zulueta, "Comparison of research team activity in two biomedical fields," *Scientometrics*, vol. 40, no. 3, pp. 423–436, 1997.
3. F. Narin and R. P. Rozek, "Bibliometric analysis of us pharmaceutical industry research performance," *Research Policy*, vol. 17, no. 3, pp. 139–154, 1988.
4. M. P. Carpenter, F. Gibb, M. Harris, J. Irvine, B. R. Martin, and F. Narin, "Bibliometric profiles for british academic institutions: An experiment to develop research output indicators," *Scientometrics*, vol. 14, no. 3, pp. 213–233, 1988.
5. G. Lewison and G. Paraje, "The classification of biomedical journals by research level," *Scientometrics*, vol. 60, no. 2, pp. 145–157, 2004.
6. A. Cambrosio, P. Keating, S. Mercier, G. Lewison, and A. Mogoutov, "Mapping the emergence and development of translational cancer research," *European journal of cancer*, vol. 42, no. 18, pp. 3140–3148, 2006.
7. G. S. McMillan, F. Narin, and D. L. Deeds, "An analysis of the critical role of public science in innovation: the case of biotechnology," *Research Policy*, vol. 29, no. 1, pp. 1–8, 2000.
8. R. J. Tijssen, "Discarding the basic science/applied sciencedichotomy: A knowledge utilization triangle classification system of research journals," *Journal of the American society for Information science and Technology*, vol. 61, no. 9, pp. 1842–1852, 2010.
9. G. Lewison and G. Dawson, "The effect of funding on the outputs of biomedical research," *Scientometrics*, vol. 41, no. 1, pp. 17–27, 1998.
10. G. Lewison and M. Devey, "Bibliometric methods for the evaluation of arthritis research." *Rheumatology*, vol. 38, no. 1, pp. 13–20, 1999.
11. K. Boyack and R. Klavans, "Multiple dimensions of journal specificity: Why journals cant be assigned to disciplines," in *The 13th conference of the international society for scientometrics and informetrics*, vol. 1, 2011, pp. 123–133.

12. K. W. Boyack, M. Patek, L. H. Ungar, P. Yoon, and R. Klavans, "Classification of individual articles from all of science by research level," *Journal of Informetrics*, vol. 8, no. 1, pp. 1–12, 2014.
13. C. Coelho, S. Das, and A. Chattopadhyay, "A hierarchical classification scheme for computationally efficient damage classification," *Proceedings of the Institution of Mechanical Engineers, Part G: Journal of Aerospace Engineering*, vol. 223, no. 5, pp. 497–505, 2009.
14. A. Kosmopoulos, I. Partalas, E. Gaussier, G. Paliouras, and I. Androutsopoulos, "Evaluation measures for hierarchical classification: a unified view and novel approaches," *Data Mining and Knowledge Discovery*, vol. 29, no. 3, pp. 820–865, 2015.
15. C. Zhai and J. Lafferty, "A study of smoothing methods for language models applied to ad hoc information retrieval," in *Proceedings of the 24th annual international ACM SIGIR conference on Research and development in information retrieval*. ACM, 2001, pp. 334–342.
16. G. Hripcsak and A. S. Rothschild, "Agreement, the f-measure, and reliability in information retrieval," *Journal of the American Medical Informatics Association*, vol. 12, no. 3, pp. 296–298, 2005.

Experimental Study of Multi-fractal Geometry on Electronic Medical Images Using Differential Box Counting

Tina Samajdar and Prasant Kumar Pattnaik

Abstract This paper focuses on the effect of fractal dimension and lacunarity to measure the roughness of digital medical images that has been contaminated with varying intensities of Gaussian noise as additive noise, Periodic noise as multiplicative noise, and Poisson noise as distributive noise. The experimental study shows that the fractal dimension successfully reveals the roughness of an image, while lacunarity reveals homogeneity or heterogeneity of the image and differential box counting method provides both the fractal dimension and lacunarity values which helps in diagnosis of diseases.

Keywords Fractal geometry · Fractal dimension · Lacunarity

1 Introduction

The Fractal Geometry is a study of physical spatial structure which is available in fragmented and irregular shape. This geometry is widely accepted by computer researchers, medical practitioners, academicians due to its ramified applications in fields like, remote image analysis, texture analysis, object detection, face detection, human eye tracking, mobile object tracking, and video processing. However, its wide influence on medical science is also very special on detection and diagnosis of various diseases. Moreover, the concept of fractal geometry was introduced by Mandelbrot in the year 1977. Pentland analyzed images of natural objects in measure of fractals. Fractal geometry is applied in most of the scenarios where Euclidean geometry formulas are the constraints with respect to shape and structure. Say examples, complex structures like landscapes, coastlines, valleys, etc., are the shapes without fixed geometrical structures. For simplification in this context, brain

T. Samajdar (✉) · P.K. Pattnaik
School of Computer Engineering, KIIT University, Bhubaneswar, India
e-mail: tinasamajdar@gmail.com

P.K. Pattnaik
e-mail: patnaikprasantfcs@kiit.ac.in

H.S. Behera and D.P. Mohapatra (eds.), *Computational Intelligence in Data Mining*, Advances in Intelligent Systems and Computing 556,
DOI 10.1007/978-981-10-3874-7_33

MRIs, dental images, ruptured skin infections, images of human eye, malignant cells, etc., also do not obey fixed geometrical structures. Hence fractal geometry is more preferable in these images. Fractal features include: Fractal dimension, self-similarity, scaling behavior, lacunarity, and hurst exponent. Fractal analysis is the study of analysis of all these fractal features. This work focuses on fractal dimension and lacunarity.

In 2015, S. Talu et al. have estimated these fractal dimension and lacunarity values on amblyopic retinal images of humans. The study motivates early diagnosis of patients with amblyopia and thus contributes in retinal disease analysis [1]. In 2013, T. Pant et al. in his paper discussed the effect of only one feature of fractal geometry, the fractal dimension on noisy digital images. Based on the experimental analysis, the roughness of the image has been estimated [2]. In 2012, P. Shanmugavadivu et al. discussed the texture analysis of digital images on the basis of fractal dimension. Fractal dimension increases with increase in noise content. The concept can be further used for noise detection in images [3]. In 2012, Andre Ricardo Backes et al. analyzed texture of color images with analysis of each color channel separately based on global estimation of fractal descriptors [4]. In 2009, M. Ivanovici et al. have estimated fractal and lacunarity values on color images of human psoriatic lesions [5]. In 2001, A.K. Bisoi et al. analyzed fractal dimension through cell counting method and discussed the necessity of lower bound on the box size to calculate the roughness of the image accurately [6].

The paper is organized as follows: Sect. 2 describes the concept of Fractal Dimension, Sect. 3 describes the concept of Lacunarity, Sect. 4 describes the proposed methodology, Sect. 5 illustrates the noise injected in the images, Sect. 6 shows the experimental studies, Sect. 7 shows the experimental outcomes, and analysis, Sect. 8 provides the conclusion.

2 Fractal Dimension

'Clouds are not spheres, mountains are not cones, coastlines are not circles, and bark is not smooth, nor does lightning travel in a straight line…' Mandelbrot coined the statement and introduced one of the vital scientific concepts of the 20th century, fractal geometry [7].

Fractal is defined as an object having two properties: self similarity and fractal dimension. Fractal grammatically means irregular and fragmented.

Self similarity refers to the fact that the object resembles itself whether scaled down or scaled up, i.e., if the object is divided into parts and again integrated, the original object is obtained. Fractal dimension is also called self similar dimension, D.

$$D = \frac{logN_r}{log\frac{1}{r}} \tag{1}$$

where, N_r is the number of self-similar objects when the object is scaled down by ratio r.

Fractals are highly important in remote image analysis since the images are rich in objects like water bodies, land covers, etc. [8]. Fractals are also highly important in fields of texture analysis. They are also used in object detection, face recognition, pattern recognition, and many more owing to a very interesting property, the fractal signature. Fractal signature is a unique and important property of fractals [9]. It is useful in almost all applications [10, 11].

3 Lacunarity

Fractal Dimension estimates the geometric complexity of the object's shape and each object has its lacunarity. Different objects can have the same value for D though having different appearance. So, Mandelbrot in 1982 proposed lacunarity to deal with the mentioned problem [12–14]. The difference in the appearance of the object is owing to the different lacunarity in the object.

Lacunarity means the gap-distribution of the object. Objects with low lacunarity means the objects are homogeneous, whereas, high lacunarity means the objects are heterogeneous. An object that is homogeneous at a small window scale can be heterogeneous at a large window scale. That means lacunarity is scale dependent [15, 16].

Lacunarity at scale r is defined as:

$$\Lambda_r(B) = \frac{E\left[(X_r^B)^2\right]}{\left(E[X_r^B]\right)^2} \tag{2}$$

where $X_r^B(n)$ denotes the histogram of the image drawn from the values from the images divided into boxes and n is the number of pixels lying in the box.

4 Proposed Methodology

In this section fractal dimension and lacunarity is estimated using Differential Box Counting (DBC) technique.

The computation follows three major steps [17, 18]:

Step 1: The quantity of the object is measured using various step sizes.

Step 2: A regression line is drawn in accordance to quantity versus. step size. The units of the axes are in log scale. This log-log plot is also known as Richardson Plot.

Step 3: The slope of the line is the fractal dimension of the object under consideration.

DBC proposed by Sarkar and Chaudhuri in 1992 estimates D quite efficiently.

$$N_r \propto \left(\frac{1}{r}\right)^{-D} \tag{3}$$

where Nr is the count of the number of boxes, r is the ratio of the size length of boxes to the side length of the image.

An image of size M * M is scaled down to a size s * s where

$$\frac{M}{2} \geq s \geq 1,$$

where s is an integer.

$$r = \frac{s}{M}$$

For different values of steps, N_r is counted.

5 Noise and Its Types

Noise is a type of disturbance which results in change in the pixel values of an image. Noise can be categorized on basis of distribution, on the nature of correlation, on the nature of noise, and on the basis of source.

Noise can be additive, multiplicative, or impulse in nature. In the experiment, Poisson noise as a category of distributive noise, varying intensities of Gaussian noise as additive noise, and varying intensities of Periodic noise multiplicative noise have been added [19]. Variances used in case of Gaussian noise are 0.01, 0.001, and 0.005 and variances used in case of Periodic noise are 0.13, 0.22, and 0.78. These values are not randomly considered, several instances are tested and the most suitable ones are used.

6 An Experimental Study of Multi-fractal Geometry on Electronic Medical Images Using Differential Box Counting

This study considers three MRI images of brain, kidney, and liver as depicted in Fig. 1. Each image of Fig. 1 is injected with varying intensities of Gaussian, Periodic, and Poisson noise individually. This means each image is contaminated with 0.001, 0.005, 0.01 of Gaussian noise, 0.13, 0.22, 0.78 of Periodic noise, and Poisson noise as shown in Fig. 2. The simulation scenario consists of boxes of size

2 * 2, 3 * 3, 4 * 4, 5 * 5, 6 * 6, and 7 * 7. The regression line is plotted using Differential Box Algorithm of Sect. 4 and hence the average D is estimated from step 3 of the same algorithm which is simulated in Matlab. For Fractal Dimension and Lacunarity we have considered Eqs. 1 and 2 and the observation of each noiseless image (Fig. 1) and noisy image (Fig. 2) are recorded in Tables 1, 2, Figs. 3, and 4. Lacunarity is computed as the ratio of standard deviation of the histogram values of the image to the square of the expectation of the histogram values of the image.

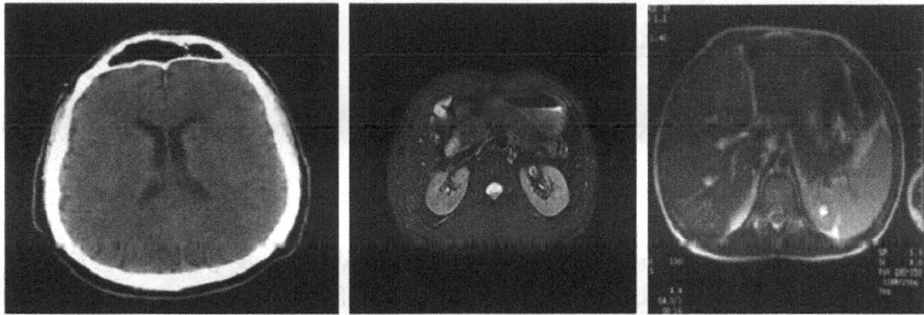

Fig. 1 Image of brain, kidney, and liver with no noise

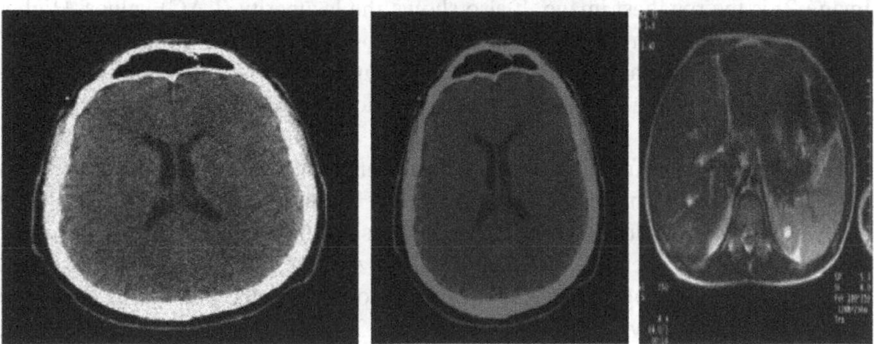

Fig. 2 Noisy images of brain (gaussian noise of variance 0.001), kidney (periodic noise of variance 0.22), and liver (poisson noise) (from *left*)

Table 1 Fractal dimension and lacunarity value of non-noisy images

	Fractal dimension	Lacunarity
Brain image (Image 1)	1.8081	**1.0664**
Kidney image (Image 2)	1.8419	1.0128
Liver image (Image 3)	**1.7411**	1.0457

Table 2 Fractal dimension values of noisy images

	Gaussian noise			Periodic noise			Poisson noise
	0.001	0.005	0.01	0.13	0.22	0.78	
Brain image (Image 1)	1.9285	1.9290	1.9316	1.8526	1.8550	1.8614	1.8682
Kidney image (Image 2)	1.9658	1.9654	1.9670	1.8779	1.8813	1.8810	1.8776
Liver image (Image 3)	1.9450	1.9456	1.9456	1.8519	1.8529	1.8493	1.8682

7 Experimental Results

Table 1 illustrates the recorded fractal dimension and lacunarity value of each noiseless image of brain, kidney, and liver, {1.8081, 1.0664}, {1.8419, 1.0128}, and {1.7411, 1.0457}, respectively. Similarly, for the noisy versions of these images the fractal dimension observations are shown in Table 2. The graphical representation of lacunarity of noisy images is shown in Figs. 3 and 4.

Table 1 shows the fractal dimension (FD) value of the three non-noisy images. It can be seen that FD (Image 3) < FD (Image 1) < FD (Image 2). As fractal dimension is a measure of roughness of an image, so it can be concluded that liver image (Image 3) is less rough than the other two images and kidney image (Image 2) is the roughest image. It also shows the lacunarity (LAC) values. Higher the lacunarity value, more the image is heterogeneous. LAC (Image 1) > LAC (Image 3) > LAC (Image 2). Brain image (Image 1) is more heterogeneous than all the other images and Image 2 is the most homogeneous image than the other images.

Table 2 shows the fractal dimension value of the noisy images. It can be seen that with the increase in the content of noise the fractal dimension also increases. It means the image with lesser noise content is less rough than the image with higher noise content. The fact can be clearly visualized for both Gaussian noise and Periodic noise. For image 1 the less rough version of it is the one that is corrupted with periodic noise of variance 0.13. For image 2 the less rough version of it is the one that is corrupted with poisson noise. For image 3 the less rough version of it is the one that is corrupted with periodic noise of variance 0.78.

Figures 3 and 4 show the lacunarity value of the noisy images. More the lacunarity, more is heterogeneity of the image. Image 1 with gaussian noise of variance 0.01 is the most heterogeneous image than all other corrupted forms of image 1. Image 2 with gaussian noise of variance 0.01 is the most heterogeneous image than all other corrupted forms of image 2. Image 3 with gaussian noise with variance 0.01 is the most heterogeneous image than all other corrupted forms of image 3.

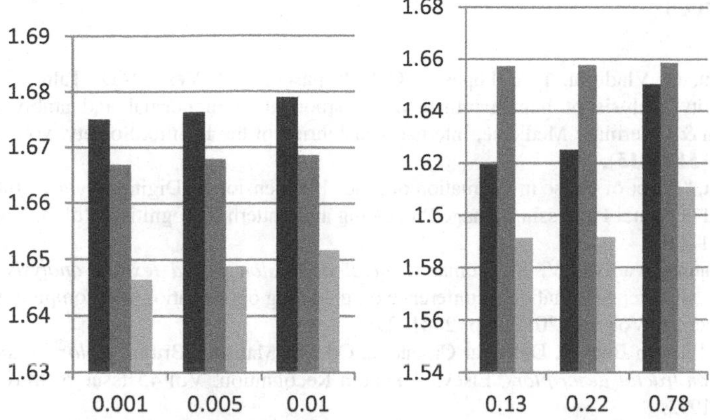

Fig. 3 Lacunarity values of images contaminated with gaussian noise and periodic noise (from *left*)

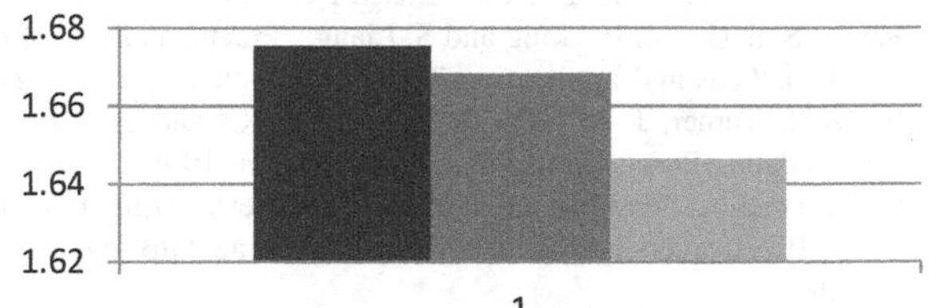

Fig. 4 Lacunarity values of images contaminated with poisson noise

8 Conclusion

The fractal dimension and lacunarity of fractal geometry is an important consideration for the noisy digital image aspects since it deals with homogeneity or heterogeneity of image with irregular shapes. The paper concludes with the following:

- The higher lacunarity value indicates for heterogeneous images in context to noise say, Gaussian, Poisson, and Periodic noise.
- Images having high noise content and its roughness are fairly indicated by its fractal dimension value.

The future scope of the work includes effect of fractal dimension and lacunarity on more fields of medical diagnosis.

References

1. S. Talu, C. Vladutiu, L.A. Popescu, C.A. Lupascu, S.C. Vesa, S.D. Talu, "Fractal and lacunarity analysis of human retinal vessel arborisation in normal and amblyopic eyes", Human & Veterinary Medicine, International Journal of the Bioflux Society, vol. 45, issue 2, pp. 45–51 **(2015).**
2. T. Pant, "Effect of Noise in Estimation of Fractal Dimension of Digital Images", International Journal of Signal Processing, Image Processing and Pattern Recognition Vol. 6, No. 5 (2013), pp. 101–116.
3. P. Shanmugavadivu, V. Sivakumar, *Fractal dimension based texture analysis of digital images*, Elsevier, International conference on modeling optimization and computing, Procedia Engineering, Vol 38, (2012), pp. 2981–2986.
4. Andre Ricardo Backes, Dalcimar Casanova, Odemir Martinez Bruno, *Color texture analysis based on fractal descriptors,* Elsevier, Pattern Recognition, Vol 45 Issue 5, May 2012, Pg. 1984–1992.
5. M. Ivanovici, N. Richard, and H. Decean, *Fractal Dimension and Lacunarity of Psoriatic Lesions*, 2nd WSEAS International Conference on biomedical electronics and biomedical informatics, ISBN: 978–960-474-110-6 **(2009).**
6. A. K. Bisoi, Jibitesh Mishra, "On Calculation of fractal dimension of images", Pattern Recognition Letters Vol. 22, Issues 6–7, pp. 631-637, Elsevier, (**2001**).
7. Alex P. Pentland, "Fractal – Based Description", SRI International, California, pp. 973–981.
8. W. Sun, G. Xu, P. Gong and S. Liang, "Fractal analysis of remotely sensed images: A review of methods and applications", Int. J. Remote Sens., vol. 27, **(2006)**, pp. 4963–4990.
9. M. J. Turner, J. M. Blackledge, and P. R. Andrews, *Fractal geometry in digital imaging*, Academic Press, Cambridge, Great Britain, 1998.
10. S. Jansson, "Evaluation of methods for estimating fractal properties of intensity images", UMEA University, Sweden. Available online: http://www.sais.se/mthprize/2007/jansson2007.pdf.
11. C. M. Hagerhall, T.Purcell, R. Taylor, "Fractal dimension of landscape silhouette outlines as a predictor of landscape preference", Journal of Environmental Psychology 24 (2004) 247–255, Elsevier.
12. A. Barcellos, "The Fractal Geometry of Mandelbrot", pp. 98–114. Available online:http://www.maa.org/sites/default/files/pdf/upload_library/22/Polya/07468342.di020711.02p00026.pdf.
13. B. B. Mandelbrot, "The Fractal Geometry of Nature", WH Freeman and Co., New York, **(1982)**.
14. T. Pant, "Noise Error Analysis in Fractal Dimension Estimation of Digital Images", *I. J. Image, Graphics and Signal Processing,* 2013, 8, pp. 55–62 Published Online June 2013 in MECS (http://www.mecs-press.org/).
15. S.W. Myint, N.Lam, "A study of lacunarity-based texture analysis approaches to improve urban image classification", Elsevier, Available online: www.sciencedirect.com.
16. Y. Quan, Y. Xu, Y.Sun, Y. Luo, "Lacunarity Analysis on Image Patterns for Texture Classification", pp. 1–8, IEEE Xplore.
17. D. Sarkar, T. Thomas, "Fractal Features based on Differential Box Counting Method for the Categorization of Digital Mammograms", International Journal of Computer Information Systems and Industrial Management Applications (IJCISIM). Available online: http://www.mirlabs.org/ijcisim ISSN: 2150–7988 Vol.2 (2010), pp. 011-019.
18. J. Li, Q. Du, C. Sun, "An improved box-counting method for image fractal dimension estimation", Available online: http://www.sciencedirect.com/science/article/pii/S0031320309000843.
19. R. C. Gonzalez and R. E. Woods, "Digital image processing", Pearson Education, New Delhi, **(2002)**.

Tsallis Entropy Based Image Thresholding for Image Segmentation

M.S.R. Naidu and P. Rajesh Kumar

Abstract Image segmentation is a method of segregating the image into required segments/regions. Image thresholding being a simple and effective technique, mostly used for image segmentation and these thresholds are optimized by optimization techniques by maximizing the Tsallis entropy. However, as the two level thresholding is extended to multi-level thresholding, the computational complexity of the algorithm is further increased. So there is need of evolutionary and swarm optimization techniques. In this paper, first time optimal thresholds are obtained by maximizing the Tsallis entropy using novel adaptive cuckoo search algorithm (ACS). The proposed ACS algorithm performance of image segmentation is tested using natural and standard images. Experiments shows that proposed ACS is better than particle swarm optimization (PSO) and cuckoo search (CS).

Keywords Image segmentation • Image thresholding • Tsallis entropy • Cuckoo search • Differential evolution

1 Introduction

Image segmentation is the pre-process step of image compression, pattern recognition, medical imaging applications, biomedical imaging, remote sensing, etc. There are many applications of image segmentation in the literature including synthetic aperture radar (SAR) image extraction, brain tumour extraction, etc. The main impartial of image segmentation is to excerpt numerous features of the image

M.S.R. Naidu (✉)
Department of Electronics and Communication Engineering,
AITAM, Tekkali, India
e-mail: msrnaidu312@rediffmail.com

P. Rajesh Kumar
Department of Electronics and Communication Engineering,
A.U. College of Engineering (A), Andhra University, Visakhapatnam, India
e-mail: rajeshauce@gmail.com

H.S. Behera and D.P. Mohapatra (eds.), *Computational Intelligence in Data Mining*, Advances in Intelligent Systems and Computing 556,
DOI 10.1007/978-981-10-3874-7_34

which can be fused or divided in order to figure objects of attention on which examination and interpretation can be accomplished. Image segmentation represents first step in image compression and pattern recognition. There are so many ways for image segmentation. The simplest and easy way of image segmentation is image thresholding. Thresholding approaches are of two types one is nonparametric and parametric. In nonparametric approach thresholding is performed based on class variance as in Otsu's method or established on an entropy criterion, such as Tsallis entropy, Fuzzy entropy and Kapur's entropy [1]. If the image is partitioned into two classes, i.e. object and background, then the threshold is termed bi-level threshold else multi-level threshold. Thresholding technique has so many real time applications like data, image and video compression, image recognition, pattern recognition, image understanding and communication. Sezgin and Sankur [2] performed comparative study on image thresholding, they classified the image thresholding into six categories. Kapur classifies the image into some classes by calculating threshold which is based on the histogram of the grey level image [3]. Otsu's method classifies the image into some classes by calculating threshold which is based on between-class variance of the pixel intensities of that class [4]. These two methods are under the category of bi-level thresholding and found efficient in case of two thresholds, but for multi-level thresholding the computational complexity is very high. Entropy may be a Shannon, fuzzy, between-class variations, Kapur's entropy, minimization of the Bayesian error and Birge–Massart thresholding strategy. The disadvantage of these techniques is that convergence time or computational time or CPU time increases exponentially with the problem. So alternative to these techniques which minimizes the CPU time for the same problem is evolutionary and swarm-based calculation techniques. Sathya and Kayalvizhi [5] applied bacterial foraging optimization algorithm (BF) for optimizing objective functions, so achieved efficient image segmentation. Further to improve convergence speed and the global searching ability of BF, they modified swarming step and reproduction step, so improved the robustness of BF and achieved fast convergence. Mbuyamba et al. [6] used Cuckoo Search (CS) algorithm for energy minimization of alternative Active Contour Model (ACM) for global minimum and exhibited that polar coordinates with CS is better than rectangular. There are so many optimization techniques available in the literature, in which a few are used for bi-level thresholding for ordinary image segmentation, Ye et al. [7] used fuzzy entropy with bat algorithm (BA) and compared the results with artificial bee colony algorithm (ABC), ant colony (ACO), PSO and Genetic algorithm (GA). Agrawal et al. [8] used Tsallis entropy with CS algorithm and compared the results with BF, PSO and GA. Horng used firefly algorithm (FA) for multilevel image thresholding [9]. Kapur's and Otsu's entropy methods are simple and effective but computationally affluent when prolonged to multilevel thresholding since they hire a comprehensive search for optimal thresholds. Bhandari et al. [10] proposed a Tsallis entropy based multilevel thresholding for coloured satellite image segmentation using high dimensional problem optimizer that is Differential Evolution (DE), WDO, PSO and Artificial Bee Colony (ABC).

In this paper, first time we applied ACS-based image thresholding for image segmentation by optimizing the Tsallis entropy and compared the results with other optimization techniques such as PSO and CS. For the performance evaluation of proposed ACS-based image thresholding we consider objective function value, Misclassification error and Structural Similarity Index (SSIM). In all parameters the proposed algorithm performance is better compared than PSO and CS.

2 Concept of Tsallis Entropy

Actually entropy is related with the amount of disorder in organization. But Tsallis prolonged the idea of entropy to an amount of uncertainty about the information gratified of the organization. It is also proved that the Tsallis entropy has additive property.

$$S(A+B) = S(A) + S(B) \tag{1}$$

By subsequent the multi-fractal theories, the Tsallis entropy extended to non-extensive organization based on a universal entropic formula:

$$S_q = \frac{1 - \sum_{i=1}^{k} (P_i)^q}{q-1} \tag{2}$$

where k is amount of options occurred in organization and q is measuring parameter of non-extensively of the organization called entropic index. This entropic form can be prolonged for a numerical self-governing organization by a pseudo-additive entropic rule:

$$S_q(A+B) = S_q(A) + S_q(B) + (1-q)S_q(A) \times S_q(B) \tag{3}$$

This type of entropy is used to maximize and to find thresholds for image thresholding. Let G is the grey levels of image and the range of these levels is {1, 2, ..., G}. Let $p_i = p_1, p_2, \ldots, p_G$ are the probability distribution of corresponding grey levels. Among these probability distribution functions some are showing background (class A) and remaining probability distribution functions are showing background (class B). The probability distributions of the foreground and background classes are given by [11].

$$p_A = \frac{p_1}{p^A}, \frac{p_2}{p^A}, \ldots, \frac{p_t}{p^A} \text{ and } p_B = \frac{p_{t+1}}{p^B}, \frac{p_{t+2}}{p^B}, \ldots, \frac{p_G}{p^B} \tag{4}$$

where $p^A = \sum_{i=1}^{t} p$ and $p^B = \sum_{i=t+1}^{G} p_i$.

Now calculate the Tsallis entropy of foreground and background which are given as:

$$S_q^m(t) = 1 - \sum_{i=t_m+1}^{G} (P_i/p^m)^q/q - 1 \tag{5}$$

$$S_q^B(t) = 1 - \sum_{i=t+1}^{G} (P_i/p^B)^q/q - 1 \tag{6}$$

Above two equations are the objective functions/fitness function which are maximized by the optimization techniques and resultant optimal thresholds are used for image thresholding. This is acquired by exploiting the objective function for bi-level thresholding:

$$\begin{aligned}[T_1, T_2, \ldots T_m] = \arg\{\max[\mathrm{S}_q^1(\mathrm{t}) + \mathrm{S}_q^2(\mathrm{t}) + \ldots \mathrm{S}_q^m(\mathrm{t}) \\ + (1-q)\mathrm{S}_q^1(\mathrm{t})\mathrm{S}_q^2(\mathrm{t}) \ldots \mathrm{S}_q^m(\mathrm{t})]\}\end{aligned} \tag{7}$$

Above equation must satisfy the following condition.

$$\left|p^m + p^{m+1}\right| - 1 < S^M < 1 - \left|p^m - p^{m+1}\right|$$

where

$$S(\mathrm{t}) = S = [\mathrm{S}_q^A(\mathrm{t}) + \mathrm{S}_q^B(\mathrm{t}) + (1-q)\mathrm{S}_q^A(\mathrm{t})\mathrm{S}_q^A(\mathrm{t})]\} \tag{8}$$

The optimal threshold value '*Th*' which is obtained by maximizing Eq. (7) and it satisfies the Eq. (8). The same equations can also extend and applicable for any multi-level thresholding. The dimensions of the optimization are equal to the number of thresholds which is to be maximized. Let m be number of thresholds with $[T_1, T_2, \ldots, T_m]$ and are maximized with Eq. (7).

$$\begin{aligned}[T_1, T_2, \ldots T_m] = \arg\{\max[\mathrm{S}_q^1(\mathrm{t}) + \mathrm{S}_q^2(\mathrm{t}) + \ldots \mathrm{S}_q^M(\mathrm{t}) \\ + (1-q)\mathrm{S}_q^1(\mathrm{t})\mathrm{S}_q^2(\mathrm{t}) \ldots \mathrm{S}_q^M(\mathrm{t})]\}\end{aligned} \tag{9}$$

To be specific, here the object is to optimize the fitness function as given in Eq. (9) using Adaptive cuckoo search algorithm.

2.1 Novel Adaptive Cuckoo Search Algorithm

The CS algorithm is projected by Yang in 2010 [12] and cuckoos step of walk follows the Levy distribution function and obeys the either Mantegna algorithm or McCulloch's algorithm. In the proposed technique, we follow a specific strategy instead of Levy distribution function. The normal CS does not have any appliance to switch the step size in the repetition process, which can lead the method to extent universal minima or maxima. Here, we try to include a step size which is relative to the suitability of the discrete nest in the search space in the present generation. The tuning parameter α is fixed in the literature. In our proposed algorithm step size follows the following equation [13]

$$step_i(\mathrm{t}+1) = (\frac{1}{t})^{|((best\,f(\mathrm{t}) - f_i|(\mathrm{t})) \div (best\,f(\mathrm{t}) - \mathrm{worst}\,f(\mathrm{t})))} \tag{10}$$

where t is the iteration search algorithm; $f_i(t)$ is the objective value ith nest in the iteration t; *bestf*(t) is the best objective in iteration t; *worstf* (t) is the worst objective value in the iteration t. Initially high value of step size is considered and is decreasing with the increment in iteration. It shows the algorithm tries to global best solution. From Eq. (11), Step size depends upon the iterations, and it shows adaptive of step size of the algorithm. From the observation step size is adaptive and chooses its value based on the fitness value. The population follows the following equation

$$X_i(t+1) = X_i(t) + randn \times step_i(t+1) \tag{11}$$

The major benefit of the naval adaptive cuckoo search is that it does not need any preliminary parameter to be distinct. It is quicker than the cuckoo search algorithm.

$$X_i(t+1) = X_i(t) + randn \times step_i(t+1) \times X_i(t) - X_{gbest} \tag{12}$$

where X_{gbest} is the universal solution amongst all X_i for I (for $i = 1, 2, \ldots, N$) at time t..

3 Results and Discussions

For the performance evolution which includes robustness, efficiency and convergence of proposed firefly algorithm, we selected "Cameraman", "Lena", "Lake" and "Goldhill" as a test images. All these images are .jpg format images and of size 225 × 225 and corresponding histograms are shown in Fig. 1. In general, perfect threshold can be selected if the histogram of image peaks is lanky, thin, symmetric,

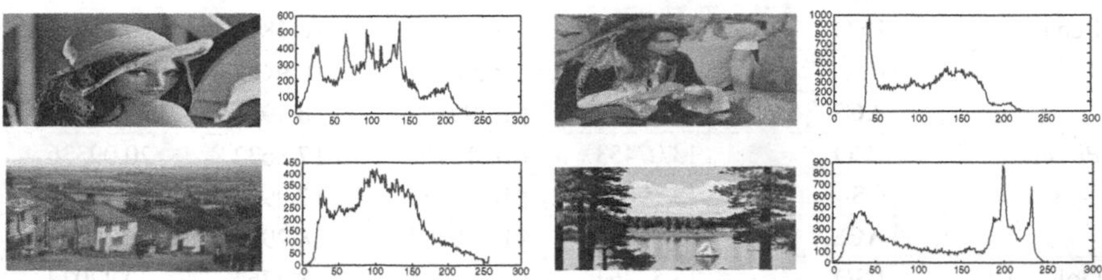

Fig. 1 Standard image and respective histograms of three methods. **a** Lena, **b** Pirate, **c** Goldhill, **d** Lake

and divided by unfathomable valleys. Cameraman, Lake, Goldhill and pirate image histograms peaks are tall, narrow and symmetric, but for Lena images histogram peaks are not tall and narrow so difficult to segment with ordinary methods. So we proposed adaptive cuckoo search algorithm-based image thresholding for effective and efficient image segmentation of above said critical images by optimizing Tsallis entropy. The performance and effectiveness of proposed adaptive CS proved better compared to other optimization techniques like PSO and CS.

3.1 Maximization of Tsallis Entropy

The ACS and other two algorithms are applied on Tsallis entropy objective function and compared the results of PSO and CS. All the algorithms are optimized to maximize the objective function. Table 1 shows the objective values CS and PSO. It is observed from Table 1 that objective values obtained with ACS using Tsallis entropy is higher than the PSO and CS for different images. Form Table 1, for Pirate image at five level thresholding, the ACS algorithm thresholds values are quite different with others.

3.2 Misclassification Error/Uniformity Measure

It is measure of uniformity in threshold image and is used to compare optimization techniques performance. Misclassification error is measured by Eq. 13

Table 1 Objective values obtained by various algorithms

Objective value					
		Th = 2	Th = 3	Th = 4	Th = 5
Images	Opt Tech	Tsallis	Tsallis	Tsallis	Tsallis
Cameraman	PSO	13.18744	16.24024	18.3295	20.01387
	CS	13.32343	16.43232	18.4544	20.23455
	ACS	13.42432	16.68494	18.4123	20.33456
Lena	PSO	13.77477	16.89825	18.7813	20.4863
	CS	13.83932	16.90002	18.9484	20.50505
	ACS	13.77199	16.99322	18.9991	20.83838
Pirate	PSO	13.07453	16.13192	17.8632	20.09336
	CS	13.19048	16.23938	17.9903	20.19393
	ACS	13.35049	16.49494	18.2939	20.39292
Goldhill	PSO	13.54264	16.63928	18.4767	20.22024
	CS	13.55933	16.89939	18.6949	20.31290
	ACS	13.59094	16.90999	18.7373	20.59406

$$M = 1 - 2 * Th * \frac{\sum_{j=0}^{Th} \sum_{i \in R_j} (\mathrm{I}_i - \sigma_j)^2}{N * (\mathrm{I}_{\max} - \mathrm{I}_{\min})^2} \tag{13}$$

where *Th* is the number of thresholds which are used to segment the image, R_j is the *j*th segmented region, I_i is the intensity level of pixel in that particular segmented area, σ_j is the mean of *j*th segmented region of image, *N* is total number of pixels in the image, I_{min} and I_{max} are the maximum and minimum intensity of image, respectively. In general misclassification errors lies between 0 and 1 and higher value of misclassification error shows better performance of the algorithm. Hence, the Uniformity measure in thresholding is measured from the difference between the maximum value, 1 (better quality of image) and minimum value, 0 (worst quality of image). Table 2 shows misclassification error of proposed and other techniques and proved proposed method have lesser misclassification error and shows better visual quality.

3.3 Structural Similarity Index (SSIM)

It estimates the visual likeness between the input image and the decompressed image/thresholded image and is calculated with below equation

$$SSIM = (2\mu_I\mu_{\tilde{I}} + C1)(2\sigma_{I\tilde{I}} + C2)/(\mu_I^2 + \mu_{\tilde{I}}^2 - C1)(\sigma_I^2 + \sigma_{\tilde{I}}^2 - C2) \tag{14}$$

where μI and μĨ are the mean value of the input image I and decompressed image Ĩ, σ_I and $\sigma_{\tilde{I}}$ are the standard deviation of original image I and reconstructed image Ĩ,

Table 2 Misclassification error (in%) for the thresholding methods

Misclassification error (in%)					
		Th = 2	Th = 3	Th = 4	Th = 5
Images	Opt Tech	Tsallis	Tsallis	Tsallis	Tsallis
Cameraman	PSO	0.984848	0.994943	0.96112	0.888909
	CS	0.949393	0.984733	0.94001	0.861111
	ACS	0.936206	0.970692	0.93981	0.858628
Lena	PSO	0.989097	0.943268	0.93443	0.820931
	CS	0.979909	0.945328	0.9198	0.815684
	ACS	0.972519	0.94903	0.90205	0.797201
Goldhill	PSO	0.983578	0.980953	0.85098	0.81987
	CS	0.978421	0.959089	0.82068	0.809168
	ACS	0.971891	0.942676	0.82816	0.831336
Lake	PSO	0.989214	0.967512	0.90089	0.848098
	CS	0.980077	0.959098	0.89165	0.82818
	ACS	0.975625	0.946025	0.84816	0.810071

Table 3 Structural similarity index (SSIM) for various algorithms

Structural similarity index					
		Th = 2	Th = 3	Th = 4	Th = 5
Images	Opt Tech	Tsallis	Tsallis	Tsallis	Tsallis
Cameraman	PSO	0.659364	0.764626	0.816141	0.770239
	CS	0.665123	0.787674	0.821234	0.818723
	ACS	0.676343	0.802143	0.842434	0.836668
Lena	PSO	0.640931	0.766572	0.781932	0.818869
	CS	0.720976	0.790435	0.790690	0.839090
	ACS	0.741289	0.809124	0.800079	0.859864
Goldhill	PSO	0.643207	0.716937	0.720002	0.768224
	CS	0.649877	0.730098	0.770987	0.790009
	ACS	0.658798	0.749642	0.790866	0.799886
Lake	PSO	0.757208	0.807438	0.819467	0.851051
	CS	0.770997	0.834567	0.846790	0.865343
	ACS	0.791235	0.858765	0.880989	0.889981

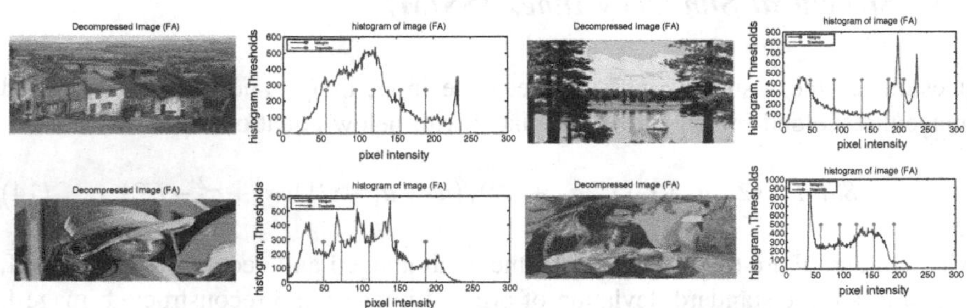

Fig. 2 Segmented images and respective optimized 5 level thresholds with ACS

$\sigma_{I\tilde{I}}$ is the cross-correlation and C1 and C2 are constants which are equal to 0.065. Table 3 shows the SSIM of various methods with Tsallis entropy and it demonstrate proposed method SSIM is higher than other methods. Figure 2 shows the segmented images and respective optimized 5 level thresholds with ACS and it shows segmentation with ACS is better than PSO and CS.

4 Conclusions

In this paper, we proposed natural inspired adaptive cuckoo search algorithm-based multilevel image thresholding for image segmentation. ACS maximizes the Tsallis entropy for efficient and effective image thresholding. The proposed algorithm is tested on natural images to show the merits of the algorithm. The results of the proposed method are compared with other optimization techniques such as PSO and CS with Tsallis entropy. From the experiments we observed that proposed algorithm has higher/maximum fitness value compared to PSO and CS. The SSIM value shows higher values with proposed algorithm than PSO and CS. It is concluded that proposed algorithm outperform the PSO and CS in all performance measuring parameters.

References

1. De Luca. A, S. Termini, A definition of non-probabilistic entropy in the setting of fuzzy sets theory, Inf. Control 20 (1972) 301–312.
2. Sezgin. M, B. Sankur, Survey over image thresholding techniques and quantitative performance evaluation, J. Electron. Imaging 13 (1) (2004) 146–165.
3. Kapur. J. N, P.K. Sahoo, A.K.C Wong, A new method for gray-level picture thresholding using the entropy of the histogram", Computer Vision Graphics Image Process. 29 (1985) 273–285.
4. Otsu. N, "A threshold selection from gray level histograms" IEEE Transactions on System, Man and Cybernetics 66, 1979.
5. Sathya. P. D and R. Kayalvizhi, "Optimal multilevel thresholding using bacterial foraging algorithm", Expert Systems with Applications, Vol. 38, pp. 15549–15564, 2011.
6. Mbuyamba. M, J. Cruz-Duarte, J. Avina-Cervantes, C. Correa-Cely, D. Lindner, and C. Chalopin, "Active contours driven by Cuckoo Search strategy for brain tumour images segmentation", Expert Systems With Applications, Vol. 56, pp. 59–68, 2016.
7. Ye. Z, M. Wang, W. Liu, S. Chen, "Fuzzy entropy based optimal thresholding using bat algorithm", Applied Soft Computing, Vol. 31, pp. 381–395, 2015.
8. Agrawal. S, R. Panda, S. Bhuyan, B.K. Panigrahi, "Tsallis entropy based optimal multilevel thresholding using cuckoo search algorithm", Swarm and Evolutionary Computation, Vol. 11 pp. 16–30, 2013.
9. Horng. M and T. Jiang, "Multilevel Image Thresholding Selection based on the Firefly Algorithm", Symposia and Workshops on Ubiquitous, Autonomic and Trusted Computing, pp. 58–63, 2010.
10. Bhandari. A. K, A. Kumar, G. K. Singh, "Tsallis entropy based multilevel thresholding for colored satellite image segmentation using evolutionary algorithms", Expert Systems With Applications, Vol. 42, pp. 8707–8730, 2015.
11. Yudong Zhang, Lenan Wu, Optimal multi-level thresholding based on maximum Tsallis entropy via an artificial bee colony approach, Entropy 13 (4) (2011) 841–859.
12. Yang. X.S, S. Deb, Cuckoo search via Levy flights, in: Proc. IEEE Conf. of World Congress on Nature & Biologically Inspired Computing, 2009, pp. 210–214.
13. M. K. Naika, R. P. Panda, "A novel adaptive cuckoo search algorithm for intrinsic discriminant analysis based face recognition", Applied Soft Computing, Vol. 38, pp. 661–675, 2016.

Efficient Techniques for Clustering of Users on Web Log Data

P. Dhana Lakshmi, K. Ramani and B. Eswara Reddy

Abstract Web usage mining is one of the essential framework to find domain knowledge from interaction of users with the web. This domain knowledge is used for effective management of predictive websites, creation of adaptive websites, enhancing business and web services, personalization, and so on. In nonprofitable organization's website it is difficult to identify who are users, what information they need, and their interests change with time. Web usage mining based on log data provides a solution to this problem. The proposed work focuses on web log data preprocessing, sparse matrix construction based on web navigation of each user and clustering the users of similar interests. The performance of web usage mining is also compared based on k-means, X-means and farthest first clustering algorithms.

Keywords Web usage mining · Sparse matrix · Clustering · Influence degree · K-means · X-means and farthest first algorithm

1 Introduction

Digitalization of information and rapid growth of information technology lead to enormous data in all domains in variety of formats and entire data may not be useful to all users as it is. Data Mining helps to extract only relevant information from these large repositories. Web is a huge repository of text documents and multimedia data. Mining useful data from the web is known as web mining and it is classified as: Web content analysis, web usage mining, and web structure analysis.

P. Dhana Lakshmi (✉) · K. Ramani
Sree Vidyanikethan Engineering College, Tirupathi, India
e-mail: mallidhana5@gmail.com

K. Ramani
e-mail: ramanidileep@yahoo.com

B. Eswara Reddy
JNTUA College of Engineering Kalikiri, Kalikiri, India
e-mail: eswarcsejntua@gmail.com

H.S. Behera and D.P. Mohapatra (eds.), *Computational Intelligence in Data Mining*, Advances in Intelligent Systems and Computing 556,
DOI 10.1007/978-981-10-3874-7_35

Web content analysis is similar to data mining technique for relational databases. Web content analysis is an extraction of useful domain knowledge or information from web documents. This information may be audio, video, text, or image. Web document may contain structured or unstructured data. Web content mining is applied to different research fields based on text and images such as Text or Image content Retrieval and artificial intelligence. Web structure analysis is used to find webpage structure or boundary extraction in WebPages accessed by the web users, which is used to project the similarities of web content and to improve the web structure. Web usage analysis is helpful to find the frequent web accessed patterns for user analysis.

From the web analytics and web application point of view, extracted domain knowledge obtained from the web usage mining could be directly applied to cross market analysis for e-market Business, e-Services, e-Education, e-Newspapers, e-Governance, Digital repositories, etc. Due to availability of large size of the web data, it is essential to analyze the web content for the e-business and online applications to avail user's recommended systems [1]. With the large number of companies using the Internet to distribute and collect information, knowledge discovery on the web has become an important research area. In web usage pattern analysis the data is extracted from the web log file [2], it contains the information such as user browsing information, who are visiting the web site, which file the user is accessing, how many bytes of data are accessed, type of Operating System, and other information. There are many techniques of web usage mining is developed by many researchers. The Sect. 2 covers a review of the web usage mining, Sect. 3 is about existing work and Sect. 4 gives the details of proposed work and Sect. 5 describes the results and Sect. 6 discussed conclusion along with future work.

2 Related Work

With rapid usage of web by a large number of customers, clustering of these people becomes an important task based on their usage or accessing of web pages. This section gives a brief review of web mining and its literature. The literature survey says that V. Vidyapriya and S. Kalaivani [3] developed an efficient clustering technique of web logs for web pages based on the hit count. It comprises data preprocessing, user and session identification, path completion and clustering is performed using farthest first technique (FFC) used for clustering. But in this method occurrence of outliers is more while clustering and not suitable for smaller datasets. B. Uma Maheswari and P. Sumathi [4] introduced a clustering mechanism of web log data which is based on many attributes such as web users, web page, etc. The clustering process includes data cleaning, user identification, session identification, and finally clustering. If the attribute is web users, clustering is done by a Basic Probability Assignment (BPA) [5]. But this method is applicable only for small repositories, therefore it lacks the scalability. Antony Selvadoss, Thanamani et al. [6] proposed a technique based on fuzzy C-means to cluster the web user activities. The mechanism involves constructing the session matrix, calculating the R index and F-score. But, this technique is highly expensive as it involves large number of computations for clustering. J. HuaXu and H. Liu [7] developed a web user clustering technique based on vector matrix: Users and URL as rows and columns and values as number of hits. Similarity measure is found by using cosine similarity function to find similarity between the vectors (rows) and then unsupervised k-means model is used to cluster the web users behavior. But, this model does not work when the variation between hits value is high. K. Santhisree et al. [8] implemented a new framework known as rough-set based DBSCAN clustering to find the user access patterns along with inter-cluster similarity identification among the clusters. The implementation is based on the set similarity, sequence similarity values. This is not valid for soft clustering. Xidong Wang et al. [9] proposed a frequent access pattern method to extract the frequent access patterns from users' accessed path for a specific website. Generation of frequent access pattern tree using this technique is difficult when more number of access records exist for each user. Wei et al. [10] implemented a multi-level model for web users clustering based on similarity interests from the user's accessed web pages. This technique is suitable only if the number of sessions is higher than the given threshold measure. No one technique is optimum in clustering the web users. Therefore, novel techniques to achieve optimum clustering and clustering with minimum time are very much necessary.

3 Existing System

Web user clustering is one of the essential tasks in web usage analysis. The goal of clustering is to group data points that are close (or similar) to each other and identify such groupings (or clusters) in an unsupervised manner [11]. Information of web user clusters has been widely used in many applications, such as solution of website structure design and optimization [12]. There are many clustering techniques employed for finding the interestingness patterns among users. One among such technique is clustering of web users through Matrix Influence Degree (MID), which contains influence degree of each web page corresponding to the user. Simple k-means is applied over the obtained MID to obtain the clusters [13]. Clustering process is pictographically represented in Fig. 1 which is a sequential process.

K-means is greedy approach for partitioning n objects into k clusters. Clusters are formed by minimizing the sum of squared distances to the cluster centers. The main task of K-means clustering is described as

- Given a set X of n points in a d-dimensional space and an integer k
- Task: choose a set of k points $\{c_1, c_2, \ldots, c_k\}$ in the d-dimensional space to form clusters $\{C_1, C_2, \ldots, C_k\}$ such that the following function (Eq. 1) is minimized.

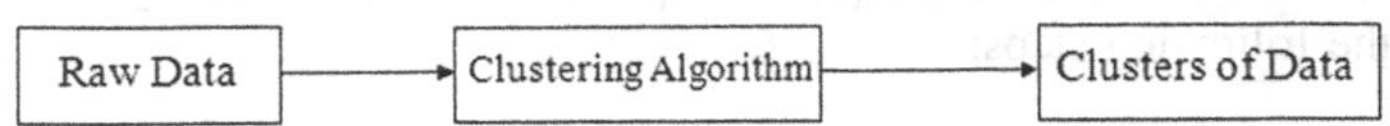

Fig. 1 Clustering of data

$$Cost(C) = \sum_{i=1}^{k} \sum_{x \in C_i} L_2^2(x - c_i) \tag{1}$$

The k-means algorithm:

- Randomly pick k cluster centers $\{c_1, \ldots, c_k\}$
- For each i, set the cluster C_i to be the set of points in X that are closer to c_i than they are to c_j for all $i \neq j$
- For each i let c_i be the center of cluster C_i (mean of the vectors in C_i)
- Repeat until convergence.

In the similar manner simple K-means operates over the generated MID obtained from web log data. J. Xiao et al. [14] developed a cluster analysis method to mine the web, this method clusters the web users based on their profiles. User profile clustering algorithm consists of three modules. First it finds the similarity measure based on Belief function and applies greedy clustering using this function then creates the common user profiles. In greedy clustering, it first randomly selects the users into common profile set and for each user calculate similarity measure, based on this choose best representative and update similarity of each point to the closest representative finally it gives the unique clusters. To generate the clusters using session representatives V. Sujatha and Punitha Valli [15] adopted a method called Hierarchical Agglomerative clustering. In this method first it considers the navigation patterns for each session as a single cluster. Clusters members are added based on similar session pages. And similar navigation patterns, it is calculated by distance similarity measure for several sessions. Density-based algorithm works [16] first searching core objects based on these objects clusters are increased and by looking for objects which are in a neighborhood with in a radius of object. The main advantage of this density-based algorithm is it can filter out noise.

Disadvantages of existing systems:

- Time consuming to build the model over large records and increases over the size of MID.
- Web user clusters are dependent on the initial point selected, which have large influence over the cluster sizes.
- More number of outliers can cause problem of ambiguity.
- Does not operate well in large variation between values of the influences degree.

4 Proposed System

In this paper, an efficient technique is implemented for clustering web users. It includes the following steps:

1. Data preprocessing of the weblog and extracting the relevant attributes for clustering.
2. Build the Sparse Matrix of every user with sessions and pages visited.
3. Generate Three Tuple matrix from sparse matrix.
4. Calculate the influence Degree of all web pages for a user.
5. Repeat the Steps 2, 3, and 4 for all web users until all sessions are completed in the web log.
6. Generate the Matrix of Influence Degree (MID) for all web pages.
7. Cluster the MID with X-means and farthest first techniques.

4.1 Data Preprocessing

The raw data of web log is as shown in Table 1, which contains set of attributes and all of these are not useful for clustering the web users. The input raw data is applied to noise removal phase, where missing values and records with error status numbers such as 400,404 are removed and only relevant attributes are selected. The required log data after cleaning is shown in Table 2, it contains relevant features such as network IP address, Time of Request, and URL are extracted to cluster the web users.

The generated three tuple matrices from log data after application of preprocessing are shown in Table 2. From Table 2 each URL is identified with unique number. This data indicates 14 web pages are accessed by web users.

4.2 Building a Sparse Matrix

Sparse matrix is a matrix in which most of the elements are zeros. Web log data is transformed into a sparse matrix. User session is considered as the criterion to group web log data. Web log data is grouped by user session for each web user and the user session is defined as 3600 s.

Let there are 'm' web users U_i, where i = 1, 2,..., m, each web user has 'n' user sessions S_j, where j = 1, 2,..., n, and there are 'p' web pages P_k, where k = 1, 2,..., p then the sparse matrix $\mathrm{SM_i}$ of U_i is shown in Eq. (2):

$$\mathrm{SM_i} = \begin{bmatrix} s1p1 & s1p2 & s1p3 \\ s2p1 & s2p2 & s2p3 \\ s3p1 & s3p2 & s3p3 \end{bmatrix}, \tag{2}$$

where, $\mathrm{S_jp_k}$ is the value equal to the times of web page P_k accessed in user session S_j by web user $\mathrm{U_i}$. 50 web log data records are considered for experimentation, the sparse matrix for the 14 users and their session usage of web pages is generated as

Table 1 Sample web log data

IP address	User ID	Time	Method/URL/protocol	Status	Size	Refer	Agent
82.117.202.158	–	[16/Nov/2009:01:12:18 +0100]	GET/galerija.php HTTP/1.0	200	4545	http://www.vtsns.edu.rs/	Moz/5.0
82.117.202.158	–	[16/Nov/2009:01:12:22 +0100]	GET/galerija.php HTTP/1.0	200	3176	http://www.vtsns.edu.rs/	Moz/5.0
82.208.207.41	–	[16/Nov/2009:01:12:43 +0100]	GET/ispit_odbijeni.php HTTP/1.0	200	826	http://www.vtsns.edu.rs/	Moz/5.0
83.136.179.11	–	[16/Nov/2009:01:22:43 +0100]	GET/index.php HTTP/1.0	200	826	http://www.vtsns.edu.rs/	Moz/5.0
83.136.179.11	–	[16/Nov/2009:01:23:43 +0100]	GET/index.php HTTP/1.0	200	826	http://www.vtsns.edu.rs/	Moz/5.0
82.208.207.41	–	[16/Nov/2009:02:12:23 +0100]	GET/ispit_odbijeni.php HTTP/1.0	200	23689	http://www.vtsns.edu.rs/	Moz/5.0
82.208.207.41	–	[16/Nov/2009:02:12:25 +0100]	GET/ispit_odbijeni.php HTTP/1.0	200	5740	http://www.vtsns.edu.rs/	Moz/5.0
82.208.207.41	–	[16/Nov/2009:02:13:43 +0100]	GET/biblioteka.php HTTP/1.1	200	3669	http://www.vtsns.edu.rs/	Moz/5.0
82.117.202.158	–	[16/Nov/2009:02:17:26 +0100]	GET/ispit_raspored_god.php HTTP/1.1	200	3669	http://www.vtsns.edu.rs/	Moz/5.0
82.117.202.158	–	[16/Nov/2009:02:17:39 +0100]	GET/ispit_raspored_god.php HTTP/1.1	200	6554	http://www.vtsns.edu.rs/	Moz/5.0
83.136.179.11	–	[16/Nov/2009:02:17:41 +0100]	GET/ispit_raspored_god.php HTTP/1.1	200	4053	http://www.vtsns.edu.rs/	Moz/5.0
82.208.255.125	–	[16/Nov/2009:02:17:44 +0100]	GET/ispit_raspored_god.php HTTP/1.1	200	1662	http://www.vtsns.edu.rs/	Moz/5.0

Moz/5.0 → Mozilla/5.0 (Windows; U)

Table 2 Web log data after cleaning

IP address	Time	URL
82.117.202.158	1:12:18	0
82.117.202.158	1:12:22	0
82.117.202.158	1:12:43	1
82.208.207.41	1:22:43	2
83.136.179.11	1:23:43	2
83.136.179.11	2:12:23	1
82.208.207.41	2:12:25	1
82.208.207.41	2:13:43	3
82.208.207.41	2:17:26	4
82.117.202.158	2:17:39	4
82.117.202.158	2:17:41	4
83.136.179.11	2:17:44	4
82.208.255.125	2:17:53	3
82.208.255.125	3:12:42	5
82.117.202.158	3:27:23	1
83.136.179.11	3:37:32	1
83.136.179.11	3:37:44	5
82.208.255.125	4:13:43	5
83.136.179.11	4:17:26	1
82.117.202.158	5:17:39	4
82.208.255.125	5:17:41	4
82.208.255.125	5:18:40	5
82.208.207.41	5:37:53	4
82.117.202.158	5:39:42	5
82.117.202.158	6:27:23	4
83.136.179.11	6:37:32	4
83.136.179.11	18:44:25	0
157.55.39.13	18:44:25	3
157.55.39.136	18:08:01	2
199.79.62.54	18:44:15	1
157.55.39.38	18:44:21	6
157.55.39.136	18:47:58	4
40.77.167.52	18:48:04	5
40.77.167.52	18:43:52	7
157.55.39.38	18:49:16	8
123.125.71.70	19:11:33	9
66.249.65.71	19:06:27	10
105.224.92.45	19:06:27	11
105.224.92.45	19:06:27	12
105.224.92.45	19:08:01	13
199.79.62.54	19:06:27	14

(continued)

Table 2 (continued)

IP address	Time	URL
105.224.92.45	19:04:32	14
157.55.39.108	19:06:26	9
105.224.92.45	19:06:27	12
105.224.92.45	19:06:29	13
105.224.92.45	19:06:30	0
105.224.92.45	19:06:27	6
105.224.92.45	19:38:05	9
66.249.65.55	19:38:27	13
66.249.65.49	19.37.27	9

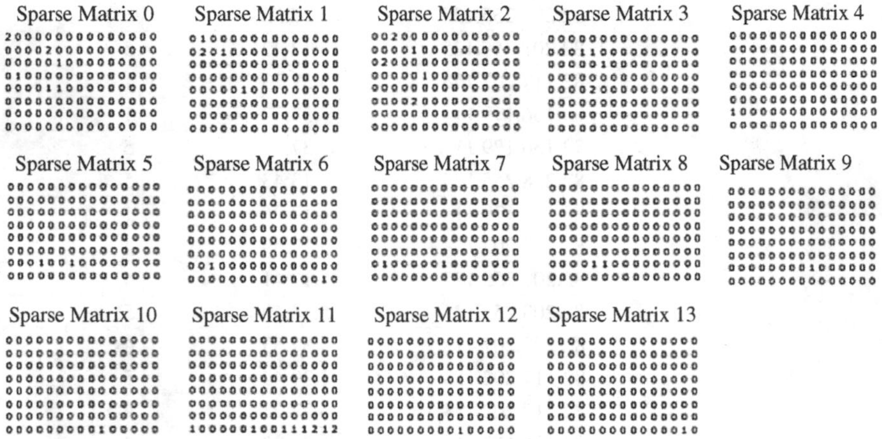

Sparse Matrix 0
```
200000000000000
000020000000000
000001000000000
010000000000000
000011000000000
000000000000000
000000000000000
000000000000000
```

Sparse Matrix 1
```
010000000000000
020100000000000
000000000000000
000000000000000
000001000000000
000000000000000
000000000000000
000000000000000
```

Sparse Matrix 2
```
002000000000000
000010000000000
020000000000000
000001000000000
000000000000000
000020000000000
000000000000000
000000000000000
```

Sparse Matrix 3
```
000000000000000
000110000000000
000001000000000
000000000000000
000020000000000
000000000000000
000000000000000
000000000000000
```

Sparse Matrix 4
```
000000000000000
000000000000000
000000000000000
000000000000000
000000000000000
000000000000000
100000000000000
000000000000000
```

Sparse Matrix 5
```
000000000000000
000000000000000
000000000000000
000000000000000
000000000000000
000000000000000
000100100000000
000000000000000
```

Sparse Matrix 6
```
000000000000000
000000000000000
000000000000000
000000000000000
000000000000000
000000000000000
001000000000000
000000000000010
```

Sparse Matrix 7
```
000000000000000
000000000000000
000000000000000
000000000000000
000000000000000
000000000000000
010000010000000
000000000000000
```

Sparse Matrix 8
```
000000000000000
000000000000000
000000000000000
000000000000000
000000000000000
000000000000000
000011000000000
000000000000000
```

Sparse Matrix 9
```
000000000000000
000000000000000
000000000000000
000000000000000
000000000000000
000000000000000
000000001000000
000000000000000
```

Sparse Matrix 10
```
000000000000000
000000000000000
000000000000000
000000000000000
000000000000000
000000000000000
000000000000000
000000000100000
```

Sparse Matrix 11
```
000000000000000
000000000000000
000000000000000
000000000000000
000000000000000
000000000000000
000000000000000
100000100111212
```

Sparse Matrix 12
```
000000000000000
000000000000000
000000000000000
000000000000000
000000000000000
000000000000000
000000000000000
000000000100000
```

Sparse Matrix 13
```
000000000000000
000000000000000
000000000000000
000000000000000
000000000000000
000000000000000
000000000000000
000000000000010
```

Fig. 2 Sparse matrixes for 14 users

displayed in Fig. 2. It is built for each user with total session count and total number of web pages in the selected website. The values in the cells of matrix indicate the visits of the web page in a particular session (s_1p_1). Value 0 in the cell indicates that page is not visited in that session with respect to the user. In each sparse matrix we can extract the useful knowledge, whether the web pages are accessed by users or not. This will give the importance or preference of accessed pages by the web users. From above diagram, consider the sparse matrix-0, the value in the cell s_0p_0 is 2, it indicates that user-0 visited the page 0 in session-0 twice. The value in cell s_4p_4 is 1, which indicates that user-0 visited the page-4 in session-4 once. The same procedure is followed for all elements of matrices. Thus, all sparse matrices indicate the activity of the user in all sessions.

4.3 Generating Three Tuple Matrix

After building the sparse matrix of each web user, we get its 3-tuple representation. 3-tuple representation of sparse matrix is a list of (i, j, v), where i is the row index of nonzero value, j is the column index of nonzero value, and v is the value of nonzero value in the sparse matrix. The three tuple matrices for the 10 users corresponding to their sparse matrices are generated as displayed in Fig. 2. Every tuple

User 0

I	J	V
0	0	2
1	4	2
2	5	1
3	1	1
4	4	1
4	5	1

User 1

I	J	V
0	1	1
1	1	1
1	3	1
4	5	1

User 2

I	J	V
0	2	2
1	4	1
2	1	2
3	5	1
5	4	2

User 3

I	J	V
1	3	1
1	4	1
2	5	1
4	4	2

User 4

I	J	V
6	0	1

User 5

I	J	V
6	3	1
6	6	1

User 6

I	J	V
6	2	1
7	13	1

User 7

I	J	V
6	1	1
6	7	1

User 8

I	J	V
6	4	1
6	5	1

User 9

I	J	V
6	8	1

User 10

I	J	V
7	9	1

User 11

I	J	V
7	0	1
7	6	1
7	9	1
7	10	1
7	11	1
7	12	2
7	13	2
7	4	2

User 12

I	J	V
7	9	1

User 13

I	J	V
7	13	1

Fig. 3 Generating three tuple matrix

Table 3 List of Influence degree for 3 tuple array

User 0	2 1 0 0 1 1 0 0 0 0 0 0 0 0 0 0
1	0 1 0 1 0 1 0 0 0 0 0 0 0 0 0 0
2	0 2 2 0 1 1 0 0 0 0 0 0 0 0 0 0
3	0 0 0 1 1 1 0 0 0 0 0 0 0 0 0 0
4	1 0 0 0 0 0 0 0 0 0 0 0 0 0 0 0
5	0 0 0 1 0 0 1 0 0 0 0 0 0 0 0 0
6	0 0 1 0 0 0 0 0 0 0 0 0 0 0 1 0
7	0 1 0 0 0 0 0 1 0 0 0 0 0 0 0 0
8	0 0 0 0 1 1 0 0 0 0 0 0 0 0 0 0
9	0 0 0 0 0 0 0 0 1 0 0 0 0 0 0 0
10	0 0 0 0 0 0 0 0 0 1 0 0 0 0 0 0
11	1 0 0 0 0 0 1 0 0 1 1 1 2 1 2
12	0 0 0 0 0 0 0 0 0 1 0 0 0 0 0 0
13	0 0 0 0 0 0 0 0 0 0 0 0 0 0 1 0

Table 4 Farthest first and X-means clustering results

K	C_1	C_2	C_3	C_4	C_5	C_6	C_7	C_8	C_9	C_{10}	C_{11}	C_{12}	C_{13}	C_{14}	C_{15}	C_{16}	C_{17}	C_{18}	C_{19}	C_{20}	C_{21}	C_{22}	C_{23}	C_{24}	C_{25}
Farthest first																									
10	165	1	5	1	1	1	2	2	4	3															
15	159	1	2	1	1	1	1	1	2	3	1	3	3	2	4										
25	149	1	1	1	1	1	1	1	1	2	1	2	1	2	3	4	2	3	2	1	1	1	1	1	1
X-means																									
10	52	8	10	7	12	6	6	7	12	65															
15	51	8	4	5	7	5	7	2	4	6	7	7	6	15	51										
25	42	2	14	8	10	9	1	1	4	5	30	12	6	5	1	6	6	4	7	5	1	6			

(row) contains three fields: i-row index of sparse matrix; j-column index of sparse matrix; v-value of the cell with index (i, j); This eliminates the zero in sparse matrix and contains only nonzero visits of the webpage with respect to the session and web page index.

From Fig. 3, three tuple for user 0 has 6 rows and three columns which have nonzero values of v (value in the cell). Consider the first row 3-tuple of user-0, it indicates 0th page visited in 0th session two times. The second row in the 3-tuple indicates in session-1 page-4 is visited twice. This is followed to all the three tuples obtained from the sparse matrix of all users.

4.4 *Influence Degree*

Influence degree of 13 web users with 15 accessed web pages is given in Table 3. Influence degree (ID) is the mean value of accessed times of one web page in one web user. We calculated influence degree of each web page from 3-tuple representation using Eq. (3).

$$ID_j = sum(j, v)/N_j, \tag{3}$$

where j is the column index of nonzero value in 3-tuple representation, *Sum* (*j*,*v*) is the sum of all values v with same j value, N_j is the number of value j. After calculation of influence degree of all web users, we get the matrix of influence degree with the row value of each influence degree of each web user. The MID obtained for the data is represented in Table 3.

Table 5 Comparison of clusters for k = 10

Algorithm	C1	C2	C3	C4	C5	C6	C7	C8	C9	C10
K-means	16	8	117	4	7	6	1	10	4	12
Farthest first	165	1	5	1	1	1	2	2	4	3
X-means	52	8	10	7	12	6	6	7	12	65

Table 6 Comparison of clusters for k = 15

Algorithm	C_1	C_2	C_3	C_4	C_5	C_6	C_7	C_8	C_9	C_{10}	C_{11}	C_{12}	C_{13}	C_{14}	C_{15}
K-means	13	8	97	2	7	3	1	8	4	13	7	5	7	4	6
Farthest first	159	1	2	1	1	1	1	1	2	3	1	3	3	2	4
X-means	51	8	4	5	7	5	7	2	4	6	7	7	6	15	51

Table 7 Comparison of clusters for k = 25

Algorithm	C_1	C_2	C_3	C_4	C_5	C_6	C_7	C_8	C_9	C_{10}	C_{11}	C_{12}	C_{13}	C_{14}	C_{15}	C_{16}	C_{17}	C_{18}	C_{19}	C_{20}	C_{21}	C_{22}	C_{23}	C_{24}	C_{25}
Farthest first	149	1	1	1	1	1	1	1	1	2	1	2	1	2	3	4	2	3	2	1	1	1	1	1	1
K-means	13	3	69	2	7	3	1	9	3	11	7	5	6	3	5	6	5	6	2	2	7	2	4	3	1
X-means	42	2	14	8	10	9	1	1	4	5	30	12	6	5	1	6	6	4	7	5	1	6			

4.5 Clustering of MID

Clustering is defined as grouping of similar type of web objects such that the web objects within a same group are similar (or related) to one another group. We performed our proposed approach MID with different clustering algorithms to analyze the effect of grouping the similar web users.

Initially farthest first, X-means clustering algorithms are applied over the obtained MID for required number of clusters. Obtained clustered instances are summarized in Table 4 as follows:

Thus from the results farthest first technique takes less time to build the model and X-means yields optimal number of cluster instances.

5 Performance Evaluation

The performance evaluation of the system gives the clustering of the web users, it is considered over the three clustering techniques: Simple K-means, farthest first and X-Means. Experiments are performed over 1000 records of web log which consist of 185 users, 40 sessions, and 25 web pages. This entire record will be converted into MID by going through many transformations from sparse matrix form. Final MID obtained is of order 185 by 25. Clustering techniques are applied over the matrix MID for desired k value. Initially, simple k-means is applied for k values 10, 15, 25, and for the same values of k, farthest first, X-means clustering techniques are applied on the same MID and the results are compared which are shown in Tables 5, 6, and 7 as follows:

From the results, compared with existing K-means approach, farthest first technique yields the clusters in less time and optimal clustered instances are obtained through X-means which is shown in Fig. 4.

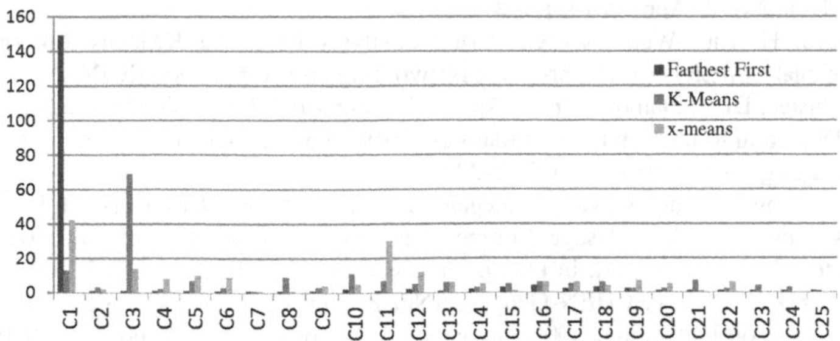

Fig. 4 Graphical representation of clustering techniques (k = 25)

6 Conclusion and Future Work

We proposed an approach to cluster the web users by transforming the web log data into a sparse matrix, then three tuple matrix is constructed by removing unused webpages. Influence degree of each web page is calculated for one user. Finally MID is formed for all web users. Proposed approach is applied on generated MID to form optimal number of clusters using X-Means and farthest first technique for clustering the web users in less time. In future our proposed approach can be extendable for finding best navigation paths using Association Rule Mining.

Acknowledgements We would like to thank our college Sree Vidyanikethan Engineering college, Tirupathi for providing valuable resources and encouragement to do research.

References

1. S. Jagan, Dr. S.P. Rajagopalan. A Survey on Web Personalization of Web Usage Mining, International Research Journal of Engineering and Technology (IRJET), Volume: 02 Issue: 01 | March-2015. pp. 6–12.
2. Rachit Adhvaryu, A Review Paper on Web Usage Mining and Pattern Discovery, Journal Of Information, Knowledge And Research In Computer Engineering, Volume – 02, Issue – 02nov 12 To Oct 13, pp. 279–284.
3. V. Vidyapriya, S. Kalaivani, An Efficient Clustering Technique for Weblogs, IJISET - International Journal of Innovative Science, Engineering & Technology, Vol. 2 Issue 7, July 2015. pp. 516–525.
4. B. Uma Maheswari, Dr. P. Sumathi, A New Clustering and Pre-processing for Web Log Mining, 2014 World Congress on Computing and Communication Technologies, IEEE, pp. 25–29.
5. Anupama D. S. & Sahana D. Gowda, Clustering Of Web User Sessions To Maintain Occurrence Of Sequence In Navigation Pattern, Second International Symposium on Computer Vision and the Internet (VisionNet'15), Elsevier 2015, pp. 558–564.
6. V. Chitraa, Antony Selvadoss Thanamani. Web Log Data Analysis by Enhanced Fuzzy C Means Clustering, International Journal on Computational Sciences & Applications (IJCSA), Vol. 4 Issue No. 2, April 2014, pp. 81–95.
7. J. HuaXu, H. Liu, Web User Clustering Analysis based on KMeans Algorithm, 2010 International Conference on Information Networking and Automation (ICINA) IEEE, pp. 6–9.
8. K. Santhisree, Dr A. Damodaram, S. Appaji, D. Nagarjuna Devi, Web Usage Data Clustering using Dbscan algorithm and Set similarities, 2010 International Conference on Data Storage and Data Engineering, IEEE, pp. 220–224.
9. Xidong Wang, Yiming Ouyang, Xuegang Hu, Yan Zhang, Discovery of User Frequent Access Patterns on Web Usage Mining, The 8th International Conference on Computer Supported Cooperative Work in Design Proceedings, 2003 IEEE, pp. 765–769.
10. LI Wei, ZHU Yu-quan, CHEN Geng, YANG Zhong, Clustering Of Web Users Based on Competitive Agglomeration, 2008 International Symposium on Computational Intelligence and Design, IEEE, pp. 515–519.
11. Xinran Yu & Turgay Korkmaz, Finding the Most Evident Co-Clusters on Web Log Dataset Using Frequent Super Sequence Mining., 2014 August, 13–15, San Francisco, California, USA. pp. 529–536.

12. T. Nadana Ravishankar & Dr. R. Shriram, Mining Web Log Files Using Self-Organizing Map and K-Means Clustering Methods, ICAREM.
13. Mohammed Hamed Ahmed Elhiber & Ajith Abraham., Access Patterns in Web Log Data: A Review. Journal of Network and Innovative Computing, ISSN 2160-2174, Volume 1 (2013), pp. 348–355.
14. J. Xiao, Y. Zhang, X. Jia & T. Li, Measuring Similarity of Interests for Clustering Web-Users, Proc. of the 12th Australian Database Conference 2001 (ADC'20OI) IEEE, Australia, 29 January - 2 February, 2001, pp. 107–114.
15. V. Sujatha & Punitha Valli, Improved User Navigation Pattern Prediction Technique From Web Log Data, International Conference on Communication Technology and System Design, 2011 Published by Elsevier Ltd, pp. 92–99.
16. J. Xiao, Y. Zhang, Clustering of Web Users Using Session-based Similarity Measures, Proc. of the 12th Australian Database Conference 2001 (ADC'20OI) IEEE, Gold Coast, Australia, 29 January - 2 February, 2001, pp. 223–228.

Critique on Signature Analysis Using Cellular Automata and Linear Feedback Shift Register

Shaswati Patra, Supriti Sinhamahapatra and Samaresh Mishra

Abstract In surge to cater the needs of modern technology and high performance computation requirement the complexity of the VLSI design increasing with complex logic design, more memory space and large test vector for testing the digital circuit. Signature analysis compresses the data. It is also known to be a compacting technique which follows the concept of cyclic redundancy checking (CRC) which in turn detects error during transmission. It is used in hardware using shift register, cellular automata, etc. as a part of VLSI design process. This paper deals with the popular mechanism of signature analysis in the context of digital system testing using LFSR and CA-based signature analysis along with its critique.

Keywords Linear feedback shift register · Cellular automata · Test vector · Signature analysis · Signature analyzer · Sequence generator · Signature checker

1 Introduction

The introduction of Signature analysis in VLSI circuit testing provides a new paradigm to the Electronics Industry. Signature analysis may be implemented by using linear feedback shift register or cellular is often represented as LFSR or CA. LFSR is faster and more efficient shift register where the input to any cell may linearly depend upon the previous states. The application of LFSR includes pseudo-random pattern generation, fast digital counters, signature analyzer [1]. However, Cellular Automata is a finite state machine whose next state of a particular cell depends upon

S. Patra (✉) · S. Mishra
School of Computer Engineering, KIIT University, Bhubaneswar, Odisha, India
e-mail: shaswati.patrafcs@kiit.ac.in

S. Mishra
e-mail: smishrafcs@kiit.ac.in

S. Sinhamahapatra
National Institute of Technology Durgapur, Durgapur, West Bengal, India
e-mail: supriti.smpatra@gmail.com

H.S. Behera and D.P. Mohapatra (eds.), *Computational Intelligence in Data Mining*, Advances in Intelligent Systems and Computing 556,
DOI 10.1007/978-981-10-3874-7_36

the present state of that cell and its neighbors [2]. It is used to develop pseudo random pattern generator, signature analyzer, cryptography [3], and error correcting code. Moreover, the LFSR and CA are important tools to implement of compression technique for Signature Analysis.

2 Signature Analysis in VLSI Design

Signature analysis compresses the data. It is also known to be a compacting technique which follows the concept of cyclic redundancy checking (CRC) which in turn detects error during transmission. It is used in hardware using linear feedback shift register, cellular automata, etc. [4]. The compression technique replaces the time taking every bit comparison of the obtained output values with the corrected values as previously computed and saved [4]. Comparing every bit requires a lot of memory since all fault-free outputs corresponding to the test vectors need to be saved [5]. On the other hand since Compression technique produces compressed data, it needs comparatively less amount of memory space for storage. Here, the reduced form of the obtained test outcome stored, is also known as a signature and the technique it follows is known as compacting or compressing. There are many compression techniques like transition counting, parity checking, syndrome checking, and signature analysis. Among which, signature analysis is the most popular that may be implemented using LFSR or CA.

2.1 Linear Feedback Shift Register

Linear feedback shift register (LFSR) is an efficient counter that generates pseudo-random pattern sequences [6]. It consists of a number of flip-flops. The XOR operation of the output of two or more flip-flops is introduced as the input of any of the flip-flops [7]. When clock pulse is given to LFSR, the signal passes bit by bit from one flip- flop to the next one [6]. The size of LFSRs can be 2 to 168, i.e., the number of registers that makes up LFSRs cannot exceed this number. These are very efficient counters, as very less number of gates is needed for its construction. Its structure is very simple and easy for the purpose of testing. It requires very little logic to implement.

Working of Linear Feedback Shift Register When the clock pulse passes through the LFSR, it generates a pseudo-random pattern, which resembles a binary number. We already know, that the XOR operation of two or more flip-flops is given as the input of any of the flip-flops [7]. An LFSR of length n produces a pseudo-random sequence of length $2^n - 1$ states if it is efficient. Generation of repeated sequence is obtained from initial state to the final till the LFSR is attached with the clock pulse. LFSR may be created using the Fibonacci or Galois configuration of gates

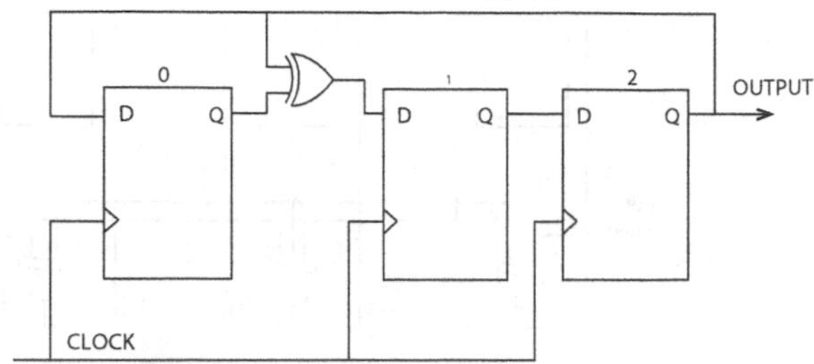

Fig. 1 3 bit internal type linear feedback shift register

and registers known as External type and Internal type LFSR [7]. In the External type of LFSR, the outputs from some of the register are combined with XOR gates and provided as input to another register. In Internal type of LFSR the gates are placed between registers. Figure 1 shows a 3-bit internal type of LFSR and is taped at stage 2 with XOR feedback.

2.2 *Cellular Automata*

Cellular Automata consists of a number of cells in a grid where each cell may be in a state on or off [8]. According to Wolfram, the CA structure is a discrete lattice of cells. Each cell in this lattice has a storage element, i.e., D flip-flop (DFF) and combinational logic (CL). The DFF may have either of the values 0 or 1. The next state of any cell may depend upon the present state of that cell and its two neighbors known as three neighborhood dependency [8]. The CL may be used to determine the rule of the CA according to some combinational logic.

Working of Cellular Automata A cellular automaton (CA) is an autonomous finite state machine (FSM). At a particular instant of time, t of each cell contain a variable known as present state (PS) [8]. At time $(t + 1)$, the next state (NS) of a cell depends on the present state of itself and its two neighbors. In its simplest form, a CA can have S different states and next state of a cell may depend on the present state of $k - 1$ neighbors and itself known k-neighborhood CA [9]. If right neighbor of the right most cell is the left most cell and vice versa, i.e., $S^t0 = S^tn$ and $S^tn + 1 = S^t1$ then the CA is known as periodic boundary CA. If the left neighbor of left most cell and right neighbor of right most cell initialized to 0 (null), i.e., $S^t0 = 0$ (null) and $S^tn + 1 = 0$ (null), then the CA is referred to as null boundary CA [9]. Finally, if the next state of the leftmost or rightmost cell depends on its right or left neighbor and the cell next to or before it then the CA is called Intermediate boundary CA [10]. Figure 2 shows the implementation of a null boundary CA. The CA has 4-cells and the output of each cell entirely depends upon the XOR operation of its present state and its neighbors, i.e., the 3-neighborhood dependency.

Based on the behavior, CA can be categorized as linear/additive CA and nonlinear CA. If the rule (i.e., next state logic function) of the CA cell is linear rule then it is

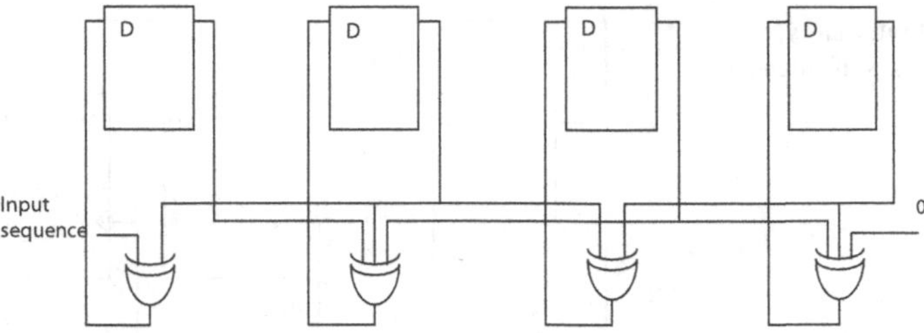

Fig. 2 Implementation of a null boundary CA

obvious that it involves only XOR operation [10] and if all the cells of a CA follow the linear rule then the CA is also linear. CA rules which involves only XNOR operation forms a complementary CA. If the CA rules involves both XOR and XNOR logics then the resulting CA is called an additive CA, but, CA with AND-OR logic is a nonadditive CA.

3 Key Mechanism of Signature Analysis

Key mechanism of signature analysis may include circuit under test (CUT), the signature analyzer and signature checker. It is because, signature analysis works with the concept of cyclic redundancy checking (CRC) using linear feedback shift register, cellular automata, etc. and compresses data [11]. Figure 3 refers to the block diagram of signature analysis. The circuit under test, generates the test vector which is passed into the signature analyzer where compression of test pattern is performed [11]. The signature analyzer can be implemented either using the LFSR or CA. The resulted signal derived from the terminals are the in reduced bit sequence than the applied test patterns [11, 12].

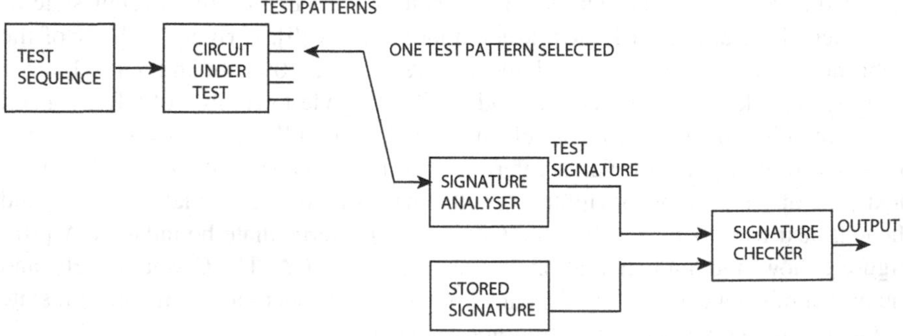

Fig. 3 Block diagram of signature analysis

The output of the compactor is known as signature. The generated signature is then checked with the fault-free signature previously calculated by a fault free circuit [12]. For all length of test sequence the length of signature is always same, and this gives a significant amount of storage required for testing the result.

4 A Critique on Signature Analysis Based on LFSR and CA

This section includes the critique through comparative study over LFSR-based signature analysis and CA-based signature analysis with help of four popular parameters namely significant reduction in amount of storage, system complexity issues, response comparison on chip, cost of test application, test time [13, 14]. Each of this discussed below:

1. *Significant reduction in amount of storage*: Bit-by-bit comparison required an abundant amount of memory for storing the correct outputs along with all the corresponding test vectors. Hence, compression technique is much simpler and requires less memory storage. For example, if Signature Analysis is performed using LFSR and CA, in both the cases the generated signature is in compressed form of the test vector hence less memory is required.
2. *System complexity issues*: Higher test generation cost is a type of complexity issue. When output is generated from the circuit, the generated result is then tested with some predefined test case. If we are to test a bigger or the whole large output, the test generation cost rises along with the system complexity. Signature Analysis may reduce generated result to be tested and hence reduce the cost. LFSR as a signature analyzer requiring least hardware as it has less XOR gates and hence less number of transistors, whereas CA requires more hardware for the complexity of its construction. Hence, the CA-based signature analysis is costlier than LFSR based signature analysis in terms of hardware requirement.
3. *Response comparison on chip*: Faster test, fewer access requirements hence the test time gets reduced in signature analysis. If the test results are bigger, then they will take more time to test and hence the total access will take place for longer times. Due to the feedback path, LFSR may require more testing time however for cascading structure and neighborhood dependency CA may take lesser testing time. Hence CA based signature analyzer responds better than that of LFSR based signature analyzer whenever the input test vector passed in to it.
4. *Cost of test application, test time*: Cost of test application gets reduced by the use of signature analysis. As the test results or 'signatures' of signature analysis are much smaller hence time for testing also gets reduced. The test time of CA-based signature analysis is less than the LFSR based signature analysis.

The comparison between LFSR and CA based on certain characteristics is referred in Table 1

Table 1 Comparison between LFSR and CA

Characteristic	LFSR	CA
Area overhead	Least-very few XOR gate	More-XOR gates are needed for every node
Performance	In internal feedback shift register it is good whereas, it is low in external feedback shift register	Due to the absence of feedback path performance is good
Randomness of parallel pattern	Less as shifting of data may cause similarity between patterns	Less as shifting of data may cause similarity
Stuck-open faults	Low: as there are less number of transitions	High: as there are more number of transitions
Computer aided design friendly	No: If pattern length change redesign required	Yes: If the pattern length changes easily new node can be cascaded
Signature aliasing	possibility is less	Possibility is lesser than LFSR. As data does not shifted from one cell to another cell
Cascading	Cascading a new cell is difficult	Cascading a new cell is easy

5 Conclusion and Future Scope

There has been tremendous growth in electronic industry mostly due to rapid advances in integration technologies. With such rapid advancements, the ability of more and more complex functions being integrated into small package has risen and modern VLSI chips contain hundreds of millions of transistors. Designing such VLSI circuits having multi-million transistors is virtually impossible without the help of computer-aided design. The VLSI design cycle contains a number of processes starting from specification followed by several different steps which ends with testing and debugging. After specification of a circuit, a functional design is created. This forms the basis of logic design. Once, logic design is done, the design of the circuit takes place. Following this, the actual circuit is physically designed. A circuit needs to be verified in order to fabricate, pack, test and if required debug. The paper concludes with followings:

- Signature analysis is a compaction technique used in the testing phase of VLSI circuit design with help of emerging tools like LFSR and CA.
- Due to the simplicity, cascade structure of additive CA it is preferable over LFSR in VLSI design.

In the future, Multi-Input Cellular Automata may be used in signature analysis to increase the efficiency.

References

1. Ravi Shankar Reddy C., Sumalatha V.: A new built in self test pattern generator for low power dissipation and high fault coverage Intelligent Computational Systems (RAICS), IEEE-2013, Pages: 19–25, doi:10.1109/RAICS.2013.6745440.
2. Abramovici Miron, Breuer Melvin A., and Friedman Arthur D.: Digital systems testing and testable design. Computer Science Press, 1990. ISBN 978-0-7167-8179-0.
3. Kumaravel A., Nickson Meetei Oinam: An application of non-uniform cellular automata for efficient cryptography. Information & Communication Technologies (ICT), IEEE conference-2013, Pages: 1200–1205, doi:10.1109/CICT.2013.6558283.
4. M. Rabaey Jan, P. Chandrakasan Anantha, and Nikoli Boriyoje: Digital integrated circuits: a design perspective. Prentice Hall electronics and VLSI series. Pearson Education, 2 edition, January 2003. ISBN-0130909963.
5. SadiqSahari Muhammad, Khari A'ain Abu, Grout Ian: A study on the effect of test vector randomness on test length and its fault coverage, Semiconductor Electronics (IEEE-ICSE), 2012. 2012, pp: 503–506, doi:10.1109/SMElec.2012.6417196.
6. Stepien Rafal, Walczac Janusz: Comparative analysis of pseudo random signals for the LFSR and DLFSR generators. Mixed Design of Integrated Circuits and Systems (MIXDES)IEEE-2013, Proceedings of the 20th International Conference. Year-2013, Pages: 598–602.
7. Xiao-chen Gu, Min-xuan Zhang: Uniform Number Generator Using Leap Ahead LFSR Architecture. Computer and Communications Security, Year-2009, Pages: 150–154, doi:10.1109/ICCCS.2009.11.
8. Saha Mousumi, K Sikdar Biplab: A self testable hardware for memory. Circuits and Systems (ICCAS), 2013 IEEE Internal Conference on year 2013, pages: 45–50, doi:10.1109/CircuitsAndSystems.2013.6671631.
9. Saha Mousumi, K Sikdar Biplab: A cellular automata based design of self testable hardware for March C-, Internal Conference on high performance computing 2013 PP: 333–338, doi:10.1109/HPCSim.2013.6641435.
10. Holdworth B. and Woods C.: Digital Logic Design, Elsevier Science, 2002. ISBN 9780080477305. Pages: 442–444.
11. Das Baisakhi, Saha Mousumi, Das Sukanta, K Sikdar Biplab: A CA based scheme of cache zone prediction for data migration In CMPs. 014 Annual IEEE India Conference (INDICON)Year: 2014 Pages: 1–6, doi:10.1109/INDICOON.2014.7030409.
12. Motavi M., Cardrilli G.C., Celliti D., Pontarelli S., Re.A Salsano M.: Design of totally self checking signature analysis checker for finite state machine. Defect and Fault Tolerance in VLSI System, 2011. Proceeding. 2001, IEEE International Symposium on, Pages: 403–411, doi:10.1109/DFTVS.2001.966794.
13. S. Mishra, P.P. Chaudhuri: Additive Cellular (CA) as a primitive structure for signature analysis. VLSI design. 1991. proceeding, Fourth CSI/IEEE International Symposium on. Year-1991, Pages: 237–242, doi:10.1109/ISVD.1991.185123.
14. Andrew B. Kahng, Jens Lienig, Igor L. Markov, and Jin Hu: VLSI Physical Design-From Graph Partitioning to Timing Closure, Springer, 2011, ISBN 978-90-481-9590-9.

SparshJa: A User-Centric Mobile Application Designed for Visually Impaired

Prasad Gokhale, Neha Pimpalkar, Nupur Sawke and Debabrata Swain

Abstract Every person has an ability of exploring the world by means of touch and this specifically becomes the core ability if a person is visually impaired. This core ability is unfolded by the mobile application SparshJa that accepts the input through touch-based Braille keypad. SparshJa is a Sanskrit word which means "Gaining Knowledge through Touch". SparshJa presents a user interface which mimics the Braille cell on touch screen phone where position of each dot is identified by means of distinct vibratory patterns. This Haptic feedback is augmented by auditory feedback that spells out the input letter. This dual feedback system makes the design of application and especially graphical interface totally user centric. The user-centric design improves utility of the application by simplifying navigation and reducing the learning curve. These claims are validated by means of on-field user trials which evaluate usability of SparshJa.

Keywords Visually impaired · User-centric design · Braille · V-Braille · Haptic feedback · User interface · Usability evaluation · Mobile application

P. Gokhale (✉) · N. Pimpalkar · N. Sawke · D. Swain
Vishwakarma Institute of Technology, Pune, Maharashtra, India
e-mail: prasad.gokhale@vit.edu

N. Pimpalkar
e-mail: neha.pimpalkar@yahoo.com

N. Sawke
e-mail: nupur_sawke@yahoo.com

D. Swain
e-mail: debabrata.swain@vit.edu

H.S. Behera and D.P. Mohapatra (eds.), *Computational Intelligence in Data Mining*, Advances in Intelligent Systems and Computing 556,
DOI 10.1007/978-981-10-3874-7_37

1 Introduction

Presently, the worldwide estimated population of visually impaired people stands at 285 million. Out of this population about 39 million people are completely visually impaired [1]. In India, the completely visually impaired population is around 15 million which is one-third of the global population. Moreover, only 5% of visually impaired children receive education in India [2]. The extent of education towards visually impaired children can be enhanced by ensuring their inclusion in the digital world. The first step in this direction is to enhance the usability of cellular phones which can plausibly connect these students to digital world. Presently the touch-based cellular phones are showing speedy growth that can potentially be customized to provide support for visually impaired. This support is not available in all cellular phones but it is slowly gaining importance and attention of the research community.

The cellular phone market presents wide range of touch screen phones based on operating systems like Android, iOS, Windows, etc. Some Android phones provide a facility of talk back which spells out the menu helping in navigation and selection of an option. But, this facility is constrained by language as the instructions are mostly in English and is inefficient during commotion. Hence, it is imperative to develop an application and specifically a user interface which is geared towards visually impaired. The user-centric design will improve the utility of smart phones.

Number of mobile applications has already been developed only for visually impaired users but these applications have problems pertaining to usability which are highlighted in [3–5]. These issues related to usability are as follows: (a) large number of on-screen targets, (b) requires lot of scrolling, (c) no source of feedback, (d) Inclusion of unsuitable gestures for visually impaired, and (e) inconsistency in user interface design. These problems increase the learning curve of an application thus making cellular phones difficult to handle. So, our main objective is to design an application that will minimize above-mentioned problems and also to concentrate on the fact that majority of visually impaired people are familiar with Braille and if they are not familiar with Braille, then this can be done with minimal efforts.

SparshJa, the mobile application presented in this paper unfolded these findings by designing a User Interface (UI) that spans entire screen and provides vibratory feedback. The UI mimics the Braille cell which is used for alphanumeric data entry. ***SparshJa*** supports dual feedback mechanism that constitutes vibratory feedback and auditory feedback. Presently, ***SparshJa*** provides basic functionalities like calling and saving contacts by employing Braille keypad as a source of input. Application supports both landscape and portrait orientation of screen. ***SparshJa*** does not require any external hardware or Internet connection. In this paper, we first present the design of ***SparshJa*** followed by the usability analysis of the application. Usability analysis refers to evaluation of an application in context of: (a) ease of use (effectiveness), and (b) degree of user satisfaction, and (c) feedback mechanism [4]. We have used on-field user trials for usability evaluation.

The rest of the paper is organized as follows: Sect. 2 presents the literature survey; Sect. 3 elaborates on design and specifically delves in User Interface (UI) design. Section 4 focuses on usability testing through user trials. Section 5 presents concluding remarks.

2 Literature Survey

A number of applications have been proposed to aid the visually impaired in the use of touch-based phones. The application No-Look Notes is an eyes-free application on iPhone in which all the 26 characters are arranged in a circular pie of eight divisions with each division representing a group of characters. Multi-touch gestures are used for selection and single-fingered interaction is used for exploring the UI and swipe gestures use minimum distance and speed for the system to recognize the gesture as an input [6]. Yet another touch-based application for visually impaired was proposed by Georgios Yfantidis and Grigori Evreinov in [7] where an adaptive interaction technique is used. Interaction is based on the gestures in any of the eight directions. The characters are arranged in a pie menu. Interpretation/acceptance of gestures is based on time. Azenkot et al. in [8] used input finger detection (IFD) in which reference points are set. Hand repositioning, touch-point inconsistency, hand drift are the three sources of noise that need to be handled when using this model for input on touch screen. All these applications are gesture based and have many on-screen targets which are against intuitive use of corners and edges by visually impaired users. Although audio feedback is rendered in all the above-mentioned applications, they make use of gestures as primary source of input and thus, the guideline of making minimum use of gestures [4, 5] is not followed. Authors of paper [9] presented two text-entry methods, namely NavTap and BrailleTap. In NavTap, the characters are grouped based on the vowels. In BrailleTap, keys 2, 3, 5, 6, 8, 9 represent six dots of the Braille cell and Braille is used for text entry. Both these applications use hardware buttons for providing input and thus violate the principle mentioned in [4] that is to obviate usage of hardware buttons.

João Oliveira et al. in [10] made a comparison of various touch-based text-entry methods: QWERTY, NavTouch, MultiTap, BrailleType. The analysis suggests that BrailleType is less error prone and easier than other methods. In BrailleType application, Braille cell is emulated on the touch screen. However, the six blocks do not span the complete screen, leaving spaces on the screen where there is no source of input. A time-out period was used for acceptance of a character. Gestures are used to delete a particular incorrect entry. SlideRule, the application implemented in [11], displays solid color when it is operating. But users with low eyesight find it difficult to use this; they instead prefer to see some objects on the screen. Hence instead of using solid colors in ***SparshJa,*** we use numbering of cell and relevant labels so that even people with poor eyesight can use this application.

Visually impaired have a penchant for familiar spatial layouts in the design of the user interface [5, 11]. Paper [12] shows a unique way of haptically representing a Braille cell on a touch screen phone. This technique replaces physically embossed dots by virtually adding different vibratory patterns over them on the screen. It allows users to read Braille characters and has shown a success rate of about 90%. Consistent navigation across the screens has shown to maximize accessibility of any application to its user. Other guidelines include reduction in location accuracy and limiting time-based gesture processing. The time out period may become a hindrance to visually impaired and may lead to increase in the rate of error [5, 10]. Furthermore, the studies indicate that having only auditory feedback may not suffice the users. This can be seen in cases where user has to interact with the device in a place of commotion. Thus tactile clues like vibratory feedback can aid the users while using the application.

The user trials in [5, 12] specified that users preferred tap than swipe gesture. Research in [5] suggests that visually impaired user prefers edges, corners, and multi-touch which enable them to judge the location of on-screen targets.

3 User Interface Design

The user interface designed of the proposed solution *SparshJa* complies with the analysis in the literature survey and inculcates the guidelines mentioned in the same.

The mobile application ***SparshJa*** mimics the Braille cell on touch screen. The dots are distinguished by means of vibratory feedback which is termed as V-Braille technology. The V-Braille proposed in [12] is used for reading. In ***SparshJa,*** we extend this concept for providing input through UI that mimics the Braille cell (refer Fig. 1). The UI uses haptic feedback along with audio feedback by means of spelling out the letter entered through touching the dots that form a particular pattern. Figure 1a. shows the UI presented to the user for accepting character input while saving contact. Figure 1b. shows the UI presented to the user for accepting numeric input while making call. The number is represented in Braille using only upper two rows hence we renamed the dot 3 as call and dot 6 as Back. The Back button is then used to correct any typographic error. But, it is worth to note that both the UI are identical and divides the screen into six blocks. The corner of cell phone can be used to identify dot 1, 3, 4, and 6 and the dot 2 and 5 in the middle row can be identified using edges and by different variation pattern from other row. This distinct vibratory pattern is known as haptics.

SparshJa provides two basic features: (1) Calling and (2) Saving Contacts, the UI presented to the user for invoking these functionalities is shown in Fig. 2. The crucial aspect of the UI design is the positioning of the main tasks at the corners of the screen as depicted in Fig. 2. The UI spans the entire screen leaving no space where there is no source of input which makes it independent from the screen size.

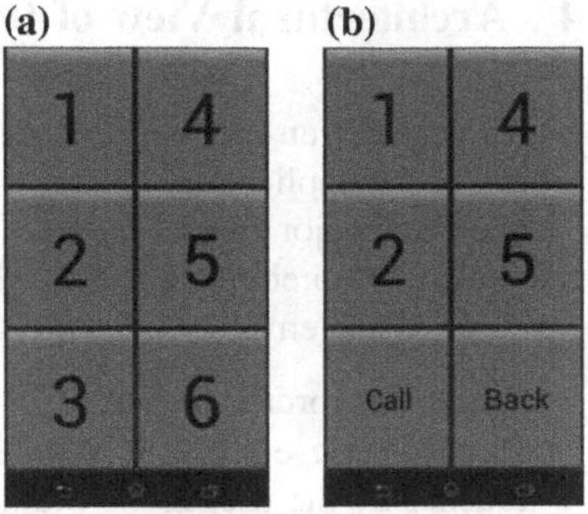

Fig. 1 Braille cell emulated on touch screen phone: **a** For alphabet entry (*left*), **b** For entering digits (*right*)

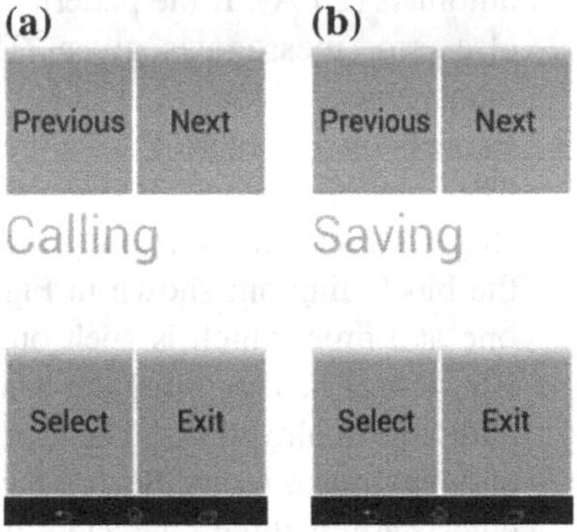

Fig. 2 Screenshots of SparshJa: **a** the UI for calling feature (*left*), and **b** the UI for saving feature (*right*)

This layout would help the users to locate different buttons on screen with ease and minimal error rate as the users use edges and corners as reference points to locate objects in spatial locality. The consistency in UI design is maintained throughout the design, thus resulting in minimal learning curve and reduced the scope for committing errors.

SparshJa makes use of clicks as primary source of input. Thus, the users would not activate unintended option accidentally. In case of any incorrect selection of option or faulty entry of a character/digit, auditory feedback will notify the users on the actions being performed and help them to rectify their errors as the application provides appropriate voice feedback which acquaints the user about which screen he/she is presently on thus guiding the user while navigating. Also haptics provide the supplementary mode of feedback. To improve upon the audio feedback, in our application, only one menu item from the menu list or one contact from the logs/phonebook is displayed on screen at a time (as shown in Fig. 2). The design of UI which covers entire screen and provides both haptic and auditory feedback is in accordance with the guidelines provided by researchers and therefore, would enhance the user experience and usability.

4 Architectural View of the Application

Fig. 3 depicted an architectural view of ***SparshJa*** elaborating various functional blocks of the application.

The four major blocks that aid the functioning of the application are: Braille Interpreter, Phonebook Handler, Sensory Feedback Management, and Handset Interface Management.

- Braille Interpreter: The user is presented with UI as shown in the Fig. 1 for alphanumeric entry. The Braille interpreter block reads the pattern of blocks touched by the user in the Braille cell and recognizes the alphabet/digit entered by the user. The interpretation of pattern is done based on the deterministic finite automata (DFA). If the pattern yields to a valid alphabet/digit then it is spelt out else error message is given. This feedback is handled by sensory feedback management block.
- Phonebook Handler: The calling using Phonebook functionality is handled by this functional block. It reads the Phonebook stored in database. This relationship between the functionalities is indicated by assigning the identical color in the block diagram shown in Fig. 3. The names in the contact list are presented one at a time which is spelt out to the user by using text to speech converter. But, if user has recorded the name of person during saving the contact then that audio file is played out. The user can navigate through the phonebook using next and previous buttons. Search facility is as well provided for which user needs to enter starting three characters of the name. The new contact can be added by entering phone number through the Braille keypad or selecting a number from call logs.

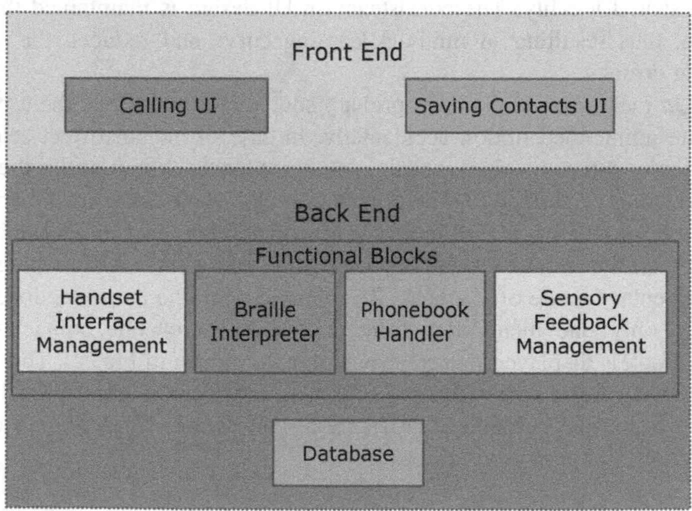

Fig. 3 Architecture diagram of SparshJa

- Sensory Feedback Management: This block provides feedback to the users' actions in order to give them information about what consequences their actions have. The application provides dual feedback mechanism in the form of haptic (vibrations) and audio. It interacts with Braille Interpreter and Phonebook Handler blocks by extending its services to them.
- Handset Interface Management: It deals mainly with the built-in services of the handset used. These include telephony manager to connect the call and audio recorder which is used to record the name of contact while saving contacts. All the four blocks interact with each other to make use of the services provided by them thus constituting the overall functioning of the application.

5 Usability Testing Through User Trials

In order to verify the credibility of the application developed and evaluate its usability, we have conducted on field user trials. The on field user trials were conducted with the participants in the age group between 14 to 16 years from a school for the visually impaired students.

5.1 User Training

In order to get users acquainted with the application, the initial training was provided. The training includes various modules as mentioned below

Train the users to use Braille keypad for text entry

This module of training made the user conversant about the way to enter Braille characters. Also the users learnt to locate various blocks of the Braille cell with the help of haptics as well as edges and corners of the screen. In case the user did not know Braille, the users can learn Braille by performing simple exercises of entering the pattern for alphanumeric using the Braille keypad which is very easy to handle due to dual feedback mechanism.

Familiarize the users with the layout of the UI

In this section of training, the users were familiarized with the UI which resembles the traditional Braille cell.

Train the users to traverse menus, phonebook, or call logs

In this module, the menu system of the application was introduced to the users where they learned to navigate through different menus with the help of audio feedback.

5.2 User Trials

The on-field user trial were conducted with 15–20 participants belonging to following three categories:

a. **Type 1**: Fully visually impaired with knowledge of Braille as well as touch screen phones.
b. **Type 2**: Partially visually impaired with knowledge of Braille as well as touch screen phones.
c. **Type 3**: Partially visually impaired with knowledge of Braille but not acquainted with touch screen phones.

All the participants performed following three tasks Fig. 4 shows two participants involved in the trial.

a. **Task 1**: To make a call using Braille Keypad
b. **Task 2**: To save a contact using Braille Keypad
c. **Task 3**: To traverse and select various options using menu

An important point of observation during the trials was that the entry of characters that involved pressing more number of buttons took slightly more time than those involving lesser number of buttons. Users were more prone to committing errors in such type of patterns.

5.3 User Trial Results and Findings

Each participant dialed five phone numbers comprising of 10 digits each, i.e., total of 50 digits were entered by each participant. Fig. 5 shows a view graph of the performance of participants evaluated based on the percentage of digits correctly entered by them. Figure 5 shows that the average accuracy entering the number

Fig. 4 Participants during user trials

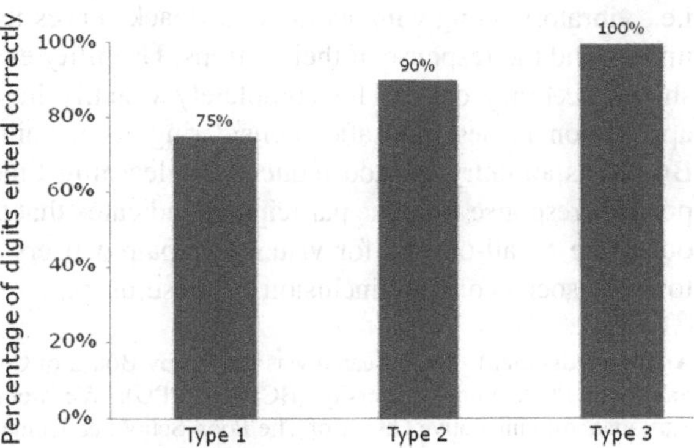

Fig. 5 Percentage of numbers entered correctly through Braille keypad

through Braille keypad stands at 75%, 90%, and 100% for Type 1, Type 2, and Type 3 users respectively. The accuracy for Type 2 and Type 3 users is very close which shows that the prior knowledge of Braille is not required. The accuracy for the completely visually impaired user stands at 75% which is fairly decent considering that they are using our application for the first time. These results indicate that the application is completely geared towards assisting the visually impaired users. Also the learning curve of the application is less. Once the participants got acquainted with the application, they were able to efficiently make use of all the features provided. One important point to be noted during the actual trials was that because of initial training, all the participants could place a call efficiently on their own, i.e., the participants made calls to a total of five phone numbers on their own without any assistance.

All the participants were enthusiastic to learn the application. They found that the application is very easy to use than the talkback phones which they are currently using. They stated that the audio feedback provided by the application was clear and simple to understand. Also they found haptic feedback useful in assisting them to recognize the button positions along with audio feedback. Participants found the navigation through the application effortless due to its consistent design. All the participants suggested that they would like to have audio input as an entry method since it would be helpful to them in case of emergency.

6 Conclusion

The application ***SparshJa*** follows complete user-centric design and we believe this is the first step taken towards assisting visually impaired users and making the touch screen devices accessible to them. The user interface of application takes into account the intuitive behavior of users of interacting with the help of corners and edges of mobile device. The application uses Braille cell structure as its foundation and keeps UI consistent across different functionalities. The UI spans entire screen which reduces error and improves the accessibility. The dual feedback mechanism, i.e., vibratory along with auditory feedback serves its purpose of helping the users understand the response of their actions. Usability evaluation of the application has shown accuracy of 75% for completely visually impaired uses indicating that the application is designed after considering its usability by its target users. Using Braille as an entry method reduced the learning time of the application. Also the positive response from the participants indicates that the application can indeed turn out to be an all-out aid for visually impaired users and prove to be the first step towards socioeconomic inclusion of these users.

Acknowledgement This research was funded by Board of College and University Development, Savitribai Phule Pune University (BCUD, SPPU). We would also like to thank Ms. Archna Sarnobat, Administrative Officer of The Poon School & Home for Blind Trust, Kothrud, Pune for allowing us to conduct on field user trials.

References

1. "World Sight Day: International Key Messages, 2015." Retrieved on June 12, 2015, http://www.iapb.org/advocacy/world-sight-day.
2. "India accounts for 20 per cent of global blind population, 2012." Retrieved June 9,2015, http://www.deccanherald.com/content/240119/indiaaccounts-20-per-cent.html.
3. Jadhav, D., Bhutkar, G., Mehta, V. Usability Evaluation of Messenger Applications for Android Phones using Cognitive Walkthrough. In APCHI '13, (Bangalore, India, 2013), ACM New York, 9–18.
4. Johnsen, A., Grønli, Tor-Morten, Bygstad, B. Making Touch-Based Mobile Phones Accessible for the Visually Impaired. In Norsk informatikkonferanse, (Bodø, Norway, 2012).
5. Kane, S., Wobbrock, J., Ladner, R. Usable Gestures for Blind People: Understanding Preference and Performance. In CHI 2011, Conference on Human Factors in Computing Systems, (Vancouver, BC, Canada, 2011), ACM New York, 413–422.
6. Bonner, M., Brudvik, J., Abowd, G., Edwards, W. "No-Look Notes: Accessible Eyes-Free Multi-Touch Text Entry." In Pervasive'10 Proceedings of the 8th international conference on Pervasive Computing, (Helsinki, Finland, May 17–20, 2010), Springer-Verlag Berlin, Heidelberg, 409–426.
7. GeorgiosYfantidis, GrigoriEvreinov, "Adaptive blind interaction technique for touchscreens". Published online: 9 July 2005 @ Springer-Verlag 2005.
8. Azenkot, S., Wobbrock, J., Prasain, S., Ladner, R., "Input Finger Detection for Nonvisual Touch Screen Text Entry in Perkinput." In Proceedings of Graphics Interface (GI '12), (Toronto, Ontario, 2012), Canadian Information Processing Society Toronto, Ont., Canada, 121–129.
9. Guerreiro, T., Lagoá, P., Santana, P., Gonçalves, D. and Jorge, J. NavTap and BrailleTap: Non-Visual Texting Interfaces. In Proceedings of RESNA (Arlington, VA, 2008).
10. Oliveira, J., Guerreiro, T., Nicolau, H., Jorge, J., Gonçalves, D., BrailleType: Unleashing Braille over Touch Screen Mobile Phones. In INTERACT'11 Proceedings of the 13th IFIP TC 13 international conference on Human-computer interaction - Volume Part I, (Springer-Verlag Berlin, Heidelberg, 2011), Pages 100–107.
11. Kane, S., Bigham, J., Wobbrock, J. Slide Rule: Making Mobile Touch Screens Accessible to Blind People Using Multi-Touch Interaction Techniques. In ASSETS'08, (Halifax, Nova Scotia, Canada, 2008), ACM New York, 73–80.
12. Jayant, C, Acuario, C., Johnson W., Hollier, J., Ladner R. VBraille: Haptic Braille Perception using a Touch-screen and Vibration on Mobile Phones. In ASSETS'10, (Orlando, Florida, USA, 2010).

A Novel Approach for Tracking Sperm from Human Semen Particles to Avoid Infertility

Sumant Kumar Mohapatra, Sushil Kumar Mahapatra, Sakuntala Mahapatra, Santosh Kumar Sahoo, Shubhashree Ray and Smruti Ranjan Dash

Abstract Now a days, thc infcrtility is a big problem for human being, especially for men. The mobility of the sperm does not depend on the number of sperm present in the semen. To avoid infertility, the detection rate of the multi moving sperms is to measured. There are different algorithms are utilized for detection of sperms in the human semen, but their detection rate is not up to the mark. This article proposed a method to track and detect the human sperm with high detection rate as compared to existing approaches. The sperm candidates are tracked using Kalman filters and proposed algorithms.

Keywords Sperm tracking · Detection rate · Background subtraction · Multi moving

---

S.K. Mohapatra (✉) · S.K. Mahapatra · S. Mahapatra · S. Ray · S.R. Dash
Trident Academy of Technology, Bhubaneswar, Odisha, India
e-mail: sumsusmeera@gmail.com

S.K. Mahapatra
e-mail: mohapatrasushil@gmail.com

S. Mahapatra
e-mail: mahapatra.sakuntala@gmail.com

S. Ray
e-mail: shree3269@gmail.com

S.R. Dash
e-mail: sranjandash33@gmail.com

S.K. Sahoo
GITA, Bhubaneswar, Odisha, India
e-mail: santosh.kr.sahoo@gmail.com

H.S. Behera and D.P. Mohapatra (eds.), *Computational Intelligence in Data Mining*, Advances in Intelligent Systems and Computing 556,
DOI 10.1007/978-981-10-3874-7_38

1 Introduction

In the emerging technological world, the infertility raises rapidly and destroys life of about 15% human couples [1]. Wenzhong et al. [2] and Menkveld et al. [3] describes the semen parameter for proper monitoring. There are different approaches are utilized to study about semen ability [4–7] to strengthen the human body. Now a days different methods are used to know the detection rate of sperms in the human semen to avoid infertility the proposed method is very efficient for tracking multiple sperm in human semen.

In the paper Sect. 2 describes the multiple sperm detection and tracking system: Sect. 3 describes the proposed algorithm. Section 4 describes experimental results and discussion. Section 5 describes conclusion part.

2 Multiple Sperm Detection and Tracking

Block diagram of our proposed multiple sperm detection and tracking system is shown in Fig. 1. Background subtraction module is given to camera.

To compute the evidence of sperm presence for each pixel on the image, segmented foreground is used. By locating storage element the detection of sperm is performed. After detection of sperm candidate, analytical models are computed for each of the candidates. To match sperms we have developed an efficient method. Each tracked sperm is represented by its analytical model and associated by Kalman filter. During matching process candidates are updated.

For background Subtraction, a pixel p represents color f(p), represented in rgl space (normalized red, normalized green and light intensity) Each pixel P_i with models, is classified as:

$$|f_n(P_i) - m_k^c| > d_{th} V_k^c \tag{1}$$

where

d_{th} decision boundary threshold
V_k^c variance in channel c
m_k^c k-th Gaussian mean vector in channel c

3 Proposed Algorithm

Step 1: Load microscopic video of human sperm.
Step 2: The line segment compute the support by Counting Sperms contained

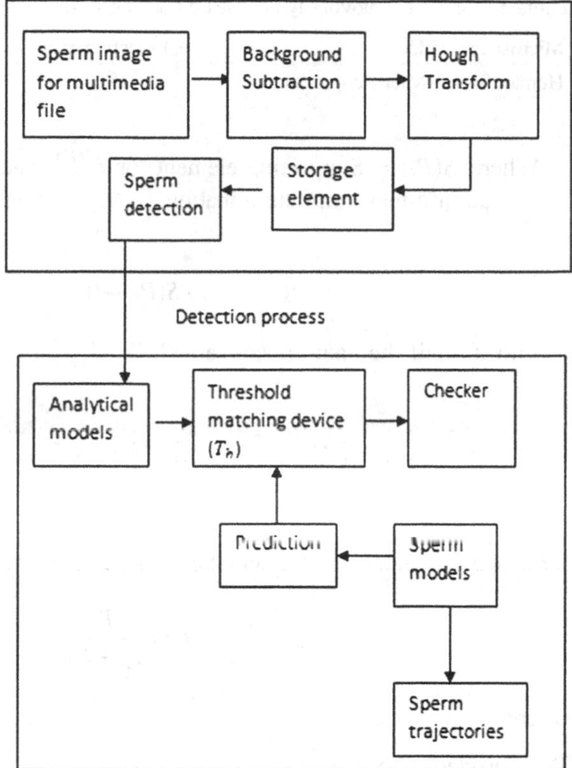

Fig. 1 The proposed sperm tracking system

(u = P_i, … P_n), $T_h^{(min)}$, $T_h^{(max)}$

Where $T_h^{(min)}$ *and* $T_h^{(max)}$ are the minimum and maximum thresholds.

Step 3: If x_i is foreground then $f_i \leftarrow f_i + 1$
Else $f_i \leftarrow f_i - 1$
end if

Step 4: $f_j = i - T_h^{(max.)} P_i$
If $f_j > 0$ $T_h^{(max.)} \leftarrow [f_i - f_j] / P_i$
else

$$T_h^{(max.)} \leftarrow f_i / P_i$$

end if

Step 5: If $T_h \geq T_h^{(min)}$
Then $S(P_i) \leftarrow T_h$

Table 1 Output for moving lyme disease tracking in rhesus macaques blood

Microscopic video	Detection rate (D_r)	Trial time in sec.
Human Sperm Microscopic Video	98.12%	0.011

Where, $S(P_i)$ = Supporting element for $T_h^{(min)}$ and $T_h^{(max)}$ for sperm dimension determination.
Else

$$S(P_i) \leftarrow 0$$

End if, until the value is converged.

4 Experimental Results and Discussion

This paper used a specific algorithm to detect multi moving sperms in human semen for proper diagnosis. The calculations of detection rate as indicated in Eq. 2.

$$D_r = \frac{T_p}{T_p + F_N} \quad (2)$$

where

T_p detected pixel
F_N undetected pixel

The Table 1 shows the output result of the proposed algorithm. From our knowledge the detection rate of proposed method is higher than the previously used approaches. The detection rate from microscopic view of semen specimen with sperms was satisfactory and for real time implementation which shown in Figs. 2, 3, 4, 5 and 6.

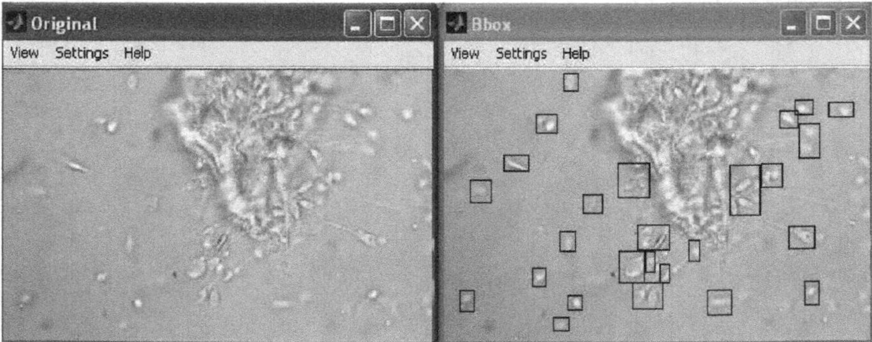

Fig. 2 The 40X detected video

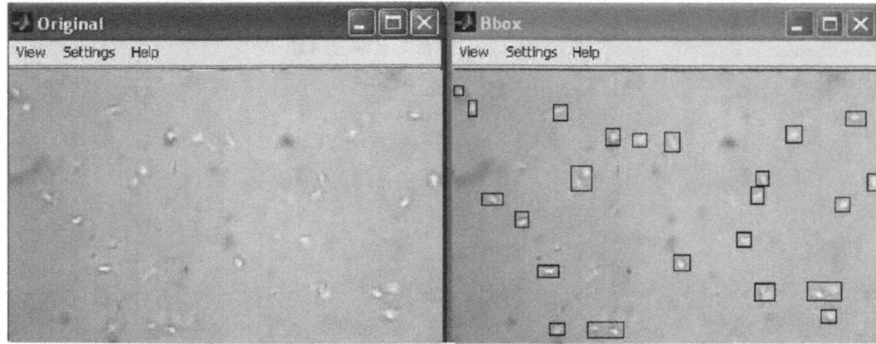

Fig. 3 The 100X detected video

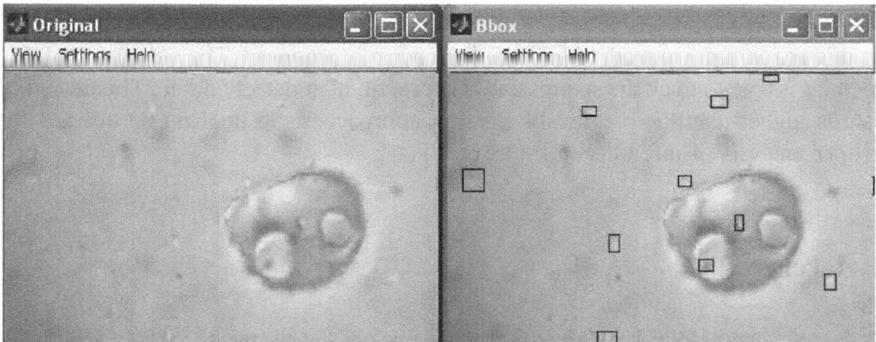

Fig. 4 The 350X detected video

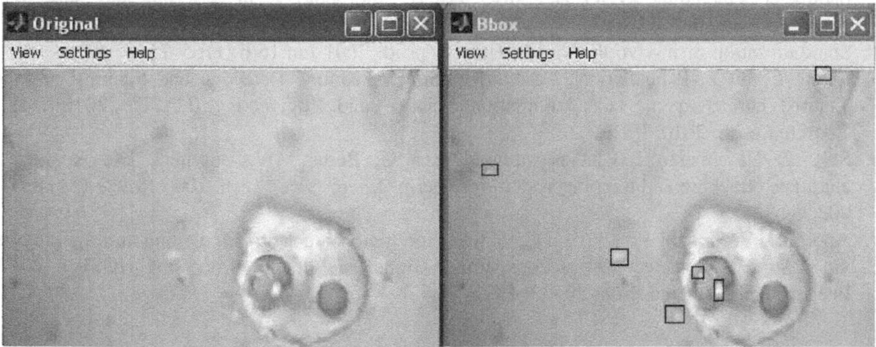

Fig. 5 The 400X detected video

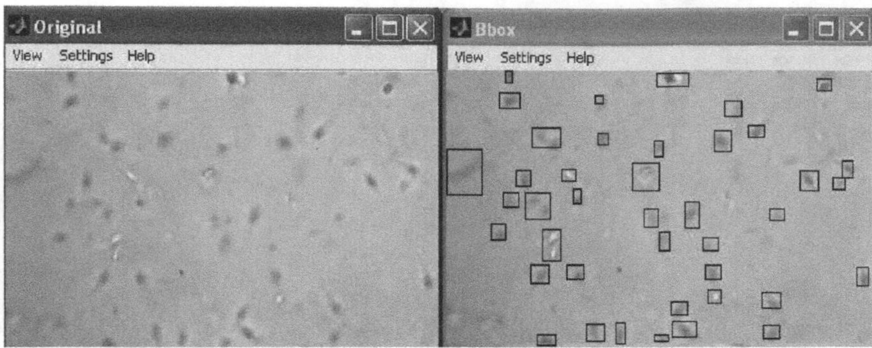

Fig. 6 The 450X detected video

5 Conclusion

In this paper, a novel method for tracking sperm is proposed. The utilization of this method is able to track the sperms and detect with high detection rate. The detection rate is higher than the previously existing approaches. So this can be utilized for proper analysis of infertility in future.

References

1. Speroff L, Fritz MA. Clinical gynecologic endocrinology and infertility: Lippincott Williams & Wilkins; 2004.
2. Wenzhong Y, Shuqun S. Automatic Chromosome Counting Algorithm Based on Mathematical Morphology. Journal of Data Acquisition & Processing. 2008;23(9):1004–9037.
3. Menkveld R, Wong WY, Lombard CJ, Wetzels AM, Thomas CM, Merkus HM, et al. Semen parameters, including WHO and strict criteria morphology, in a fertile and subfertile population: an effort towards.
4. Standardization of in-vivo thresholds. Hum Reprod. 2001 Jun;16(6):1165–71.
5. Leung C, Lu Z, Esfandiari N, Casper RF, Sun Y, editors. Detection and tracking of low contrast human sperm tail. Automation Science and Engineering (CASE), 2010 IEEE Conference on; 2010: IEEE.
6. Shi LZ, Nascimento J, Chandsawangbhuwana C, Berns MW, Botvinick EL. Real-time automated tracking and trapping system for sperm. Microsc Res Tech. 2006 Nov;69(11):894–902.
7. Abbiramy V, Shanthi V, Allidurai C, editors. Spermatozoa detection, counting and tracking in video streams to detect asthenozoospermia. Signal and Image Processing (ICSIP), 2010 International Conference on; 2010:IEEE.

The Use Robotics for Underwater Research Complex Objects

Sergei Sokolov, Anton Zhilenkov, Anatoliy Nyrkov and Sergei Chernyi

Abstract In the present article, the authors give an overview of the historical development of underwater technical vehicles and dwell on the modern representatives of this class of vehicles-autonomous (Unmanned) underwater vehicles (AUV). The authors touch upon the structure of this class of vehicles, problems to be decided, and perspective directions of their development, cover scientific materials basing mainly on the data about the benefits of the application of the vehicles in a number of tasks of narrow focus.

Keywords Autonomous · Robot · Underwater · Global positioning system

1 Introduction

In the scientific surroundings, the statements that Oceanographic researches are, in many aspects, more complex and challenging, even compared to space research are not rare. Now, in the middle of the second decade of the twenty-first century, there is no doubt that the study of the oceans has become an issue of global importance, covering economic, industrial, social, defense and many other activities and interests of the society in the modern world. We are seeing the necessity of expanding the boundaries of Oceanographic research, increasing the number of types and growth of quality of measurements in the water column as well as their

S. Sokolov · A. Nyrkov · S. Chernyi (✉)
Admiral Makarov State University of Maritime and Inland Shipping, Dvinskaya st., 5/7, 198035 Saint-Petersburg, Russia
e-mail: sergiiblack@gmail.com

A. Zhilenkov
Saint Petersburg National Research University of Information Technologies, Mechanics and Optics, Saint Petersburg, Russia

S. Chernyi
Kerch State Maritime Technological University, 298309 Kerch, Russia

H.S. Behera and D.P. Mohapatra (eds.), *Computational Intelligence in Data Mining*, Advances in Intelligent Systems and Computing 556,
DOI 10.1007/978-981-10-3874-7_39

systematization, increasing of the depths of research, which is caused by the growing necessity of sea bottom studies, etc.

Most projects of similar research are applied and become known mainly due to their commercial implementation. Undoubtedly, in the modern realities, it is more and more difficult to conduct scientific research regardless of the urgent production problems [1, 2]. Thus, cooperation of commercial structures, research institutions, military departments, etc. is required for the successful implementation of most projects. Certainly, the results of such projects can be effectively used by all the organizations listed above. For example, almost any innovation in the surveillance and search, under-ice and other studies in one form or another can be used for strategic or tactical military purposes.

2 Modern Technologies of AUV

The latest developments of AUV in Russia can present a small apparatus "MT-2012 (Fig. 1) and so-called solar AUV or SAUV [3, 4] (Fig. 2), designed by Far East branch of the Russian Academy of Sciences.

We can also mention the device "Bluefin-21" worked out by the company "Bluefin Robotics" and used to search the crash site of the plane MH370 of Malaysian airlines.

The peculiarities of modern AUV are as follows: lack of functional dependence on the support vessel, high speed of searching, large coverage, wide range of depths of immersion, accuracy of the target coordinate determination, accuracy of holding its own place in the area of deployment, underwater autonomy of the vehicle, stealth capability (physical fields), universality for performing a wide range of tasks, gathering data in close proximity to the object.

From the overview above you can see how the emergence of new ideas and tasks in constructing of the AUV has shifted the focus of current technological research.

Fig. 1 Compact offline unmanned underwater vehicle MT-2012

Fig. 2 Solar Autonomous underwater vehicle (SAUV)

It is difficult to list the entire list of technologies used in the development of the AUV. The main directions of technological development manifested in recent decades are:

- autonomy;
- power supply;
- navigation;
- sensorics;
- communication.

Despite all the advances made in these areas, they remain priorities to this day. Any limitation of development of these technological areas are the direct limitations of using AUV.

3 Intellectualization

Working out every new project of AUV researchers have to solve the question of the necessary level of intellectualization of the AUV, developed for specific tasks, and ways of realization of this level. Such features of this area, as the architecture of intelligent systems, mission planning, situation recognition, and making decisions are difficult tasks. Some successful solutions of these tasks are embodied in hardware, in other cases, developers tend to insist that to reach the stated goals they do not need a high level of intellectualization and to provide a list of preprogrammed instructions is quite enough. This statement is rather controversial, in reality there are only few significant developments in the area and most developers prefer to focus on the issue of informational autonomy of AUV. But work in this direction is going on and it is not uncommon to meet a navigation system based on the artificial neural networks or similar systems among functioning models. Existing systems provide some recognition of emergencies, recognition of the real-time set of some simple geometric shapes or standard signals on the basis of data received from the vision systems. There are some successful solutions in the selection of the target images or signals in the presence of masking noise and false targets. A particular issue here is the need for processing in real time, as well as the difficulty of obtaining high-quality informative signal from the vision systems, as underwater conditions for fulfilling these tasks are particularly difficult. Similar challenges can be seen in providing AUV with intelligent control.

Another direction of research is creation of simulators and models for training and testing of intelligent systems of developed AUV. This is not a trivial task and successful implementation of it will allow researchers to design more effective AUV and explore their behavior, debugging algorithms, architecture, and systems prior to costly field tests.

4 Informational Autonomy

The number of successes in this area, obtained for other Autonomous vehicles, do not affect the AUV. There are only few programs aimed at solving the problems of autonomy of the AUV. However, with the observed increase of missions performed by the AUV, the necessity of such programs is obvious. Some researchers believe that the problems of autonomy must be addressed in conjunction with the development of intellectualization of control systems for AUV. This would allow AUV to adapt to the environment to the best of their resources. It is worth mentioning that the vast majority of modern AUV require a vessel or vessels of technical support to be near their place of their work. Moreover, these vessels must carry special equipment, both electronic and constructive. All this seriously increases the cost of AUV using and thus reduces the economic attractiveness of their using.

As it was noted in the historical overview, the use of the mode of cooperative operation of several AUV was first proposed in the 80s of the last century. Some work has been done in this direction but even a half of potential of this direction has not been implemented. In recent years, the number of tasks where the use of the mode of cooperation of AUV is recognized as perspective grows. The greatest success was achieved in using this mode for the search and clearance of mines. However, this is only a small part and there are still a lot of unresolved problems left.

5 The Autonomy of Energy

This figure in the history of the evolution of AUV has grown from several hours to tens of hours of Autonomous run. Some samples are able to work for days and few samples are able to work for years. However, this increase in autonomy has become possible due to reducing of sensitivity of the onboard systems, as well as the speed of movement of these devices. The researchers try to increase this indicator by means of lots of measures, which include the development of new energy sources: batteries or alternative sources; reduction of consumption of electric power by control systems and power units of AUV. The latter, in particular, include measures to reduce losses of electric power spent on the hull and its parts resistance to water.

Currently, the batteries that operate on the basis of fuel cell like aluminum-oxygen, hydrogen-oxygen, and lithium batteries have the highest capacity. They provide, respectively, a battery life of AUV from 36 to 60 h.

In some applications, the solution may be in the use of solar energy, as it was done in the domestic SAUV (Fig. 2). The devices of this class require the provision of phases of charge with rising to surface and phases of work. However, this approach is not universal. For example, it is unsuitable for under-ice studies. So, in AUV of the ALTEX project, designed to study currents in under-ice space, the researchers use batteries on the aluminum and oxygen elements.

Another solution is the use of Glider type AUV (Fig. 3) that uses, as a rule, the energy of heating to change the buoyancy of AUV so that it slides on a parabolic or sinusoidal trajectory in the water column. The potential autonomy of these devices is measured in years.

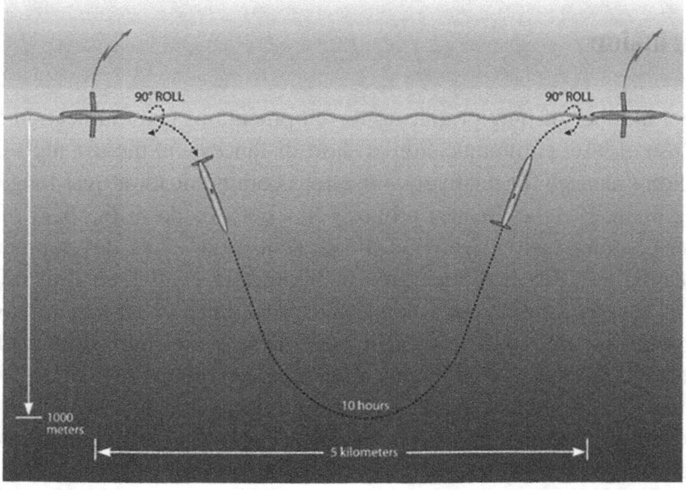

Fig. 3 Glider type AUV

6 Navigation Systems

Navigation systems continue to improve in terms of increasing of the accuracy of positioning. In recent years, even relatively cheap AUV received the global positioning system (GPS). However, for positioning such apparatus must periodically rise to the water surface, which is acceptable not for all tasks of AUV. So now, the research continues in the field of navigation systems that use the environment "habitat" of AUV. Usually such studies are based on the use of gravimetric or similar near-bottom characteristics.

Sensor systems, data processing, and surround vision. Most researches in this area are devoted to increase of the resolution capability of systems of visual and acoustic vision of AUV. The development of microprocessor technology allows researchers to obtain high-resolution images from increasingly long distances. This is evidenced by, for example, a successful project LENS of acoustic vision of high-resolution based on a movable acoustic lenses [4, 5]. But such developments become impractical and unclaimed due to the lack of appropriate systems of processing, analysis, recognition of images in offline mode for decision-making on maneuvering and control of AUV. For example, the above-mentioned project AUSS that possesses the energy autonomy of approximately 10 h of work, required up to 10 h on automatic processing and recognition of pictures obtained in the process. And in the end, the final decision on the results of processing had to be made by a human operator. Another part of development in this direction is the use of modes of cooperation between groups of AUV touch for large-scale studies and collaborative information processing. These simple examples show how close the interconnections of priority technical directions of the development of modern AUV mentioned above are.

7 Conclusion

There are a number of new promising developments in communication technologies. The use of laser communication at short distances and the use high-frequency fields (radiofrequency) for relatively noise-free communication over long distances are among them. But the greatest progress has been made in the field of acoustic underwater communication, which at the present stage of its development allows communicating at a distance, measured in kilometers, while speed of information exchange is several kilobits per second, and probability of error is relatively low. Another promising area of development in the field of communications of AUV is the prospect of realization of communication between groups of ANPA via network infrastructure. Such network AUV can communicate with the vessel of technical support by a buoy or other object on the surface that is able to communicate with ground networks like the Internet. The problem here is not only in the organization of the communication between AUV, but also in ensuring of its efficacy.

From technical point of view, the most demanded progress in the development of AUV is the progress in the field of increase of their autonomy, particularly energy autonomy. And although the rest mentioned above problems are also a priority, it is obvious that the possibilities of ANPA are primarily limited by their low autonomy.

High modularity of modern AUV and path of their development toward multipurpose use, leading developers to use the analogue of the standard "Plug and Play" in AUV, requires a corresponding powerful software for AUV. As for the relevance in accordance with their intended purposes, we are now seeing a rapid formation and development of the oil and gas industry at sea, and of sea communications made by means of the underwater optical fiber cable lines. The development of these areas requires regular surveys of underwater communications, lying, as a rule, at large and medium depths. Some success in this field can be presented in the form of a few layouts of devices with mixed (remote and Autonomous) control that have passed the shallow water test.

References

1. Chernyi, S.: Use of Information Intelligent Components for the Analysis of Complex Processes of Marine Energy Systems. Transport and Telecommunication Journal, 17(3), (2016) 202–211. DOI: 10.1515/ttj-2016-0018
2. Nyrkov, A., Sokolov, S., Zhilenkov, A., Chernyi, S.: Complex modeling of power fluctuations stabilization digital control system for parallel operation of gas-diesel generators. 2016 IEEE NW Russia Young Researchers in Electrical and Electronic Engineering Conference (EIConRusNW), St. Petersburg (2016) 636–640. doi: 10.1109/EIConRusNW.2016.7448264
3. Kamgar-Parsi, G., B. Johnson. D.L. Folds, Belcher, E.O.: High-resolution underwater acoustic imaging with lens-based systems, Int. J. Imaging Sys. Technol. 8 (1997) 377–385
4. Welsh, R., et. al.: "Advances in Efficient Submersible Acoustic mobile Networks" International UUV Symposium, Newport RI, 2000
5. Blidberg, D.R., Jalbert, J.C., and Ageev, M.D.: A Solar Powered Autonomous Underwater Vehicle System, International Advanced Robotics Program, (IARP) (1998)

Elicitation of Testing Requirements from the Selected Set of Software's Functional Requirements Using Fuzzy-Based Approach

Mohd. Sadiq and Neha

Abstract Software requirements elicitation is employed to find out different types of software requirements. In literature, we find out that goal-oriented requirements elicitation (GORE) techniques do not underpin the identification of testing requirements from the functional requirements (FR) in early phase of requirements engineering. Therefore, to tackle this research issue, we proposed an approach for the elicitation of the testing requirements from FR. In real-life applications, only those requirements are implemented which are selected by stakeholders; and tested by testers after implementation during different releases of software. So in the proposed method we used fuzzy-based technique for FR selection on the basis of nonfunctional requirements (NFR). Finally, an example is given to explain the proposed method.

Keywords Functional requirements · Nonfunctional requirements · Testing requirements · Fuzzy-based approach · Triangular fuzzy numbers

1 Introduction

The objective of requirements elicitation method is to find out the need of the stakeholders for the successful development of software [16]. Sadiq and Jain [15] classify the requirements engineering (RE) processes into five subprocesses, i.e., identification (i.e., elicitation), representation (i.e., modeling), analysis, verification and validation, and management of requirements. Different people are involved in

Mohd. Sadiq (✉)
Computer Engineering Section, UPFET, Jamia Millia Islamia
(A Central University), New Delhi 110025, India
e-mail: sadiq.jmi@gmail.com

Neha
Department of Computer Science and Engineering, Al-Falah University
(A NAAC- UGC 'A' Grade University), Dhauj, Faridabad, Haryana, India
e-mail: silent.angel02@gmail.com

H.S. Behera and D.P. Mohapatra (eds.), *Computational Intelligence in Data Mining*, Advances in Intelligent Systems and Computing 556,
DOI 10.1007/978-981-10-3874-7_40

RE process and testing process, and therefore, they have different mindset during software development [11–13]. For example, the mindset of RE people is to identify, analyse, and model the need of the software. On the other hand side, the mindset of the testing people is to identify as many errors as possible in the proposed system [2, 3].

Several researchers discuss the need to link the requirements with testing [2, 4, 7, 9, 10, 18]. For example, in 2002, Graham [7] discussed the need to link the requirements with testing. Therefore, keeping in view the observation of [7], in 2008, Uusitalo et al. [18] discussed the need to bring RE and tester closer to each other by proposing some set of guidelines. In 2009, Kukkanen et al. [10] integrate the RE process and testing to increase the quality of research and development. In their work, they also observed that when requirements engineer and tester's work together then it form a solid basis for the successful development of the software product. In 2009, Post et al. [14] focused on linking the functionality of the software requirements and verification. In 2011, Barmi et al. [2] suggest that by coupling functional requirements (FR) and testing requirements, a more accurate testing plan can be developed; and this testing plan helps to improve the cost and schedule of the project. In another study, Wnuk et al. [19] shared some experiences from software industry and they identify that test cases can be considered as the model of the software requirements.

In literature, we find out that GORE techniques [15–17] do not underpin the elicitation of testing requirements from the functional requirements (FR) in early phase of requirements engineering. Therefore, to tackle the above research issue, we proposed a method for the elicitation of the testing requirements from FR. In real-life applications, there could be N number of requirements in the system, i.e., N = 100 or 1000, and only those requirements are implemented which are selected by the stakeholders. Selected requirements are delivered during different release of software after implementation. Therefore, the selection of requirements is an important issue during RE process. In our work, we used fuzzy-based technique for software requirements selection (SRS).

The remaining part of this paper is structured as follows: In Sect. 2, we present an insight into fuzzy set theory. Proposed method is given in Sect. 3. Case study is given in Sect. 4. Finally, the conclusions and future work are given in Sect. 5.

2 Fuzzy Set Theory

Professor Lotfi A. Zadeh introduced the view of fuzzy set theory in 1965 to deal the vagueness and impreciseness in decision-making process [20]. There are different types of fuzzy numbers which are used in real-life applications like triangular fuzzy number, trapezoidal fuzzy numbers, etc [8]. The representations of these numbers are used to rank the fuzzy numbers or to rank the alternatives and requirements [4, 5, 16]. In our method, we used the graded mean integration representation (GMIR) method for representing the fuzzy numbers, proposed by Chen and Hsieh in 1998 [4].

Suppose W = (m, q, s) is a TFN whose membership function is defined as follows:

$$f_{W(x)} = \begin{cases} \frac{x-m}{q-m}, & m \le x \le q, \\ \frac{x-s}{q-s}, & q \le x \le s, \\ 0, & otherwise \end{cases} \quad (1)$$

Since

$$L_W(x) = \frac{x-m}{q-m}, m \le x \le q$$

$$R_W(x) = \frac{x-s}{q-s}, q \le x \le s$$

and

$$L_W^{-1}(h) = m + (q-m)h \quad 0 \le h \le 1$$

$$R_W^{-1}(h) = s + (q-s)h \quad 0 \le h \le 1$$

$L_W(x)$ and $R_W(x)$ are the function L and R of the fuzzy number W, respectively. $L_W^{-1}(x)$ and $R_W^{-1}(x)$ are the inverse functions of $L_W(x)$ and $R_W(x)$ at level h. The GMIR of W can be written as [4, 5]:

$$P(W) = \frac{\int_0^1 \frac{h\{L_W^{-1}(h) + R_W^{-1}(h)\}}{2} dh}{\int_0^1 h\, dh} \quad (2)$$

$$P(W) = \frac{\int_0^1 \frac{h\{m+(q-m)h+s+(q-s)h\}}{2} dh}{\int_0^1 h\, dh} \quad (3)$$

$$P(W) = \int_0^1 \frac{h\{m+(q-m)h+s+(q-s)h\}}{2} dh / \int_0^1 h\, dh$$

$$P(W) = \int_0^1 \frac{mh + qh^2 - mh^2 + sh + qh^2 - sh^2}{2} dh / \int_0^1 h\, dh$$

$$P(W) = \left\{ \frac{\left[\frac{mh^2}{2} + \frac{qh^3}{3} - \frac{mh^3}{3} + \frac{sh^2}{2} + \frac{qh^3}{3} - \frac{qh^3}{3}\right]_0^1}{2} \right\} / \{\frac{1}{2}\}$$

$$P(W) = \frac{m + 4q + s}{6} \quad (4)$$

Let us assume that W_1 and W_2 are two TFNs. Using (4), we can write the fuzzy multiplication operation of W1 and W2 as:

$$W_1 \odot W_2 = \frac{1}{6}(m_1 + 4q_1 + s_1) \odot \frac{1}{6}(m_2 + 4q_2 + s_2) \quad (5)$$

where $\odot$ is a fuzzy multiplication operator.

3 Proposed Method

In this section, we present the proposed method for the elicitation of the testing requirements from the selected set of software's requirements. There are three steps in the proposed method, i.e., Step 1: Elicitation of requirements; Step 2: Applying GMIR method for SRS; Step 3: Eliciting testing requirements from the selected requirements.

Step 1: Elicitation of requirements

In this step, the need of the stakeholders is identified. For the elicitation of requirements, we use the goal-oriented methods because these methods can refine and decompose the requirements by using AND/OR graph; and present a visual representation of the requirements clearly [6, 16, 17].

Step 2: Applying GMIR method for SRS

Before applying the GMIR method, we classify the FR and NFR [1]. Then hierarchical representation for software requirements selection (HR-SRS) would be constructed. After that we apply the Eqs. 4 and 5 for SRS, as discussed in Sect. 2.

Step 3: Eliciting testing requirements from the selected requirements

In this step, testing requirements are elicited from the selected set of requirements. Requirements which have highest priority are selected for the elicitation of testing requirements. Requirements engineer and tester will work together for the generation of the test cases in early phase of RE.

4 Case Study

In this section, we apply the proposed method on Institute Examination System (IES).

Step 1: Elicitation of requirements

We apply the goal-oriented method for the identification of FR and NFR of IES. As a result, we identify the following requirements, as discussed in our previous work [16, 17].

R1: Login part; R2: Record of the student's fee; R3: Semester result; R4: Seating arrangement; R5: Online examinations; R6: Examination form; R7: activities related to examination"; R8: hall ticket of students; R9: Approve_examination form; R10: examination fee_online_payment; R11: Security; R12: Cost effectiveness; R13: Usability.

Here, from requirements R1 to R10 are the FRs; and from R11 to R13 are the NFRs.

Now, we find out that which requirement would be implemented during software releases. Therefore, the objective of the next step is how to select the software requirements [4, 5].

Step 2: Applying GMIR method for SRS

The HR-SRS problem is given in Fig. 1. The linguistic variables (LV) for the importance weight of each NFR are defined in Table 1. The importance weight of each NFR is given in Table 2. The LV for the evaluation of FR on the basis of NFRs are given in Table 3.

The LV of Table 3 would be used by stakeholders to compute the ranking values of FR on the basis of NFRs; and the results are shown in Table 4.

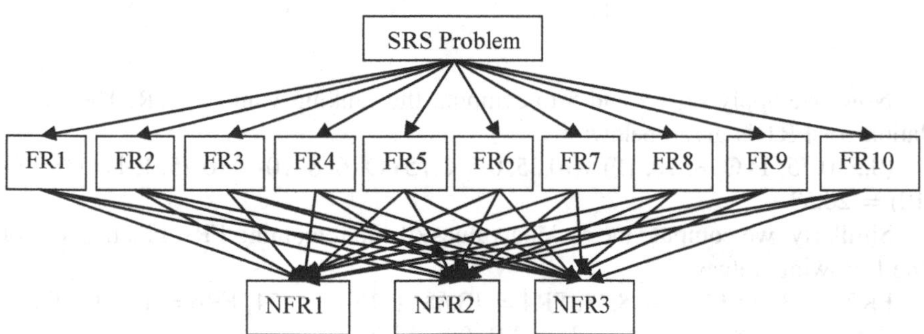

Fig. 1 HR-SRS problem

Table 1 LV for the importance weight of each NFR

Linguistic terms	TFN
Very Low (VL)	(0, 0, 0.25)
Low (L)	(0, 0.25, 0.5)
Medium (M)	(0.25, 0.5, 0.75)
High (H)	(0.5, 0.75, 1)
Very High (VH)	(0.75, 1, 1)

Table 2 Importance weight of each NFR

NFRs	TFN
Security (NFR1)	H
Economic (NFR2)	M
Usability (NFR3)	VH

Table 3 LV for the evaluation of FR

Linguistic terms	TFN
Very Weak (VL)	(2, 2, 4)
Weak (W)	(2, 4, 6)
Fair (F)	(4, 6, 8)
Strong (S)	(6, 8, 10)
Very Strong (VS)	(8, 10, 10)

Table 4 Evaluation of each FR with respect to NFRs

FRs	NFR1	NFR2	NFR3
FR1	VS	S	VS
FR2	F	F	M
FR3	S	F	VS
FR4	W	F	S
FR5	VS	W	VS
FR6	S	S	S
FR7	W	S	F
FR8	S	F	VS
FR9	F	F	M
FR10	S	W	M

Now, we apply Eqs. 4 and 5 to compute the ranking values of FR. The calculation for FR1 is given below:

(0.5, 0.75, 1) ⊙ (8, 10, 10) + (0.25, 0.5, 0.75) ⊙ (6, 8, 10) + (0.75, 1, 1) ⊙ (8, 10, 10) = 20.53

Similarly, we compute the ranking values of the remaining FRs. Finally, we got the following values:

FR2 = 13.26; FR3 = 18.28; FR4 = 13.68; FR5 = 18.54; FR6 = 17.84; FR7 = 12.76; FR8 = 18.28; FR9 = 13.26; FR10 = 13.76

The ranking values of FR indicate that FR1 has the highest ranking values, i.e, Login part. This module is important because it has the highest priority; therefore, it would be implemented first. In the next step, we elicit the testing requirements for FR1.

Step 3: Eliciting testing requirements from the selected requirements

Based on the ranking values of the previous step, we identify that FR1 has the highest priority. The snapshot of the Login module is given in Fig. 2. To identify the testing requirements, we first refine and decompose the Login Module into

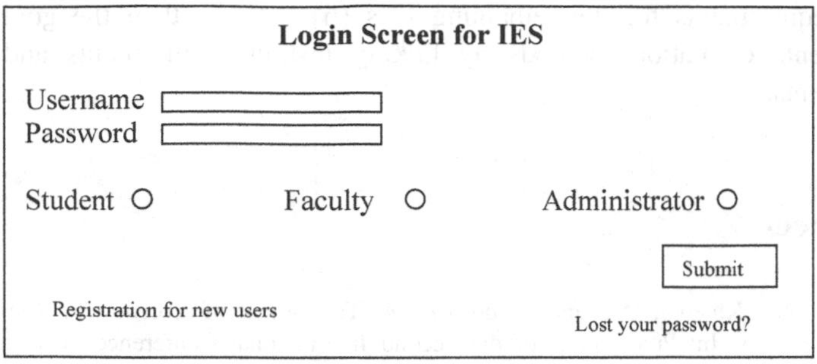

Login Screen for IES

Username
Password

Student O Faculty O Administrator O

Submit

Registration for new users

Lost your password?

Fig. 2 Login part

subrequirements, i.e., Username; Password; selection of the types of users, i.e., Student, Faculty, and Administration; Submit button; Registration for new users; and Forgot password, as shown in Fig. 2. The testing requirements (TR) for FR1 are given below:

TR1: Username should contain letters, numbers, and periods; TR2: Username should not be contain left blank; TR3: The length of username should not exceed 40 characters; TR4: Username should not start with any special symbol; TR5: Password should be at least six characters. The first character should be capital letter. In the password, there should be any three numeric values; TR6: Password should not contain spaces and period; TR7: Password should not be more than 40 characters; TR8: Does the link for the registration for new users' works correctly?; TR9: Does the "Lost your password" link work properly?; TR10: Do the forward and backward buttons work properly?; TR11: How errors are handled and displayed?

By linking the testing with the requirements, we have elicited 11 testing requirements for one FR, i.e., Login Part. Similarly, we can elicit the TR for the remaining FRs.

5 Conclusions and Future Work

This paper presents a method for the elicitation of testing requirements from the selected set of functional requirements. Proposed method includes three steps, i.e., (i) elicitation of requirements, (ii) applying fuzzy-based GMIR method for the selection of software requirements (SRS), and (iii) eliciting testing requirements from the selected requirements. We have applied proposed method for the elicitation of testing requirements for one of the FR of Institute Examination System which have highest ranking value, i.e., Login Part. As a result, we have elicited 11 testing requirements for one FR, i.e., Login Part. By linking testing with the requirements we can develop a successful software product because of the knowledge of the requirements engineer and tester in early phase of requirements engineering. In future, we will try to work on the following issues: (i) to elicit the testing requirements for the remaining FRs (ii) to strengthen the goal oriented requirements elicitation methods by linking testing requirements and security requirements.

References

1. Afreen N., Khatoon A., and Sadiq M.: A Taxonomy of Software's Non-Functional Requirements. In: Proceedings of the Second International Conference on Computer and Communication Technologies, Vol. 379 of Advances in Intelligent Systems and Computing, Springer, (2015) 47–53.
2. Barmi Z.A., Ebrahimi A. H., and Feldt R.: Alignment of Requirements Specification and Testing: A Systematic Mapping Study. In: 4th International conference on Software Testing, Verification, and Validations Workshop (2011) 476–485.
3. Bjarnason E., Unterkalmsteiner M., Borg M., and Engstrom E.: A Multi-case Study of Agile Requirements Engineering and the Use of Test Cases as Requirements. Information and Software Technology (2016).
4. Chen S. H., Hsieh C. H. Graded Mean Integration Representation of Generalized Fuzzy Numbers. In: Sixth International Conference on Fuzzy Theory and Its Applications, Taiwan (1998) 1–6.
5. Chou C.C.: The Representations of Multiplication Operation on Fuzzy Numbers and Application to Solving Fuzzy Multiple Criteria Decision Making Problems. In: PRICAI, LNAI, Springer (2006) 161–169.
6. Garg N., Sadiq M., and Agarwal P.: GOASREP: Goal Oriented Approach for Software Requirements Elicitation and Prioritization using Analytic Hierarchy Process. In: 5th International Conference on Frontiers in Intelligent Computing Theory and Applications, Springer (2016).
7. Graham D., "Requirements and Testing: Seven Missing Link Myths", IEEE Software, Vol. 19, No. 5, pp. 15–17, 2002.
8. Ishizaka A. and Labib A.: Review of the main Development in the Analytic Hierarchy Process. Expert Systems with Applications, 38 (11) (2011), 14336–14345.
9. Kamata I. M. and Tamai T.: How Does Requirements Quality Relate to Project Success or Failure? In: 15th IEEE International Requirements Engineering Conference (2007) 69–78.
10. Kukkanen J., Vakevainen K., Kauppinen M, and Uusitalo M.: Applying Systematic Approach to Link Requirements and Testing: A Case Study. In: Asia Pacific Software Engineering Conference (2009), 482–488.
11. Li J., Zhang H., Zhu L., Jeffery R., Wang Q., and Li M.: Preliminary Results of a Systematic Literature Review on Requirements Evolution. In: ACM-EASE (2012) 12–21.
12. Lindstrom D. R.: Five Ways to Destroy a Development Project. IEEE Software, 10(5) (1993) 55–58.
13. Merz F., Sinz C., Post H, Georges T., and Kropf T.: Bridging the Gap between Test Cases and Requirements by Abstract Testing. Innovations in Systems and Software Engineering, 11 (4) (2015), 233–242.
14. Post H., Sinz C., Merz F., Gorges T., and Kropf T.: Linking Functional Requirements and Software Verification. In: IEEE International Requirements Engineering Conference, (2009) 295–302.
15. Sadiq M. and Jain S. K.: An Insight into Requirements Engineering Processes. In: 3rd International Conference on Advances in Communication, Network, and Computing LNCSIT-Springer, (2012) 313–318, Chennai, India.

16. Sadiq M. and Jain S. K.: Applying Fuzzy Preference Relation for Requirements Prioritization in Goal Oriented Requirements Elicitation Process. International Journal of Systems Assurance Engineering and Maintenance, Springer, 5(4), (2014) 711–723.
17. Sadiq M., Jain S. K.: A Fuzzy Based Approach for Requirements Prioritization in Goal Oriented Requirements Elicitation Process. In: 25th International Conference on Software Engineering and Knowledge Engineering, Boston, USA, June 27-June 29, 2013.
18. Uusitalo E., Komssi M., Kauppinen M., and Davis A. M.: Linking Requirements and Testing in Practice. In: International Requirements Engineering Conference, 2008.
19. Wnuk K., Ahlberg L., and Persson J.: On the Delicate Balance between RE and Testing-Experiences from a Large Company. International Requirements Engineering Conference, (2014) 1–3.
20. Zadeh L. A.: Fuzzy Logic = Computing with Words. IEEE Transactions on Fuzzy Systems, 4 (2) (1996) 103–111.

Analysis of TCP Variant Protocol Using Active Queue Management Techniques in Wired-Cum-Wireless Networks

Sukant Kishoro Bisoy, Bibudhendu Pati, Chhabi Rani Panigrahi and Prasant Kumar Pattnaik

Abstract In this work, we analyzed the performance of TCP variant protocols in wired-cum-wireless networks considering active queue management (AQM) techniques such as random exponential marking (REM) and adaptive virtual queue (AVQ) along with Droptail. For analysis, we consider Reno, Newreno, Sack1, and Vegas as TCP variants and proposed a network model for wired-cum-wireless scenario. Then, the performance of TCP variants is analyzed using delayed acknowledgement (DelACK) technique. The simulation results show that Newreno performs better than others when DelACK is not used. However, when DelACK is used, the performance of Vegas is better than others irrespective of AQM techniques.

Keywords AQM · REM · AVQ · DelACK · TCP variants · NS2

1 Introduction

Due to rapid growth of Internet and popularity of mobile devices during the last decade, several new techniques have been discovered, which allow different researchers to improve the performance of wireless applications [1]. In such scenario, the transmission media become more heterogeneous (wired-cum-wireless).

S.K. Bisoy (✉) · B. Pati · C.R. Panigrahi
Department of Computer Science and Engineering, C.V. Raman College of Engineering, Bhubaneswar, India
e-mail: sukantabisoyi@yahoo.com

B. Pati
e-mail: patibibudhendu@gmail.com

C.R. Panigrahi
e-mail: panigrahichhabi@gmail.com

P.K. Pattnaik
School of Computer Engineering, KIIT University, Bhubaneswar, India
e-mail: patnaikprasantfcs@kiit.ac.in

H.S. Behera and D.P. Mohapatra (eds.), *Computational Intelligence in Data Mining*, Advances in Intelligent Systems and Computing 556,
DOI 10.1007/978-981-10-3874-7_41

A few works have been proposed which are basically developed based on the wireless technology [2–5]. But, the characteristics of wired networks are different from wireless networks. Even through TCP [6] works well for wired network, it is not working well for wired-cum-wireless network. It assumes that any loss of packet is an indication of congestion in the network. However, it may not be true for wireless network due to many reasons. It may be node disconnection due to high mobility, frequent bandwidth changing, and high bit error rate which causes huge packet loss and unnecessary throughput degrades.

When a packet arrives in the queue and if the queue is full, then it is forced to drop any new incoming packets. Whenever any packet drop occurs, the sender interprets that there is congestion in the network. To overcome congestion in the network, all the TCP sources reduce the window size to control the transmission rate. Based on the perception of congestion, the TCP estimates the congestion and able to control the congestion in the Internet [7].

In order to improve link utilization, the router may accommodate more packets in their queue. However, different queue management follows different dropping policy to manage the queue. Droptail mechanism follows passive queue management. It drops the incoming packets once the queue gets filled. Due to passive nature, it suffers from global synchronization problem and long delay. To overcome the above problems, AQM [8] techniques have been developed in recent years. AQM technique can be used to improve high link utilization and reduce delay and packet loss. In this work, we analyze the performance of TCP variants using a proposed network model for wired-cum-wireless network. Different AQM techniques are used to analyze TCP variants.

The rest of the paper is organized as follows: Sect. 2 describes the related work. Sections 3 and 4 present the AQM techniques and TCP variants, respectively. Section 5 provides the simulation setup required for this work. The analysis of results is described in Sect. 6. Finally, Sect. 7 concludes the work.

2 Related Work

TCP has many variants such as Tahoe, Reno, Newreno, Sack1, and Vegas. The congestion control mechanisms and performance of each protocol are different from others. The performance of TCP variants is analyzed in wireless network using delayed acknowledgement strategy [9]. The simulation results show that Vegas achieve higher throughput, lower retransmissions, and lower delay than Newreno. The authors in [10] analyzed the intra-protocol fairness between Reno and Vegas using some routing protocols. Then, delayed acknowledge strategy is used to analyze the fairness between TCP Reno and TCP Vegas. The results show that the fairness between Reno and Vegas is better using DSDV protocols than AODV and DSR. Use of delayed acknowledgement can improve the fairness further. In [11], authors analyzed the interaction between TCP variants and ad hoc routing protocol in wireless network. The results show that OLSR is a suitable ad hoc routing protocol for TCP variants to achieve higher throughput and lower packet loss ratio than DSDV and OADV.

3 Active Queue Management Techniques

AQM [8] technique can be used to notify the incipient congestion before the queue become fill. It adopts the dynamic behavior of network and minimizes the congestion. The router using AQM technique provides congestion information by dropping/marking the packets. For marking the packets, explicit congestion notification (ECN) [12, 13] mechanism can be used in the IP header. Further, faster congestion notification (FCN) [14] mechanism is used to manage queue efficiently and to notify congestion faster.

The most popular widely used AQM technique is random early detection (RED) [15]. It is recommended by Internet Engineering Task Force (IETF) [16] as default AQM scheme. It prevents global synchronization problem faced by Droptail. Random exponential marking (REM) [17] is a rate and queue-based technique and adopts congestion measure differently from others. End user is notified about congestion by marking the packets. In RED, queue length is used to measure the congestion. In REM, the congestion is measured by price. It calculates the aggregate link price and notifies the sender through the parking probability. Irrespective of the number of users, it is trying to stabilize queue and input rate. In AVQ input traffic rate is used as congestion measure to notify the sender. It maintains a virtual queue whose capacity is less than actual queue. Whenever there is an overflow in the virtual queue, the packets in real queue is marked. To achieve desired utilization of link capacity, the virtual queue is updated for each packet arrival. Originally, it was proposed as rate-based technique [18].

4 TCP Variants

Most of the Internet traffic is carried by TCP protocol to transport data over Internet. It is utilized in many applications of Internet such as World Wide Web (WWW), E-mail, file transfer protocols (FTP), Secure Shell, peer-to-peer file sharing, and some streaming media applications. Several versions of TCP have been developed and implemented in the literature [19, 20]. In wireless network, performance of TCP relies on routing mechanisms used to transmit the packets. The congestion control and loss recovery mechanism of TCP variant is different from each other.

4.1 Reno

Reno is one of the variants of TCP protocol. It increases their window size by one per RTT until the packet loss occurs. It reduces the window size to half for each packet loss detected. It follows additive increase and multiplicative decreases (AIMD) [21] policy for window management.

4.2 Newreno

Newreno is developed with small modification in Reno at the sender side [22]. New improved fast recovery algorithm is implemented in Newreno, which helps to recover multiple packet losses in a single window and minimizes as many as transmission timeout events [23]. This is suitable for a wireless environment where packet loss occurs in burst.

4.3 Sack1

This technique specifies the segment which has received correctly. It overcomes the problem of multiple packets dropping in each window. Instead of sending cumulative acknowledge, Sack1 provides information of lost packets selectively in the acknowledgement, which helps to send multiple lost segment in each Round Trip Time (RTT) [24].

4.4 Vegas

Unlike Reno, Vegas does not reply on packet loss for window management. Vegas follows delay-based approach and Reno follows loss-based approach. Vegas is able to detect congestion before any segment loss occurs [25]. To detect congestion in the network, it finds measured RTT of each packet. Vegas estimates the level of congestion and updates the window size (CWND) using following:

Diff = (expected − actual) * BaseRTT;
Expected = current window size (CWND)/minimum round trip time (BaseRTT);
Actual = current window size (CWND)/actual round trip time (BaseRTT);

Based on the *Diff* value, source can update the window size. If Diff $< \alpha$, TCP Vegas increases the CWND by 1 during the next round trip time and decreases the

CWND by 1 for Diff > β and remains unchanged, otherwise. The alpha (α) and beta (β) are the two threshold values of Vegas. Vegas always try to regulate its queue length between α and β packets in the queue. It is important for TCP Vegas to have an accurate estimation of available bandwidth because it adjusts window size based on the estimation of propagation delay and BaseRTT. Therefore, the performance of TCP Vegas is very sensitive to the accuracy of BaseRTT.

5 Simulation Setup

We analyzed the performance of different TCP variant protocols using a proposed network model for wired-cum-wireless scenario. For analysis, we used Network Simulator-2 (NS-2) [26]. The proposed network model is shown in Fig. 1 is having 6 nodes. S1 and S2 are associated with TCP source and D1 and D2 is the TCP sink. S1 and S2 are connected with R1 through wirelessly. Others are connected with wired links. The common link (R1 to R2) is passed by two TCP flows from wireless sources S1 and S2 to wired destination D1 and D2 and all the flows have same RTT. The bandwidth of bottleneck link R1–R2 is 2 Mb and delay is 16 ms. The bandwidth and delay of side links from R2 to D1 and R2 to D2 are 20 Mb, 1 ms, respectively. TCP variants such as Reno, Newreno, Sack1, and Vegas are used as transport layer protocols and for buffer management Droptail, REM, and AVQ techniques are used at R1.

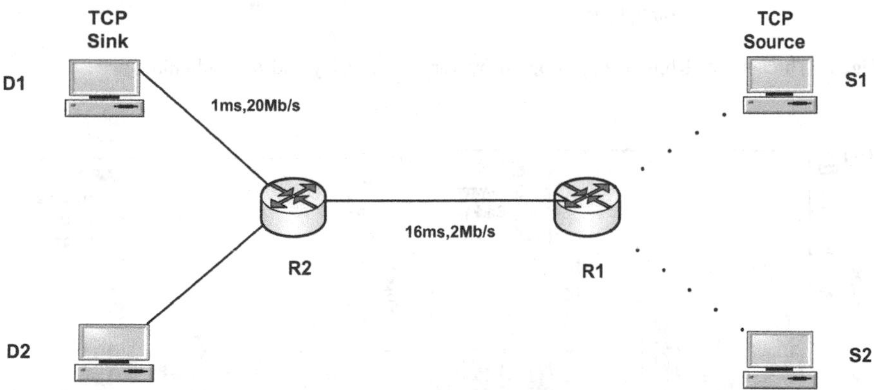

Fig. 1 The proposed network model

6 Results Analysis

In this section, the performances of TCP variants are provided in terms of throughput (Kbps) in two parts. In first part, performances of different TCP Variants are analyzed using different AQM techniques and in second part DelACK technique is used along with different AQM techniques.

6.1 Using AQM Techniques

In this section, the results obtained for each AQM technique are shown in Figs. 2, 3, and 4 by varying delay and bandwidth between R1 and R2. Using Droptail as

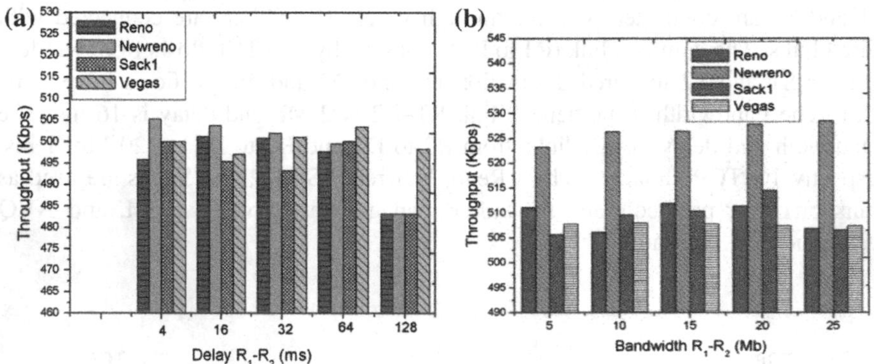

Fig. 2 Throughput (Kbps) using Droptail by varying **a** Delay and **b** Bandwidth

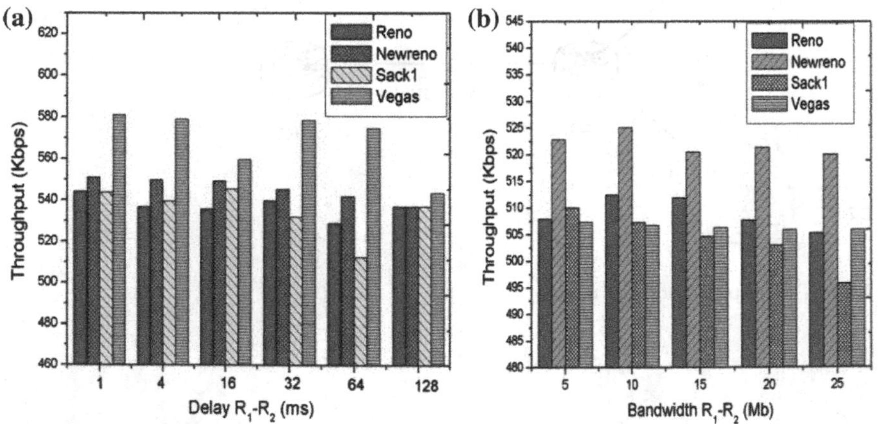

Fig. 3 Throughput (Kbps) using REM by varying **a** Delay and **b** Bandwidth

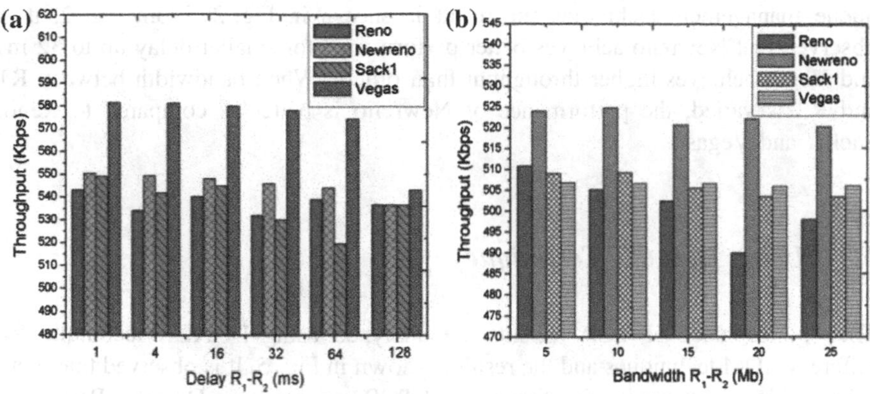

Fig. 4 Throughput (Kbps) using AVQ by varying **a** Delay and **b** bandwidth

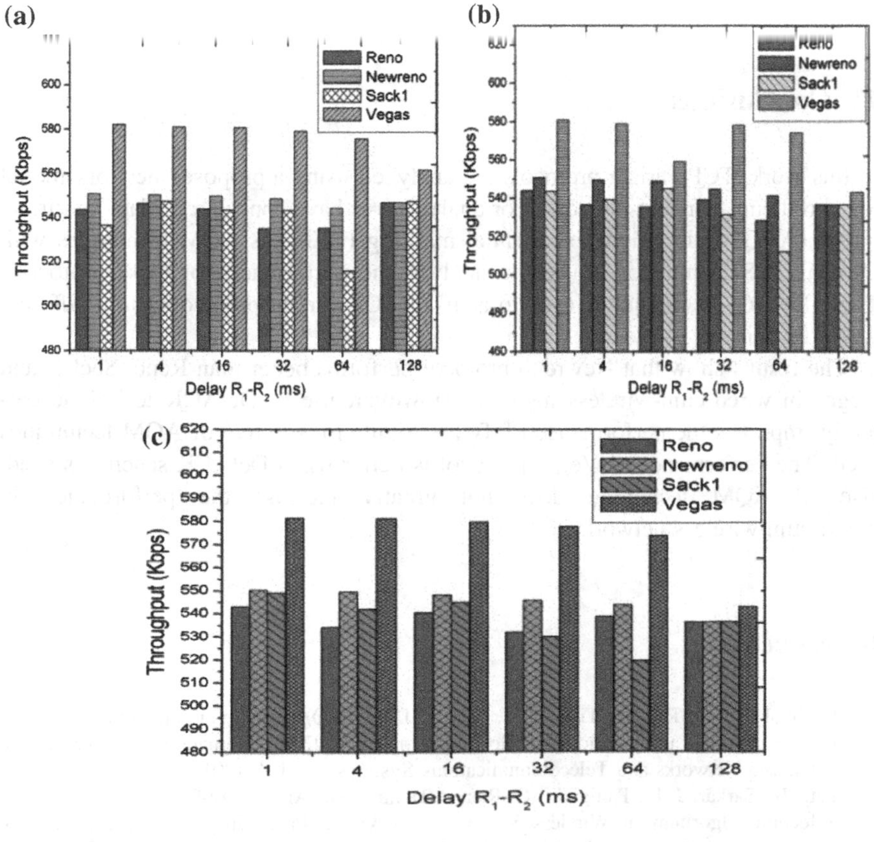

Fig. 5 Throughput (Kbps) using DelACK for **a** Droptail, **b** REM and **c** AVQ

queue management technique, the result is shown in Fig. 2. From Fig. 2, it is observed that Newreno achieves better performance for smaller delay up to 32 ms and Vegas achieves higher throughput than others. When bandwidth between R1 and R2 is varied, the performance of Newreno is better as compared to Reno, Sack1, and Vegas.

6.2 Using DelACK Technique

Then performance of TCP variants are analyzed using DelACK technique for different AQM techniques and the result is shown in Fig. 5. It is observed that there is insignificant difference in performance of TCP variants using Droptail, REM, and AVQ. The TCP performance remains same irrespective of AQM techniques used. However, when DelACK is used, Vegas performs better in achieving higher throughput as compared to Reno, Newreno, and Sack1.

7 Conclusions

In this work, TCP variant protocols are analyzed using a proposed network model for wired-cum-wireless network. For analysis, we have considered adaptive virtual queue (AVQ) and random exponential marking (REM) as AQM techniques with Droptail. NS2 simulator is used to analyze the performance of these protocols. Then, DelACK technique is used to analyze TCP variant protocols using different AQM techniques.

The results show that Newreno protocol performs better than Reno, Sack1, and Vegas in wired-cum-wireless network. However, use of DelACK technique certainly improves the performance of TCP variants irrespective of AQM techniques used. The performance of Vegas protocol is better when DelACK scheme is used. Use of AQM technique does not greatly increase the performance in wired-cum-wireless networks.

References

1. Panigrahi, C. R., Pati, B., Tiwary, M., Sarkar, J. L.: EEOA: Improving energy efficiency of mobile cloudlets using efficient Offloading Approach. IEEE International Conference on Advanced Networks and Telecommunications Systems, pp. 1–6, (2015).
2. Pati, B., Sarkar, J. L., Panigrahi, C. R. and Tiwary, M.: An Energy-Efficient Cluster-Head Selection Algorithm in Wireless Sensor Networks. In Proceedings of 3rd International Conference on Mining Intelligence and Knowledge Exploration, pp. 184–193, (2015).

3. Sarkar, J. L., Panigrah, C. R., Pati, B., Das, H.: A Novel Approach for Real-Time Data Management In Wireless Sensor Networks. In Proc. of 3rd International Conference on Advanced Computing, Networking and Informatics, vol. 2, pp. 599– 607, (2015).
4. Panigrahi, C. R., Sarkar, J. L., Pati, B., Das, H.: S2S: A Novel Approach for Source to Sink Node Communication in Wireless Sensor Networks. In Proceedings of 3rd International Conference on Mining Intelligence and Knowledge Exploration, pp. 406–414, (2015).
5. Pati, B., Sarkar, J. L., Panigrahi, C. R. and Verma R. K.: CQS: A Conflict-free Query Scheduling Approach in Wireless Sensor Networks. In Proc. of 3rd International Conference on Recent Advances in Information Technology, pp. 13–18, (2016).
6. Postel, J.: Transmission Control Protocol. RFC 79 (1980).
7. Kurose, J., Ross, K.: Computer Networks: a top-down approach featuring the Internet, 2nd ed. Addison Wesley (1996).
8. Braden, B., Clark, D., Crowcroft, J., Davie, B., Deering, S., Estrin, D., Floyd, S., Jacobson, V., Minshall, G., Partridge, C., Peterson, L., Ramakrishnan, K., Shenker, S., Wroclawski, J., and Zhang, L.: Recommendations on Queue Management and Congestion Avoidance in the Internet, RFC Editor (1998).
9. Bisoy, S. K., Das, A., Pattnaik, P. K., and Panda, M. R.: The Impact of Delayed ACK on TCP Variants Protocols in Wireless Network. In: proceedings of IEEE International Conference on High Performance Computing & Application (ICHPCA), pp. 1–6, 22–24 December (2014).
10. Bisoy, S. K., P. K. Pattnaik.: Throughput of a Network Shared by TCP Reno and TCP Vegas in Static Multi-hop Wireless Network. In: proceedings of Springer ICCIDM, Vol. 1, pp. 471–481, December (2015).
11. Bisoy, S. K. and Pattnaik, P. K.: Interaction between Internet Based TCP Variants and Routing Protocols in MANET. In: Int. Conf. on Frontiers of Intelligent Computing: Theory and Applications (FICTA), pp. 423–433 (2013).
12. Floyd, S.: TCP and explicit congestion notification. ACM Computer Communication Review, 24 (5), pp. 10–23 (1994).
13. Ramakrishnan, K., Floyd, S.: A proposal to add Explicit Congestion Notification (ECN to IP.), RFC 2481(1999).
14. Mohammed, M., Kadhum, M., Hhassan, S.: A new congestion management mechanism for next generation routers. Journal of engineering science and technology (JESTEC), 3(3), pp. 265–271 (2008).
15. Floyd, S. and Jacobson, V.: Random early detection gateways for congestion avoidance. IEEE Transactions on Networking, 1(4), pp. 397–413 (1993).
16. Feng, W., Kandlur, D., Saha, D., and Shin, K.: Blue: A new class of queue management algorithms", IEEE Transactions on Networking, 10(4), pp. 513–528 (2002).
17. Athuraliya, S., Li, V. H., Low, S. H., and Yin, Q.; Rem: Active queue management", pp. 48–53, Vol. 15, Issue 3, IEEE Network (2001).
18. Kunniyur, S., Srikant, R.: Analysis and design of an adaptive virtual queue (AVQ) algorithm for active queue management, Proc. ACM SIGCOMM, pp. 123–134 (2001).
19. Jacobson, V.: Congestion avoidance and control. Computer Communication Review, 18(4), pp. 314–29 (1988).
20. Jacobson, V.: Modifed TCP Congestion Avoidance Algorithm. Technical report (1990).
21. Chiu, D. and Jain, R.: Analysis of the Increase and Decrease Algorithms for Congestion Avoidance in Computer Networks. Computer Networks and ISDN Systems, 17, pp. 1–14 (1989).
22. Hoe, J.: Start-up Dynamics of TCP's Congestion Control and Avoidance Scheme, Master's thesis, MIT (1995).
23. Floyd, S., Henderson, T., and Gurtov, A.: The NewReno Modification to TCP's Fast Recovery Algorithm. RFC 3782 (2004).
24. Fall, K. and Floyd, S.: Simulation-based comparison of tahoe, reno, and sack tcp. Computer Communication Review, vol. 26, pp. 5–21 (1996).

25. Brakmo, L., O'Malley, S., and Peterson, L.: TCP Vegas: New Techniques for Congestion Detection and Avoidance. In: Proc. of ACM SIGCOMM, pages 24–35 (1994).
26. Information Sciences Institute, The Network Simulator NS-2, http://www.isi.edu/nanam/ns/, University of Southern California.

Image Texture-Based New Cryptography Scheme Using Advanced Encryption Standard

Ram Chandra Barik, Suvamoy Changder and Sitanshu Sekhar Sahu

Abstract Encapsulation of information using mathematical barrier for forbidding malicious access is a traditional approach from past to modern era of information technology. Recent advancement in security field is not restricted to the traditional symmetric and asymmetric cryptography; rather, immense security algorithms were proposed in the recent past, from which biometric-based security, steganography, visual cryptography, etc. gained prominent focus within research communities. In this paper, we have proposed a robust cryptographic scheme to original message. First, each message byte, the ASCII characters ranging from Space (ASCII-32) to Tilde (ASCII-126), is represented as object using flat texture in a binary image which is decorated as n by n geometrical-shaped object in images of size N × N. Create a chaotic arrangement pattern by using the prime number encrypted by Advanced Encryption Standard (AES). The sub-images are shuffled and united as rows and columns to form a host covert or cipher image which looks like a grid-structured image where each sub-grid represents the coded information. The performance of the proposed method has been analyzed with empirical examples.

Keywords Grid structure · Cryptography · Texture · Shuffling pattern · AES (Advanced Encryption Standard) · Steganography · Visual cryptography

R.C. Barik (✉) · S. Changder
Department of Computer Applications, National Institute of Technology, Durgapur, West Bengal, India
e-mail: ramchbarik@gmail.com

S. Changder
e-mail: suvamoy.nitdgp@gmail.com

S.S. Sahu
Department of Electronics & Communication Engineering, Birla Institute of Technology, Mesra, Ranchi, Jharkhand, India
e-mail: sitanshusekhar@gmail.com

H.S. Behera and D.P. Mohapatra (eds.), *Computational Intelligence in Data Mining*, Advances in Intelligent Systems and Computing 556,
DOI 10.1007/978-981-10-3874-7_42

1 Introduction

Mathematical barrier plays a prominent role for protecting information in the security arena. The uses of electronic media and technology are thriving day by day in this new era of communication where information security is a major challenge for multi-dimensional applications since many years. Cryptography provides an opportunity for secure communication in the presence of third parties and plays a significant role of abstracting data into an unreadable format. Although the existing cryptography techniques provide good security, but at the same time it increase the computational complexity. Hence, there is a need of an efficient as well as simple cryptography approach. Generally, messages are being encrypted using various algorithms, and the sending and receiving mode is Internet through online login, secure service login, Internet banking, database login, military secret code interchange, etc. Encryption of plaintext using public key and private key is a significant area of attention among many researchers and mathematicians in the field of information security and cryptography. Cryptography applied to a digital image is a new dimension of information security field known as steganography, visual cryptography, etc. Concealing or embedding text in an image requires the intensity modification.

Freeman [1] proposed chain codes for description of digitized curves, contours, and drawings to represent the data. Chain codes represent the digital contour by a sequence of line segments of specified length and direction. Islam et al. [2] proposed a new dimension of message encryption where edges in the cover image have been used to embed messages. It performs better or at least at par with the state-of-the-art steganography techniques but provides higher embedding capacity. Yicong et al. [3] proposed a novel encryption algorithm using a bit plane of a source image as the security key to encrypt images. Chang and Yu [4] proposed an encryption scheme, which is better as compared to its predecessors, but it is still vulnerable to attacks if it uses the same key to encrypt different images. But again, Chang provides a different approach to show that their scheme can be broken into some pairs of plain image and cipher image for secure communication. Wang et al. [5] presented a new method of optical image encryption with binary Fourier transform computer-generated hologram (CGH) and pixel scrambling technology. The orders of the pixel scrambling, as well as the encrypted image, are used as the keys to decrypt the original image and thereby claimed to provide higher security. Furthermore, the encrypted image is binary, so it is easy to be fabricated and robust against noise and distortion. Kuang Tsan-Lin [6] proposed a new type of encoding methods to encrypt hidden (covert) information in host images. The encrypted information can be plot, fax, word, or network, and it must be encoded with binary codes. Li et al. [7] proposed a typical binary image scrambling/permutation algorithm exerting on plaintext of fixed size and a novel optimal method to break it with some known/chosen plaintexts. Yang and Kot [8] proposed a novel blind data-hiding method for binary images authentication by preserving pixel neighborhood which locates the embeddable pixels in a block of different block schemes.

Wange et al. [9] proposed a new dimension of hiding pattern by improved embedding capacity with embedding distortion using authenticated and annotated scanned images. Jung et al. [10] proposed a new data-hiding method for binary images that relies on block masking to distribute keys to two parts and then authenticates the right authorized part.

Most of the existing security schemes are computationally expensive and introduce inherent difficulty in the processing of data. Cryptography applied in the form of image is emerging as a new dimension to information security field. It minimizes the computational load as well as improves the security by introducing some difficulties to third parties.

The aim of the proposed algorithm is to divide the original image into n * n pixels of blocks or sub-images and then encrypt the message inside those blocks in binary image. The geometrical smooth texture which is based on binary provides a form of security corresponding to the original message. The smooth textures are generated and arranged in a binary image form to make it robust from external malicious access. Then this image is transmitted to receiver by any kind of communication mode; after that in receiver side, the image is decrypted and the data is decoded to get back the result. This paper gives the overall cryptography process in a digital image, which makes it a unique approach.

2 Materials and Methods

The detailed methodology of the proposed cryptographic scheme is shown in Fig. 1, which describes the flow for encryption and decryption part of the approach.

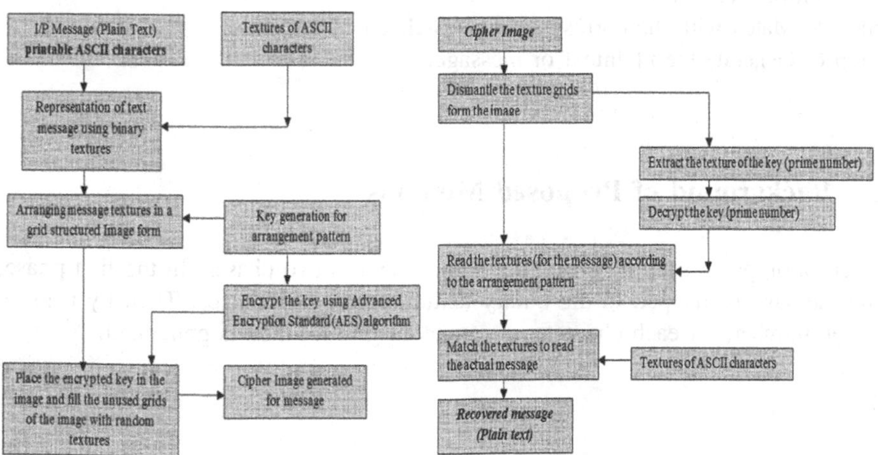

Fig. 1 Flow diagram of the proposed method for encryption and decryption

2.1 Algorithm

The procedure for implementing the proposed approach is described as follows:

2.1.1 Encryption

Step 1: Map the plaintext or message into its corresponding sub-binary image.
Step 2: Generate a random number series based on the size of message using 2 prime numbers (within the range of message size).
Step 3: Embed the sub-image holding the objects into the grids based on the random number leaving few grids.
Step 4: Concatenate two prime numbers and encrypt by AES using a key.
Step 5: Embed the AES-encrypted text and the key in the remaining grid. Pad the vacant grid by arbitrary texture sub-image.
Step 6: Save the grid-structured covert image.

2.1.2 Decryption

Step 1: Dismantle the grid-structured cipher image into a series of 16 × 16 sub-images.
Step 2: Read the last 24 grid sub-images and recognize the pattern by direct matching. Find the prime key.
Step 3: Based on the prime key generate same series of random number representing grid sub-image.
Step 4: Recognize each grid by direct matching or using any other image texture recognition techniques.
Step 5: Match with the corresponding ASCII character value.
Step 6: Generate the plaintext or message.

3 Background of Proposed Methods

Encryption process of the proposed method depicts two phases. In the first phase, the message is mapped to the binary texture of the sub-images. Then by using a prime number, for each character one random grid location is generated.

3.1 Texture Creation for ASCII Character

The proposed encryption method is based on the grid-structured binary image. The message is concealed in an image having certain number of grids or blocks. The message in the form of ASCII character is mapped for certain smooth texture onto a sub-image made of binary pixels. Table 1 contains the texture for 95 printable ASCII characters starting from Space (32) to tilde (126). Equation 1 describes the formation of binary image with the smooth texture.

First, a matrix of N × N pixels with black background is created, i.e., the intensity is zero, which can be realized as an image having black foreground as well as background. Over the black background, the foreground object is represented as white symbolic lines. The image Im(x, y) can be mathematically expressed in Eq. 1.

$$\mathrm{Im}(x,\ y) = \mathrm{Mat}_{ij}\begin{cases} =1 & \text{for} \quad \text{points on the object} \\ =0 & \text{for} \quad \text{background points} \end{cases} \tag{1}$$

Table 1 The texture for ASCII character Space (32) to tilde (126)

In the binary image, zero (0) represents black and one (1) represents white. The texture of the objects is represented in the form of binary images of N × N pixels and stored in both sender and receiver sides. The original binary matrix of N × N pixels is divided into small blocks/grids of the size of the sub-images (n × n), and these sub-images are to be contained within it in a judicial way.

A series of such sub-image texture arranged in row and column formats to form a host image. Equation 2 describes the concatenation. The texture of the original message is represented in the form of images of N × N pixels and stored in both sender and receiver sides. The original matrix of N × N pixels is divided into small blocks/grids of the size of the sub-images (n × n).

$$\sum_{i=1}^{n} \mathrm{subIm}[i]_{16\times16} \rightarrow \mathrm{Im}_{512\times512} \qquad (2)$$

where $\mathrm{Img}_{512\times512}$ is the original matrix which is divided into n sub-matrix or sub-image $\mathrm{subIm}[i]_{16\times16}$ $\forall i = 1, 2, \ldots n$. Equation 2 describes the merging of sub-images or matrices to form a host image. Plaintext could be any alphabets, numbers, alphanumeric characters (ASCII and Unicode family, etc.).

3.2 Random Grid Location Generator

The proposed method uses random number generator to generate the random number grid location between ranges of the grid. Based on the size of the message and the two other prime or random numbers P_1 and P_2

$$X_{n+1} = P_1 X_n + P_2 (\mathrm{mod}\, N) \quad n = 0, 1, 2 \qquad (3)$$

Here, we are assuming that both P1 and P2 both are the prime numbers. For example, in Fig. 2, the Lena gray image (256 × 256) is being divided into

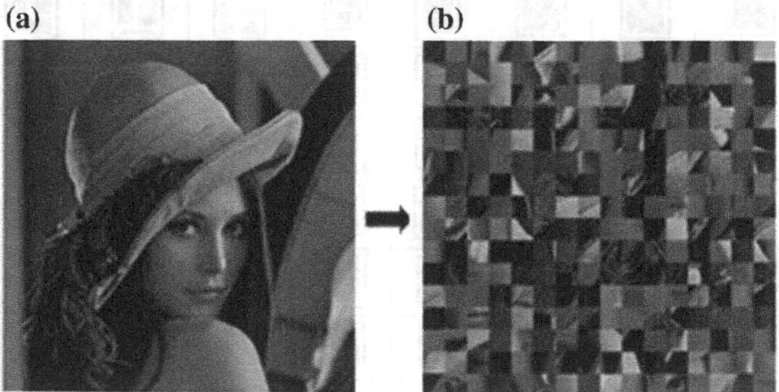

Fig. 2 Lena as the source image is divided into 16 × 16 blocks, which is shuffled to form grid-structured image. Grayscale Lena image, **b** 16 × 16 pixels block shuffled

16 × 16 blocks comprising as a whole 256 grid points or sub-matrices. Then, it is shuffled to make it as a confusing pattern.

The two prime numbers used in random number to generate a series of random number is equal with the size of message.

3.3 Advanced Encryption Standard (AES)

National Institute of Standards and Technology (NIST) has announced a dominant encryption technique over other popular technique as Advanced Encryption Standard (AES) in 2001, which is a symmetric-key block cipher [11]. 128 bits data blocks can be encrypted and decrypted by AES using key size as 128, 192, or 256 bits depending upon the number of encryption rounds 10, 12, or 14.

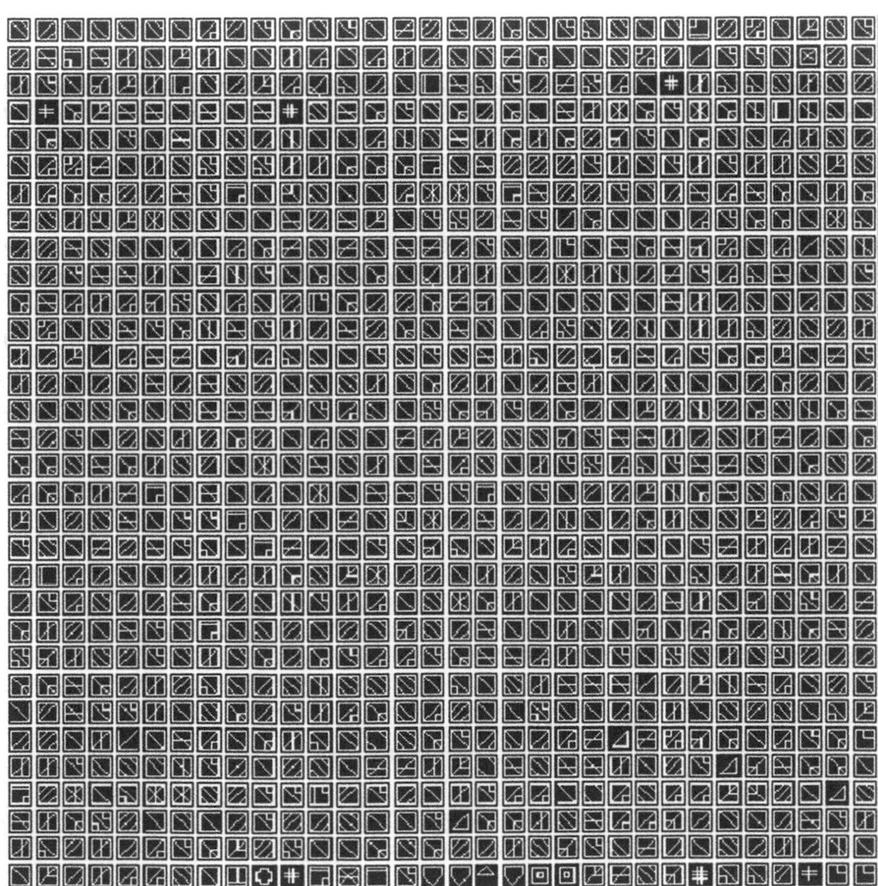

Fig. 3 Binary grid-structured cipher or covert image

4 Result and Discussion

First, the texture creation is performed in binary matrix format having size 16 × 16 for universal ASCII character starting from Space (ASCII = 32) to Del (ASCII = 126). Then, 95 sub-images in the form of binary image with smooth texture are created.

4.1 Empirical Analysis

4.1.1 Encryption

Let us take a plaintext having 1000 character to be encrypted as "*NICACI ssfueoc no btho ryoeth nda opsntcailaip ni teh orbda rsaea fo mtcnooacumnii nooceytglh, mrctupeo cesncei & frimotniano yrcesuit. sThi foececnner msai ot ingrb heregott miadaecc, sciinstste psfsrr, soeo aecrrseh ashrlsoc adn nsdtetsu to haesr and msnastediei nnmaitofoir on odeelgknw nad fnicsiteci ahcserer wsrok eaetdlr to citpgnmu, o trkinnoewg, adn socifamntri to uiscssd eth aalrpitcc snlcheaegl rctuenodene and eth iolusstno ad.edotp The erconeefnc wlli dprvoie het oustrha dna esieltsrn with teuisnopprtoi fro aitoalnn dna neanotailitnr aotlblnoairco nad nrtwkienog aongm eiitsivnuesr dan nsiniuottsit rfom indIa dan adbroa fro mtorp-nigo rreaesch nad vilodgnepe gtoeo.snclhie iTsh imas ot opremot slroiatnatn fo isabc eareshcr iton pliedap aotsiniivtgen dna rencvot aipldep ivgiestontian iotn etp.caric hisT eceonnferc lwli sloa eeract resaeaswn atobu hte enmratciop fo bacsi nicisicfet rchrasee ni ndeterfif elidsf nda cginatmh ihts htwi teh protucd eamkrt.(b) a + 22^2a2b + ^b^ = +a itxar,{ }1.*"

Using the proposed algorithm, the encrypted cipher image looks like Fig. 3, which depicts a chaotic pattern.

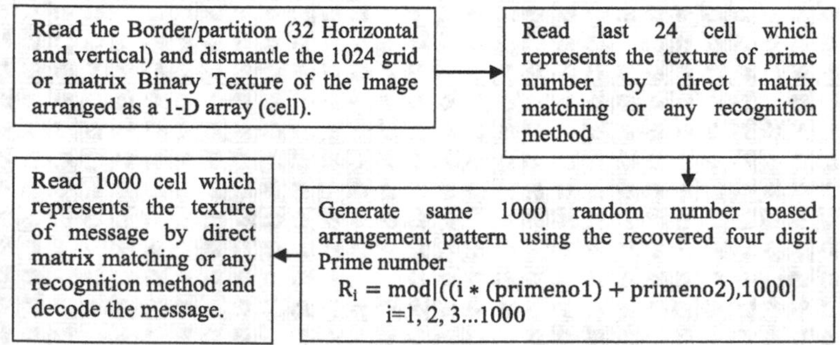

Fig. 4 The steps to decrypt the message from cipher image

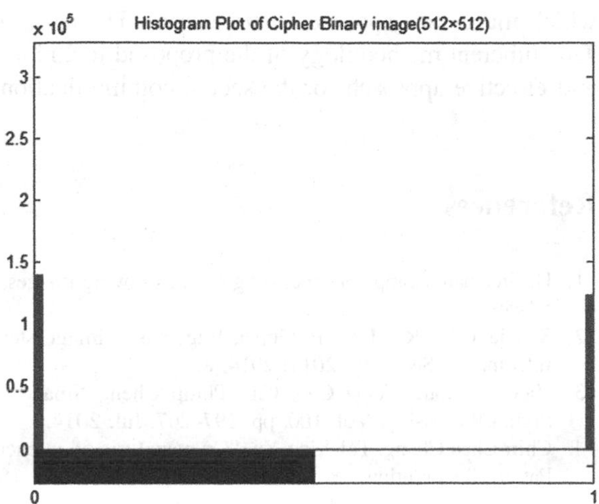

Fig. 5 The histogram plot of encrypted binary image

4.1.2 Decryption

The step to decrypt information from cipher image is shown as a flow diagram in Fig. 4.

4.2 Histogram and Entropy Analysis

The histogram of the cipher image embedding encoded smooth texture-based object is shown in Fig. 5.

Entropy is a texture analysis of the image for measuring the statistics of randomness. It evaluates the percentage of disorder. The entropy level of the cipher image in the proposed algorithm is high. The percentage of disorder or entropy level of the grid-structured image without the objects embedded is 0.5325 but entropy level of the grid-structured image after embedding objects as sub-images with entropy 0.9971 which makes it a robust method.

5 Conclusion

Image-based security gains immense popularity in the field of information security. In the last two decades, various algorithms were proposed for information security. In this paper, we have proposed a new dimension of image level encryption using popular Advanced Encryption Standard (AES). The proposed method uses earlier standard technique AES as well as a new form grid structure-based sub-image which makes this approach a unique one. Huge size of message can be encrypted. This inherent methodology of the proposed technique makes it a simple, low-cost, and effective approach for the secure communication.

References

1. H. Freeman, Computer processing of line drawing images, ACM Computer Surveys 6 (1974) 57–59.
2. S. Islam, M. R. Modi, P. Gupta, Edge-based image steganography, EURASIP Journal on Information Security, (2014) 2014, 8.
3. Yicong Zhou, Weijia Cao, C.L. Philip Chen, "Image encryption using binary bitplane", Signal Processing, Vol. 100, pp. 197–207, Jul. 2014.
4. Chin-Chen Chang, Tai-Xing Yu, "Cryptanalysis of an encryption scheme for binary images", Pattern Recognition Letters, Vol. 23, no. 14, pp. 1847–1852, Dec. 2002.
5. Yong-Ying Wang, Yu-Rong Wang, Yong Wang, Hui-Juan Li, Wen-Jia Sun, "Optical image encryption based on binary Fourier transform computer-generated hologram and pixel scrambling technology", Optics and Lasers in Engineering, Vol. 45, no. 7, pp. 761–765, Jul. 2007.
6. KuangTsan Lin, "Digital information encrypted in an image using binary encoding", Optics Communications, Vol. 281, no. 13, pp. 3447–3453. Jul. 2008.
7. Chengqing Li, Kwok-Tung Lo, "Optimal quantitative cryptanalysis of permutation-only multimedia ciphers against plaintext attacks", Signal Processing, Vol. 91, no. 4, pp. 949–954, Apr. 2011.
8. Huijuan Yang and Alex C. Kot, "Pattern-Based Data Hiding for Binary Image Authentication by Connectivity-Preserving" IEEE Transactions on Multimedia Vol. 9, no. 3, pp. 475–486, Apr. 2007.
9. Chung-ChuanWange, Ya-Fen Changd, Chin-Chen Changc, Jinn-Ke Jana, Chia-Chen Linb, "A high capacity data hiding scheme for binary images based on block patterns", Journal of Systems and Software, Vol. 93, pp. 152–162, Jul. 2014.
10. Ki-Hyun Jung, Kee-Young Yoo, "Data hiding method in binary images based on block masking for key authentication", Information Sciences, Vol. 277, pp. 188–196, Sep. 2014.
11. NIST "Advanced Encryption Standard (AES)", Federal Information Processing Standards Publication 197. 2001.

MusMed: Balancing Blood Pressure Using Music Therapy and ARBs

V. Ramasamy, Joyanta Sarkar, Rinki Debnath, Joy Lal Sarkar, Chhabi Rani Panigrahi and Bibudhendu Pati

Abstract Recently, increase or decrease in blood pressure level is one of the main problems for human all over the world which causes heart attack, Brain stroke, and many other diseases. There are many reasons for blood pressure and is increasing day-by-day and controlling blood pressure is one of the difficult tasks for every patient. For that patients take different kinds of medicines according to doctors suggestion. But, effects of these medicines gradually decrease at the time passes by. In this work we propose a treatment named as MusMed which is the combination of music therapy and medicine which helps to control blood pressure of human body. We used Indian classical raga by instrumental guitar as a music therapy and olmesartan molecule as a medicine. The results indicate that the combination of music therapy and medicine works well as compared to only medicine. Our results are validated through mercury sphygmomanometer.

Keywords Music therapy · ARBs · Blood pressure

V. Ramasamy (✉)
Department of Computer Science and Engineering, Park College of Engineering and Technology, Coimbatore, India
e-mail: researchrams@gmail.com

J. Sarkar
Department of Instrumental Music, Rabindra Bharati University, Kolkata, India
e-mail: joyanta35032@gmail.com

R. Debnath
International Institute of Rehabilitation Science and Research, Bhubaneswar, India
e-mail: rinki0381@gmail.com

J.L. Sarkar · C.R. Panigrahi
Department of Computer Science, Central University of Rajasthan, Ajmer, India
e-mail: joylalsarkar@gmail.com

C.R. Panigrahi
e-mail: panigrahichhabi@gmail.com

B. Pati
Department of Computer Science and Engineering,
C.V. Raman College of Engineering, Bhubaneswar, India
e-mail: patibibudhendu@gmail.com

H.S. Behera and D.P. Mohapatra (eds.), *Computational Intelligence in Data Mining*, Advances in Intelligent Systems and Computing 556,
DOI 10.1007/978-981-10-3874-7_43

1 Introduction

The rate of increase of Blood Pressure (BP) patients creates huge difficulties for every human life and also to the medical science. From 40–45 years, the chances of high or low BP increase. There are different causes for BP like age, family history, life style [1]. But, most of the risks can be controlled like body weight, high salt, alcohol intake etc. [1] by reducing risk factors people can save their life from heart attack, brain stroke etc. [2]. The reason behind this is that high BP damage our life that means it can damage our blood vessels and also make them narrower and more rigid [3] and for that overall work done by heart will increase because to push the blood of our body heart has to work more. So, if a patient does not take any serious action for reducing high BP, the patient suffers from many diseases like angina, heart attack, kidney failure, and stroke etc. [4]. For controlling BP, patients take various kinds of medicines on a regular basis. There are several kinds of medicines used for Calcium Channel Blockers (CCB), $K+$ Channel activators, Angiotensin (AT1) blockers etc. But, in all medicines there are certain limitations because the half life of every medicine is not 24 h except telmesartan and also the effects of those medicines reduce gradually [5]. Music plays a vital role for controlling BP which is basically related to the human emotion, relaxation, joy, and sadness etc. [6]. There are various domains like immunity, stress etc. where music can influence our health by neuro-chemical changes. In this paper, we present a combination of Music Therapy (MT) with medicine named as MusMed which can balance our BP. We used slow rhythmic for high BP patients and high rhythmic for low BP patients and we used Indian classical raga as a MT and olmesartan as an Angiotensin Receptor Blockers (ARB) (according to doctors prescription) as a medicine.

The rest of the paper is organized as follows: Sect. 2 describes the related work. Section 3 presents the problem formulation, Sect. 4 presents our proposed approach. Section 5 presents mathematical analysis and in Sect. 6 we present the results obtained along with the analysis of results. Finally, we conclude the paper in Sect. 7.

2 Related Work

The application of MT is increasing day-by-day [7]. MT plays a vital role in human life. In [7], authors analyzed how MT can improve immunity, stress etc. through neuro-chemical systems. When people listen music it can help in improving feelings or euphoria [8, 9]. In [10–12] authors used positron emission tomography (PET) for investigating the regional cerebral blood flow (rCBF). Where, own selection of music which reliably induces chills- down-the-spine [10]. Nucleus Accumbens (NAc) can be introduced when people listen music which is very unfamiliar [11] and also when people sing as compared to speech [12]. There are different studies which are used for finding the neural correlates [13]. By listening music, people can relief from postoperative pain [14]. The reason behind is that music can stimulate the release of endogenous opioid peptides within the brain.

MT which is a combination of the relaxation process and classical music which is also called as Guided Imagery and Music (GIM) was introduced to decrease the activation of Hypothalamic-Pituitary-Adrenal (HPA) [15, 16]. In [17], authors proposed a technique which can lowering of cortisol levels by MT as compared to silent. In case of surgery patient when they select their own music that reduces the cortisol levels and is more effective then the experimenter selected music [18]. In [19], authors proposed an approach for reducing the anxiety of people by using MT using HRV. In [20], authors studied that slow rhythmic music effects well in case of high BP patients whereas the high rhythmic music effects well in case of low BP patients. The rosary prayer and yoga mantras can balance human cardiovascular system [21]. There are different kinds of medicines which can reduce high BP but these do not work for 24 h except telmesartan. But, the effect is less in case of MDA (serum Malondialdehyde) etc.

3 Problem Formulation

There are several reasons for which BP can fluctuate from it's normal range. Different patients take different medicines based on doctors suggestions and some patients can not able to take medicine due to its high. MT is one of the great solutions which can jointly work with medicine as well as without medicine. The following condition shows MT takes care of when BP fluctuates from it's normal range.

$$Exec(MT, ARBs) = \begin{cases} MT, & \text{if } BP < NR \text{ or } BP > NR \\ ARBs, & \forall day \end{cases}$$

where, MT works when BP fluctuates from it's Normal Range (NR) and ARBs to be taken by the patient for all day according to the doctor's suggestion.

4 Proposed Approach

Balancing BP is one of the difficult tasks now-a-days. There are various ARBs which can balance BP. We used the combination of MT as well as different ARBs according to doctor's prescription. From juxtaglomerular cells of kidney, *Renin* is produced which is basically a prototype enzyme and is also called as an *angiotensinogenase*. The angiotensinogen produces *Angiotensin-I* and *Angiotensin-I* first is converted into *Angiotensin- II* and later *Angiotensin- II* is again converted into the *Angiotensin- III* Both *Angiotensin- II* and *Angiotensin- III* are responsible for Aldosterone secretion

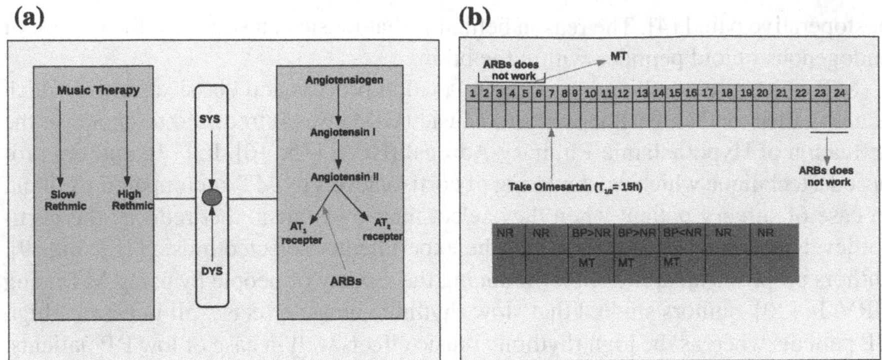

Fig. 1 **a** Music therapy and ARBs, **b** Music therapy works when effects of ARBs are reduced

from Adrenal Cortex which is basically cased to increase BP. The ARBs work on *Angiotensin- II* and block *angiotensin- II.*

Thus ARBs work to reduce high BP. But, there are several reasons such as stress, life style, and also rapid change in the environment for which BP fluctuates from it's normal range. We introduce MT along with ARBs which can balance BP and basically works for synchronization as well as improved cardiovascular rhythms by improving variability [21]. Figure 1a shows how MT and ARBs jointly work to balance BP. In case of MT, it mainly focuses on two types: *slow rhythmic* and *high rhythmic*. The slow rhythmic is used to reduce BP and high rhythmic just works to increase the BP. Sensors monitor BP in each time and based on the different readings it selects the rhythmic whether it is slow or first.

Figure 1b shows that when a patient takes medicine (e.g. Olmesertan), the half life of that medicine is over before 6 h from the time the medicine will be taken in the next day. So, in between 6 h the range of BP may be high and the patient may suffer from serious heart attack or brain stroke. The MT can be used for controlling the high BP. The example is true for high BP patients whereas the effects may be same in case of low BP also.

5 Mathematical Formulation

We assume that there are *i* number of patients suffer from high BP and is given in Eq. (1).

$$T_H^{BP} = p_1^H + p_2^H + p_3^H + \cdots + p_i^H \tag{1}$$

Now, there are *j* number of patients suffer from low BP and is given in Eq. (2).

$$T_L^{BP} = p_1^L + p_2^L + p_3^L + \cdots + p_j^L \tag{2}$$

So, total $(i + j)$ number of patients suffer from BP unbalance and is given in Eq. (3).

$$T_{H,L}^{BP} = (p_1^H + p_2^H + p_3^H + \cdots + p_i^H) + (p_1^L + p_2^L + p_3^L + \cdots + p_j^L) \quad (3)$$

Now, high BP patients should reduce their BPs upto normal range. So, Eq. (1) can be rewritten as:

$$Norm(T_H^{BP}) = H_{Norm}(p_1^H + p_2^H + p_3^H + \cdots + p_i^H) \quad (4)$$

Now, the low BP patients should increase their BP upto the normal range so the Eq. (1) can be rewritten as:

$$Norm(T_L^{BP}) = L_{Norm}(p_1^L + p_2^L + p_3^L + \cdots + p_j^L) \quad (5)$$

Now after taking ARBs high BP can be minimized. We consider $M_1^H, M_2^H, \ldots, M_i^H$ are the values for reduced BP. So, from Eq. (1) we get:

$$R(T_H^{BP}) = (p_1^H - M_1^H) + (p_2^H - M_2^H) + (p_3^H - M_3^H) + \cdots + (p_i^H - M_i^H) \quad (6)$$

But, to control high BP, R.H.S. of Eq. (6) should be $p_1^H - M_1^H, p_2^H - M_2^H, p_i^H - M_i^H \rightarrow NR$

After combining MT, high BP should meet it's *NR* and is given as in Eq. (7).

$$NR(T_H^{BP}) = (p_1^H - (M_1^H + m_1^H)) + (p_2^H - (M_2^H + m_2^H)) + (p_3^H - (M_3^H + m_3^H)) + \cdots + (p_i^H - (M_i^H + m_i^H)) \quad (7)$$

Similarly, low BP can be maximized by taking ARBs. We consider $M_1^L, M_2^L, \ldots, M_i^L$ as the values of reduced BP. So, from Eq. (2) we get:

$$R(T_L^{BP}) = (p_1^L + M_1^L) + (p_2^L + M_2^L) + (p_3^L + M_3^L) + \cdots + (p_i^L + M_i^L) \quad (8)$$

But, to controlling low BP R.H.S. of Eq. (8) should be $p_1^L + M_1^L, p_2^L + M_2^L, p_i^L + M_i^L \rightarrow NR$

After combining MT, low BP should meet it's *NR* and is given as in Eq. (8).

$$NR(T_L^{BP}) = (p_1^L + (M_1^L + m_1^L)) + (p_2^L + (M_2^L + m_2^L)) + (p_+^L - (M_3^L + m_3^L)) + \cdots + (p_i^L + (M_i^L + m_i^L)) \quad (9)$$

When a patient takes medicine at t_a time, the life of the medicine gradually decreases and it does not work after certain amount of time. We assume that medicine will work in a body for next t_b time that means for $(t_t - t_b)$ time the medicine does not effect. Where, t_t indicates the time between first dosage and the next dosage. We assume that due to decrease of the half life of medicine, the BP also gradually increases and is given in Eq. (10).

$$R(T_H^{BP}) = (p_1^H - M_1^H + a_1^H) + (p_2^H - M_2^H + a_2^H) + (p_3^H - M_3^H + a_3^H) + \cdots + (p_i^H - M_i^H + a_i^H) \tag{10}$$

And for low BP will gradually decrease when half life decreases and is given in Eq. (11).

$$R(T_L^{BP}) = (p_1^L + M_1^L + a_1^L) + (p_2^L + M_2^L + a_2^L) + (p_3^L + M_3^L + a_3^L) + \cdots + (p_i^L + M_i^L + a_i^L) \tag{11}$$

Theorem 1 *The combination of music therapy and ARBs can balance BP with respect to ARBs alone.*

Proof For high BP, patients can take medicine (Olmeserten) at morning 8.00 a.m. The half life of that medicine is 13–15 h. So rest 11 or 9 h medicine does not work and BP will gradually increases and the effects of medicine in a body can decreases (From Eqs. (10) and (11)). So in such situation music therapy can be taken care of. Music therapy which is basically works for syncronisation as well as improved cardiovascular rhythms by improving variability [21]. So overall BP can be controlled when the effects of medicine minimized. Basically at last 6 h is very crucial because patients can not take same medicine within that period. Because duration of to take medicine like Olmeserten is 13–15 h. So in that duration patients can face heart attack or another strong diseases and the results cause death. So according to the Eq. (6), if patients take music therapy within that period the overall BP can be minimized and save the life also.

6 Results and Analysis

For analysis of our approach that is how MT produces effects on human body, we took 10 patients who suffer from hypertension and took ARBs as well. Out of 10 patients, we took 5 females and 5 male patients and their ages range from 45–65 years and all are married.

For MT we used instrumental music with *Desh* raga. Although there are different kinds of music we took only Indian classical music (Instrumental) and also we only choose raga desh because it effects well for high BP patients. We used *Stethoscopes* and also *mercury sphygmomanometer* for analysis of BP (Systolic and diastolic). Every patient took medicines for BP at morning 8 a.m. according to doctors suggestion. Initially, we only measured BP after taking only medicine (where all patients took molecule Olmesertan) and for other days we analyzed same patients with same medicine along with MT. We took BP of each patient separately where, all patients

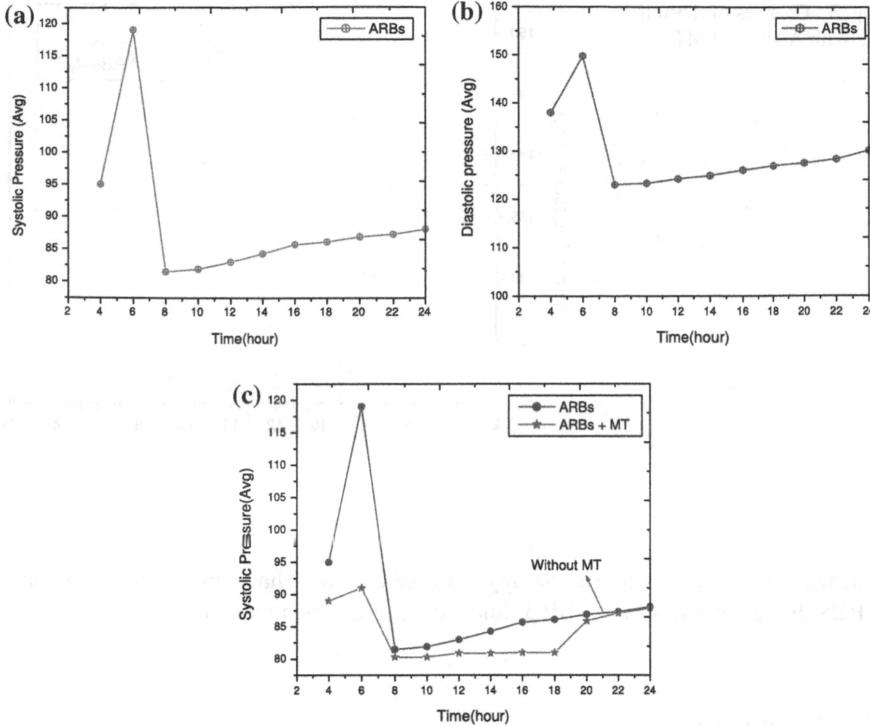

Fig. 2 **a** Changes of systolic BP with ARBs, **b** Changes of diasstolic BP with ARBs, **c** Changes of systolic BP with ARBs and MT

listen to a piece of music for 35 min. Before we test, we took post test BP readings for all patients.

Figure 2a, b show the changes of systolic and diastolic pressure when all patients take molecule Olmesertan. From Fig. 2a, b it is clear that Olmesertan produces effects for 18 h in our body and last 6 h BP gradually increases. Figures 2c and 3 show the comparison of results for changes of BP when patients take only ARBs and both ARBs and music therapy. From Figs. 2c and 3, it is clear that when patients take music therapy the BP can be controlled. Interesting case occurs In Fig. 2c during the duration of 20–24 P.M the change of BP almost same as compared to ARBs and the reason behind is that during this period patients do not take any MT.

So, the effects almost same. Although there are large number of molecules which doctors suggest we only take those patients which are taking same medicine like olmesartan which is basically used for hypertension patients to reduce the high blood pressure. We only analyzed the results for high BP patients which forms one group and after getting all readings from all patients we take average where room temperature range from 20–24 cc. To ensure that all patients do not suffer from any strong

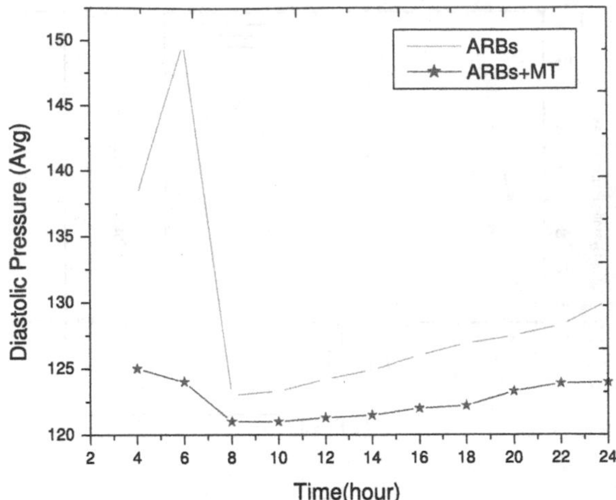

Fig. 3 Changes of diastolic BP with ARBs and MT

diseases, initially we performed blood test and ECG test. From Fig. 2b, c, it is clear that when patients take music therapy with ARBs it will have more effect than only ARBs. For analysis, we took BP 3 times reading for each patient.

7 Conclusion

In this paper, we investigated how music therapy works well for controlling BP of our body which can helps to avoid heart attack or brain stroke. We see that music therapy works well with the combination of ARBs because some of the ARBs can not show effects 24 h of our body. In such scenario music therapy performs well. The results indicate that if patients can take music therapy regularly along with ARBs that can help for controlling BP. In future, we would like to investigate the effect of MT without ARBs in human body and also with low BP patients.

References

1. Cidad, P., Novens, L., Garabito, M., Batlle, M., Dantas, A.P., Heras, M., Lpez-Lpez, J.R., Prez-Garca, M.T., Roqu, M.: K+ Channels Expression in Hypertension After Arterial Injury, and Effect of Selective Kv1.3 Blockade with PAP-1 on Intimal Hyperplasia Formation. Cardiovascular Drugs and Therapy, **28**(6), 501–511 (2014).
2. Wang, Y., Xie, F., Kong, M.C., Lee, L.H., Ng. H.J., Ko, Y.: Cost-effectiveness of Dabigatran and Rivaroxaban Compared with Warfarin for Stroke Prevention in Patients with Atrial Fibrillation. Cardiovascular Drugs and Therapy, **28**(6), 575–585 (2014).
3. Basciftci, F., Eldem, A.: Using reduced rule base with Expert System for the diagnosis of disease in hypertension. Int. federation for medical and biological engineering, **51**(12), 1287–1293 (2013).

4. Diaz, K.M., Tanner, R.M., Falzon, L., Levitan, E.B., Reynolds, K., Shimbo, D., Muntner, P.: Visit-to-visit variability of blood pressure and cardiovascular disease and all-cause mortality: a systematic review and meta-analysis. Hypertension, **64**(5) 965–982 (2014).
5. http://www.medicalnewstoday.com/releases/91285.php, Last Accessed 25 June, 2016.
6. Konecni, Vladimir, J.: Does music induce emotion? A theoretical and methodological analysis. Psychology of Aesthetics, Creativity, and the Arts, **2**(2), 115–129 (2008).
7. Chanda, M.L. and Levitin, D.J.: The neurochemistry of music. Trends in Cognitive Sciences, **17**(4), 179–193 (2013).
8. Harrison, L., Loui, P.: Thrills, chills, frissons, and skin orgasms: toward an integrative model of transcendent psychophysiological experiences in music. Frontiers in Psychology, **5**(790), 1–6 (2014).
9. Sloboda, J.A.: Music structure and emotional response: some empirical findings. Psychol. Music, 19, 110–20 (1991).
10. Blood, A.J., Zatorre, R.J.: Intensely pleasurable responses to music correlate with activity in brain regions implicated in reward and emotion. Proc. Natl. Acad. Sci, 98, 11818–11823 (2001).
11. Brown, S., Martinez, M.J., and Parsons, L.M.: Passive music listening spontaneously engages limbic and paralimbic systems. Neuroreport., 15, 2033–2037 (2004).
12. Jeffries, K.J., Fritz, J.B., Braun, A.R.: Words in melody: an h(2)15o pet study of brain activation during singing and speaking. NeuroReport, 14, 749–754 (2003).
13. Koelsch, S. Towards a neural basis of music-evoked emotions. Trends Cogn. Sci, 14, 131–137 (2010).
14. Dehcheshmeh, F.S., Rafiei, H.: Complementary and alternative therapies to relieve labor pain: A comparative study between music therapy and Hoku point ice massage. Complementary Therapies in Clinical Practice, 21, 229–232 (2015).
15. McKinney, C.H., Antoni, M.H., Kumar, M., Tims, F.C., McCabe, P.M.: Effects of guided imagery and music (GIM) therapy on mood and cortisol in healthy adults. Health Psychol, **16**(4), 390–400 (1997).
16. McKinney, C.H. The effect of selected classical music and spontaneous imagery on plasma beta-endorphin. J. Behav. Med, 20, 85–99 (1997).
17. Khalfa, S. et al. Effects of relaxing music on salivary cortisol level after psychological stress. Ann. N. Y. Acad. Sci, 999, 374–376 (2003).
18. Wallston, K.A.: The validity of the multidimensional health locus of control scales. J. Health Psychol, 10, 623–631 (2005).
19. http://www.cancerresearchuk.org/about-cancer/cancers-in-general/treatment/complementary-alternative/therapies/music-therapy, last accessed 29 June, 2016.
20. http://www.dailymail.co.uk/health/index.html, last accessed 29 June, 2016.
21. Bernardi, L., Sleight, P., Bandinelli, G., Cencetti, S., Fattorini, L., Wdowczyc-Szulc, J., Lagi, A.: Effect of rosary prayer and yoga mantras on autonomic cardiovascular rhythms: Comparative study. British Medical Journal, 325, 1446–1449 (2005).

Interprocedural Conditioned Slicing

Madhusmita Sahu and Durga Prasad Mohapatra

Abstract A technique, named *Node Marking Conditioned Slicing* (NMCS) algorithm, has been proposed to compute conditioned slices for interprocedural programs. First, the *System Dependence Graph* (SDG) is constructed as an intermediate representation of a given program. Then, NMCS algorithm selects the nodes satisfying a given condition by marking process and computes the conditioned slices for each variable at each statement during marking process. A stack has been used in NMCS algorithm to preserve the context in which a method is called. Some edges of SDG have been labeled to signify which statement calls a method.

Keywords Conditioned slicing · Conditioned slice · Node Marking Conditioned Slicing (NMCS) algorithm

1 Introduction

Program slicing is a decomposition approach used to segregate programs based on the control flow and data flow between various program statements. A slicing criterion is used to compute a slice. The slice can be *static* or *dynamic* based on the input to the program. *Conditioned slicing* is a generalized form of *static slicing* and *dynamic slicing*. A *conditioned slice* is a tuple <*C, s, v*>, where *C* is some condition, *s* is a desired point in the program, and *v* is a variable. Conditioned slicing discards those portions of the pilot program that cannot influence the variables at the desired point on the satisfaction of conditions. Details about conditioned slicing can be found in [1–3].

M. Sahu (✉) · D.P. Mohapatra
Department of CSE, National Institute of Technology, Rourkela 769008, Odisha, India
e-mail: madhu_sahu@yahoo.com

D.P. Mohapatra
e-mail: durga@nitrkl.ac.in

H.S. Behera and D.P. Mohapatra (eds.), *Computational Intelligence in Data Mining*, Advances in Intelligent Systems and Computing 556,
DOI 10.1007/978-981-10-3874-7_44

The objective of this work is to compute conditioned slices for interprocedural programs. To the best of our knowledge, no method has been proposed for computing conditioned slices for interprocedural programs. In this paper, a method has been proposed for the said purpose. First, the *System Dependence Graph* (SDG) for the given program is constructed. Then, the nodes satisfying the given condition are marked, and the slices are computed during marking process.

The remainder of the paper is structured as follows: Section 2 presents some background details of our technique. In Sect. 3, we discuss our proposed *Node Marking Conditioned Slicing* (NMCS) algorithm to compute conditioned slice. Section 4 discusses some works related to conditioned slicing. Section 5 renders conclusion and future work.

2 Background

The conditioned slice can be efficiently computed using a *System Dependence Graph* (SDG) as an intermediate program representation. The technique of Horwitz et al. [4] is adopted to construct the system dependence graph (SDG). The details can be found in [4].

An example program written in C is shown in Fig. 1. We have adopted the program from Canfora et al. [1], but with slight modifications. The modifications are due to the presence of multiple procedures and the absence of a variable *test0*. The integer *c* and a sequence of *c* integers, *x*, are given as input to the program; the value of the integers *psum*, *pprod*, *nsum*, and *nprod* are computed. The integers *psum* and *nsum* calculate the sum of the positive and negative numbers, respectively. Similarly, the integers *pprod* and *nprod* calculate the product of the positive and negative numbers, respectively. The program returns the greatest sum and product computed. The checking of a number for positiveness and negativeness is done through a function *test*. The greatest sum and product for positive and negative numbers are computed through a function *check*. The static slice of the example program shown in Fig. 1 for slicing criterion *<20, sint>* consists of statements numbered as 1, 3, 4, 5, 7, 9, 11, 12, 13, 15, 17, 18, 20, 22, 23, 24, 25, 26, 27, 28, and 29. The dynamic slice of the example program shown in Fig. 1 for slicing criterion *<20, sint>* for the input $n = 3$ and $x = \{8, -3, 11\}$ consists of statements numbered as 1, 3, 4, 5, 7, 9, 11, 12, 13, 15, 17, 18, 20, 22, 23, 24, 25, 26, 27, and 28.

3 Node Marking Conditioned Slicing (NMCS) Algorithm

The system dependence graph (SDG) for the given program is constructed statically only once. For a given condition in the slicing criterion, our NMCS algorithm marks the nodes satisfying the condition. During the marking process, the conditioned slices are also computed. Whenever there is a method call, the corresponding

Fig. 1 Example C program

```
int test(int);
int check(int,int);
void main()
{
int c,i,psum,pprod,nsum,nprod,x,sint,pint;
printf("How many numbers:");
scanf("%d",&c);
i=1;
psum=0;
pprod=1;
nsum=0;
nprod=1;
while(i<=c)
{
printf("Enter a number:");
scanf("%d",&x);
if(test(x))
{
psum+=x;
pprod*=x;
}
else
{
nsum+=x;
nprod*=x;
}
i++;
}
sint=check(psum,nsum);
pint=check(pprod,nprod);
printf("Sum is %d\n",sint);
printf("Product is %d",pint);
}
int test(int p)
{
if(p>0)
return 1;
else
return 0;
}
int check(int p,int q)
{
if(p>=q)
return p;
else
return q;
}
```

node is recorded in a variable, CS_e. We maintain a stack, *SMC*, to keep track of call context. Whenever a parameter-out edge is traversed, the label of the edge is pushed to the stack; whenever a parameter-in edge or call edge is encountered, the stack is popped. The top of the stack is compared with CS_e and if both are same, then the conditioned slice is updated. In the remainder of the paper, the terms *vertex* and *node* are used interchangeably (Fig. 2).

Fig. 2 Conditioned slice of example program shown in Fig. 1 for slicing criterion *<test(x)! = 0, 20, sint>*

```
     int test(int);
     int check(int,int);
1    void main()
     {
     int c,i,psum,pprod,nsum,nprod,x,sint,pint;
2    printf("How many numbers:");
3    scanf("%d",&c);
4    i=1;
5    psum=0;
6    pprod=1;
7    nsum=0;
8    nprod=1;
9    while(i<=c)
     {
10   printf("Enter a number:");
11   scanf("%d",&x);
12   if(test(x))
     {
13   psum+=x;
14   pprod*=x;
     }
     else
     {
15   nsum+=x;
16   nprod*=x;
     }
17   i++;
     }
18   sint=check(psum,nsum);
19   pint=check(pprod,nprod);
20   printf("Sum is %d\n",sint);
21   printf("Product is %d",pint);
     }
22   int test(int p)
     {
23   if(p>0)
24   return 1;
     else
25   return 0;
     }
26   int check(int p,int q)
     {
27   if(p>=q)
28   return p;
     else
29   return q;
     }
```

Let *cslice(u)* indicates the conditioned slice for slicing criterion <*C, s, var*>, where *var* is a variable, *C* is a condition, and *s* is a statement corresponding to node *u*. Let $s_1, s_2, \ldots, s_k$ indicate all the marked predecessor nodes of *u* in the dependence graph. Then, the conditioned slice for slicing criterion <*C, s, var*> is given by:

$$cslice(u) = \{u, s_1, s_2, \ldots, s_k\}\ U\ cslice(s_1)\ U\ cslice(s_2)\ U \ldots\ U\ cslice(s_k)$$

The proposed NMCS algorithm is presented in the pseudocode form below.

Node Marking Conditioned Slicing (NMCS) Algorithm

1. **SDG Construction**: Construct the SDG for the given program *P* only once as follows:

(a) For each statement in *P*,

i. Create a node in the graph.

ii. If node *x* is a method call node, then

A. Create actual-in parameter nodes corresponding to the number of parameters passed in *x*.

B. Create actual-out parameter nodes corresponding to the number of global variables modified in *x*.

iii. If node *x* is a method entry node, then

A. Create formal-in parameter nodes corresponding to the number of parameters passed in *x*.

B. Create formal-out parameter nodes corresponding to the number of global variables modified in *x*.

(b) If node *x* controls the execution of node *y*, then

i. Add a *control dependence edge* from *x* to *y*, $x \rightarrow y$.

(c) If node *x* defines a variable *v* and node *y* uses *v*, then

i. Add a *data dependence edge* from *x* to *y*, $x \rightarrow y$.

(d) If node *x* calls a method which is defined at node *y*, then

i. Add a *call edge* from *x* to *y*, $x \rightarrow y$.

ii. Label the edge $x \rightarrow y$ with *x*.

(e) If node y returns a value to the call node x, then

i. Add a *data dependence edge* from y to x, $y \rightarrow x$.

ii. Label the edge $y \rightarrow x$ with x.

(f) If node y is actual-in and node z is formal-in parameter nodes for a call node x, then

i. Add a *parameter-in edge* from y to z, $y \rightarrow z$.

ii. Label the edge $y \rightarrow z$ with x.

(g) If node y is actual-out and node z is formal-out parameter nodes for a call node x, then

i. Add a *parameter-out edge* from z to y, $z \rightarrow y$.

ii. Label the edge $z \rightarrow y$ with x.

(h) If there is a path from actual-in node x to actual-out node y, then

i. Add *summary edge* from x to y, $x \rightarrow y$.

(i) If there is a path from actual-in node x to corresponding call node y, then

i. Add *summary edge* from x to y, $x \rightarrow y$.

2. **Initialization**: Do the following before finding the conditioned program *CP* for given program *P*.

(a) Unmark all the nodes of the SDG.

(b) Set *cslice(u)* = φ for every node u of the SDG satisfying a condition *C*.

3. **Updations**: For the slicing criterion <*C, s, var*>, carry out the following for each statement s, corresponding to node u of the program *P* satisfying the condition *C* till the end of the program.

(a) For every variable v used at u

i. Update *cslice(u)* to *cslice(u)* = $\{u, s_1, s_2, \ldots, s_k\}$ *U cslice*(s_1) *U cslice*(s_2) *U*... *U cslice*(s_k), where $s_1, s_2, \ldots, s_k$ are the marked predecessor nodes of u in the SDG.

(b) Mark the node u.

(c) Let CS_e be the call site for edge *e*. If *u* is a method call vertex, then do the followings:

i. Mark the node *u*.

ii. Mark the associated actual-in and actual-out vertices at *u*.

iii. Mark the corresponding method entry vertex.

iv. Mark the formal-in and formal-out vertices associated with the method entry vertex.

v. Set $CS_e = u$.

(d) Let *SMC* be a stack to keep track of call context. If *u* returns a value to a method call node *v*, then do the following:

i. Push label of the edge $u \rightarrow v$ to the stack *SMC*.

ii. Update *cslice(v) = cslice(v) U cslice(u)*.

(e) If predecessor *p* of *u* is a formal-in parameter node, then

i. Pop the stack *SMC*.

ii. If top of the stack *SMC* is equal to CS_e, then

A. Set $cslice(p) = \{p, s_1, s_2, \ldots, s_k\}\ U\ cslice(s_1)\ U\ cslice(s_2)\ U \ldots U\ cslice(s_k)$, where $s_1, s_2, \ldots, s_k$ are the marked predecessor nodes of *p* in the SDG.

4. **Slice Look Up**: For the slicing criterion <*C, s, var*>, carry out the following on the SDG.

(a) Look up *cslice(u)* for variable *var* to get the slice.

//node *u* corresponds to the statement *s*.

(b) Unmark the remaining marked nodes.

(c) Display the resulting slice.

Working of the Algorithm

An example is used to elucidate the functioning of our NMCS algorithm. Consider the example C program given in Fig. 1. Figure 3 shows its SDG.

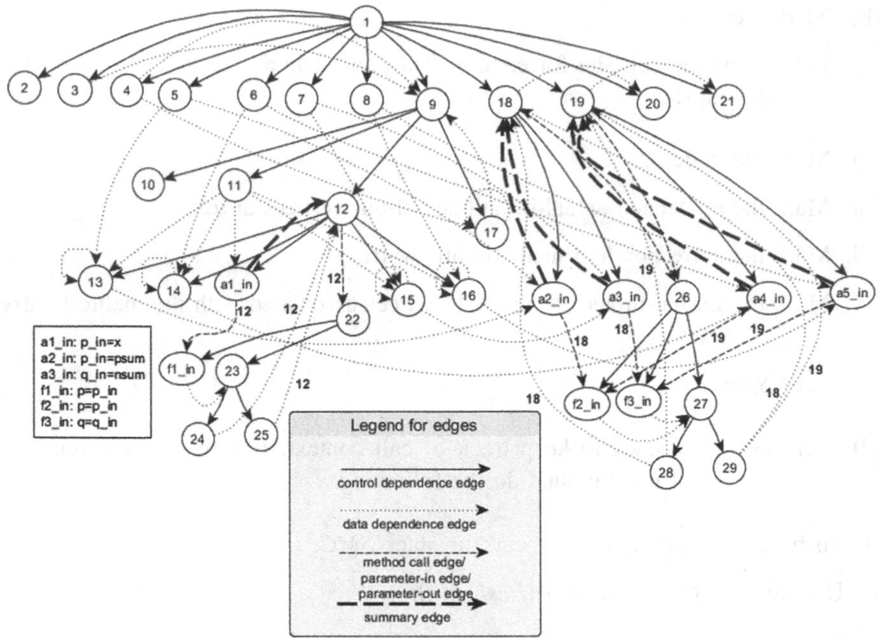

Fig. 3 SDG of the example C program depicted in Fig. 1

In the initialization step of our NMCS algorithm, all the nodes of the SDG are first unmarked. For every node *u* of the SDG satisfying a condition *C*, $cslice = \varphi$ is set. Consider the slicing criterion <*C, s, var*>, where *C* = "*test(x)! = 0*", *s* = "20", and *var* = "*sint*". This means that we have to find conditioned slice for the variable *sint* at statement number 20 for the given condition that *test(x)* is positive. Our NMCS algorithm first marks the nodes 1, 2, 3, 4, 5, 6, 7, 8, 9, 10, 11, 12, 22, 23, 24, 13, 14, 17, 18, 26, 27, 28, 19, 26, 27, 28, 20, and 21 in order. The actual parameter nodes at the calling method and the formal parameter nodes at the called method are also marked. During marking process, NMCS algorithm also computes the conditioned slice for each marked node in the SDG. When a method call node 12 is marked, $CS_e = 12$ according to Step 3(c). When node 24 is marked, the label of the edge $24 \rightarrow 12$ is pushed to the stack *SMC* using Step 3(d), i.e., label 12 is pushed to the stack *SMC*.

Here, $u = 24$ and $v = 12$. Thus, *cslice(12)* is updated as *cslice(12)* = *cslice(12) U cslice(24)*.

When node 23 is marked, Step 3(e) is applied. One of the predecessors of node 23 is node *f1_in* and *p* = *f1_in*, $u = 23$. Stack *SMC* is popped, and the popped item is 12. It is equal to CS_e as $CS_e = 12$. Thus, *cslice(f1_in)* is updated as *cslice (f1_in) = {f1_in, a1_in, 22} U cslice(a1_in) U cslice(22)*.

The same procedures, i.e., Step 3(c), Step 3(d), and Step 3(e), are applied for method call nodes 18 and 19.

Now, we shall determine the conditioned slice computed for variable *sint* at the statement number 20 for the given condition that *test(x)* is positive. According to NMCS algorithm, the conditioned slice for variable *sint* is given by:

$$cslice(20) = \{20, 1, 18\}\ U\ cslice(1)\ U\ cslice(18).$$

The above expression is evaluated recursively to obtain final conditioned slice at statement 20. The final conditioned slice contains the statements corresponding to the following set of nodes:

$$\{1, 3, 4, 5, 7, 9, 11, 12, 13, 17, 18, 20, 22, 23, 24, 26, 27, 28\}$$

The bold nodes in Fig. 4 show the statements that are included in the conditioned slice. Also these statements are shown bold in Fig. 2.

Table 1 shows the complete list of *cslice(u)* for all nodes satisfying the condition *C* = *"test(x)! = 0"*.

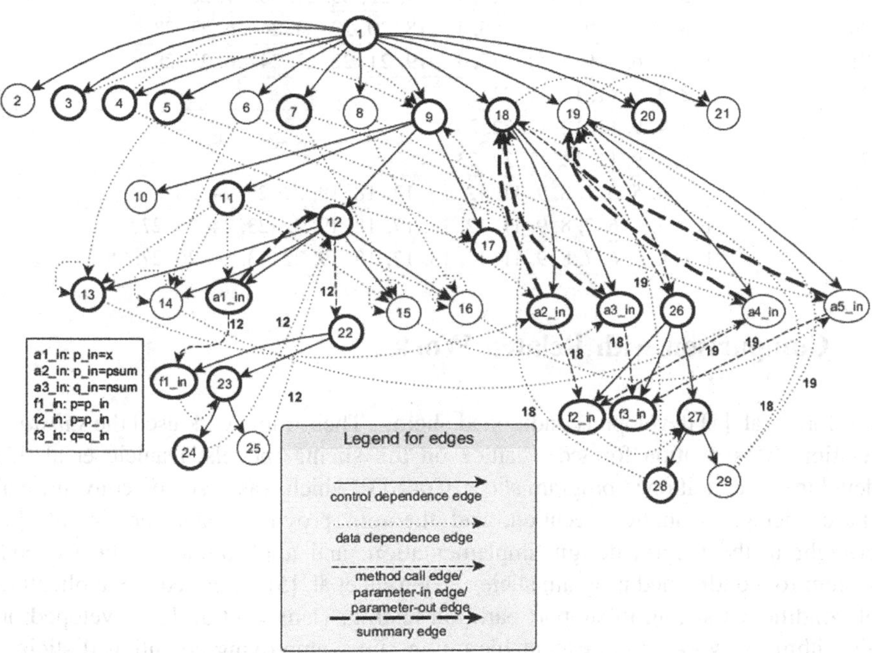

Fig. 4 Conditioned slice of example C program given in Fig. 1 for slicing criterion *<test(x)! = 0, 20, sint>* are depicted as bold nodes

Table 1 *cslice(u)* for all nodes satisfying the condition C = *"test(x)! = 0"*

u	*cslice(u)*
1	1
2	1, 2
3	1, 3
4	1, 4
5	1, 5
6	1, 6
7	1, 7
8	1, 8
9	1, 3, 4, 9, 17
10	1, 3, 4, 9, 10, 17
11	1, 3, 4, 9, 11, 17
12	1, 3, 4, 9, 11, 12, 17, 22, 23, 24
13	1, 3, 4, 5, 9, 11, 12, 13, 17, 22, 23, 24
14	1, 3, 4, 6, 9, 11, 12, 14, 17, 22, 23, 24
17	1, 4, 9, 17
18	1, 3, 4, 5, 7, 9, 11, 12, 13, 17, 18, 22, 23, 24, 26, 27, 28
19	1, 3, 4, 6, 8, 9, 11, 12, 13, 17, 19, 22, 23, 24, 26, 27, 28
20	1, 3, 4, 5, 7, 9, 11, 12, 13, 17, 18, 20, 22, 23, 24, 26, 27, 28
21	1, 3, 4, 6, 8, 9, 11, 12, 13, 17, 19, 21, 22, 23, 24, 26, 27, 28
22	1, 3, 4, 9, 11, 12, 17, 22
23	1, 3, 4, 9, 11, 12, 17, 22, 23
24	1, 3, 4, 9, 11, 12, 17, 22, 23, 24
26	1, 3, 4, 5, 6, 7, 8, 9, 11, 12, 13, 17, 18, 19, 22, 23, 24, 26
27	1, 3, 4, 5, 6, 7, 8, 9, 11, 12, 13, 17, 18, 19, 22, 23, 24, 26, 27
28	1, 3, 4, 5, 6, 7, 8, 9, 11, 12, 13, 17, 18, 19, 22, 23, 24, 26, 27, 28

4 Comparison with Related Work

Canfora et al. [1] developed conditioned slicing. Their framework used the *subsume* relation. This relation formed a lattice on the slicing models. Danicic et al. [2] developed a conditioned program slicer, ConSIT, which was based on conventional static slicing, symbolic execution, and theorem proving. Later, Fox et al. [3] brought in the theory, design, implementation, and applications of the ConSIT system for conditioned program slicing. Hierons et al. [5] described the application of conditioned slicing to support partition testing. Danicic et al. [6] developed an algorithm for working out executable union slices employing conditioned slicing. Cheda et al. [7] developed a technique for finding conditioned slices to be applied to first-order functional logic languages. Silva [8] surveyed some work on program slicing based techniques. He described the features and applications of each technique utilizing example, and established the relations between them.

All these works [1–3, 5–7] have been dealt with computing conditioned slices for intraprocedural programs. They have not considered the interprocedural aspects. The work carried out for computing conditioned slices for interprocedural programs is scarce. We have computed conditioned slices for interprocedural programs.

5 Conclusion

We have presented a technique to compute conditioned slices for interprocedural programs. We have named the technique as Node Marking Conditioned Slicing (NMCS) algorithm. First, we construct the system dependence graph (SDG). Then, we mark the nodes satisfying the given condition in slicing criterion. We have also computed the slices during marking process using only the marked nodes. In future, we will implement our technique and find the slice computation time. We will also develop techniques to compute conditioned slices for object-oriented softwares, Web applications, aspect-oriented software, etc.

References

1. Gerardo Canfora, Aniello Cimitile and Andrea De Lucia. Conditioned Program Slicing, *Information and Software Technology*, Vol. 40, No. 11–12, pp. 595–607, December 1998.
2. Sebastian Danicic, Chris Fox, Mark Harman and Rob Hierons. ConSIT: A Conditioned Program Slicer. In *proceedings of International Conference on Software Maintenance (ICSM '00)*, pp. 216–226, 2000.
3. Chris Fox, Sebastian Danicic, Mark Harman and Robert M. Hierons. ConSIT: A Fully Automated Conditioned Program Slicer. *Software-Practice and Experience-SPE*, Vol. 34, No. 1, pp. 15–46, 2004.
4. Susan Horwitz, Thomas Reps and David Binkley. Interprocedural Slicing using Dependence Graphs. *ACM Transactions on Programming Languages and Systems*, Vol. 12, No. 1, pp. 26–60, January 1990.
5. Rob Hierons, Mark Harman, Chris Fox, Mohammed Daoudi and Lahcen Ouarbya. Conditioned Slicing Supports Partition Testing, *Software Testing, Verification and Reliability*, Vol. 12, No. 1, pp. 23–28, 2002.
6. Sebastian Danicic, Andrea De Lucia and Mark Harman. Building Executable Union Slices using Conditioned Slicing. In *proceedings of 12th IEEE International Workshop on Program Comprehension*, pp. 89–97, 24–26 June 2004.
7. Diego Cheda and Salvador Cavadini. Conditioned Slicing for First Order Functional Logic Programs. In *proceedings of 17th International Workshop on Functional and (Constraint) Logic Programming (WFLP '08)*, pp. 1–14, Elsevier Science, Vol. 34, July 2008.
8. Josep Silva. A Vocabulary of Program Slicing Based Techniques. *ACM Computing Surveys (CSUR)*, Vol. 44, No. 3, June 2012.

Face Biometric-Based Document Image Retrieval Using SVD Features

Umesh D. Dixit and M.S. Shirdhonkar

Abstract Nowadays, a lot of documents such as passport, identity card, voter id, certificates contain photograph of a person. These documents are maintained on the network and used in various applications. This paper presents a novel method for the retrieval of documents using face biometrics. We use trace of singular matrix to construct face biometric features in the proposed method. K-nearest neighbor approach with correlation distance is used for similarity measure and to retrieve document images from the database. Proposed method is tested on the synthetic database of 810 document images created by borrowing face images from face94 database [1]. Results are compared with discrete wavelet transform features (DWT), which is counterpart of singular value decomposition (SVD). Proposed features in combination with correlation similarity measure provided mean average precision (MAP) of 75.73% in our experiments.

Keywords Face biometrics • Document image retrieval • Singular value decomposition • Face detection

1 Introduction

Huge number of documents such as identity cards, certificates, and passports contain photograph of a person. Retrieval of such documents based on the photograph (face image) will help to search and access all documents belonging to a particular person. Such work will find its importance in offices, government

U.D. Dixit (✉)
Department of Electronics & Communication Engineering, B.L.D.E.A's,
V.P.Dr. P. G. Halakatti CET, Bijapur 586103, India
e-mail: uddixit@rediffmail.com

M.S. Shirdhonkar
Department of Computer Science & Engineering, B.L.D.E.A's,
V.P.Dr. P. G. Halakatti CET, Bijapur 586103, India
e-mail: ms_shirdhonkar@rediffmail.com

H.S. Behera and D.P. Mohapatra (eds.), *Computational Intelligence in Data Mining*, Advances in Intelligent Systems and Computing 556,
DOI 10.1007/978-981-10-3874-7_45

organizations, as well as in crime branches to collect and verify information about a person.

In this paper, we propose a method for retrieving of documents based on photograph or face images. In the proposed system, initially we separate out the photograph (face image) from the query document image and then based on features extracted from the face image, other documents in the database that contain similar face image are retrieved.

The main contributions of this paper include proposing new idea for extracting features from face image employing SVD. When an image is decomposed using SVD, the singular matrix generated consists of only diagonal elements and these elements will be unique in nature. We suggest using trace of this singular matrix as a feature for the retrieval of documents based on face biometrics. The proposed method is tested using an artificial database consisting of 810 documents, and the results are also compared with discrete wavelet transform (DWT) features, which are counterpart of SVD. The rest of the paper is organized as follows: Sect. 2 discuss about the literature related to this work; in Sect. 3, we explain the proposed work in detail; Sect. 4 provide discussion about the results; and finally Sect. 5 concludes the paper.

2 Related Work

Cao [2] presented theory on singular value decomposition and its application for digital image processing. They also introduced a method for face recognition using SVD features. They treated a set of known faces as vectors in subspace called "face space," spanned by a small group of "base faces." Projection of a new image on to the base face is compared with known face images for recognition.

Nefian and Hayes [3] presented a new approach in using PCA. They applied PCA on wavelet sub-bands. They employed wavelet transform for decomposing an image into different frequency bands, and mid-range frequency subband is used for PCA representation.

Vikram et al. [4] proposed a technique for person-specific document image retrieval. They used principal component analysis (PCA) and linear discriminant analysis subspace method for face recognition. In their work, they recommended calculating the average of the face images, by normalizing the size and tagging all the documents to this average face image.

Jang [5] proposed a novel approach for face detection that employs a user-oriented language model for face detection. The proposed method works by projecting face images which represents the significant variations among known faces. They were motivated by information theory and used Eigen vectors, which are significant feature values called as the Eigen faces.

A novel Walshlet pyramid-based face recognition technique was presented by Kekre et al. [6], in which features are extracted from Walshlets applied on various levels of image decomposition. Sadek [7] presented a work describing the applications of SVD for image processing with research challenges. This paper provides a survey on SVD-based image processing applications. It also includes new contributions using the properties of SVD for image processing applications. Keyvanpour and Tavoli [8] presented a survey on document image retrieval based on signature, logo, and layout structure of the documents. It also proposes a framework for the classification of document image retrieval approaches.

In our proposed method, initially we segment the face image from the document and then we extract features from this segmented image. For segmentation of face image, we compute the energy contributed by the connected components, and the component with highest energy is considered as face image or photograph in the document. We used the trace of the singular matrix of decomposed face image as a set of features for matching query with database documents, correlation distance metric for similarity measure for ranking and retrieval of documents.

3 Proposed Method for Face Biometric-Based Document Retrieval

We divide the process of face biometric-based document image retrieval into three tasks: In the first task, we detect and separate out photograph or face image from the query document; the second task is to extract features; and in the third, we retrieve the documents based on similarity. Following sections describe these three tasks in detail.

3.1 Photograph Detection from the Document Image

In the documents such as identity cards, passports, and certificates, portion that consists of photograph will have more energy compared to any of its other parts. This idea is to detect and separate out the face image from the document. Algorithm 1 shows the steps used for the detection of face image from the query document.

Algorithm 1: Face image detection from query document image

1. **Input:** Query document image,
   Output: Face image.
2. **Begin**
3. Convert color image to grayscale.
4. Find connected components in the image.
5. Compute energy of each connected component using Eq. (1)

$$EN_{CC} = \frac{1}{M \times N} \sum_{i=1}^{M} \sum_{j=1}^{N} |CC(i,j)| \quad (1)$$

Where, EN_{CC} is energy of connected component CC[M:N] with size M × N.

6. Photograph or Face image = Connected component with Max (EN_{CC})
7. **End**

Document containing face or photograph of person is used as an input to the algorithm. Normally, these documents are stored as color images. In preprocessing step, color document images are converted to gray scale and median filter is applied to remove impulse noise. Then, we identify connected components in the document image and compute energy of each component using Eq. (1). Connected component with highest energy is assumed as part of document image containing a photograph or face image. This component is then extracted from the document image to compute its features in the next step. Figure 1 shows sample results of face detection and extraction from document image.

3.2 *Feature Computation for Extracted Face or Photograph from Document*

Features of an image decide the performance of image analysis, matching, and retrieval results. Here, we propose SVD-based features for the retrieval of document images based on face biometrics. SVD is an algebraic tool, which can be used in various applications of image processing ranging from representation to compression of an image [4]. It allows decomposing of an image into three matrices that include a singular matrix containing diagonal elements. A square matrix is called singular, only when one of its singular values is zero. Singular values contain huge information about the image and will be unique in nature. This motivated us to use singular values as features in the proposed method. Algorithm 2 shows the steps used for constructing the feature vector.

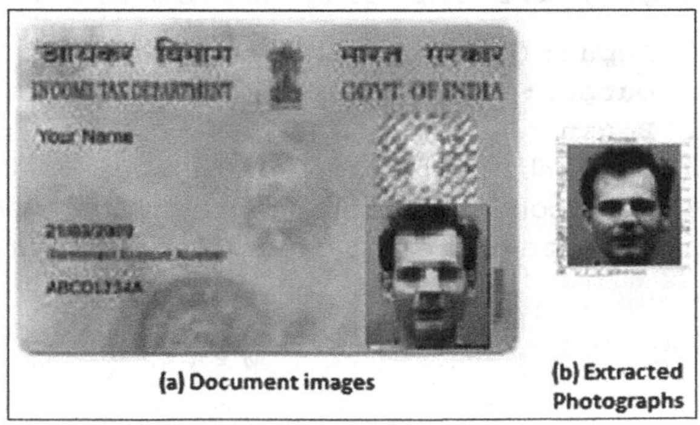

Fig. 1 Sample result of face detection and extraction from document image

Algorithm 2: Feature computation for extracted face or photo from document

1. **Input:** Face image, **Output:** Feature Vector (FV)
2. **Begin**
3. Resize face image to 64 × 64 pixels.
4. Divide face image into four blocks as F_1, F_2, F_3 and F_4 with size 32 × 32.
5. Decompose F_1, F_2, F_3, F_4 using SVD and obtain set of singular values $\{SF_1\}$, $\{SF_2\}$, $\{SF_3\}$ and $\{SF_4\}$.
6. Store diagonal elements of four image blocks

$$Diag[1:128] = \{SF_1\}\ U\ \{SF_2\}\ U\ \{SF_3\}\ U\ \{SF_4\} \quad (2)$$

   Where Diag[1:128] holds singular values of SF_1, SF_2, SF_3, and SF_4
7. Compute Feature Vector

$$FV[1:32] = \left\{ \sum_{i=1}^{4} Diag(i,i),\ \sum_{i=5}^{8} Diag(i,i),\ \ldots,\ \sum_{i=125}^{128} Diag(i,i) \right\} \quad (3)$$

   Where, each element of 'FV' is trace of singular matrix obtained after decomposition.
8. Return FV
9. **End**

Let "F" be an extracted face image resized to 64 × 64 pixels. We divide image "F" into four blocks that gives rise to F1, F2, F3, and F4 each of size 32 × 32 pixels for obtaining features. Application of SVD on these blocks can be represented using Eq. (2).

$$F_i[32:32] = U_i \times D_i \times V_i^T,\ for\ i = 1\ to\ 4 \quad (4)$$

In Eq. (2), U_i, D_i, and V_i^T are the decomposed matrices of F_i, respectively, each with size 32 × 32. Columns of matrices U_i will be Eigen vectors of $F_i \times F_i^T$, columns of matrices V_i will be Eigen vectors of $F_i^T \times F_i$, and D_i the diagonal matrices that contains singular values. The singular values are obtained as a square root of Eigen values of $F_i^T \times F_i$, and these represent unique information about the image. Each diagonal matrix D_1, D_2, D_3, and D_4 contain 32 singular values after decomposition. Now to obtain the feature vector, trace of the matrix is computed by dividing each diagonal matrix D_i into 8 parts. The sum of diagonal elements of a matrix is called its trace and in Eq. (3), it is computed as $\sum_{i=1}^{N} Diag(i,i)$. Thus, we created a feature vector FV with 32 elements.

3.3 Retrieval of Documents

Initially we extract features using Algorithm 2 from all the preprocessed documents in the database and store these features in a file to create indexed documents. Similarly, feature vector from query document is obtained while processing the query. Then, we employ following steps for the retrieval of documents.

- Let FVDB[1:N] be the array of feature vectors of all the documents stored in the database, and FVQ is the feature vector obtained from the query document.
- Compute correlation distance value between FVQ and each of FVDB vector using (5) for document matching.

$$CorrelationDist[1:N] = 1 - \frac{(X-M_x)\left(Y-M_y\right)^T}{\sqrt{(X-M_x)(X-M_x)^T \times \left(Y-M_y\right)\left(Y-M_y\right)^T}} \quad (5)$$

where X and Y are the feature vectors, M_x and M_y the mean of vectors X and Y.

- Apply K-nearest neighbor approach to retrieve top 10 documents using correlation similarity distance measure. Here, the documents are sorted based on computed distance values, such that document corresponding to lowest distance at the top, to retrieve top K number of documents for the user.

4 Results and Discussion

Due to unavailability of public database, we created our own database consisting of 810 document images for testing the proposed method. Database stores 30 documents for 27 persons, leading to a total of (27 Persons × 30 Documents) 810 documents with various sizes, including identity cards, passports, and certificates. Face images in the database are borrowed from publicly available database face94 [1].

Precision, recall, and F-measure are used as evaluation metrics. Precision is a measure indicating the fraction of retrieved documents that are relevant to the query; recall provides fraction of relevant documents retrieved out of total relevant documents stored in the database; and F-measure is a combined metric given by (6).

$$F-measure = \frac{2 \times Precision \times Recall}{Precision + Recall} \quad (6)$$

Figure 2 shows the sample result of document retrieval using face biometrics. The sample result shown have successfully retrieved relevant documents yielding a precision of 70% and 23.33% of recall.

For comparing proposed method with DWT, we executed 27 randomly selected queries and calculated average precision and recall considering Top 1, Top 5, Top

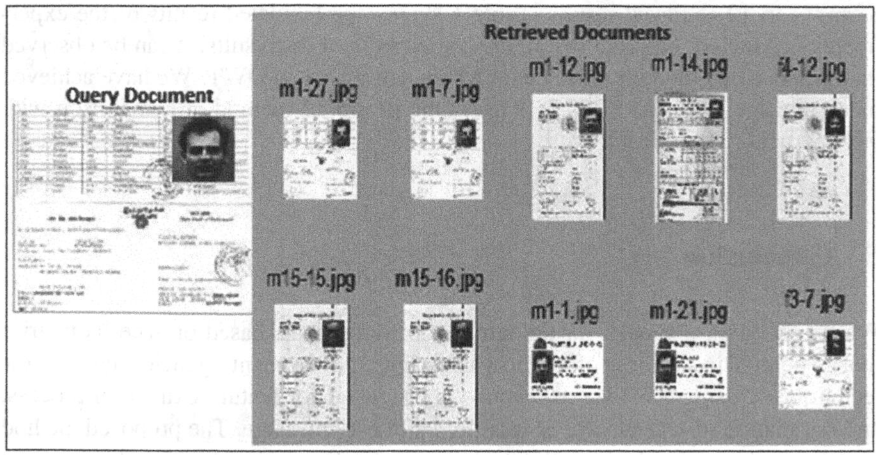

Fig. 2 Sample query document and retrieved documents

Table 1 Experimental results

Top matches	Proposed method (SVD-Based features)		Using DWT features	
	Average precision	Average recall	Average precision	Average recall
Top 1	100	3.3	100	3.3
Top 5	77.78	12.96	65.93	10.99
Top 8	69.91	18.64	51.39	13.7
Top 12	55.25	22.1	41.98	16.79
Mean values	**75.73**	**14.25**	**64.82**	**11.19**

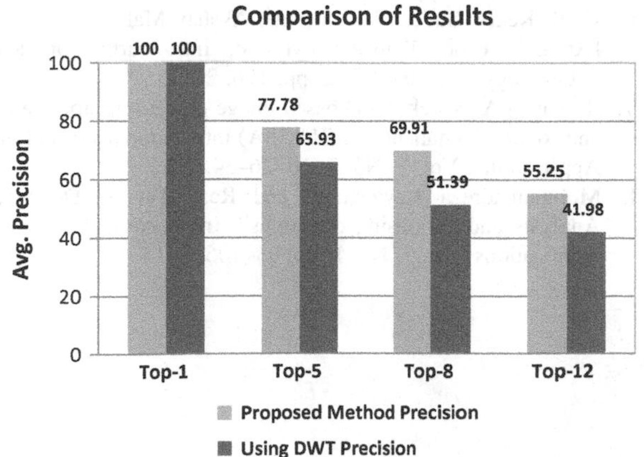

Fig. 3 Sample query document and retrieved documents

8, and Top 12 retrieved results. Table 1 shows the tabulated results of the experiments conducted. Figure 3 shows the comparison of the results. It can be observed that proposed method provides better results compared to DWT. We have achieved mean average precision of 75.73% using the proposed method in comparison with 64.82% using DWT features.

5 Conclusion

We proposed an approach for the retrieval of documents based on face biometrics that is helpful in business offices, organizations, government agencies, etc. A new technique of using SVD-based features is proposed for feature extraction process, and correlation distance metric is used for similarity measure. The proposed method is tested and compared with DWT features on database created with 810 document images. With the proposed method, we achieved a mean average precision of 75.73%.

References

1. Dr. Libor Spacek, Faces Directories, Faces 94 Directory, http://cswww.essex.ac.uk/mv/allfaces.
2. Lijie Cao: Singular Value Decomposition Applied To Digital Image Processing, Division of Computing Studies, Arizona State University Polytechnic Campus, Mesa, Arizona 85212, pp. 1–15.
3. Ara V. Nefian and Monsoon H. Hayes: Hidden Markov Models For Face Detection And Recognition, IEEE Transactions On Pattern Analysis And Machine Intelligence,Vol. 1, pp. 141–145,1999.
4. Vikram T.N, Shalini R. Urs, and K. Chidananda Gowda: Person specific document retrieval using face biometrics, ICADL-2008, LNCS 5362, pp. 371–374, Springer-Verlag Berlin Heidelberg 2008.
5. M. Daesik Jang: User Oriented Language Model for Face Detection, IEEE workshop on person oriented vision, pp. 21–26, 2011.
6. H. B. Kekre, Sudeep D. Thepade, Akshay Maloo: Face Recognition Using Texture Features Extracted From Walshlet Pyramid, Int. Journal on Recent Trends in Engineering & Technology, Vol. 05, No. 2, pp. 186, 2011.
7. Rowayda A. Sadek: SVD based image processing applications: State of the art, Contributions and Research challenges", (IJACSA) International Journal of Advanced Computer Science and Applications, Vol. 3, No. 7, pp. 26–34, 2012.
8. Mohammadreza Keyvanpour and Reza Tavoli: Document image retrieval: Algorithms, Analysis and Promising directions", International Journal of Software Engineering and Its Applications Vol. 7, No. 1, pp. 93–106, 2013.

Learning Visual Word Patterns Using BoVW Model for Image Retrieval

P. Arulmozhi and S. Abirami

Abstract Bag of visual words (BoVW) model is popularly used for retrieving relevant images for a requested image. Though it is simple, compact, efficient, and scalable image representation, one of its major drawbacks is the visual words formed by this model are noisy that leads to mismatched visual words between two semantically irrelevant images and thus the discriminative power gets reduced. In this paper, a new pattern is learnt from the generated visual words for each image category (group) and the learnt pattern is applied to the numerous images of each category. The uniqueness and correctness of the learnt pattern are verified leading to the reduction of false image matches. This pattern learning is experimented using Caltech 256 dataset and obtained higher precision values.

Keywords Bag of visual words · Visual word pattern · Quantization error · Discriminative power

1 Introduction

The voluminous amount of multimedia information is progressively flourishing due to the technology development in digital world. Adding to this, the growing social media such as facebook, twitter, etc., influence the over abundance of various kinds of data such as text, image, video, and audio. In this context, to retrieve the relevant images, content-based image retrieval (CBIR) [1] is a kind of study that uses the content of images such as color, texture, and shapes for efficient image search and this representation should be compact, scalable, and invariant to geometrical transformable features. Shape-based image representation is popularly used, and

P. Arulmozhi (✉) · S. Abirami
Department of Information Science and Technology, College of Engineering, Anna University, Chennai, India
e-mail: arulmozhikec@gmail.com

S. Abirami
e-mail: abirami_mr@yahoo.com

H.S. Behera and D.P. Mohapatra (eds.), *Computational Intelligence in Data Mining*, Advances in Intelligent Systems and Computing 556,
DOI 10.1007/978-981-10-3874-7_46

this paper also focuses on this shape-based image representation. As a result, many techniques are progressing for shape-based image retrieval. Bag of visual word (BoVW) is one among the extensively used representation model.

BoVW [2] model representation is an orderless arrangement of histograms which carries the frequencies of visual words and hence causes low discriminative power; availability of noisy data makes false positive matches among the two entirely different images and thus causing quantization error [3]. In order to improve the false positive matches, many methods [4, 5, 6, 7, 8, 9] are formed. The authors [9] created vocabulary by utilizing both the class-specific codebook and universal codebook, and they used randomized clustering forest for developing discriminative visual word vocabularies. In [10], the authors created a model for clustering, that clusters ambiguous visual words and suggested to get an effective codebook.

In this paper, with a view to reduce the false matches that are caused due to quantization error, learning patterns from the generated visual words are carried out (learnt pattern) and used in this learnt pattern to test with the images to improve the precision. This learnt pattern can be used either directly for image matching or doing post processing that helps to increase the precision and there by false matches can be substantially reduced. The contributions of this paper are

- Learning new visual word pattern from the existing codebook using frequency summation and occurrence count based on the individual visual words.
- Assign a category for images (belonging to the learnt category) by applying learnt visual word patterns.

In this regard, the rest of the paper is structured in the following way. The related works are discussed in Sect. 2. The basic working principle of BoVW and learning visual word pattern and verifying with the test images are detailed in Sect. 3. Experimentation of the proposed method is discussed in Sect. 4. In Sect. 5, conclusion and possible future works that can be carried out are specified.

2 Related Works

Lots of work is performed in order to improve the retrieval accuracy and efficiency of image retrieval. In [11], the authors have assumed visual words as nouns and useful contents namely color, shape, locations as adjectives, and they incorporated multiple features that capture the relationship between multiple features. In [12, 13], weightage are given based on the importance of the features and considered as a way of improving discrimination power of images. In [12], they proposed a method by including more query-based latent visual word spatial co-occurrences, reweighted by giving more importance to the related visual words. In [14], the authors took contextual information to provide weightage by exploring its spatial relations and the dominant orientation of local features for identifying near duplicate images. In [15], the authors proposed a new coding strategy that integrates the relationship involved with visual features and visual words and taking pairwise relations that exist between visual words. In order to have code-free feature quantization, [16] proposed method that generated discriminative bit vector namely binary SIFT (BSIFT) by quantizing SIFT descriptor. In [17], the authors take the advantages of both hamming-based and lookup-based methods to form strong hash function. In [18], the authors proposed to perform geometric verification by obtaining binary code using spatial context information around each feature by forming a circle around it.

Furthermore, one of the emerging techniques is deep learning [19, 20, 21], which produce high-level features from low-level features. In [19], the authors have proposed a method which uses the visual bag of words for generating high-level feature by using deep belief network and thus they applied for 3D shape retrieval and recognition that preserves good semantic ability. Thus, numerous researches in image retrieval is based on BoVW model makes us to choose this BoVW representation for learning patterns apart from its simplicity and compact representation.

3 Proposed Methodology

Retrieving similar images should possess high accuracy and efficiency, and in order to support this, many image representations are developed and BoVW representation is one among them. It is already a popular representation and proved as efficient method in text-based retrieval. In this paper, pattern learning is performed for different image category and verified by applying the pattern learnt for the images belonging to different categories.

3.1 BoVW Representation

In order to reduce semantic gap, a middle-level representation namely BoVW representation is becoming famous in image retrieval, where after extracting features [22], they are grouped based on their similarity by using any clustering methods and a representative of each group is formed and termed as codeword/visual word. This collection of codewords is coined as vocabulary. For any given images, the features are encoded using this codebook, and a histogram is formed by finding the visual word occurrences and thus image representation is formed.

3.2 *Proposed Learning of Visual Word Patterns*

Even though BoVW visual search is fruitful, it causes false positive images. As a remedy to this problem, many research works are done as discussed in related works. As our contribution, patterns are learnt from visual words for each image category. The idea is each image belonging to a category will have a set of visual words that is different from the visual words of another image category. With this idea, the patterns of visual words are learnt for different categories of images and applied for the images belonging to the image categories to verify the learnt patterns.

3.2.1 Learning Visual Word Pattern

The codewords generated using BoVW representation are the key point for learning the pattern. The Fig. 1 shows how the patterns are learnt from visual words. For a set of images belonging to a category, frequencies of individual visual words are counted, and it is repeated for other category images also. Then a frequency matrix for each image category is formed having individual images as rows and visual words as columns. To learn patterns, two methods are followed, one is by taking column wise frequency summation of visual words and the second way is to count column wise visual words occurrence and then arrange in decreasing order of frequency summation/occurence count VWs. To find patterns, for example, to get 3 patterns, first 3 visual words are taken, thus for 'n' patterns first 'n' visual words are considered. The algorithm is described as follows:

1. For an image category $catgy_i$ where $i = \{1, 2 \ldots c\}$, c is the number of category

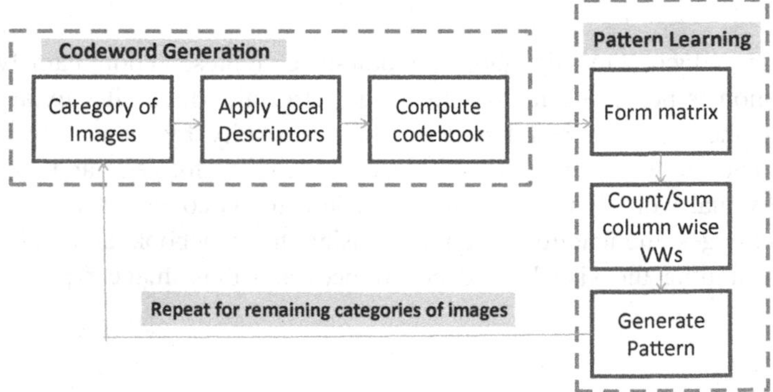

Fig. 1 Learning patterns from visual words

(a) Find the interesting features $\overset{v}{\underset{i=1}{}} f_i$ here v is the number of features of an image and it varies depending on the image homogeneity.
(b) Cluster the features to find codewords $CW = \{v_1, v_2 \ldots v_x\}$ where x is the number of cluster centers/visual words.

2. Count the frequency occurrence of the visual words for the given image category.
3. Form a matrix for all the images belonging to that category

$$\begin{array}{c} \\ I_1 \\ I_2 \\ \vdots \\ I_m \end{array} \begin{array}{c} \begin{array}{cccc} V_1 & V_2 & \ldots & V_2 \end{array} \\ \begin{pmatrix} W_{11} & W_{12} & \ldots & W_{1n} \\ W_{21} & W_{22} & \ldots & W_{2n} \\ \vdots & \vdots & \ddots & \vdots \\ W_{m1} & W_{m2} & \ldots & W_{mn} \end{pmatrix} \end{array}$$

where W_{mn} are v_x frequencies in case of frequency summation and for occurrence count procedure

$$W_{mn} = \begin{cases} 1 & \text{if } v_x \text{ is present} \\ 0 & \text{otherwise} \end{cases}$$

4. Find the sum/count of each column of the matrix belonging to individual visual Words.
5. Arrange the count of all visual words in descending order and based on the final weightage, first ‘n’ visual words are taken as pattern for that category.
6. Repeat step 1 to 5 for all the remaining image categories.

3.2.2 Verifying the Learnt Pattern

Once the pattern is learnt, using this pattern, the test images (belonging to the trained categories) have to fix a category. For a considered test image, after performing image quantization, the frequencies of visual words are counted and the visual words are arranged in descending order based on their frequencies. Here, position of the pattern is considered as an important factor for calculating the score. Thus, based on the visual word position the score is calculated as follows: Assume ‘n’ learnt patterns are placed in positions 1, 2, 3.... n. The learnt visual pattern of a category is compared with the arranged visual words generated for the test images and if both matches, its matching position is noted and if not matched, the position value is taken as zero. Once the positions are finalized, the order of the non-zero positions are checked and if they are in ascending order, they follow with the learnt pattern position and hence assumed as “positions are same” else assumed as “positions are not same.” For zero positions the zero weightage is given. The “wtg” for each position are generated using the Eq. 1

$$\text{wtg} = \begin{cases} wt, & \text{if positions are same} \\ \frac{\text{wt}}{2}, & \text{if positions not same} \\ 0, & \text{otherwise} \end{cases} \quad (1)$$

where

$$\text{wt} = \frac{1}{\text{n}}, \text{ n is the number of pattern considered}$$

As an example, Fig. 2 shows score calculation for 2 test images, assuming the pattern size as 5. Here, fourth row shows the matched positions of the test image. First two positions '3' and '7' are in ascending order, and hence, "wtg" is given as 0.2(1/5 = 0.2) as they fit into the category as 'positions are same"; for zero position, zero is assigned as "wtg"; finally, the fifth and sixth position values are fitting into "positions are not same" and thus giving 0.1(0.2/2) as "wtg". The score is the summation of the "wtg" of the 5 patterns. In similar way, the wtg and score computation for another test image are presented in the same example.

The final score is calculated by summing the wtg of the entire pattern as Eq. 2

$$\text{Score} = \textstyle\sum_{i=1}^{n} \text{wtg}. \quad (2)$$

Similarly, the score for the other learnt pattern category are calculated and finally the category of maximum score is assigned as the image's category. Repeat the procedure for all the test images that belong to the learnt category. The algorithm for deciding a category for a given test image is depicted as follows:

1. For a given category test image, find the features using local descriptor.
2. Quantize each feature to its nearest neighbors.
3. Count the frequencies of each visual word and sort them in decrement order.
4. Compare each pattern learnt with the sorted image visual words.

Description	**First 5 patterns**					**Score**
Pattern learnt	56	401	40	345	57	
Pattern position	1	2	3	4	5	
a. Test Image matched position	3	7	0	167	59	
Weights	.2	.2	0	.1	.1	**=0.6**
b. Test Image matched position	148	32	59	8	2	
Weights	.1	.1	.1	.1	.1	**= 0.5**

Fig. 2 Example for score calculation for 5 patterns

a. Perform

$$indx = \begin{cases} pos_i, & \text{if matched,} \\ 0, & \text{otherwise} \end{cases}$$

where pos_i is position of matched pattern

b. Find the positions that apply to "positions are same," "positions are not same," and assign "wtg" using the Eq. 1.
c. Compute the score as specified by Eq. 2.
d. Find score for all the patterns belonging to each category and assign the max (score) value as its category.

5. Repeat from step 1 to step 4 for the remaining test images.

4 Experimental Results and Discussion

For learning patterns and assigning category for images, CalTech-256 dataset [23] is experimented using MATLAB 2013. The dataset has 256 different category images, and approximately, the total images it contains are 29,700. For experimental purpose only airplane, butterfly, horse, hibiscus, and mushroom class images are taken. To learn patterns from visual words [24], 50 and 100 images per category are considered. To assign category for images, different 100 images belonging to each category are considered. Here to test the category assignments 5, 10, 25, and 50 patterns are considered and their results are tabulated. Table 1 shows the first 10 patterns learnt from visual words for different categories based on frequency summation and occurrence count of column wise visual words. It can be seen that the patterns of each category are differing from each other.

Considering the airplane category for the patterns of 5, 10, 25, and 50, the category assigned using frequency summation process and occurrence count processes is shown in Tables 2 and 3, respectively. Here img1, img2…img10 are the test images belonging to the airplane category. It can be seen that for some images

Table 1 List of first 10 learnt pattern for 500 vocabulary with 50 images/category belonging to 5 different category images based on frequency summation and occurrence count process

Image	Frequency summation process	Occurrence count process
Airplane	421,14,88,472,34,139,90,334,125,473	88,34,14,469,90,334,357,271,472,421
Butterfly	139,471,14,226,131,267,238,34,425,47	34,111,229,267,159,270,301,336,80,121
Horse	139,267,34,14,232,226,238,316,125,318	318,267,90,357,462,406,34,424,79,53
Hibiscus	34,226,356,445,57,178,316,336,392,267	356,34,316,226,270,392,484,79,178,301
Mushroom	470,22,466,34,180,252,178,459,158,473	470,22,252,466,34,178,180,347,473,316

Table 2 List of category assignment for 10 airplane images for the patterns 5, 10, 25, and 50 using frequency summation process (1-airplane; 2-butterfly; 3-horses; 4-hibiscus and 5-mushroom)

No. of patt.	Categories assigned to the airplane test images										Precision
	Img1	Img2	Img3	Img4	Img5	Img6	Img7	Img8	Img9	Img10	
5	3	[1;5]	[1;3;5]	[1;3;5]	1	[1;3]	[1;3]	[1;5]	3	5	0.7
10	3	1	[1;3;5]	5	3	3	[2;3]	[1;5]	5	5	0.3
25	3	1	3	1	5	3	3	[1;5]	[1;3]	[1;5]	0.5
50	3	3	1	1	[1;3]	3	1	1	1	[1;5]	0.7

Table 3 List of category assignment for 10 airplane images are listed for the patterns 5, 10, 25 and 50 using occurrence count process (1-airplane; 2-butterfly; 3-horses; 4-hibiscus and 5-mushroom)

No. of patt.	Categories assigned to the airplane test images										Precision
	Img1	Img2	Img3	Img4	Img5	Img6	Img7	Img8	Img9	Img10	
5	1	1	5	[1;3]	1	1	1	1	[1;2;5]	[1;3]	0.9
10	1	1	5	1	1	1	1	1	1	1	0.9
25	2	1	1	1	1	1	1	1	1	1	0.9
50	1	1	1	1	1	[1;5]	1	1	1	[1;5]	1

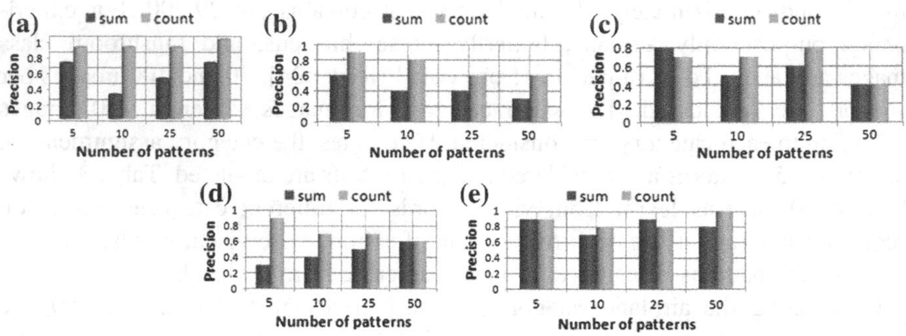

Fig. 3 Chart showing precision for both frequency summation and occurrence count process used for learning pattern and Precision calculation for **a** airplane **b** butterfly **c** horse **d** hibiscus and **e** mushroom

more than one category is assigned. As a evaluation metric, precision is considered where precision is the fraction of retrieved images that are relevant as in Eq. 3

$$\text{Precision} = \frac{\text{No. of relevant images retrieved}}{\text{No. of retrievedimages}} = \text{P(relevant/retrieved)} \quad (3)$$

Figure 3 shows the precision calculated for the 5 image categories using frequency summation and occurrence count process. From the graph, it is clear that almost for all the test images, occurrence count-based pattern learning is giving better results, and it is good to choose lower number of patterns than higher number of patterns. According to this graph, it is proven that almost 5 patterns show better results. As a future work, the learnt pattern is to be used either for image matching process or applied as the post-processing step to remove false positive matches and thus more relevant images are retrieved by improving the retrieval accuracy.

5 Conclusion

Bag of visual word representation is one of the popularly used methods for image retrieval and image classification. In order to reduce the false matches among the images, the patterns are learnt from the visual words that belongs to different categories by two ways; one by taking the frequency summation and the second way by mere occurrence count of the column wise visual words. This is validated for the set of images using Caltech 256 dataset by finding their categories, and precision values are calculated to show that the occurrence count-based pattern learning performs better and 5 number of pattern is enough for providing better precisions, and thereby it reduces the false matches. As future works, weight calculation using neural network and applying learnt pattern for image retrieval are to be carried out.

References

1. Datta R, Joshi D, Li J, Wang JZ. Image retrieval: Ideas, influences, and trends of the new age. ACM Computing Surveys (CSUR). (2008) 40(2):5.
2. Sivic J, Zisserman A. Video Google: A text retrieval approach to object matching in videos. In Computer Vision, Proceedings. Ninth IEEE International Conference (2003), 1470–1477.
3. Philbin J, Chum O, Isard M, Sivic J, Zisserman A: Lost in quantization: Improving particular object retrieval in large scale image databases. In Computer Vision and Pattern Recognition, CVPR 2008. IEEE Conference (2008), 1–8.
4. Wu Z, Ke Q, Isard M, Sun J.,: Bundling features for large scale partial-duplicate web image search. In Computer Vision and Pattern Recognition, CVPR, (2009), 25–32.
5. Philbin J, Chum O, Isard M, Sivic J, Zisserman A., Object retrieval with large vocabularies and fast spatial matching. In Computer Vision and Pattern Recognition. (2007), 1–8.
6. Yang YH, Wu PT, Lee CW, Lin KH, Hsu WH, Chen HH.,: ContextSeer: context search and recommendation at query time for shared consumer photos. In Proceedings of the 16th ACM international conference, (2008), 199–208.
7. Turcot P, Lowe D.,: Better matching with fewer features: The selection of useful features in large database recognition problems. In ICCV workshop, (2009).
8. Perronnin F, Dance C, Csurka G, Bressan M.,: Adapted vocabularies for generic visual categorization. In Computer Vision–ECCV, (2006), 464–475.
9. Moosmann, F., Triggs, W.,: Randomized clustering forests for building fast and discriminative visual vocabularies. Article in Advances in neural information processing systems, (2006).

10. Van Gemert, J. C., Veenman, C. J., Geusebroek, J. M.,: Visual word ambiguity. Pattern Analysis and Machine Intelligence, IEEE Transactions on, 32(7), (2010), 1271–1283.
11. Xie, L., Wang, J., Zhang, B., & Tian, Q.,: Incorporating visual adjectives for image classification. Neurocomputing, (2015).
12. Chen, Y., Dick, A., Li, X., & Van Den Hengel, A., Spatially aware feature selection and weighting for object retrieval. Image and Vision Computing, 31(12), (2013) 935–948.
13. Wang, J., Li, Y., Zhang, Y., Wang, C., Xie, H., Chen, G., Gao, X.,: Bag-of-features based medical image retrieval via multiple assignment and visual words weighting. IEEE transactions on medical imaging, 30(11), (2011), 1996–2011.
14. Jinliang Yao, Bing Yang, and Qiuming Zhu,: Near-Duplicate Image Retrieval Based on Contextual Descriptor, IEEE Signal Processing Letters, . 22, No. 9, (2015), 1404–1408.
15. Yang, Y. B., Zhu, Q. H., & Pan, L. Y.,: Visual feature coding for image classification integrating dictionary structure. Pattern Recognition, 48(10), (2015), 3067–3075.
16. Zhou, W., Li, H., Hong, R., and Tian, Q.,: BSIFT: toward data-independent codebook for large scale image search. Image Processing, IEEE Transactions, 24(3), (2015), 967–979.
17. Zhao, H., Wang, Z., Liu, P., & Wu, B.,: A fast binary encoding mechanism for approximate nearest neighbor search. Neurocomputing, 178, (2016), 112–122.
18. Zhen Liu, Houqiang Li, Wengang Zhou, and Qi Tian,: Contextual hashing for the large scale Image search, IEEE Transactions on Image Processing, . 23, No. 4, (2014), 1606–1614.
19. Liu, Z., Chen, S., Bu, S., & Li, K.,: High-level semantic feature for 3D shape based on deep belief networks. In Multimedia and Expo (ICME), IEEE International Conference (2014).
20. Li, Q., Li, K., You, X., Bu, S., & Liu, Z.,: Place recognition based on deep feature and adaptive weighting of similarity matrix. Neurocomputing, 199, (2016), 114–127.
21. Guo, Yanming, Yu Liu, Ard Oerlemans, Songyang Lao, Song Wu, and Michael S. Lew.,: Deep learning for visual understanding: A review. Neurocomputing (2015).
22. Lowe, D. G.,: Distinctive image features from scale-invariant keypoints. International journal of computer vision, 60(2), (2004), 91–110.
23. Griffin, G., A. Holub, and P. Perona.: Caltech-256 object category dataset California Inst. Technol., Tech. Rep. 7694, (2007).
24. Vedaldi, A., & Fulkerson, B.,: VLFeat: An open and portable library of computer vision algorithms (2008).

Test Scenario Prioritization Using UML Use Case and Activity Diagram

Prachet Bhuyan, Abhishek Ray and Manali Das

Abstract Software testing mainly aims at providing softwarc quality assurancc by verifying the behavior of a software using a finite set of test cases. The continuous evolution of software makes it impossible to perform exhaustive testing. The need for regression testing is to uncover new software bugs in existing system after some changes have been made to ensure that the existing functionalities are working fine. Re-executing the whole test suite is time-consuming as well as expensive. Hence, this issue can be handled by test case prioritization technique. Prioritization helps to organize the test suites in an effective manner where high-priority test cases are executed earlier than the low priority test cases based on some criteria. In this paper, a new prioritization approach is proposed using UML use case diagram and UML activity diagram. We have applied our technique to a particular of a case study which indicates the effectiveness of our proposed approach in prioritizing test scenarios.

Keywords Regression testing · UML · Activity diagram · Test scenarios · Control flow graph · Test case prioritization

1 Introduction

Regression testing is needed to uncover new software bugs in existing system after some changes have been made to ensure that the existing functionalities are working fine. Regression testing is checking the entire system using the old test suits. It also ensures that the changes incorporated does not affect the other modules of

P. Bhuyan (✉) · A. Ray · M. Das
School of Computer Engineering, KIIT University, Bhubaneswar
751024, Odisha, India
e-mail: pbhuyanfcs@kiit.ac.in

A. Ray
e-mail: arayfcs@kiit.ac.in

M. Das
e-mail: manali.das1991@gmail.com

H.S. Behera and D.P. Mohapatra (eds.), *Computational Intelligence in Data Mining*, Advances in Intelligent Systems and Computing 556,
DOI 10.1007/978-981-10-3874-7_47

the software and the customer's business requirements. Due to the increase in size and complexity of the software, re-executing the whole test suite is time-consuming as well as expensive. It is impractical and inefficient to perform exhaustive testing. To handle this issue, an optimized way of testing is needed that would identify the maximum errors in less time and effort. Test case prioritization helps to organize the test case in an effective manner such that the beneficial test cases are executed first. Elbaum et al. [1] defined prioritization as follows:

Given: T, a test suite; PT, the set of possible permutations of T : f is a function from PT to the real numbers.

Problem: Find $T' \in PT$ such that $(\forall T'')$ $(T'' \in PT)$ $(T'' \neq T')$ $[f(T') \geq f(T'')]$.

Here, PT represents the set of all possible prioritizations of T and f is a function that, applied to any such ordering, yields an award value for that ordering.

The advantage of prioritization is to detect faults as early as possible which help the testers to understand and resolve the errors at an early stage. Test case prioritization can be classified into two categories: code-based test prioritization and model-based test prioritization [2]. Code-based prioritization techniques generally use information such as statement coverage, branch coverage, and number of faults discovered. Model-based prioritization is based on system models. Many different models such as extended finite-state machine (EFSM), specification and description language (SDL), and UML models have been used. Experimental comparisons have proved that prioritization techniques based on system models detects fault early compared to code-based techniques and is less expensive.

The Unified Modeling Language (UML) is a standardized notation for modeling object-oriented systems. Use case diagram corresponds to high-level functional requirements and represents the different ways in which a system can be used by the users. The event flow in an use case diagram can be categorized into Basic flow of events or main scenario and Alternate flow of events or alternate scenario [3]. The dynamic behavior of a system is captured by the activity diagram. A particular use case can be illustrated using one or more than one activity diagrams [4]. This paper focuses on a new prioritization technique based on UML use case and activity diagram. We have prioritized the test scenarios obtained from the activity diagram depending on activity coverage, decision node coverage, weights assigned to the activity nodes, weights assigned to the edges, and complexity value of each test path.

The rest of the paper is structured as follows: Sect. 2 gives an overview of the literature survey in this field where various test case prioritization approaches have been described. Section 3 describes the proposed methodology to prioritize test scenarios, and Sect. 3.1 discusses the working of our proposed work using a case study. Conclusion and future work are presented in Sect. 4.

2 Related Work

In this section, various works done on prioritization techniques have been discussed.

Korel et al. [2] gave an overview of prioritization techniques based on source code and system models. They experimentally proved that model-based prioritization techniques are inexpensive compared to code-based prioritization techniques. Results have proved that model-based prioritization is more effective in detecting faults early as compared to code-based prioritization.

Mohanty et al. [5] proposed a prioritization technique for component-based software. They have used state chart diagram to generate a component interaction graph (CIG). CIG describes the interactive scenarios within the components. Athira et al. [6] proposed a model-based test case prioritization approach using UML activity diagram.

Sapna et al. [4] used UML activity diagram for prioritization. They have converted the activity diagram into tree structure and identified main and alternate scenarios. Sapna et al. [7] used UML use case diagram to prioritize scenarios. Prioritization takes into account the constructs such as actor priority, use case priority, along with customer assigned priority of use case diagram. Use case priority is calculated as sum of the weights assigned to customer priority and technical priority.

Bhuyan et al. [8] used UML use case and activity diagram to address regression testing method. They have identified some SOA testing challenges such as the heterogeneous nature of Web services, unavailability of source code, and structure of services. Prioritization is done based on activity coverage. Test case covering the maximum number of activities gets the highest priority.

3 Proposed Methodology

In this section, we have presented our proposed framework of prioritizing test scenarios. The proposed framework is shown in Fig. 1. The proposed framework is described as follows:

1. *Draw the use case diagram*: The different functions performed by the system are represented by the use case diagram.

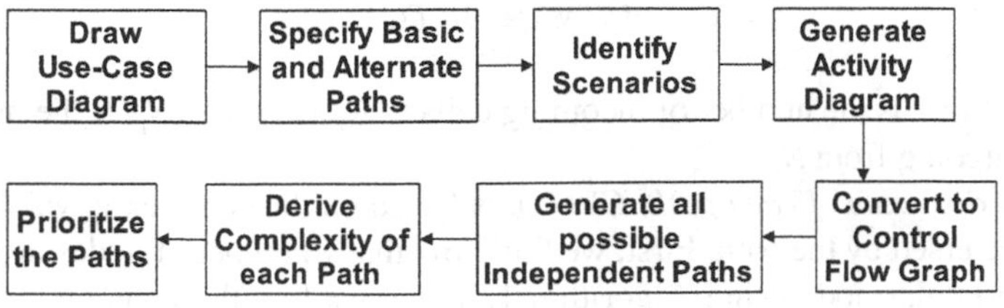

Fig. 1 Proposed framework of prioritizing test scenarios

2. *Specify the basic and alternate flow of events*: The main functions performed by the system represent the basic scenario. The exceptional behavior due to wrong inputs is represented by alternate scenario.
3. *Identify Scenarios*: A scenario may comprise of a basic flow or a combination of basic flow and alternate flows.
4. *Generate Activity Diagram*: Activity diagram represents the sequence of interactions or various actions performed by the users. Each activity is shown as a node and transition from one activity to another is represented as edges.
5. *Conversion of activity diagram to control flow graph*: Each activity in the activity diagram is mapped as a node in the control flow graph. The start node is a node having outgoing edges but no incoming edge. The end node is having no outgoing edge. The nodes having one incoming edge and two outgoing edges represent decision nodes.
6. *Conversion of activity diagram to control flow graph*: The control flow graph is traversed using depth-first search (DFS) algorithm to generate all possible independent paths such that each node is executed at least once. Each independent path represents a scenario.
7. *Prioritization Procedure*: To prioritize the test scenarios, we calculate the following:

   (a) Weight of each path based on the information complexity of each node traversed by the path.
   (b) Path complexity.
   (c) Node coverage by each path.
   (d) Decision node coverage by each path.
   (e) Weight of each path based on the weights assigned to the edges traversed by the path which takes into account the type of coupling existing between nodes connecting the edges.
   (f) Total weight of each path.

The calculation of these five factors is explained below:

- *Weight of each test path, $W(P_i)$* is calculated using Eq. 1.

$$W(P_i) = \sum (\textit{Information complexity of each node traversed by } P_i) \tag{1}$$

  Information complexity of each node is calculated using Eq. 2

$$IN(N_i) \times OUT(N_i) \tag{2}$$

  where IN(N_i) is the number of incoming calls to N_i and OUT(N_i) is the number of calls outgoing from N_i.
- *Complexity value of each path, $CV(P_i)$*, is based on the complexity value of each node traversed by the path. First, we find how many number of nodes are affected by a particular node using Algorithm 1. Then, we find the complexity value of each node using Eq. 3.

$$CV(N_i) = \frac{Amount\ of\ node\ contribution}{Total\ number\ of\ affected\ nodes} \tag{3}$$

where amount of node contribution is the number of times the node N_i is encountered out of the total number of affected nodes.

- *Node coverage by each path, $N(P_i)$*: is calculated depending on the total number of nodes traversed by the path P_i.
- *Decision node coverage by each path, $D(P_i)$*: is calculated depending on the number of decision nodes encountered by the path P_i.
- *Weight of each path, $W1(P_i)$*: is calculated using Eq. 4 which add the weights assigned to the edges traversed by the path. Weights are assigned to edges by taking into account the type of coupling such as data coupling, stamp coupling, and control coupling between the nodes connecting the edges. We have assumed the weights of data coupling, stamp coupling, and control coupling as 1, 2, and 3, respectively.

$$W1(P_i) = \sum(weights\ of\ the\ edges\ traversed\ by\ P_i) \tag{4}$$

- *Total weight of each path, $TW(P_i)$*: is the sum of all the above factors and is presented in Eq. 5.

$$TW(P_i) = W(P_i) + CV(P_i) + N(P_i) + D(P_i) + W1(P_i) \tag{5}$$

The paths are prioritized based on the total weight of each path. Path with highest weight is given maximum priority.

Algorithm 1 Reachable Nodes

Input: Control Flow Graph, Entrypoint
Output: All the nodes reachable from Entrypoint.
Initialization: Set *entrypoint*=node, Hashset *affected nodes*=null, Linked list *nodes* stores the adjacent nodes.

1: $affectednodes.put(entrypoint, newHashset < String > ())$ {Put entrypoint in Hashset *affectednodes*}
2: $nodes \leftarrow getadjacentnodes(node)$ {Adjacent nodes are stored in Linked list *nodes*}
3: **for** $x \in nodes$ **do**
4: $affectednodes.put(entrypoint, x)$ {Enter corresponding adjacent nodes of the *entrypoint* to the Hashset}
5: call $REACHABLENODE(graph, x)$ {Recursive function call}
6: **end for**

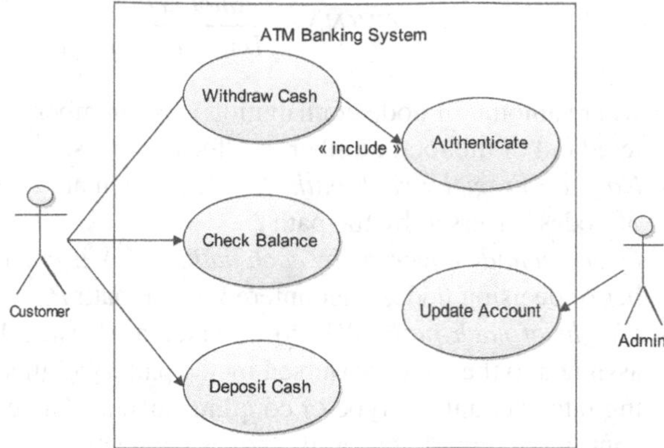

Fig. 2 Use case diagram of 'ATM Banking System'

3.1 A Case Study: ATM Banking System

In this section, our prioritization approach is illustrated with the help of a case study 'ATM Banking System'.

1. *Draw the use case diagram*: The UML use case diagram has two actors, namely customer and admin. 'Withdraw cash,' 'Check balance,' and 'Update account' are the different use cases represented in the use case diagram in Fig. 2.
2. *Specify the basic and alternate flow of events*: The Basic flow of events are identified as follows: *System displays Welcome screen, User inserts valid card, User enters valid PIN, User enters valid amount, System checks available balance, User withdraws cash, System prints receipt, System updates account, System ejects card, User takes card.* The alternate flow of events is identified as follows: *User inserts invalid card, User inserts invalid PIN, Insufficient balance.*
3. *Identify Scenarios*: Basic flow or combination of Basic flow and Alternate flow represents a scenario as shown in Table 1.
4. *Generate Activity Diagram*: The activity diagram representing the sequence of interactions is shown in Fig. 3.
5. *Converting activity diagram to control flow graph*: The activity diagram is converted into control flow graph as shown in Fig. 4. Each activity in the activity diagram is represented by a node in the control flow graph. The mapping of activity and the node number is given below:
   Node 1: Start
   Node 2: System displays Welcome screen
   Node 3: User inserts card
   Node 4: Check valid card
   Node 5: User enters PIN
   Node 6: Check valid PIN
   Node 7: User enters amount
   Node 8: System gets balance

Table 1 Scenario table

Scenario 1 (Main scenario)	Withdraw cash
Scenario 2 (Alternate Scenario 1)	Invalid card
Scenario 3 (Alternate Scenario 2)	Invalid PIN
Scenario 4 (Alternate Scenario 3)	Insufficient balance

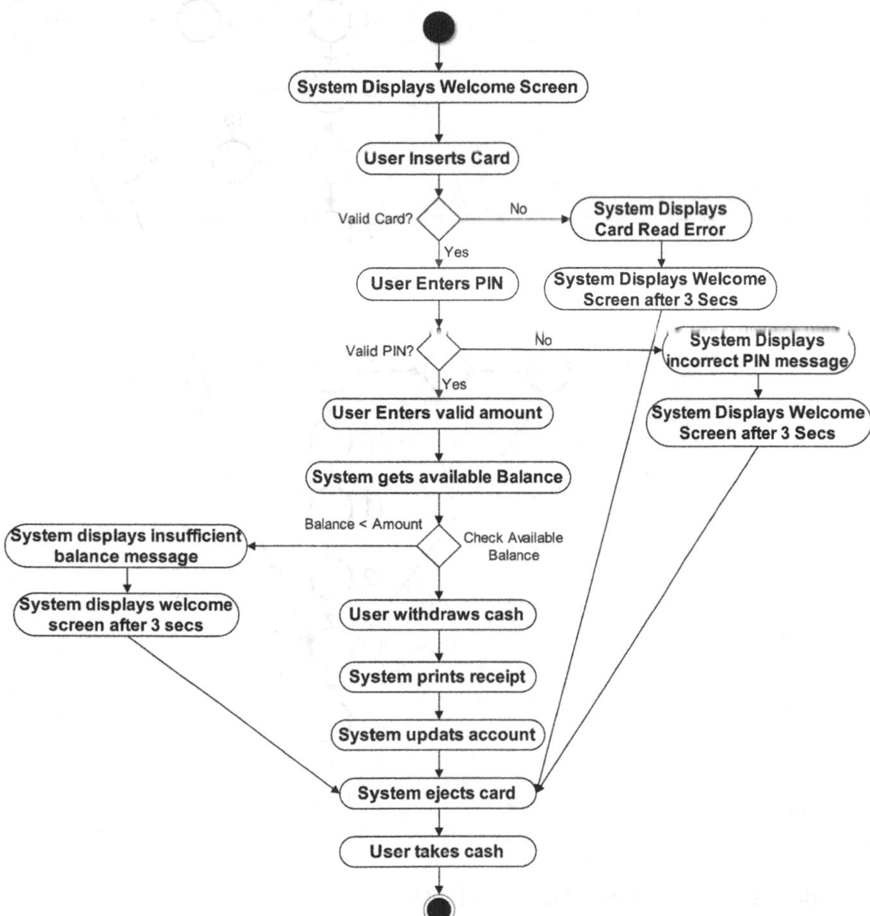

Fig. 3 Activity diagram of 'ATM Banking System'

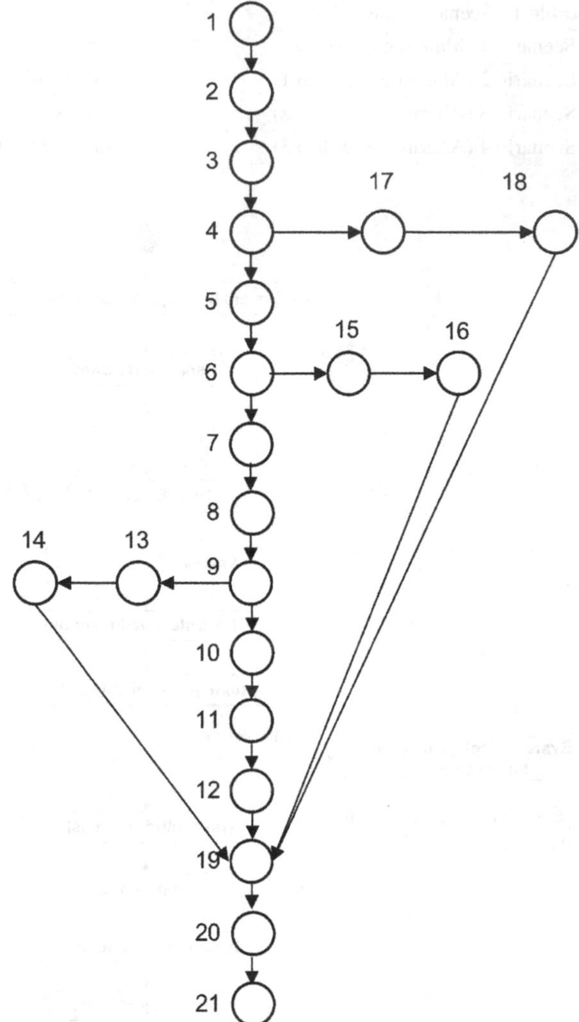

Fig. 4 Control flow graph for activity diagram of ATM Banking System

Node 9: Check whether balance available
Node 10: User withdraws cash
Node 11: System prints receipt
Node 12: System updates account
Node 13: System displays insufficient balance
Node 14: System displays Welcome screen after 3 s
Node 15: System displays invalid PIN
Node 16: System displays Welcome screen after 3 s
Node 17: System displays card read error
Node 18: System displays Welcome screen after 3 s

Node 19: System ejects card
Node 20: User takes card
Node 21: Finish

6. *Generate all possible independent paths*: After applying DFS algorithm, the following independent paths are generated from the control flow graph as shown below:
   P0: 1→2→3→4→5→6→7→8→9→10→11→12→19→20→21
   P1: 1→2→3→4→17→18→19→20→21
   P2: 1→2→3→4→5→6→15→16→19→20→21
   P3: 1→2→3→4→5→6→7→8→9→13→14→19→20→21
   Path P0 can be represented as Scenario 1 that is basic path which corresponds to 'withdraw cash.' Path P1, P2, and P3 can be represented as alternate scenarios corresponding to 'invalid card,' 'invalid PIN,' and 'insufficient balance.'
7. *Prioritization Procedure to prioritize test scenarios*:
   (a) Calculation of weight of each path, W(P_i): First, we calculate the information complexity of each node in the control flow graph as shown in Table 2. The weight of each test path, W(P_i), is the summation of information complexity of each node traversed by the path given in Table 3.
   (b) Calculation of path complexity, CV(P_i): Three main activities in an ATM Banking System are insert card, enter pin, and enter amount to withdraw which are represented as node 3, 5, and 7 in the graph. We will find out the nodes that are affected by the nodes 3, 5, and 7 applying Algorithm 1 as shown below:
   Nodes Affected by 3: 19, 17, 18, 15, 16, 13, 14, 11, 12, 21, 20, 10, 7, 6, 5, 4, 9, 8
   Nodes Affected by 5: 19, 15, 16, 13, 14, 11, 12, 21, 20, 10, 7, 6, 9, 8
   Nodes Affected by 7: 21, 20, 10, 19, 9, 8, 13, 14, 11, 12
   The complexity values of each node is calculated by applying Eq. 1. The detailed description is shown in Table 4.
   The complexity value of each path, CV(P_i) is the summation of complexity value of each node traversed by the path. It is shown in Table 5.
   (c) Calculation of node coverage, N(P_i): The number of nodes that each independent path encounters is described as node coverage by each path as calculated in Table 6.
   (d) Calculation of decision node coverage, D(P_i): The number of decision nodes each independent path encounters is described as decision node coverage by each path as calculated in Table 7.
   (e) Calculation of weight of each path, W1(P_i): Different types of coupling exist between the nodes in the control flow graph. For example, 'insert card' in node 3 and 'check valid card' in node 4 communicate via data items; hence, they are data coupled. The detailed description in shown in Table 8. The weight of each path is the sum of the weights of the edges traversed by it as shown in Table 9.

Table 2 Information complexity table

Node (N_i)	IN (N_i)	OUT (N_i)	IN (N_i) × OUT (N_i)
1	0	1	0
2	1	1	1
3	1	1	1
4	1	2	2
5	1	1	1
6	1	2	2
7	1	1	1
8	1	1	1
9	1	2	2
10	1	1	1
11	1	1	1
12	1	1	1
13	1	1	1
14	1	1	1
15	1	1	1
16	1	1	1
17	1	1	1
18	1	1	1
19	4	1	4
20	1	1	1
21	1	0	0

Table 3 Weight of paths based on information complexity

Independent paths (P_i)	Path description	Weight of each path W(P_i)
P0	1-2-3-4-5-6-7-8-9-10-11-12-19-20-21	19
P1	1-2-3-4-17-18-19-20-21	11
P2	1-2-3-4-5-6-15-16-19-20-21	14
P3	1-2-3-4-5-6-7-8-9-13-14-19-20-21	18

(f) Calculation of total weight of each path, TW(P_i): The total weight of each path is calculated by adding all the above-calculated factors taken together as shown in Table 10. The paths are prioritized by arranging the total weights of each independent path in descending order such that the path showing higher weight is the most complex path and needs to be tested at an early stage as given in Table 11.

Table 4 Complexity value of nodes

Node	Amount of node contribution	Complexity value of each node
1	0	0
2	0	0
3	0	0
4	1	0.05
5	1	0.05
6	2	0.12
7	2	0.12
8	3	0.17
9	3	0.17
10	3	0.17
11	3	0.17
12	3	0.17
13	3	0.17
14	3	0.17
15	2	0.12
16	2	0.12
17	1	0.05
18	1	0.05
19	3	0.17
20	3	0.17
21	3	0.17

Table 5 Path complexity based on node complexity value

Independent paths (P_i)	Path description	Complexity value of each path CV(P_i)
P0	1-2-3-4-5-6-7-8-9-10-11-12-19-20-21	1.67
P1	1-2-3-4-17-18-19-20-21	0.67
P2	1-2-3-4-5-6-15-16-19-20-21	0.94
P3	1-2-3-4-5-6-7-8-9-13-14-19-20-21	1.5

Table 6 Node coverage values

Independent paths (P_i)	Path description	Node coverage of each path N(P_i)
P0	1-2-3-4-5-6-7-8-9-10-11-12-19-20-21	15
P1	1-2-3-4-17-18-19-20-21	9
P2	1-2-3-4-5-6-15-16-19-20-21	11
P3	1-2-3-4-5-6-7-8-9-13-14-19-20-21	14

Table 7 Decision node coverage values

Independent paths (P_i)	Path description	Decision node coverage of each path, D(P_i)
P0	1-2-3-4-5-6-7-8-9-10-11-12-19-20-21	3
P1	1-2-3-4-17-18-19-20-21	1
P2	1-2-3-4-5-6-15-16-19-20-21	2
P3	1-2-3-4-5-6-7-8-9-13-14-19-20-21	3

Table 8 Weights assigned to edges

Edge	Type of coupling	Weight
1–2	No	0
2–3	Control	3
3–4	Data	1
4–5	Control	3
4–17	Control	3
5–6	Data	1
6–7	Control	3
6–15	Control	3
7–8	Data	1
8–9	Stamp	2
9–10	Control	3
9–13	Control	3
10–11	Stamp	2
11–12	Stamp	2
12–19	Control	3
17–18	Control	3
18–19	Control	3
15–16	Control	3
16–19	Control	3
13–14	Control	3
14–19	Control	3
19–20	No	0
20–21	No	0

Table 9 Weight of paths based on coupling factor

Independent paths (P_i)	Path description	Weight of each path W1(P_i)
P0	1-2-3-4-5-6-7-8-9-10-11-12-19-20-21	24
P1	1-2-3-4-17-18-19-20-21	13
P2	1-2-3-4-5-6-15-16-19-20-21	17
P3	1-2-3-4-5-6-7-8-9-13-14-19-20-21	23

Table 10 Total weight of each path

Independent paths (P_i)	Path description	Total weight TW(P_i)
P0	1-2-3-4-5-6-7-8-9-10-11-12-19-20-21	62.73
P1	1-2-3-4-17-18-19-20-21	34.78
P2	1-2-3-4-5-6-15-16-19-20-21	45.10
P3	1-2-3-4-5-6-7-8-9-13-14-19-20-21	59.57

Table 11 Prioritized scenario table

Scenario	Path	Path description	Total weight TW(P_i)
Scenario 1	P0	1-2-3-4-5-6-7-8-9-10-11-12-19-20-21	62.73
Scenario 4	P3	1-2-3-4-5-6-7-8-9-13-14-19-20-21	59.57
Scenario 3	P2	1-2-3-4-5-6-15-16-19-20-21	45.10
Scenario 2	P1	1-2-3-4-17-18-19-20-21	34.78

4 Conclusion and Future Work

In this paper, a new test scenario prioritization approach is proposed using UML use case diagram and activity diagram. In our work, model-based technique is used where system models are used to prioritize test scenarios. A particular use case is selected and elaborated as an activity diagram. The effectiveness of our approach is shown using a case study of 'ATM Banking System'. We can extend our approach to accommodate other UML diagrams and non-functional features for test case identification and prioritization.

References

1. Sebastian Elbaum, Alexey G. Malishevsky,Gregg Rothermel, Test Case Prioritization: A Family of Empirical Studies, IEEE Transactions on Software Engineering, VOL. 28, NO. 2 (2002)
2. Bogdan Korel, George Koutsogiannakis, Experimental Comparison of Code-Based and Model-Based Test Prioritization, IEEE International Conference on Software Testing Verification and Validation Workshops (2009)
3. Rajani Kanta Mohanty, Binod Kumar Pattanayak and Durga Prasad Mohapatra, UML Based Web Service Regression Testing Using Test Cases: A Case Study,ARPN Journal of Engineering and Applied Sciences,VOL. 7, NO. 11 (2012)
4. Sapna P.G., Hrushikesha Mohanty, Prioritization of Scenarios based on UML Activity Diagrams,First International Conference on Computational Intelligence, Communication Systems and Networks (2009)
5. Sanjukta Mohanty, Arup Abhinna Acharya, Durga Prasad Mohapatra, A Model Based Prioritization Technique for Component Based Software Retesting Using UML State Chart Diagram, IEEE (2011)

6. Athira B, Philip Samuel, Web Services Regression Test Case Prioritization, IEEE (2010)
7. Sapna P.G., Hrushikesha Mohanty,Prioritizing Use Cases to aid ordering of Scenarios, Third UKSim European Symposium on Computer Modeling and Simulation (2009)
8. Prachet Bhuyan, Abhishek Kumar, Model Based Regression Testing Approach of Service Oriented Architecture (SOA) Based Application: A Case Study, International Journal of Computer Science and Informatics (2013)

Conditioned Slicing of Aspect-Oriented Program

Abhishek Ray and Chandrakant Kumar Niraj

Abstract The different variants of slicing techniques of aspect-oriented programs (AOPs) are used in software maintenance, reuse, debugging, testing, program evolution, etc. In this paper, we propose a conditioned slicing algorithm for slicing AOPs, which computes more precise slice in comparison with dynamic slice. First, we have constructed an intermediate representation named *conditioned aspect-oriented dependence graph* (CAODG) to represent the aspect-oriented programs. The construction of CAODG is based on execution of aspect-oriented program with respect to pre-/post-condition rule, which is defined in aspect code. Then, we have proposed a conditioned slicing algorithm for AOP using the proposed CAODG.

Keywords Aspect-oriented programming · Pre-/post-condition · Conditioned aspect-oriented dependence graph · Conditioned slicer

1 Introduction

Program slicing, first introduced by Weiser [7], is a decomposition technique with regard to some slicing criterion, which is employed in several fields of software engineering such as debugging, testing, program comprehension, and component reuse. To achieve these applications, several program slicing techniques [1–3, 9–12, 14, 15] were proposed to compute precise slices in various programming constructs such as structured/procedural programming, object-oriented programming (OOP), and aspect-oriented programming. In this paper, we have proposed an algorithm to compute the conditioned slice of AOPs. We have chosen AOP as a programming construct because it addresses the major concern of OOP called as *crosscutting concern*. First, we have defined pre-/post-condition rules in aspect code and utilize them

A. Ray (✉) · C.K. Niraj
KIIT University, Bhubaneswar, Odisha, India
e-mail: arayfcs@kiit.ac.in

C.K. Niraj
e-mail: ckniraj@gmail.com

H.S. Behera and D.P. Mohapatra (eds.), *Computational Intelligence in Data Mining*, Advances in Intelligent Systems and Computing 556,
DOI 10.1007/978-981-10-3874-7_48

through non-aspect code. Then, we have constructed an intermediate representation named *conditioned aspect-oriented dependence graph* (CAODG) by using program execution trace [6] method. Next, We have proposed a conditioned slicing algorithm to compute the conditioned slice of AOP. Lastly, we have implemented the proposed algorithm for computing conditioned slices.

The rest of the paper is organized as follows. Section 2 presents few definitions used in the paper. Section 3 discusses the construction of intermediate representation for AOP. Section 4 describes the proposed algorithm for conditioned slicing of aspect-oriented programs and also discusses the working of our proposed algorithm. Section 5 presents the comparison with related work. Finally, Sect. 6 concludes the paper.

2 Basic Concepts and Definitions

This section illustrates the definitions of intermediate program representations and the various types of dependencies associated with them.
Conditioned-Weaving (K_n) **node**: In a directed graph $G(N, E)$, a Condition-Weaving (K_n) node holds the pre-condition rules in aspect code and add them to a non-aspect code in aspect-oriented programming.
Condition_Slice (CS_G): It is used for storing the computed conditioned slice statements, according to pre-condition rule of the program, and will satisfy the post-condition rule of program P. Otherwise, program is not suitable for condition slicing.
ExecuTrace(T): It is used to trace and store the flow of program execution, which will utilize for conditioned slicing.
DisplayResult(): It is used for displaying the result after conditioned slicing of aspect-oriented program.

3 CAODG: Our Proposed Intermediate Representation

In this section, we describe the construction of *conditioned aspect-oriented dependence graph* (CAODG).

The CAODG is a *directed graph* $G(N, E)$, where N is the set of vertices that correspond to the statements and E is the set of edges between vertices N. In the proposed CAODG, we have considered the following types of nodes and dependency edges exist:

- **Statement Node**: It is used to represent the executed program lines according to their numbers.
- **Start/Stop Node**: It shows the start and stop of program execution.

- **Control Dependence Edge**: It represents the control flow relationships of a program during execution.
- **Data Flow Edge**: It represents the relevant data flow relationships of a program.
- **Weaving Edge**: It is used for joining of aspect code and non-aspect code at appropriate join points.
- **Error Message Edge**: It shows the presence of error in a program P during runtime.

Node Creation: The node creation is done statically only once with respect to the statement number of the program. First, we write non-aspect code and aspect code, which consist pre-/post-condition rule for the execution of variable X and Y. Then, we initialized the statements of non-aspect code from starting with numbers $(1, 2, 3, \ldots n)$ and with aspect code from starting with numbers $(n + 1, n + 2, \ldots m)$.
Graph Construction: Our CAODG construction is based on dynamic aspect-oriented dependence graph (DADG) [6] of aspect-oriented programming. The DADG [6] uses execution trace method to know the execution of the program and graph is constructed on their execution. Similarly, the CAODG uses execution flow method and construction of the graph is done behalf of AspectJ program execution.

Conditioned-Weaving (K_n) node
Procedure K_n node selection:

i. int ExecuTrace(T);
ii. for (int K_n: T)
iii. from node M to node N in a directed graph G;

   (a) Procedure K_n-node-search (G,M)
      1. if node M = K_n node, then marked it.
      2. else if node M ≠ K_n node, then again call K_n-node-search (G,N)
      3. else exit;

Non-Aspect Code

```
import java.util.Scanner;
1.  public class Client {
2.  public static void main (String arg [])
    {
3.  Point p = new Point();
4.  Scanner sc = new Scanner(System.in);
5.  System.out.println("Plz enter the value of X");
6.  int x = sc.nextInt();
7.  System.out.println("Plz enter the value of y");
8.  int y = sc.nextInt();
9.  p.moveBy(x, y);
10. System.out.println(" X = " + x +" and Y = " + y);}}
11. class Point
    {
12. int X, Y;
13. public int getX()
14. { return X;   }
15. public int getY()
16. { return Y;   }
17. public void setX(int x)
18. { X=x; }
19. public void setY(int y)
20. { Y=y; }
21. public void moveBy(int dx, int dy)
    {
22. setX(2*dx);
23. setY(2*dy);
    }
  }
```

Aspect Code

```
24. public aspect PointCondition {
25. int Min_X = 0, Min_Y = 0;
26. int Mid_X = 50, Mid_Y = 50;
27. int Max_XY= 100;
28. before(int newX): call(void Point.setX(int)) && args(newX) {
29.   if ( newX <= Min_X || newX >= Mid_X )
30.   throw new IllegalArgumentException( " X is Less than MinX or greater than."); }
31. after(Point p, int newX) returning:call(void Point.setX(int))&&target(p)&&args(new){
32.   if (newX > Max_XY)
33.   throw new IllegalArgumentException( " X should be Less than Max X.");
34.   System.out.println(" X Co-ordinate : " + newX ); }
35. before(int newY): call(void Point.setY(int)) && args(newY){
36.   if ( newY <= Min_Y || newY >= Mid_Y )
37.   throw new IllegalArgumentException(" Y is Less than MinY or greater than Max."); }
38. after(Point p, int newY) returning:call(void Point.setY(int))&&target(p)&&args(newY){
39.   if (newY > Max_XY)
40.   throw new IllegalArgumentException(" Y should be Less than Min Y.");
41.   System.out.println(" Y Co-ordinate : " + newY );
      }
}
```

Fig. 1 An AspectJ program

Determination of (K_n) node: Conditioned-Weaving (K_n) node represents the joining of non-aspect code and aspect code at appropriate join point with pre-condition rule.

Consider the AspectJ program as shown in Fig. 1. The main objective of the program is to set the value of *X* coordinate and *Y* coordinate of a graph during run-time. For the conditioned slicing, pre-/post-condition rule is defined in aspect code. In Fig. 1, the non-aspect code or base code has two classes (Point and Client). The pre-/post-condition rules are defined with the help of *before* and *after* advice. At the time of implementation of AspectJ program, first main class of base code (*Client class*) requires an input from user for *X* and *Y* coordinates and then determines the pre-condition rule before setting the value of *X* or *Y* coordinate in *moveBy()* method of base code and afterward adjusting the value of *X* or *Y* in base code *moveBy()* method again it satisfies the post-condition rule.

Program Execution

```
2.  public static void main (String arg []){   {
3.  Point p = new Point();
11. class Point     {
4.  Scanner sc = new Scanner(System.in);
5.  System.out.println("Plz enter the value of X");
6.  int x = sc.nextInt();
7.  System.out.println("Plz enter the value of y");
8.  int y = sc.nextInt();
9.  p.moveBy(x, y);
21. public void moveBy(int dx, int dy)  {
22. setX(2*dx);
28. before(int newX): call(void Point.setX(int)) && args(newX) {
17. public void setX(int x)
18. { X=x; }
29. if ( newX <= Min_X || newX >= Mid_X )
30. throw new IllegalArgumentException( " X is Less than MinX or greater than."); }
18. { X=x; }
31. after(Point p, int newX) returning: call(void Point.setX(int))&&target(p) && args(newX){
17. public void setX(int x)
18. { X=x; }
32. if (newX > Max_XY){
33. throw new IllegalArgumentException( " X should be Less than Max X.");
22. setX(2*dx);
34. System.out.println(" X Co-ordinate :  " + newX ); }
23. setY(2*dy);
35. before(int newY): call(void Point.setY(int)) && args(newY){
19. public void setY(int y)
20. { Y=y; }
36. if ( newY <= Min_Y || newY >= Mid_Y )
37. throw new IllegalArgumentException(" Y is Less than MinY or greater than Max."); }
20. { Y=y; }
38. after(Point p, int newY) returning: call(void Point.setY(int))&&target(p) && args(newY)
19. public void setY(int y)
20. { Y=y; }
39. if (newY > Max_XY)
40. throw new IllegalArgumentException(" Y should be Less than Min Y.");
23. setY(2*dy);
41. System.out.println(" Y Co-ordinate :  " + newY );
10. System.out.println(" X = " + x +" and Y = " + y);}}
```

Fig. 2 Execution trace of the program given in Fig. 1 for pre-condition rules = ($X \geq 0; Y \geq 0$)

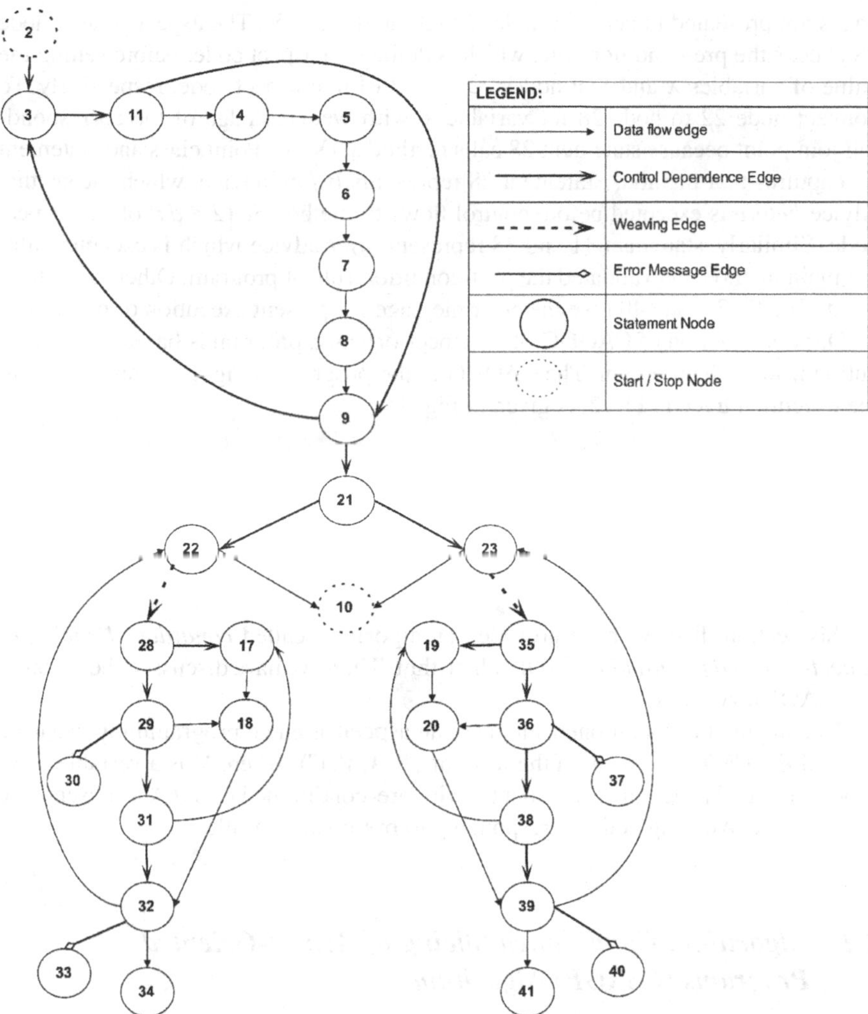

Fig. 3 CAODG for the execution trace of the program given in Fig. 2

Figure 2, represents the flow of program execution, which is based on the defined *pre-condition* to satisfy the *post-condition* rule. In this example, we have defined the pre-condition rule as *MinX or MinY* ≤ 0 and post-condition rule as *MaxXY* > 100 for program *P*.

In Fig. 3, there are four cases of dependence edges. There is a control dependence edge between node 2 and 3. Node 4 and 5 are represented by a data dependence edge. In between node 29 and node 30, there has an error message dependence edge. Another pair of nodes is represented by the control, data, and error message dependence edges which based on the relation between program instructions. Weaving edges are presented in between node (22–28) and (23–35). The aspect code at node 28 checks the pre-condition rule, which is defined in aspect code, before setting the value of variables *X* and *Y* at nodes 22 and 23 of non-aspect code, respectively. To connect node 22 to node 28 for variable *X* with weaving edge of the corresponding join point because statement 28 calls method *setX()* of Point class and statement 22 captures that method. Statement 28 represents *before* advice, which means that advice *before* is executed before control flows to method *Set(2 * dx)* of non-aspect code. Similarly, statement 31 and 38 represent *after* advice which is executed after the main method and validated the post-condition rule of program. Otherwise, statements 30, 33, 37, and 40 give the error message for present execution of program.

Our construction of CAODG of an aspect-oriented program is based on the execution flow of the program. The CAODG of the program in Fig. 1, corresponding to the execution trace in Fig. 2, is given in Fig. 3.

4 Conditioned Slicing of Aspect-Oriented Programs (CSAOP) Algorithm

In this section, first we have proposed an algorithm called *conditioned slicing of aspect-oriented programs* (CSAOP) algorithm. Then, we have discussed the working of CSAOP algorithm.

To compute the conditioned slicing of an aspect-oriented programming, we have defined the *slicing criterion* in the form of $\langle P, A, V, C \rangle$, where V is a variable in an aspect-oriented program at statement P with pre-conditioned C and A represents the Conditioned-Weaving node corresponding to pre-condition rule.

4.1 Algorithm: Conditioned Slicing of Aspect-Oriented Programs (CSAOP) Algorithm

1. **Stage 1**: CAODG construction:

   (a) Create node *Sn* in the CAODG for the each statement *s* in the trace file.
   (b) Add following dependence edges to these nodes.
      i. Two node *n* and *X* have following dependence, do
         If node *n* calls node *X* for variables, then
            Add data dependence edge *(n, X)*.
         else, if node *n* calls node *X*
          for control the variable, then
            Add control dependence edge *(n, X)*.
         else, if node *n* calls node *X* of aspect code, then
            Add weaving dependence edge *(n, X)*.

else,
Add error message dependence edge *(n, X)*.

2. **Stage 2**: Managing Run-time Updation:

   (a) Initialize: Before traversing the CAODG, do
      i. Set Condition_Slice(CS_G) = ϕ
      ii. Set ExecuTrace(T) = [1, 2, 3, …, m − 1, m]
      iii. Unmark all the nodes
   (b) Run-Time Updation: Traverse the CAODG with slicing criterion $\langle P, A, V, C\rangle$
      a. Call Procedure Conditioned-Weaving (K_n) node selection.
      b. Mark all control dependence edges corresponding to K_n node.
   (c) DisplayResult(CS_G): Display the computed conditioned slice.

4.2 Working of CSAOP Algorithm

In this section, we illustrate the working of our proposed algorithm with the help of the AspectJ program as shown in Fig. 1. First, we have constructed CAODG statically once by considering the execution trace of the program as shown in Fig. 1. The constructed CAODG is shown in Fig. 3. Let the slicing criterion for computing the conditioned slice is $\langle 10, 28, X, X \geq 0\rangle$. This slicing criterion requires us to find the backward conditioned slice for the variable (*X* with respect to call at statement 10. The execution of the program is based on the pre-/post-condition rule of program. For the input value of variable *X* = *2 and Y* = *4,* which satisfy the pre-/post-condition rule, the program will execute the following statements:
P = 2→ 3→ 11→ 4→ 5→ 6→ 7→ 8→ 9→ 21→ 22→ 28→ 17→ 18→ 29→ 30→ 18→ 31→ 17→ 18→ 32→ 33→ 22→ 34→ 23→ 35→ 19→ 20→ 36→ 37→ 20→ 38→ 19→ 20→ 39→ 40→ 23→ 41→ 10.

So, our CSAOP algorithm marks the nodes: 2, 3, 11, 4, 5, 6, 7, 8, 9, 21, 22, 28, 17, 18, 29, 30, 31, 32, 33, 34, 23, 35, 19, 20, 36, 37, 38, 39, 40, 41, 10. These statements are stored in flow of program execution file named *ExecuTrace(T)*. After the execution of program, we have to manage the changes during run-time, which is defined in Stage-2 of CSAOP algorithm. But, before traversing the CAODG, we have to initialize the pre-/post-condition rule for conditioned slicing.

Consider, pre-condition for *X* and *Y* coordinates is
$X \geq 0; Y \geq 0$
and post-condition for *X* and *Y* coordinate is
$X \leq 100; Y \leq 100$

In the considered slicing criterion $\langle 10, 28, X, X \geq 0\rangle$, from node 10, variable *X* with pre-condition rule (X ≥ 0) backward trace the *Conditioned-Weaving* node through a procedure *Conditioned-Weaving* (K_n)*-node-search* in the CAODG. In

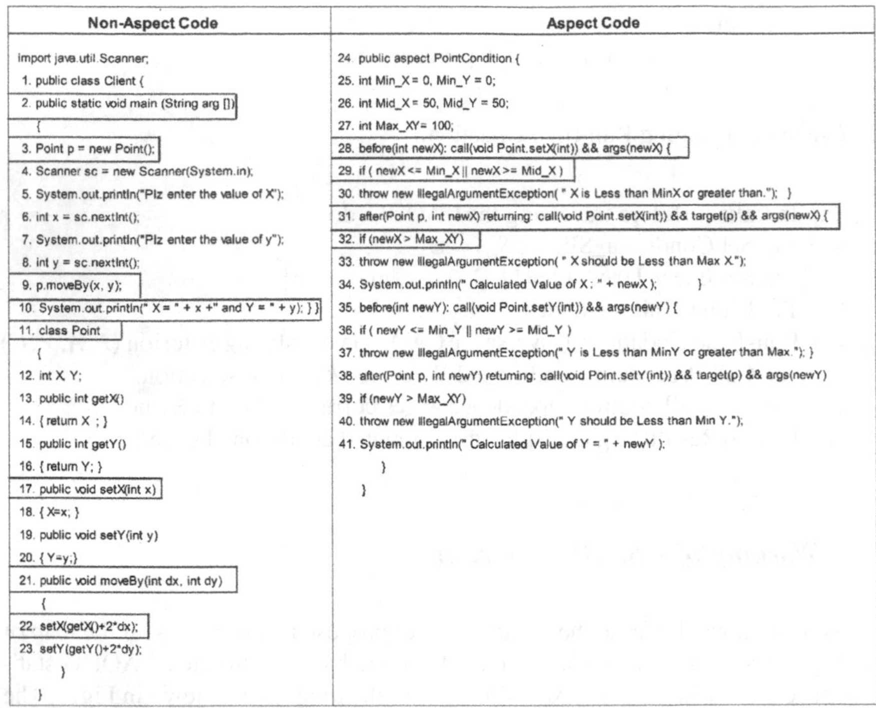

Non-Aspect Code

```
import java.util.Scanner;
 1. public class Client {
 2. public static void main (String arg [])
    {
 3. Point p = new Point();
 4. Scanner sc = new Scanner(System.in);
 5. System.out.println("Plz enter the value of X");
 6. int x = sc.nextInt();
 7. System.out.println("Plz enter the value of y");
 8. int y = sc.nextInt();
 9. p.moveBy(x, y);
10. System.out.println(" X = " + x +" and Y = " + y); } }
11. class Point
    {
12. int X, Y;
13. public int getX()
14. { return X ; }
15. public int getY()
16. { return Y; }
17. public void setX(int x)
18. { X=x; }
19. public void setY(int y)
20. { Y=y;}
21. public void moveBy(int dx, int dy)
    {
22. setX(getX()+2*dx);
23. setY(getY()+2*dy);
        }
    }
```

Aspect Code

```
24. public aspect PointCondition {
25. int Min_X = 0, Min_Y = 0;
26. int Mid_X = 50, Mid_Y = 50;
27. int Max_XY= 100;
28. before(int newX): call(void Point.setX(int)) && args(newX) {
29. if ( newX <= Min_X || newX >= Mid_X )
30. throw new IllegalArgumentException( " X is Less than MinX or greater than.");  }
31. after(Point p, int newX) returning: call(void Point.setX(int)) && target(p) && args(newX) {
32. if (newX > Max_XY)
33. throw new IllegalArgumentException( " X should be Less than Max X.");
34. System.out.println(" Calculated Value of X : " + newX );        }
35. before(int newY): call(void Point.setY(int)) && args(newY) {
36. if ( newY <= Min_Y || newY >= Mid_Y )
37. throw new IllegalArgumentException(" Y is Less than MinY or greater than Max."); }
38. after(Point p, int newY) returning: call(void Point.setY(int)) && target(p) && args(newY)
39. if (newY > Max_XY)
40. throw new IllegalArgumentException(" Y should be Less than Min Y.");
41. System.out.println(" Calculated Value of Y = " + newY );
        }
    }
```

Fig. 4 The conditioned slice of the program given in Fig. 1 with slicing criterion $\langle 10, 28, X, X \geq 0 \rangle$

Fig. 3, for variable *X*, we get first *Conditioned-Weaving* node at 28, which is connected to node 22. This signifies that the *Conditioned-Weaving* node is node 28, which is a *before advice* in aspect code and likewise it is satisfying the pre-condition rule and computed conditioned slicing on given slicing criterion will satisfy post-condition rule at node 31 for variable *X*. Finally, the computed conditioned slicing of aspect-oriented program with slicing criterion $\langle 10, 28, X, X \geq 0 \rangle$ is: $\{10, 2, 3, 11, 9, 21, 22, 28, 17, 29, 31, 32\}$. The statements included in the conditioned slice are represented in dotted rectangular box in Fig. 4 for slicing criterion $\langle 10, 28, X, X \geq 0 \rangle$.

5 Comparison with Related Work

In the absence of any existing algorithm for conditioned slicing of aspect-oriented programs, we compare our proposed algorithm with existing algorithms for computing conditioned slicing of procedural program as well as object-oriented programs.

Danicic et al. [5] have introduced a technique for computing executable union slices by the help of conditioned slicing, which is useful in program comprehension and component reuse. They have only considered object-oriented programs for a slice. Hierons et al. [8] have introduced the partition testing through conditioned slicing. They have used ConSIT [4] tool for the partition testing through conditioned slicing of procedural programs, which helps to find the fault in test partition. Ray et at. [13] have proposed test case design technique using conditioned slicing of activity diagram. Mainly, their approach is used for test case coverage with models.

Our proposed algorithm computes the conditioned slice for aspect-oriented programs. Our approach is based on the *pre-/post*-condition rule, which is defined in aspect code rather than non-aspect code of AspectJ programs.

6 Conclusion

In this paper, we have proposed a conditioned slicing algorithm for aspect-oriented programs. First, we have constructed a dependence-based intermediate program representation *conditioned aspect-oriented dependence graph* (CAODG) on defined *pre-/post-condition rule in aspect code* for aspect-oriented program. The CAODG is constructed based on the program execution trace. Then, we have proposed an algorithm called *conditioned slicing of aspect-oriented programming* (CSAOP) to compute conditioned slicing of aspect-oriented programs.

References

1. BINKELY, D. W., AND GALLAGHER, K. B. Program Slicing, Advances in computers. Tech. rep., Academic Press, San Diego, CA.
2. B. KOREL, AND J. LASKI. Dynamic Program Slicing. *Information Processing Letters 29*, 3 (1988).
3. CANFORA, G., CIMITILE, A., AND LUCIA, A. Conditioned program slicing. *Information and Software Technology 40* (1998), 595–607.
4. C. FOX, M. HARMAN, AND R.M. HIERONS. ConSIT: A Conditioned Program Slicer. In *International Conference on Software Maintenance (ICSM)* (2000), pp. 216–226.
5. DANICIC, S., LUCIA, D. A., AND HARMAN, M. Building Executable Union Slices using Conditioned Slicing, 2004.
6. D.P. MOHAPATRA, M. SAHU, R. KUMAR, AND R. MALL. Dynamic Slicing of Aspect-Oriented Programs. *Informatica* (2008), 261–274.
7. HARMAN, M., AND DANICIC, S. Amorphous Program Slicing. Master's thesis, School of Computing, University of North London, 1997.
8. HIERONS, R., HARMAN, M., FOX, C., DAOUDI, M., AND OUARBYA, L. Conditioned Slicing Supports Partition Testing. *The Journal of Software Testing, Verification and Reliability* (2002), 23–28.
9. J. FERRANTE, K. OTTENSTEIN, AND J. WARREN. The Program Dependence Graph and Its use in Optimization, 1987.
10. M. HARMAN, R.M. HIERONS, C. FOX, AND J. HOWROYD. Pre/Post conditioned Slicing. In *International Conference on Software Maintenance (ICSM)* (Nov 2001), pp. 138–147.
11. MUND, G., AND MALL, R. A Efficient inter-procedural Dynamic Slicing Method. *Information and Software Technology* (2006), 791–806.

12. MUND, G., MALL, R., AND S. SARKAR. Computation of Intra-procedural Dynamic Program Slices. *Information and Software Technology* (2003), 499–512.
13. RAY, M., BARPANDA, S., AND MOHAPATRA, D. Test Case Design using Conditioned Slicing of Activity Diagram. *International Journal of Recent Trends in Engineering 1*.
14. SIKKA, P., AND KAUR, K. Program Slicing Techniques and their Need in Aspect Oriented Programming. *International Journal of Computer Applications 70*, 3 (2013), 11–14.
15. WEISER, M. Programmers Use Slices When Debugging. *Communications of the ACM* (July 1982), 446–452.

Comparative Analysis of Different Land Use–Land Cover Classifiers on Remote Sensing LISS-III Sensors Dataset

Ajay D. Nagne, Rajesh Dhumal, Amol Vibhute, Karbhari V. Kale and S.C. Mehrotra

Abstract Determination and identification of land use–land cover (LULC) of urban area have become very challenging issue in planning a city development. In this paper, we report application of four classifiers to identify LULC using remote sensing data. In our study, LISS-III image dataset of February 2015, obtained from NRSC Hyderabad, India, for the region of Aurangabad city (India) has been used. It was found that all classifiers provided similar results for water body, whereas significant differences were detected for regions related to residential, rock, barren land and fallow land. The average values from these four classifiers are satisfactory in agreement with Toposheet obtained from the Survey of India.

Keywords Remote sensing · Land use–land cover (LULC) · Supervised classification · Maximum likelihood classifier (MLC) · LISS-III

1 Introduction

Life and developmental stir are completely based on land, so it is absolutely one of the most important natural resources. The land use and land cover pattern of a region is a consequence of natural and socioeconomic factors, utilized by human in

A.D. Nagne (✉) · R. Dhumal · A. Vibhute · K.V. Kale · S.C. Mehrotra
Department of Computer Science & IT, Dr. Babasaheb Ambedkar Marathwada University, Aurangabad, MS, India
e-mail: ajay.nagne@gmail.com

R. Dhumal
e-mail: dhumal19@gmail.com

A. Vibhute
e-mail: amolvibhute2011@gmail.com

K.V. Kale
e-mail: kvkale91@gmail.com

S.C. Mehrotra
e-mail: mehrotra.suresh15j@gmail.com

H.S. Behera and D.P. Mohapatra (eds.), *Computational Intelligence in Data Mining*, Advances in Intelligent Systems and Computing 556,
DOI 10.1007/978-981-10-3874-7_49

time and space. Land use refers to the type of utilization to which man has put the land. In developing countries, urban development makes changes on land use and land cover in many areas across the world. Urban development has expanded the exploitation of natural sources, so it makes effects on the patterns of LULC. Now, LULC has become a main central factor in current strategies for managing natural resources and also for monitoring the environmental changes. Rapid urbanization caused serious losses of land, forest area and water bodies, so it creates and demands more occasions for new urban developments. Globally, LULC plays an important role in international policy issues [1, 2].

Remote sensing and GIS techniques for LULC application are very cost-effective methods. The basic premise in using satellite images for change detection is that changes in land cover result in changes in radiance values that can be remotely sensed. The satellite imageries are useful to accurately compute various LULC categories. It also helps in managing the spatial data infrastructure. This spatial information is very important for observing urban expansion and also in change-detections studies. Land use–land cover classifications are performed either by using supervised classifiers or by unsupervised classifiers, and these techniques are available in all the image-processing tools [2–6].

Remote sensing technology provides earth surface coverage with a temporal resolution. This is an important benefit of remote sensing systems, and it can be used to perform analysis and inventory of surroundings and its resources. This innovation gives appropriate distinction between different objects where as its temporal resolution gives dynamic changes on the earth surface and atmosphere. It required many aspects and it has to be unified into one system for accurate analysis [7, 8].

2 Anderson Classification (LULC) Levels

The term land use is utilized extensively to connote the human exercises ashore comprising of adjustments of the common habitat into a constructed or non-assembled environment. The Anderson classification scheme of LULC comprises the different classes into the levels [9, 10].

3 Study Area

Our study was done on Aurangabad municipal corporation region (AMC), which was located at N 19° 53′ 47″ - E 75° 23′ 54″. The place is well known for its tourist places such as Ajanta caves, Ellora caves, Panchakki, Daulatabad fort and Bibi-ka-Makbara. The city is also in competition to be made as smart city. The study will be used to provide relevant information for making it smart [11–13] (Fig. 1)

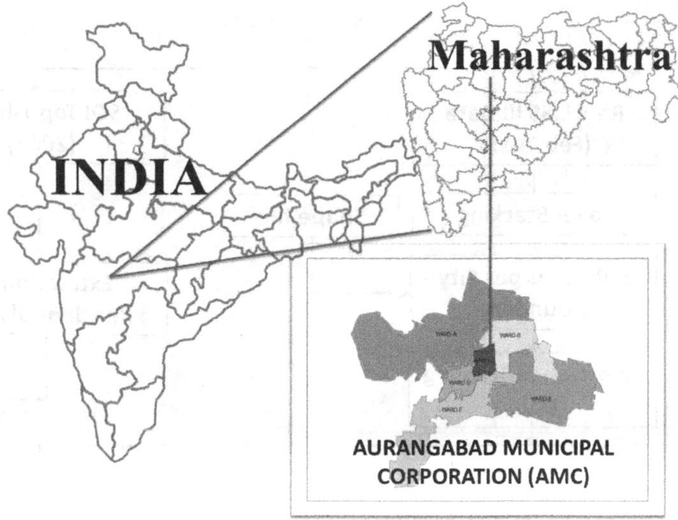

Fig. 1 Study area (*Aurangabad Municipal Corporation*)

4 Methodology and Data Used

In this study, we have used multispectral LISS-III data (Feb 2015) and Survey of India's Toposheet. The Toposheet (No. E43D5) was procured from the SoI, having a 1:50000 scales and which was surveyed and prepared in 2005–2006. The map was digitized by scanning method and then we have performed the georeferencing on it with longitude and latitudes using the ArcGIS software. The obtained LISS-III data were geometrically and radiometrically corrected by the provider having four spectral bands with 23.5 m spatial resolution. The detailed methodology is shown in Fig. 2. The primary objective of this study is to identify the urban land use–land cover with integrating remote sensing and GIS.

A raw LISS-III data, there is a separate file for every band and its file format is tiff. We have performed the layer stacking by using ENVI tool, so a single image with all bands was obtained. After performing a layer stacking, we had extracted Aurangabad municipal corporation (AMC) area using the shape file of the region as shown in Fig. 3.

4.1 Classification

Every land use–land cover classification uses an Anderson classification schemes, and we have considered a Level 1 classification of Anderson classification schemes. Because we have used a LISS-III dataset and its spatial resolution is 23.5 m, by

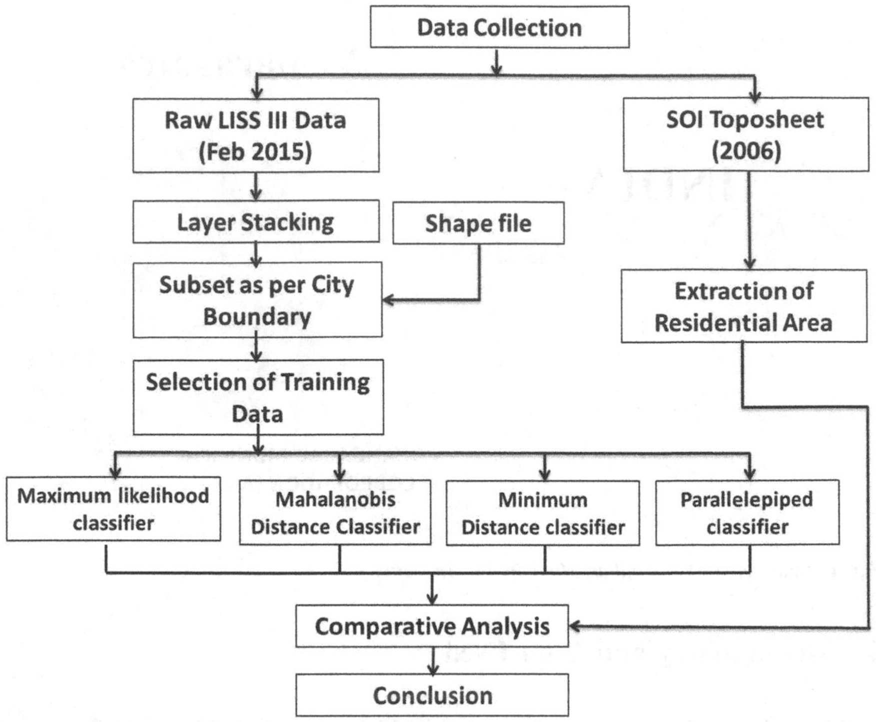

Fig. 2 Proposed methodology

Fig. 3 Aurangabad municipal corporation areas

Table 1 Anderson Level 1 and 2 land cover codes

Sr. no	Level 1 classes	Level 2 classes
(1)	Urban/settlement area/built-up	(11) Residential
		(12) Commercial
		(13) Industrial
		(14) Transportation/communications
		(15) Industrial/commercial complexes
		(16) Mixed urban
		(17) Other urban
(2)	Agricultural area	(21) Crops and pastures
		(22) Orchards, vineyards, Nurseries
		(23) Confined feeding operations
		(24) Other agricultural
(3)	Rangeland	(31) Herbaceous rangeland
		(32) Shrub and brush rangeland
		(33) Mixed rangeland
(4)	Forests	(41) Deciduous forests
		(42) Evergreen forests
		(43) Mixed forests
(5)	Water body	(51) Streams and canals
		(52) Lakes
		(53) Reservoirs
		(54) Bays and estuaries
(6)	Hills/rock	(61) Hills with vegetation
		(62) Hills without vegetation
(7)	Barren land	(71) Dry salt flats
		(72) Beaches
		(73) Sandy areas (non-beach)
		(74) Bare exposed rock
		(75) Strip mines, quarries
		(76) Transitional areas
		(77) Mixed barren

using this spatial resolution, we can perform only Level 1 classification scheme. To identify or perform a Level 2 classification, it is required a to have a data with spatial resolution less than 5 m [9, 10].

We have applied four classification techniques by assuming six types of objects in the extracted the images of Aurangabad city. These four classifiers were maximum likelihood [14–16], Mahalanobis distance [14, 15], minimum distance [14, 15] and parallelepiped classifier [14, 15]. The types of objects were categorized

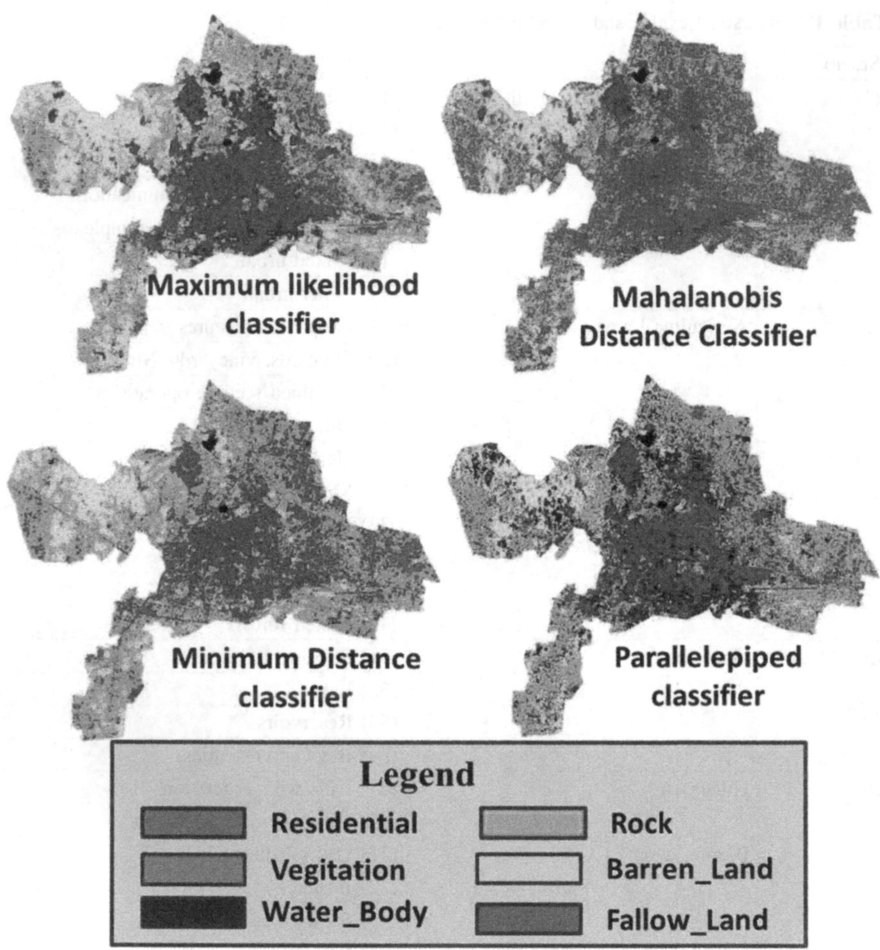

Fig. 4 LISS-III (*2015*) classified images of Aurangabad region

as residential, vegetation, water body, rock, barren land and fallow land. This object classification is subset of the Anderson scheme as given in the Table 1. The results are shown in Fig. 4 for all four classifiers with these six classes and given in Table 2.

Table 2 Classification results

Anderson scheme code	Class name	Color	Minimum distance (%)	Maximum likelihooc (%)	Mahalanobis (%)	Parallelepiped (%)	Average (%)
1	Residential	Magenta	21.06	28.50	18.27	32.90	25.18
6	Rock	Cyan	9.84	21.62	16.99	14.81	15.82
5	Water body	Blue	0.36	0.30	0.28	0.33	0.32
2,4	Vegetation	Green	4.07	8.89	3.51	5.30	5.44
7	Barren land	Yellow	51.75	35.15	31.65	43.82	40.59
3	Fallow land	Red	12.92	5.54	29.31	2.84	12.65

Table 3 Comparison of computed algorithms

Algorithm name	Overall accuracy (%)	Kappa coefficient	Remark
Minimum distance	88.41	0.84	High
Maximum likelihood	93.43	0.90	Highest
Mahalanobis	83.01	0.76	Less
Parallelepiped	81.46	0.75	Lowest

5 Result and Discussion

The results of the study indicate that the result will follow the same trend irrespective of classifiers. This study clearly indicates that these technologies have the ability to produce or prepare base maps of Aurangabad city. Comparing the results of different classifiers, it is observed that there is no significant difference using different types of classifiers.

Accuracy assessment is a quantitative assessment of the performance of the particular classifier and it also determines a quality of information from the remotely sensed data. To perform an accuracy assessment of a particular classification technique, it required two sources of information: first one is the classified image and the second one is the ground truth or referenced image. Accuracy assessment was performed by comparing the classified image with a ground truth. For this purpose, ten ground truth samples were taken for each class, so a total of sixty ground truths were taken [16, 17].

It can be seen from the Table 3 that performance of the maximum likelihood is better than all other methods, whereas the performance of parallelepiped is worst. However, the performance of all classifier technique is found to be satisfactory, varying in the range of 93.43–81.46% and Kappa coefficient values from 0.76 to 0.90. The average values given in Table 2 are good estimate of different types of land in Aurangabad city.

6 Conclusion

The LISS-III multispectral image datasets of February 2015 were used to identify LULC of the Aurangabad municipal corporation area. In this study, four classifiers were used to identify LULC of the AMC. By comparing all the four classifiers, results clearly indicated that there are no significant changes in water body; a minor change has been identified in vegetation area. There is a major difference between residential, rock, barren land and fallow land, because every classification technique uses different algorithms and methods; for example, Mahalanobis distance classifier is more useful to determine similarity of an unknown sample set to a known one, whereas maximum likelihood classifier classifies every pixel into its corresponding class. Since there are some variations in major classes, an average of each class was calculated. It provides a satisfactory value for each types of land. Maximum likelihood classifier has highest accuracy of 93.43% with a Kappa coefficient of 0.90 as compared to other classifiers and parallelepiped classifier has lowest accuracy of 81.46% with 0.75 Kappa coefficients. These results were also compared with the Survey of INDIA's Toposheet and it is found to be in agreement with the analysis of every classes. The results reported are very useful for future planning of AMC, as it provides information about various types of land available.

Acknowledgements The authors would like to acknowledge (1) UGC–BSR fellowships (2) DST_FIST and (3) UGC-SAP(II)DRS Phase-I and Phase-II F.No.-3-42/2009 and 4-15/2015/DRS-II for Laboratory Facility to Department of CS & IT, Dr.
B.A.M. University, Aurangabad(MS), India.

References

1. Bhagawat, Rimal.: Application Of Remote Sensing And GIS, Land Use/Land Cover Change In Kathmandu Metropolitan City, Nepal, Journal Of Theoretical And Applied Information Technology (2011)
2. Innocent, Ezeomedo., Joel, Igbokwe.: Mapping and Analysis of Land Use and Land Cover for a Sustainable Development Using High Resolution Satellite Images and GIS, Remote Sensing for Landuse and Planning (2013)
3. A., A., Belal., F., S., Moghanm.: Detecting urban growth using remote sensing and GIS techniques in Al Gharbiya governorate, Egypt, The Egyptian Journal of Remote Sensing and Space Sciences (2011) 14, 73–79
4. Nayana, Ratnaparkhi., Ajay, Nagne., Bharti, Gawali.: Analysis of Land Use/Land Cover Changes Using Remote Sensing and GIS Techniques in Parbhani City, Maharashtra, India. International Journal of Advanced Remote Sensing and GIS (2016), Volume 5, Issue 4, pp. 1702–1708, Article ID Tech-573 ISSN 2320 – 0243
5. Nayana, Ratnaparkhi., Ajay, Nagne., Bharti, Gawali.: A Land Use Land Cover classification System Using Remote Sensing data, International Journal of Scientific & Engineering Research, Volume 5, Issue 7, (2014)
6. Govender, M., K., Chetty., H., Bulcock.: A review of hyperspectral remote sensing and its application in vegetation and water resource studies, Water Sa 33.2 (2007)
7. Navalgund, R., R., K., Kasturirangan.: The Indian remote sensing satellite: a programme overview, Proceedings of the Indian Academy of Sciences Section C: Engineering Sciences 6.4 (1983): 313–336
8. Ajay, Nagne., Rajesh, Dhumal., Amol, Vibhute., Yogesh, Rajendra., Karbhari, Kale., Suresh, Mehrotra.: Suitable Sites Identification for Solid Waste Dumping Using RS and GIS Approach: A Case Study of Aurangabad, (MS) India, The 11th IEEE INDICON 2014 Conference, Pune, (2014). 978-1-4799-5364-6/14/$31.00, 2014 IEEE
9. James, R., Anderson., Ernest, E., Hardy., John, T., Roach., Richard, E., Witmer.: A Land Use And Land Cover Classification System For Use With Remote Sensor Data, Geological Survey Professional Paper 964 (1976)
10. Ajay, Nagne., Bharti, Gawali.: Transportation Network Analysis by Using Remote Sensing and GIS a Review, International Journal of Engineering Research and Applications (2013) ISSN: 2248-9622 Vol. 3, Issue 3, pp. 070-076
11. Ajay, Nagne., Amol, Vibhute., Bharti, Gawali., Suresh, Mehrotra.: Spatial Analysis of Transportation Network for Town Planning of Aurangabad City by using Geographic

Information System, International Journal of Scientific & Engineering Research, Volume 4, Issue 7, (2013)

12. Kashid, Sumedh., Ajay, Nagne., Karbhari, Kale.: Solid Waste Management: Bin Allocation and Relocation by Using Remote Sensing & Geographic Information System. International Journal of Research in Engineering and Technology, Volume: 04 Issue: 12 (2015)
13. Dhananjay, Nalawade., Sumedh, Kashid., Rajesh, Dhumal., Ajay, Nagne., Karbhari, Kale.: Analysis of Present Transport System of Aurangabad City Using Geographic Information System. International Journal of Computer Sciences and Engineering Vol.-3, 124–128 (2015), E-ISSN: 2347-2693
14. Amol, Vibhute., Rajesh, Dhumal., Ajay, Nagne., Yogesh, Rajendra., Karbhari, Kale., Suresh, Mehrotra.: Analysis, Classification, and Estimation of Pattern for Land of Aurangabad Region Using High-Resolution Satellite Image, Proceedings of the Second International Conference on Computer and Communication Technologies, Advances in Intelligent Systems and Computing AISC Series 11156, Vol. 380, pp 413–427, Springer India. DOI 10.1007/978-81-322-2523-2_40, 04 (2015)
15. Amol, Vibhute., Ajay, Nagne., Bharti, Gawali., Suresh, Mehrotra.: Comparative analysis of different supervised classification techniques for spatial land use/land cover pattern mapping using RS and GIS, International Journal of Scientific & Engineering Research, Volume 4, Issue 7, (2013)
16. Rajesh, Dhumal., Amol, Vibhute., Ajay, Nagne., Yogesh, Rajendra., Karbhari, Kale, Suresh, Mehrotra.: Advances in Classification of Crops using Remote Sensing Data, Cloud Publications International Journal of Advanced Remote Sensing and GIS (2015), Volume 4, Issue 1, pp. 1410–1418, Article ID Tech-483 ISSN 2320 – 0243
17. Foody, Giles, M.: Status of land cover classification accuracy assessment, Remote sensing of environment 80.1 (2002): 185–201

Review on Assistive Reading Framework for Visually Challenged

Avinash Verma, Deepak Kumar Singh and Nitesh Kumar Singh

Abstract Objective of this paper is to review various approaches used for providing the assistive reading framework for the visually challenged persons. Visually challenged persons are the persons who are either blind or having any kind of difficulty in reading any printed material to acquire the domain knowledge. A lot of research is being done for visually challenged to make them independent in their life. We want to study the existing assistive technology for visually challenged and then propose a robust reading framework for visually challenged. The discussed reading framework will help the visually challenged person to read normal printed books, typed documents, journals, magazines, newspapers and computer displays of emails, Web pages, etc., like normal persons. It is a system based on image processing and pattern recognition using which visually challenged person carries or wears a portable camera as a digitizing device and uses computer as a processing device. The camera captures the image of the text to be read along with the relevant image and data. An optical character recognition (OCR) system segregates the image into text and non-text boxes, and then, the OCR converts the text from the text boxes to ASCII or text file. The text file is converted to voice by a text-to-speech (TTS) converter. Thus, blind person would 'hear' the text information that has been captured. So this technology can help the visually challenged to read the captured textual information independently which will be communicated as voice signal output of a text-to-speech converter.

Keywords Reading framework · Assistive technology · Visually challenged · OCR and TTS

A. Verma (✉)
BBD University, Lucknow, India
e-mail: avinash.verma93@yahoo.com

D.K. Singh
Integral University, Lucknow, India
e-mail: deepak.iiita@gmail.com

N.K. Singh
BBDNITM, Lucknow, India
e-mail: niteshboss87@gmail.com

H.S. Behera and D.P. Mohapatra (eds.), *Computational Intelligence in Data Mining*, Advances in Intelligent Systems and Computing 556,
DOI 10.1007/978-981-10-3874-7_50

1 Introduction

The World Blind Union (WBU) [1] on behalf of blind and partially sighted persons of the world estimates 258 million blind and visually challenged persons in about 190 countries and of this, about one-third of them are living in India. Conventionally, it is accepted that the visually challenged people, especially the born blind, can do very limited types of jobs. The reasons for this restricted range of jobs are twofold—firstly, society finds that they are not suitable for certain types of jobs because of the need to protect the worker and the people around the person and the second reason being that the visually challenged could acquire very limited knowledge and skills. Most of the effort of the visually challenged individual and society goes to make out that the person lives in the world comfortably with the help of devices such as Braille, canes, and dogs. The support cost for the visually challenged is considerably high. The existing assistive technologies [2–7] have helped the visually challenged in different ways at different point in time. However, most of these technologies tried to provide minimal support so that the challenged person is able to survive rather than leading a fuller life. However, the current ICTs have the means and capabilities of restoring the fullness to life by making the challenged person to live life like any one of us. The moment we provide the capabilities that would make the visually challenged to behave like normal human being, the set of problems associated with the challenged would disappear. The person can learn any discipline or activity and contribute to the society and have the same quality of life as anyone else.

2 Review of Literature

Most of the technology developed for visually challenged person is based on image acquisition and then extraction of the text from the image with the help of optical character recognition technology also known OCR [8]. Then, text file output of the OCR engine is converted to speech with the help of text-to-speech (TTS) synthesis [9] technology. Thus, the visually challenged person gets to listen whatever the text is being captured by him. Figure 1 shows the overall system classified as:

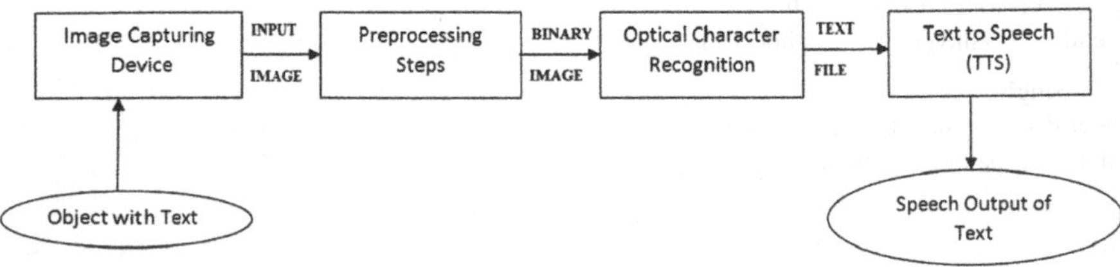

Fig. 1 Architecture of reading framework for visually challenged

2.1 *Image Acquisition by Visually Challenged*

In image acquisition, visually challenged person has to capture the image of document, book, warning sign, computer display, magazine, journals, and newspaper which is done with the help of capturing device such as portable mobile camera or any kind of specially prepared goggle that can capture the image of the document to read. As the image will be captured by a visually challenged person, the image will contain the document text as well as the surrounding. Input image will also be affected with the effect of noise caused due to different lighting condition, effect of shadow, skew, and blur. This will make the localization of the text a challenging problem, and various text extraction techniques [10, 11] have been proposed for the text localization in natural scene [12] which can localize the text present in a captured image. The geometric rectification of captured image [13] will require removing the effects of skew which makes the recognition of text difficult.

2.2 *Preprocessing Steps Involved Before OCR*

As the image is acquired by the visually challenged person, it will be associated with a higher degree of complexities such as effect of shadow, different lighting conditions, effect of skew, and blur. If the input image is affected by the skew and the blur, then we first take care of the blur in the image. Blur in the image is mainly due to the motion of either the object or the capturing device mainly camera at the time of image capture, and this type of blur is known as motion blur.

$$g(x,y) = f(x,y) * h(x,y) + n(x,y) \tag{1}$$

In expression (1), blurred image can be modeled as the convolution of the ideal image $f(x,y)$ with the blur kernel $h(x,y)$ to get the blurred image $g(x,y)$ with additive noise $n(x,y)$ which is induced at the time of image capture. Motion blur kernel also known as point spread function (PSF) is the function of two parameters which is length of blur L and angle of blur θ. Therefore, motion deblurring tries to estimate the blur kernel $h(x,y)$, and then, it performs the deconvolution operation which is the inverse of convolution operation to recover original image $f(x,y)$ also known as latent image from the blurred image. When the text document to be captured translates with a relative velocity V with respect to the camera, the blur length L in pixels is $L = \mathrm{VT}_{\mathrm{exposure}}$, where $\mathrm{T}_{\mathrm{exposure}}$ is the time duration of the exposure. The expression for motion blur is given as,

$$h(x,y) = \begin{cases} \frac{1}{L}, & \textit{if } \mathbf{0} \leq |x| \leq \boldsymbol{L}\cos\boldsymbol{\theta}; \boldsymbol{y} = \boldsymbol{L}\sin\boldsymbol{\theta} \\ \mathbf{0}, & \textit{otherwise} \end{cases}$$

When the angle of blur $\theta = 0$, it is called horizontal motion blur. In deblurring, the original image $f(x, y)$ requires the deconvolution of the PSF $h(x, y)$ with the observed image $g(x, y)$. In most of the cases, the PSF is known prior to the deblurring, and this type of deblurring is done using the classical well-known techniques such as inverse filtering, wiener filtering, least square filtering, and recursive Kalman filtering that are available. If the deblurred image or the resultant image $f(x, y)$ after deblurring operation is affected with the skew, then the image is deskewed by finding the angle of skew, and then, the image is rotated to get the image which is free from the effect of skew. To make the recognition of the text from the image an effective process using an OCR we require various standard preprocessing steps before applying the OCR engine.

2.3 Various Applications Proposed for the Visually Challenged

In mid-1990s, Xerox launched a device called reading edge, which scans printed materials and reads it out load to a user. It also provides a Braille interface so that a blind person may read the content in Braille. The reading edge devices incorporate a scanner, speech generation software system, and a Braille keypad for editing. Users could adjust the reading speed and have a choice of nine speaking voices. While the device was a handy aid for a blind person, still its usage required significant effort. A book has to be laid in proper orientation for it to be scanned. Also the unit, due to its size and weight, cannot be freely carried. One of the Android applications proposed is R-Map [14], it uses the camera to capture the image, and optical character recognition and text-to-speech engine are used to provide read-out loud service. Mobile phone has lower processing power than a desktops or notebooks. However, providing a processing power that can work in real-life environment in mobile phone is difficult. Mobile camera may lead to various problems in the image such as skew, blur, curved base lines, and autofocus mechanism, thereby causing the best available OCR to fail. Tesseract OCR [15] engine is used which provides text segmentation and recognition. The recognized text is send to text-to-speech (TTS) engine and then further to text-to-speech synthesis. Due to limited mobile screen size, it is hard for a visually challenged person to take the image of the long printed material. Akshay Sharma, Abhishek Srivastava, and Adhar Vashishth proposed a reading system for visually challenged [16] that uses a scanner to scan the document image which is given as a input to the OCR module which performs the text and non-text segmentation and recognition and generates a text file as output which is converted to the speech by text-to-speech module. The difficulty with this is that a visually challenged person cannot put the document into the scanner in proper orientation. Chucai Yi, Yingli Tian, and Aries Arditi proposed a portable camera-based text and product label reading for handheld objects for blind persons [17], and this helps blind person to read the text label and product packaging from handheld object in their day-to-day life. A motion-based region of interest (ROI) [12] is proposed by asking the user to shake the object. This method extracts the moving object from its complex background, and then, text extraction and recognition are performed on the object. The recognized text is outputted to the blind as voice using text-to-speech mechanism. Trinetra [18] by Patrick E. Lanigan, Aaron M. Paulos, Andrew W. Williams, Dan Rossi, and Priya Narasimhan is a cost-effective assistive technologies developed for visually challenged person to make them independent in their life. The objective of the Trinetra system is quality improvement of the life of visually challenged with the help of different networked devices to support them in navigation, shopping, and transportation. Trinetra uses barcode-based solution consisting of combination of off-the-shelf components, such as Internet and Bluetooth-enabled mobile phone, text-to-speech software, and a portable barcode reader.

3 Review of Optical Character Recognition (OCR) Algorithms

Most of the above-used reading framework used optical character recognition [2, 19, 20]. OCR is an image processing technology in which we extract the text in an image and convert it into a text file. Input to the optical character recognition module is the acquired bitmap image. On the input image, we have to perform various preprocessing operations such as binarization, blur removal, skew [13] correction, and noise removal step [21]. Then, text and non-text segmentations are performed to isolate text from graphics, and text is then segmented into various lines. From each line, we segment different words, and then, these words are segmented as individual characters. Character recognition of the individual character is performed, and output will be the text file containing the recognized text. Tesseract [15] is an open-source OCR engine that was developed at HP between 1984 and 1994. Tesseract was a PhD research project in HP Labs, Bristol, and became famous as commercial OCR engines due to the failure of previous OCR engines on best quality print. Tesseract work concentrated more on improving rejection efficiency than on base-level accuracy. Tesseract was sent for the Annual Text of OCR Accuracy at UNLV in 1995, where it overpowered different commercial OCR engines of that time. After which it was released as open-source software by HP in 2005.

3.1 Architecture of Tesseract OCR Engine

Tesseract [15] assumes binary image as input with polygonal text regions defined optionally. Initially, the outline of the components is extracted with connected

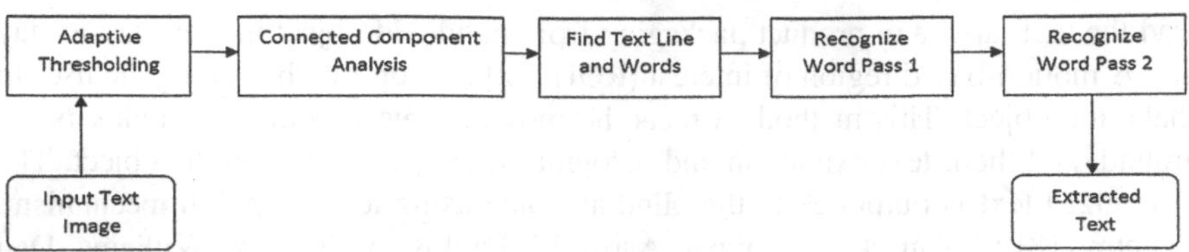

Fig. 2 Overall architecture of Tesseract OCR engine

component analysis. This was expensive step at that time, but it had the advantage of recognizing the inverse text easily like black-on-white text. At this stage, outlines are collected together, purely by nesting into Blobs. Blobs are organized into text lines, and the lines and regions are analyzed for fixed pitch or proportional text. On text lines, word-level segmentation is performed according to the type of character spacing. Fixed pitch text is broken immediately by character cells. Proportional text is broken into words using definite spaces and fuzzy spaces. Recognition is a two-pass process. In the first pass, each word was recognized. Satisfactory recognized words are passed to an adaptive classifier as training data. The adaptive classifier then gets a chance to more accurately recognize text lower down the page. Since the adaptive classifier may have learned few useful words too late to make a contribution near the top of the page, second pass is run over the page for better result, in which words that were not recognized well enough are recognized again. Figure 2 shows the overall architecture of Tesseract OCR engine.

4 Review of Text-to-Speech Engine

After the optical character recognitions of the image, we get the text file as output. Now, text has to be converted into speech or Braille so that the visually challenged person can understand it. Text-to-speech engine converts the text output from OCR engine into its corresponding speech output.

4.1 Architecture of Text-to-Speech Engine

Text-to-speech engine [22–24] works by initially performing the preprocessing on the text, and then, the speech generation is performed. The preprocessing module prepares the input text for further processing the operations such as text analysis and text normalization, and then, translating into a phonetic or some other linguistic representation is performed. In preprocessing, the spell checking is performed on the text based on the punctuation marks, and the formatting of the paragraph is done. In text normalization, the abbreviations and acronyms are handled. Text normalization enhances the speech output. In linguistic analysis, morphological

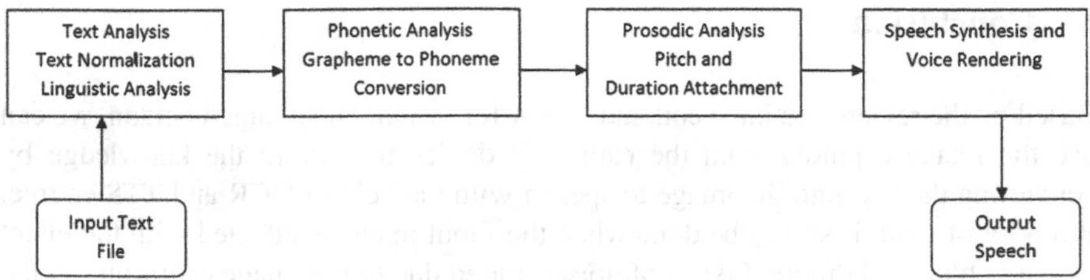

Fig. 3 Overall architecture of text-to-speech engine

operations for proper pronunciation of word and syntactic analysis to facilitate in handling ambiguities in written text are performed. Speech generation process consists of phonetic analysis which is used to find the phone level within the word. Each phone is tagged with the information about the sound to produce and how to be produced. This is followed by grapheme to phoneme conversion based on dictionary. Next step is prosody analysis which attaches the pitch and duration information. Then, after the speech synthesis that is rendering of voice, we get the speech out form the text-to-speech engine. Figure 3 shows the overall architecture of text-to-speech engine.

4.2 Various Text-to-Speech (TTS) Engine Developed

Considerable research has been done for the task of text-to-speech conversion. A screen reader in Indian languages has been designed by speech and Vision Lab in Language Technologies research center at IIIT Hyderabad named reading aid for visually impaired (RAVI) [25]. The screen reading software (RAVI) is an assistive technology to help visually impaired people to use or access the computer and Internet. RAVI uses text-to-speech technology to speak out the contents on the current screen (display) of a computer system. This is a PC-based software system. For wide adoption of the envisaged system, support must be provided for Indian languages. Due to the presence of large number of Indian languages and scripts, this task is going to be huge. Efforts have been made to convert Indian text into speech at various institutes, such as 'ACHARYA' at IIT Madras, TTS for multiple Indian languages at CDAC, KGP-Talk at IIT Kharagpur, and TTS for Oriya language at Utkal University. Today, many mobile phone companies provide text-to-speech facility. Google has introduced TTS for Android-based phones available in English, Spanish, French, and Italian languages. All these research identify host of research issues, and a new system can be build upon all this existing work to develop capability to generate speech from multiple Indian language texts.

5 Conclusion

Based on the review, we have concluded that for visually challenged person, we can use the image captured from the capturing device to acquire the knowledge by converting the text into the image to speech with the help of OCR and TTS engine. But a lot of work is still to be done when the input image is affected with the effect of skew, blur, and different types of noises added due to the image capturing on the move. In order to increase the efficiency of the system, we need to handle the following areas:

- Adaption of different text sizes and autofocus.
- Stabilization of page image against vibrations (in cases where the person is holding the book and reading).
- Elimination of the effect of skewed and blurred image.
- Elimination of effects of shadows and the contrast changes.
- Interpretation of the text in terms of subheadings, paragraphs, and emphasis on words or sentences through bold, italics, etc.
- Conversion of text to voice.
- Integrating the system into a workable and portable system.

References

1. World Blind Union www.worldblindunion.org.
2. Marion A. Hersh, Michael A. Johnson "Assistive technology for Visually Impaired and Blind people" Book by Springer ISBN 978-1-84628-866.
3. Roberto Netoa, Nuno Fonsecaa "Camera Reading For Blind People" HCIST 2014 International Conference on Health and Social Care Information Systems and Technologies, Procedia Technology 16 (2014) 1200–1209 published by ELSEVIER.
4. Zeng Fanfeng North China University Of Technology NCUT Beijing, China "Application research of Voice control in Reading Assistive device for visually impaired persons" 2010 International Conference on Multimedia Information Networking and Security, 2010 IEEE.
5. Maria M. Martins, Cristina P. Santos, Anselmo Frizera. Neto, Ramón Ceres "Assistive mobility devices focusing on Smart Walkers: Classification and review" Robotics and Autonomous Systems 2011 published by ELSEVIER.
6. Nihat Kocyigita, Pinar Sabuncu Artara "A challenge: Teaching English to visually-impaired learners" Procedia - Social and Behavioral Sciences 199 (2015) Elsevier.
7. Itunuoluwa Isewon, Jelili Oyelade, Olufunke Oladipupo "Design and Implementation of Text To Speech Conversion for Visually Impaired People" International Journal of Applied Information Systems Foundation of Computer Science FCS, New York, USA Volume 7– No. 2, April 2014.
8. Sukhpreet Singh "Optical Character Recognition Techniques: A Survey" CIS Journal of emerging trends in Computing and Information Science Vol 4, June 6, 2013.
9. Suhas R. Mache, Manasi R. Baheti, C. Namrata Mahender "Review on Text-To-Speech Synthesizer" International Journal of Advanced Research in Computer and Communication Engineering Vol. 4, Issue 8, August 2015.

10. Rainer Lienhart, Member, IEEE, and Axel Wernicke "Localizing and Segmenting Text in Images and Videos" IEEE TRANSACTIONS ON CIRCUITS AND SYSTEMS FOR VIDEO TECHNOLOGY, VOL. 12, NO. 4, APRIL 2002.
11. Honggang Zhang, Kaili Zhao,Yi-ZheSong, Jun Guo "A Text extraction from natural scene image: A survey" Neurocomputing 122 (2013) ELSEVIER.
12. Lukas Neumann and Jiri Matas "A method for text localization and recognition in real-world images" published at the 10th Asian Conference on Computer Vision, Queenstown, New Zealand.
13. Jian Liang, Daniel DeManthon "Geometric rectification of Camera-Captured Document Image" IEEE Transaction on Pattern Analysis and Machine Intelligence Vol 30 No 4 April 2008.
14. Akbar S. Shaik, G Hossain "Desgin and Development and performance Evaluation of Reconfigured Mobile Android Phone for People Who are blind or Visually Impaired" In Proceedings of the 28th ACM International Conference on Design of Communication" New York, NY, USA, 2010. ACM. October 2010.
15. Ray Smith Google Inc. theraysmith@gmail.com "An Overview of the Tesseract OCR Engine" Google Inc OSCON 2007.
16. Akshay Sharma, Abhishek Srivastava, Adhar Vashishth "An Assistive Reading System for Visually Impaired using OCR and TTS" International Journal of Computer Applications, Vol 95, No. 2 June 2014.
17. Chucai Yi, Yingli Tian and Aries Arditi "Portable Camera-Based Assistive Text and Product Label Reading from Hand Held Objects for Blind Persons" IEEE/ASME Transaction on Mechatronics.
18. Patrick E. Lanigan, Aaron M. Paulos, Andrew W. Williams, Dan Rossi and Priya Narasimhan "Trinetra: Assistive Technology for Grocery Shopping for the Blind" Published by Cylab in 2006.
19. U. Pal, B.B. Chaudhuri "Indian script character recognition: a survey" Pattern Recognition 37 (2004) 1887–1899 published by ELSEVIER.
20. Sangheeta Roy, Palaiahnakote Shivakumara, Partha Pratim Roy, Umapada Pal, Chew Lim Tan,Tong Lu "Bayesian classifier for multi-oriented video text recognition system" Expert Systems with Applications 42 (2015) ELSEVIER.
21. C. Gonzalez Richard E. Woods "Digital Image Processing" Book Third Edition Rafael Interactive Pearson International Edition prepared by Pearson Education PEARSON Prentice Hall.
22. Suhas R. Mache, Manasi R. Baheti, C. Namrata Mahender "Review on Text-To-Speech Synthesizer" International Journal of Advanced Research in Computer and Communication Engineering Vol. 4, Issue 8, August 2015.
23. Gore Megha B., Surwase S. V., Kailash. J. Karande "Technical Aspects in Development of Marathi Calculator using Text to Speech Synthesizer" 2014 IEEE Global Conference on Wireless Computing and Networking (GCWCN).
24. Kaveri Kamble, Ramesh Kagalkar "A Review: Translation of Text to Speech Conversion for Hindi Language" International Journal of Science and Research (IJSR) Volume 3 Issue 11, November 2014.
25. www.ravi.iiit.ac.in.
26. Bindu Philips and R.D. Sudhaker Samuel "Human Machine Interface – A smart OCR for the visually challenged", International Journal of Recent Trends in Engineering, Vol 2, No. 3 November 2009.

Performance Evaluation of the Controller in Software-Defined Networking

Suchismita Rout, Sudhansu Shekhar Patra and Bibhudatta Sahoo

Abstract Classical Internet architecture plays major obstacles in IPV6 deployment. There exits different reasons to extend classical approach to be more polished. Next generation of future Internet demands for routing not only within the same network domain but also outside the domain. Along with this, it offers many attributes such as network availability, end-to-end network connectivity, QoS management dynamically, and many more. The application area extends from small network size to big data center. To take in hand all these concerns, software-defined networking (SDN) has taken the major role in current situations. SDN separates the control plane and data plane to make the routing more versatile. We model the packet-in message processing of SDN controller as the queueing systems $M/M/1$.

1 Introduction

There is a tremendous evolution in the mechanism to exchange information over the past few years. This process is evolved significantly. The evolution of networks transits from traditional public switched telephone to network with cellular mobile and IP [1]. All mechanisms require a supporting infrastructure that allows sending and receiving information. For the traditional packet switched telephone network, Internet protocol provides the infrastructure. But it is very difficult to manage these IP networks [2]. In IP network, routing is the major role. The route is found to transfer

S. Rout (✉)
School of Computer Science and Engineering, KIIT University, Bhubaneswar, Odisha, India
e-mail: suchismita.rout28@gmail.com

S.S. Patra
School of Computer Application, KIIT University, Bhubaneswar, Odisha, India
e-mail: sudhanshupatra@gmail.com

B. Sahoo
Department of Computer Science and Engineering, National Institute of Technology, Rourkela, Odisha, India
e-mail: bibhudatta.sahoo@gmail.com

H.S. Behera and D.P. Mohapatra (eds.), *Computational Intelligence in Data Mining*, Advances in Intelligent Systems and Computing 556,
DOI 10.1007/978-981-10-3874-7_51

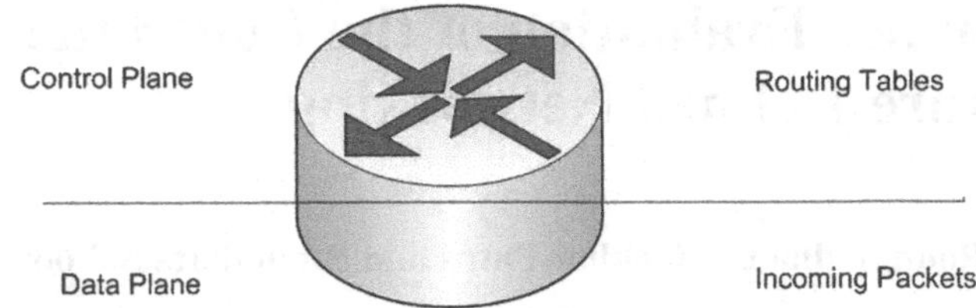

A router has a control plane that handles routing tables and data plane that handles incoming packets

Fig. 1 The routing tables are handled by control plane, and the incoming packets are handled by data plane in a router

packet from source to destination. It is called routing. The other nodes play the role of routers and gateways. The packet is transferred through a number of nodes to reach at the desired point. Different routing algorithms have been developed from past several years to barter information between faraway nodes. The hardware devices that are used to execute these routing algorithms are known as routers. The desktop computers, laptops, and PDAs are the example of routers. Each router makes it own decisions for routing as the routing mechanism is distributed. In order to centralize routing mechanism, a controller is added in the networks which will take the routing decision for the entire networks. With the introduction of central controller in network to manage all routing decision, the router is not concerned on complex and expensive computational algorithm. These routers which are controlled by the central controller are known as the programmable routers. This type of network is known as software-defined networking (SDN) [3].

In a network, router has to perform two main tasks. The former is to make the routing tables updated regularly with changing network topology. This is called control plane of routers. The latter is called data plane of router. It forwards the data packet in support of these tables. The control plane concerned with routing tables and data plane with forwarding the packets based on the information of the control plane. SDN is a novel networking archetype that is used to detach the two planes for network management simplification and enable improvement and advancement as given in Fig. 1 [4].

2 Architecture of SDN

SDN mainly supports two types of controller models: centralized and distributed. Each model has its different speciality to be considered based on the infrastructure and requirements. Finally, there is a hybrid controller model that combines the benefits of both these models. But in our analysis, we only consider the centralized model.

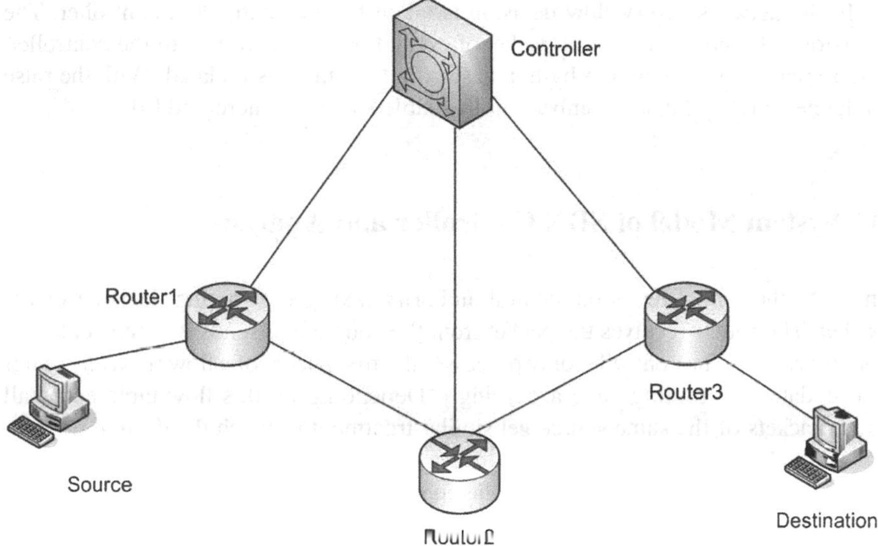

Fig. 2 An architecture of SDN

In centralized model, a single central controller has to manage and supervise the network. This architecture is officially supported by ONF. Network states and intelligence depend on a single decision point. Controller must be able to visualize the load of a switch globally across the routing path. All the routing-related information such as statistics, errors, and faults are collected and communicated to the controller with the help of OpenFlow. Some analytic algorithms can be used to detect switch overload and detect future load of the network [5].

In Fig. 2, source sends the packet to the router and then the router sends it to the controller. Then, the packet is processed, and the routing table of the router is updated. This revised routing table of the router is known as programming in SDN. The controller manages the flow table in such a manner that every single packet of the source gets analogous accomplishment in the network. The controller receives only the first packet and does the required processing for that packet one. The router front-wards every other packets depending on the instructions of the controller. In SDN, the router acts as a dumb device. Though the entire router has the flow tables, it has no ability to update it without the direction of the controller. One of the benefits of the SDN is that the routers are free from any computation-related tasks. All the routing decisions are taken by the controller centrally. However, this central controller has certain restrictions as limitations in terms of scalability and reliability issues. An alternate approach is to make control plane in a distributed fashion to execute applications of centralized controller. In a network, more than one controller can be present, so it is difficult to make the control action to be centrally controlled. But SDN can be centrally controlled. Such an architecture is called logically centralized [6–8].

In the network, every flow decision is taken by the centralized controller. The decision is based on the initial packet of every flow that is routed to the controller. After creating a path on hop-by-hop basis, the flow table is updated. With the raise in the network traffic, the numbers of flow table states are increased [9].

3 System Model of SDN Controller and Analysis

In SDN, the controller is the central authority taking the routing decision of the packet. The router receives the packet from the source and sends it to the controller for processing. The controller only processes the first packet of a flow from any router and updates the routing table accordingly. Depending on this flow table entry, all other packets of the same source get similar treatment to reach the desired destination. This flow rule is set by the controller, and the router has no ability to update them without any instructions from the controller. A router acts as a dumb device [10].

We represent the overall functionality into a system model and consider the system contains a single controller. The message arrival process of SDN controller is shown in Fig. 3. Due to abruptness and persistence of network traffic, the packet arrival process is not a Poisson stream, but the flow surfacing process conforms to the Poisson distribution. The Poisson process fits into the sequence of the first packets in all flows. In SDN architecture, the router sends the first packet of a flow to the controller conforming Poisson distribution and the flow of packets in controller is a hybrid of multiple Poisson streams.

Suppose SDN controller manipulates k routers. It receives a packet from the ith one with a Poisson distribution parameter λ. A queuing system based on this process is in state E_n at time t if the number of packets in the system is n if N(t) = n. Here, N(t) gives the length of packet-in message queue at a SDN controller in the interval [0, t], which is a birth and death process on the countable infinite state set E = $\{0, 1, 2, \ldots\}$. A birth occurs when a packet arrives from the source, and death occurs when packet reached the destination. Here, the steady-state solution for queuing model considered for birth rate is λ_n and death rate is μ_n. On arrival of each packet, the controller determines the respective flow rule by looking in its internal forwarding table. If there is no path, the controller finds its path and processes the packet. Therefore, the processing time of the packet in the controller satisfies the negative exponential distribution with parameter μ. In summary, we are using concise queuing model $M/M/1$ to characterize the packet-in message processing of a SDN controller.

This model assumes a random (Poisson) arrival pattern λ and random (exponential) service time distribution μ. The arrival rate does not depend on the number of packets in the system. For mathematical analysis, we use Markovian process to model the system behavior. The number of packets in an $M/M/1$ system forms a continuous-time Markovian chain (CTMC) where the state of the system corresponds to the number of packets in the system. The state transition diagram is illustrated in Fig. 4, where the state n denotes the number of packets in the controller.

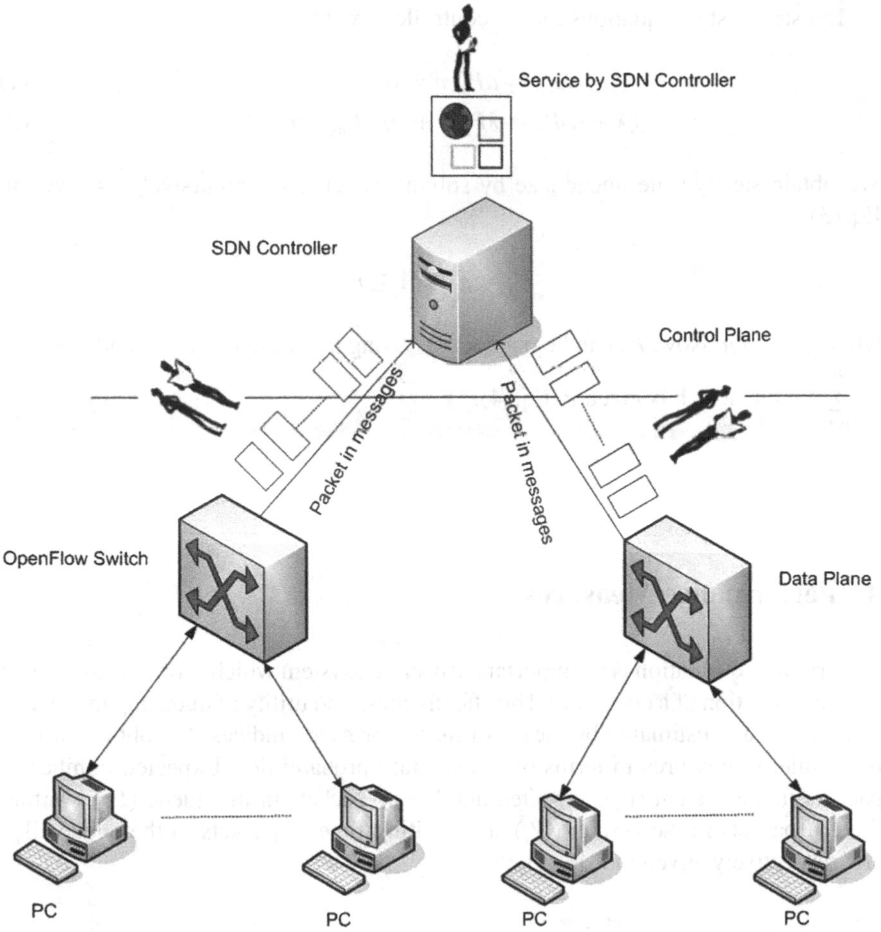

Service by SDN Controller

SDN Controller

Control Plane

Packet in messages

Packet in messages

OpenFlow Switch

Data Plane

PC

PC

PC

PC

Fig. 3 The message arrival process of SDN Controller

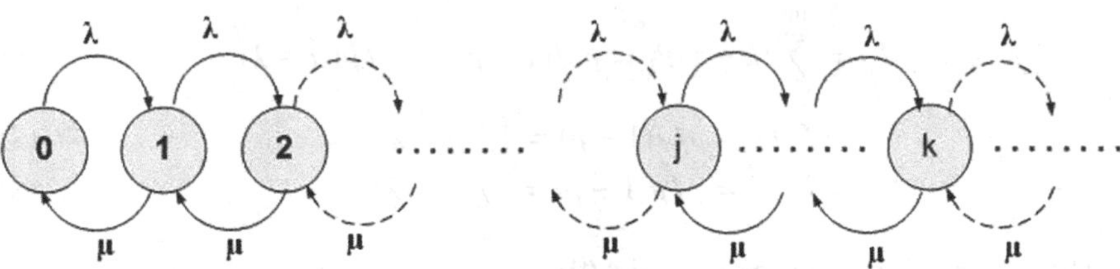

Fig. 4 State transition diagram for a $M/M/1$ Queue

The steady-state equations for the controller are given by

$$\lambda P_0 = \mu P_1, n = 0 \tag{1}$$

$$(\lambda + \mu)P_n = \lambda P_{n-1} + \mu_{j+1} P_{n+1}, n \geq 1 \tag{2}$$

We obtain steady-state queue size by solving Eqs. (1)–(2) recursively as given in Eq. (3).

$$P_n = \rho_1^n P_0, 1 \geq n \tag{3}$$

where $\rho = \lambda/\mu$. Now, P_0 can be obtained by using the normalization condition $\sum_{i=0}^{\text{inf}} P_i = 1$, which is given in Eq. (4).

$$P_0 = 1 - \rho. \tag{4}$$

4 Performance Measures

Performance evaluation is an important aspect of a system which is of crucial interest for the evaluation of a controller. The effectiveness and utility of queueing model can be depicted and estimated by means of its performance indices. We obtain various performance measures in terms of steady-state probabilities. Expected number of packets in the system (L_s), expected number of packets in the queue (L_q), waiting time of packets in the system (W_s), and waiting time of packets in the queue (W_q) are, respectively, given in Eqs. (5)–(8).

$$L_s = \sum_{k=0}^{\text{inf}} kP_k = \rho/(1-\rho) = \lambda/(\mu - \lambda) \tag{5}$$

$$L_q = \sum_{k=1}^{\text{inf}} (k-1)P_k = \rho^2/(1-\rho) = \lambda^2/\mu(\mu - \lambda) \tag{6}$$

$$W_s = L_s/\lambda = \rho/\lambda(1-\rho) = 1/(\mu - \lambda) \tag{7}$$

$$W_q = L_q/\lambda = \rho/\mu(1-\rho) = \rho/(\mu - \lambda) \tag{8}$$

The server utilization is given in Eq. (9).

$$U_s = 1 - P_0 = \lambda/\mu = \rho \tag{9}$$

5 Numerical Illustrations

In this section, we illustrate the numerical tractability that shed light on the performance aspect of controller which is of crucial interest for the SDN. Figure 5 graphically shows the relationship between the average system size L_s and the queue utilization ρ. As seen, L_s is an increasing function of ρ. For low values of ρ, L_s increases somewhat slowly, and then, as ρ gets closer to 1, L_s increases very rapidly, eventually growing to an infinite limit as $\rho -> 1$. The same situation is also for the average queue size. It is depicted in Fig. 6. The impact of λ on W_s and W_q is shown in Figs. 7 and 8, respectively, for different μ. It has been observed that W_s increases as λ increases and finally reaches to infinite. The same is the scenario for W_q. Figure 9 shows the impact of ρ on server utilization U_s.

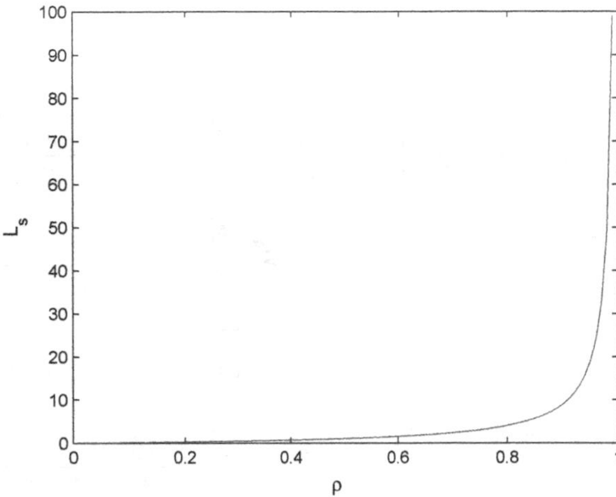

Fig. 5 Impact of ρ on L_s

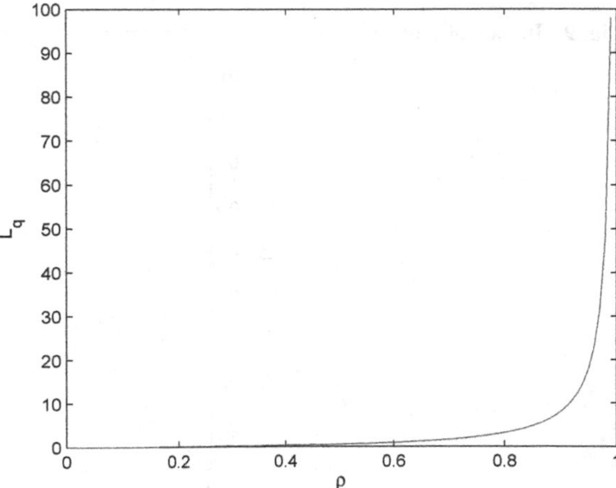

Fig. 6 Impact of ρ on L_q

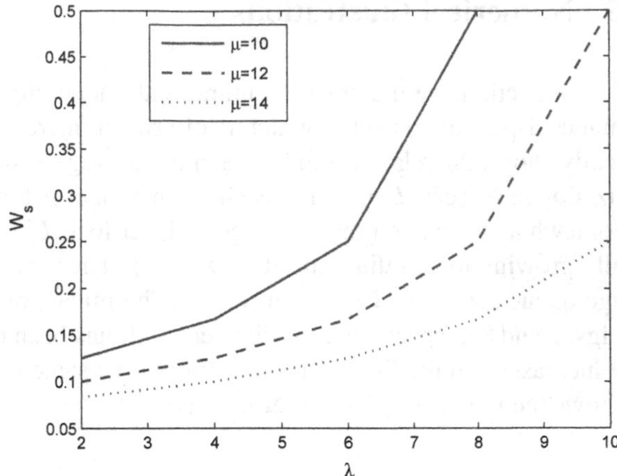

Fig. 7 Impact of λ on W_s

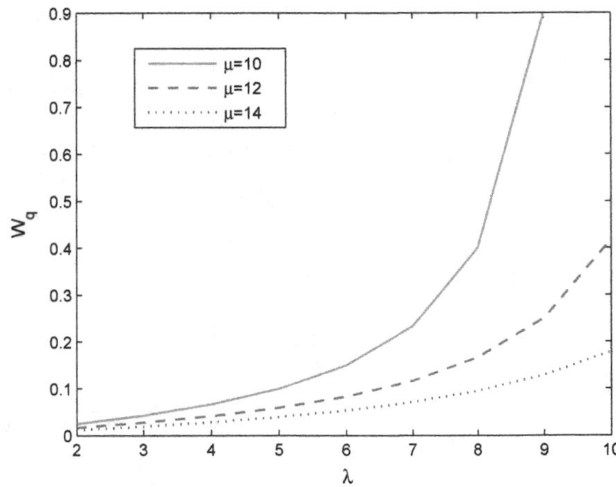

Fig. 8 Impact of λ on W_q

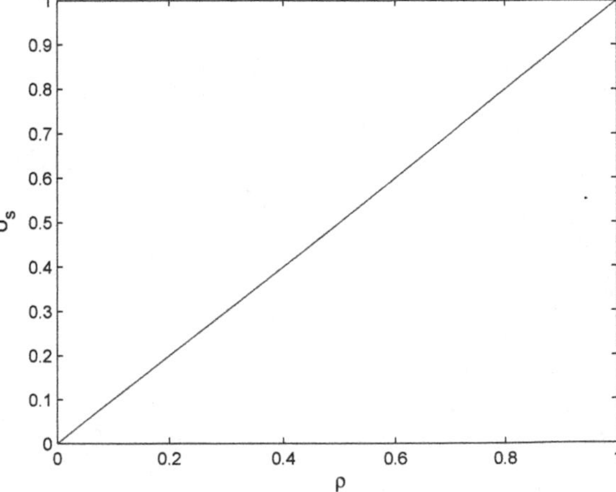

Fig. 9 Impact of ρ on U_s

6 Conclusion

Software-defined networking is a pioneering trend in communication networks. Many researchers focus on the SDN architecture as it logically divides the data and the control plane of a router. Thus, this new technology is redesigning the architecture of traditional network by breaching the rigid assimilation of two planes. In SDN, all the routing decisions are taken by the controller. Various factors are involved in taking the decisions regarding the number of controllers and their replacements. An analytical Markovian model was developed and steady-state system size distribution obtained using recursive method. Various performance measures of the system indicate the performance of the SDN controller.

References

1. Hasan, Syed Faraz. Emerging trends in communication networks. Springer, 2014.
2. Benson, Theophilus, Aditya Akella, and David A. Maltz. "Unraveling the Complexity of Network Management." In NSDI, pp. 335–348. 2009.
3. Hu, Fei, Qi Hao, and Ke Bao. "A survey on software-defined network and openflow: from concept to implementation." IEEE Communications Surveys and Tutorials 16, no. 4, pp. 2181–2206, 2014.
4. McKeown, Nick, Tom Anderson, Hari Balakrishnan, Guru Parulkar, Larry Peterson, Jennifer Rexford, Scott Shenker, and Jonathan Turner. "OpenFlow: enabling innovation in campus networks." ACM SIGCOMM Computer Communication Review 38, no. 2, pp. 69–74, 2008.
5. Hakiri, Akram, Aniruddha Gokhale, Pascal Berthou, Douglas C. Schmidt, and Thierry Gayraud. "Software-defined networking: Challenges and research opportunities for future internet." Computer Networks 75, pp. 453–471, 2014.
6. Li, Chung-Sheng, and Wanjiun Liao. "Software defined networks." IEEE Communications Magazine 51, no. 2 (2013): 113–114.
7. Casado, Martin, Teemu Koponen, Scott Shenker, and Amin Tootoonchian. "Fabric: a retrospective on evolving SDN." In Proceedings of the first workshop on Hot topics in software defined networks, pp. 85–90. ACM, 2012.
8. Kanaumi, Yoshihiko, Shuichi Saito, and Eiji Kawai. "Toward large-scale programmable networks: Lessons learned through the operation and management of a wide-area openflow-based network." In 2010 International Conference on Network and Service Management, pp. 330–333. IEEE, 2010.
9. Raghavendra, Ramya, Jorge Lobo, and Kang-Won Lee. "Dynamic graph query primitives for sdn-based cloudnetwork management." In Proceedings of the first workshop on Hot topics in software defined networks, pp. 97–102. ACM, 2012.
10. Levin, Dan, Andreas Wundsam, Brandon Heller, Nikhil Handigol, and Anja Feldmann. "Logically centralized?: state distribution trade-offs in software defined networks." In Proceedings of the first workshop on Hot topics in software defined networks, pp. 1–6. ACM, 2012.

Task-Scheduling Algorithms in Cloud Environment

Preeta Sarkhel, Himansu Das and Lalit K. Vashishtha

Abstract Cloud computing has increased its popularity due to which it is been used in various sectors. Now it has come to light and is in demand because of amelioration in technology. Many applications are submitted to the data centers, and services are given as pay-per-use basis. As there is an increase in the client demands, the workload is increased, and as there are limited resources, workload is moved to different data centers in order to handle the client demands on as-you-pay basis. Hence, scheduling the increasing demand of workload in the cloud environments is highly necessary. In this paper, we propose three different task-scheduling algorithms such as Minimum-Level Priority Queue (MLPQ), MIN-Median, Mean-MIN-MAX which aims to minimize the makespan with maximum utilization of cloud. The results of our proposed algorithms are also compared with some existing algorithms such as Cloud List Scheduling (CLS) and Minimum Completion Cloud (MCC) Scheduling.

Keywords Cloud computing · Task scheduling · Makespan · Cloud utilization

1 Introduction

In the era of modern technological advancement, cloud computing is an emerging platform to deliver high-scale services to all the users [1]. The cloud resources are provided to the end user on the pay-per-use basis. In a cloud environment, usually, there are many modes to rent all the resources from a cloud provider in different

P. Sarkhel (✉) · H. Das · L.K. Vashishtha
School of Computer Engineering, KIIT University, Bhubaneswar, India
e-mail: preetasarkhel@gmail.com

H. Das
e-mail: das.himansu2007@gmail.com

L.K. Vashishtha
e-mail: lalitkvashishtha@gmail.com

H.S. Behera and D.P. Mohapatra (eds.), *Computational Intelligence in Data Mining*, Advances in Intelligent Systems and Computing 556,
DOI 10.1007/978-981-10-3874-7_52

ways such as advanced reservation (AR), best effort (BE), immediate, deadline sensitive (DS). In advanced research, immediate mode resources are non-preemptable, whereas in BE and DS mode resources are preemptable. The two algorithms for federated IaaS cloud system proposed by Li et al. [2] are Cloud Min-Min Scheduling (CMMS) and Cloud List Scheduling (CLS). In these algorithms, the authors have solved the task-scheduling problem for the heterogeneous cloud environment. In a heterogeneous environment, all the clouds have different properties and are independent of each other. A directed acyclic graph (DAG) is used to represent the tasks in an application. Tasks are independent of each other and have different arrival time. The algorithms such as Median Max Scheduling (MEMAX), Minimum Completion Cloud Scheduling (MCC), Cloud Min-Max Normalization Scheduling (CMMN) are proposed by Panda and Jana [3]. The main goal of this algorithm is to improve the makespan. Virtualization is used to allocate resources based on their needs and supports green computing. Skewness concept is introduced to measure the variable utilization of the server. Overload avoidance is achieved to lead good performance. Here, a hotspot and cold spot are used to explain green computing algorithm. The contributions made by the author are to avoid the overload in large scale [4]. The main objectives of the paper were overload avoidance and green computing [5]. O.M. Elzeki et al. [6] had proposed an improved version on Max-Min algorithm and had discussed an algorithm based on RASA algorithm. Improved Max-Min algorithm is based on the expected execution time instead of completion time on the basis of selection. This algorithm achieves lower makespan than RASA and original Max-min algorithm. We have addressed three different task scheduling problem such as Minimum-Level Priority Queue (MLPQ), MIN-Median (MINM), and Mean-MIN-MAX using DAG for a cloud environment and have proposed three different algorithms to solve the problem. The experimental results show that its performance is better than other proposed algorithm in terms of makespan.

The rest of the paper is organized as follows: The system model and the task-scheduling problem are considered in Sect. 2. The proposed algorithms are explained by a working example in Sect. 3. The experimental results are shown in Sect. 4. Finally, we conclude this paper in Sect. 5.

2 System Model and Problem Definition

Consider clouds having different datacenters. These clouds are used to provide services. These services are made available by setting up VMs in the datacenters. So these VMs are the cloud that is formed in the datacenters [2], and these VMs have different characteristics. There is a manager server that keeps on tracking the current status of all VMs present in the datacenter. A manager server communicates with other manager servers when an application arrives to distribute the tasks.

Note that while distributing the tasks, the manager server must be well informed of the resource availabilities in other cloud systems. The VMs and the resources associated with it are set free once the execution is completed.

2.1 *Problem Definition*

A number of requests arrive in cloud ever time. The requests may be of an application or a service. These applications and services are composed of several tasks. Tasks can be of two types—independent and dependent. Dependency among tasks is represented in the form of directed acyclic graph (DAG). The tasks are allocated on virtual machines residing on hosts, which is called scheduling. This is also a NP-complete problem. Scheduling of all the tasks in a huge environment such as cloud is very complex and critical. Most of the complex problems [7–10] are solved by near optimal solutions. The data transferred between such large applications are very huge and massive. So, the cost produced will also be high. Therefore, the cost needs to be minimized by minimizing the scheduling length. We have proposed some algorithms based on different parameters.

2.2 *System Model*

There are 'N' of clouds $C = \{C1, C2, C3, \ldots, C_N\}$ and 'M' of applications in the form of $A = \{A1, A2, A3, \ldots, A_M\}$, where each application is the set of task T with precedence among the tasks. The applications are represented in the form of directed acyclic graph (DAG) $G = (V_i, E_i)$, where V_i is the set of vertices (nodes) and E_i is the set of edges (links). The node represents a task, and each edge represents the dependency among the tasks. The task B can only be scheduled after A for execution if task A has successfully completed its execution and has released all the resources associated with it. The arrival time of each application is same. The main goal is to assign applications to clouds in such a way that maximum utilization of cloud is achieved with minimum makespan.

3 Proposed Algorithms

3.1 *Minimum-Level Priority Queue (MLPQ)*

The objective of this algorithm is to minimize the makespan with maximum utilization of cloud. The basic concept of this algorithm is as follows: At first, all the tasks that are ready for execution are added to the queue q. Once the tasks are added to the queue, we can now consider these sorted tasks for their allocation of resources and we can allocate a resource to the task which has minimum execution time and will be allocated to that cloud which can finish it at an earlier time. The task being assigned starts its execution if and only all its predecessor tasks are completed and the cloud that is being allocated is available at that time. Once the tasks have been assigned, it is removed from the queue and its child nodes are added to the queue. The process repeats until the queue is empty. We implement MLPQ algorithm in cloud which is not available in the literature. The algorithm is shown in Fig. 1. The DAG representation is given in Fig. 4 and its execution time of the tasks of the different clouds are given in Table 1. The approach is described as follows: The tasks that are ready for execution are A, J, and M, and they take minimum execution time of 3, 2, 2. The task J has minimum execution time in cloud 4, i.e., 2 unit of time. So, J is assigned first to cloud 4. Now task J is removed from the queue, and all its child nodes are added to the queue and the queue is updated. Next, the queue has tasks A, M, I, and H. All the tasks present in the updated queue are again compared among each other with their minimum execution time. The procedure is repeated until the queue is empty. The total makespan of the cloud is 13 units of time.

Algorithm: Minimum Level Priority Queue (MLPQ)

Input: A DAG with different tasks, n different clouds, Execution time.

Output: An assignment of the tasks to the cloud such that the overall processing time (makespan) is minimized.

Step 1: The tasks that are ready for execution are added to the minimum priority queue Q.

Step 2:

for i: task V_i € Q

2.1: Find the minimum execution time for each task.

2.2: Select the task which have the minimum execution time.

2.3: Find the cloud C_{min} having the minimum execution time.

2.4: Assign the task to cloud C_{min}.

end

Step 3: Remove the task from queue Q.

Step 4: Add the successor of that task to Q.

Step 5: Repeat step 2-4, until the queue is empty.

Step 6: **Stop.**

Fig. 1 MLPQ algorithm

Table 1 Execution time of 15 tasks in 4 clouds

Tasks	A	B	C	D	E	G	H	I	J	K	L	M	N	O	P
Cloud 1	9	3	4	5	6	9	5	2	8	6	10	2	9	11	6
Cloud 2	3	6	8	6	5	6	7	4	4	3	9	3	10	3	8
Cloud 3	5	7	9	2	2	8	6	6	12	7	4	4	2	8	7
Cloud 4	6	2	10	4	3	3	2	3	2	4	11	12	8	10	4

Algorithm: MIN-Median (MINM)

Input: A DAG with different tasks, n different clouds, Execution time.

Output: An assignment of the tasks to the cloud such that the overall processing time (makespan) is minimized.

Step 1: The tasks that are ready for execution are added to the list R.

Step 2:

for i: task Vi € R

2.1: Calculate the medians of execution time of the tasks.

2.2: Select the task which have the minimum median value.

2.3: Find the cloud C_{min} having the minimum execution time.

2.4: Assign the task to cloud C_{min}.

end

Step 3: Remove the task from list R

Step 4: Add the successor of that task to R.

Step 5: Repeat step 2-4, until the list is empty.

Step 6: Stop.

Fig. 2 MINM algorithm

3.2 *MIN-Median (MINM)*

The MIN-Median algorithm is designed to minimize the makespan. The basic concept of the algorithm is as follows: First, it calculates the median of the execution time of the all the tasks present in the list named R that are ready for execution over all the clouds. Then the task with minimum median value is selected and the task is assigned to the cloud having the minimum execution time. The algorithm is shown in Fig. 2. The DAG representation is given in Fig. 4 and its execution time of the tasks of the different clouds are given in Table 1. The ready tasks are A, J, and M which have the minimum median values as 5.5, 6, and 3.5. So the task M has a minimum median value of 3.5 with minimum execution time of 2 units of time, and it is allocated to cloud 1 as minimum completion time is achieved here. Then, the

Table 2 Scheduling sequence in minimum-level priority queue

Time	0–2	2–4	4–8		
Cloud 1	M	I	C		
Time	0–3	3–6	6–9		
Cloud 2	A	O	K		
Time	0–2	2–4	4–6	6–8	8–12
Cloud 3		N	D	E	L
Time	0–2	2–4	4–6	6–9	9–13
Cloud 4	J	H	B	G	P

child nodes N and O of the tasks are ready for execution. The median values of the tasks over the clouds are 8.5 and 9. So the tasks now present in the list are A, J, N, and O. Task A has minimum value so it is allocated to minimum completion time cloud 2. The process is continued until the list is empty. The processing time/makespan is 16 units.

3.3 Mean-MIN-MAX

In this algorithm, we proposed a method in which the mean of the execution time of the ready tasks are calculated. Then, it selects the tasks which have minimum mean value and allocates it to the cloud which has the maximum execution time. The algorithm is shown in Fig. 3. The DAG representation is given in Fig. 4 and its execution time of the tasks of the different clouds are given in Table 1. The tasks which are ready for execution are A, J, and M which have the minimum mean values as 5.75, 6.5, and 5.25. So the task M has a minimum mean value of 5.25 with minimum execution time of 2 units and it is allocated to cloud 4 as maximum completion time is achieved here. Now, the child nodes N and O of the task are ready for execution and their mean values are 7.25 and 4.5, respectively. So the tasks now present in the list are A, J, N, and O. Task O has minimum mean value with minimum execution time of 3 units, and it is allocated to the cloud 4 because maximum completion time is achieved here. Similarly, the process continues until the list is empty. The makespan is 15 units.

3.4 Case Study

Consider the DAGs as shown in Fig. 4. There are three applications $A1 = (T1, E1)$, $A2 = (T2, E2)$, and $A3 = (T3, E3)$, where $A1 = (T1 = \{A, B, C, D, E, G\}$ and $E1 = \{AB, AC, AD, AE, BG\})$, $A2 = (T2 = \{J, I, H, K, L\}$ and $E2 = \{JI, JH, JK, IL\})$, and $A3 = (T3 = \{M, N, O, P\}$ and $E3 = \{MN, MO, NP\})$. Suppose the arrival time of the $A1$, $A2$, and $A3$ is 0. The applications are listed into four different clouds.

Algorithm: Mean-MIN-MAX (MMIMX)

Input: A DAG with different tasks, n different clouds, Execution time.
Output: An assignment of the tasks to the cloud such that the overall processing time (makespan) is minimized.

Step 1: The tasks that are ready for execution are added to the list R.
Step 2:
  for i: task $V_i \in R$
  2.1: Calculate the mean of the execution time of the tasks.
  2.2: Select the task which have the minimum mean value.
  2.3: Find the minimum execution time of the task.
  2.4: Assign it the cloud C_{max}.
  end
Step 3: Remove the task from list R
Step 4: Add the successor of that task to R.
Step 5: Repeat step 2-4, until the list is empty.
Step 6: Stop.

Fig. 3 MMIMX algorithm

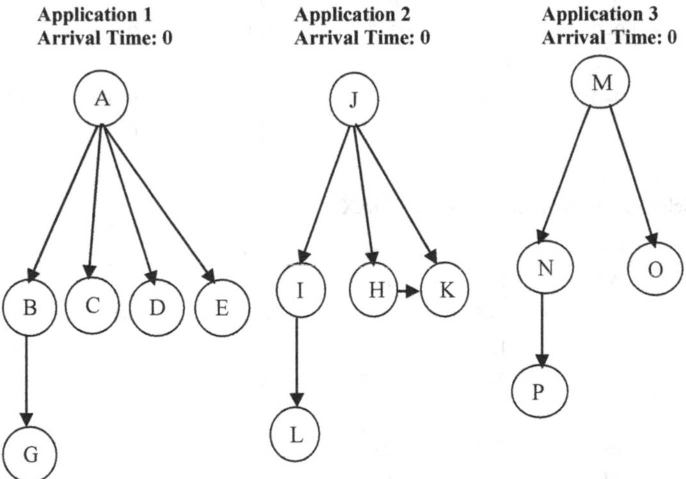

Fig. 4 DAG of three applications

Task dependency is maintained in this problem. Table 1 shows the execution time of each and every task on each and every cloud environments. Note that in each cloud, the execution time of the task is different.

4 Experimental Results

In order to implement the proposed approach, we have used the experiment on the system which is having Intel(R) core(TM) i3 Processor 1.80 GHz, Windows 8 platform using Dev-C++5.11 IDE as a development kit. We compare our proposed algorithm with Cloud List Scheduling (CLS) and Minimum Completion Cloud Scheduling (MCC) [2, 3]. Table 2 shows the scheduling sequence of our proposed algorithm MLPQ, Table 3 shows the scheduling sequence in MINM, and Table 4 shows the scheduling sequence of Mean-MIN-MAX. The task assignment in MLPQ, MINM, and MMIMX is shown in Tables 3, 4, and 5, respectively. The comparison of algorithms based on cloud makespan is shown in the Fig. 5. From the figures, we can observe that our proposed algorithm has the minimum makespan and gives better performance compared to MCC and CLS (Tables 6 and 7).

Table 3 Scheduling sequence in MIN-Median

Time	0–2	2–7	7–9	9–13		
Cloud 1	M		I	C		
Time	0–3	3–9	9–12	12–15		
Cloud 2	A		K	O		
Time	0–3	3–5	5–7	7–9	9–13	
Cloud 3		E	D	N	L	
Time	0–3	3–5	5–7	7–9	9–12	12–16
Cloud 4		B	J	H	G	P

Table 4 Scheduling sequence Mean-MIN-MAX

Time	0–3	3–5	5–8	8–11		
Cloud 1	A	E	G	O		
Time	0–3	3–5	5–7	7–9	9–11	11–15
Cloud 2		D		H	N	P
Time	0–3	3–5	5–7	7–9	9–12	
Cloud 3		B	J	I	K	
Time	0–2	2–3	3–7	7–9	9–13	
Cloud 4	M		C		L	

Table 5 Task assignments in minimum-level priority queue

Tasks	A	B	C	D	E	G	H	I	J	K	L	M	N	O	P
Clouds	2	4	1	3	3	4	4	1	4	2	3	1	3	2	4

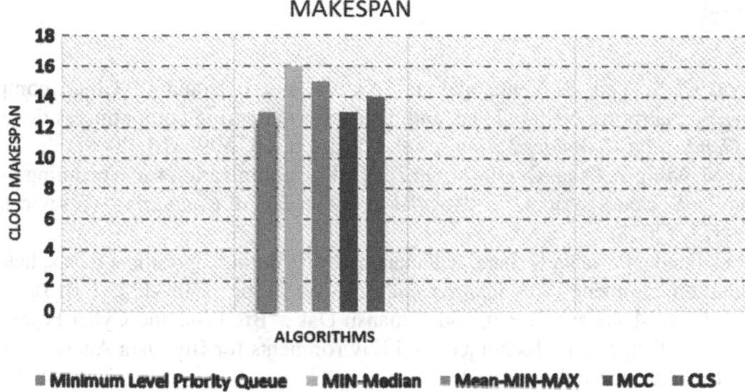

Fig. 5 Comparison chart of cloud makespan

Table 6 Task assignments in MIN-Median

Tasks	A	B	C	D	E	G	H	I	J	K	L	M	N	O	P
Clouds	2	4	1	3	3	4	4	1	4	2	3	1	3	2	4

Table 7 Task assignments in Mean-MIN-MAX

Tasks	A	B	C	D	E	G	H	I	J	K	L	M	N	O	P
Clouds	1	3	4	2	1	1	2	3	3	3	4	4	2	1	2

5 Conclusion and Future Work

We have studied about different cloud computing environment and about the task-scheduling problem in a cloud environment. In a cloud environment, a large number of applications are submitted regularly to the cloud data centers to obtain various services on a pay-per-use basis. But due to limited number of resources in cloud, data centers are heavily loaded with tasks; some workloads of overloaded cloud are transferred to other loaded cloud data centers to handle the end-user demands. Therefore, scheduling is important. We have proposed an algorithm in which all the applications have the same arrival time and we have used a DAG to represent it. Here, we have implemented the proposed algorithm and have found that it shows a better result than other existing algorithms such as Minimum Completion Cloud Scheduling (MCC) and Cloud List Scheduling (CLS). The comparison results based on makespan and utilization shows that the proposed algorithms provides better results than the existing algorithms in terms of cloud makespan. In future, we would try to implement the proposed algorithm considering the other parameters such as energy consumption and fault tolerance.

References

1. R. Buyya, C. S. Yeo, S. Venugopal, J. Broberg, and I. Brandic, "Cloud computing and emerging IT platforms: Vision, hype, and reality for delivering computing as the 5th utility," *Future Generation Computer Systems*, vol. 25, no. 6, pp. 599–616, 2009.
2. Li J, Qiu M, Ming Z, Quan G, Qin X, Gu Z, "Online optimization for scheduling preemptable tasks on IaaS cloud system," *J Parallel & Distributed Computing* (Elsevier), Vol. 72, pp. 666–677, (2012).
3. Sanjaya K. Panda, Prasanta K. Jana, "Efficient task scheduling algorithms for the heterogeneous multi-cloud environment," *J of Supercomputing,* Vol. 71, pp. 1505–1533 (2015).
4. Panigrahi, C R, M Tiwary, B Pati, and Himansu Das., "Big Data and Cyber Foraging: Future Scope and Challenges." In Techniques and Environments for Big Data Analysis, pp. 75–100. Springer International Publishing, 2016.
5. L. Dhivya, Ms. K. Padmaveni, "Dynamic Resource Allocation Using Virtual Machines for Cloud Computing Environment", *IJREAT International Journal of Research in Engineering Advanced Technology*, Vol. 2, No.1, pp. 1-4, (2014).
6. O M Elzeki, M Z Reshad and M A Elsoud, ''Improved Max-Min Algorithm in Cloud Computing,'' *International Journal of Computer Applications* Vol.50, pp. 22–27, (2012).
7. Das, Himansu, D.S. Roy, "A Grid Computing Service for Power System Monitoring", in *International Journal of Computer Applications* (IJCA), 2013, Vol. 62 No. 20, pp 1–7.
8. Das, Himansu, D.S. Roy, "The Topological Structure of the Odisha Power Grid: A Complex Network Analysis", in *International Journal of Mechanical Engineering and Computer Applications (IJMCA)*, 2013, Vol.1 Issue 1, pp 12–18.
9. Das, Himansu, A K Jena, P K Rath, B Muduli, S R Das, "Grid Computing Based Performance Analysis of Power System: A Graph Theoretic Approach", in *Advances in Intelligent Systems and Computing*, Springer India, 2014, pp. 259–266.
10. Das, Himansu, G S Panda, B Muduli, and P K Rath. "The Complex Network Analysis of Power Grid: A Case Study of the West Bengal Power Network." In *Intelligent Computing, Networking, and Informatics*, Springer India, 2014, pp. 17–29.

Skin-Colored Gesture Recognition and Support Vector Machines-Based Classification of Light Sources by Their Illumination Properties

Shreyasi Bandyopadhyay, Sabarna Choudhury, Riya Ghosh, Saptam Santra and Rabindranath Ghosh

Abstract The illumination characteristics of light sources can determine whether the light source is a normal or a faulty one. The proposed work is based on moving a platform containing a light source in both horizontal and vertical directions by gesture recognition. The gesture recognition done by Fuzzy C means and snake algorithm-based skin color detection makes the recognition more accurate. The illumination values of the light source are obtained by a webcam. The set of data helps in classification of the state of an unknown light source (normal or faulty) by support vector machines with radial basis function as kernel with a yield of an error rate of about 0.6% marking the efficacy of the system and making the system a novel and sophisticated one.

Keywords Illumination characteristics · Gesture recognition · Skin color detection · Fuzzy C means · Snake algorithm · Support vector machines · Radial basis function

---

S. Bandyopadhyay (✉) · S. Choudhury · R. Ghosh · S. Santra · R. Ghosh
Electronics and Communication Department, St. Thomas' College of Engineering and Technology, Kolkata, India
e-mail: s.bando.93@gmail.com

S. Choudhury
e-mail: sabarna.choudhury@gmail.com

R. Ghosh
e-mail: ghriya4@gmail.com

S. Santra
e-mail: saptamsantraece@gmail.com

R. Ghosh
e-mail: rnghosh@gmail.com

H.S. Behera and D.P. Mohapatra (eds.), *Computational Intelligence in Data Mining*, Advances in Intelligent Systems and Computing 556,
DOI 10.1007/978-981-10-3874-7_53

1 Introduction

A number of light sources such as incandescent, compact fluorescent lamp, LED, high-intensity discharge, neon lights, lasers, and so on have inundated our workplace as well as our homes. The different light sources have different illumination parameters such as brightness (measured in lux), luminous flux (measured in lumen), color temperatures, and luminous intensity (measured in candela). Lux meters used for measuring brightness or other spectrometers are often error-prone and so integrating sphere method or goniophotometer method [1, 2] is recommended. It is basically a closed sphere method where a light source is placed at the center of the sphere. The spectral response of this photodetector provides total luminous response of the light source under test and the method provides a relative measurement. The methods are time-consuming and costly. Therefore, there is a scope for a simpler and low-cost alternative for these tests. A similar semi-automated test setup can be made; however, using few computer attachments, a dedicated algorithm is introduced in this paper. The prototype of intensity profile measurement system presented here works on the basis of capturing the images of the bulb at different orientation with a simple webcam connected to a computer and creating a polar profile of intensity in azimuth (φ) and elevation (θ). The orientations to azimuth and elevation are controlled by gesture-recognition principles, initially posed before the webcam. In other words, rotations of the bulbs about axes can be sequenced according to the user and only one angular motion can simply be performed. In many cases, the azimuth profile only may be necessary to pass the sources under consideration, as found out from data sheet of the light source [3]. The system reduces the duration of previous tests and makes it automated yet retaining the rigor of the illumination engineering fundamentals. The inclusion of gesture-recognition principles to this setup is used to ease the operations and to reduce duration of tasks. Posture-based approaches, for example, utilize static images, concentrating only on the shape of the hand to extract features such as hand contours, fingertips, and finger directions [4–7]. The proposed system creates a virtual interaction between human and computer without any manual device, where the recognized gestures can be used to control a system or convey meaningful information. The skin color can also be detected by various algorithms [8, 9]. The real time Fuzzy C means and snake algorithm, described later, are used to provide the gestures to the proposed system.

The outline of the paper is as follows: Section 2 deals with the experimental setup and Sect. 3 illustrates the gesture-recognition process. Section 4 gives an idea of the proposed methodology while Sect. 5 shows the classification by SVM. The results and conclusion are discussed in Sect. 6.

2 Experimental Setup

The block diagram of the proposed setup is as shown in Fig. 1. The designed platform is capable of moving in both horizontal and vertical directions. The system is operated by using a laptop or a computer. Two webcams are needed, one for capturing and analyzing the illumination values of the bulb and the other for processing the gesture recognition. Moreover, the setup is painted black to avoid the reflection. From the images captured, the intensity profile and the polar diagram can be obtained as a result. The alignment of the bulb with webcam should be done with utmost care to obtain the intensity profile at exact values of angle [10].

Four different types of bulb are demonstrated in the context for complete understanding of the different illumination patterns. They are incandescent, compact fluorescent lamp (CFL), and green and white light emitting devices (LED). Here, LED bulbs are taken into consideration. The algorithm after extracting the images in RGB mode processes them to obtain the Y-component (commonly termed as luma) from the YCbCr mode as it provides much better compatibility in real-life situations. Y-component can be directly correlated with brightness, and since the images are obtained at a perfect perpendicular direction, the Y-component can be comparable to luminous intensity I [10].

3 Gesture Recognition

Gesture recognition, one of the significant tools of automation plays a vital role in developing the sense of automation in this work. Various posture movements, color-based recognition are some of the types of gesture recognition. Movement of postures using certain colored caps and recognizing them make the process a simpler one, though it is not a generalized case and may vary a lot from place to place, from time to time and is also region based. However, it solves the purpose. Here, we have developed a sophisticated system of gesture recognition to exploit

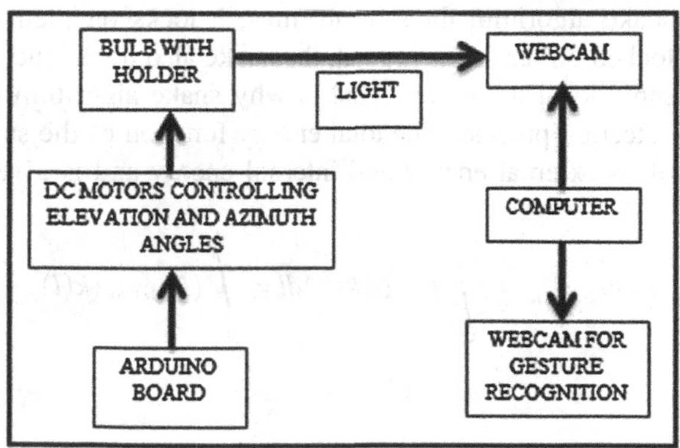

Fig. 1 Block diagram of the proposed system

the skin color detection rather than using and depending on colored stuffs. This sophisticated system is comprised of a method which computes, compares, and draws the optimal detection of gesture simultaneously, which makes the proposed system of detection a novel one. This method is based on Fuzzy C Means and Snake algorithm, which have been discussed in the following sections.

3.1 Skin Color Detection

Skin color detection using Fuzzy C Means has been performed. The following algorithm illustrates the computation of the clusters of Fuzzy C Means and extracting the skin color out of it.

1. A video is started.
2. A snapshot of an image containing the hand is obtained.
3. Fuzzy C means segmentation technique in that image is performed.
4. Morphological techniques are used to remove the noise and other connected components that are much smaller in area.
5. Hand (skin color) is detected.
6. Go back to step 2. This operation is repeated till the video is on.

Here, we have taken the exponential factor (m) to be as 2 and the change of error to be 0.001. The maximum number of iterations is taken to be 300, and number of clusters is taken as 6. The results so obtained contain the extracted skin color along with some background-detected portion giving the same response as that of the skin. To get hold of an accurate estimation of the skin-colored region, we pass this result through a snake algorithm. Here, we are not concerned about the accuracy of detection of skin, rather we are concerned about getting the best possible approximation, which eliminates background noise and shows the presence or absence of skin color based on which the entire system works. The choice of snake algorithm is based on the relevance of using it in this particular system development. Some of the principal advantages of snake algorithm are that the snakes are actively used in tracking the dynamic objects, which is the movement of hand in this case. Once the snake algorithm finds its features, it locks on them spontaneously and when the locked on feature is moved, the snake also tracks the moving object by tracking the same local minimum. That is why snake algorithm has been chosen here for the detection process. The total energy function of the snake contour model is the sum of its external energy and internal energy and is given as follows:

$$Energy^*_{snake} = \int_0^1 E_{snake}(k(l))dl = \int_0^1 (E_{\text{internal}}(k(l)) + E_{image}(k(l)) + E_{con}(k(l)))dl, \tag{1}$$

where E_{image} is called the image energy and E_{con} is the constraint energy.

α(continuity), β(curvature), and γ(Image) values are set depending on the image gradient, edges of the image, and contrast between background and the feature.

The choice of the following key parameters makes the detection an efficient one. $\alpha = 0.7$; $\beta = 0.5$; $\gamma = 0.3$. Stop criteria = 0.0001; Maximum number of iterations = 200. Figures 2 and 3 show the results of the skin color detection.

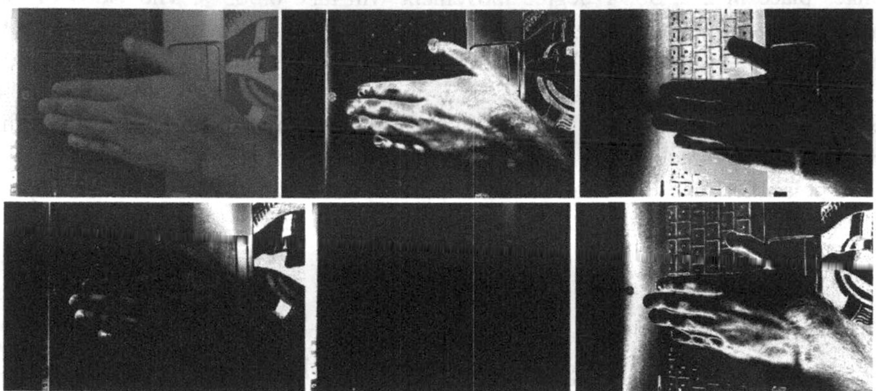

Fig. 2 Clusters of Fuzzy C Means showing the extraction of the

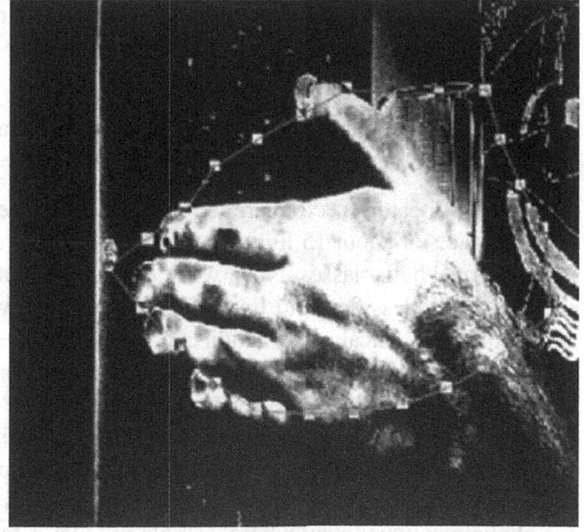

Fig. 3 Approximate estimation marked by a boundary after the computation of snake algorithm

4 Proposed Methodology

The proposed system comprises of a platform capable of holding a light source and moving in both horizontal and vertical axes. The platform is moved with a help of a dc motor, which is operated by gesture recognition as discussed earlier. The video of the moving platform with the light source is captured by a webcam and analyzed simultaneously. The dc motor calibration, an important aspect of this work, is calculated as follows: It has been observed that a full rotation (360°) of the dc motor takes place in 2 s. So, 1 degree movement will take 0.005 s. The video of the rotating platform with the light source is disintegrated into 771 frames, which is 100 s long. Then, 1 frame takes 0.129 s. Accordingly in 0.129 s, dc motor will move for 25.8°. Thus, to get all the samples in a single rotation, we will have to take 14 samples. The analysis of the video gives the illumination characteristics of the light source in both directions from which it is possible to plot the characteristics curves in a polar plot. The illumination characteristics are recorded which further help in classification of an unknown light source before it is used. Here, the properties of LED bulbs are shown.

5 Classification Using Support Vector Machines (SVM)

Support vector machine (SVM) has been used here in order to analyze the extracted features and find an optimal way to classify the images into their respective classes, normal and faulty light sources. SVM generally is a 2-class classifier. It classifies a set of data into two clusters. Here, we want the SVM to classify the data points into two classes: normal and faulty or abnormal bulbs. The two feature vectors used here are as follows:

1. Illumination coefficient of a bulb in horizontal plane.
2. Illumination coefficient of a bulb in vertical plane.

These two feature vectors are fed to the SVM trainer system. We have tested and taken features of about 15 light sources. SVM trainer classifies data into the normal and abnormal bulb classes. Now, when one sample bulb is taken, its features are fed into the SVM classifier, and finally, we get to know which class it belongs to. The horizontal and vertical illumination coefficients are measured, and it is observed that a bulb is only called faulty if its illumination value falls below the threshold value as obtained from other. The relevance of SVM here is of significant importance. Usually, in a manufacturing industry, testing is done in a GO or NO-GO way. This process is widely used in various industries. All the products are not tested as it is really a time-taking issue when it comes to test lakhs of products. So, they choose the above-mentioned way where a single product is taken out from a lot and then it is tested. If it appears to be normal then the lot is considered to be normal, and if it appears faulty, then the entire lot comes under screening. So, SVM is quite relevant in testing one of the bulbs by its illumination characteristics.

6 Results and Conclusion

Some of the recorded data for LED bulb have shown in Tables 1 and 2. These data are trained and classified by the support vector machines. To avoid saturation of illuminance values and also to bring down the values above saturation (if any) an X-ray plate has been used which behaves as a neutral density filter. Tables 1 and 2 show some of the results that are recorded. Figure 4 shows it graphically in a polar plot.

The following image, Fig. 5, shows the training and classification of the illumination properties of LED light sources.

The various classifiers used for the classification process have different error rates. The following graph, Fig. 6, shows the error rates of the classifiers used. The best of them is the radial basis function. Using this as a kernel in SVM yields a result of only 0.6% error rate as compared to linear, quadratic, multi-layer perceptron, and polynomial kernels.

The proposed system makes a huge impact in automation fields with its dynamic gesture-recognition system. The classification of the state of light sources using support vector machines (SVM) can help to determine whether the light source is usable (normal) or unusable (faulty).

Table 1 Data recorded for LED bulb at horizontal axis

Angle (Pi Radian)	Illuminance level (cd)
0.152	154
0.22	183
0.41	190
0.716	165
0.769	187

Table 2 Data recorded for LED bulb at vertical axis

Angles (Pi Radian)	Illuminance level (cd)
0	56
0.143	57
0.286	55
0.429	55
0.572	58

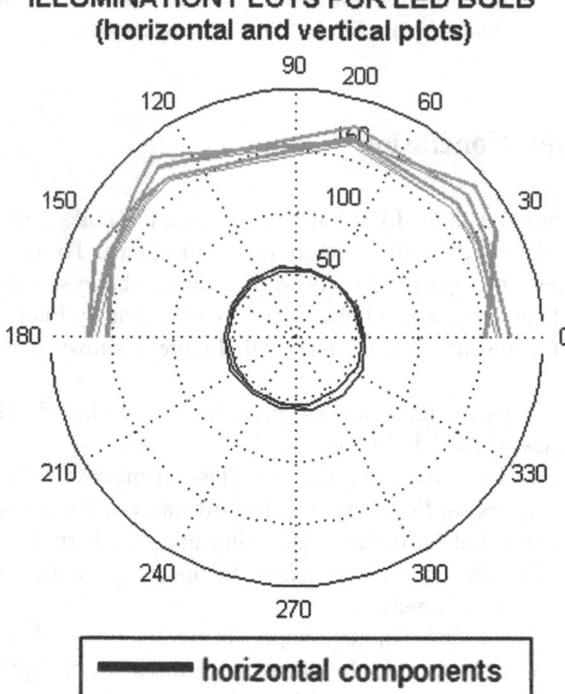

Fig. 4 Illumination plots for led bulbs in horizontal and vertical axes

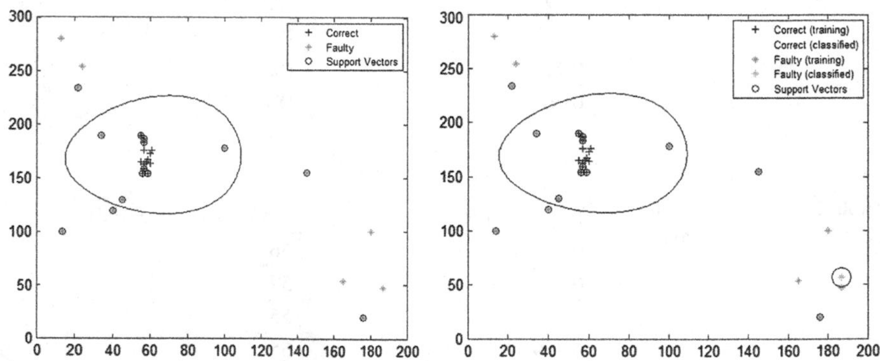

Fig. 5 SVM training and classification of LED light sources

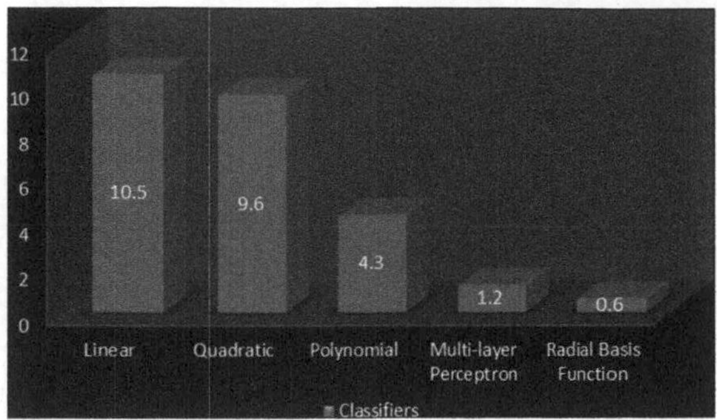

Fig. 6 Error rates of classifiers used

References

1. Jack L. Lindsey, The Fairmont Press Inc., 1991.
2. Labsphere, Technical guide - integrating sphere, 2016.
3. Luxembright, Technical datasheet of Luxembright LED, 2016.
4. Markos Sigalas, Haris Baltzakis, Panos Trahanias. "Gesture Recognition Based on Arm Tracking for human robot interaction", The 2010 IEEE/RSJ International Conference on Intelligent Robots and Systems October 18–22, 2010, Taipei, Taiwan.
5. Rafiqul Zaman Khan, Noor Adnan Ibraheem. "Hand Gesture Recognition: A Literature Review", International Journal of Artificial Intelligence & Applications (IJAIA), Vol 3, No 4, July 2012.
6. Yikai Fang, Kongqiao Wang, Jian Cheng, Hanqing Lu. "A Real-Time Hand Gesture Recognition Method". 2012.
7. Alsheakhali, Ahmed Skaik, Mohammed Aldahdouh, Mahmoud Alhelou. "Hand Gesture Recognition System", ICICS 2011, Jordan.
8. Chieh Li, Yu Hsiang Liu, Huang-Chia Shih, "Adaptive skin color tone detection with morphology-based model refinement", Information, Communications and Signal Processing (ICICS), 2013, 9th International Conference, Taiwan, pp 1 – 4.
9. Abdul Rahman Hafiz, Md Faijul Amin, Kazuyuki Murase, "Using complex-valued Levenberg-Marquardt algorithm for learning and recognizing various hand gestures", The 2012 International Joint Conference on Neural Networks (IJCNN), pp 1 – 5.
10. S. Choudhury, S. Bandyopadhyay, S. Santra, R.Ghosh, A. Ray, K. Palodhi, "Gesture recognition based relative intensity profiler for different light sources", submitted in IEEE proceedings, ICACCI 2016.

Fig. [illegible]

References

[illegible]

Test Case Prioritization Using UML State Chart Diagram and End-User Priority

Namita Panda, Arup Abhinna Acharya, Prachet Bhuyan and Durga Prasad Mohapatra

Abstract The intangible behaviour of software has given rise to various challenges in the field of testing software. One of the major challenges is to efficiently carry out regression testing. Regression testing is performed to ensure that any modifications in one component of the software do not adversely affect the other components. But, the retesting of test cases during regression testing increases the testing time and leads to delayed delivery of the software product. In this paper, a dynamic model, i.e. UML state chart diagram, is used for system modelling. Further, the UML state chart diagram is converted into an intermediate representation, i.e. State Chart Graph (SCG). The SCG is traversed to identify the affected nodes due to certain modification in the software. This information, about the affected nodes, is periodically stored in a historical data store across different versions of the software. Next time, when regression testing is carried out for any change, the stored data decides the pattern of frequently affected nodes for prioritizing the test cases and further it decides the criticality value (CV) of a test case. Along with this, to strengthen the prioritization the test sequence, two more criteria, i.e. priority set by the end-user for different functions and browsing history of the end-user, are also added up. This approach is found to be very efficient as we are able to model dynamic nature of applications, maintain a historical data store of the test cases and track the complete life of an object.

N. Panda (✉) · A.A. Acharya · P. Bhuyan
School of Computer Engineering, KIIT University,
Bhubaneswar 751024, India
e-mail: npandafcs@kiit.ac.in

A.A. Acharya
e-mail: aacharyafcs@kiit.ac.in

P. Bhuyan
e-mail: pbhuyanfcs@kiit.ac.in

D.P. Mohapatra
Department of Computer Science & Engineering,
National Institute of Technology,
Rourkela 769008, India
e-mail: durga@nitrkl.ac.in

H.S. Behera and D.P. Mohapatra (eds.), *Computational Intelligence in Data Mining*, Advances in Intelligent Systems and Computing 556,
DOI 10.1007/978-981-10-3874-7_54

Keywords State chart diagram · State chart graph · Regression testing · UML model · Prioritization

1 Introduction

Software testing plays a vital role in measuring the functional and non-functional quality attributes, which shows the path to improve upon the quality of the software. Maintaining quality is a tedious task due to exponential growth in complexity of the software. Quality can be ensured by adopting a good testing methodology. Therefore, more than half of the development effort and time are spent on testing [1, 2]. According to IEEE [3, 4], the quality can be ensured over the lifespan of the software, if different defects present in the modified product can be identified and the software testers are able to define necessary steps to fix all the defects in due time. The act of testing performed to ensure that the modified components of the software are not going to adversely affect the existing product is called ***Regression Testing*** [5]. In reality, it is not feasible to retest all test cases, for fixing all the bugs identified in the modified product, due to limitation of time. Test case selection, test suite minimization and test case prioritization are the three major techniques to carryout regression testing. As per the analysis done by different researchers done over time, test case prioritization plays a vital role in increasing the average percentage of fault detection during regression testing [6–8]. Test case prioritization is the process of rearrangement of test cases present in the test suit. This rearrangement is done based on some test criteria such as statement coverage, branch coverage and mark transition coverage, so that the test cases with higher priority will be executed first [6–8]. Test case prioritization technique improves the average percentage of fault detection (APFD) and makes the debugging process easier.

In this paper, we have proposed an approach for test case prioritization using the concept of historical data store and the priority of different features set by the end-user. The proposed approach takes the system model as input, which is modelled using UML, i.e. State Chart Diagram (SCD) [9]. Then an intermediate graph, i.e. state chart graph (SCG), is generated from SCD. The graph is traversed to find out all the linearly independent paths, and accordingly test scenarios are generated, which are stored in a test case repository. Then, depending upon the modification done to different components of the software, the respective modified and affected nodes are identified using forward slicing algorithm and the node details are stored in the historical data store. Frequent pattern of the affected nodes are generated by applying association rule mining (ARM) [10] on the historical data store. Then, priority values are assigned to each node depending upon their occurrence in the frequent pattern. Further, priority values are assigned to each test case by summing up the priority values of each node the test case traverses. Finally, two new factors are considered, i.e. priority of the functions (represented through nodes of SCG) set by the user and browsing history of the end-user, to prioritize the test cases.

The rest of the paper is organized as follows: Sect. 2 discusses the related work and its analysis, and the proposed approach is discussed in Sect. 3. Section 4 discusses the conclusion and future work.

2 Related Work

All the research work discussed in this section aims at increasing the average percentage of fault detection through test case prioritization. Khandai et al. [11] proposed a model-based test case prioritization technique using business criticality value (BCV). The authors have maintained a repository of the affected functions which helped them in setting up the criteria for test case prioritization. For a new version the software, the repository of the affected function is analyzed and each affected function is assigned with the BCV. Then, the business criticality test value (BCTV) for all test cases are calculated and the prioritization is done. But finding the business criticality value of the non-functional requirements is the biggest challenge.

A novel approach of test case prioritization using ARM has been proposed by Acharya et al. [12]. The authors have mainly focused on the structural data and actual data and termed them as graph data and observation data, respectively. This information is stored in a data store and acts as an input to ARM. The resultant frequent pattern works as the input to the test case prioritization process. The process assigns priority value to the nodes present in the pattern. The author has not considered the non-functional aspects of the system while finding the graph or observation data. The authors have considered the UML activity diagram, which is a semi-formal model, and the correct mapping of the system design with code is the real challenge. This problem can be avoided to a greater extend by using formal models.

Muthusamy et al. [13] have considered four practical weight factors such as priority allotted by the customer, code execution complexity observed by the developer, requirement change, impact of fault, test case completeness and traceability. These factors helped to identify the severe faults.

Huang et al. [14] proposed a method of cost-cognizant test case prioritization based on the use of historical records. As per their method, historical records were gathered from the latest regression testing. Then they proposed a new genetic algorithm which was applied to determine the most effective order. Test case prioritization was done based on their historical information, which provides high test effectiveness during testing.

Khalilian et al. [15] used test case performance technique for test case prioritization. In that technique, they are directly computing the priority of each test case using a mathematical equation. But, they have proposed a new prioritization equation with variable coefficients gained according to the available historical performance data. The historical performance data of a program acts as a feedback from the previous test sessions. Finally, they have validated their approach using APFD metric.

The existing research methodologies discussed in this section have not considered factors such as priority set by the end-user, the browsing history of the end-user and the modifications that done to the sub-functions where these factors play a crucial role in deciding the prioritization sequence of the test cases.

3 Proposed Approach

In this section, the proposed approach for test case prioritization using association rule mining and criticality of test cases is discussed. In the proposed approach, the test case prioritization is done through system models, which validate the model-based prioritization. In this paper, the system behaviours are modelled using UML state chart diagram [9].

Figure 1 depicts the proposed framework for test case prioritization and Sect. 4 discusses the working of the approach. The detailed data flow name for the proposed framework is given in Table 1.

The proposed approach comprises of four major steps. The phases are summarized below:

- Test cases generation from State Chart Diagram (SCD).
- Maintaining the historical data store and generating frequent pattern of nodes.

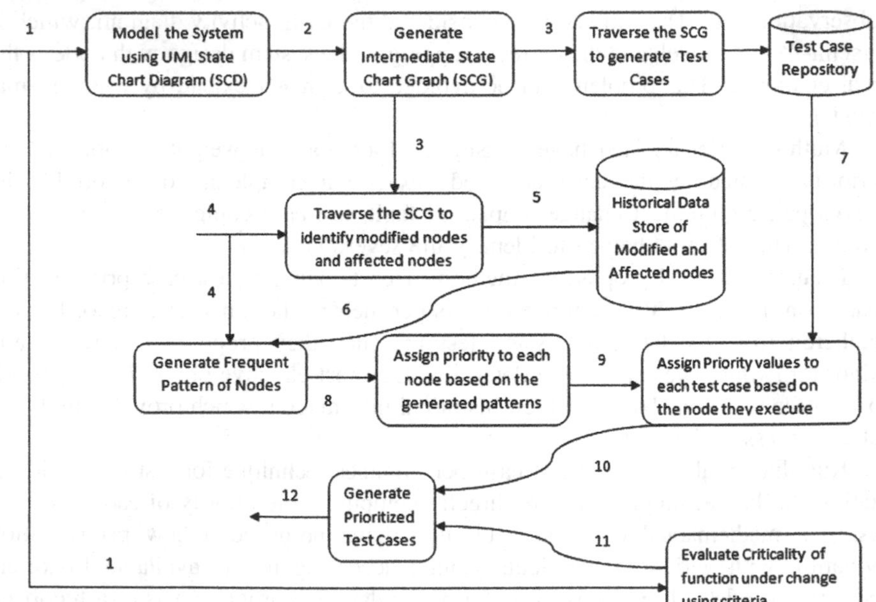

Fig. 1 Proposed framework for test case prioritization

Table 1 Detailed data flow name table for the proposed framework

Sl. no. data flow in the proposed framework	Name of the data flow
1	Requirement specification document
2	UML state chart diagram
3	State chart graph
4	Functional modification detail
5	Graph data on modified and affected node
6	Historical data on modified and affected node
7	Test cases
8	Frequent pattern of nodes
9	Prioritized node sequence
10	Set of test cases with priority value
11	Priority of test cases based on the criticality
12	Prioritized test cases

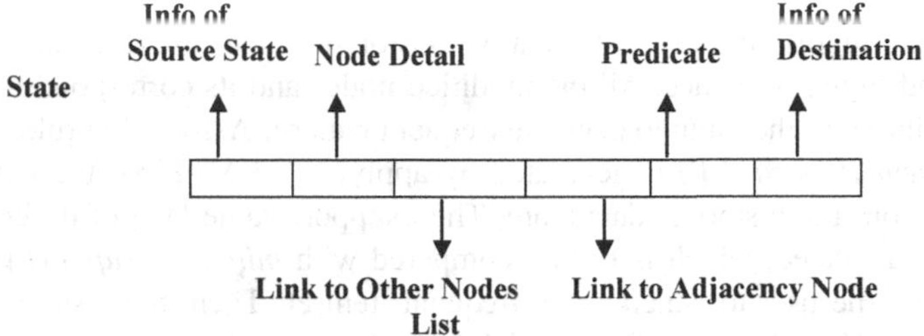

Fig. 2 Node structure of state chart graph

- Evaluating the priority value of each test case.
- Generating the prioritized test cases.

3.1 Test Cases Generation from State Chart Diagram (SCD)

First we discuss the test case generation phase from State Chart Graph. The system is modelled using UML *State Chart Diagram* (SCD) after getting information from the requirement specification documents. Then, the SCD is converted into a intermediate graph known as *State Chart Graph* (SCG). Each transition is an event and can be mapped to a function in the application. So, each transition is represented as a node in the SCG and two different transitions (nodes) are connected by an edge in the graph. The predicate for the state change for each transition is also stored in the graph. The node structure of the graph is given in Fig. 2. During traversal of the graph, the predicates will be a part of the test scenario. Then, SCG is traversed to find all the linearly independent paths present in the graph. This will ensure path coverage through path testing. The node sequence present in each path will lead to one test scenario. These test scenarios are stored in a test repository for further references.

3.2 Maintaining Historical Data Store and Generating Frequent Pattern of Nodes

The second phase deals with maintaining historical data store of modified and affected nodes across the various versions of the applications. Over the period of time, the identified modified node present in the SCG and the corresponding affected nodes are found out and stored in the historical data store. Further, we discuss the generation of frequent pattern of nodes for a given recent modification detail using *ARM* [10]. Whenever there is a modification in the application, then all the modified and corresponding affected nodes are collected from SCG using forward slicing algorithm. This information is stored in the historical data store for current and future reference. All the modified nodes and its corresponding affected node details act as the input to generate frequent pattern. Association rules depicting the *Frequent Pattern* (FP) is generated by applying ARM. First, the itemsets are collected from the historical data store. Then support value [10] of different set of items are calculated, which is further compared with *minimum support* [10] value specified by the user to generate the frequent itemset. Then, the resultant itemset, i.e. *Frequent Node Pattern* (FNP), which leads to generation of an association rule, is generated after comparing the confidence value of the frequent itemset with the *minimum confidence* value [10] specified by the user.

3.3 Evaluating the Priority Value of Each Test Case

This phase discusses the process followed to assign the priority values to the nodes present in FNP. First, the priority value of nodes is calculated based on the FNP. Now, the nodes are assigned with a priority values using the formula given in Eq. 1. The priority value is decided based on the summation of the total number of sub-functions affected in each node across the different version of the software. The evaluated value is termed as *Criticality Value* (CV).

$$\textbf{Criticality Value (CV) of Node N} = \sum_{i=1}^{n} N_i \qquad (1)$$

where n = total number of version change,
N_i = Number of sub-functions affected during *i*th version change.

The second add-on to the priority value is calculated by considering two criteria, i.e. (i) ***priority set by the user for each function*** and (ii) ***browsing history of the end-user***. For the first criterion, the user will set a priority value in the range <1, N> in the specification document where N is the total number of nodes present. This value is represented by *End User Priority* (EUP). The second criterion is calculated by capturing the browsing history of the end-user. The browsing history will contain the total number of access made to a function from the inception of the function in the software. This is represented by the term *Total Access* TA (N) i.e. Total Access made to a function represented by a node N in the SCG. The *Total Priority* (TP) of a node is calculated by adding all the three factors, which is given in Eq. 2.

$$\textbf{Total Priority(TP) of Node N} = \textbf{CV(N)} + \textbf{EUP(N)} + \textbf{TA(N)} \quad (2)$$

Then, for each test case, the priority value is calculated by adding the TP of each node that is present in the test scenario of the corresponding test case. The priority of a test case is represented by *Priority Value of Test Case* (PVT).

3.4 Generating the Prioritized Test Cases

Finally, the test cases are prioritized depending upon the priority value of test cases (PVT). A test case having a higher PVT value is kept earlier in the test suit than a test case with comparatively lower PVT value.

4 Conclusion and Future Work

The proposed approach encompasses different activities such as generation of test cases from UML state chart diagram, maintaining historical data store and generating frequent pattern, evaluating priority value of each test case and finally generating prioritized test cases, respectively. Maintaining a historical data store and considering criteria such as function-wise priority set by the end-users and browsing history of the end-user are going to give efficient and accurate prioritized sequences to do regression testing. This approach is found to be very efficient as we are able to model dynamic nature of applications, maintain a historical data store of the test cases and track the complete life of an object. But in this approach, the non-functional aspects of the system are not taken into account while doing prioritization. The real challenge in considering non-functional requirements is quantifying it.

References

1. Kundu, Debasish, Samanta, Debasis.: "A novel approach to generate test cases from uml activity diagram". Journal of Object Technology. 8(3) (2009) 65–83.
2. Sarma, M., R. Mall.: Automatic test case generation from uml models. 10th International Conference on Information Technology. (2007). 196–201.
3. Swain, S. K., Mohapatra, D. P.: Test case generation from behavioural uml models. International Journal of Computer Applications. 6(8) (2010). 5–11.
4. "IEEE Standard Glossary of Software Engineering Terminology." http://www.ieeexplore.ieee.org/IEEE Standard Glossary of Software Engineering Terminology.
5. Mall, R.: Fundamental of Software Engineering. 3rd edn. PHI Learning Private Limited, New Delhi (2009).
6. Tyagi, Manika, Malhotra, Sona.: An approach for Test Case Prioritization Based on Three Factors. International Journal of Information Technology & Computer Science. 4 (2015) 79–86.
7. D. D. Nardo, N. Alshahwan, L. Briand, and Y. Labiche. Coverage based regression test case selection, minimization and prioritization: a case study on an industrial system. Software Testing, Verification and Reliability, 25(4):371–396, June 2015.
8. Haidry, S. e Zehra, Miller, T.: Using dependency structures for prioritization of fundamental test suites. IEEE Transactions on Software Engineering. February 39(2) (2013).
9. "UML 2.4 Diagrams Overview." http://www.uml-diagrams.org/uml-24diagrams.html.
10. Han, J., Kamber, M.: Data Mining: Concepts and Techniques. 2nd edn. Morgan Kaufmann Publishers, 500 Sansome Street, Suite 400, San Francisco, CA 94111(2010).
11. Khandai, S., Acharya, A. A., Mohapatra, D. P.: Prioritizing test cases using business test criticality value. International Journal of Advanced Computer Science and Applications. 3(5) (2011) 103–110.
12. Acharya, A. A., Mahali, P., Mohapatra, D. P.: Model Based Test Case Prioritization using Association Rule Mining. Computational Intelligence in Data Mining. 3(2015) 429–440.
13. Muthusamy, Thillaikarasi, Seetharaman., K.: A new effective test case prioritization for regression testing based on prioritization algorithm. International Journal of Applied Information Systems (IJAIS). January 6(7) (2014) 21–26.
14. Y.-C. Huang, K.-L. Peng, and C.-Y. Huang. A history-based cost cognizent test case prioritization in regression testing. The Journal of Systems and Software, Elsevier, 85:626–637, 2012.
15. A. Khalilian, M. A. Azgomi, and Y. Fazlalizadeh. A improved method for test case prioritization by incorporating historical test data. Science of Computer Programming, Elsevier, 78:93–116, 2012.

Performance Analysis of Spectral Features Based on Narrowband Vegetation Indices for Cotton and Maize Crops by EO-1 Hyperion Dataset

Rajesh K. Dhumal, Amol D. Vibhute, Ajay D. Nagne, Karbhari V. Kale and Suresh C. Mehrotra

Abstract The objective of this paper is to estimate and analyze the selected narrowband vegetation indices for cotton and maize crops at canopy level, generated by using EO-1 Hyperion dataset. EO-1 Hyperion data of the date 15th October 2014 has been collected from United States Geological Survey (USGS) Earth Explorer by data aquisition request (DAR). After performing atmospheric corrections by using Quick Atmospheric Correction (QUAC), we have applied selected narrowband vegetation indices specifically those which are based on greenness/leaf pigments namely NDVI, EVI, ARVI, SGI, and red-edge indices such as RENDVI and VOG-I. Statistical analysis has been done by using the statistical t-test, it is found that there is a more significant difference in the mean of the responses of cotton and maize to NDVI, ARVI and VOG-1 than EVI & RENDVI, whereas, the response to SGI for both the crops is very close to each other.

Keywords Narrowband vegetation indices · EO-1 Hyperion data · Spectral features · NDVI · EVI · ARVI

R.K. Dhumal (✉) · A.D. Vibhute · A.D. Nagne · K.V. Kale · S.C. Mehrotra
Department of Computer Science & Information Technology, Dr. Babasaheb Ambedkar Marathwada University, Aurangabad 431004, MS, India
e-mail: dhumal19@gmail.com

A.D. Vibhute
e-mail: amolvibhute2011@gmail.com

A.D. Nagne
e-mail: ajay.nagne@gmail.com

K.V. Kale
e-mail: kvkale91@gmail.com

S.C. Mehrotra
e-mail: mehrotra.suresh15j@gmail.com

H.S. Behera and D.P. Mohapatra (eds.), *Computational Intelligence in Data Mining*, Advances in Intelligent Systems and Computing 556,
DOI 10.1007/978-981-10-3874-7_55

1 Introduction

The recent literature says that the narrow bands are essential for providing additional information with significant enhancements than the broad bands in reckoning biophysical and biochemical characteristics of agricultural crops [1]. Various narrowband vegetation indices used in this work are explained in this section, and these narrowband indices have been used to generate data which represents amount of greenness and chlorophyll red-edge values of cotton and maize crops. To improve vegetation signal from multispectral or hyperspectral data and to provide an imprecise measure of green vegetation amount, various indices reported in the literature, which can be proposed by combining values from multiple bands into a single value, or some of them associated with biophysical characteristics and others with biochemical characteristics of the vegetation.

Some authors have suggested the numerous best vegetation indices of different kinds to distinguish the crop types those are greenness/leaf pigment indices namely ARVI, EVI, NDVI, and SGI; chlorophyll red-edge indices such as RENDVI and VOG-1; light use efficiency indices such as SIPI and PRI; and leaf water indices such as DWSI and NDWI [2, 3]. Here firstly concentrated on NDVI (Normalized Difference Vegetation Index) presented by Rouse et al. (1973), which is most commonly used to highlight vegetated areas and their condition [4], then subsequently EVI, ARVI, SGI, RENDVI and VOG –I have been examined. The equations for these indices are as given in Table 1.

1.1 Enhanced Vegetation Index (EVI)

This index was proposed by Huete et al. to improve the vegetation signal with superior sensitivity in high biomass regions and better vegetation monitoring through a decoupling of the canopy background signal and to reduce the impacts of atmosphere [5]. The equation for EVI for hyperspectral data is as given in Eq. (1).

$$EVI = G([\rho_{864} - \rho_{671}]/[\rho_{864} + C1 \times \rho_{671} - C2 \times \rho_{467} + L]) \quad (1)$$

where ρ is atmospherically corrected reflectance (center wavelength in nanometers). L = 1 is used as canopy background adjustment, C1 = 6, C2 = 7.5 (coefficients of the aerosol resistance term), and G = 2.5 (gain factor) [5].

1.2 Atmospherically Resistant Vegetation Index (ARVI)

The main advantage of this index is its resistance to atmospheric effects, as compared to the normalized difference vegetation index which is accomplished by a self-adjustment procedure for the atmospheric influence on the red band, using the

Table 1 Narrowband indices used in this work defined for Hyperion bands

Sr. No.	Index	Abbreviation	Equation	Reference
1	Normalized difference vegetation index	NDVI	$(B_{51} - B_{32})/(B_{51} + B_{32})$	Rouse et al. (1973)
2	Enhanced vegetation index	EVI	$2.5\,([B_{51} - B_{32}]/[B_{51} + 6 \times B_{32} - 7.5 \times B_{12} + 1])$	Huete et al. (2002)
3	Atmospherically resistant vegetation index	ARVI	$(B_{51} - [2 \times B_{32} - B_{12}])/(B_{51} + [2 \times B_{32} - B_{12}])$	Kaufmann and Tenre (1992)
4	Sum green index	SGI	$(B_{16} + B_{17} + B_{18} + B_{19} + B_{20} + B_{21} + B_{22} + B_{23} + B_{24} + B_{25})/10$	Lobell and Asner (2003)
5	Red-edge normalized difference vegetation index	RENDVI	$(B_{40} - B_{35})/(B_{40} + B_{35})$	Gitelson et al. (1996)
6	Vogelmann red-edge index	VOG-I	B_{39}/B_{37}	Vogelmann et al. (1993)

B is the reflectance at corresponding Hyperion band (atmospherically corrected reflectance) to the original wavelength formulations difference in the radiance between the blue and the red bands to correct the radiance in the red band. Simulations using radiative transfer computations on arithmetic and natural surface spectra, for numerous atmospheric conditions, show that ARVI has an analogous dynamic range to the NDVI, but it is less sensitive to atmospheric effects than the NDVI [6]. The formula for ARVI for narrowband data is as given in Eq. (2).

$$\mathrm{ARVI} = (\rho_{864} - [2 \times \rho_{671} - \rho_{467}])/(\rho_{864} + [2 \times \rho_{671} - \rho_{467}]) \tag{2}$$

1.3 Sum Green Index (SGI)

Sum green index is used to observe changes in vegetation greenness. It is highly sensitive to vegetation canopy opening. It is computed by taking the average of reflectance from 500 nm to 600 nm range of the electromagnetic spectrum, the common range for green vegetation is 10–25 (in a unit of % reflectance) [7, 8] the equation for SGI is as given in Eq. (3).

$$\mathrm{SGI} = (\rho_{508} + \rho_{518} + \rho_{528} + \rho_{538} + \rho_{549} + \rho_{559} + \rho_{569} + \rho_{579} + \rho_{590} + \rho_{600})/10 \tag{3}$$

1.4 Red-Edge Normalized Difference Vegetation Index (RENDVI)

The reflectance near 0.70 um has found to be very sensitive to red-edge position as well as chlorophyll concentration [9] RENDVI gives normalized value at the wavelength of 752 and 701 nm, and the equation for this index is as given in Eq. (4).

$$\mathrm{RENDVI} = (p_{752} - p_{701})/(p_{752} + p_{701}) \tag{4}$$

1.5 Vogelmann Red-Edge Index (VOG-I)

It is simple red edge index invented by Vogelmann, various studies in literature has shown that red edge region to be useful for accessing the variety of condition in vegetation, whereas several studies proved that red edge portion is the indicator of plants phonological status in that point of view, here we have examined the Vogelmann Red Edge Index (VOG-I) which can be obtained by ratio of p742/p722 (p is the reflectance at closest Hyperion band) [10].

$$\mathrm{VOG} - 1 = (p_{742}/p_{722}) \tag{5}$$

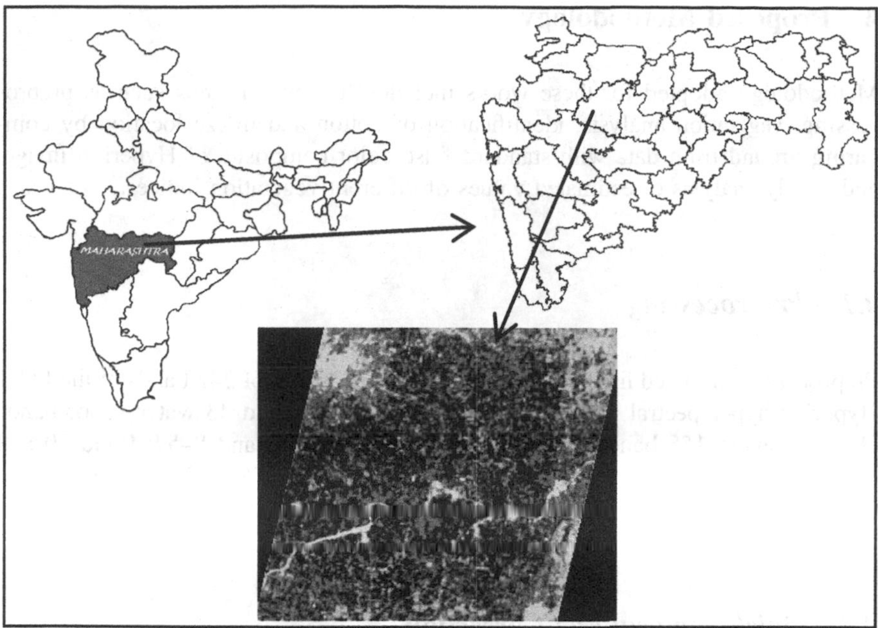

Fig. 1 Study area represented by false color composite of EO-1 Hyperion image

2 Study Area

Study area covers the part of Kanhori, Pimpalgaon Walan, Pal, and Wanegaon villages in Phulambri Taluka in Aurangabad District of Maharashtra State, India. It is part of Marathwada region and Aurangabad division. These villages are about 35 km toward north from district headquarters Aurangabad, 5 km from Phulambri, 350 km from state capital Mumbai, and located at 20°07'13.5"N 75°23'05.3"E and surrounded by Khultabad and Kannad Taluka toward west, Aurangabad Taluka towards south, and Sillod Taluka towards east [11] (Fig. 1).

3 Dataset and Fieldwork

The data was acquired from Hyperion sensor of United States Geological Survey (USGS) Earth observing-1 (EO-1) satellite. The Hyperion Hyperspectral data having 242 spectral bands with approximately 10 nm bandwidth, 30 m spatial resolution and 7.75 km swath. The range of spectral bands of this data is from 0.4 to 2.5 μm [12]. The data used in this work was projected to the Universal Transverse Mercator (UTM) zone 43 north using World Geodetic System (WGS)-84 datum. Ground truth points have been collected by using GPS (Global Positioning System)-enabled smartphone.

4 Proposed Methodology

Methodology adopted for these works includes four major steps such as preprocessing, vegetation analysis, identification of cotton and maize location by comparing ground truth data with standard false color composite of Hyperion image, and finally analysis of estimated values of different vegetation indices.

4.1 Preprocessing

Preprocessing initiated firstly by removing bad bands, out of 242 bands of the EO-1 Hyperion hyperspectral image, there are 44 bad bands and 43 water vapor bands [13], following 155 bands were utilized for this study: Band 8–57, Band 79 and 83–119, Band 133–164, Band 183 and band 184, Band 188–220. For atmospheric correction, QUAC (Quick Atmospheric Correction) tool in ENVI was used.

4.2 Quick Atmospheric Correction

The Quick Atmospheric Correction (QUAC) model is used for atmospheric correction of hyperspectral images in visible, near infrared to short wave infrared wavelength region. As compared to other methods, it requires only imprecise specification of sensor band locations or central wavelengths with radiometric calibration rather than other metadata. Quick Atmospheric Correction is a scene-based realistic approach and based on the radiance values of the image used for the removal of atmospheric effects. It takes atmospheric compensation factors from the image scene directly while ignoring secondary information with high computational speed. It provides better repossession of reasonable reflectance spectra while an image did not have proper wavelength or radiometric calibration or solar illumination intensity is not known. QUAC model of ENVI was used to perform this task. QUAC will give improved results for further processing. QUAC does not involve first-principle radiative transfer calculations. QUAC is considerably faster than other physics-based methods, and it is likewise more fairly accurate [14, 15].

After performing the atmospheric correction, vegetation analysis has been applied on preprocessed image by using image analysis tool which evaluates different vegetation indices, and then, we focused on above discussed indices. As spatial resolution of Hyperion image is 30 m, so it is quite challenging to identify ground truth points on the same image, and we have identified those sites by using pixel locator tool and observed the indices values for cotton and maize crops with the help of spatial pixel editor.

5 Results and Discussion

As Hyperion image used for this study is of October 15, 2014, at that time, the age of cotton crop is 15–20 weeks, which would be the boll opening stage of cotton life cycle. Similarly, in October, maize is also well matured. After performing the vegetation analysis, we have focused on six indices as given in Figs. 2 and 3, and then, we studied the basic statistics of obtained indices values as shown in Table 2.

We found that for Greenness/leaf pigment indices, mean of NDVI for cotton is 0.4340, which is greater than maize 0.3492, similarly for EVI mean for cotton is 0.4725 which is also greater than the maize 0.3656. ARVI for cotton is 0.4006 and for maize is 0.3040 but for SGI mean for cotton is 0.1566 which is close to maize 0.1505. Similar things replicated in t-stats that NDVI, ARVI has significant difference than EVI. The response of SGI was found to be partially similar in both the crops. For Red Edge indices VOG-1 is having more significant difference than the RENDVI. The graphical representation is as given in Fig. 4.

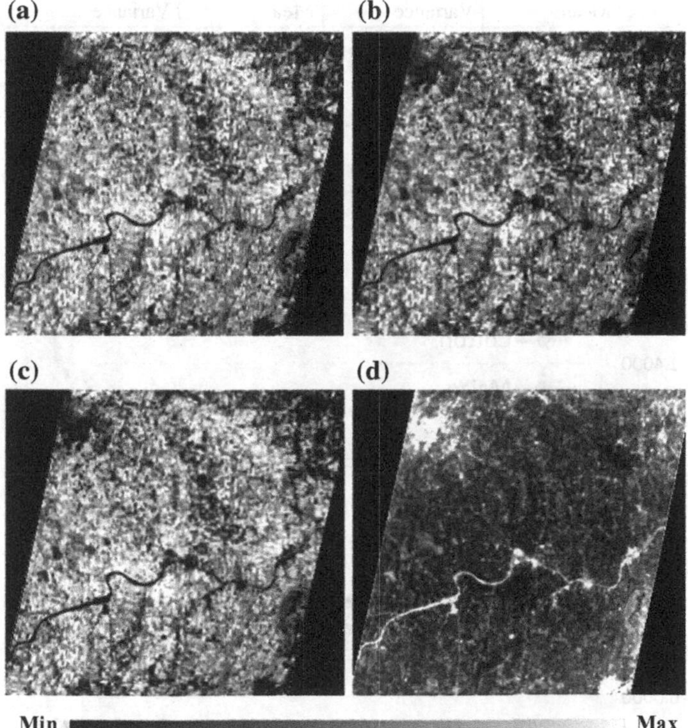

Fig. 2 Output images of greenness/leaf pigment indices **a** Normalized difference vegetation index, **b** enhanced vegetation index, **c** atmospherically resistant vegetation index, **d** sum green index

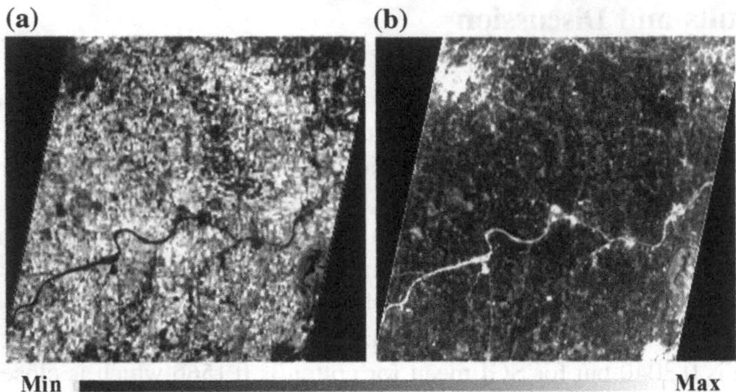

Fig. 3 Output images of chlorophyll red-edge indices **a** Red-edge normalized difference vegetation index, **b** Vogelmann red-edge index

Table 2 Statistics observed for studied vegetation indices

Indices	Cotton		Maize		T-stat
	Mean	Variance	Mean	Variance	
NDVI	0.4340	0.0068	0.3492	0.0032	3.5932
EVI	0.4725	0.0164	0.3656	0.0067	2.9869
ARVI	0.4006	0.0110	0.3040	0.0059	3.1534
SGI	0.1566	0.0000	0.1505	0.0012	0.7411
RENDVI	0.3817	0.0061	0.3218	0.0025	2.7478
VOG-I	1.5236	0.0285	1.3925	0.0054	3.0229

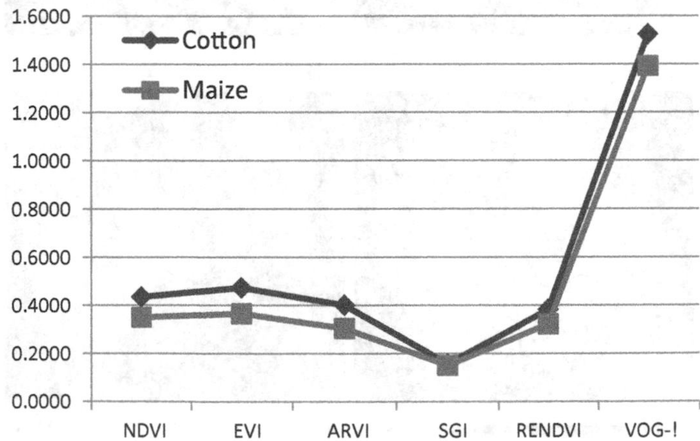

Fig. 4 Graph showing difference in mean of cotton and maize for six indices

6 Conclusion

After analyzing these indices, we conclude that, in this data of October 2014, the performance of greenness/leaf pigment and red-edge indices is greater for cotton than the maize. SGI is the exception for that, this may happen because it focuses on the greenness of vegetation, i.e., average reflectance between 500 and 600 nm region of electromagnetic spectrum. The dataset used for this study is of October 2014, and the result may change in temporal datasets because phonological status of crops gets changed throughout their life cycle. In greenness/leaf pigment indices NDVI and ARVI are having the more significant difference in the mean, than the mean of EVI for studied crops. In the case of Red edge indices, VOG-1 has significant difference in mean than RENDVI for both the crops. This study is very useful toward identifying unique spectral features at canopy level for crops. This work can be extended by considering other narrowband vegetation indices and to identify distinct spectral features for better crop discrimination.

Acknowledgements The authors would like to acknowledge to Deity, Government of India, for providing financial assistance under Visvesvaraya PhD Scheme and also thanks to DST-FIST program and UGC for laboratory facilities under UGC_SAP (II) DRS Phase-I F.No.3-42/2009, Phase-II 4-15/2015/DRS-II to Department of CS & IT, Dr. B. A. M. University, Aurangabad, M. S., India. The authors are thankful to USGS (United States Geological Survey) for providing EO-1 Hyperion data for this study.

References

1. Thenkabail, Prasad, S., Ronald B. Smith, and Eddy De Pauw.: Hyperspectral vegetation indices and their relationships with agricultural crop characteristics. Remote sensing of Environment 71.2 (2000): 158–182.
2. Galvão, L. S.: Crop Type Discrimination Using Hyperspectral Data. In Hyperspectral Remote Sensing of Vegetation, Thenkabail, P. S., Lyon, J. G. and Huete, A. (Eds.) Boca Raton, London, New York: CRC Press/Taylor and Francis Group. (2011). 17; 397–422.
3. Dhumal, R. K., Vibhute, A. D., Nagne, A. D., Rajendra, Y. D., Kale, K. V., & Mehrotra, S. C.: Advances in Classification of Crops using Remote Sensing Data. International Journal of Advanced Remote Sensing and GIS, 4(1), (2015): pp-1410.
4. Rouse, J. W., Haas, R. H., Schell, J. A., and Deering, D. W.: Monitoring vegetation systems in the Great Plains with ERTS. Proceedings of Third ERTS-1 Symposium, Washington, DC, December 10–14, NASA, SP-351, Vol. 1, pp. 309–317, (1973).
5. Huete, A. R., Didan, K., Miura, T., Rodriguez, E. P., Gao, X., and Ferreira, L. G.: Overview of the radio-metric and biophysical performance of the MODIS vegetation indices. Remote Sensing of Environment, 83, 195–213, 2002.
6. Kaufman, Y. J. and Tanré, D.: Atmospherically resistant vegetation index (ARVI) for EOS-MODIS. IEEE Transactions on Geoscience and Remote Sensing, 30, 261–270, 1992.
7. Lobell, D.B. and Asner, G.P.: Comparison of Earth Observing-1 ALI and Landsat ETM + for crop iden-tification and yield prediction in Mexico. IEEE Transactions on Geoscience and Remote Sensing, 41, 1277–1282, 2003.
8. Dutta, R.: Review of Vegetation Indices for Vegetation Monitoring. Proceeding of Nay pyi, ACRS, (2014).

9. Gitelson, A. A., Merzlyak, M. N., and Lichtenthaler, H. K., Detection of red edge position and chlorophyll content by reflectance measurements near 700 nm, Journal of Plant Physiology, 148, 501–508, 1996.
10. Vogelmann, J. E., Rock, B. N., and Moss, D. M.: Red edge spectral measurements from sugar maple leaves. International Journal of Remote Sensing, 14, 1563–1575, 1993.
11. http://www.onefivenine.com/india/villages/Aurangabad-District/Phulambri/Kanhori accessed on 27-March-2016.
12. Beck, Richard. "EO-1 user guide v. 2.3." Department of Geography University of Cincinnati (2003).
13. Vibhute, A. D., Kale, K. V., Dhumal, R. K., & Mehrotra, S. C. (2015, December). Hyperspectral imaging data atmospheric correction challenges and solutions using QUAC and FLAASH algorithms. In 2015 International Conference on Man and Machine Interfacing (MAMI) (pp. 1–6). IEEE.
14. Bernstein, Lawrence S., et al.: Quick atmospheric correction code: algorithm description and recent upgrades. Optical engineering 51.11 (2012): 111719–1.
15. Pervez, W., and S. A. Khan. "Hyperspectral Hyperion Imagery Analysis and its Application Using Spectral Analysis." The International Archives of Photogrammetry, Remote Sensing and Spatial Information Sciences 40.3 (2015): 169.

Measuring Hit Ratio Metric for SOA-Based Application Using Black-Box Testing

A. Dutta, S. Godboley and D.P. Mohapatra

Abstract In our proposed work, we discuss how to generate test cases automatically for BPEL processes to compute Hit Ratio percentage of an SOA application. First, we design an SOA-based application using OpenEsb tool. That application is supplied to code converter to get XML code of the designed application. Then, we have supplied this XML code to Tcases tool to generate test cases according to black-box testing technique. These test cases are supplied to Hit Ratio Calculator to compute Hit Ratio percentage. On an average of four SOA-based applications, we achieved Hit Ratio percentage as 63.94%.

Keywords SOA · Hit Ratio · Black-box testing

1 Introduction

Service-oriented architecture (SOA) is a design for building business applications as an arrangement of loosely coupled black-box components organized to provide a very much characterized level of service by connecting together business processes. Application components are orchestrated and loosely coupled. The technology is based on services and communication among them. A service is an independent unit of programming that performs a particular task. It has three parts: an interface, an agreement, and execution. The interface characterizes how a service supplier will perform demands from a service customer, the agreement characterizes how the service supplier and the service purchaser ought to connect, and the implementation is the genuine service code itself. Since the interface of a service is partitioned

A. Dutta · S. Godboley (✉) · D.P. Mohapatra
DOS Lab, Department of CSE, NIT Rourkela, Odisha, India
e-mail: sanghu1790@gmail.com

A. Dutta
e-mail: arpitad10j@gmail.com

D.P. Mohapatra
e-mail: durga@nitrkl.ac.in

H.S. Behera and D.P. Mohapatra (eds.), *Computational Intelligence in Data Mining*, Advances in Intelligent Systems and Computing 556,
DOI 10.1007/978-981-10-3874-7_56

from its implementation, a service supplier can execute a solicitation without the service buyer knowing how it does as such, the service purchaser just stresses over consuming services. Software applications created on top of SOA are progressively prominent, yet testing them remains a test. So, testing is mandatory for SOA-based applications. There are some current work performed for generating automatic test cases for SOA-based services. But, coverage of XML elements is missing.

Service-Oriented Architecture (SOA) is a method of coordinating software [1]. If commercial enterprises projects stick to the standards of SOA, the result will be a stock of particular units called services, which consider a brisk reaction to change. SOA is an application of cloud computing in view of the request and replay model for asynchronous and synchronous applications. Service interface is independent from the implementation of application. SOA-based application developers are using services. They are unaware of the fundamental concept of services. A service can be developed in any languages such as .Net or J2EE.

SOA have the accompanying characteristics as follows: Services of SOA contain interfaces in XML documents which are platform independent. WSDL is the standard used to depict the services. Services communicate with messages formally characterized by means of XSD (also called XML Schema). Customers and suppliers communicate among themselves using services which happens in heterogeneous environment. There is no practical information about the supplier.

Software testing is the technique which distinguishes the bugs, errors, and faults of a system to produce a good quality product. Its main objective is to test most extreme number of test cases to reach every feasible solution with an expectation to discover most extreme bug inside the system. Software testing is of two types: Black-box testing and White-box Testing. In our proposed work, we have used test cases generated using Block-box testing.

In this paper, we propose an approach to calculate Hit Ratio percentage using test cases generated by black-box tester Tcases for Service-Oriented Applications. Using OpenEsb tool, we design SOA-based application. OpenEsb provide an XML code as an output. It is based on designed framed in SOA. This XML code is supplied to Tcases to generate test cases for XML code. Using the information from test cases generated, we compute Hit Ratio percentage. To measure Hit ratio, we have developed Hit Ratio Calculator. For this paper, we experimented four SOA applications.

The rest of the article is organized as follows: Sect. 2 presents the literature survey. Section 3 shows our proposed approach. Section 4 discusses about experimental results. Section 5 concludes the proposed approach and suggesting some future work.

2 Literature Survey

In this section, we discuss some related work.

Shamsoddin-Motlagh et al. [2] addressed a method for automatic generation of test cases for SOA services. First step is tocreate a control flow graph of BPEL and to creat subgraphs of services by WSIG file. Their proposed algorithm called Random Generation and Genetic Algorithm generates random test cases for the graph. The proposed algorithm is compared against standard genetic algorithm.

Ma et al. [3] illustrated a method to automatically generate test cases for BPEL process using Stream X-machine-based testing methods. It automatically generates test cases for business process unit testing of BPEL. Due to asynchronous property and distributed nature test cases, generation for integrated web services is difficult.

Mayer et al. [4] introduced a layer-based approach for developing framework for reusable, white-box BPEL unit testing. These are used for developing a new framework for testing. This new framework utilizes specific BPEL, to describe communications in BPEL process. The new framework provides automatic test execution in a standardized way.

Kumar et al. [5] found some various kind of methodologies to solve the problems encountered during SOA testing. From various perspectives, the authors viewed the challenges faced in SOA testing and improved testability by detecting the problems.

Kumar et al. [6] introduced testing model for SOA services. The new test model generates automatic test cases for SOA services. They have developed test case generation algorithm and test case selection algorithm. The proposed algorithms give coverage for XML Schema (XSD) elements of WSDL.

Nonchot et al. [7] focuses on generating test cases from a Business Process Model and Notation (BPMN) along with a BPEL. They developed a tool that generates test cases from a BPMN.

Godboley et al. [8–12] presented an automated approach to generate test data that helps to achieve an increase in MC/DC coverage of a program under test. Transformation techniques are used for transforming program. This transformed program is inserted into the CREST TOOL to generate test suite and increase the MC/DC coverage.

3 Our Proposed Work

In this section, we discuss our proposed work in detail. Firstly, we present the overview of our proposed work, followed by the schematic representation and at last the implementation.

3.1 Overview of Framework

Figure 1 shows the schematic representation of our proposed work. First of all, we used OpenEsb framework to develop SOA-based application. Then, supplying this developed application into code converter to generate XML code. Supplying this

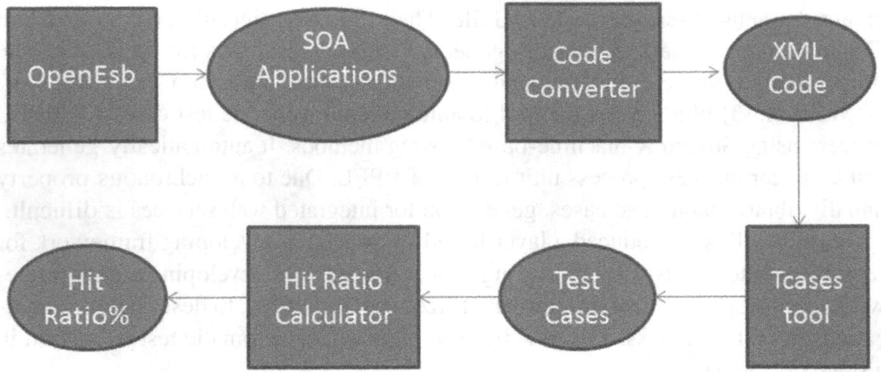

Fig. 1 Schematic representation of our proposed approach

XML code into Tcases to generate test cases according to black-box testing. We have developed a Hit Ratio Calculator which uses the generated test cases and compute Hit Ratio%.

3.2 Description of Our Proposed Work with an Example

In this section, we explain our proposed approach with the help of an example. Let us take the example of *Quadratic Equation*. Quadratic Equations requires total five services to complete the operation. Figure 2 shows the development of SOA-based

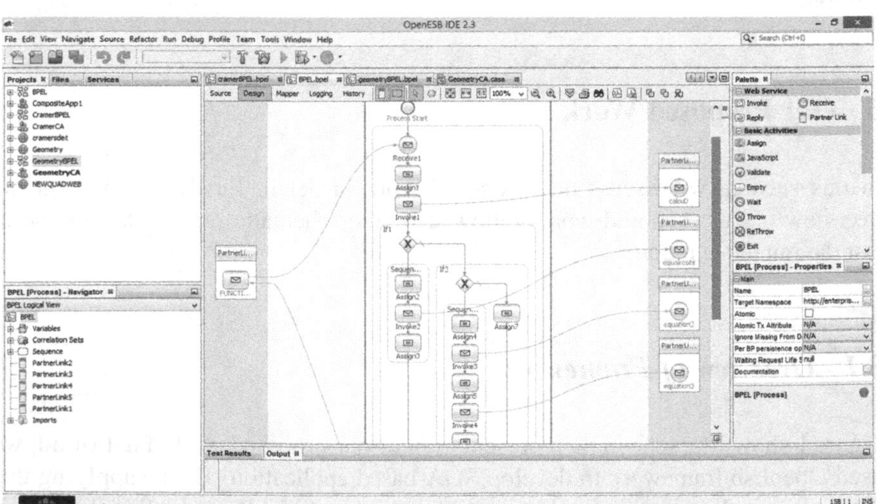

Fig. 2 BPEL architecture for Quadratic Equation using openEsb

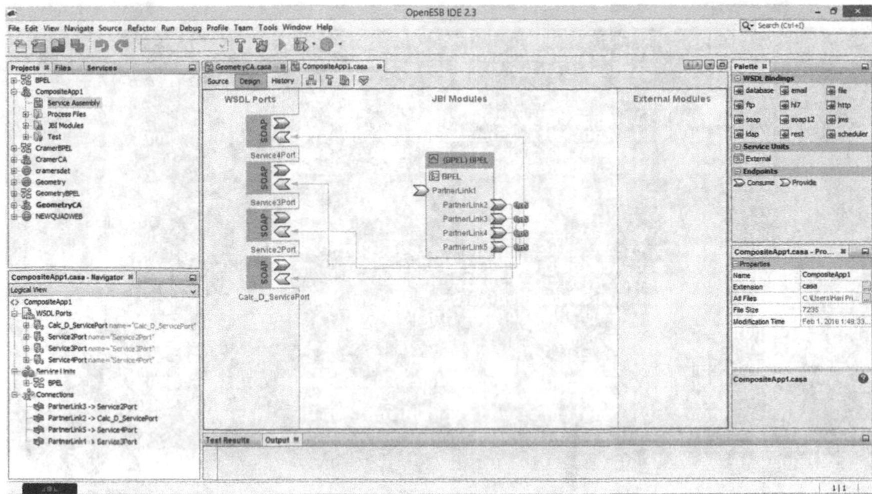

Fig. 3 Composite view of SOA-based application

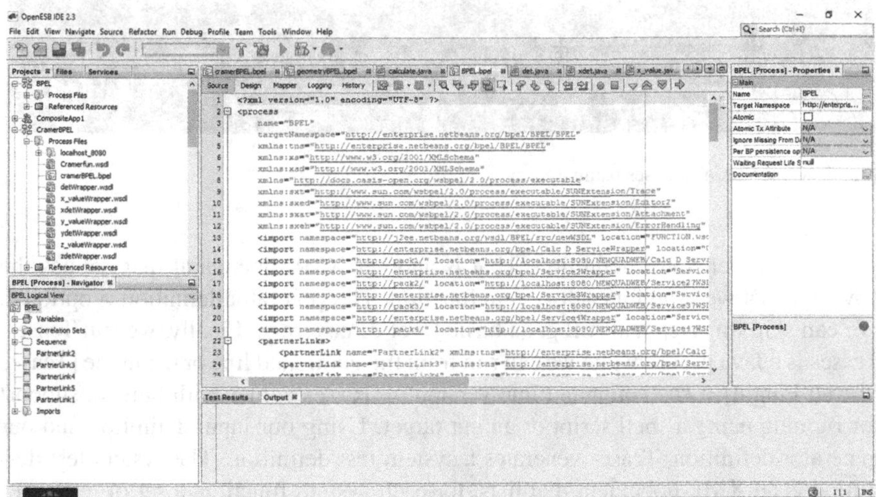

Fig. 4 Generated XML code

application for Quadratic Equation using OpenEsb. OpenEsb provides the composite view as shown in Fig. 3 for Quadratic Equation. Now, we required the code to generate the test cases. Code Converter provides the XML code for the developed application for Quadratic Equation. Figure 4 shows the converted code in the form of XML. Now, this XML code is supply to Tcases tool to generate test cases according black-box concept. First, we create a system input definition, an XML document that defines our system as a set of functions. For each system function, the system input definition defines the variables that characterize the function input space. Then, we

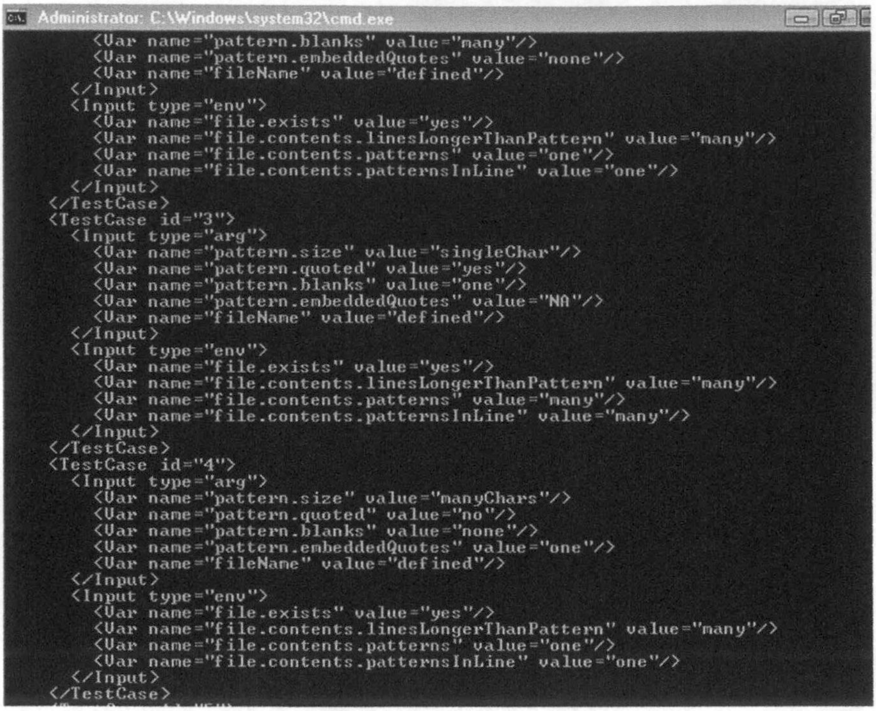

Administrator: C:\Windows\system32\cmd.exe


```
    <Var name="pattern.blanks" value="many"/>
    <Var name="pattern.embeddedQuotes" value="none"/>
    <Var name="fileName" value="defined"/>
  </Input>
  <Input type="env">
    <Var name="file.exists" value="yes"/>
    <Var name="file.contents.linesLongerThanPattern" value="many"/>
    <Var name="file.contents.patterns" value="one"/>
    <Var name="file.contents.patternsInLine" value="one"/>
  </Input>
</TestCase>
<TestCase id="3">
  <Input type="arg">
    <Var name="pattern.size" value="singleChar"/>
    <Var name="pattern.quoted" value="yes"/>
    <Var name="pattern.blanks" value="one"/>
    <Var name="pattern.embeddedQuotes" value="NA"/>
    <Var name="fileName" value="defined"/>
  </Input>
  <Input type="env">
    <Var name="file.exists" value="yes"/>
    <Var name="file.contents.linesLongerThanPattern" value="many"/>
    <Var name="file.contents.patterns" value="many"/>
    <Var name="file.contents.patternsInLine" value="many"/>
  </Input>
</TestCase>
<TestCase id="4">
  <Input type="arg">
    <Var name="pattern.size" value="manyChars"/>
    <Var name="pattern.quoted" value="no"/>
    <Var name="pattern.blanks" value="none"/>
    <Var name="pattern.embeddedQuotes" value="one"/>
    <Var name="fileName" value="defined"/>
  </Input>
  <Input type="env">
    <Var name="file.exists" value="yes"/>
    <Var name="file.contents.linesLongerThanPattern" value="many"/>
    <Var name="file.contents.patterns" value="one"/>
    <Var name="file.contents.patternsInLine" value="one"/>
  </Input>
</TestCase>
```


Fig. 5 Generated test cases using Tcases

can create a generator definition. That is another XML document that defines the coverage that we want for each system function. The generator definition is optional. We can skip this step and still get a basic level of coverage. Finally, we run Tcases. Tcases is a Java program that we can run from the command line or using the [Tcases Maven Plugin].[1] The command line version of Tcases comes with built-in support for running using a shell script or an ant target. Using our input definition and our generator definition, Tcases generates a system test definition. The system test definition is an XML document that lists, for each system function, a set of test cases that provides the specified level of coverage. Each test case defines a specific value for every function input variable. Tcases generates not only valid input values that define successful test cases, but also invalid values for the tests cases that are needed to verify expected error handling. Figure 5 shows the test cases generated through Tcases. Tcases also provides the report of all test cases. Figure 6 shows the test cases report generated through Tcases. Here, we can observe the capability of Tcases to identify the valid and failure test cases as shown in Fig. 6. For, this Quadratic Equation Tcases generated total ten test cases. Tcases identified four of valid test cases and six as failure test cases. Therefore, our developed Hit Ratio Calculator computes

[1] http://www.cornutum.org/tcases/docs/tcases-maven-plugin/.

```
11:42:31.257 INFO  org.cornutum.tcases.Tcases - Tcases 1.5.4 (2016-02-18)
11:42:31.319 INFO  o.c.t.generator.TupleGenerator - FunctionInputDef[find]: Preparing constraint info
11:42:31.335 INFO  o.c.t.generator.TupleGenerator - FunctionInputDef[find]: Generating test cases
11:42:31.366 INFO  o.c.t.generator.TupleGenerator - FunctionInputDef[find]: Created 6 valid test cases
11:42:31.381 INFO  o.c.t.generator.TupleGenerator - FunctionInputDef[find]: Created 4 failure test cases
11:42:31.381 INFO  o.c.t.generator.TupleGenerator - FunctionInputDef[find]: Completed 10 test cases
```

Fig. 6 Test cases report to identify valid and failure test cases

Hit Ratio% as 60%. Hit Ratio% is calculated using following formula:

$$\textit{Hit Ratio\%} = \frac{\textit{Valid test cases}}{\textit{Total number of testcases}} \times 100 \quad (1)$$

4 Experimental Results

In this section, we discuss experimental result of our proposed work.

Table 1 shows the characteristics of application developed using OpenEsb. Column 2 shows the name of applications. Column 3 shows the total services used for developing application. Column 4 shows the classes when this application converted in skeleton Java program.

Table 2 shows the experimental results of our proposed approach. Column 3 shows the total generated test cases. Column 4 shows the valid test cases. This valid test cases also defined the meaning of "Hit". Column 5 shows the failure test cases. Column 6 shows the Hit Ratio%. This metric is computed using Eq. 1. On an average of four SOA-based applications, we computed Hit Ratio is 63.94%.

Table 1 Characteristics of applications

Sl.No.	Application name	Services	Classes
1	Quadratic	4	3
2	Cramer	7	3
3	Geometry	5	4
4	Calculator	5	0

Table 2 Experimental results

Sl.No.	Application name	Test cases	Valid test cases	Failure test cases	Hit Ratio%
1	Quadratic	10	6	4	60%
2	Cramer	17	7	10	41.2%
3	Geometry	3	3	0	100%
4	Calculator	11	6	5	54.55%

5 Conclusion and Future Work

In this paper, we have proposed and developed an approach to compute Hit Ratio% for SOA-based applications. First, we design an SOA-based application using OpenEsb tool. That application is supplied to code converter to get XML code of the designed application. Then, we have supplied this XML code to Tcases tool to generate test cases according to black-box testing technique. These test cases are supplied to Hit Ratio Calculator to compute Hit Ratio percentage. We have also described the approach using an example with some screenshots. For this work, we have experimented our proposed work for four applications. Through results, we conclude that on an average, the computed Hit Ratio percentage for four SOA-based application is 63.94%.

In our future work, we will extend our work to compute test cases for white-bax testing, so that we can compute many coverage criteria, such as statement coverage, branch coverage, and MC/DC coverage.

References

1. Karimi, O., and Modiri, N., Enterprise Integration using Service Oriented Architecture. *Advanced Computing*, 2(5), p. 41, (2011).
2. Ebrahim Shamsoddin-Motlagh. Automatic Test Case Generation for Orchestration Languages at Service Oriented Architecture. *International Journal of Computer Applications*, Volume 80, Number 7, October 2013.
3. Ma, C., Wu, J., Zhang, T., Zhang, Y. and Cai, X., Automatic test case generation for BPEL using stream x-machine. *International Journal of u- and e-Service, Science and Technology*, 1(1), pp. 27–35, (2008).
4. Mayer, P., and Lübke, D., Towards a BPEL unit testing framework. In *Proceedings of the 2006 workshop on Testing, analysis, and verification of web services and applications* (pp. 33–42). ACM, (July 2006).
5. Kumar, A., and Singh, M., An Empirical Study on Testing of SOA based Services. *International Journal of Information Technology and Computer Science (IJITCS)*, 7(1), p. 54, (2014).
6. Kumar, A., A Novel Testing Model for SOA based Services. *International Journal of Modern Education and Computer Science (IJMECS)*, 7(1), p. 31, (2015).
7. Nonchot, C., and Suwannasart, T., A Tool for Generating Test Case from BPMN Diagram with a BPEL Diagram. In *Proceedings of the International MultiConference of Engineers and Computer Scientists* (Vol. 1), (2016).
8. Godboley S., Improved Modified Condition/Decision Coverage using Code Transformation Techniques. *Thesis (MTech) NIT Rourkela* (2013).
9. Godboley S., and Mohapatra DP., Time Analysis of Evaluating Coverage Percentage for C Program using Advanced Program Code Transformer *7th CSI International Conference on Software Engineering* Pages 91–97, (Nov 2013).
10. Godboley S., Prashanth GS., Mohapatra DP, and Majhi B., Increase in Modified Condition/Decision Coverage using program code transformer. *IEEE 3rd International Advance Computing Conference (IACC)* Pages 1400–1407, (Feb 2013).

11. Godboley S., Prashanth GS., Mohapatra DP, and Majhi B., Enhanced modified condition/decision coverage using exclusive-nor code transformer. *2013 International Multi-Conference on Automation, Computing, Communication, Control and Compressed Sensing (iMac4s)* pages 524–531, (March 2013).
12. Godboley, S., Mohapatra, D.P., Das, A., and Mall, R., An improved distributed concolic testing approach. *Software: Practice and Experience*, 47(2), pp. 311–342, (2017). DOI:10.1002/spe.2405.

Circularly Polarized MSA with Suspended L-shaped Strip for ISM Band Application

Kishor B. Biradar and Mansi S. Subhedar

Abstract The broadband circular polarized (CP) microstrip patch antenna (MSA) for ISM band (2.4 GHz) applications is proposed. The proposed antenna consists of a suspended square ring along with corner-chopped square shape slot and a suspended horizontal L-shaped strip line. Two cylindrical probes which connect L-shaped strip and radiating patch are used to feed patch at two different positions. This feed network of two probes excited two orthogonal signals of equal magnitude which generated CP radiation. The protocol of proposed antenna is simulated and fabricated, and experimental results shows that 10 dB impedance bandwidth of 21.75% (2.13–2.65 GHz), 3 dB axial ratio bandwidth of 19.6% (2.16–2.60 GHz), and measured simulated gain over 3 dB axial ratio (AR) are 7.16 dBi.

Keywords Circular polarization · Axial ratio · Broadband antenna · Suspended L-strip

1 Introduction

Increasing growth of wireless communication technologies, many satellite system, and GPS applications demands design of high-gain, broadband CP antenna. Circularly polarized antenna allows more adjustable adaptation between transmitter and receiver antennae. CP can be retrieved when asymmetry is introduced in patch geometry or exciting two orthogonal modes in radiating patch of equal amplitude and 90° phase difference. The antenna with single feed, it is mostly observed that thickness of the substrate affects the axial ratio bandwidth [1]. Horizontal microstrip feed line technique is used to achieve good impedance bandwidth in [2]. In dual-feed and

K.B. Biradar (✉) · M.S. Subhedar
Pillai HOC College of Engineering & Technology, Khalapur, Raigad 410206, Maharashtra, India
e-mail: biradarkishor@gmail.com

M.S. Subhedar
e-mail: msubhedar@mes.ac.in

H.S. Behera and D.P. Mohapatra (eds.), *Computational Intelligence in Data Mining*, Advances in Intelligent Systems and Computing 556,
DOI 10.1007/978-981-10-3874-7_57

four-feed antenna structures, modifications in patch are not required. By using 9° hybrid or power divider circuit, wide axial ratio bandwidth can be achieved as in [3]. Drawback of such technique is complicated structure and large antenna size.

L-shaped strip feed structure is also used for good impedance matching technique [4]. To obtain CP, symmetrical slot can be feed using suspended horizontal L-shaped coupling strip [5, 6]. Using double line strip feed technique and ring slot patch antenna, 3 dB AR bandwidth is achieved to be 10.5% [7]. Chen et al. proposed antenna with the four probes and microstrip feed line [8]. It is an effective way to excite CP, and axial ratio bandwidth of 16.4% is observed.

In this paper, using single SMA probe which is connected between ground plane and L-shaped strip and two cylindrical shaped probes connected to L-shape strip and suspended square-ring radiating patch form dual-feed CP patch antenna operating at ISM band. By optimizing various dimensions of L-strip and selecting the position of two probes, good CP is achieved. In addition, because of simple structure, antenna is very easy to fabricate. This paper is assembled as follows: Section 2 discusses on antenna geometry, Sect. 3 contains experimental results, and Sect. 4 concludes paper. All antenna simulation is carried out using IE3D software.

2 Design and Geometrical View of Antenna

Figure 1 shows the geometry of proposed CP antenna. A square-shaped FR4 substrate (thickness (h) = 1.6 mm, $tan\delta = 0.002$, relative permittivity (ϵ_r) = 4.4) of length L_2 is printed with a square-ring patch of length L_1, and whole assembly is suspended on ground plane of length L_3. Figure 1c shows the L-shaped strip with two arm, one along x-axis and another along y-axis having length P_2 and P_1 and

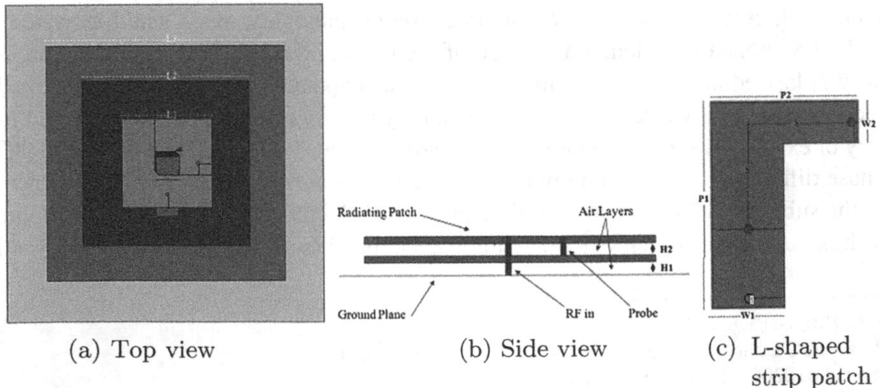

(a) Top view (b) Side view (c) L-shaped strip patch

Fig. 1 Proposed antenna geometry

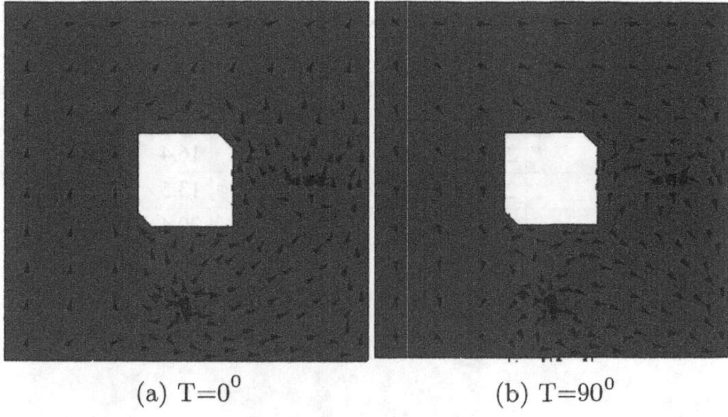

Fig. 2 Current distributions on the radiating patch

width W_2 and W_1, respectively. The height between ground and L-strip is H_2 and L-strip and radiating ring patch is H_1. Figure 1b shows side view of proposed antenna geometry. The input port is given along y-axis, while other arm is along the x-axis. Two probes, on x and y axes, are connected to radiating patch at length s from the center of the patch. Value of the s should be approximately $P_2 - W_1/2 - 0.6$ mm [9]. Square of length l with corner-chopped length d_1 is slotted on radiating patch. The radius of the transmission line and two shorting pins is 0.6 mm. To achieve CP and good impedance matching, two probes should be equidistant from the center.

The revised parameters of the proposed antenna are $L_1 = 38$ mm, $L_2 = 70$ mm, $L_3 = 100$ mm, $P_1 = 9$ mm, $P_2 = 5$ mm, $W_1 = 9$ mm, $W_2 = 5$ mm, $s = 13$ mm, $l = 10$ mm, $d_1 = 1.5$ mm, and $H_1 = H_2 = 3$ mm. This antenna operates with a center frequency of 2.45 GHz.

The two probes are used in proposed structure responsible for increasing the current distribution on suspended square-ring radiating patch. From Fig. 2, it observed that the radiating patch is excited with two orthogonal currents. As the phase of the current distribution along y-axis has a 900° phase difference with that on the x-axis, current distribution is clockwise, and hence, right-hand circular polarization (RHCP) characteristics is achieved. Calculated current value at two-feed points approximately equals to 2.45 GHz. Also, there is very small difference in amplitude at this mode, and both frequencies are orthogonal to each other that confirms circular polarization of antenna. Table 1 indicates the comparison of the proposed antenna structure with existing structures; from this, it is observed that proposed antenna structure has wide AR bandwidth and size of the antenna is much smaller than listed ones.

Table 1 Comparison with existing antenna structures

Reference	Reported bandwidth	
	Input impedance bandwidth (%)	Axial ratio bandwidth (%)
Chen and Chung [8]	23.9	16.4
Wang et al. [2]	25.8	13.5
Wang et al. [3]	30.4	20.4
Wu et al. [9]	14.3	13.1
Proposed antenna	21.75	19.6

3 Measured Results

Square ring with a corner-chopped square shape slot and a horizontal suspended L-shaped strip line is fabricated on different FR4 substrates. The comparison between measured and simulated S_{11} bandwidth is shown in Fig. 5, and simulated radiation pattern is shown in Fig. 6. Prototype of antenna is fabricated and experimentally demonstrated as shown in Fig. 7. It shows the below 10 dB impedance bandwidth of 21.75% (2.13–2.65 GHz) is shown in Fig. 3 and below 3 dB axial ratio bandwidth of 19.6% (2.16–2.60 GHz) is shown in Fig. 4. Measured simulated gain over 3 dB axial ratio is 7.16 dBi.

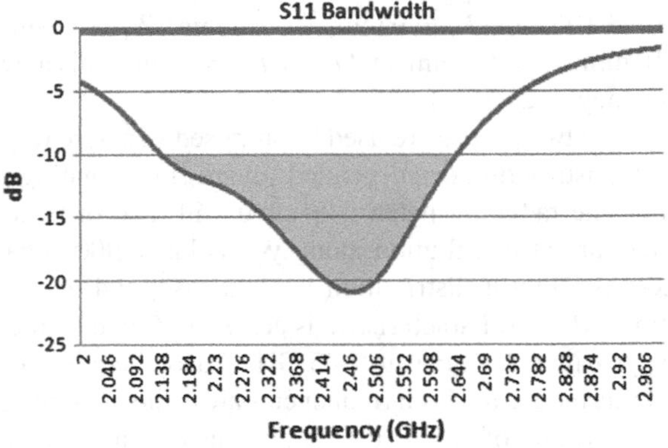

Fig. 3 Simulated S_{11} bandwidth

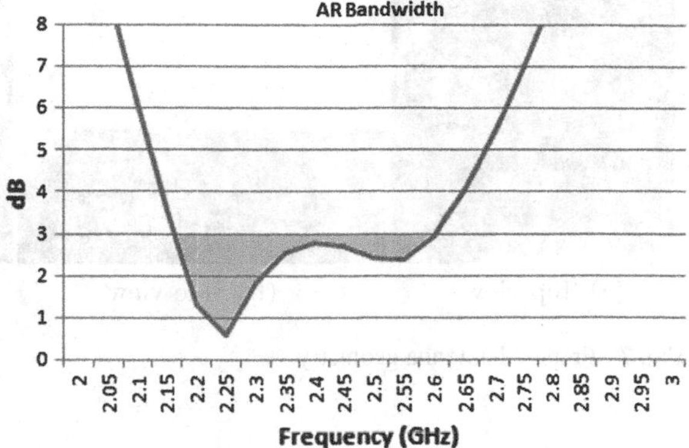

Fig. 4 Simulated AR bandwidth

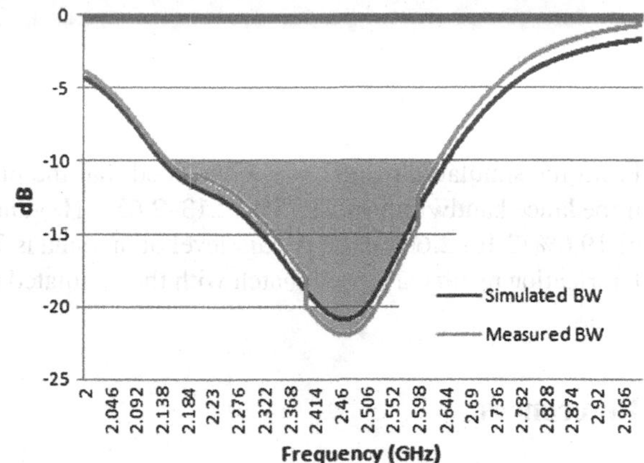

Fig. 5 Comparison of simulated and measured bandwidth

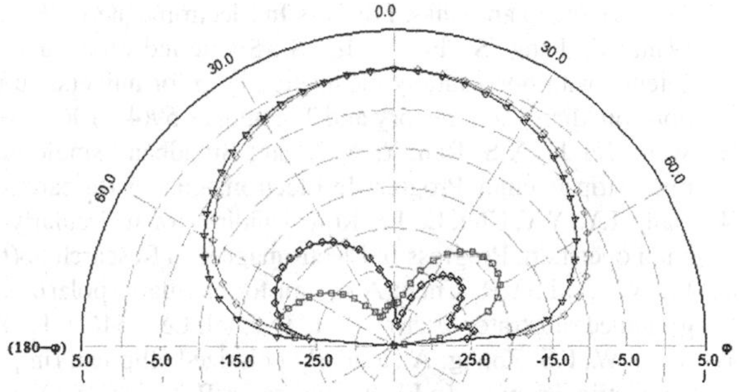

Fig. 6 Simulated radiation Pattern

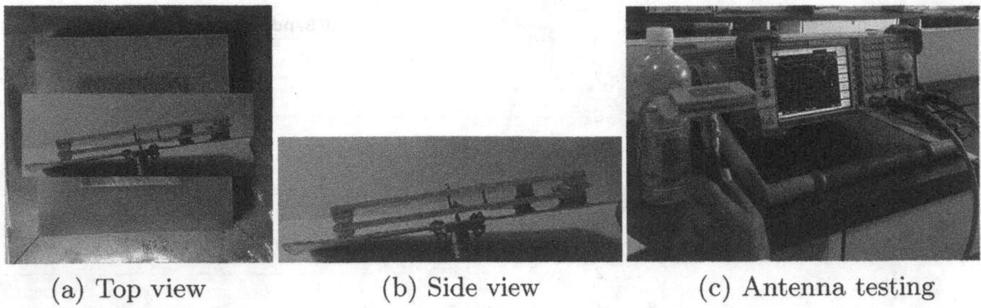

(a) Top view (b) Side view (c) Antenna testing

Fig. 7 Proposed antenna geometry

4 Conclusion

A suspended square ring with a corner-chopped square shape slot patch antenna feed by two cylindrical probes and suspended horizontal L-shaped strip has been presented. Proposed feeding structure gives broadband circular polarized bandwidth. From the simulated results, it is concluded that the proposed antenna has an input impedance bandwidth of 21.75% (2.13–2.65 GHz) and 3 dB axial ratio bandwidth of 19.6% (2.16–2.60 GHz). A gain level of antenna is 7.16 dBi over AR bandwidth. Fabrication results are good match with the simulated result.

References

1. Yang S.S., K.F. Lee, A., K.M. Luk.: Design and study of wideband single feed circularly polarized microstrip antennas. Progress In Electromagnetics Research 80, 45–61 (2008)
2. Wang, Z., Fang, S., Fu, S., Jia, S.: Single-fed broadband circularly polarized stacked patch antenna with horizontally meandered strip for universal uhf rfid applications. IEEE Transactions on Microwave Theory and Techniques 59(4), 1066–1073 (April 2011)
3. Wang Xu. F., X.S. Ren, Y., S.T. Fan: Broadband single-fed single-patch circularly polarized microstrip antenna. Progress In Electromagnetics Research 34, 203–213 (2013)
4. Hang T.Y., Y.Y. Chu, L., J.S. Row: Unidirectional circularly-polarized slot antennas with broadband operation. Progress In Electromagnetics Research 56(6), 1777–1780 (2008)
5. Liu W.L., M., T.R. Chen: A design for circularly polarized ring slot antennas backed by thin grounded substrate. Microw. Opt. Technol. Lett. 54(7), 1768–1770 (2012)
6. M.Z., W., F.S. Zhang: A circularly polarized elliptical ring slot antenna using an l-shaped coupling strip. Progress In Electromagnetics Research 35(6), 29–35 (2012)
7. Sze J.Y. Hsu C.I., Ho M.H., O.Y., M.T., W.: Design of circularly polarized annular-ring slot antennas fed by a double-bent microstripline. IEEE Transactions on Antennas and Propagation 55(11), 3134–3139 (Nov 2007)
8. Z.N. Chen, X.Q., Chung, H.L.: A universal uhf rfid reader antenna. IEEE Transactions on Microwave Theory and Techniques 57(5), 1275–1282 (May 2009)
9. Jian-Jun Wu, Xue-Shi Ren, Y.Z. Y., Wang, Z.D.: A broadband circularly polarized antenna fed by horizontal l-shaped strip. Progress In Electromagnetics Research 41, 175–184 (2013)

A Genetic Algorithm with Naive Bayesian Framework for Discovery of Classification Rules

Pooja Goyal and Saroj

Abstract Genetic algorithms (GAs) for discovery of classification rules have gained importance due to their capability of finding global optimal solutions. However, building a rule-based classification model from large datasets using GAs is very time-consuming task. This paper proposes an efficient GA that seeds the initial population with gain ratio-based probabilistic framework. The gain ratio is normalized information gain of attributes and is not biased toward multi-valued attributes. In addition, the proposed approach computes the fitness of individuals from a pre-computed matrix of posterior probabilities using Bayesian framework instead of making repeated training database scans. This approach for fitness computation increases the efficiency of the proposed GA by eliminating a large number of database scans. The enhanced GA is validated on ten datasets from UCI machine learning repository. The results confirm that the proposed approach performs better than a GA with randomly initialized population (GARIP) and Naïve Bayesian classifier.

Keywords Genetic algorithm · Classification rule discovery · Gain ratio · Naïve bayesian classifier

1 Introduction

Classification is a well-studied data mining task for predictive modeling. A classification model is learnt from training data and, subsequently, it is used to predict the class labels of data instances unseen during the learning phase. There are several algorithms such as Decision Trees, Neural Networks (NNs), Support Vector

P. Goyal (✉) · Saroj
Department of Computer Science and Engineering, Guru Jambheshwar University of Science and Technology, Hisar 125001, Haryana, India
e-mail: poojagoyal681992@gmail.com

Saroj
e-mail: ratnoo.saroj@gmail.com

H.S. Behera and D.P. Mohapatra (eds.), *Computational Intelligence in Data Mining*, Advances in Intelligent Systems and Computing 556,
DOI 10.1007/978-981-10-3874-7_58

Machines (SVMs) and Rough Sets to build classification models [1]. Some of them like NN and SVM are known for their high accuracy but models are not very comprehensible. Rough Sets produce a large number of classification rules that reduces comprehensibility. Decision trees generate accurate and comprehensible classifiers in the form of "*If-Then*" rules; however, these build a model by selecting one attribute at a time following some greedy heuristic approach [2, 3].

GAs are stochastic algorithms for solving optimization problems that involve large search space. GAs are based on the natural phenomenon of genetic inheritance and Darwin theory of evolution [4]. Building a classification model is an optimization problem in the sense that it searches for a global optimal classifier (made of rules in "*If-Then*" form). A genetic algorithm is able to learn good classifiers because of their robust search mechanism in the candidate rules' search space. The main purpose of using GAs for discovering classification rules is that they perform a global search and deal with attribute interactions better than the other greedy rule induction algorithms. However, GAs have some limitations too. Usually, a GA starts its evolution from a randomly initialized population. A randomly initialized population will have classification rules of extremely low fitness. Such an approach may either fail or it may take long time to discover optimal classification model. Moreover, a GA has to repeatedly scan the training database to compute the fitness of classification rules [5]. Several methods such as reducing data, preventing duplicate fitness computations, and seeding the initial population have been used to address this problem. This paper suggests an efficient genetic algorithm for discovery of classification models. The efficiency is enhanced on two fronts: (i) By seeding the initial population to have better-fit rules and (ii) By using Bayesian framework for fitness computations that eliminates the need for repeated training database scans. This approach estimates fitness of candidate solutions using Bayesian framework instead of computing True Positive rate (TP), False Positive rate (FP), False Negative rate (FN), and True Negative rate (TN) via database scans. The suggested approach computes a matrix of posterior probabilities of alleles of all the attributes of a dataset. It scans the training data once for every attribute to compute these posterior probabilities.

The rest of the paper is organized as follows: Sect. 2 presents the related work to enhance the performance of GA for discovery of classification rules. Section 3 describes Naive Bayesian framework briefly. Section 4 portrays the novel GA design which includes gain ratio-based probabilistic initialization and fitness evaluation using Naïve Bayesian framework. This section also describes the genetic operators used. The experimental design and results are presented in Sect. 5. Section 6 concludes the paper and gives the future scope of this work.

2 Related Work

Various attempts have been made to apply GAs for the discovery of classifier. Some of the earlier and well-known Genetic rule miners are Genetic-based Inductive Learning (GIL), High Dimensionality Pattern Discovery and Classification Systems (HDPDCS), and Coverage-based Genetic Induction (COGIN) algorithm [6]. Some of the recent successful genetic algorithm approaches in the domain of classification rule discovery are suggested in [7–11]. A commonly used genetic approach starts with a randomly initialized population of classification rules. These randomly initialized rules have either zero or very low fitness and hardly cover any examples in training data. Hence, fitness proportionate selection becomes ineffective and a GA takes a very large number of generations to evolve an optimal classification model. Moreover, GAs in the domain of discovery of classification rules make thousands of training database scans while computing the fitness of individual rules during their evolution [5]. Two approaches to address these issues are seeding initial population with domain knowledge and reducing database scans to compute fitness.

2.1 Seeding Initial Population

One approach to deal with randomly initialized population is to seed it with the examples from the training data itself [12]. This approach randomly selects some of the examples from training data. These examples are extremely specialized rules that will cover at least one example. The examples can be used to create generalized rules by dropping some of the attributes value pairs [13, 14]. In this context, Sharma and Saroj (2014) have recently proposed a bottom-up Pittsburgh approach for discovering classification models [15]. Liu and Kwok (2000) have used a non-random selection of examples to serve as seeds. Their seeding method was based on the idea that a seed should be a data instance lying at the center of the cluster of instances belonging to the same class as that of seed's class. Kapila and Saroj (2010) and Vashishtha et al. (2011) have proposed genetic algorithms that seed the initial population on the basis of entropy of attributes [5, 16]. Because entropy is biased toward multi-valued attributes, we have seeded the initial population on the basis of probabilities computed from gain ratio of attributes.

2.2 Reducing Database Scans

A database scan is a computationally expensive operation and, therefore, it is important to reduce the database scans to enhance the execution time of GAs. Some researchers have attempted to augment a GA with a long-term memory in the form of data structures such as Binary Search Tree and heap to reduce the database scans. These approaches use a data structure to store the fitness of the rules generated during the evolution [5, 17]. In [5], the authors have used a heap to organize the candidate solutions (Classification rules in *If-Then* form) and their fitness. During evolution whenever a rule is generated, the heap is searched for the rule. If the rule is found, then its fitness is not re-evaluated but retrieved from the heap. If search fails, then the fitness of the rule is evaluated and it is inserted in the heap. However, such approaches need extra amount of memory and additional computational effort. In this work, we have computed a matrix of posterior probabilities and used a Bayesian framework for fitness evaluations.

3 The Naive Bayesian Classifier

A Bayesian classifier is a statistical classifier, i.e., it predicts class membership probabilities for a given instance [1]. It is simple but has comparable performance to decision tree and neural network classifiers. The naive Bayesian classifiers work as follows:

Let us assume a dataset D of training examples with m classes, C_1, C_2, … C_m. Given a test example X to classify, a naïve Bayesian classifier computes the posterior probabilities $P(C_i|X)$ and it predicts the class C_i for the test example X if $P(C_i|X) > P(Cj|X)$ where $1 \leq j \leq m$. A naïve Bayesian classifier is based on Bayes formula given in Eq. 1.

$$P(Ci|X) = \frac{P(X|Ci) * P(Ci)}{P(X)}. \quad (1)$$

The class prior probability $P(C_i)$ is measured as $|C_{i,D}|/|D|$; where $|C_{i,D}|$ refers to number of training examples that belong to class C_i. The denominator $P(X)$ (prior probability) can be avoided, as it is constant for all the classes. As attributes are assumed conditionally independent, thus

$$P(X|Ci) = \prod_{k=1}^{n} P(x_k|Ci) = P(x_1|C_i) * P(x_2|C_i) * \cdots * P(x_n|C_i). \quad (2)$$

In above Eq. 2, x_k represents an allele of an attribute.

4 The Proposed GA Design

We have designed a GA using Michigan approach for encoding rules. Each individual represents a single classification rule in the GA population [18]. We have devised GA operators and fitness function especially suitable for discovery of classification rules.

4.1 Population Initialization

A randomly initialized population usually does not have good rules to direct the search toward globally optimal rule set. Therefore, the initial population is seeded with gain ratio of attributes. The gain ratio of an attribute indicates its predictive power. Higher the gain ratio an attribute has, more relevant it is for predicting class labels. The stepwise initialization process is as follows:

Computing Information Gain. The attributes with higher information gain are able to produce pure partitions with respect to their values and class labels. The information gain of an attribute (A) is computed as given in Eq. 3.

$$\text{Gain}\,(A) = \text{Entropy}\,(D) - \sum_{j=1}^{v} \frac{|Dj|}{|D|} * \text{Entropy}\,(D_j). \tag{3}$$

where $Entropy(D) = -\sum_{k=1}^{m} P_k * (\log_2 P_k)$; m is the number of classes, P_k is the probability of kth class, and v is number of values/alleles of attribute A. The value $|Dj|/|D|$ refers to the weight of the jth partition.

Computing Gain Ratio. The Gain Ratio is normalized information gain. Gain Ratio and initialization probability for an attribute (A) are computed as given in Eqs. 4 and 6.

$$\text{Gain Ratio}(A) = \frac{\text{Gain}(A)}{\text{SplitInfo}(A)}. \tag{4}$$

$$\text{Where, SplitInfo}\,(A) = \sum_{j=1}^{v} \frac{|Dj|}{|D|} * \log_2 \left(\frac{|Dj|}{|D|}\right). \tag{5}$$

$$\text{Initialization Prob}\,(A) = \frac{GainRatio(A) - MinGainRatio}{MaxGainRatio - MinGainRatio}. \tag{6}$$

Encoding. An individual chromosome in the proposed GA population represents a classification rule in "*If* A *Then* C" form; where A (antecedent part) containing conjunction of predicting attributes and C is the predicted class. An individual rule is encoded as an alphanumeric string of fixed length. The length of a rule is equal to the number of attributes in a given dataset. The locus of each attribute in a rule is fixed. An attribute is initialized by one of its alleles in proportion to its initialization probability. If an attribute has high initialization probability, it gets initialized frequently. If an attribute has low initialization probability, it is more often initialized with a "#" character. The "#" symbol represents a "don't care state" which stands for absence of an attribute in a candidate rule. For illustrating encoding scheme, consider the Mushroom dataset with 22 predicting attributes and two classes ("e" = edible and "p" = poisonous). Therefore, each chromosome will be

A_1	A_2	A_3	A_4	A_5	A_6	A_7	A_8	A_9	$A_{10\text{-}19}$	A_{20}	A_{21}	A_{22}	A_{23}
#	#	#	#	n	#	#	#	#	# … #	#	#	#	e

Fig. 1 Encoding scheme for a rule

of length 23. The fifth attribute named as "odor" is initialized by its alleles more frequently because it has the highest gain ratio and hence the highest probability of initialization. A classification rule "*If* odor = none, *Then* class = edible" is encoded as depicted in Fig. 1.

4.2 Computing Fitness

In this work, a Bayesian framework is applied to compute fitness of classification rules. The first part of Eq. 7 is the posterior probability of an allele x_k conditioned on class C_i occurring in a rule and the second part is the prior probability of the class of the rule.

$$\text{Fitness}\,(R) = \prod_{k=1}^{n} P(x_k|C_i) * P(C_i). \tag{7}$$

Computing a matrix of count and posteriori probabilities. A frequency matrix is created to store the frequency count of each of the alleles with respect to each class by scanning training data. The first column contains the names and alleles of the attributes. The second and third columns store the count of the corresponding alleles and posteriori probability under the edible and poisonous classes, respectively. For illustration purpose, a part of matrix containing only two attributes of Mushroom dataset is given in Table 1. A Laplace correction needs to be applied to avoid getting zero probability for the alleles with zero count in the frequency matrix [18]. The last row contains the prior probabilities for the class labels.

The fitness of the rules is computed by multiplying the entries of posterior probabilities from Table 1 for respective alleles occurred in the rule antecedent. It is further multiplied by prior probability of the respective class as given in the last row of Table 1. The fitness computed for two sample rules is depicted in the Table 2.

4.3 Genetic Operators

Selection. The selection process selects individuals from current generation to breed for next generation. In present work, roulette wheel selection method is used. The roulette wheel selects number of parents determined by a parameter of GA named as percentage replacement.

Table 1 Count and posterior probabilities for the alleles of selected attributes

Attributes		Class "e = edible"		Class "p = poisonous"	
Name	Alleles	$\lvert x_k \wedge {''e''} \rvert$	$P(x_k \mid C_i) = \frac{\lvert x_k \wedge {''e''} \rvert}{\lvert {''e''} \rvert}$	$\lvert x_k \wedge {''p''} \rvert$	$P(x_k \mid C_i) = \frac{\lvert x_k \wedge {''p''} \rvert}{\lvert {''p''} \rvert}$
Odor	a (almond)	352	0.1119	0	0.0005
	c (creosote)	0	0.0003	172	0.0889
	f (fishy)	0	0.0003	1428	0.7384
	l (anise)	360	0.1144	0	0.0005
	m (musty)	0	0.0003	31	0.0160
	n (none)	2433	0.7736	79	0.0408
	p (pungent)	0	0.0003	224	0.1158
	y (yellow)	0	0.0003	7	0.0036
Spore-print-color	h (chocolate)	0	0.0003	1416	0.7310
	k (black)	1499	0.4719	203	0.1048
	n (brown)	1522	0.4792	203	0.1048
	r (green)	0	0.0003	60	0.0309
	u (purple)	40	0.0125	0	0.0005
	w (white)	90	0.0283	46	0.0237
P(class) = $\lvert C_i \rvert / \lvert D \rvert$		$\lvert {''e''} \rvert / \lvert D \rvert = 3145/5079 = 0.62$		$\lvert {''p''} \rvert / \lvert D \rvert = 1934/5079 = 0.38$	

Table 2 Fitness computation of sample rules

Rule	Fitness
If odor = none, *Then* class = edible	0.7736 * 0.62 = 0.479
If odor = fishy and spore-print color = chocolate, *Then* class = poisonous	0.7384 * 0.731 * 0.38 = 0.205

Crossover. The crossover (or recombination) operator swaps genetic material between two "parents" creating two new offspring. Here, one-point simple crossover is used. As interspecies rule breeding would result in bad rules, a crossover is permitted within the rules of same species.

Mutation. The mutation operator randomly changes the value/allele of an attribute with another value/allele belonging to the domain of that attribute. Mutation operator may specialize a candidate rule by replacing "#" (don't care state) with one of the alleles and generalize a rule by replacing an allele with a "#" symbol. In our GA implementation, an attribute having more gain ratio is mutated with greater probability.

4.4 Maintaining Diversity in GA Population

A crowding scheme is employed to discover a set of diverse classification rules. It uses overlapping GA populations. In GAs with overlapping populations, only a certain percentage of population is replaced with newly generated offspring. An offspring replaces the worst but most similar individual in the GA population to keep different species of rules. An offspring is compared with some of the worst fit members of the GA population using a similarity function. The similarity between two rules ($R1$ and $R2$) is measured as given in Eq. 8.

$$\text{Similarity}\ (R1,\ R2) = \frac{|R1 \cap R2|}{|R1 \cup R2|}. \tag{8}$$

where $|R1 \cap R2|$ is the count of training tuples covered by both rules and $|R1 \cup R2|$ refers to training tuples covered by either rule $R1$ or rule $R2$. This similarity function depends on the fact that different species of rules will occupy different and mutually exclusive niches in the training data. Using above formula for measuring similarity, an offspring will replace a worst individual of its own species. This plan preserves the diversity in the population and prevents the convergence to the single best solution.

4.5 Building Classifier Model

The classifier model is built incrementally by updating it at the end of every run of the proposed GA. The rules obtained in a GA run are sorted in decreasing order of fitness, and then, the best rule is added to the classifier. Subsequently, sequential covering strategy is employed to remove the examples covered by this rule from the training dataset. This process continues until almost all the examples of training set have been covered. Some examples can be left uncovered to avoid over fitting of the model. The number of examples left uncovered is decided by a given parameter.

5 Experimental Setup and Results

The proposed GA has been implemented in R software and evaluated on ten datasets from the UCI Machine Learning repository. The characteristics of these data sets are summarized in Table 3. The values for GA parameters are given in Table 4.

To evaluate the performance of the proposed GA, it is compared to a GA with randomly initialized population (GARIP) and Naïve Bayesian classifiers (Table 5).

Table 3 Description of datasets used in our experiment

Dataset	# instances	# predicting attribute		# class
		# nominal	# numerical	
Mushroom (MS)	5644	22	0	2
Vote (VT)	232	16	0	2
Nursery (NS)	12960	8	0	5
Car (CR)	1728	6	0	4
Tic-tac-toe (TTT)	958	9	0	2
Credit-a (Ca)	690	8	6	2
Breast cancer (BC)	683	0	10	2
Pima (PM)	768	0	8	2
Iris (IS)	150	0	4	3
Chess (CS)	3196	36	0	2

Table 4 Values of GA parameters

Sr. no.	Name of the parameter	Value
1	Population size	60
2	No. of generations	20
3	Crossover probability	0.25
4	Mutation probability	0.05
5	Replacement	8
6	Number of uncovered examples for sequential covering strategy	1% of the training dataset size

Table 5 Predictive Accuracy obtained from GARIP, Naive Bayesian and the proposed GA

Dataset	GARIP	Naïve bayesian	The proposed GA
Mushroom	85.2	95.56	99.8
Vote	87.0	90.94	99.1
Nursery	66.3	90.32	83.4
Car	69.2	82.46	97.9
Tic-tac-toe	68.4	69.62	92.8
Credit-a	82.5	84.92	84.9
Breast cancer	73.9	97.42	94.6
Pima	67.0	67.70	91.6
Iris	94.0	90.00	94.3
Chess	79.0	87.89	96.8

In GARIP, all the attributes are initialized randomly. A tenfold cross validation was used to evaluate the performance of these algorithms. The resulting predictive accuracies have been averaged over 10 folds of the sampling technique.

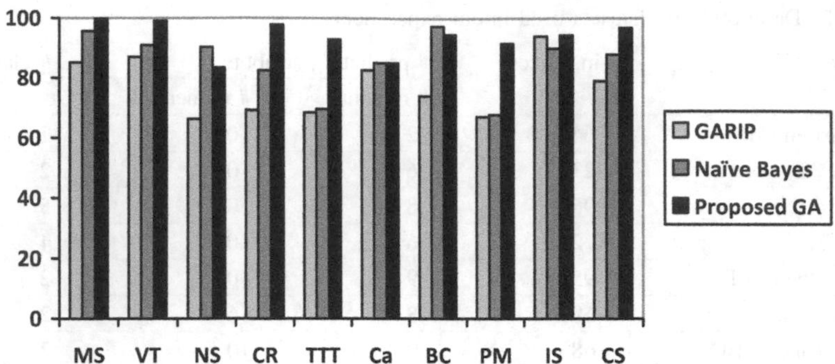

Fig. 2 Comparison of the proposed GA with other classification algorithm

Table 6 A comparison of runtime on three datasets (time measured in second)

Name of the dataset	Runtime of a traditional GA (with database scans) [T_1]	Runtime of the proposed GA [T_2]	Speed gain [T_1/T_2]
Nursery	7.24	0.17	42.59
Mushroom	4.12	0.17	24.24
Car	1.78	0.12	14.83

It is clearly visible from Table 5 and the graph shown in Fig. 2 that the proposed genetic algorithm is definitely better than GARIP. The proposed GA-based classifier is either better or comparable to Naïve Bayes Classifier for most of the datasets. The pure Naïve Bayesian approach performs better only for two datasets (Nursery and Breast Cancer). Moreover, the proposed GA does not make large number of database scans for computing fitness of classification rules over successive generations. Table 6 compares runtimes of a traditional GA (with database scans) and the proposed algorithm for the three largest datasets out of the ten datasets used for experimentation. This Table shows that the proposed genetic algorithm runs several times faster.

6 Conclusion and Future Scope

This paper has proposed an efficient genetic algorithm for discovering classification rules. The proposed algorithm seeds the initial population on the basis of gain ratio of attributes. The proposed genetic algorithm gets a better start with the presence of better-fit rules in the initial population. This type of initialization leads to evolution of a better-fit classifier in lesser number of generations.

The proposed genetic algorithm eliminates a number of training database scans for fitness computations. It makes a matrix of posterior probabilities from training data and uses this matrix for computing fitness of rules during GA runs. The database scans take significant amount of time for large datasets. This work has enhanced the efficiency of genetic algorithms in the domain of classification rule discovery. One of the limitations of the suggested approach is that it does not work well if dependencies exist between attributes and needs to incorporate attribute dependency in fitness computation mechanism.

References

1. Han, J., Kamber, M., Pei J.: Data Mining: Concepts and Techniques. Morgan Kaufmann Publishers, Burlington, MA (2011).
2. Zhou, C., Xiao, W., Tirpak, T.M., Nelson, P.C.: Evolving Accurate and Compact Classification Rules with Gene Expression Programming. IEEE Transactions on Evolutionary Computation. 7, 519–531 (2003).
3. Maimon, O., Rokach, L.: Introduction to Knowledge Discovery and Data Mining. In: Maimon, O. and Rokach, L. (eds.) Data Mining and Knowledge Discovery Handbook. pp. 1–15. Springer US (2009).
4. Michalewicz, Z.: Genetic Algorithms + Data Structures = Evolution Programs. Springer Berlin Heidelberg, Berlin, Heidelberg (1996).
5. Saroj, Kapila, Kumar, D., Kanika: A Genetic Algorithm with Entropy Based Probabilistic Initialization and Memory for Automated Rule Mining. In: Meghanathan, N., Kaushik, B.K., and Nagamalai, D. (eds.) Advances in Computer Science and Information Technology. pp. 604–613. Springer Berlin Heidelberg (2011).
6. Vashishtha, J., Kumar, D., Ratnoo, S., Kundu, K.: Mining Comprehensible and Interesting Rules: A Genetic Algorithm Approach. International Journal of Computer Applications. 31, 39–47 (2011).
7. Sarkar, B.K., Sana, S.S., Chaudhuri, K.: A Genetic Algorithm-based Rule Extraction System. Applied Soft Computing. 12, 238–254 (2012).
8. Juan Liu, J., Kwok, J.T.-Y.: An Extended Genetic Rule Induction Algorithm. In: Proceedings of the 2000 Congress on Evolutionary Computation, 2000. pp. 458–463 vol. 1 (2000).
9. Fidelis, M.V., Lopes, H.S., Freitas, A.A.: Discovering Comprehensible Classification Rules with a Genetic Algorithm. In: Proceedings of the 2000 Congress on Evolutionary Computation, 2000. pp. 805–810 vol. 1 (2000).
10. Noda, E., Freitas, A.A., Lopes, H.S.: Discovering Interesting Prediction Rules with a Genetic Algorithm. In: Proceedings of the Congress on Evolutionary Computation. CEC 99. p. 1329 (1999).
11. Shi, X.J., Lei, H.: A Genetic Algorithm-Based Approach for Classification Rule Discovery. In: 2008 International Conference on Information Management, Innovation Management and Industrial Engineering. pp. 175–178. IEEE, Taipei (2008).
12. Freitas, A.A.: Data Mining and Knowledge Discovery with Evolutionary Algorithms. Springer Science & Business Media (2013).
13. Walter, D., Mohan, C.K.: ClaDia: A Fuzzy Classifier System for Disease Diagnosis. In: Proceedings of the 2000 Congress on Evolutionary Computation, 2000. pp. 1429–1435 vol. 2 (2000).
14. Kwedlo, W., Krętowski, M.: Discovery of Decision Rules from Databases: An Evolutionary Approach. In: Żytkow, J.M. and Quafafou, M. (eds.) Principles of Data Mining and Knowledge Discovery. pp. 370–378. Springer Berlin Heidelberg (1998).

15. Sharma, P., Ratnoo, S.: Bottom-up Pittsburgh Approach for Discovery of Classification Rules. In: International Conference on Contemporary Computing and Informatics (IC3I). pp. 31–37. IEEE, Mysore (2014).
16. Kapila, Saroj, Kumar, D., Kanika: A Genetic Algorithm with Entropy based Initial bias for Automated Rule Mining. In: 2010 International Conference on Computer and Communication Technology (ICCCT). pp. 491–495 (2010).
17. Gantovnik, V.B., Anderson-Cook, C.M., Gürdal, Z., Watson, L.T.: A Genetic Algorithm with Memory for Mixed Discrete–Continuous Design Optimization. Computers & Structures. 81, 2003–2009 (2003).
18. Freitas, A.A.: A Survey of Evolutionary Algorithms for Data Mining and Knowledge Discovery. In: Ghosh, D.A. and Tsutsui, P.D.S. (eds.) Advances in Evolutionary Computing. pp. 819–845. Springer Berlin Heidelberg (2003).

Study of a Multiuser Kurtosis Algorithm and an Information Maximization Algorithm for Blind Source Separation

Monorama Swain, Rachita Biswal, Rutuparna Panda and Prithviraj Kabisatpathy

Abstract An attempt is made in this study to use two distinct algorithms to inspect blind source separation (BSS). In this paper, we have used multiuser Kurtosis (MUK) algorithm for BSS and an information maximization algorithm for separation and deconvolution of voice signals. Among the various criteria available for evaluating the objective function for BSS, we have considered information theory principles and Kurtosis as a measure for statistical independence. The MUK algorithm uses a combination of Gram–Schmidt orthogonalization and a stochastic gradient update to achieve non-Gaussian behavior. A correlation coefficient is used as an evaluation criterion to analyze the performance of both the algorithms. Simulations results are presented at the end of the study along with the performance tabulation for discussion and analysis.

Keywords Blind source separation · Information maximization · Kurtosis · Correlation coefficient

M. Swain (✉) · R. Biswal
Department of Electronics and Tele-communication, SIT Bhubaneswar, Bhubaneswar, India
e-mail: mswain@silicon.ac.in

R. Biswal
e-mail: biswal.sanjana770@gmail.com

R. Panda
Department of Electronics and Tele-communication, VSSUT, Burla, India
e-mail: r_ppanda@yahoo.co.in

P. Kabisatpathy
Department of ETC, CVRCE, Bhubaneswar, India
e-mail: pkabisatpathy@gmail.com

H.S. Behera and D.P. Mohapatra (eds.), *Computational Intelligence in Data Mining*, Advances in Intelligent Systems and Computing 556,
DOI 10.1007/978-981-10-3874-7_59

1 Introduction

The basic use of blind source separation (BSS) technique is to extract the underlying signal of interest from a set of mixture. The recent trends in BSS and component analysis are based on matrix factorization to get specific knowledge about the signal. The goal of BSS can be considered as estimation of true physical sources of mixing systems. It is based on various unsupervised learning methods and assumes only some statistical knowledge about sources and systems. Assumption is made that a set of independent sources are present in the data whose actual properties are unknown and all the source signals are mutually independent. From the central limit theorem, if the non-Gaussian behavior of a set of signals is maximized, independent signals that have maximum entropy are recovered [1, 2], and some measure of independence is then defined. The data are subsequently decorrelated by maximizing this measure. Here, the source separation is based on statistical methods, namely second-order statistics information maximization algorithm and fourth-order statistics Kurtosis maximization.

The paper consists of the following sections. In Sect. 2, the problem formulation has been briefly described. In Sect. 3, we learn about the information maximization approach. In Sect. 4, we describe about the different algorithms used for BSS. In Sect. 5, we explain all the experimental procedures and the corresponding results. Section 6 winds up the paper.

2 Methodology

The model consists of a statistically independent signal S, mixed with matrix A resulting a vector signal Y. Then, a demixing matrix W is estimated that is acclimated to restore original signal S.

The BSS signal model is given in Eq. (1)

$$X(r) = AS(r) + v(r) \quad (1)$$

where

$S(r) = [s_1(r), s_2(r), \ldots, s_p(r)]^T$	is a $p \times 1$ vector of transmitted source signals;
$Ap \times q$	mixing matrix
$X(r)\, q \times 1$	vector of acquired signal
$v(r)\, q \times 1$	vector of add-on noise samples at time instant r
T	matrix/transpose vector.

Figure 1 gives the basic overview of BSS model, Here, we studied the BSS problem taking three audio sources (2 female, 1 male) using both the multiuser Kurtosis and information maximization algorithms for estimating the demixing matrix W. Subsequently, the cross-correlation coefficient is calculated to prove a

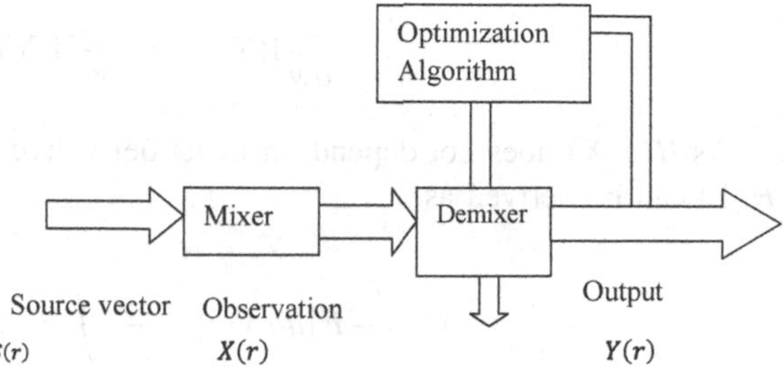

Fig. 1 System overview of the BSS process

match among the input and output data. The three sources $s_1(r), s_2(r), s_3(r)$ are mixed through a mixing matrix A and give observed mixed signals $X_1(r), X_2(r), X_3(r)$ [3, 4]. The intent of the source separation algorithms is to evaluate all source signals. The plan is to find the demixing matrix W^{TV} considering the components of the reconstructed signals are mutually independent without knowing A, and distribution of the source signal $S(r)$ is shown in the following equation

$$Y(r) = W^T(r)X(r) \tag{2}$$

3 Information Maximization Approach

Infomax is considered as an information maximization algorithm, formulated by Anthony J. Bell and T.J. Sejnowski [5]. It is one of the leading techniques used to separate instantaneous mixtures.

This method consists of minimizing the mutual information as well as augmenting output entropy. According to Laughlin principles, it states that when inputs crossed through a sigmoid function, maximal information transmission takes place. It takes place when sloping part of the sigmoid is optimally aligned with the high-density parts of the inputs [5]. This can be resolved by using an adaptive logic, which uses a stochastic gradient ascent rule. The process of maximizing the mutual information such that the output Y of neural network processor contains about its input X is given by

$$I(Y,X) = H(Y) - H(Y|X) \tag{3}$$

where $H(Y)$ is the entropy of the output and $H(Y|X)$ is that entropy of the output not considering from the input. Here, the leveling between X and Y is deterministic, and $H(Y|X)$ has its lowest possible value which deviates to $-\infty$. Equation (3) can be attributed as follows, with respect to a parameter w, considering the calibration from X to Y:

$$\frac{\partial}{\partial \mathrm{w}} \mathrm{I}(\mathrm{Y}, \mathrm{X}) = \frac{\partial}{\partial \mathrm{w}} \mathrm{H}(\mathrm{Y}) \tag{4}$$

As $H(Y|X)$ does not depend on w, its derivative is zero. The entropy of output $H(Y)$ can be derived as

$$(Y) = -E\left[lnf_y(y)\right] = -\int_{-x}^{x} f_y(y) lnf_y(y) dy \tag{5}$$

$E[.]$ gives excepted value and $f_y(y)$ is the probability density function of the output. For multiple input and output system according to A.J. Bell and T.J. Sejnowski, given an input x weight matrix W, a bias vector w_0, and a monotonically transformed output vector

$$y = g(W_x + w_0) \tag{6}$$

The probability function of y can be computed as

$$f_y(y) = \frac{f_x(x)}{|J|} \tag{7}$$

where $|J|$ is the complete value of the Jacobian transformation. To maximize entropy, we have to maximize $ln|J|$. Considering logistic function as the nonlinearity, i.e., $g(u) = (1 + e^u)^{-1}$, $u = W_x + w_0$, the resulted learning rules will be

$$\frac{dH(y)}{dW} = W^{-T} + (1 - 2y)x^T \tag{8}$$

$$\frac{dH(y)}{dw_0} = 1 - 2y \tag{9}$$

4 Multiuser Kurtosis (MUK) Maximization Algorithm

Over a decade, different researchers for blind source separation have proposed higher-order statistics-based optimization algorithms. In this study, a MUK algorithm is implemented for solving the BSS problem. The usual MUK assumptions are made—the transmitted signals are reciprocally independent likewise distributed and transmitted signals share the same non-Gaussian distribution [6, 7]. From the BSS model, we know that $Y(r) = W^T(r)X(r)$

which can be written as

$$Y(r) = W^T(r)AS(r) = G^T(r)S(r) \tag{10}$$

where $W(r)$ is $n \times m$ matrix and $G^T(r)$ $m \times m$ global response matrix.

This algorithm is implemented for communication signal, which is typically considered to be sub-Gaussian, but the separation of speech signal is ended by recognizing the speech signal to be super-Gaussian [6]. If each $s_j(r)$, $(j = 1, 2, \ldots, p)$ is a non-Gaussian, zero-mean sequence $\{s_j(r)\}$, $\{s_l(r)\}$ are statistically independent for $j \neq l$, the necessary conditions are for recovery of source or transmitted signals at the outputs:

1. $|KY_j(r)| = |K_s|, j = 1, 2, \ldots, p$
2. $\mathrm{K}|Y_j(r)|^2 = \sigma_s^2, j = 1, 2, \ldots, p$
3. $E\left(Y_j(r)Y_l^*(r)\right) = 0; j \neq 1$

Where $E(x) = E\left(|x|^4\right) - 2E^2\left(|x|^2\right) - |E(x^2)|$ is the (un-normalized) Kurtosis of x, σ_s^2, K_s are the variance and Kurtosis of each $s_j(r)$, respectively, and * marks complex conjugate.

The BSS prototype is derived from conditions (1), (2), and (3).

$$\begin{cases} \max\limits_{\mathrm{G}} \mathrm{F(G)} = \sum_{\mathrm{j}=1}^{\mathrm{p}} |\mathrm{K(Y_j)}| \\ \text{subject to } \mathrm{G^H G = I_p} \end{cases} \tag{11}$$

where $\mathrm{I_p}$ is the $\mathrm{p} \times \mathrm{p}$ identity matrix and H marks Hermitian transpose. The constraint comes from the evidence that, according to (2) and (3), E $(\mathrm{YY^H}) = \sigma_s^2 \mathrm{I_p}$ (we have also simulated that $\sigma_s^2 = 1$). We call this equation to be the multiUser Kurtosis (MUK) maximization basis for blind source separation. The following steps are involved for the implementation of MUK algorithm

1. First, update $W(r)$ in the direction implementation of the instantaneous gradient (controlled by a step size μ):

$$W'(r+1) = W(r) + \mu\, sign(K_s) X^*(r) Y(r) \tag{12}$$

2. Continuing the iteration to satisfy the orthogonality basis: $G^H(r+1)G(r+1) = I_p$. Considering H as unitary, it gives

$$W^H(r+1)W(r+1) = I_p \tag{13}$$

3. For

$$j = 1, \ldots, p: \tag{14}$$

where δ_{lj} is Kronecker delta, Eq. (13) can also be written as

$$\begin{cases} \min_{W_j} \Delta(W_j) = \left(W_j - W_j^{'}\right)^H \left(W_j - W_j^{'}\right) \\ \text{subject to: } W_l^H W_j = \delta_{lj}, l = 1, \ldots, j \end{cases} \tag{15}$$

4. In order to compute the problem (15), the Lagrangian of $\Delta(W_j)$ is as follows:

$$L_{\Delta\left(w_j, \lambda_j, \mu_j, v_j\right)} = \Delta W_j - \lambda_j (W_j^H W_j - 1) - \sum_{l=1}^{j-1} \mu_{lj} Re\,(W_l^H W_j) - \sum_{l}^{j-1} v_{lj} Im\,(W_l^H W_j) \tag{16}$$

Table 1: The MUK algorithm

1. r = 0: Initialize W(0) = W_0
2. for r > 0
3. Obtain $W^{'}(r+1)$ from (12)
4. Obtain $W_1(r+1) = W_1^{'}(r+1)/W_1^{'}(r+1)$
5. for j = 2: p
6. Calculate $W_j(r+1)$ from (21)
7. Go to 5
8. $W(r+1) = [W_1(r+1), \ldots, W_p(r+1)]$
9. Go to 2

5. Where $\lambda_j, \mu_{lj}, v_{lj}$ are real scalars. The gradient of $L_{\Delta\left(w_j, \lambda_j, \mu, v\right)}$ with respect to W_j to zero $\left(\frac{dL_\Delta}{dW_j^{'}} = 0\right)$, so for each j the equation will be:

$$W_j - W_j^{'} - \lambda_j W_j - \int_{l=1}^{j-1} \beta_{lj} W_l = 0 \tag{17}$$

where $\beta_{lj} = 1/2\left(\mu_{lj} + i v_{lj}\right)$ and $i = \sqrt{-1}$ from (17)

$$\begin{cases} \lambda_j = 1 - W_j^H W_j^{'} \\ \beta_{lj} - W_j^H W \end{cases} \tag{18}$$

which gives

$$(1-\lambda_j)W_j = W_j' - \sum_{l=1}^{j-1} W_l^H W_j' W_l \tag{19}$$

According to (19),

$$W_j \propto \left(W_j' - \sum_{l=1}^{j-1} W_l^H W_j' W_l \right) \tag{20}$$

where ∝ denotes "proportional."

Now

$$W_j(r+1) = \frac{w_j'(r+1) - \sum_{l=1}^{j-1} (w_l^H(r+1)w_j'(r+1))w_l(r+1)}{\left\| w_j'(r+1) - \sum_{l=1}^{j-1} (w_l^H(r+1)w_j'(r+1))w_l(r+1) \right\|} \tag{21}$$

The Gram–Schmidt orthogonalization of matrix $W_j'(r+1)$ is implemented in steps 4–8 of the algorithm. An important property associated with that of the MUK algorithm is that it can efficiently and effectively handle both the sub-Gaussian and super-Gaussian inputs. Furthermore, the MUK algorithm can also be altered to handle non-symmetrical input. The algorithm is tested and implemented.

5 Experimental Methods and Analysis of Results

(A) BSS and Blind Deconvolution Using Information Maximization

1. Blind Source Separation Results

In the infomax approach, we generate a mixing matrix A random in nature with values within the range of -1 and 1 that are uniformly distributed. The matrixes S and X are both $m \times n$ matrices, and X is calculated from S by permutating its time index to produce $S^{\dagger}$. The mixture X is a combination of mixing matrix $X = AS\dagger$. The demixing matrix W and the bias vector w_0 were then trained. Here, in this experiment, we have taken 3 sources (2 females and 1 male) that were successfully separated and show cross-corelation coefficient values. The learning rate value is 0.001 that has been taken for the experiment. The strength and the direction of a linear relationship between the estimated and the actual value of the variable can be easily estimated by the cross-correlation coefficient. If the coefficient is closer to either -1 or 1, then the correlation between the variables is considered to be stronger. Figures 2, 3, and 4 show the source signals, the first two from female subjects and the last one from a male subject. Figure 5 shows the mixing signal obtained by procedures mentioned earlier. Figure 6a–c shows the retrieved signals derived from the mixed signal.

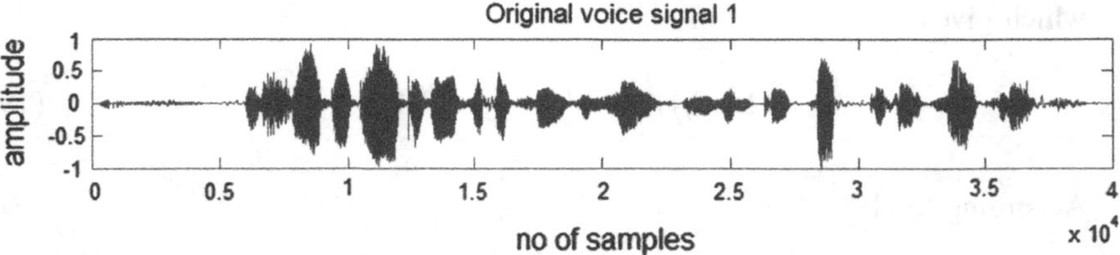

Fig. 2 First source signal

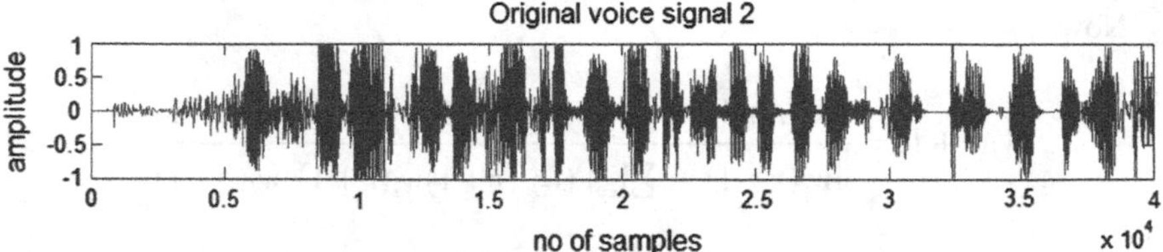

Fig. 3 Second source signal

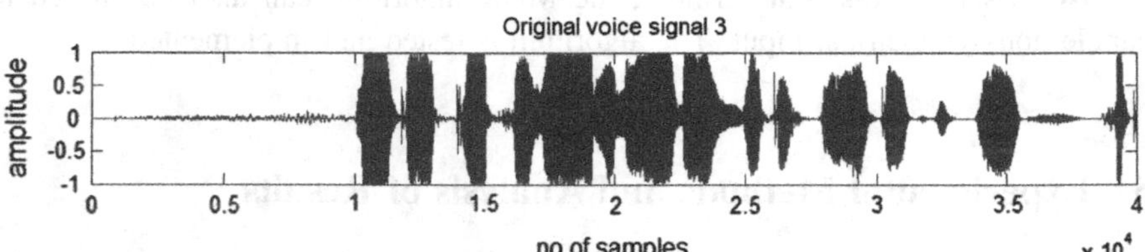

Fig. 4 Third source signal

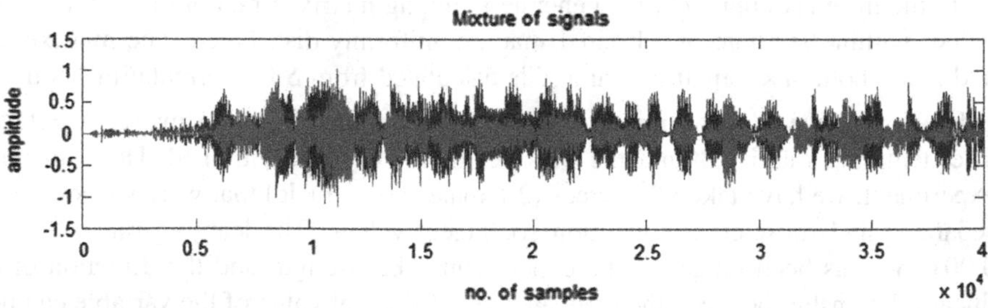

Fig. 5 Mixing signal

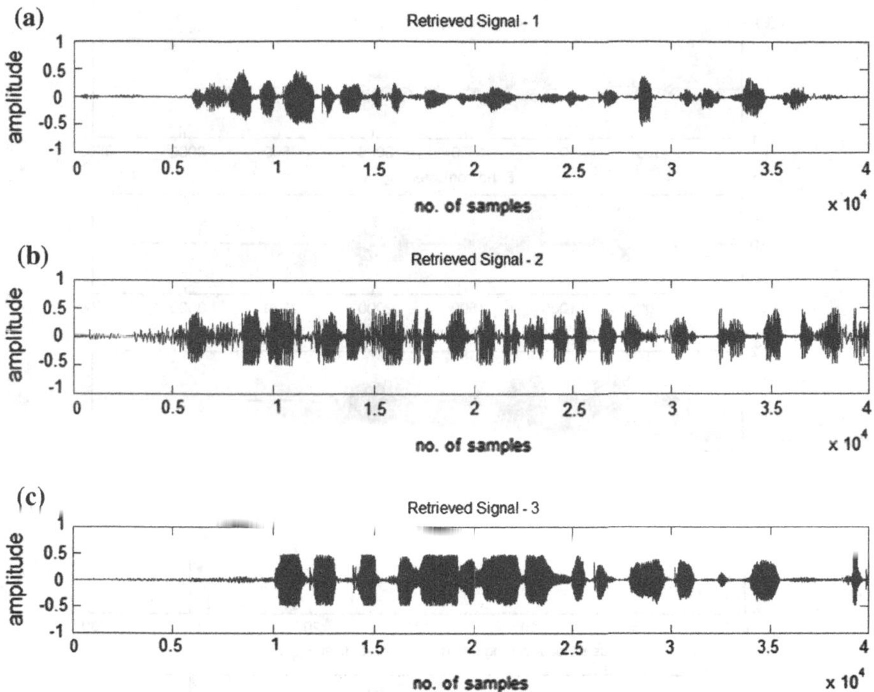

Fig. 6 **a** First retrieved signal. **b** Second retrieved signal. **c** Third retrieved signal

The cross-correlation coefficient values for infomax are as follows:

$$r11 = \begin{bmatrix} 1.0000 & 0.9992 \\ 0.9992 & 1.0000 \end{bmatrix}, r22 = \begin{bmatrix} 1.0000 & 0.9995 \\ 0.9995 & 1.0000 \end{bmatrix}$$
$$r33 = \begin{bmatrix} 1.0000 & 0.9978 \\ 0.9978 & 1.0000 \end{bmatrix}$$

2. Blind Deconvolution Results

In blind deconvolution method, we have taken a single unknown speech signal S(r) which is convolved with an unknown tapped delay line filter $a_1, a_2, \ldots, a_N$ giving a corrupted signal $x(r) = a(r) * s(r)$, i.e., $a(r)$ is the impulse response of the filter. The task is to revert S(r) by convolving $X(r)$ with a known filter $w_1, w_2, \ldots, w_L$ which switches the effect of the filter $a(r)$. Speech signals were convolved with various filters, and learning rules were used to execute blind deconvolution [8, 9]. Here, we have taken the learning rate 0.0001 for better evaluation. From the experiment, we have displayed the convolving filter and deconvolving filter and learnt filter and the spectrogram as shown in the Fig. 7a–c. Here, a 6.25 ms echo is added to the signal that sounds a mild audible barrel effect.

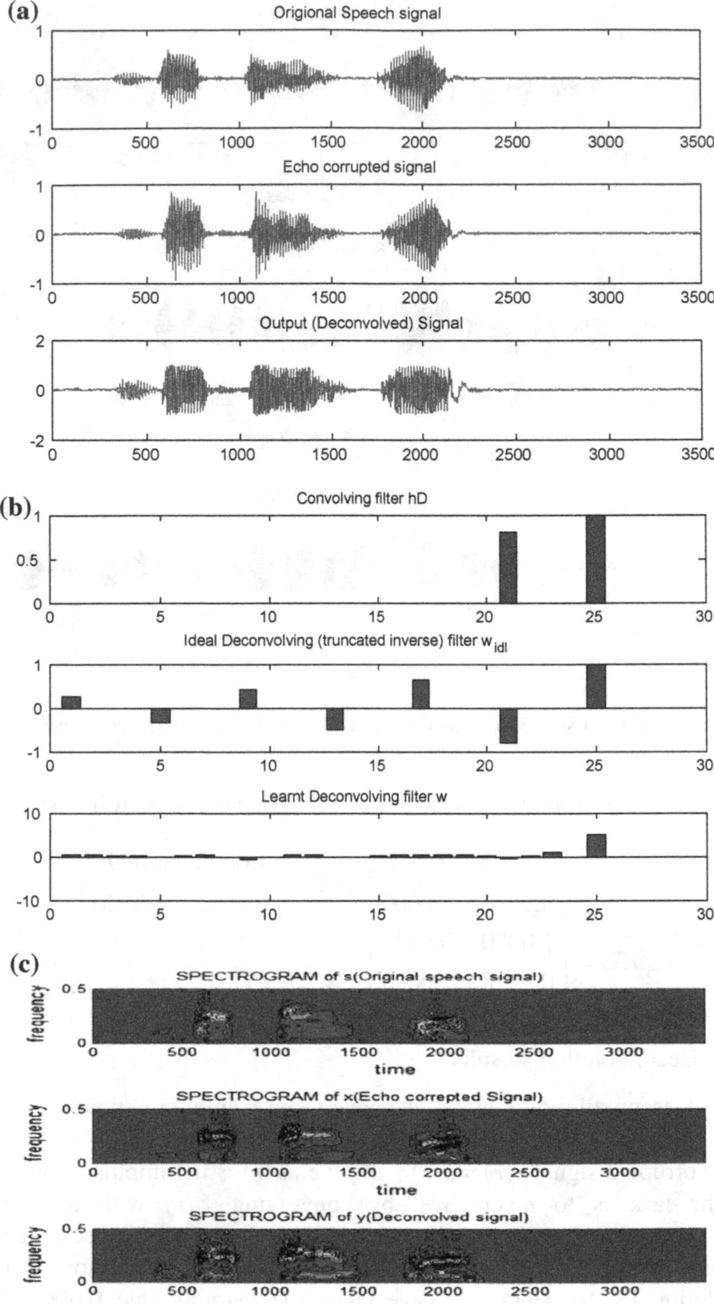

Fig. 7 **a–c** Blind deconvolution results

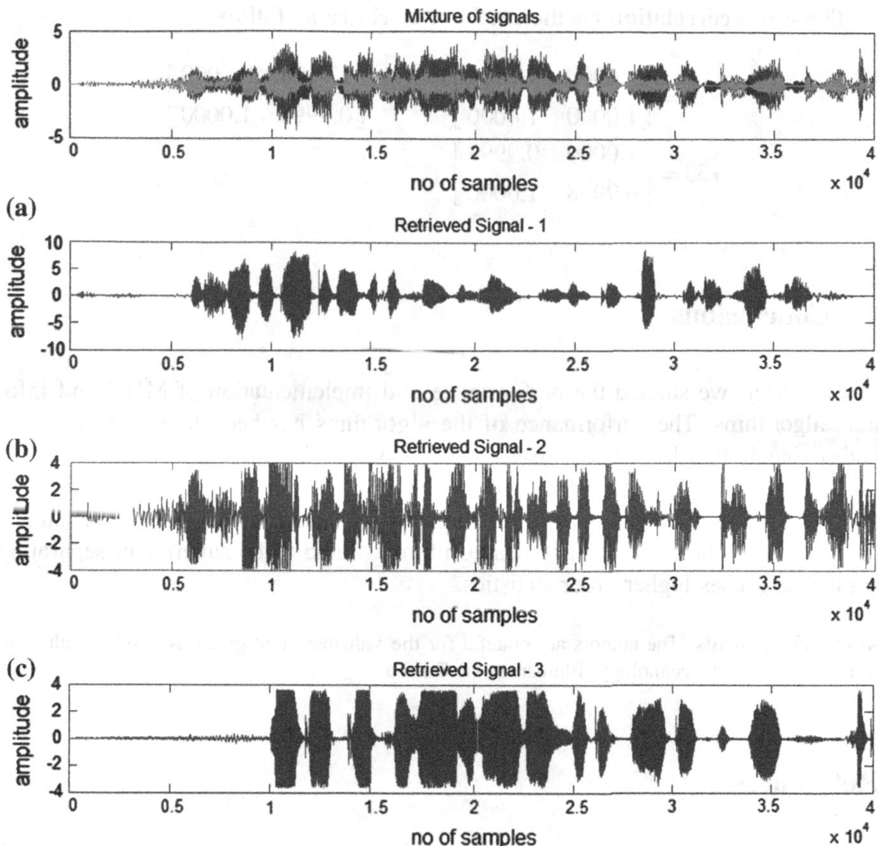

Fig. 8 Mixing signal. **a** First retrieved signal. **b** Second retrieved signal. **c** Third retrieved signal

(B) Multiuser Kurtosis Maximization Algorithm

The performance of the MUK algorithm shows us that there is successful separation between the different sources. The constrained multiuser optimization criterion gives us the pathway for the derivation of the algorithm. At each iteration, the algorithm followed by a Gram–Schmidt orthogonalization that projects the updated settings on the constraint [10, 11] performs stochastic gradient adaptation. It was observed from the figures that the cross-correlation coefficient is nearly equal to 1 as compared to infomax. Hence, the second-order statistics of X (i.e., the autocorrelation matrix R_{XX}) do not provide enough information, so higher statistics are most preferable [12]. Figure 8 shows the mixing signal, and Fig. 8a–c shows the retrieved signals.

The cross-correlation coefficient for MUK are as follows:

$$r11 = \begin{bmatrix} 1.0000 & 1.0000 \\ 1.0000 & 1.0000 \end{bmatrix}, r22 = \begin{bmatrix} 1.0000 & 0.9998 \\ 0.9998 & 1.0000 \end{bmatrix}$$

$$r33 = \begin{bmatrix} 1.0000 & 0.9998 \\ 0.9998 & 1.0000 \end{bmatrix}$$

6 Conclusions

In this paper, we studied the performance and implementation of MUK and infomax algorithms. The performance of the algorithms has been tested on real-time audio signals that have been mixed artificially. An investigation of blind deconvolution on three source signals and one unknown echo source signal was conducted. The robustness of the algorithms is evaluated in terms of normalized correlation coefficients. The MUK algorithm is found to be superior in separating signals as it uses higher-order statistics.

Acknowledgements The authors are grateful for the valuable input given by Prof. J. Talukdar, Silicon Institute of Technology, Bhubaneswar, Odisha.

References

1. P. Comon and L. Rota, "Blind separation of independent sources from convolutive mixtures," IEICE Trans. on Fundamentals, vol. E86-A, no. 3, pp. 542–549, Mar 2003.
2. Hyvärinen and E. Oja, "Independent Component Analysis: Algorithms and Applications", Neural Networks, vol 13, pp. 411–430, 2000.
3. E. Weinstein, M. Feder, and A. Oppenheim., "Multichannel signal separation by decorrelation." IEEE Trans. Speech Audio Proc., vol. 1, no. 4, pp. 405–413, Oct 1993.
4. A. Hyvarinen, et al., "Independent Component Analysis", John Wiley & Sons Co, 2001.
5. Bell and T. Sejnowski. An Information Maximization Approach to Blind Separation and Blind Deconvolution. Neural Computation, 7:1129–1159, July 1995-97.
6. Constantinos B. Papadias, Member, IEEE, Globally Convergent Blind Source Separation Based on a Multiuser Kurtosis Maximization Criterion IEEE Transaction on Signal Processing, vol. 48, no. 12, pp 3508–3519, Dec., 2000.
7. Shahab Faiz Minhas, Patrick Gaydecki "A hybrid algorithm for blind source separation of a convolutive mixture of three speech sources" Eurasip journal on advances in signal processing, page 1–15,2014.
8. Konstantinos Diamantaras, Gabriela Vranou, Theophilos Papadimitriou, "Multi Input Single Output Nonlinear Blind separation of Binary Sources" IEEE Transactions on signal processing, Vol. 61, No. 11, June 1,2013.
9. C. B. Papadias. "Kurtosis-based criteria for adaptive blind source separation". In IEEE International Conference on Acoustics, Speech, and Signal Processing, pages 2317–2320, Seattle, WA, USA, May 12–15, 1998.

10. A. Cichocki, R. Unbehauen, and E. Rummert, "Robust learning algorithm for blind separation of signals," Electronics Letters, vol. 30, no. 17, pp. 1386–1387, 18 August 1994.
11. D. Yellin and E. Weinstein Multichannel signal, "separation: Methods and analysis," IEEE Trans. Sig. Proc., vol. 44, no. 1, pp. 106–118, Jan 1996.
12. M. Kadou, K. Arakawa, "A Method of Blind Source Separation for Mixed Voice Separation in Noisy and Reverberating Environment", IEICE Tech. Rep., vol. 108, no. 461, SIS2008-81, pp. 55–59, March 2009.

Use of Possibilistic Fuzzy C-means Clustering for Telecom Fraud Detection

Sharmila Subudhi and Suvasini Panigrahi

Abstract This paper presents a novel approach for detecting fraudulent activities in mobile telecommunication networks by using a possibilistic fuzzy c-means clustering. Initially, the optimal values of the clustering parameters are estimated experimentally. The behavioral profile modelling of subscribers is then done by applying the clustering algorithm on two relevant call features selected from the subscriber's historical call records. Any symptoms of intrusive activities are detected by comparing the most recent calling activity with their normal profile. A new calling instance is identified as malicious when its distance measured from the profile cluster centers exceeds a preset threshold. The effectiveness of our system is justified by carrying out large-scale experiments on a real-world dataset.

Keywords Call detail records · Clustering · Possibilistic fuzzy c-means · Fraud detection

1 Introduction

In recent years, the usage of mobile phones for communication has revolutionized the telecom industry along with the increase in the mobile phone subscriptions. This results in the rise of telecom fraud, which occurs whenever a fraudster performs deceptive methods to get the telephonic services free of charge or at a reduced rate. This problem leads to the loss of subscriber's faith in the service provider company as well as the revenue losses for the organization. According to a study [1] done by Financial Fraud Action United Kingdom (FFA UK), £23.9 million was lost in 2014 in the UK due to various fraudulent activities, which is three times more than the

S. Subudhi · S. Panigrahi (✉)
Department of Computer Science & IT, Veer Surendra Sai University of Technology, Burla 768018, India
e-mail: spanigrahi_cse@vssut.ac.in

S. Subudhi
e-mail: sharmilasubudhi1@gmail.com

H.S. Behera and D.P. Mohapatra (eds.), *Computational Intelligence in Data Mining*, Advances in Intelligent Systems and Computing 556,
DOI 10.1007/978-981-10-3874-7_60

previous year. The figure shown in the study reflects the growing trend of losses resulting due to rise in fraudulent activities in the telecom industry. Therefore, there is a need to address the mobile phone fraud problem in a quick manner for minimizing the financial losses.

The most common type of telecom fraud is known as the superimposed fraud, which can only be detected by the analysis of a genuine user's account for the presence of any kind of fraudulent activities made by the fraudster by exploiting the genuine account. This type of fraud can remain undetected for a long time as the presence of fraudulent activities is comparatively small in the overall call volume [2]. In this work, we aim at detecting the superimposed mobile phone fraud by applying possibilistic fuzzy c-means (PFCM) clustering on the subscriber's call detail records (CDRs). We have demonstrated the effectiveness of our proposed system by performing extensive experiments on reality mining dataset [3]. To the best of our knowledge, this is the first ever attempt to develop a mobile phone FDS by using PFCM clustering technique.

The rest of the paper is organized as follows: Section 2 discusses the previous work done in mobile phone fraud detection. Section 3 focuses on the fundamental concept of PFCM clustering. The next section elaborates the proposed FDS along with its working methodology. In Sect. 5, we have discussed the results obtained from the experimental analysis. Finally, in Sect. 6, we conclude the paper with some future enhancements of the proposed model.

2 Related Work

In this section, some published works have been reviewed that are relevant to the mobile phone fraud detection problem. In paper [4], the authors have suggested the usage of Dempster–Shafer theory and Bayesian inferencing for information fusion from various sources for the detection of fraudulent activities. The authors of [5] present the application of feed forward neural network and hierarchical agglomerative clustering for fraud detection with five different user profiles for each user. They have applied each technique independently on those user profiles for discriminating illegitimate calls from the legitimate ones by visualizing different aspects of the model. The work in [6] proposes an FDS for the detection of malicious activities present in a subscriber account by data visualization using self-organizing map (SOM).

Another recent work [7] suggests the building of five different user profiles from the features selected by applying four different feature selection algorithms. A genetic programming (GP)-based classifier is then used for the detection of fraudulent patterns. The usefulness of K-means clustering and hierarchical agglomerative clustering algorithms has been presented in [8] for fraud detection. The discrimination of fraudulent signatures from the genuine ones are done by applying these two methods independently on five different user profiles built from the CDRs of each user.

Although several methodologies have been suggested for developing an efficient mobile phone FDS, one of the major issues in the above-mentioned systems is the limited applicability of various hard clustering methods for solving such type of real-world problem in which there is no crisp boundary for segregating the normal user profile and intrusive patterns. Moreover, an individual data point may belong to more than one cluster with different membership values. For improving the accuracy of fraud detection, we have therefore applied the possibilistic fuzzy c-means clustering algorithm in the current work. Besides, this method is superior to two other fuzzy clustering algorithms, namely fuzzy c-means (FCM) and possibilistic c-means (PCM) as it solves the outlier sensitivity problem of FCM and the overlapped cluster issue of PCM.

3 Background Study

In this section, we briefly describe the working principle of PFCM for demonstrating the training and fraud detection methodologies of our proposed system.

3.1 Possibilistic Fuzzy C-Means Clustering

Possibilistic fuzzy c-means (PFCM) [9] clustering is a hybrid of two most widely used fuzzy clustering algorithms, namely FCM [10] and PCM [11]. PFCM overcomes the inefficiency of handling noisy instances of FCM and the coincident cluster problem of PCM simultaneously. PFCM takes unlabeled instances of a dataset and attempts to form clusters by finding the most appropriate point as centroid in each cluster. A membership value and typicality value is then assigned to every point in the clusters. This is attained by minimizing the *objective function* as stated below:

$$\begin{aligned} MinJ_{m,\eta}(U,T,V;D) = & \sum_{j=1}^{n}\sum_{i=1}^{c}(a_{ij}^{m}+bt_{ij}^{\eta})\left\|d_j-v_i\right\|_A^2 \\ & +\sum_{i=1}^{c}\gamma_i\sum_{j=1}^{n}(1-t_{ij})^{\eta} \end{aligned} \tag{1}$$

subject to constraints $\sum_{i=1}^{c} u_{ij}=1\forall j$ and $0\le u_{ij},t_{ij}\le 1$ where $J_{m,\eta}$ is the objective function, m is the fuzzifier weighting exponent, and η is the scale parameter. $D=\{d_1, d_2,\ldots,d_n\}$ is the dataset with n points on which PFCM is to be performed, $U=[u_{ij}]$ is the membership matrix, $T=[t_{ij}]$ is the typicality matrix, $V=\{v_1, v_2,\ldots, v_c\}$ is a matrix of c cluster centers, and $\left\|d_j-v_i\right\|_A$ is the inner product norm used to compute the distance between cluster center v_i and the data point d_j, $\gamma_i>0$ is a user defined constant value and $a > 0$ is the significance of the membership value, $b > 0$ is the significance of typicality value, $m > 1$ and $\eta > 1$.

PFCM exhibits the FCM properties when $b = 0$ and displays PCM characteristics when $a = 0$. The clustering output becomes more favorable towards PCM as the value of b increases with respect to a and vice versa. For PFCM, a larger value of b is required to be considered as compared to a in order to reduce the effect of outlier instances. Likewise, the effects of noisy points can be reduced for a higher value of m than η. However, a very large value of m can cause the clustering model to be more receptive toward PCM. On giving a dataset as input to PFCM, it produces three different outputs—fuzzy membership matrix (U), typicality matrix (T), and a set of cluster centers (V) computed by using Eq. (1).

4 Proposed Approach

The proposed mobile phone FDS monitors the calling activities of the subscribers by analyzing their CDRs and identifies any fraudulent patterns by applying the PFCM clustering technique. The flow of events in our FDS is partitioned into two phases—training phase and fraud detection phase.

4.1 Training Phase

The training phase deals with the construction of behavioral profile of each user. We have considered the following relevant features for the representation of CDR of a user:

$$\langle user\_id, timedt, dur, type\_call \rangle$$

The feature *user_id* is used to uniquely identify each user by taking the anonymous interpretation of the IMEI (International Mobile Equipment Identity) number of the user's mobile device. The *timedt* denotes the date (ddmmyyyy) and time (hh:min:sec in 24-h format) of a call when it is made. Similarly, *dur* signifies the call duration measured in seconds and *type_call* refers to the type of calls made, which has been mapped to numerals as: 0 for local calls, 1 for national calls and 2 for international calls in our approach. For instance, suppose <7, *29042004171119, 50, 1*> represents a CDR of a user. This example indicates a call record having *user_id* as 7, *timedt* is 29-04-2004 and 17:11:19, *dur* is 50 s and *type_call* = 1 (national).

Initially, we perform normalization on the CDRs by converting all points in the range of [0, 1], since the high-valued attribute fields can cause bias while clustering. The CDRs of a subscriber are partitioned into training and testing sets. Once the segmentation of the dataset is complete, the parameter setting of PFCM is carried out by conducting experiments. The user behavioral profiles are then obtained by employing PFCM clustering technique on the training dataset based on the attributes: *dur* and *type_call*.

4.2 Fraud Detection Phase

After the profile building of a user is successfully completed, the testing set is used in the clustering model for the detection of any kind of fraudulent patterns present in the CDRs. This is accomplished by initially measuring the Euclidean distance (*d*) of the incoming call record with each cluster centroid and finally comparing *d* with a preset threshold value (*th*). The threshold value has been determined through rigorous experimentation as discussed in Table 3 of Sect. 5. If the distance value is higher than or equal to the threshold, then the call is marked as a fraudulent one. On the other hand, if the distance is smaller than the threshold value, then the call is identified as genuine. Upon detecting any illegitimate activities, the service provider company can obtain confirmation regarding the call from the respective subscriber.

5 Experimental Results and Discussions

The proposed FDS has been implemented in MATLAB 8.3 on a 2.40 GHz i5-4210U CPU system. The usefulness of our FDS has been presented by testing with reality mining dataset [3]. Initially, we have performed tests for the determination of optimal parameter values needed for PFCM clustering. Once the required parameters are obtained, we then conduct the fraud detection experiments.

The reality mining dataset consists of call records, messaging records, and much more information of 106 users gathered over a 9-month period from September 2004 to April 2005. We have used the following standard performance metrics—true positive rate (TPR), false positive rate (FPR), accuracy, precision, and F-score to measure the effectiveness of our proposed system. *TPR* denotes the ratio of truly positive samples that are correctly classified by the classifier. *FPR* measures the fraction of rejected genuine samples that incorrectly identify as fraudulent by the classifier. *Accuracy* estimates the correctness of a classifier. *Precision* can be depicted as the proportion of correct classification made by the system, and *F-score* refers to the harmonic mean of precision and TPR.

5.1 Determination of PFCM Parameters

In this section, we discuss the estimation of the correct combination of the clustering parameters required for the working of PFCM. Initially, we perform a set of

Table 1 Determination of optimal number of clusters

c	1	2	3	4
PC	1	*0.9824*	0.9456	0.9346

experiments for finding out the required number of clusters. We have considered partition coefficient (PC) index [12] for measuring the clustering validity as shown in Table 1.

$$PC = \frac{1}{n} \sum_{i=1}^{c} \sum_{j=1}^{n} u_{ij}^{2} \tag{2}$$

where PC denotes the average relative quantity of membership sharing done between the fuzzy subset pairs in the $U = [u_{ij}]$ matrix, c is the number of clusters, and n refers to the data points on which clustering is to be performed. The optimal cluster number (c^{+}) is chosen as follows:

$$c^{+} = max_{2 \leq c \leq n-1} PC \tag{3}$$

which has been presented in italics for better visualization in Table 1. The PC values are computed using Eq. (2), which produces the maximum value, i.e., PC = 0.9824 at $c^{+} = 2$ after satisfying with Eq. (3). Hence, we have chosen the optimal number of clusters $c^{+} = 2$.

After the correct number of clusters is determined, we then find the optimal combination of the other four PFCM parameters—significance of the membership value (*a*), significance of typicality value (*b*), weighting exponent (*m*), and scale parameter (η). The clustering output of PFCM is presented in Table 2 by taking different combinations of parameters with $c^{+} = 2$ along with a Fuzziness Performance Index (FPI) value. The FPI [13] can be defined as a measurement of the degree to which different classes share membership values. The optimal partition of fuzzy clustering can be found by minimizing the FPI value as this implies that the cluster elements have minimum overlapping between themselves. The FPI value can be calculated as follows:

$$FPI = 1 - (c^{*}PC - 1)/(c - 1) \tag{4}$$

where c is the number of clusters and PC is the partition coefficient index. From Table 2, it is quite clear from the cluster centroids $\{v_1, v_2\}$ that except at run 5 and run 6, all other runs produce overlapped clusters. The FPI values are calculated by using Eq. (4). However, the FPI value of run 6 is lesser than the FPI value of run 5. Therefore, we chose the parameter values of a, b, m, and η of Run 6 as an optimal combination of PFCM parameters, which has been italicized in Table 2 for better visualization.

The effectiveness of our proposed system also depends on the threshold value (*th*). The variations in TPR, FPR, accuracy, precision, and F-score over different threshold values are depicted in Table 3. It is clear from Table 3 that for

Table 2 Results produced by PFCM with different parameter values

Run	a	b	m	η	v_1	v_2	FPI
1	1	1	2	2	0.0082	0.0082	−1.878e+03
					0.5286	0.5286	
2	1	3	2	2	0.0090	0.0090	−1.878e+03
					0.5259	0.5259	
3	1	6	2	2	0.0098	0.0098	−1.878e+03
					0.5190	0.5190	
4	1	7	2	2	0.0099	0.0099	−1.878e+03
					0.5168	0.5168	
5	1	1	5	1.5	0.0097	0.0036	−3.2893 e+03
					0.4437	0.8428	
6	*1*	*1*	*7*	*1.5*	*0.0097*	*0.0018*	*−3.4189 e+03*
					0.4175	*0.9509*	
7	1	5	5	1.5	0.0083	0.0083	−1.878e+03
					0.5065	0.5065	
8	1	5	5	10	0.0087	0.0087	−1.878e+03
					0.5096	0.5096	
9	1	1	2	7	0.0079	0.0079	−1.878e+03
					0.5282	0.5282	
10	1	4	3	2	0.0094	0.0094	−1.878e+03
					05123	0.5123	

Table 3 Variation in different performance metrics over different threshold values

Threshold (*th*)	TPR (in %)	FPR (in %)	Accuracy (in %)	Precision (in %)	F-score (in %)
0.001	90.15	9.83	90.16	91.50	90.82
0.003	*95.07*	*9.25*	*93.09*	*92.34*	*93.69*
0.005	93.30	10.44	91.49	90.50	91.88
0.007	90.50	10.23	90.16	90.95	90.73
0.009	91.00	11.36	89.89	90.10	90.55

th = 0.003, our proposed system exhibits maximum TPR = 95.07% and minimum FPR = 9.25%. Hence, we choose the *th* = 0.003 for efficient fraud detection.

Table 4 presents a comparative performance analysis of various clustering methods on different performance metrics by taking the cluster number $c = 2$. It can be clearly seen that PFCM outperforms other clustering techniques by yielding better performance in terms of all performance metrics while keeping FPR = 9.25% at the lowest level. This selection is essential as the failure to detect a fraud causes direct loss to the service provider while the actions required to handle the false alarms also tend to be costly.

Table 4 Performance analysis of various clustering techniques

Clustering method	TPR (in %)	FPR (in %)	Accuracy (in %)	Precision (in %)	F-score (in %)
PFCM	*95.07*	*9.25*	*93.09*	*92.34*	*93.69*
FCM	75.68	12.34	80.59	89.94	82.15
PCM	84.68	11.69	86.17	91.26	87.85
K-means	69.51	15.03	75.80	87.08	77.31

6 Conclusions

In this work, a novel mobile phone fraud detection system has been suggested by employing PFCM clustering technique. The fraud detection procedure is segmented into two phases—training phase and fraud detection phase. For measuring the efficiency of our system, the reality mining dataset has been used. PFCM clustering algorithm is used for behavioral profile construction of mobile phone subscribers as well as for identification of any intrusive signatures present in their profiles. The experimental results show the ability of PFCM in detecting fraudulent activities of various users. Based upon the outcomes, it can be concluded that by using PFCM clustering technique, this kind of real-world problematic scenario can be addressed effectively.

Acknowledgements The authors are highly grateful to the Department of Computer Science & Engineering and IT, Veer Surendra Sai University of Technology, Burla, Sambalpur, Odisha, India, for providing the required amenities and support for making this successful investigation.

References

1. Kosmides, M,: Telephone fraud on rise in UK, study finds (2014) http://www.counterfraud.com/fraud-types-n-z/telecoms-fraud/telephone-fraud-on-rise-inuk-study-finds–1.htm, accessed: 30 January, 2016.
2. Cox, Kenneth C., et al.: Brief application description; visual data mining: Recognizing telephone calling fraud. Data Mining and Knowledge Discovery1.2 (1997) 225–231.
3. Eagle, N., Pentland, A.S.: Reality mining: sensing complex social systems. Personal and ubiquitous computing 10.4 (2006) 255–268.
4. Panigrahi, S., et al.: Use of dempster-shafer theory and Bayesian inferencing for fraud detection in mobile communication networks. Australasian Conference on Information Security and Privacy. Springer Berlin Heidelberg, (2007).
5. Hilas, C.S., Paris A.M..: An application of supervised and unsupervised learning approaches to telecommunications fraud detection. Knowledge-Based Systems 21.7 (2008) 721–726.
6. Olszewski, D.: Fraud detection using self-organizing map visualizing the user profiles." Knowledge-Based Systems 70 (2014) 324–334.
7. Hilas, C.S., et al.: A genetic programming approach to telecommunications fraud detection and classification. (2014).

8. Hilas, C.S., Paris A.M., Ioannis T.R.: Clustering of Telecommunications User Profiles for Fraud Detection and Security Enhancement in Large Corporate Networks: A case Study. Applied Mathematics & Information Sciences 9.4 (2015) 1709.
9. Pal, N.R., et al.: A possibilistic fuzzy c-means clustering algorithm. IEEE transactions on fuzzy systems 13.4 (2005) 517–530.
10. Bezdek, J.C., Ehrlich, R., Full, W.: FCM: The fuzzy c-means clustering algorithm. Computers & Geosciences 10.2–3 (1984) 191–203.
11. Krishnapuram, R., Keller, J.M.: A possibilistic approach to clustering. IEEE transactions on fuzzy systems 1.2 (1993) 98–110.
12. Wang, W., Zhang, Y.: On fuzzy cluster validity indices. Fuzzy sets and systems 158.19 (2007) 2095–2117.
13. Odeh, I. O. A., Chittleborough, D. J., McBratney, A. B.: Soil pattern recognition with fuzzy-c-means: application to classification and soil-landform interrelationships. Soil Science Society of America Journal 56.2 (1992) 505–516.

A Classification Model to Analyze the Spread and Emerging Trends of the Zika Virus in Twitter

B.K. Tripathy, Saurabh Thakur and Rahul Chowdhury

Abstract The Zika disease is a 2015–16 virus epidemic and continues to be a global health issue. The recent trend in sharing critical information on social networks such as Twitter has been a motivation for us to propose a classification model that classifies tweets related to Zika and thus enables us to extract helpful insights into the community. In this paper, we try to explain the process of data collection from Twitter, the preprocessing of the data, building a model to fit the data, comparing the accuracy of support vector machines and Naïve Bayes algorithm for text classification and state the reason for the superiority of support vector machine over Naïve Bayes algorithm. Useful analytical tools such as word clouds are also presented in this research work to provide a more sophisticated method to retrieve community support from social networks such as Twitter.

Keywords Zika · Twitter analysis · Twitter classification · Support vector machines · Naïve Bayes algorithm

1 Introduction

The Zika virus is responsible for causing the Zika disease and is primarily carried by the Aedes species mosquito. The incubation period of the disease lasts for at most a week and has symptoms such as fever, rashes, headache, and conjunctivitis. Zika virus was declared as a Public Health Emergency of International Concern (PHEIC) by World Health Organization (WHO) on February 1, 2016. At present,

B.K. Tripathy (✉) · S. Thakur · R. Chowdhury
School of Computing Science and Engineering, VIT University, Vellore, India
e-mail: tripathybk@vit.ac.in

S. Thakur
e-mail: saurabh.chandrakantthakur2013@vit.ac.in

R. Chowdhury
e-mail: rahul.chowdhury2013@vit.ac.in

H.S. Behera and D.P. Mohapatra (eds.), *Computational Intelligence in Data Mining*, Advances in Intelligent Systems and Computing 556,
DOI 10.1007/978-981-10-3874-7_61

there are no cures such as vaccines or any other form of treatment for this disease and thus makes it a serious global health issue.

Social networking such as Twitter and Facebook has often been treated as useful sources of information for community support on social outbreaks, especially on the global spectrum [9]. Twitter is a popular microblogging Web site where users interact socially by posting messages or the so-called 'tweets' on the Twitter platform. Twitter data have previously been used for various data analysis such as sentiment analysis [3, 8] and event detection and can be easily accessed by the publicly available Twitter API (application program interface). Twitter is highly popular in mobile application throughout the world and the users can post tweets that can be considered as precise sources of information as they have a 140-character limit [4]. Moreover, there are many verified accounts of reputed people, organizations, and communities and thus add more credibility to the tweets.

Preprocessing of the tweets: The Twitter Streaming API was used to collect the most recent tweets. The tweets collected by the API are then preprocessed initially to make the later analysis easier. The URLs, hashtags, and user mentions are separated from the text in the original tweet. We also provide an initial analysis of the tweets such as showing graphically the countries from where tweets related to Zika are being tweeted the most.

2 Preprocessing of Tweets

A Python script was written with the help of Twitter Streaming API which collects the most recent raw tweets with keywords such as 'Zika,' 'Zika virus,' and 'Aedes'—the species name of the mosquito causing Zika in a text file and then each tweet is converted into JSON (JavaScript Object Notation) for easy manipulation and handling of data. Pandas, an open source library for data manipulation in Python, is then used to store the data in a data frame with columns such as Twitter ID, created-at, text, and favorite-counts.

A total of 4751 tweets were collected, and after removing the re-tweets, we were left with 1471 unique tweets. The original tweet contains many elements other than the original text such as hashtags, external links, and user mentions. Thus, for proper analysis of the tweet, from the text we separated (1) stop words such as 'a', 'an,' and 'the'; (2) user mentions; (3) hashtags; (4) URLs or external Web site links; and (5) special characters such as emoticons. This process of segregation left us with the tweet containing only the main words. A special type of analytical methodology called the word clouds was then used, which when given an array of words, gives us insight into what words have the highest frequency and are important for the analysis. Therefore, word clouds were generated for main text, hashtags, and user mentions.

The training tweets were then given class label according to the three classes—(1) tweets related to fight and prevention against Zika; (2) tweets related to cure for Zika; and (3) tweets related to damage caused by the Zika virus, mainly the infected areas and the death caused. Word clouds were also generated for each of the three classes.

3 Empirical Model

In this section, we propose a novel system for classifying tweets related to 'Zika.' The system architecture is shown in Fig. 1.

Building the Classification Model: After the preprocessing of the collected tweets, we divide our initial data set into training data set and testing data set having 67% and 33% number of tweets, respectively. All the tweets in the training data set belong to any of the three classes—'fight and prevention,' 'cure,' and 'infected and death' since each class has separate set of keywords, hence preventing overlapping of the classes.

As previous research [8, 9] suggests the efficiency of support vector machine (SVM) algorithm [1, 6] and Naïve Bayes algorithm [7] for text classification, we use them to train our data and evaluate the accuracy of our methodology using the training data set.

Comparisons: The accuracy of the SVM and Naïve Bayes algorithm is compared, and then, we justify why SVM was chosen as the final classification algorithm for the empirical model.

Analyzing Tweets and Community Support: After building the intelligent model and determining the accuracy of the empirical model, we have demonstrated how social networks such as Twitter can be used to gather community support about diseases like Zika by analyzing the classified tweets.

Few research has been done in building intelligent models for community support on social networking sites, and thus, our approach demonstrates one such novel method.

As mentioned, we decided to use the support vector machine (SVM) algorithm and Naïve Bayes algorithm for our classification as it has been seen earlier that both SVM and Naïve Bayes algorithm are suitable algorithms for text classification [10, 12].

The support vector machine (SVM) algorithm [11] is a non-probabilistic binary linear classifier. The model represents data entities as points on a sample coordinate plane in such a way that there is a clear gap between the groups of entities of different classes. The reason SVM work very well for text categorization is that text categorization involves many features (sometimes more than 5000) and SVM handles large feature space.

Fig. 1 Model architecture

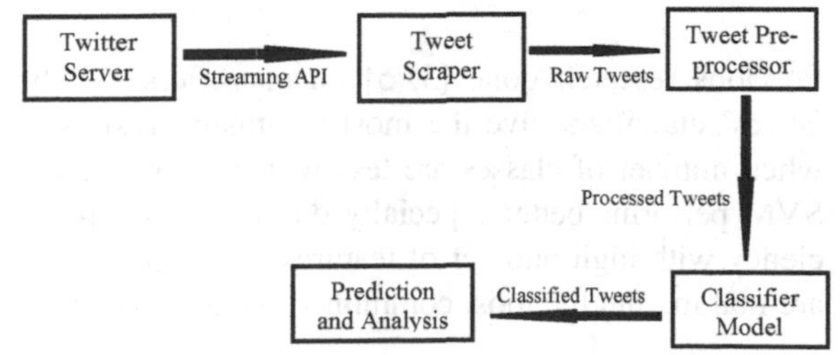

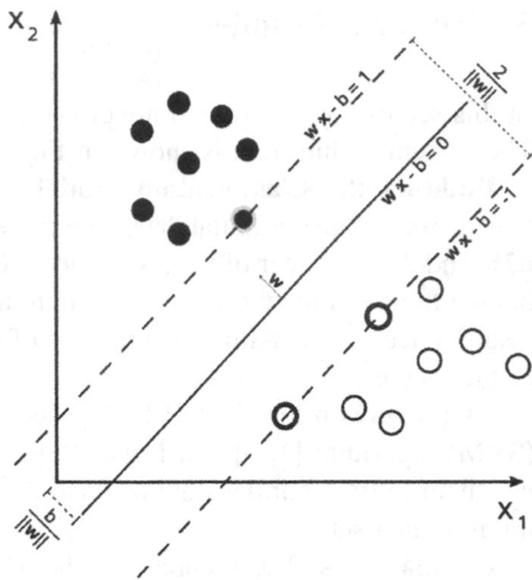

Fig. 2 Maximum margin hyperplane and margins for an SVM from 2 classes

If there are *n* points in the form (x_1, y_1)... (x_n, y_n) where y_i is the class for x_i, then it is possible to draw a maximum margin hyperplane between groups having $y_i = 1$ and $y_i = -1$ and the hyperplane can be expressed in the following form:

$$w.x - b = 1 \text{ and } w.x - b = -1 \; w.x - b = 1 \text{ and } w.x - b = -1 \tag{1}$$

where w is the vector normal to the hyperplane and is shown in Fig. 2.

Naïve Bayes' classifiers [2] are simple probabilistic linear classifiers and are based on Bayes' theorem. All Naïve Bayes' classifiers assume that the features are independent of each other.

If there are n entities in the feature space represented by a vector $x = (x_1,\ldots,x_n)$, class C_k then using Bayes' theorem, the conditional probability can be expressed as follows:

$$p(C_k|x) = \frac{p(C_k)p(x|C_k)}{p(x)} \tag{2}$$

4 Comparison

Previous research done [3, 5] on classifiers clearly states that SVM and Naïve Bayes' classifiers give the most promising results when compared to others. But when number of classes are less which in our case is 3, it is highly probable that SVM performs better especially due to its non-probabilistic approach and its efficiency with high number of features which in our case was around 2804. Features are nothing but the most commonly occurring words in the corpus of all the tweets.

To ensure no overfitting takes place, we have used a fivefold stratified cross-validation methodology to get our accuracy score. The data were divided into 5 subsamples maintaining the ratio of tweets belonging to each class. For accuracy measures, we used simple classification accuracy metric which shows the fraction of correct predictions over total predictions. $\hat{y}$ in Eq. 3 is predicted target variable and y is actual target variable.

$$\text{accuracy}(y, \hat{y}) = \frac{1}{\eta_{\text{samples}}} \sum_{i=0}^{\eta_{\text{samples}}-1} 1(\hat{y}_i = y_i) \tag{3}$$

Table 1 presents the mean accuracy score for both SVM and Naïve Bayes' classifier. Clearly, SVM offers more scalability due to its linear modeling approach and is much more accurate.

5 Analyzing Tweets and Community Support

The word clouds generated are depicted in this section. Following are the figures and their description:

Figure 3 shows the word cloud for all the hashtags retrieved from the tweets. As depicted, words in bigger font such as 'Zika' and 'rio2016' are the most frequent hashtags.

Table 1 Table depicting the comparison between Naïve Bayes' classifier and support vector machine

Classifier name	Fivefold stratified crossvalidation score
Multinomial Naïve Bayes' classifier	0.84695
Support vector machine (linear kernel)	0.89452

Fig. 3 Hashtag word cloud

The word cloud in Fig. 4 is related to the twitter accounts which have tweeted related to Zika most frequently.

Word clouds were also generated for the three classes:

(1) Figure 5 for class 'fight and prevention,'
(2) Figure 6 for class 'cure,' and
(3) Figure 7 for class 'infected and death.'

Fig. 4 Twitter mention word cloud

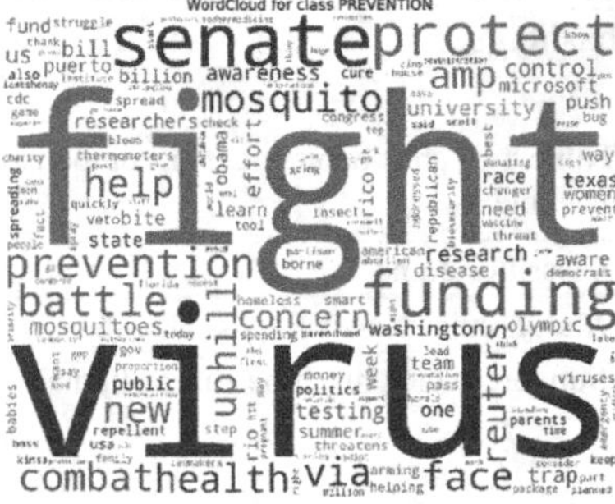

Fig. 5 Fight and prevention

Fig. 6 Cure

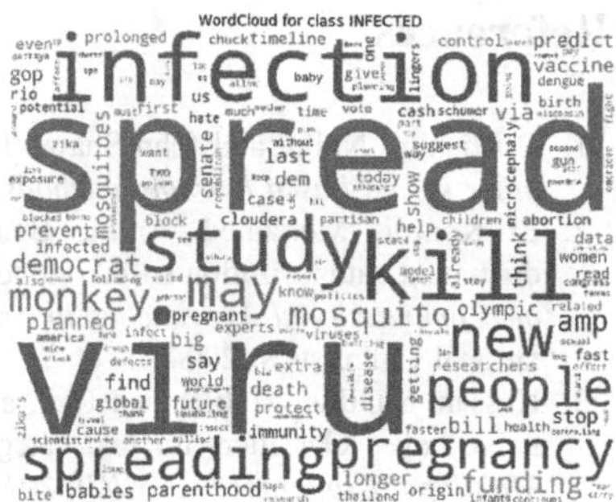

Fig. 7 Infected and death

6 Conclusion

In this paper, we present a detailed analytics of most recent tweets collected over a span of two days. We have tried to extract some information out of the tweets that is presented succinctly in the form of word clouds. Finally, a model is proposed to scrape and gather most recent tweets using twitter stream API and analyze them in real time and finally classifying them into one of the three broad categories as stated above. We were successful in achieving an accuracy of almost 90% with support vector machine classification. According to the analysis, it was inferred that in the tweets gathered 36.50%, 24.94%, and 38.54% of tweets belonged to 'fight and prevention,' 'cure,' and 'infected and death,' respectively. These values also provide a statistical evidence of social community support and awareness available for Zika presently.

7 Applications and Future Work

This model might find application in community support analysis where one can easily predict what is community more worried about whether it is the cure or the spread. Model can also help in making sense of data over time when a huge amount of tweets spanning over several months is available.

For future work, we can try observing the behavior of the model on a relatively larger data set spanning over a period of 3–4 months. A time-based analysis and geospatial analysis are some major branches of development for this model. It would really be interesting to observe how the community support changes over time. Such model can easily be tweaked to later on analyzing various other events and not just epidemics like Zika.

References

1. Cristianini, Nello, and John Shawe-Taylor. *An introduction to support vector machines and other kernel-based learning methods*. Cambridge university press, 2000.
2. El Kourdi, Mohamed, Amine Bensaid, and Tajje-eddine Rachidi. "Automatic Arabic document categorization based on the Naïve Bayes algorithm." *Proceedings of the Workshop on Computational Approaches to Arabic Script-based Languages*. Association for Computational Linguistics, 2004.
3. Hassan, Sundus, Muhammad Rafi, and Muhammad Shahid Shaikh. "Comparing SVM and naive bayes classifiers for text categorization with Wikitology as knowledge enrichment." *Multitopic Conference (INMIC), 2011 IEEE 14th International*. IEEE, 2011.
4. Joachims, Thorsten. "Text categorization with support vector machines: Learning with many relevant features." *European conference on machine learning*. Springer Berlin Heidelberg, 1998.
5. Khan, Aamera ZH, Mohammad Atique, and V. M. Thakare. "Combining lexicon-based and learning-based methods for Twitter sentiment analysis." *International Journal of Electronics, Communication and Soft Computing Science & Engineering (IJECSCSE)* (2015): 89.
6. Lerman, Kristina, and Rumi Ghosh. "Information contagion: An empirical study of the spread of news on Digg and Twitter social networks." *ICWSM* 10 (2010): 90–97.
7. McCallum, Andrew, and Kamal Nigam. "A comparison of event models for naive bayes text classification." *AAAI-98 workshop on learning for text categorization*. Vol. 752. 1998.
8. Pak, Alexander, and Patrick Paroubek. "Twitter as a Corpus for Sentiment Analysis and Opinion Mining." *LREc*. Vol. 10. 2010.
9. Sakaki, Takeshi, Makoto Okazaki, and Yutaka Matsuo. "Earthquake shakes Twitter users: real-time event detection by social sensors." *Proceedings of the 19th international conference on World wide web*. ACM, 2010.
10. Sebastiani, Fabrizio. "Machine learning in automated text categorization." *ACM computing surveys (CSUR)* 34.1 (2002): 1–47.
11. Shmilovici, Armin. "Support vector machines." *Data Mining and Knowledge Discovery Handbook*. Springer US, 2005. 257–276.
12. Tong, Simon, and Daphne Koller. "Support vector machine active learning with applications to text classification." *Journal of machine learning research* 2. Nov (2001): 45–66.

Prediction of Child Tumours from Microarray Gene Expression Data Through Parallel Gene Selection and Classification on Spark

Y.V. Lokeswari and Shomona Gracia Jacob

Abstract Microarray gene expression data play a major role in predicting chronic disease at an early stage. It also helps to identify the most appropriate drug for curing the disease. Such microarray gene expression data is huge in volume to handle. All gene expressions are not necessary to predict a disease. Gene selection approaches pick only genes that play a prominent role in detecting a disease and drug for the same. In order to handle huge gene expression data, gene selection algorithms can be executed in parallel programming frameworks such as Hadoop Mapreduce and Spark. Paediatric cancer is a threatening illness that affects children at age of 0–14 years. It is very much necessary to identify child tumours at early stage to save the lives of children. So the authors investigate on paediatric cancer gene data to identify the optimal genes that cause cancer in children. The authors propose to execute parallel Chi-Square gene selection algorithm on Spark, selected genes are evaluated using parallel logistic regression and support vector machine (SVM) for Binary classification on Spark Machine Learning library (Spark MLlib) and compare the accuracy of prediction and classification respectively. The results show that parallel Chi-Square selection followed by parallel logistic regression and SVM provide better accuracy compared to accuracy obtained with complete set of gene expression data.

Keywords Parallel gene selection · Chi-Square · Parallel logistic regression · Parallel SVM · Hadoop map reduce · Spark MLlib

---

Y.V. Lokeswari (✉) · S.G. Jacob
Department of CSE, Sri Sivasubramaniya Nadar College of Engineering,
Kalavakkam, Chennai, India
e-mail: lokeswariyv@ssn.edu.in

S.G. Jacob
e-mail: shomonagj@ssn.edu.in

H.S. Behera and D.P. Mohapatra (eds.), *Computational Intelligence in Data Mining*, Advances in Intelligent Systems and Computing 556,
DOI 10.1007/978-981-10-3874-7_62

1 Introduction

Deoxyribonucleic acid (DNA) of a cell has coding and non-coding segments. Coding segments are celled genes. Genes specify structure of proteins. DNA is transcribed into messenger Ribo Nucleic Acid (mRNA). mRNA is translated into proteins. Gene expression analysis is the process of monitoring the expression levels of thousands of genes simultaneously under a particular condition. Microarray technology makes this possible and the quantity of data generated from each experiment is enormous. Microarray gene expression data is huge in volume, and it is of interest to discover the set of genes that can be used as class predictors. Gene selection algorithms are necessary for selecting the most informative genes. Moreover, microarray data consists of small number of samples, while number of attributes corresponding to number of genes is typically in thousands. This creates high likelihood of finding false positives both in determining differentially expressed genes and in building predictive models. The vast amount of raw gene expression data leads to statistical and analytical challenges including classification of data set into correct classes. The goal of classification is to identify the differentially expressed genes that may be used to predict class membership for new samples. Gene (Feature) selection methods aims at selecting important genes and they are categorized as Filter, Wrapper and Embedded methods. Filter method selects features based on discriminating criteria that are relatively independent of classification. Wrapper method utilizes classifier as a black box to select subset of features based on their predictive power. Embedded method combines both of the above-mentioned approaches. Some of the feature selection algorithms are Fisher Filtering, ReliefF, Step-wise Discriminate Analysis, Chi-Square, Attribute subset selection, Principal Component Analysis (PCA) and Singular Value Decomposition (SVD) [1].

The motivation of Attribute subset selection is to find a minimum set of attributes such that resulting probability distribution of data classes is as close as possible to the original distribution obtained using all attributes. Few attribute subset selection methods are discussed here. Step-wise forward selection starts with empty set of attributes, determines best of original attributes and adds them to set. Step-wise backward elimination starts with full set of attributes and removes the worst attribute remaining in the set at each step. Combined forward selection and backward elimination combines both approaches. Decision Tree Induction in which attributes that do not appear in tree are assumed to be irrelevant for classification. Set of attributes appearing in the tree form the reduced subset of attributes. In this paper, the authors have implemented parallel Chi-Square gene selection approach with the objective of scaling gene selection algorithm for huge number of genes. Chi-Square method is used to find correlation between genes (attributes). These attribute subset selection algorithms can also be run in parallel to improve the performance on large data sets.

In parallelization of algorithms, the data set will be split into chunks of n/p by the master, where n is number of samples and p be number of processors. Each data split will be allocated to slaves and master also contributes in processing.

After processing, slaves have to report to master, master have to aggregate and produce final result. This parallelization of algorithms gives rise to certain critical issues such as synchronization, communication overhead, load imbalance and latency in accessing files from disk. In order to address these issues, Hadoop Map Reduce—a parallel programming model has been used in literature [2, 3]. Hadoop is a parallel programming model used to process very large data sets. Hadoop consists of a Mapreduce engine which takes care of parallel computing and a Hadoop Distributed File System (HDFS) for distributed storage. Mapreduce consists of a Job Tracker and a set of Task Trackers under each Job Tracker. The Job Tracker allots tasks to Task Trackers. Task Trackers run Map and Reduce functions as stipulated by the user. Figure 1 depicts the major modules in Mapreduce framework. Mapreduce splits the input file into multiplc small filcs and are assigned to each task tracker. Task tracker that runs Map function will take a split file as input and produces a *<key, value>* pair. The combine function performs local sorting and grouping to produce an intermediate *<key, list (values)> pair.* The partitioning function maps the output of mapper to one of the *R* regions, where *R* equals the number of reducers. The *i-th* reducer takes intermediate key value pairs from *i-th* region. The intermediate key value pairs are again sorted and grouped in Reducer. Sorted *<key, value>* pair is given as input to Reduce function which aggregates and produces the final values to an output file. The results of task trackers will get stored in HDFS. HDFS consists of a Name Node and a set of Data Nodes under a Name Node. Name Node contains metadata and file system namespace. Data Nodes consists of number of data blocks of 64 MB size and stores original data and replicas. As a fault tolerant mechanism, Hadoop replicates each data block and stores in different locations. The drawback of Hadoop Mapreduce is latency in accessing files or results from HDFS. Spark is an Apache framework which runs on top of YARN (Yet Another Resource Negotiator), and it has the advantage of in-memory computation. Spark provides Machine Learning library (Spark MLlib) which helps to handle execution of machine learning algorithms on huge amount of data.

The paper has been structured to present the work as follows. Section 2 narrates the related work in microarray gene data classification and clustering. Section 3 briefs about proposed system with parallel Chi-Square gene selection, parallel logistic regression and parallel SVM classifier. Section 4 explains experimental analysis. Section 5 gives results and discussion. Section 6 concludes the paper on

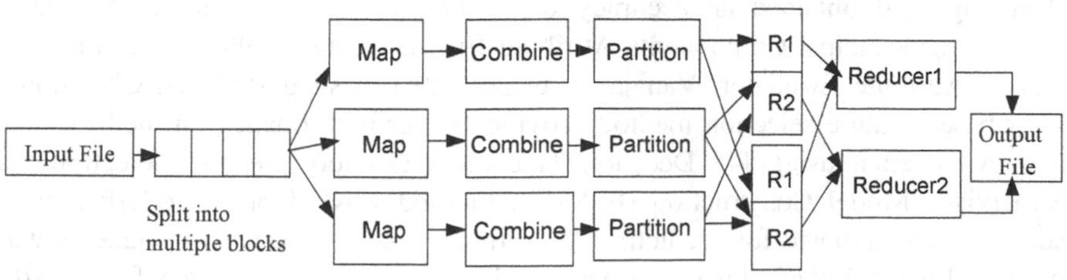

Fig. 1 Mapreduce framework

which machine learning technique is more suitable for learning from microarray gene data set.

2 Related Work

Data Mining in Clinical data sets: A review by Jacob et al. [4] reviewed about data mining techniques applied on medical data sets and conveyed that Quinlan's C4.5 and Random Tree algorithms provided 100% accuracy. The paper discussed that machine learning and statistical techniques applied to gene expression data helps to distinguish tumour morphology, predicting post treatment outcome and to find molecular markers for diseases [5]. Shomona et al. [1, 6] provided 16 classification algorithms that are applied for classifying breast cancer along with feature selection methods. Jirapech-Umpai et al. [7] presented the importance of gene selection from Microarray gene data set and discussed multiclass classification of leukaemia and NCI60 data sets. Evolutionary algorithm was proposed to identify near-optimal set of predictive genes for classification. Leave one out Cross Validation (LOOCV) method was used to build the classifier and validate it. Ben-Dor et al. [8] insisted that gene expression data aid in development of efficient cancer diagnosis and classification. Measuring gene expression levels under different conditions is important for understanding gene function and how various gene products interact. Gene expression data help in better understanding of cancer. Individual gene expression levels and scoring methods were used for separation of tissue types. They have presented results of performing LOOCV with classifiers such as K-Nearest Neighbor (K-NN), Support Vector Machine (SVM), AdaBoost and a novel clustering-based classification. Piatetsky-Shapiro et al. [9] discussed about challenges for microarray data mining and importance of gene selection for classifying cancer types, searching for new drugs and treatments. Lavanya et al. [10] had used attribute selection methods such as SVM Attribute Evaluator, Principal Components Attribute Evaluator and Symmetric Uncertainty Attribute Evaluator that yielded an accuracy of 73.03%, 96.99% and 94.72%, respectively, for Wisconsin breast cancer data set. They suggest that the best feature selection method for a particular data set depends on number of attributes, attribute type and instances. Lavanya et al. [11] had used above-mentioned attribute selection methods for breast cancer data set with ensemble of decision trees using Bootstrap Aggregation (Bagging) and obtained an accuracy of 74.47% for SVM Attribute Evaluator, 97.85% for Principal Components Attribute Evaluator and 95.96% for Symmetric Uncert Attribute Evaluator. Vanaja, S. et al. [12] had suggested that Chi-Square filter-based feature selection method provides better performance on medical data and evaluated it using J48 Decision Tree on Weka tool. Feature Selection via Supervised Model Construction (FSSMC) method was adopted. ReliefF feature selection was adopted for predicting Diabetics. Presence/absence of Diabetes was predicted using ReliefF feature selection with Naive Bayes and C4.5 for classification. Selective Naive Bayes was used for filtering and ranking the attributes of medical data set. Stroke disease was predicted using Naive Bayes, Neural Network and Decision Tree classifiers by reducing dimensions using Principal Component Analysis (PCA). Breast cancer patients are classified into carcinoma in situ and malignant potential of SEER (Surveillance, Epidemiology, and End Results) data set using C4.5. PCA with Fuzzy K-NN classifier was used for predicting Parkinson's disease. Non Parallel Plan Proximal Classifier (NPCC) ensemble was used for binary and multiclass classification of cancer gene analysis. Feature selection algorithm that works on both binary and multiclass data sets, sometimes generate high accuracy on binary data sets, but give low accuracy on multiclass data sets [13]. Rajeswari K. et al. [14] proposed a novel approach for feature selection using correlation and by generating association rules. They provided a statement that removing irrelevant attributes will improve the performance of classification and cost of classification may get reduced so that patients need not take unnecessary tests that are not required for predicting the disease. Devi, M.A. et al. [15] provided a comparison of clustering algorithms with feature selection on breast cancer data set. Hierarchical clustering takes additional time for model construction compared to K-means and density-based clustering. Principal component filter had improved the accuracy in execution of K-means and density-based clustering. Wang et al. [16] proposed a 'α' depended degree-based feature selection approach for robust feature selection and compared with Chi-Square, Information Gain, ReliefF and Symmetric Uncertainty methods. The results revealed that proposed method was superior to the canonical dependent degree of attribute-based method in robustness and applicability. Hassanien, A.E. [17] proposed techniques which are based on inductive decision tree learning algorithm that has low complexity with high transparency and accuracy. Decision tree considers only subset of features which leads to best performance and the ratio of correct classification of new sample is high. Zhang, H. et al. [18] introduced a new method for gene selection called Chi-Square test-based integrated rank gene and direct classifier. Initially weighted integrated rank of gene importance was obtained from Chi-Square tests of single and pair-wise gene interactions. Then, ranked genes were introduced sequentially and redundant genes were removed using leave one out cross validation of Chi-Square test-based direct classifier. This was performed within the training set to obtain informative genes. Finally, accuracy was measured for independent test data with learned Chi-Square-based direct classifier. Nguyen, C. et al. [19] provided a machine learning method based on Random Forest and feature selection for Wisconsin breast cancer data set. The results show that classification accuracy was 100% for best case and around 99.8% for average case. Peralta, D. et al. [20] discussed about feature selection algorithm based on evolutionary computation that uses Hadoop Mapreduce to obtain subset of features from big data sets. Feature selection method was evaluated using SVM, logistic regression and Naive Bayes with Spark framework. Islam, A.T. et al. [21] pointed out that scalable gene selection methods are required for Microarray data analysis due to rapidly increasing volume of microarray data. Parallel gene selection using Mapreduce programming model was used. The proposed method uses K-NN classifier algorithm for evaluating classification accuracy. Begum, S. et al. [22] proposed an ensemble-based SVM called AdaBoost Support Vector Machine (ADASVM) for classifying cancer from microarray gene expression data. Leukaemia data set was used to compare the performance of ADASVM, K-NN and SVM classifiers. Jeyachidra, J. et al. [23] compared feature selection algorithms such as Gini Index, Chi-Square and MRMR (Maximum Relevance and Minimum Redundancy) and evaluated J48 and Naive Bayesian classifier using leave one out cross validation. It was identified that Chi-Square and Gini Index yield a classification accuracy of approximately 85%.

Based on the review of previous work in the area of gene selection on microarray gene expression data, it was identified that (i) there is a clear scarcity of efficient parallel algorithms to mine from voluminous gene data (ii) Parallelization of gene selection approaches has not been explored on clinical data, although its importance has been emphasized in the literature. (iii) Paediatric tumours are one of the major causes for child mortality, especially in low- and middle-income countries. Yet, not much work has been devoted to unearthing the genetic variants contributing to the disease.

This being the rationale behind this research, the authors have made an investigation on the possibility of mining optimal genes contributing to paediatric tumours through computational methods based on parallelization of gene selection and classification techniques. The proposed methodology and experimental results are discussed in the following sections.

3 Proposed Parallel Gene Selection and Parallel Classification for Detection of Child Tumours on Spark

The proposed parallel gene selection algorithm is parallel Chi-Square on Spark MLlib. Child tumour is the life-threatening disease that affects children at the age of 0–14 years. Childhood leukaemia is most commonly occurring type of tumour in children. Such deadly disease can be predicted using microarray gene expression data. Microarray gene data has thousands of genes with very less number of samples. It is necessary to apply gene selection approaches for selecting most optimal genes that play vital role in causing cancer. The selected genes are to be evaluated using classifiers to identify the sub-types of cancer. In order to scale for huge number of genes in a sample, parallel gene selection algorithm called parallel Chi-Square was used. Selected genes are evaluated using parallel logistic regression and support vector machine on Spark MLlib.

3.1 Chi-Square Gene Selection Algorithm

Chi-Square correlation coefficient was utilized for finding correlation between genes (attributes) [31]. Chi-Square value is computed using the Eq. 1.

$$\chi^2 = \sum_{i=1}^{c} \sum_{j=1}^{r} \frac{(o_{ij} - e_{ij})^2}{e_{ij}} \quad (1)$$

where $O_{i\,j}$ is observed (actual) frequency of joint event of genes (A_i, B_j) and $e_{i\,j}$ is expected frequency of (A_i, B_j) which is computed using Eq. 2. '*r*' and '*c*' are number of rows and columns in contingency table.

$$e_{ij} = \frac{Count(A = a_i) \times Count(B = b_j)}{N} \quad (2)$$

where N is number of data tuples. *Count* ($A = a_i$) is number of tuples having value a_i for A (*True Positive*). *Count* ($B = b_j$) is number of tuples having value b_j for B (*True Negative*). A and B are the gene pairs considered for correlation. The sum is computed over all of $r \times c$ cells in a contingency table. The χ^2 value needs to be computed for all pair of genes. The χ^2 statistics test the hypothesis that genes A and B are independent. The test is based on significance level with $(r - 1) \times (c - 1)$ degrees of freedom. If Chi-Square value is greater than the statistical value for given degree of freedom, then the hypothesis can be rejected. If the hypothesis can be rejected, then we say that genes A and B are statistically related or associated.

Chi-Square test was executed parallel on Spark MLlib in order to scale for enormous amount of genes in a sample. This greatly reduces the training time of a model and selects optimal genes in very less time.

3.2 Framework for Parallel Gene Selection and Classification on Spark

Gene selection algorithms will improve accuracy and training time of a model. Chi-Square was used to select genes, and it was evaluated on parallel classification and prediction algorithms. Figure 2 depicts the framework of the proposed system for classifying and predicting paediatric tumour.

1. Microarray gene expression data for paediatric tumour was collected from Orange laboratories.
2. Parallel Chi-Square selector (ChiSqSelector) was executed on Apache Spark for selecting most important genes for cancer classification or prediction
3. Genes were selected in subsets of varying sizes starting from 25, 50, 100, 200 and so on till 1000.
4. Selected genes were evaluated using prediction and classification algorithms such as logistic regression and SVM, respectively, on Spark MLlib [24].
5. Accuracy of prediction and classification of child tumour was obtained.

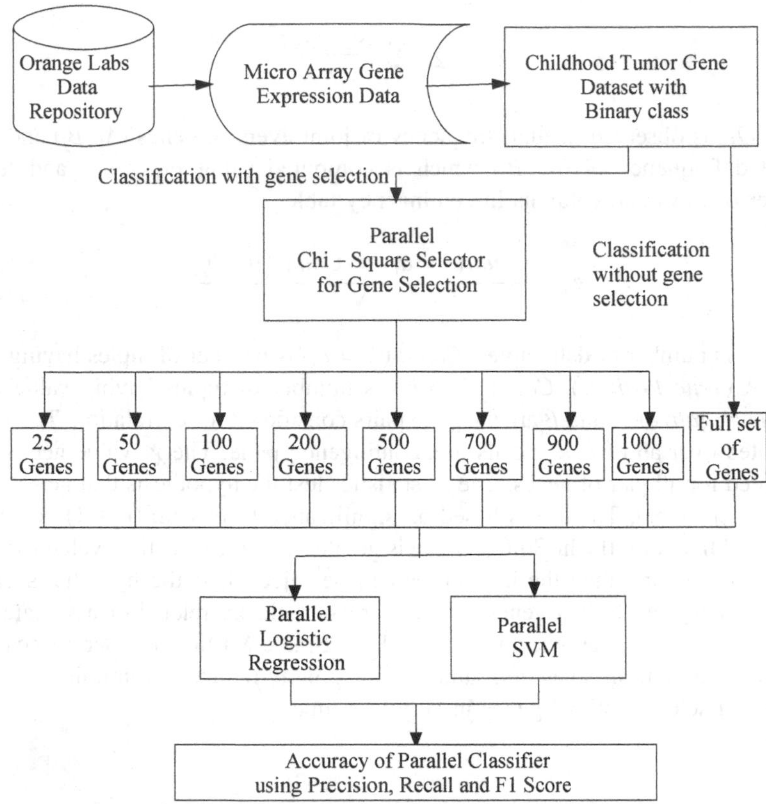

Fig. 2 Parallel gene selection and classification on spark

4 Experimental Analysis

Parallel programming framework called Apache Spark was installed on Ubuntu 14.04 box. Microarray gene data set for paediatric tumour was collected from Orange laboratory [25], and pre-processing was done manually. Spark MLlib can process data only in libSVM format, and it cannot handle missing values and categorical values for attributes. Hence, gene data set was converted into libSVM format. The selected paediatric tumour gene data set does not contain any missing values and all gene values were numerical. Initially, parallel logistic regression and SVM were run on original gene data set, and accuracy was computed. Parallel Chi-Square selector was executed on Spark MLlib in order to select optimal genes. Gene sets of 1000, 900, 700 till 25 were selected using parallel Chi-Square selector as indicated in Fig. 3. Finally, the selected genes were evaluated on parallel logistic regression and parallel SVM. Accuracy of selected genes for prediction and classification was computed. The gene data set consisted of two sub-types of leukaemia

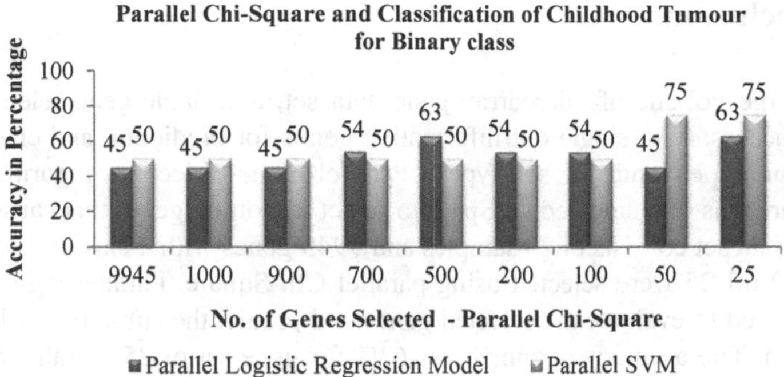

Fig. 3 Accuracy of parallel Chi-Square with parallel logistic regression and parallel SVM

cancer that occur in children. They are Ewing's sarcoma (EWS) and Rhabdomyosarcoma (RMS) cancer.

5 Results and Discussions

The original gene data set consists of 9945 genes with 23 samples. Random sampling was done to split 60% of data set for training and 40% for testing. The accuracy of predicting and classifying the correct cancer sub-type was only 45% from parallel logistic regression and 50% from parallel SVM. When selected genes are evaluated on parallel logistic regression and SVM, the results show that SVM performs better with 75% classification accuracy based on Area under Curve metric as shown in the below chart in Fig. 3.

The results show that accuracy was improved when genes were selected in different ranges. There was slight increase in accuracy with parallel logistic regression, when selected genes were of 700 and 500. With 500 genes, parallel logistic regression provided an accuracy of 63% based on Precision, Recall and F1 score metrics. On the other hand, parallel SVM maintained an accuracy of 50% from 1000 genes to 100 genes as shown in Fig. 3.

On selecting 50 genes, the accuracy of parallel logistic regression decreased to 45% whereas with parallel SVM, it was 75% of accuracy based on Area under Curve. On selecting 25 genes, the accuracy improved to 63% with logistic regression and 75% with SVM. So, it was identified that parallel Chi-Square with logistic regression provided prediction accuracy of only 63%, and SVM provides 75% with 25 selected genes. These variations in accuracy on selecting different gene sets were mainly because of genetic information in the input data set. When most informative genes were selected, it results in higher accuracy, which was not the same when less informative genes were selected. The gene selection algorithm varies according to number of samples and type of attributes (genes).

6 Conclusion

Due to large volume of microarray gene data set, a scalable gene selection algorithm is necessary to single out informative genes for prediction and classification of tumour types and its sub-types. Parallel gene selection algorithm called Chi-Square was implemented on Spark to select important genes that causes cancer. Original data set consists of 23 samples and 9945 genes with 2 classes. Gene sets of 1000, 900 till 25 were selected using parallel Chi-Square. Parallel logistic regression was used to evaluate the selected genes and predict the sub-types of leukaemia in children. The accuracy obtained was 63% for gene set of 25. Parallel SVM was used to evaluate the selected genes and classify the tumour sub-type. The accuracy obtained was 75% for gene set of 25. From the results, it was identified that SVM performs better, but still the accuracy can be improved by analyzing other gene selection algorithms. In future, the work has to be extended for more number of classes, pre-processing the noisy gene data set and explore other classification algorithms such as Decision Tree Induction and Random Forest in improving the accuracy of classification.

References

1. Shomona Gracia Jacob, Dr.R.Geetha Ramani, P.Nancy: Feature Selection and Classification in Breast Cancer Datasets through Data Mining Algorithms, In Proceedings of the IEEE International Conference on Computational Intelligence and Computing Research (ICCIC'2011), Kanyakumari, India, IEEE Catalog Number: CFP1120J-PRT, ISBN: 978-1-61284-766-5. (2011). 661–667.
2. Masih, Shraddha, and Sanjay Tanwani: Data Mining Techniques in Parallel and Distributed Environment-A Comprehensive Survey. In International Journal of Emerging Technology and Advanced Engineering (March 2014), Vol. 4, Issue 3, (2014) 453–461.
3. Pakize., Seyed Reza and Abolfazl Gandomi: Comparative Study of Classification Algorithms Based on MapReduce Model. In International Journal of Innovative Research in Advanced Engineering (2014), ISSN (2014): 2349–2163.
4. Jacob, S.G. and Ramani, R.G.: Data mining in clinical data sets: a review. training, 4(6). (2012).
5. Yeh, J.Y: Applying data mining techniques for cancer classification on gene expression data. In Cybernetics and Systems: An International Journal, 39(6), (2008). 583–602.
6. Shomona Gracia Jacob, Dr.R.Geetha Ramani, Nancy.P: Classification of Splice Junction DNA sequence data through Data mining techniques, ICFCCT, 2012, held at Beijing, China, May 19–20, ISBN:978-988-15121-4-7, (2012). 143–148.
7. Jirapech-Umpai, T. and Aitken, S.: Feature selection and classification for microarray data analysis: Evolutionary methods for identifying predictive genes. In BMC bioinformatics, 6(1), (2005).148.
8. Ben-Dor, A., Bruhn, L., Friedman, N., Nachman, I., Schummer, M. and Yakhini, Z.: Tissue classification with gene expression profiles. In Journal of computational biology, 7(3–4), (2000). 559–583.
9. Piatetsky-Shapiro, G. and Tamayo, P.: Microarray data mining: facing the challenges. In ACM SIGKDD Explorations Newsletter, 5(2), (2003).1–5.

10. Lavanya, D. and Rani, D.K.U.: Analysis of feature selection with classification: Breast cancer datasets. Indian Journal of Computer Science and Engineering (IJCSE), 2(5), (2011), 756–763.
11. Lavanya, D. and Rani, K.U:. Ensemble decision tree classifier for breast cancer data. In International Journal of Information Technology Convergence and Services, 2(1), (2012).17.
12. Vanaja, S. and Kumar, K.R.: Analysis of feature selection algorithms on classification: a survey. In International Journal of Computer Applications, 96(17) (2014).
13. Ramaswamy, S., Tamayo, P., Rifkin, R., Mukherjee, S., Yeang, C.H., Angelo, M., Ladd, C., Reich, M., Latulippe, E., Mesirov, J.P. and Poggio, T:. Multiclass cancer diagnosis using tumour gene expression signatures. In Proceedings of the National Academy of Sciences, 98 (26), (2001).15149–15154.
14. Rajeswari K, Vaithiyanathan, V. and Pede, S.V:. Feature selection for classification in medical data mining. In International Journal of Emerging Trends and Technology in Computer Science (IJETTCS), 2(2), (2013). 492–7.
15. Devi, M.A. and Sarma, D.D., Comparison of Clustering Algorithms with Feature Selection on Breast Cancer Dataset. In Journal of Innovation in Computer Science and Engineering, (2015).59–63.
16. Wang, X. and Gotoh, O:. A robust gene selection method for microarray-based cancer classification. In Cancer informatics, 9, (2010).15–30.
17. Hassanien, A.E: Classification and feature selection of breast cancer data based on decision tree algorithm. In Studies in Informatics and Control, 12(1), (2003). 33–40.
18. Zhang, H., Li, L., Luo, C., Sun, C., Chen, Y., Dai, Z. and Yuan, Z:. Informative gene selection and direct classification of tumour based on chi-square test of pairwise gene interactions. In BioMed research international, (2014).
19. Nguyen, C., Wang, Y. and Nguyen, H.N.: Random forest classifier combined with feature selection for breast cancer diagnosis and prognostic. In Journal of Biomedical Science and Engineering, 6(5), (2013).551.
20. Peralta, D., del Río, S., Ramírez-Gallego, S., Triguero, I., Benitez, J.M. and Herrera, F". Evolutionary Feature Selection for Big Data Classification: A MapReduce Approach. Mathematical Problems in Engineering, 501, (2015), 246139.
21. Islam, A.T., Jeong, B.S., Bari, A.G., Lim, C.G. and Jeon, S.H: MapReduce based parallel gene selection method. Applied Intelligence, 42(2), (2015), 147–156.
22. Begum, S., Chakraborty, D. and Sarkar, R: Cancer classification from gene expression based microarray data using SVM ensemble. In 2015 International Conference on Condition Assessment Techniques in Electrical Systems (CATCON) IEEE (2015), 13–16.
23. Jeyachidra, J. and Punithavalli, M: February. A comparative analysis of feature selection algorithms on classification of gene microarray dataset. In Information Communication and Embedded Systems (ICICES), 2013 IEEE International Conference on (2013), 1088–1093.
24. http://spark.apache.org/mllib/.
25. http://www.biolab.si/supp/bi-cancer/projections/info/EWSGSE967.htm.

Handover Decision in Wireless Heterogeneous Networks Based on Feedforward Artificial Neural Network

Archa G. Mahira and Mansi S. Subhedar

Abstract In heterogeneous networks, vertical handover decision is a significant issue due to increasing demand of customers to access various service features among them. In order to provide a seamless transfer between various technologies, the effect of various user preference metrics and network conditions needs to be considered. This paper proposes a multilayer feedforward artificial neural network algorithm for handover decision in wireless heterogeneous networks. Neural network aids in taking the handover and selection of best candidate based on data rate, service cost, received signal strength indicator (RSSI) and velocity of mobile device. Experimental results show an improvement in reducing number of handover effectively as compared to other existing systems. It is found that probability of handover decision is also improved.

Keywords Wireless heterogeneous networks · Seamless mobility · Self-adaptive handover · Artificial neural networks

1 Introduction

The customer's needs for accessing to high-speed data services anywhere at anytime led to the development of next-generation wireless technologies. The forthcoming wireless networks are capable of providing high-speed data services with seamless connectivity and integration among various mobile heterogeneous networks (HetNets). The goal of seamless connectivity is to provide continuity of access to any kind of desired data services at instant of time. Hence, mobility management is a very challenging task while maintaining service continuity.

A.G. Mahira (✉) · M.S. Subhedar
Department of Electronics & Telecommunication, Pillai HOC College of Engineering & Technology, Rasayani, Tal. Khalapur, Dist. Raigad 410206, Maharashtra, India
e-mail: archageeson@gmail.com

M.S. Subhedar
e-mail: mansi_subhedar@rediffmail.com

H.S. Behera and D.P. Mohapatra (eds.), *Computational Intelligence in Data Mining*, Advances in Intelligent Systems and Computing 556,
DOI 10.1007/978-981-10-3874-7_63

When a mobile station (MS) gets transferred from the coverage area of its serving base station (BS) to another, there occurs either a vertical handover or a horizontal handover. In 4G networks, vertical handover occurs due to the integration of different wireless technologies. In this paper, we proposed an artificial neural network (ANN)-based algorithm to decrease the effect of handover counts in forthcoming heterogeneous networks.

The rest of the paper is organised as follows: Sect. 2 reviews previous work related to this article. Section 3 presents artificial neural network-based handover decision algorithm. Experimental results are presented in Sect. 4. Section 5 concludes the paper.

2 Related Work

Recently, numerous research works have been proposed for the handover mechanism to provide an uninterrupted service to customers. Kustiawan et al. proposed a fuzzy logic-based handover decision algorithm with a channel equalisation technique called Kalman filter to reduce handover initiations effectively [1]. Simulation results show a 88.88% reduction in number of handover. Calhan et al. proposed a handoff decision system based on ANN in order to reduce the handoff latency. Simulation results show that a reduction in handoff delay and the number of handoff is satisfactory [2]. Alsamhi et al. discussed an intelligent handoff algorithm to enhance quality of service (QoS) in high-altitude platforms using neural networks [3]. Results show that handoff rate and blocking rate are enhanced. A machine learning scheme based on neural networks for vertical handover is presented to achieve seamless connectivity and always best connected call status [4]. The performance results showed an enhancement of QoS perceived by both voice and data services. A new network selection algorithm is proposed based on particle swarm optimisation [5]. Results show that performance of the system significantly reduces computational complexity and time. Xiaohuan Yan et al. presented a comprehensive survey of the vertical handover algorithms developed to satisfy user requirements [6]. A variety of vertical handover decision algorithms are also discussed that include RSSI-based, cost function-based and QoS-based schemes. A solution to the challenging task of handoff is proposed using artificial neural networks [7]. The proposed method is able to distinguish the best existing network by matching predefined user preferences. A new neural network prediction system that is able to capture some of the patterns exhibited by users moving in a wireless environment is presented [8]. A neural network-based method to model access network and an adaptive parameter adjustment algorithm is presented in [9]. The simulation study shows that the scheme allows user to access the destination network quickly and variation in throughput can be ignored efficiently. Another algorithm based on particle swarm optimisation for vertical handoff decision in HetNets is presented in [10]. This method reduces call blocking probability. Mubarak et al. introduced a self-adaptive handover (FuzSAHO) algorithm based on fuzzy logic [11]. The simulation results indicate a reduction in handover delay and ping-pong handover effectively.

3 Proposed System

A multilayer feedforward artificial neural network algorithm is employed to find the best target network accessible to user equipment to handover the ongoing call process. Figure 1 depicts architecture of proposed handover algorithm using artificial neural networks. The proposed algorithm consists of four layers. First layer is composed of four input neurons such as MS velocity, monetary cost, data rate, RSSI and output layer as handover decision. The two hidden layers are composed of several nodes which use hyperbolic tangent sigmoid transfer functions.

The proposed vertical handover criterion uses several input parameters to provide seamless connectivity among various wireless heterogeneous technologies such as received signal strength, monetary cost, data rate and mobile station velocity. RSSI

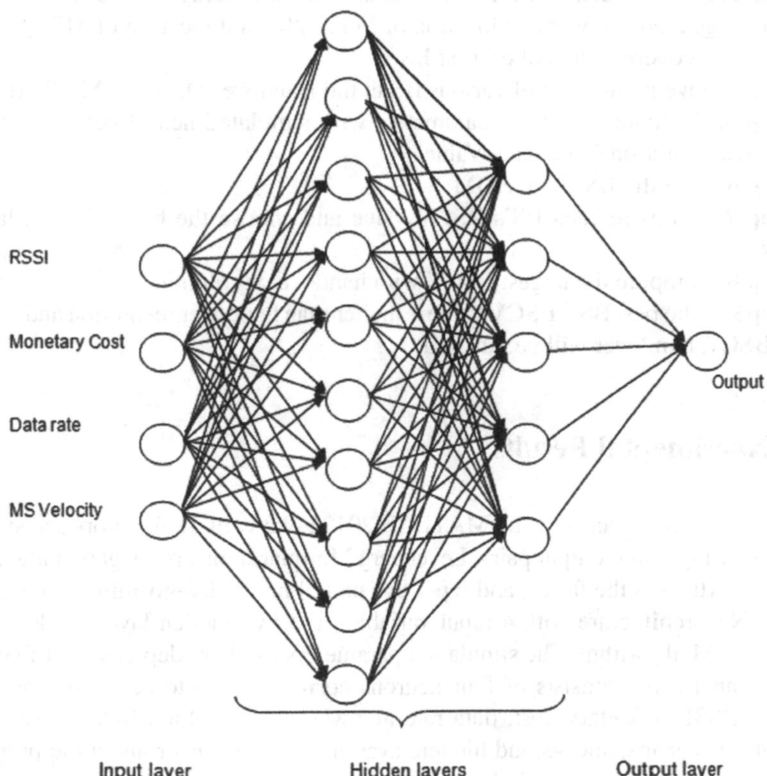

Fig. 1 Structure of proposed artificial neural network

Table 1 Range of input parameters

Inputs	Range
RSSI (dBm)	−90 to −65
Monetary cost	0.5–2.7
Data rate (mbps)	1–4
MS velocity (m/s)	0–40

(received signal strength indicator) indicates the availability of a network, and velocity indicates the movement of mobile terminal in network coverage area. These are the most important criterion for both the horizontal and vertical handover mechanisms. Network bandwidth conditions are provided with the help of data rate. Monetary cost indicates different charging policies provided by different networks to the user. The range of input variables is depicted in Table 1.

The proposed algorithm for handover decision can be summarised as follows.

Step 1: Define simulation parameters and measure the received signal strength using cost231hata propagation model

Step 2: Load trained ANN structure and get the parameters for all neighbouring base stations for handover

Step 3: Decide whether the mobile station is in the coverage area of current wireless heterogeneous network. If location of MS $\leqq 200$ and location of MS $\geqq 1$, then MS is in the coverage area of current BS

Step 4: Save parameters of various BS in the Handover Decision Matrix (HDM)

Step 5: Evaluate each BS's parameters with simulated neural network and get BSCV (Base Station Candidacy Value)

Step 6: Save the BSCV in HDM

Step 7: Compare each BS's BSCV value and choose the best BS with largest BSCV

Step 8: Compare the largest BSCV with handover resolution

Step 9: If the best BS's BSCV value is higher than handover resolution and current BS's BSCV, handover will be initiated.

4 Experimental Results

Experiments are obtained using MATLAB 2012a platform. In the proposed system, to train the input and output pairs, Levenberg Marquardt backpropagation algorithm is chosen which is the fastest and repetitive neural network algorithm. The feedforward ANN architecture with 4 input variables and two hidden layers is developed based on LM algorithm. The simulation parameters used are depicted in Table 2.

The input layer consists of four neurons corresponding to each input variable such as RSSI, monetary cost, data rate and MS velocity. First hidden layer consists of 10 neurons, and second hidden layer consists of 5 neurons in the proposed ANN architecture. The hyperbolic tangent sigmoid transfer functions are selected for first two hidden layers. Hyperbolic tangent is used as activation function since it is

Table 2 Parameters used for simulation

Simulation parameters	Value
Propagation model	cost231hata
Effective BS antenna height	50 m
Effective receiver antenna height	2 m
Distance between BS to BS	1 km
Handover resolution	0.2

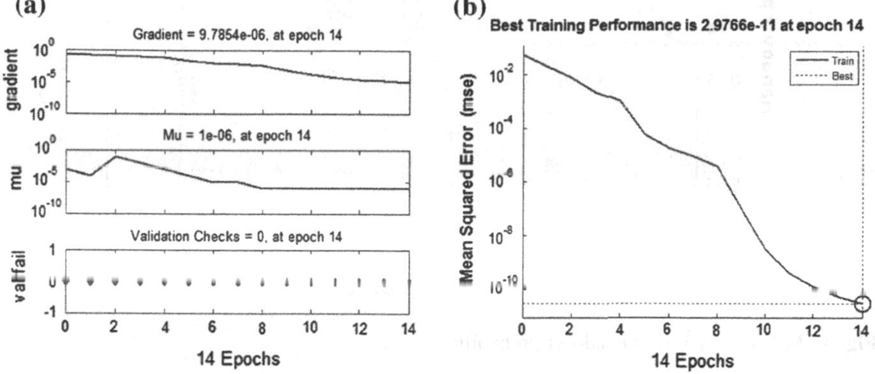

Fig. 2 **a** Training process of ANN and **b** performance plot of proposed ANN

completely symmetric, and its derivative can easily be obtained [2]. Output layer gives handover decision value. It is a binary signal that varies between 0–1 where 1 represents urgently required handover and 0 represents no handover is required. The linear function is selected as transfer function for output layer. The value of RSSI varies from −90 to −65 dBm, and the range of velocity varies between 0 and 40 m/s. The learning rate and maximum number of epochs to train ANN have been set to 0.01 and 1000, respectively.

Figure 2a depicts the training process of multilayer feedforward system for vertical handover decision (MFVHO). It shows that error between desired output and actual output reduces by the correction of network weights. The error value obtains minimum value 9.7854e-06 at epoch 14. The best training performance is 2.9766e-11 at epoch 14 as shown in Fig. 2b. As shown in Fig. 3, handover decision probability increases as velocity increases. Hence, the system is capable of transferring the access to the neighbouring BS quickly and efficiently, whenever the user reaches the boundary of coverage region of current base station (BS).

We compared handoff initiation scenarios with several existing schemes in the literature. We have 4, 14, 27 and 36 number of handoffs using combination of Kalman filter and fuzzy logic scheme [1], Mamdani fuzzy logic, Kalman filtered RSSI and traditional fixed RSSI approaches. As shown in Fig. 4, the proposed MFVHO system reduces the number of handover considerably as compared to other existingalgorithms with a relative reduction percentage of 93.75. The calculation of relative

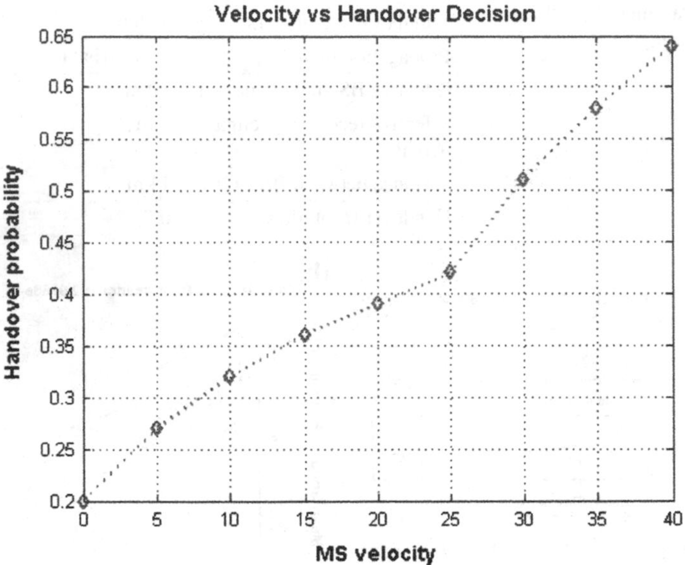

Fig. 3 MS velocity versus handover probability

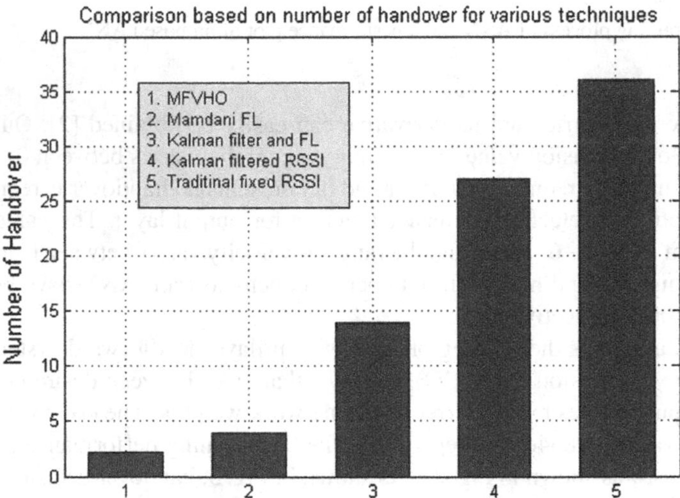

Fig. 4 Comparison based on number of handover for various existing techniques with MFVHO

reduction is given as follows: $(36 - 2.25)/36 * 100 = 93.75\%$. During an ongoing call process, if the number of handovers is more, then it results in a greater probability of call dropping, a delay in processing of handover requests. Thus, RSSI will get reduced to a level of unacceptable quality over a longer period of time due to higher number of handover.

5 Conclusion

A feedforward neural network-based handover decision algorithm has been presented in this paper for the selection of best available network while transfer of ongoing call or data services takes place. The input set of MFVHO algorithm considered several factors such as data rate, monetary cost, velocity of user equipment and received signal strength. The simulation study shows that the proposed multicriterion improves reduction percentage by 93.75% and reduces the number of handover effectively. The reduction in number of handover reduces call dropping probability as well as delay in processing of handover initiation requests effectively. Thus, it is endorsed that our MFVHO criterion outperforms various existing handover decision techniques for intersystem roaming scenarios between heterogeneous networks.

References

1. I. Kustiawan, K. H. Chi.: Handoff Decision Using a Kalman Filter and Fuzzy Logic in Heterogeneous Wireless Networks. IEEE Communication Letters 19, 1–4 (2015)
2. A. Calhan, C. Ceken.: Artificial neural network based vertical handoff algorithm for reducing handoff latency. Wireless Personal Communications 71, 2399–2415 (2013)
3. S. H. Alsamhi, N. S. Rajput.: An intelligent hand-off algorithm to enhance quality of service in high altitude platforms using neural network. Wireless Personal Communications 82, 2059–2073 (2015)
4. N. M. Alotaibi, S. S. Alwakeel.: A neural network based handover management strategy for heterogeneous networks. 14th IEEE International Conference on Machine Learning and Applications, 1210–1214 (2015)
5. K. Ahuja, B. Singh, R. Khanna.: Particle swarm optimization based network selection in heterogeneous wireless environment. Optik-International Journal for Light and Electron Optics 125, 214–219 (2014)
6. Yan, X., Sekercioglu, Y. A., Narayanan, S.: A survey of vertical handover decision algorithms in fourth generation heterogeneous wireless networks. Computer Networks 54, 1848–1863 (2010)
7. Nassr, N., Guizani, S., Al-Masri, E.: Middleware vertical handoff manager: a neural network-based solution. IEEE International Conference on Communications, 5671–5676 (2007)
8. Capka, J., Boutaba, R.: Mobility prediction in wireless networks using neural networks. Management of Multimedia Networks and Services 3271, 320–333 (2011)
9. Chai, R., Cheng, J., Pu, X., Chen. Q.: Neural network based vertical handoff performance enhancement in heterogeneous wireless networks. Wireless Communications, Networking and Mobile Computing (WiCOM), 1–4, (2011)
10. Nan, W., Wenxiao, S., Shaoshuai, F., Shuxiang, L.: PSO-FNN based vertical handoff decision algorithm in heterogeneous wireless networks. 2nd International Conference on Challenges in Environmental Science and Computer Engineering 11, 55–62 (2011)
11. Ben-Mubarak, M., Ali, B. M., Noordin, N. K., Ismail, A., Ng, C. K.: Fuzzy logic based self-adaptive handover algorithm for mobile WiMAX. Wireless Personal Communications 71, 1421–1442 (2013)

Tweet Cluster Analyzer: Partition and Join-based Micro-clustering for Twitter Data Stream

M. Arun Manicka Raja and S. Swamynathan

Abstract Data stream mining is the process of extracting knowledge from continuously generated data. Since data stream processing is not a trivial task, the streams have to be analyzed with proper stream mining techniques. In many large volume of data stream processing, stream clustering helps to find the valuable hidden information. Many works have concentrated on clustering the data streams using various methods, but mostly those approaches lack in some core tasks needed to improve the cluster accuracy and quick processing of data streams. To tackle the problem of improving cluster quality and reducing the time for data stream processing time in cluster generation, the partition-based DBStream clustering method is proposed. The result has been compared with various data stream clustering methods, and it is evident from the experiments that the purity of clusters improves 5% and the time taken is reduced by 10% than the average time taken by other methods for clustering the data streams.

Keywords Tweets · Micro-clustering · Macro-clustering · Cosine similarity · DBStream · Partition · Join

1 Introduction

In many real-world applications related to stock trading, social network data streams, and sensor networks, a large volume of data are generated. In particular, social media data streams are widely generated at a rapid rate. Twitter is the most probably used social media application for instantly sharing the message of size

M. Arun Manicka Raja (✉) · S. Swamynathan
Department of Information Science and Technology, College of Engineering Guindy, Anna University, Chennai 600025, Tamil Nadu, India
e-mail: arunmanick@auist.net

S. Swamynathan
e-mail: swamyns@annauniv.edu

H.S. Behera and D.P. Mohapatra (eds.), *Computational Intelligence in Data Mining*, Advances in Intelligent Systems and Computing 556,
DOI 10.1007/978-981-10-3874-7_64

140 characters. Twitter generates large volume of data streams pertaining to various topics, posted by different users across the world.

It is essential to apply some mining tasks on these twitter data streams such as classification, clustering, and frequent item set mining. Data stream clustering is the process of grouping or categorizing the live data streams based on its relevance. In this work, initially, the arrived data streams are invoked to create subgroups. These subgroups are then used to create the final macro-clusters. Data stream clustering techniques mainly focus on creating clusters with high purity in a less time. The clustering process is performed in both online and offline modes. When the streams are generated, in the online mode itself, the streams are invoked to create almost closely relevant clusters. The time taken for generating these micro-clusters is low.

The partition and join-based DBStream clustering is proposed for clustering the evolving twitter data streams. The proposed method uses the micro-clustering tasks for generating clusters. Initially, the newly arriving twitter data streams are either grouped or merged into the existing micro-clusters. Then, the non-growing micro-clusters are pruned. Finally, after pruning has been done, the macro-cluster is formed using the available micro-clusters.

The main contributions of partition and join-based DBStream clustering are presented as follows: (1) In the partition and join-based DBStream clustering, the newly arriving twitter data streams are mapped into the suitable available initial subgroups created out of the received data streams. Thus, the cluster search time is reduced for assigning the data into the appropriate micro-cluster. (2) In the partition and join-based DBStream, if the arriving data streams are not assigned to any micro-clusters, then those clusters are considered as outlier, and over a period, those outlier data are expired. (3) The experimental results have proved that the partition and join-based DBStream clustering outperformed DStream and DenStream clustering techniques in terms of purity of clusters and the data stream processing time.

The paper is organized as follows: Section 2 provides the related work of data stream clustering. Section 3 explains concepts and overview of partition and join-based DBStream clustering. Section 4 describes the implementation details and experimental results with detailed discussion. The conclusion and possible future enhancements are presented in Sect. 5.

2 Related Work

Social media application generates large volume of data as continuous stream information. These data streams have to be processed continuously for generating meaningful outcomes. The parallel data stream clustering has been incorporated from online k-means clustering and outlier mechanisms. The data streams are processed using apache storm stream processing engine for adopting the clustering speed when parallel execution is performed [1]. It is ultimately important for identifying the appropriate stream clustering algorithms. StreamK++ is one of the data stream clustering algorithms, and it is widely used in most of the data stream clustering applications. When compared to BIRCH and StreamLS, Stream++ produces clusters with better quality [2]. The points in the same cluster are having higher similarity among them and having much lesser similarity with the data points in the other clusters. The method is suitable for recognizing the clusters of any arbitrary shapes [3].

Density-based clustering is able to detect any arbitrary shape clusters, and also, it does not require the number of clusters in advance. A multi-density clustering algorithm is used for clustering the data streams. A pruning technique helps to remove the noisy data streams from the real-time data [4]. DDenStream is the modified version of DenStream. It applies the decaying technique as fading function in the evolving data streams and improves the quality of the clusters [5]. The HSWStream is used for clustering high-dimensional data streams. It eliminates the old points while clustering and only considers the new data points received for further clustering process [6]. Haar wavelet transform is used to extract feature from every data block and preserves the feature detail of recent data. Recent biased distance function is used to apply the recent biased clustering algorithm [7].

Expectation–maximization (EM) is used effectively to cluster the data stream in a distributed environment. A test and cluster strategy is used for reducing the processing cost, and it is also effective for online clustering over large volume of data streams [8]. Stream ensemble fuzzy C-means is used for data stream clustering in the immense data and analyzes the data and combines results to gain more accurate analysis than the clustering performed by the individual algorithm [9]. Expectation–maximization algorithm is used to initialize the compound Gaussian clustering algorithm [10]. The statistical change point detection test is combined with the affinity propagation. The clustering model is rebuilt whenever the test detects change in the underlying data distribution [11]. The algorithm handles detailed correlations among finite points for producing high-quality clusters [12].

A support vector-based clustering is applied to cluster the data streams. The data elements of a stream are mapped into the kernel. The support vectors are used as the summary information of the historical data elements for constructing cluster boundaries with arbitrary cluster shapes. Multiple spheres are dynamically maintained for adapting both abrupt and gradual changes [13]. Index structure clustering data stream tree is used with improved space partition. By applying this, it is possible to handle large volume of data streams with arbitrary cluster shapes [14]. Data stream clustering with improved similarity search tree, buffer, speed processing, and local aggregation strategy leads to the adaptation of data streams with different speeds. The noises in the data stream are processed using core micro-cluster and outlier micro-cluster buffer [15, 16].

In spite of various works that have been carried out for data stream clustering, still there is a lack in the performance of the data stream clustering. In this work, the data stream clustering is performed with a novel idea of subset group generation for efficient online micro-clustering of twitter data streams with decaying functionality-based macro-cluster generation. It comparatively reduces the time taken for clustering the data streams since the micro-clustering in online improves clustering speed and efficiently performs macro-clustering in offline mode.

The system also adopts to the scalability and flexibility of processing large data streams in a real-time setup.

3 Overview of the Partition and Join-based DBStream Clustering

The tweet stream clustering framework consists of various components as shown in Fig. 1 such as data stream collector, pre-processing unit, subgroup creator, micro-cluster generator, cluster partitioning and joining, and macro-cluster generation.

3.1 Twitter Data Stream Collector

The streaming API is mainly used for getting the live data streams from the twitter. Tweets collected using the tweet stream collector are passed to the tweet pre-processing component.

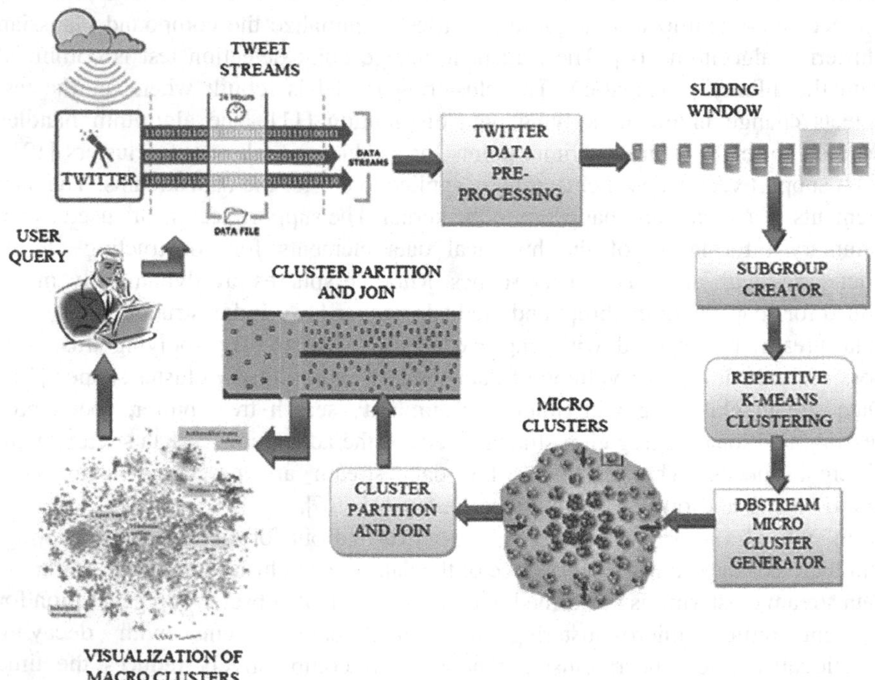

Fig. 1 Overview of the partition and join-based DBStream clustering

3.2 Pre-processing

In some cases, the tweets are only smileys or emoticons. Many times, the tweets are posted in the regional languages but typed using English letters. All these kinds of tweets do not contain any rich information pertaining to any subject. Thus, those tweets are considered as invalid tweets. It is considered to remove those tweets in the pre-processing stage.

3.3 Subgroup Creator

The twitter data are continuously received and stream information have to be processed while generated. In this sub group creation stage, the stream of data is passed along with the sliding window. Each window of data is of particular size. In the window of data, the subset of data is identified and stored separately.

3.4 Micro-cluster Creator

Once the subgroups have been created, then the similar subgroups are identified based on the data available in each subset. The similarity is measured, and the micro-clusters are generated. The micro-cluster is implemented with the DBStream clustering algorithm available in the massive online analysis framework. As long as the data streams are received, the micro-clusters are generated and the summary statistics of these clusters are used for further final cluster generation.

3.5 Partition and Join

After the creation of the micro-clusters, those are stored as a repository for creating the macro-cluster using the partition and join process. This process helps to partition the cluster with less similarity data streams and make them as new cluster. In addition, when new data streams are arrived, those can be compared with the already available clusters. If it is similar, then it can be combined with the existing cluster or a new cluster can be created. When two or more micro-clusters are same with the data, then those micro-clusters are joined together for creating a new cluster.

3.6 Macro-cluster Creator

The result of partition and join step is the creation of macro-clusters. The macro-clusters are the combination of the micro-clusters. All the micro-clusters are formed as separate macro-clusters. Few micro-clusters are not reachable to the macro-clustering process level, since those clusters are expired during the cluster formation.

4 Experimental Results and Discussion

The twitter data streams were collected using twitter REST API for over a period of last four months from the date of experiment conducted. The twitter data streams were collected by focusing on the assembly election 2016 of Tamil Nadu in India. Mainly, the data collection focused on the tweets representing the major political parties and their leaders and various public issues targeted by these political parties. Totally, around 1 million tweets have been collected on these topics. All the tweets are stored in the JSON format with various parameters of tweets.

Since twitter data streams are continuously received, those tweets have to be simultaneously processed. For that, subgroup creator is implemented by identifying the number of subgroups needed for the arriving data streams dynamically. The sliding window is used for receiving the streams and process a window of data in the subgroup creator. The k-means clustering process is used for identifying the number of subgroups needed over the available twitter data streams. Around five times, the k-means clustering has been performed on these tweets with the average of 2,00,000 tweets per every time.

Consider b(i) as the lowest average dissimilarity of i to any cluster. But i is not a member in that cluster. The cluster having lowest average dissimilarity is considered as the neighboring cluster of i since it is the successive best-fit cluster for point i. A silhouette coefficient is defined as represented in Eq. 1.

$$s(i) = \frac{b(i) - a(i)}{\max\{a(i), b(i)\}} \tag{1}$$

If s(i) needs to be very close to 1, then it must be like a(i)¡¡b(i). Since a(i) is the measure of the dissimilarity of i to the own cluster, even a small value of it is appropriately matched. In case, if b(i) is very large, then it is considered that i is not appropriately matched with the neighboring cluster. If s(i) is close to +1 means, the stream datum is suitably clustered. If s(i) is close to −1, then it is appropriate that the stream datum must have clustered with the neighboring cluster. If s(i) is near 0 means, it is considered that the stream datum is in the middle of two clusters. The average s(i) over the entire data in a cluster can be a measure of tightly grouped characteristics of the data in that cluster.

In the micro-clustering model, each cluster is considered as a spherical unit of data with dimension d. With the use of feature clusters, the i-th cluster data streams are represented as points Xi1; Xi2;...; Xini and the time stamp values are Ti1; Ti2; Tini by (CF2xi; CF1xi; CF2ti; CF1ti; ni).

CF2xi keeps track of the sum of the squares of the data streams, and its j-th data is represented as mentioned in Eq. 2.

$$CF2_{i_j}^{x} = \sum_{k=1}^{n_i} (X_{i_{k_j}})^2 \tag{2}$$

CF1xi stores the sum of data stream elements, and its j-th data arrival is denoted as mentioned in Eq. 3.

$$CF1_{i_j}^{\infty} = \sum_{k=1}^{n_i} X_{i_{k_j}} \tag{3}$$

CF2ti records the sum of the squares of the time values, and it is written as in Eq. 4.

$$CF2_i^t = \sum_{k=1}^{n_i} T_{i_k}^2 \tag{4}$$

CF1ti keeps the sum of time values, and it is described as in Eq. 5.

$$CF1_i^t = \sum_{k=1}^{n_i} T_{i_k} \tag{5}$$

ni represents the number of data stream elements available in the i-th cluster. The centroid and radius, relevance time value of a micro-cluster is represented in Eq. 7. The j-th data arrival of the centroid of the i-th micro-cluster is denoted as in Eq. 6.

$$Centroid_{i_j} = \frac{CF1_{i_j}^x}{n_i} \tag{6}$$

The radius of the micro-cluster shows the sparsity of the data streams in the micro-cluster, and it is measured using the root-mean-square (RMS) variation of the data stream elements from the centroid that is as mentioned in Eq. 7.

$$Radius_i = \frac{1}{d}\sum_{j=1}^{d} \sqrt{\frac{CF2_{i_j}^x}{n_i} - \left(\frac{CF1_{i_j}^x}{n_i}\right)^2} \tag{7}$$

The time values of the data stream elements contained in a micro-cluster are denoted as m = (2/ni) for the arriving data stream elements. It is assumed that the data stream elements follow the normal distribution and the j-th data arrival for which the mean is represented as mentioned in Eqs. 8 and 9 with its standard deviation. Many algorithms also capture the dispersion of data points by storing the variance. When the re-clustering is performed, only the distances between the micro-clusters are used. Mostly, the micro-clusters, which are closer to each other, may end up in a same cluster. The micro-cluster generation algorithm has been described as follows:

$$\mu_{i_j} = \frac{CF1^t_{i_j}}{n_i} \tag{8}$$

$$\sigma_{i_j} = \sqrt{\frac{CF2^t_{i_j}}{n_i} - \left(\frac{CF1^t_{i_j}}{n_i}\right)^2} \tag{9}$$

Input: Data subgroups from the sliding window
Output: Microclusters
for j = 1 to number of subgroups
1. find the appropriate k microclusters from the sub
groups depending upon the mean silhouette value
2. generate microclusters
3. if the silhouette value <0.25
re-cluster microclusters
end
end

The newly arriving tweet streams are received and checked for the similarity of the content, and it is assigned to the respective clusters.

Every micro-cluster is assigned with a minimum weight value. The distance metric is also specified for calculating the distances between the cluster centers.

The cluster density with Gaussian parameter as the entire density weight of the data stream elements is represented in Eq. 10.

$$\rho_i = \sum_{j=1}^{n} \frac{d_{ij}}{d_c} \exp(-\frac{d_{ij}}{d_c}) \tag{10}$$

where n represents the total data stream elements, dij denotes the Euclidean distance among the i-th and j-th micro-clusters, and dc represents the cutoff measure distance. The local cluster distance is the distance between the ith micro-cluster and the other neighbor clusters, which is represented in Eq. 11.

$$\psi_i = \min_{j, \rho_j > \rho_i}(d_{ij}) \tag{11}$$

The macro-clustering algorithm for the data stream is represented as follows:

Input: microclusters, mic-clu
Output: macroclusters, mac-clu:
function Macrolustering(mic-clu)
for i = 0 to micronum - 2 do

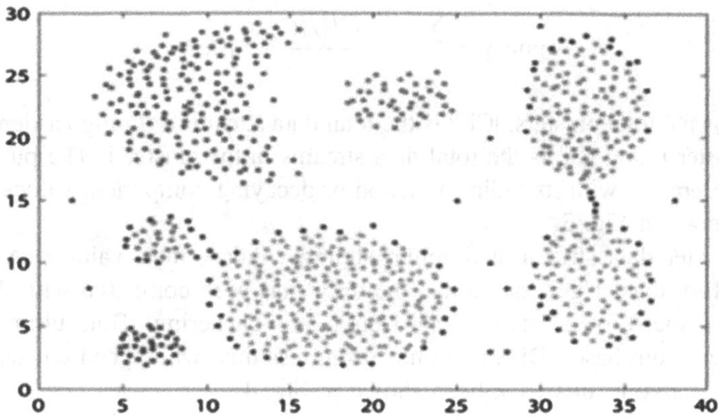

Fig. 2 Generated macro-clusters

```
for j = i to micronum - 1 do
calculate CentreDistanceMat[i, j]
end for
end for
receive dc by arranging the CenterDistanceMat
for i = 0 to micronum - 1 do
calculate ρ[i]
end for
for i = 0 to micronum - 1 do
calculate ψ[i]
end for
for i = 0 to micronum - 1 do
create and assign cluster data labels
end for
return mac-clu
end function
```

The macro-clusters have been generated, and it has been visualized in Fig. 2.

4.1 Cluster Evaluation

The evaluation of cluster is a prominent task in the cluster analysis process. The main objective of the cluster evaluation process is to evaluate the clustering results of the newly implemented algorithm with the previously available data stream clustering algorithm. Though various measures are used for the cluster evaluation, the cluster purity is widely used in many implementations. The quality of the clustering process is evaluated by using the average cluster purity. It is defined as represented in Eq. 12.

$$\text{purity} = \frac{\sum_{i=1}^{k}(|C_i^d|/|C_i|)}{K} * 100\% \tag{12}$$

where K is the total clusters, |Cid| is the total data streams with higher density label in the cluster i, and |Ci| is the total data streams in the cluster i. The purity of the cluster generation with its fading function or decaying component values has been clearly shown in Fig. 3.

The twitter data stream at different sliding window inter value size has been processed for cluster generation and the result has been compared with the stream processing speed of DStream and DenStream clustering. But, ultimately, the partition and join-based DBStream has taken less time with speed clustering time. The comparative results have been shown in Fig. 4.

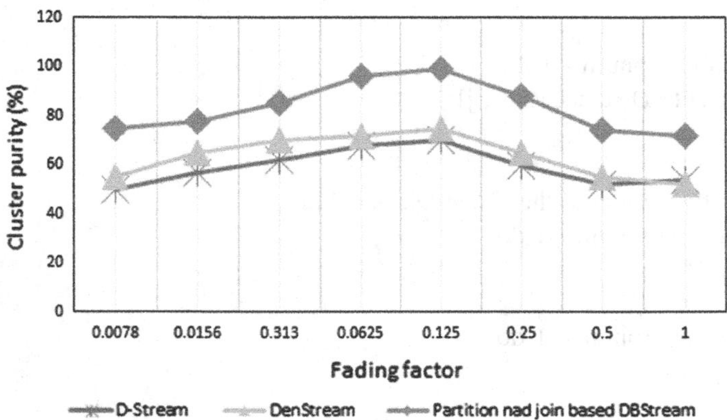

Fig. 3 Purity of the cluster versus decaying component value

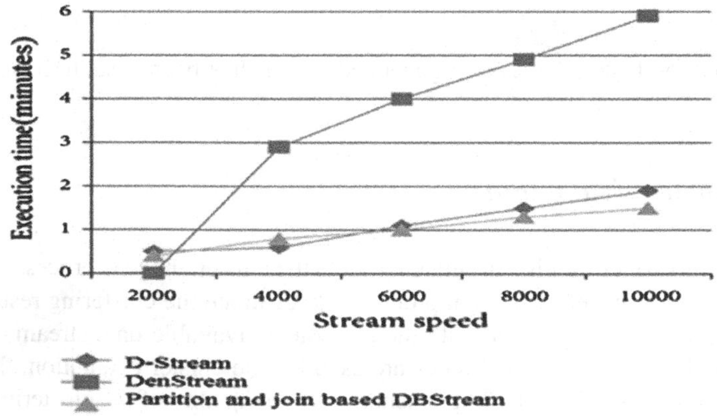

Fig. 4 Processing time versus arrival of data streams

5 Conclusion and Future Work

In this work, the partition and join-based DBStream clustering has been proposed for evolving twitter data streams. The proposed partition and join-based DBStream clustering method has achieved high-quality clusters and low computational time when compared to the existing methods. The main unique process between the existing methods and the proposed method is the cluster assignment. The join process has reduced the time complexity of macro-cluster formation. The subgroup data structure makes the update of the data streams very fast. Further, the implemented method has increased the cluster purity improves 5% with the micro-cluster generation and expiry tasks. The future possible enhancement includes the improvement of partition and join-based DBStream clustering as a distributed DBStream clustering method.

References

1. X. Gao, E. Ferrara and J. Qiu, "Parallel Clustering of High-Dimensional Social Media Data Streams," 15th IEEE/ACM International Symposium on Cluster, Cloud and Grid Computing (CCGrid), Shenzhen, 2015, pp. 323–332.
2. A. Kaneriya and M. Shukla, "A novel approach for clustering data streams using granularity technique," International Conference on Advances in Computer Engineering and Applications (ICACEA), Ghaziabad, 2015, pp. 586–590.
3. G. Lin and L. Chen, "A Grid and Fractal Dimension-Based Data Stream Clustering Algorithm," International Symposium on Information Science and Engineering, Shanghai, 2008, pp. 66–70.
4. A. Amini, H. Saboohi and T. Y. Wah, "A Multi Density-Based Clustering Algorithm for Data Stream with Noise," IEEE 13th International Conference on Data Mining Workshops, Dallas, 2013, pp. 1105–1112.
5. M. Kumar and A. Sharma, "Mining of data stream using DDenStream clustering algorithm," IEEE International Conference in MOOC Innovation and Technology in Education (MITE), Jaipur, 2013, pp. 315–320.
6. W. Liu and J. OuYang, "Clustering Algorithm for High Dimensional Data Stream over Sliding Windows," IEEE 10th International Conference on Trust, Security and Privacy in Computing and Communications, Changsha, 2011, pp. 1537–1542.
7. Qian Zhou, "A recent-biased clustering algorithm of data stream," Second International Conference on Mechanic Automation and Control Engineering (MACE), Hohhot, 2011, pp. 3803–3808.
8. A. Zhou, F. Cao, Y. Yan, C. Sha and X. He, "Distributed Data Stream Clustering: A Fast EM-based Approach," IEEE 23rd International Conference on Data Engineering, Istanbul, 2007, pp. 736–745.
9. R. Fathzadeh and V. Mokhtari, "An ensemble learning approach for data stream clustering," 21st Iranian Conference on Electrical Engineering (ICEE), Mashhad, 2013, pp. 1–6.
10. M. m. Gao, J. z. Liu and X. x. Gao, "Application of Compound Gaussian Mixture Model clustering in the data stream," International Conference on Computer Application and System Modeling (ICCASM 2010), Taiyuan, 2010, pp. V7-172-V7-177.

11. X. Zhang, C. Furtlehner, C. Germain-Renaud and M. Sebag, "Data Stream Clustering With Affinity Propagation," IEEE Transactions on Knowledge and Data Engineering, vol. 26, no. 7, 2014, pp. 1644–1656.
12. H. Zhu, Y. Wang and Z. Yu, "Clustering of Evolving Data Stream with Multiple Adaptive Sliding Window," Data Storage and Data Engineering (DSDE), International Conference on, Bangalore, 2010, pp. 95–100.
13. C. D. Wang, J. H. Lai, D. Huang and W. S. Zheng, "SVStream: A Support Vector-Based Algorithm for Clustering Data Streams," IEEE Transactions on Knowledge and Data Engineering, vol. 25, no. 6, 2013, pp. 1410–1424.
14. Huanliang Sun, Ge Yu, Yubin Bao, Faxin Zhao and Daling Wang, "CDS-Tree: an effective index for clustering arbitrary shapes in data streams," 15th International Workshop on Research Issues in Data Engineering: Stream Data Mining and Applications (RIDE-SDMA'05), 2005, pp. 81–88.
15. Kehua Yang, HeqingGao, Lin Chen and Qiong Yuan, "Self-adaptive clustering data stream algorithm based on SSMC-tree," 4th IEEE International Conference on Software Engineering and Service Science (ICSESS), Beijing, 2013, pp. 342–345.
16. Charu C. Aggarwal, "A Framework for Clustering Evolving Data Streams" Proceedings of the 29th VLDB Conference, Berlin, Germany, Vol. 29, 2003, pp. 81–92.

Contour-Based Real-Time Hand Gesture Recognition for Indian Sign Language

Rajeshri R. Itkarkar, Anilkumar Nandi and Bhagyashri Mane

Abstract Gesture recognition system is widely being developed recently as gesture-controlled devices are on a large scale used by the consumers. The gesture may be in static or in dynamic form, typically applied in robot control, gaming control, sign language recognition, television control etc. This paper focuses on the use of dynamic gestures for Indian sign language recognition. The methodology is implemented in real time for hand gestures using contour and convex hull for feature extraction and Harris corner detector for gesture recognition. The accuracy results are obtained under strong, dark, and normal illumination. The overall accuracy achieved for Indian sign language recognition under dark illumination is 81.66. With Indian sign language application, the recognized gesture can also be applied for any machine interaction.

Keywords Gesture recognition · Hand gestures · Harris corner detector

1 Introduction

Gesture recognition in real time is a continuous research. There are millions of Indian people using sign language, as gesture is the simplest way for interaction. Gestures are strongest means of communication all around the world [1]. In sign language, each gesture is assigned a meaning and strong rule of grammar is applied to trace the gesture [2]. Gestures are recognized by using different algorithms which allows the machine to interact easily and efficiently with any human. Physically handicapped

R.R. Itkarkar (✉) · A. Nandi
BVB COE, Hubli, India
e-mail: itkarkarrajashri@yahoo.com

A. Nandi
e-mail: anilnandy@bvb.edu

B. Mane
JSPMs RSCOE, Pune, India
e-mail: mane.bhagyashree5@gmail.com

H.S. Behera and D.P. Mohapatra (eds.), *Computational Intelligence in Data Mining*, Advances in Intelligent Systems and Computing 556,
DOI 10.1007/978-981-10-3874-7_65

people can interact naturally and effectively to control any device in 3D virtual environment. Basically, hand gestures are classified as static and dynamic gestures. Static gestures are the gestures which are still or not changing with time, and dynamic gestures are those which change with time [3]. Hand gestures are recognized using gloves or vision technique. Glove-based recognition uses sensors to detect the position and orientation of the gesture. Vision-based gestures make use of camera, computer vision, and image processing for gesture recognition [4]. Reza Hassanpour et al. presented a survey on human machine interaction for analyzing, modeling, and recognizing hand gestures with different algorithms and their approaches [5]. Le Tran Nguyen et al. proposed a real-time recognition model which used Moore Neighborhood algorithm for feature extraction. The Hu invariant moments are the features extracted, and the classifier used is memory-based logistic classifier with an accuracy of 90.4 [6]. The advantage of logistic classifier is that they are able to self tune according to the noise level of the data points in the training set. Feng-Sheng Chen et al. developed a real-time system in which Fourier descriptor was used for hand tracking and extraction in spatial domain, and motion analysis was used in temporal domain. The combined features in spatial and temporal domain were further applied to an HMM classifier for recognition and achieved an accuracy of 90% [7]. The limitation of this article was that they did not had enough training data to make a good estimate of the HMM model parameters. Kam Lai et al.proposed a real-time gesture recognition system using kinect camera. The skeleton-based extraction and nearest-neighbor classifier was applied with a close accuracy of 100% The kinect camera gave the higher accuracy, but it offers a poor performance for outdoor applications [8]. Shen Wu et al. in their work used kinect sensor to record the cloud points, from which skeleton features were extracted by using laplacian contraction method. A partition-based classifier was applied to improve the accuracy and robustness of the system [9]. Bretzner L et al. have developed a prototype system in real time, in which a multi-scale color image features are extracted in terms of scale, orientation, and hand position [10]. Choi J et al. developed a hand gesture-controlled power point presentation model, distance transform and circular profiling method was applied [11]. Zhu X et al. extracted common patch-level features and fused them using kernel descriptors and SVM classifier to recognize a single frame hand gesture [12]. The accuracy was less for the complex gestures.

The flow of the paper is followed by introduction. Section 2 presents real-time hand gesture recognition. Section 3 discusses the implementation and results with analysis.

2 Real-Time Hand Gesture Recognition Method

The proposed system uses a simple web camera for video sequence acquisition from which a frame is captured in the YUY2 format. In YUY2 format, the data is represented in an array of unsigned values, where the first byte contains the Y sample i.e., Luminance, the second byte contains the U sample (Cb) i.e., Color difference

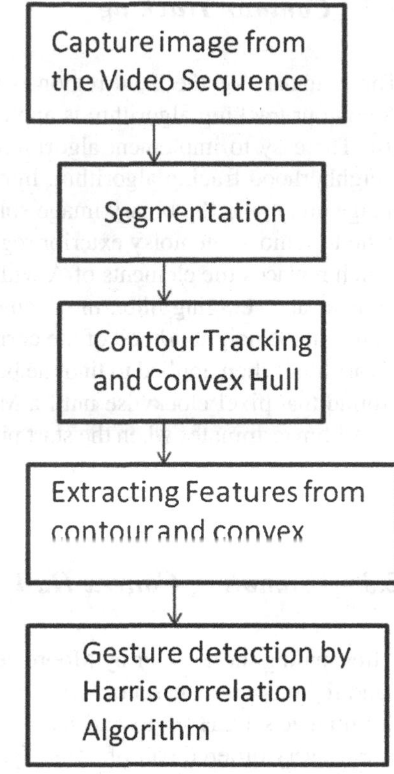

Fig. 1 Proposed methodology for real-time indian sign language

for blue, the third byte contains the second Y sample, and the fourth byte contains the V sample value (Cr). YUY2 format is usually preferred in the 4:2:2 pixel format for Microsoft. Thus, the frame captured from the video sequence, in YUY2 format, is converted first in the RGB format. The RGB format is further converted into gray scale. The proposed work is implemented in four modules: segmentation, contour tracking and estimating the convex hull, feature extraction, and classification. Figure 1 shows the flow for the proposed methodology.

2.1 Segmentation

Segmentation is the process in which the image is divided into multiple segments. In gesture recognition, segmentation is usually used to locate objects and boundaries in the image. Segmentation in the proposed method is used to convert gray scale image into binary image, so that only two segments are present in the image, one is the hand and other is the background.

2.2 Contour Tracking

The boundary of the hand region is required to be estimated to find the contour. A contour tracking algorithm is applied to track the contour in clockwise direction [6]. The easy to implement algorithm are the square tracing algorithm and Moore Neighborhood tracing algorithm. In contour tracking algorithm, first threshold the image and make the binary image solid by filling it with holes. Median filtering is done to remove the noisy exterior regions. We have used the function medfilt2 (A) which replaces the elements of A with median of the neighbors. Median filtering is same as an averaging filter, in which each output pixel is set to an average of pixel values in the neighborhood of the corresponding output pixel. Moore neighborhood algorithm is then applied to find the boundary. The idea is to hit a black pixel and go around that pixel clockwise until a Moore neighborhood black pixel is found. The algorithm terminates when the start pixel is once again located. The black pixel over which we have moved is the contour of the pattern.

2.3 Estimating Convex Hull

After finding the contour by Moore neighborhood algorithm, the convex hull of the hand region or for the set of P points is estimated as follows: Set P of 2D points

0. Initialize set CH (P) to null set
1. For every ordered pair of points (p, q) from P
2. If all OTHER points lie to the right of directed line through p and q
3. Include directed edge (p, q) to CH (P)
4. Sort the convex hull edges in clockwise order.

2.4 Feature Extraction

When the input data to an algorithm is too large to be processed and contains redundant bits, then this input data can be converted or represented into reduced set of features called as feature extraction. In the proposed method, the hand region properties such as centroid, minor axis length, major axis length, and area are calculated. Figure 2 shows the hand region properties.

2.5 Classification

The classifier used in this method is Harris correlate algorithm. In this classifier, to recognize the gesture, a small window is used. By shifting the window in any

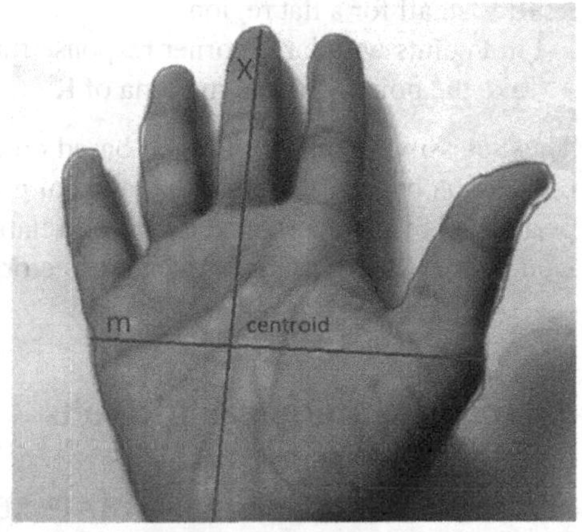

Fig. 2 Hand region showing centroid, minor axis, and major axis

direction, it gives a large change in intensity. Example change of intensity for a shift of [u, v] is given by

$$E(u, v) = \Sigma w(x, y)[I(x + u, y + v) - I(x, y)]^2 \tag{1}$$

$$E(u, v) = [u, v]\, M \begin{bmatrix} u \\ v \end{bmatrix} \tag{2}$$

$$M = w(x, y) \begin{bmatrix} I_x^2 & I_x I_y \\ I_x I_y & I_y^2 \end{bmatrix} \tag{3}$$

E (u, v) is the change of intensity for shift [u, v] as given in Eq. 1. Equation 2. It is represented in terms of matrix M which is a 2×2 matrix evaluated from image derivative; from the intensity change in shifting window, Eigen values are calculated. M is evaluated as per Eq. 3. Classification of hand region depends on the Eigen values of M. The measure of the corner response is given by R as per Eq. 4, and detected matrix by Eq. 5 and traced M by Eq. 6.

$$R = \det M - k(\operatorname{trace} M)^2, k - \text{empirical constant} = 0.04\text{–}0.06 \tag{4}$$

$$\det M = \lambda_1 \lambda_2 \tag{5}$$

$$\operatorname{trace} M = \lambda_1 + \lambda_2 \tag{6}$$

R depends only on Eigen values of M.

- R is large for a corner
- R is negative with large magnitude for an edge

- |R| is small for a flat region
- Find points with large corner response function R (R > threshold)
- Take the points of local maxima of R

Thus, classification is performed based on correlation, auto-correlation, and cross-correlation by verifying the fundamental matrix M. In this method, the detection is performed by estimating the Harris correlation-based corner detectors. The comparison between the training and testing vectors is done by measuring the distance.

3 Implementation and Results

The method is implemented in MATLAB using web camera, with image size 120 × 160. Indian sign language is recognized in real time for numbers 0–9 and alphabets A to L. The total training vectors used are 1300, approximately 60 images for each sign.

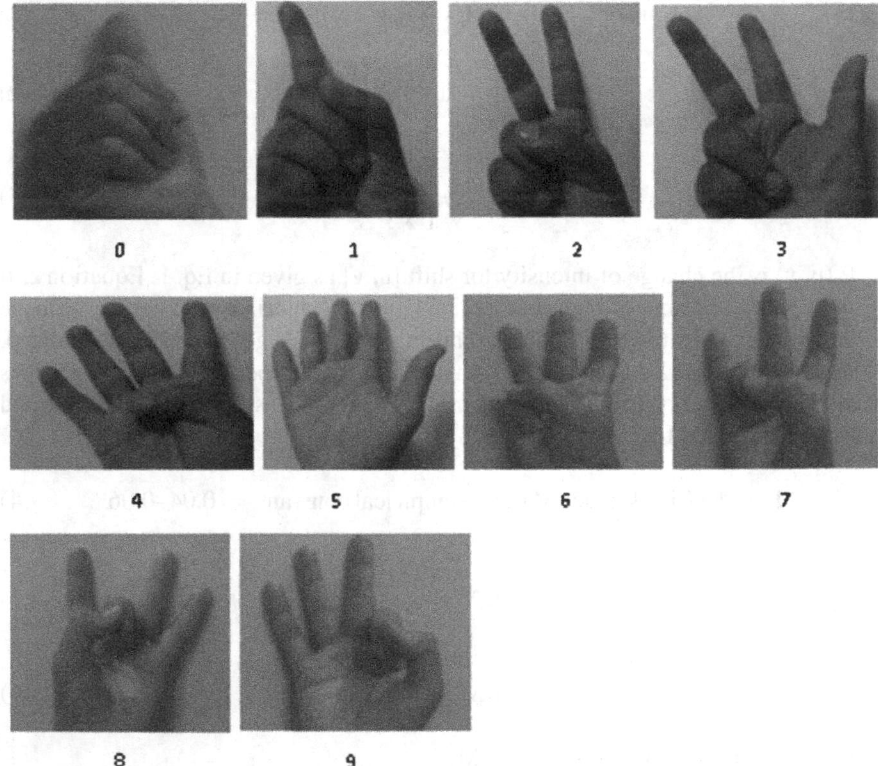

Fig. 3 Indian sign language

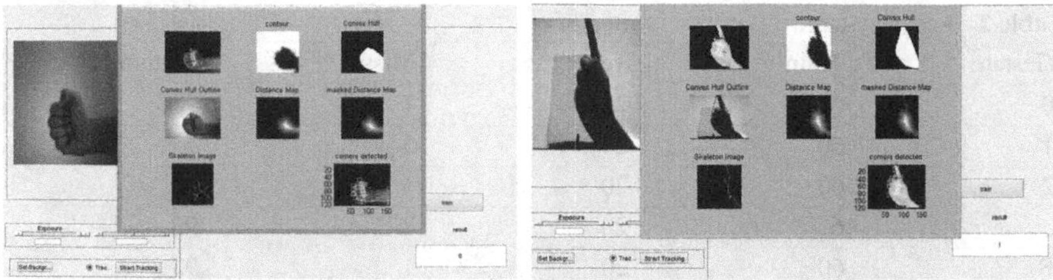

Fig. 4 Result for number system 0 and 1

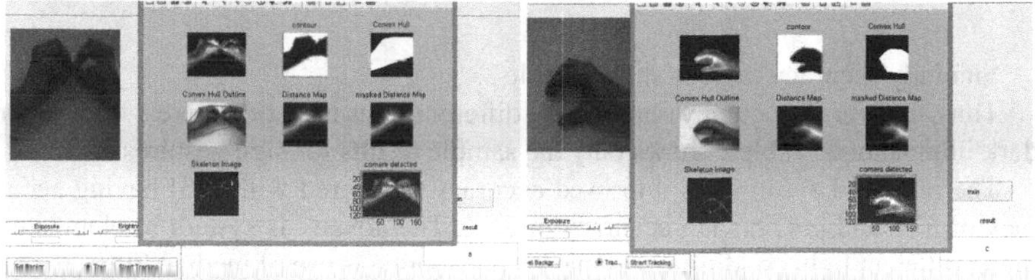

Fig. 5 Result for alphabet 'B' and 'C'

3.1 Results

Database is created with fix white background with the web camera for different light conditions. The Indian sign language is created as shown in Fig. 3. The results are obtained for almost all Indian signs and some sample results are as shown in Figs. 4 and 5.

3.2 Result Analysis

The results are analyzed for different light conditions, such dark, strong, and normal. Table 1 shows the analysis for different sign languages in three light conditions.

Table 1 Sample results under different light condition

Light condition	Gesture	Accuracy (%)
Normal illumination	0	80
Dark illumination	0	95
Strong illumination	0	70
Normal illumination	C	70
Dark illumination	C	75
Strong illumination	C	60

Table 2 Accuracy table

Gesture	Train	Test	Miss	Accuracy (%)
0	60	20	1	95
1	60	20	5	75
2	60	20	6	70
3	60	20	2	90
B	60	20	1	95
C	60	20	5	75

Similarly accuracy table is shown Table 2.

Thus, the overall accuracy obtained for different signs is 81.66% in real time under dark illumination. Table 2 shows only the sample results for sign gestures.

The proposed system has improved accuracy compared with [13] having accuracy of 87% and [14] having accuracy of 76%. The novelty is system implemented under different light condition, and major contribution is use of median filters which improves robustness toward noise.

4 Conclusion

Thus Indian sign language detection in real time is a challenging task like other sign languages. The methodology implemented is using contour and convex hull based with the simple detection algorithm used. The system can improve the accuracy with dark illumination. The future work is going, considering the light conditions and background. The accuracy obtained is 81.66% and also be implemented and analyzed using other classification techniques.

Acknowledgements The Author1 would like to wish thanks to Dr. D. S. Bormane Principal, JSPMs Rajarshi shahu college of Engineering, Pune, where she is presently working as assistant professor, for permitting her to work as research scholar at BVB Hubli, Karnataka.

References

1. Henrik Birk and Thomas Baltzer Moeslund, Recognizing Gestures From the Hand Alphabet Using Principal Component Analysis, Masters Thesis, Laboratory of Image Analysis Aalborg University, Denmark, 1996
2. Siddharth S. Rautaray Anupam Agrawal, Vision based hand gesture recognition for human computer interaction: a survey, Artif Intell Rev doi:10.1007/s10462-012-9356-9
3. McNeill D (1992) Hand and mind: what gestures reveal about thought. University of Chicago Press. ISBN: 9780226561325

4. G.R.S Murthy and R S Jadon, A review of vision based hand gestures recognition, International Journal of Information Technology and Knowledge Management July - December 2009, Volume 2, No. 2, pp. 405–410
5. Reza Hassanpour, Asadollah Shahbahrami, Human Computer Interaction Using Vision Based Hand Gesture Recognition, Journal of Advances in Computer Research 2(2010)
6. Le Tran Nguyen, Cong Do Thanh, Tung Nguyen Ba, Cuong Ta Viet, Ha Le Thanh, Contour Based Hand Gesture Recognition Using Depth Data, Advanced science and technology letters, Vol 29. (SIP- 2013) pp. 60-65, ISSN: 2287- 1233ASTL
7. Feng-Sheng Chen, Chih-Ming Fu, Chung-Lin Huang, Hand gesture recognition using a real-time tracking method and hidden Markov models, Image and Vision Computing 21 (2003) 745758, 2003 Elsevier Science B.V
8. Kam Lai, Janusz Konrad, and Prakash Ishwar, A gesture-driven computer interface using Kinect, 978-1-4673-1830-3/12/ IEEE
9. Shen Wu, Feng Jiang and Debin Zhao, Hand Gesture Recognition based on Skeleton of Point Clouds, 2012 IEEE fifth International Conference on Advanced Computational Intelligence (ICACI) October 18–20, 2012 Nanjing, Jiangsu, China
10. Bretzner L, Laptev I, Lindeberg T (2002) Hand gesture recognition using multi-scale colour features, hierarchical models and particle filtering. In: Fifth IEEE international conference on automatic face and gesture recognition, pp. 405410. doi:10.1109/AFGR.2002.1004190
11. Choi J, Park H, Park, J-I (2011) Hand Shape Recognition Using Distance Transform and Shape Decomposition. 18th IEEE International Conference on Image Processing (ICIP), September 2011; pp. 36053608
12. Zhu X, Wong KK (2012) Single-Frame Hand Gesture Recognition Using Color and Depth Kernel Descriptors. 21st International Conference on Pattern Recognition (ICPR)
13. Ebrahim Aghajari, Damayanti Gharpure Real Time Vision-Based Hand Gesture Recognition for Robotic Application, International Journal of Advanced Research in Computer Science and Software Engineering, Volume 4, Issue 3, March 2014 ISSN: 2277 128X
14. Quentin De Smedt, Hazem Wannous, Jean-Philippe Vandeborre Skeleton-based Dynamic hand gesture recognition, CVPR open access workshop paper provided by computer vision foundation

Firefly Algorithm for Feature Selection in Sentiment Analysis

Akshi Kumar and Renu Khorwal

Abstract Selecting and extracting feature is a vital step in sentiment analysis. The statistical techniques of feature selection like document frequency thresholding produce sub-optimal feature subset because of the non-polynomial (NP)-hard character of the problem. Swarm intelligence algorithms are used extensively in optimization problems. Swarm optimization renders feature subset selection by improving the classification accuracy and reducing the computational complexity and feature set size. In this work, we propose firefly algorithm for feature subset selection optimization. SVM classifier is used for the classification task. Four different datasets are used for the classification of which two are in Hindi and two in English. The proposed method is compared with feature selection using genetic algorithm. This method, therefore, is successful in optimizing the feature set and improving the performance of the system in terms of accuracy.

Keywords Sentiment analysis · Feature selection · Swarm intelligence · Firefly algorithm · Genetic algorithm · Support vector machine (SVM)

1 Introduction

With the advent of social media, opinion-rich data resources such as microblogging sites, personal blogs, and online review sites have proliferated enormously. People express their views or opinions/attitudes on a variety of issues, discuss current issues, complain, and provide feedback and suggestions for the products and policies they use in their daily life or which concern them. This unstructured social media data is used to mine the overall attitude of the writer toward a specific issue.

A. Kumar · R. Khorwal (✉)
Department of Computer Science & Engineering,
Delhi Technological University, Delhi, India
e-mail: thekhorwal@gmail.com

A. Kumar
e-mail: akshikumar@dce.ac.in

H.S. Behera and D.P. Mohapatra (eds.), *Computational Intelligence in Data Mining*, Advances in Intelligent Systems and Computing 556,
DOI 10.1007/978-981-10-3874-7_66

Sentiment analysis [1], as an intelligent mining technique, helps to capture and determine opinions, emotions, and attitudes from text, speech, and database sources, which correspond to how users retort to a particular issue or event. Sentiment mining from social media content is a tedious task, because it needs in-depth knowledge of the syntactical and semantic, the explicit and implicit, and the regular and irregular language rules.

Sentiment analysis is a multi-step process encompassing various sub-tasks such as sentiment data collection, feature selection, sentiment classification, and sentiment polarity detection [2]. Feature selection in sentiment analysis has a very significant role in enhancing accuracy of the system as the opinionated documents usually have high dimensions, which can adversely affect the performance of sentiment analysis classifier. Effective feature selection technique recognizes significant and pertinent attributes and improves the classification accuracy, thereby reducing the training time required by classifier. Due to the high-dimensional, unstructured characteristics of the social media content, this problem of text classification manifolds, thus fostering the need to look for improved and optimized techniques for feature selection.

The traditional methods for feature selection such as chi-square, information gain, and mutual information [3] are successful in reducing the size of the corpus but with a compromised accuracy. These produce sub-optimal feature subsets as feature subset selection problem lies in the category of non-polynomial (NP)-hard problems. Evolutionary algorithms have been successful at coming up with good solutions for complex problems, when there is a way to measure quality of solutions [4]. Algorithms such as nature-inspired algorithms [5], genetic algorithms [6], simulated annealing [7] have been explored much in the literature for improved classification. Swarm intelligence is a distributed system whereby self-cooperating global behavior is produced by anonymous social agents interacting locally having local perception of its neighboring agents and the surrounding environment [8].

Firefly algorithm, developed by Xin-She Yang [9] in the year 2008, is a biologically inspired algorithm centered on flashing patterns of fireflies. FA is a relatively new algorithm and outperforms other swarm intelligence algorithms. It is a population-based metaheuristic algorithm based on pattern of fireflies where each firefly represents potential solutions to the problem in the search space. The proposed system uses firefly optimization algorithm for feature selection. The proposed technique is compared with feature selection using genetic algorithm and the baseline model and is validated for four different datasets.

2 Related Work

SA has received a lot of focus from researchers and analysts in recent years. T. Sumathi et al. [10] (2013) introduced ABC algorithm for feature selection technique for selecting optimum feature set, thereby improving classification accuracy and reducing computational complexity. The ABC technique is a

Table 1 Analysis of feature selection optimization techniques

Technique	Dataset	Classifier	Accuracy without optimization	Accuracy with optimization	Year	
ABC	Product reviews	SVM	55	70	2015	[11]
ABC	Internet movie database (IMDb)	Naïve Bayes	85.25	88.5	2014	[10]
		FURIA	76	78.5		
		RIDOR	92.25	93.75		
Hybrid PSO/ACO2	Product reviews, government data	Decision tree	83.66	90.59	2014	[15]
PSO	Twitter data	SVM	71.87	77	2012	[16]
PSO	Restaurant review	CRF	77.42	78.48	2015	[12]

powerful optimization technique and is widely used in optimizing NP-hard problems. Opinion mining is used for classifying movie reviews where features are optimized using ABC algorithm. This method improved the classification accuracy in tune of 1.63–3.81%. Ruby Dhruve et al. [11] investigated ABC algorithm for calculating weight of sentiment. The proposed technique incorporated ABC algorithm for the classification to improve the accuracy of the classifier with BOW and BON features for optimizing the best result of the classification. Experimental results show an accuracy improvement from 55 to 70%.

Particle swarm optimization was used in [12] for feature selection in aspect-based sentiment analysis. The proposed technique here could automatically determine the most relevant features for sentiment classification and also these works focus on the extraction of the aspect terms. Experiments revealed that the system is able to achieve better accuracy with a feature set having lower dimensionality. A comparative analysis of work done in feature selection using evolutionary algorithms is given in Table 1. Distance-based discrete firefly algorithm is used in [13] for optimal feature selection using mutual information criterion for text classification.

The system proposed in this work uses mutual information-based criterion which can measure the association between two features selected by the fireflies and determine corrections of features. The work done in this system produces results having greater accuracy.

3 Methodology

The proposed system selects optimum feature subset from high-dimensional feature set for sentiment analysis using firefly algorithm and also the results are compared with genetic algorithm. The fundamental idea of firefly algorithm is that the fireflies use information about brightness from its neighbors to assess themselves. Each firefly is attracted toward its brighter neighbor based on distance. In the standard firefly algorithm, the search strategy is reliant on its control parameters, i.e., absorption and randomness. Optimality degree and the time required to obtain the optimality are two important evaluation aspects of the feature selection problem. The current methods of feature subset selection are successful in achieving either of the two criteria but not both. So, firefly algorithm is used here to tackle both the problems simultaneously.

3.1 *Feature Selection Using Firefly Optimization Algorithm*

A discrete FA is proposed in this section to solve the feature selection problem. Pseudo code for the proposed system is presented in Algorithm 1. The fitness function f is used to measure the fitness based on the accuracy of particular solution χ_i.

Step 1 The first step is to initialize the firefly parameter size for the firefly population (N), α—the randomness parameter, γ—absorption coefficient, and t_{max}—maximum number of generations required for termination process.

Step 2 Next, initial firefly position is initialized. Initially, a random position is allocated to fireflies. $X_i = [x_{i1}; \ldots; x_{in}]$, where n represents number of features in total and m solutions are considered as candidate solutions initially.

Algorithm 1 Pseudo code of the Firefly algorithm

```
Feature Selection Using Firefly Algorithm
Input : N number of fireflies, T_max Maximum number of iterations
        γ Absorption parameter, α Randomness Parameter
Output : Optimal firefly position and its fitness
 1. Initialize parameters N, T_max, γ and α
 2. Initialize χ_i = ϕ, subset of feature selected by i^th firefly
 3. Initialize x_i, position of each firefly subjecting to Σx_ij =s
 4. Calculate fitness f(χ_i)
 5. Sorting the fireflies in accordance with f(χ_i)
 6. While t < T_max
      for i = 1 to N (for each firefly)
        for j = 1 to N (for each firefly)
          if f(χ_j) > f(χ_i)
            move firefly i towards firefly j using equation
          end if
         update γ and corresponding attractiveness
        end for
       end for
      evaluate the position of fireflies
        t=t+1
     end while
 7. Output the feature subset
```

Step 3 In the next step, the fitness of the population is calculated. For this case, the function of fitness is the accuracy of the classifier model.

Step 4 Firefly position modification: The firefly having less brightness value would move toward a firefly with more brightness. The new position is based on the modification in each dimension of firefly.

Step 5 The new solution produced by modifying firefly position is examined using the fitness value of the SVM classifier. Solutions with very low accuracy are discarded.

Step 6 Store the best solution attained till now and increment the generations counter.

Step 7 If the termination criteria are satisfied, then we stop the search, otherwise, go to step 3. The termination criteria used in this method is either maximum number of generations exausts or if classification error is negligible.

3.1.1 Initial Population and Encoding of Fireflies

Initial population is generated randomly as array of binary bits, with length equal to the total number of features. The potential solution has the form $\vec{X}(i) = (x_{i1}, x_{i2}, \ldots, x_{in})$ where $x_{ij} \in \{0, 1\}, \mathrm{i} = 1, 2, \ldots, \mathrm{N}$, where N specifies the number of feature set, i.e., the population size and $\mathrm{j} = 1, 2, \ldots \mathrm{n}$ where n is the number of features. Firefly length is equal to the total number of features. Suppose we have a feature set $\mathrm{F} = (\mathrm{f1}, \mathrm{f2}, \mathrm{f3}, \ldots, \mathrm{fn})$, then a firefly is represented as a binary vector of length n. In $\vec{X}(i)$, if any bit holds a value of "0," then the particular feature is not used for training the classifier, and a value "1" indicates that the corresponding feature is used for the classification. We define the number of features we want to use for the classification. For example, considering a set of features, $\mathrm{F} = (\mathrm{f}_1, \mathrm{f}_2, \mathrm{f}_3, \mathrm{f}_4, \mathrm{f}_5, \mathrm{f}_6, \mathrm{f}_7, \mathrm{f}_8, \mathrm{f}_9, \mathrm{f}_{10})$, and taking N the total population size as 3, the firefly population can be symbolized as follows:

$$\vec{X}(1) = (1, 1, 0, 1, 0, 1, 0, 1, 0, 1)$$
$$\vec{X}(2) = (0, 1, 1, 1, 0, 0, 1, 1, 0, 1)$$
$$\vec{X}(3) = (1, 0, 0, 1, 1, 1, 1, 0, 1, 0)$$

The initial population is generated randomly for N solutions of feature sets. In the initial feature set, the bit positions are randomly assigned as 1 or 0. This is done by generating a uniform random number c in range [0, 1] for every bit position of the feature vector string, i.e., X_{id} of *X(i).* Based on the random value, every firefly *X(i)* is created as follows:

$$X_{id} = \begin{cases} 1 & \text{if } c \text{ is} < 0.5 \\ 0 & \text{otherwise} \end{cases} \quad (1)$$

3.1.2 Updating the Global and Best Firefly Position Value

The best particle from the initial population is first selected based on the fitness value, i.e., the accuracy of the classifier. Now, the firefly having low value moves toward the firefly having more value of the brightness, i.e., fitness function (accuracy).

$$p_{ij} = \frac{1}{1 + e^{\vartheta_{ij}}} \quad (2)$$

Initially, the local best and global best are set as same which is the best firefly position from the initial population. The terms in Eq. (3) are for continuous optimization, to use it for discrete optimization the terms are required to be converted into discrete form. So they are converted into discrete form using function given in Eq. (2). P_{ij} in Eq. (2) is the probability that jth bit is set in x_i. The movement of a firefly i toward firefly j based on attraction is computed using Eq. (3). Here, rand is a random number generated in range [0, 1]. The second term in equation is because of the attraction between the two fireflies due to brightness variance, and the third term used is for bringing randomization to make exploration.

$$\vartheta_{ij} = \beta_o e^{-\gamma r_{ij}^2}(x_{kj} - x_{ij}) + \alpha\left(\text{rand} - \frac{1}{2}\right) \quad (3)$$

$$x_{ij}^{t+1} = \begin{cases} 1 & \text{if } p_{ij} \geq \text{rand} \\ 0 & \text{otherwise} \end{cases} \quad (4)$$

The ith firefly uses the update rules that are given by Eq. (4). In this equation, rand is used to denote a random number generated on interval (0, 1). Specific value could also be used instead of the rand value used in Eq. (4). Figure 1 shows the flowchart of firefly alorithm for feature selection.

3.1.3 Parameters Controlling Exploration and Exploitation

The two parameters, randomness and attractiveness, control the quality of solutions produced and also the convergence rate.

The α, i.e., the randomness parameter affects the light transmission, and it can be varied for producing variations in the solution and providing more diversity in candidate solution space. It basically represents the noise existing in the system, $\alpha \in [0, 1]$. The parameter γ also known as absorption coefficient characterizes the

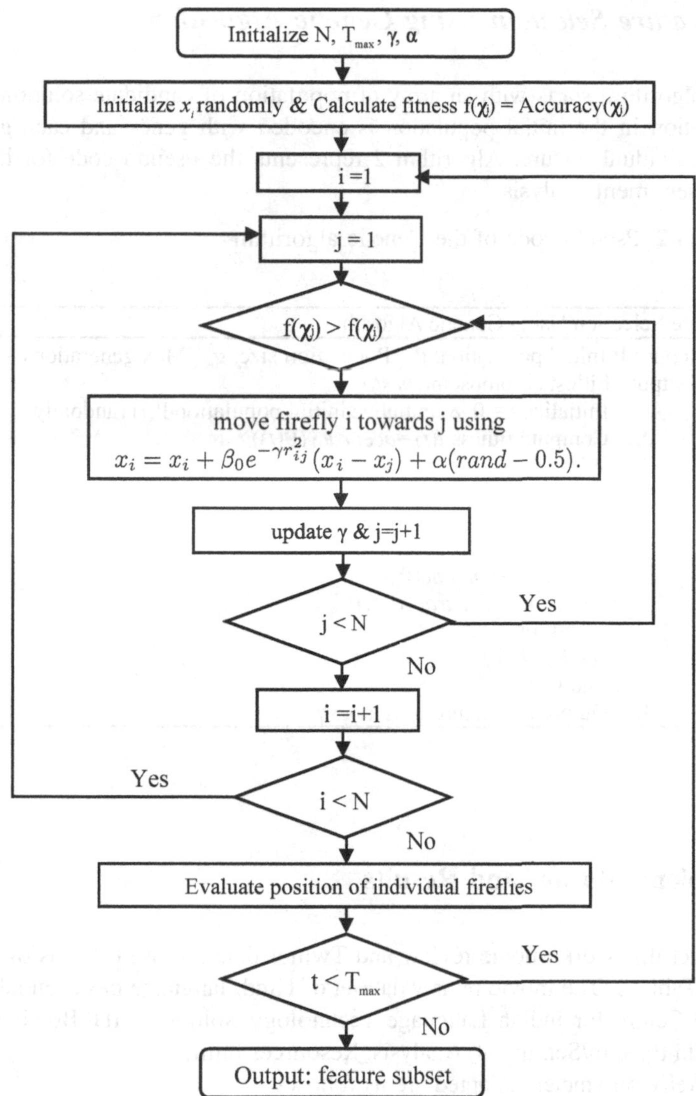

Fig. 1 Flowchart of firefly algorithm for feature selection

variation of the attractiveness, and its value is vitally important for evaluating the convergence speed of algorithm [14]. It controls exploration and exploitation of algorithm and is generally selected in range [0, ∞]. If $\gamma \rightarrow \infty$, the brightness and attractiveness will go down significantly, which bases the fireflies getting lost in the search process, whereas, if $\gamma \rightarrow 0$, the brightness and the attractiveness will be constant. Proper setting of α and γ enhances the performance of the algorithm.

3.2 Feature Selection Using Genetic Algorithm

Genetic algorithm starts with an array of population of candidate solutions, where each solution in the initial population is encoded with genes and each gene represents individual feature. Algorithm 2 represents the pseudo code for this algorithm in sentiment analysis.

Algorithm 2 Pseudo code of the Genetic algorithm

```
Feature Selection Using Genetic Algorithm
Input : P Initial population, P_size Population size, g_max Max generations
Output : Fittest chromosome bestp
  1.  Initialize t = 0  & initialize initial population P(t) randomly
  2.  Compute fitness f(t) =accuracy(P(t))
  3.  While t < g_max do
        t = t + 1
        for individual (i) in P (for each individual)
            P_i(t) = crossover(P_i(t - 1))
            P_i(t) = mutate(P_i(t))
            F_i(t)= accuracy(P_i(t))
        end for
       update bestp
      end while
  4.  Output the feature subset bestp
```

4 Implementation and Results

To conduct this work, movie review and Twitter data are used details of which is given in Table 2. The movie review dataset of Hindi language has been taken from Resource Centre for Indian Language Technology Solutions, IIT Bombay (http://www.cfilt.iitb.ac.in/Sentiment_Analysis_Resources.html).

The firefly parameters adapted are as follows:

(1) Firefly population size: 50
(2) Firefly length = Total number of features
(3) Number of generation: 50
(4) Absorption coefficient α: 0.5, and attractiveness parameter γ: 0.9

The result of the classification of SVM classifier, SVM-genetic classifier, and SVM-firefly classifier for the four datasets used in this work is given below in Table 2. Comparisons with baseline SVM system and optimized SVM systems shows that the optimized system produces more accuracy as compared to the

Table 2 Dataset used

Dataset	Language	Number of samples	Positive reviews	Negative reviews
Movie review	English	1000	500	500
Movie review	Hindi	302	127	125
Twitter data keyword—"Delhi government"	English	800	400	400
Twitter data keyword—"दलिली सरकार"	Hindi	500	250	250

Table 3 Average percentage improvement in accuracy

Model	Movie review (english)	Movie review (hindi)	Twitter data keyword—{"Delhi," "government"}	Twitter data keyword—{"दलिली," "सरकार"}
SVM	79.55	74.12	80.3	73.5
Genetic-SVM	82.15	77.31	83	77
Firefly-SVM	85.29	79.6	86.71	78.46

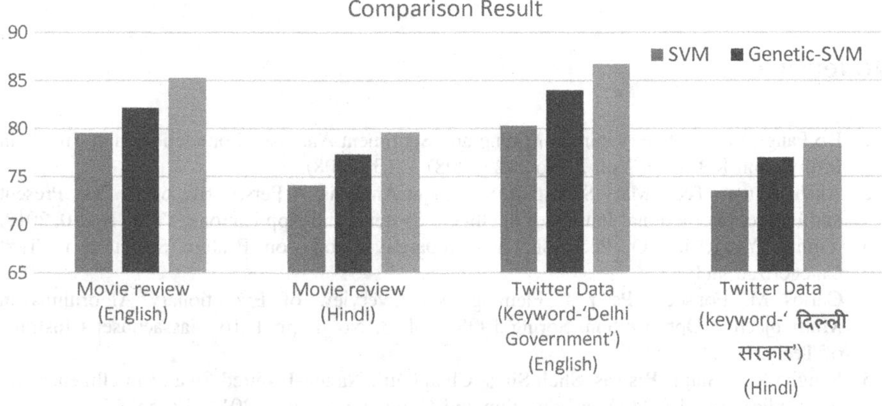

Fig. 2 Comparison results of firefly method, genetic method, and baseline SVM

baseline model. Table 3 shows accuracies of different datasets with different classification methods. This shows that promising accuracies with much reduced feature set can be achieved using FA.

The firefly algorithm is a powerful optimization algorithm and produces great improvement in accuracy in our model as can be seen from Fig. 2. The genetic algorithm which is also an evolutionary algorithm model is better than the baseline model but less efficient than the firefly model of feature extraction. The system is validated with four different datasets of which two are in Hindi language and other two in English language.

5 Conclusion

Sentiment analysis systems are used in almost every domain be it social, business, or political as opinions are key influencers of almost all human activities. Formed by perception, these opinions guide our conducts in various domains. Sentiment analysis finds many applications, few of them being product perception: evaluating customers' sentiments toward particular product, evaluating trend change over time, and defining new targets for marketing. The accuracy of sentiment analysis systems is very important for these applications. Reduction of the feature vector size considerably leads to improvements in accuracy as there are features which are noisy and redundant and not required for the classification. In this work, feature subset reduction is achieved using firefly algorithm. It improves the accuracy of sentiment analysis considerably by reducing the feature set size. The hybrid firefly-SVM model brings an accuracy improvement of 5.64 on average which is significant improvement. The experiment results reveal that the classification accuracy increases on an average by 3% in the case of genetic algorithm and by 5.64% in the case of firefly algorithm. Also, the hybrid method works very well for languages other than English as shown in this work whereas hybrid method works well for Hindi language also.

References

1. Bo Pang., Lilliam Lee.: Opinion Mining and Sentiment Analysis. Foundations and Trends in Information Retrieval. Vol. 2, No 1–2 (2008) 1–13 (2008).
2. Akshi Kumar, Teeja Mary Sebastian, Sentiment Analysis: A Perspective on its Past, Present and Future, International Journal of Intelligent Systems and Applications, Vol.4, No.10, 2012.
3. Yiming Yang, Jan O. Pederson, A Comparative study on Feature Selection in Text Categorization(1997).
4. Carlos M. Fonseca, Peter J. Fleming. An Overview of Evolutionary Algorithms in Multiobjective Optimization. Spring 1995, Vol. 3, No. 1, pp 1–16 Massachusetts Institute of Technology.
5. Sangita Roy, Samir Biswas, Sheli Sinha Chaudhuri, Nature-Inspired Swarm Intelligence and Its Applications, I.J. Modern Education and Computer Science, 2014, 12, 55–65.
6. Ekbal, A., Saha, S., and Garbe, C. S. "Feature selection using multi objective optimization for named entity recognition" 20th International Conference on Pattern Recognition, IEEE, pp. 1937–1940,2010.
7. William L. Goffe, Gary D. Ferrier, John Rogers, Global optimization of statistical functions with simulated annealing, Journal of Econometrics Volume 60, Issues 1–2, 1994, pp 65–99.
8. Mehdi Hosseinzadeh Aghdam *, Nasser Ghasem-Aghaee, Mohammad Ehsan Basiri, Text feature selection using ant colony optimization, Expert Systems with Applications 36 (2009) 6843–6853.
9. Xin-She Yang. Firefly algorithm, stochastic test functions and design optimization. International Journal of Bio-Inspired Computation, 2(2):78–84, 2010.
10. T. Sumathi, S. Karthik, M.Marikkannan, "Artificial Bee Colony Optimization for Feature Selection in Opinion Mining", Journal of Theoretical and Applied Information Technology, 2014. vol. 66 no.1.

11. Ruby Dhurve, Megha Seth, " Weighted Sentiment Analysis Using Artificial Bee Colony Algorithm", International Journal of Science and Research (IJSR), ISSN (Online): 2319–7064.
12. Deepak Kumar Gupta, Kandula Srikanth Reddy, Shweta, Asif Ekbal, "PSO-ASent: Feature Selection Using Particle Swarm Optimization for Aspect Based Sentiment Analysis", Natural Language Processing and Information Systems, Volume 9103 pp 220–233.
13. Long Zhang, Linlin Shan, Jianhua Wang, Optimal feature selection using distance-based discrete firefly algorithm with mutual information criterion, Neural Computing and Applications, ISSN 1433-3058, Springer, 2016.
14. Xin-She Yang, Xingshi He, Firefly Algorithm: Recent Advances and Applications, International Journal of Swarm Intelligence, 2013 Vol.1, No.1, pp. 36–50.
15. George Stylios, Christos D. Katsis, Dimitris Christodoulakis, " Using Bio-inspired Intelligence for Web Opinion Mining", International Journal of Computer Applications Vol 87 – No.5, 2014.
16. Abd. Samad Hasan Basari, Burairah Hussin, I. Gede Pramudya Ananta, Junta Zeniarja, "Opinion Mining of Movie Review using Hybrid Method of Support Vector Machine and Particle Swarm Optimization".

Adaptive Dynamic Genetic Algorithm Based Node Scheduling for Time-Triggered Systems

B. Abdul Rahim and K. Soundara Rajan

Abstract Nowadays, there has been tremendous increase in the use of reliable systems for safety critical applications such as avionics and automotives. As the systems become reliable, fault-tolerant design is must often involving strict timings. The implementation of such systems with traditional event-triggered approach is inappropriate; consequently, time-triggered approach is taking control. The time-triggered architectures are hard real-time embedded systems. Similarly, the scheduling process has to be redefined for optimality in resource allocation. The schedulability of tasks in such systems is analysed with meta-heuristic approach; genetic algorithm for optimization of processing nodes. Further, an adaptive approach has been arrived called adaptive dynamic genetic algorithm which allocates tasks to available nodes in a better optimized way for multiprocessor architecture.

Keywords Safety critical systems · Time-triggered systems · Genetic algorithms

1 Introduction

In automotive and aerospace systems, there has been exponential increase in the use of electronics and computer-based control, replacing predominantly mechanical systems [1]. These paradigm shifts from mechanical control to x-by-wire systems encompass safety critical requirements. The critical system is so-called if its malfunction may lead to economic or loss of life. Thus, these systems should possess

B. Abdul Rahim (✉)
Department of ECE, Annamacharya Institute of Technology and Sciences,
Rajampet 516126, AP, India
e-mail: abraheem@rediffmail.com

K. Soundara Rajan
Department of ECE, JNTUA College of Engineering,
Ananthapuramu 515002, AP, India
e-mail: soundararajan_jntucea@yahoo.com

H.S. Behera and D.P. Mohapatra (eds.), *Computational Intelligence in Data Mining*, Advances in Intelligent Systems and Computing 556,
DOI 10.1007/978-981-10-3874-7_67

highest level of safety integrity and reliability [2]. The reliability is nothing but in a special interval there should be probability of no failures and also this is the measure of performance for fault-tolerant systems. Thus, fault-tolerant design is must for applications involving high level of reliability and safety [3]. The objective of the development of time-triggered computer architecture is its application in distributed real-time fault-tolerant systems [4]. In [5], it has been verified that time-triggered architectures provide clear advantage over the present design approaches for safety critical systems, focused on fault tolerance especially in case of transport sector [5].

The electronic control units (ECUs) typically get their inputs from sensors and switches, compute them, and then enforce the output to actuators. If all these devices and sensors were to be connected point-to-point, then the cable length may grow to several kilometres. This would add to overall cost and also reliability problems. To overcome this problem, several communication network protocols such as LIN, CAN, TTP, Byteflight, and most recently Flexray were developed. In present-day high-end vehicles, approximately 70 ECUs are interconnected to exchange nearly 2500 signals in between them and these may grow if the application requires short latencies, demand for determinism and possessing high accuracy of operation [6, 7].

Depending on communication and processing activities between the two nodes, event- or time-triggered approaches are selected for the design. The event-triggered process is based on well-known interrupt mechanism, where activities are initiated with the significant change in the state is recognized whereas in time-triggered structure the communication process and the sequence of activities are put into action at a predetermined time [8]. The event-triggered architecture works on anticipatory scheduling and at a precise synchronization of the tasks, whereas time-triggered architectures have set activation period and can be scheduled offline. The constraint is a prior knowledge of worst-case situation should be known for consideration in development process [9]. The major setback of event-triggered system is non-determinism; that is, there is no guarantee of message transfer and conflict of access whereas in time-triggered systems as only one node or ECU is authorized for communication, collision is eliminated, and the tasks are executed in allotted time period. That is the network works on time-division multiple access (TDMA) concept [10].

Tasks have important estimations to make such as cost, duration, and resource usage for performance measures. Task requires use of single or set of resources for execution and the resource utilization may change during the task in operation. The estimations of cost and duration may depend on the resources employed to the task or used by the task. The performance measures found to be probabilistic or deterministic, depending upon the resources employed to complete the task, which can be executed in many modes. Various scheduling problems having resource constraints are proposed, implemented, and evaluated. The solution processes are organized in two noticeable classes: heuristic methods and exact methods. Further, these classes are classified into deterministic and stochastic approaches. Exact methods find solution for an existing problem otherwise provides some indication, whereas heuristic optimization solutions do not have such assurance. However, some assurance can be furnished analytically that some strength of optimality in the solutions is achieved. Stochastic process consists of probabilistic operations where the operations may not be executed in the same way, but the results obtained may be same for two or more similar operations while the deterministic methods work in the similar way each time, whenever the problem is applied.

Many methods concentrated completely on message scheduling and decided to be static. In static scheduling, the data have to be prefetched and can be pipelined for assignment in different stages during the task execution while the dynamic scheduling process uses the dynamic components/job tasks. Here, the execution time of the task cannot be estimated in advance so the task allocation is performed on the fly. Thus, the dynamic scheduling process consumes more runtime overhead compared to static scheduling process.

Static cyclic scheduling is considered for most of the hard real-time systems, often it is list scheduling where order of selection of tasks is important as given in [11–15]. However, the alternatives proposed are evolutionary approaches as used in [16]. A feasible solution to multiprocessor, distributed hard real-time systems was proposed in [17] was simulated annealing approach, but it requires large computation time. In this work, we have investigated with genetic algorithm for better optimal solution to overcome the problems stated earlier, i.e. to achieve best solution.

Different scheduling problems such as travelling salesman problem and job shop assignment, have been studied in the past. In this work, genetic algorithm is applied to the travelling salesman problem (TSP) involves implementation of crossover function, calculation of fitness function, and then applying mutation operator for generating best solution. A good calculation of fitness function obtained depends on the actual length of the solution. Various methods of applying the crossover and mutation function were discussed in [18]. Since, lot of research work has been performed with simple genetic algorithm, in multiobjective scheduling it is required to implement with modifications in genetic algorithm. This be able to perform better than the simple genetic algorithm in many instances and the work presented here reflects a clear move towards making modifications based on works carried out in [19, 20].

In [19], the standard task graph (STG) is used as benchmark function for the evaluation of multiprocessor scheduling algorithms. In STG, all the task graphs are generated randomly without consideration of communication cost. As it is hard to evaluate on all the task graphs, they have randomly chosen several task graphs from the standard STG and performed the evaluation. The tests have been carried out with small number of nodes. For large number of nodes, this method becomes complex, hence suggested, TSP combined with genetic algorithm may result in optimum solution. In [20], nearest-neighbour and genetic algorithm using travelling salesman problem for scheduling in multiprocessor system was compared and suggested that genetic algorithm with TSP yields best solution. But they have not visited for multiple TSP which can be used in multiprocessor applications. Simulation results show the algorithms suitability if the number of tasks is more and available ECUs is also large in number compared to the existing algorithms. Thus, schedulability of tasks in such system is analysed with proposed genetic algorithm (named as adaptive dynamic genetic algorithm).

The next section discusses the system architecture over which the scheduling of tasks is obtained. The third section deals with problem description of the scheduling with genetic algorithm approach for the said architecture and the fourth section presents the proposed algorithm. Finally after the results and discussion, the paper is concluded.

2 The System Architecture

The system architecture is developed over time-triggered paradigm systems which is fully synchronous and built around time-triggered network used for message transmission and also for synchronization [21]. Figure 1a shows four electronic modules or nodes connected to common bus. Each electronic module consists of a communication controller which disassociates the communication subsystem with the host subsystem. The communication process between the nodes is governed by time-triggered protocol TTP/C [22]. The nodes of cluster exchange its messages in between by TTP/C communication protocol. According to the static schedule, the communication subsystem decides message transmission and reception of a particular node is relevant or not. The TTP/C network consists of two channels, replicating each other. The host subsystem of electronic module executes the application and may possess an activator or a sensor. The communication subsystem in the node is communication controller designed to execute the features of TTP/C protocol. The nodes can be interchangeable in case of faults, hence specified as smallest replaceable unit (SRU). As said earlier, the fault-tolerant unit (FTU) is the combination of more than two nodes as shown in Fig. 1a. The node and the network is called the cluster [23–25]. A particular service is delivered by the fault-tolerant unit even if a node fails from the group.

The bus accessing scheme is derived from the global time notion controlled by cyclic time-division multiple access, where every active module has its own TDMA slot. Each node in its allocated TDMA slot sends or receives the message. The sequence of TDMA slots which form a TDMA round is shown in Fig. 1b. Every TDMA round carries the messages to the nodes in the allotted slots. After completion of a TDMA round, another TDMA round is started but maybe with a different or in sequence message in the same pattern as the previous one. The number of TDMA rounds varies depending upon the messages in for transfer and

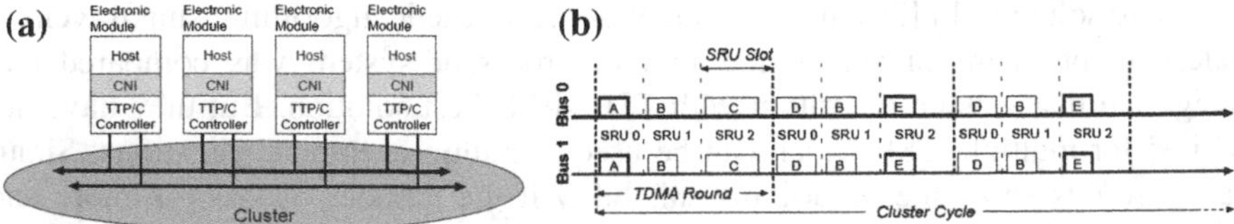

Fig. 1 **a** TTA scheme with electronic modules connected by replicated communication channels and **b** cluster cycle representation

this determines the length of the cluster cycle. When the cluster completes all operations, the transmission pattern starts again in new cluster cycle. The system does two types of schedules; which are communication schedule and scheduling of individual node. All the nodes in the system share same network for message transmission and reception which is basically of broadcast type. As different tasks are performed by the node, the node scheduling is different and it should be framed optimistically. The operating system executes this schedule in time-triggered manner. The propagation of errors due to malfunctioning of node or collection of malicious data from other nodes is prevented by communication network interface (CNI), i.e., babbling idiot problem can be prevented. Any single hardware failure is tolerated, and if any malicious host produces, erroneous data can never interfere in correct operation of the cluster. That is, fail silence condition is guaranteed by the communication controller. Thus, system becomes fault-tolerant and hence applicable to safety critical systems such as automotives, avionics, and railways [1].

3 The Genetic Algorithm

The genetic algorithm is a search-based heuristic, employing the evolution processes found in natural biology, introduced by John Holland in 1970 [26]. Reasonable results have been regularly produced with the basic application of a genetic algorithm to simple problems, whereas for the larger problems, application of genetic algorithms often results in poor performance. This is largely due to natural characteristics of genetic search operation and also due to the relationships between a genetic representation and genetic operators. Each evolutionary step is called generation, according to their level of fitness, from this new set of approximations are created, these are propagated through operators borrowed from natural genetics. The genetic algorithm is based on fundamental genetic operations on chromosomes such as selection, crossover, and mutation, to modify the solution for obtaining appropriate resultant offspring. The best-suited individuals which are selected for evolution process create the next generation.

In the search space, it selects many points at time and rapidly converges to possible solution or optimal solution. If the problem is too large, the genetic algorithm suffers from complex computations. It works on group of individuals for possible group of solutions instead of a single solution. This gives scope for parallel processing by genetic algorithm for complex and large problems. It works on iterative procedure having all possible solutions for the current optimization problem. At every generation, the chosen individuals are decoded and are assessed based on fitness function. The performance of the individual in the population depends on the number of times; it is chosen in a generation. The crossover is performed at single point or at multipoints to form new individuals from the selected individuals. The newly formed individuals are subjected to mutation that is some changes are caused by mistake during the process of gene copying from parents and this happens at random.

It is required to create initial population over which the new generations are possible. Certain important considerations have to be looked for in task scheduling of multiprocessor or distributed embedded systems. The prominent are *dependency* and *resource sharing*. The dependent tasks have to be scheduled for execution on the same node or ECU. This will eliminate communication cost. The resource sharing by some tasks is preferably on priority basis, and unfinished tasks may be shared among all nodes or ECUs. The scheduler should monitor and store in one queue accessed by all the processors or nodes.

```
Psuedo-code for Genetic Algorithm
 t := 0; initialize time
  Initialize population N(t);
  Evaluate fitness of population N(t);
  While not complete do
    t := t + 1;
    N'(t) := select parents N(t);
    recombine N'(t);
    mutate N'(t);
    evaluate fitness of N'(t);
    N := survive N,N'(t);
 End while
```

4 The Proposed Algorithm

The above algorithms present task allocation on an ECU considering uniprocessor architecture whereas the TTA is a multiprocessor architecture having constraints of fault-tolerant design. Instead of single crossover and single mutation, a multiple crossover and multiple mutation is performed to meet the above requirements. This is said to be dynamic genetic algorithm (DGA). In addition, a multiple warehouse (*Task generator*) condition is added here to solve the above problem more efficiently. Each salesman visits a set of cities starting from a warehouse and returning back and also exactly one salesman visits each city. Thus, a simple genetic algorithm is combined with multiwarehouse multiple travelling salesmen problem (MWMTSP) to schedule on multiprocessor system like TTA. MWMTSP with variable number of salesmen using genetic algorithm finds a near optimal solution referred in our work as adaptive dynamic genetic algorithm (ADGA).

The MWMTSP is a more challenging problem than simple MTSP; also it is NP-hard, which means it is difficult to solve with exact procedures. This has to be dealt more efficiently and effectively. Figure 2 illustrates the block diagram of the proposed method. Here, grouping is done with respect to nodes and warehouses in

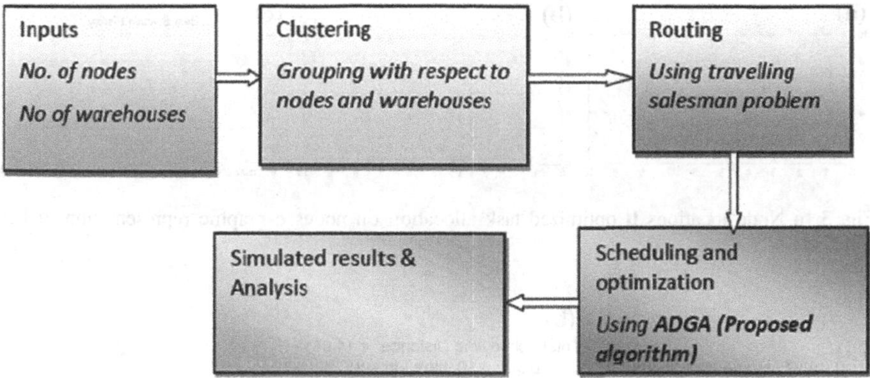

Fig. 2 Block diagram of proposed method

random fashion with Euclidean distance calculation. The routing problem is considered from travelling salesman problem. The main aim was to implement parallelism and minimize timing cost. The scheduling and optimization is done using proposed adaptive dynamic genetic algorithm. The results are evaluated with respect to number of nodes, number of clusters or routes, and best solution. In this work, we follow the functionality that allows finding best solution of task allocation on ECUs (minimum sum of all tour lengths). MWMTSP solution is such that there are no two routes based at the same warehouse.

```
Pseudo-code for proposed Algorithm
Initialization
t=0;
N(t)= ECUs;
Wi…n = Warehouses;
  Evaluate availability of Wi…n  and N (t);
  For
      Wi ≠ Wn
  Create a population
  While not complete do
  Run GA         % GA = Genetic Algorithm
  t := t + 1;
    N'(t)  := select best of available ECUs N(t);
      crossover N'(t);
      mutate N'(t);
      select N'(t);
      Ns := allocate task N,N'(t);
      Remove Scheduled Task
  End while
 End For
```

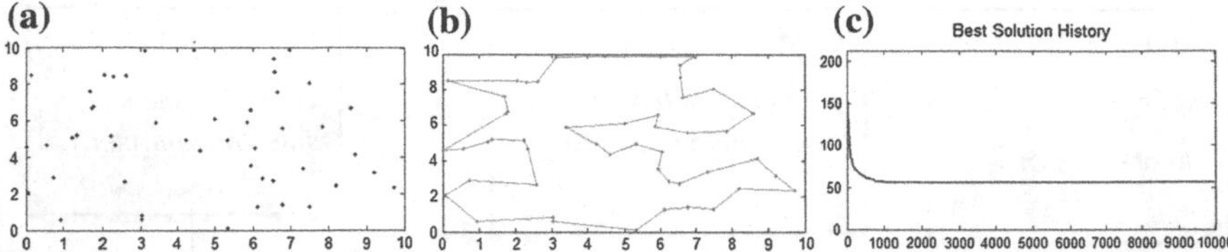

Fig. 3 a Node locations b optimized task allocation on nodes c graphic representation of best solution

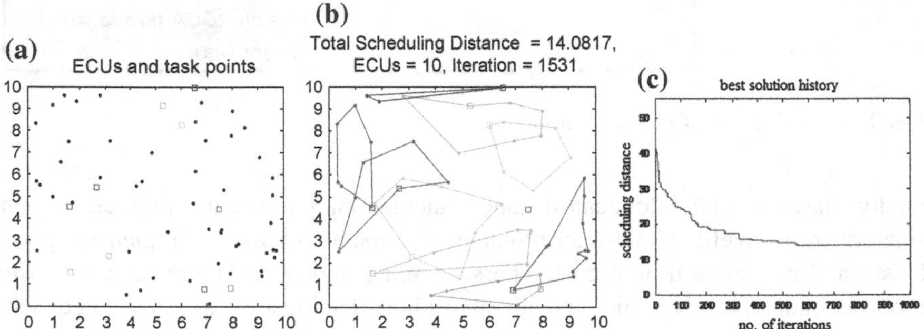

Fig. 4 a Node locations and task points; b optimum scheduled paths of 10 nodes; c graph representing scheduling distance versus iterations

5 Results and Discussions

In a time-triggered system, there are large sets of ECU's present for performing huge number of tasks at minimum time with a speed of 10 MB/S for each channel. Time-triggered system continuously checks all sensor point at which the task is performed. For real-time embedded applications, it is required to recognize for limited resource utility the algorithm has to run several times to yield optimal results. The results obtained show that approximately 15% scheduling time is reduced as shown in Fig. 3. A larger population size slows down the genetic algorithm run, whereas a smaller value leads to travelling around of a small search space. This is due to the crossover and mutation operators inherent with genetic algorithm.

In ADGA approach, 10 nodes or ECUs are used for allocating 50 tasks to the nearest ECU which is optimized by genetic algorithm approach. Random fixing of nodes is shown in Fig. 4a indicated with square points, and the dots indicate the actuators or sensors with tasks that require to be executed. By crossover operation, the nodes and tasks exchange their routes. By mutation process with a suitable fitness function final optimized schedule is obtained shown in Fig. 4b. Figure 4c is a graph depicting the scheduling distance after different iterations. It is seen that approximately 60% communication distance is reduced after large number of iterations. Table 1 shows the sequence of tasks scheduled over 10 nodes or ECUs after final optimized schedule.

Table 1 Sequence of tasks allocated to ECUs

ECU	Sequence of scheduled tasks which are allocated to ECUs											
E 1	34	40	32	35	11	E 1						
E 2	49	42	41	5	19	E 2						
E 3	46	8	48	17	33	30	E 3					
E 4	7	14	37	44	15	25	E 4					
E 5	27	50	26	31	28	E 5						
E 6	21	23	2	43	29	E 6						
E 7	47	38	3	6	22	E 7						
E 8	45	16	E 8									
E 9	1	36	9	13	39	12	20	24	10	11	18	E 9
E 10	E 10											

6 Conclusions

The time-triggered system has been developed for the fast-growing aeronautic and automotive applications. The scheduling of tasks over electronic control units of the system becomes difficult if they are large in number. Here, we have investigated meta-heuristic genetic algorithm to overcome the complexity problems by conventional list scheduling. Compared to genetic algorithm, the proposed algorithm for travelling salesman problem gives better optimized scheduling results. Although genetic algorithm is simple and for multiobjective routing, modified version is to be adopted. The results are quite comparable, and the computing time has reduced considerably. Further, the scheduling can be investigated with other evolutionary algorithms for hard real-time embedded systems.

References

1. C Scheidler, et.al, Time Triggered Architectures, EMMSEC'97, Advances in information Technologies: The business challenge, IOS Press, (1997), 758-765.
2. Mike Falla (edt), Advances in safety critical Systems-Results and achievements from the DTI/EPSRC R&D Program (1997)1–2.
3. Martin L Shooman, Reliability of Computer Systems and Networks. John Wiley (2000) 14–15.
4. H kopetz, Real Time Systems, Kluwer Academic Publishers, Boston, (1997).
5. M Sparchmann, Modeling a controller for a time triggered protocol, PhD Thesis, Vienna, University of Technology, (1997).
6. A Albert, Comparison of event triggered and time triggered concepts with regard to distributed control systems, in Embedded World, (2004).
7. N Navet, et.al, Trends in automotive communication systems, Proceedings of IEEE, Vol.93 (6), (2005) 1204–1224.
8. H kopetz, Event triggered versus Time triggered real time systems, Technical report 8/91, Insitut fur Technische Informatik TU Vienna, Austria (1991).

9. L Sha, R Rajkumar, J P Lehoczky, Priority Inheritance Protocols: An Approach to real time synchronization, IEEE Transactions on Computers, Vol 39 (9), (1990) 1175–1185.
10. S Poledna, Tolerating Sensor timing faults in highly responsive hard real time systems, IEEE Transactions on Computers, Vol 44 (2), (1995).
11. D Ullman, NP-Complete scheduling problems, journal of computer systems & science, Vol.10(3), (1975) 384–393.
12. J A Stankovic, et.al, implications of classical scheduling results for real time systems, Technical report UM-CS-94–089, computer science dept. University of Massachusetts, (1994).
13. N Audsley, et.al, Fixed priority preemptive scheduling: An Historical perspective, Real time systems, Vol.8(3), (1995).
14. F Balarin, et.al, Scheduling for embedded real time systems, IEEE Design and Test of Computers, Jan-Mar, (1998).
15. J Xu and D L Parnas, On satisfying timing constraints in hard real time systems, IEEE Transactions on Software Engineering, Vol 19 (1), (1993).
16. M Schwehm and T Walter, Mapping and Scheduling by Genetic Algorithms, Conference on Algorithms and Hardware for parallel processing, (1994) 832–841.
17. K Tindell, A Burns and A J Wellings, Allocating Hard real time tasks (An NP-Hard problem made easy), Journal of Real time systems, Vol. 4(2), (1992) 145–165.
18. Johnson D.S. & McGeoch L.A., "The Traveling Salesman Problem: A Case Study in Local Optimization", in: E.H.L. Aarts, J.K. Lenstra (Eds.), Local Search in Combinatorial Optimization, Wiley, New York, (1997), 215–310.
19. Probir Roy, Md. Mejbah Ul Alam and Nishita Das, "Heuristic Based Task Scheduling In Multiprocessor Systems With Genetic Algorithm By Choosing The Eligible Processor", International Journal of Distributed and Parallel Systems (IJDPS) Vol.3(4), (2012).
20. Besan Al Salibi, M B Jelodar and Ibrahim Venkat, "A Comparative study between the nearest neighbor and genetic algorithms: A revisit to the TSP", IJCSEE, Vol 1(1), (2013) 34–38.
21. H kopetz and G Gruenstiedl, TTP- A Protocol for Fault Tolerant Real Time Systems, IEEE Computer, Vol 24(1), (1994), 14–23.
22. H kopetz, et.al, A Synchronization strategy for a TTP/C Controller, SAE paper 960120, SAE press Warrendale, (1996), 19–27.
23. SAE: Class C Application Requirements – J2056/1, SAE Handbook, SAE Press Waarendale, (1994), 23.366–23.372.
24. H kopetz and R Nossal, The Cluster Compiler – A Tool for the Design of Time Triggered RTS, ACM SIGPLAN Workshop, (1995).
25. C Scheidler, L J Schafers and O K Fuhrmann, Software Engineering for parallel systems: The TRAPPER Approach, HICCS-28, IEEE CS Press, (1995), 349–358.
26. G R Harik, F G Lobo and D E Goldberg, The compact Genetic Algorithm, IEEE Transactions on Evolutionary Computation, Vol 3(4), (1999) 287–297.

Deformation Monitoring of Volcanic Eruption Using DInSAR Method

P. Saranya and K. Vani

Abstract The high-resolution images provided by TerraSAR-X satellite is very much useful for the displacement monitoring of earth phenomenon. In this paper, a model is developed for the deformation monitoring of Kilauea volcano in Hawaii Island region. This deformation monitoring is done with the help of differential interferometric synthetic aperture radar (DInSAR) algorithms. For this, a two-pass TerraSAR-X dataset of Hawaii region is used. The input master and slave images are preprocessed to remove speckle from the image, and then atmospheric correction is made to get the noise-free data. The preprocessed data are co-registered and then interferogram is calculated from this co-registered master and slave images. Phase unwrapping is performed to remove the multiple phase in the images. After that the vertical, horizontal, and LOS displacement is calculated. The results of the proposed model are validated with existing DInSAR method.

Keywords SAR interferometry · Differential SAR interferometry · Deformation · Monitoring · Volcano · Coherence · Filtering · Phase unwrapping

1 Introduction

Differential interferometric SAR is one of the powerful techniques for monitoring the displacement information of earth phenomenon such as earthquake, land slide, and volcanic eruption. But one-dimensional information is not enough to monitor

P. Saranya (✉) · K. Vani
Department of Information Science and Technology, College of Engineering,
Anna University, Chennai, India
e-mail: saranyasivam@auist.net

K. Vani
e-mail: vani@annauniv.edu

H.S. Behera and D.P. Mohapatra (eds.), *Computational Intelligence in Data Mining*, Advances in Intelligent Systems and Computing 556,
DOI 10.1007/978-981-10-3874-7_68

these activities. Therefore, three-dimensional information is used for the monitoring of deformation. Lot of research is going on for displacement modelling using differential SAR interferometry. DInSAR has become an important tool in remote sensing over the past years for the estimation of ground surface displacement based on time evolution. This technique is able to analyze large areas periodically at low cost and to monitor the changes even in the millimeter level [1–5]. DInSAR technique records change in distance between ground and sensor. It is based on the comparison of INSAR phases. The contribution of this work is to develop a DInSAR model for the deformation monitoring. In this paper, two interferograms are generated based on the difference in phases. One is between master and slave images and the other from digital elevation model (DEM) to reduce topographical error. After that the filtered interferogram is phase unwrapped to give a continuous measure of displacement in LOS region across the image by the global phase unwrapping algorithm [6]. Finally, differential interferogram is generated with respect to time series. Then the deformation is calculated in horizontal and vertical directions. The results of the proposed model are analyzed by comparing the results with existing method.

2 Related Works

Differential SAR interferometry is used to study the evolution of the surface displacement pattern due to as many main shocks [7]. The ascending of magma through volcano system causes earthquake and deformation of the ground. Volcanic deformation is nothing but the change in shape and dimensions of volcanoes [8]. The change in dimension is due to the change in geometry. The parameters which are considered for monitoring the deformation are the distance between the points, the position of the points, the height difference between the points, and the magnitude of tilting. From this, deformation is nothing but change in position or movement of point in volcano. There are different types of measuring volcanic deformation such as GPS data, ground measurements, radar data. [9]. Ground measurement will provide high accuracy data, but it is difficult to get the information in practical due to various factors such as safety and human resources. But synthetic aperture radar gives solution to the problem. Because of its capability of getting data in all weather conditions and also time-independent, it can provide data in both day and night time. Previously deformation is calculated by applying analytical method [10]; it gives solution but it needs data with no topographic effect and also it works in small pressure region. Nowadays, DInSAR method is used for deformation monitoring in all earth hazards phenomenon from time to time, i.e., periodical monitoring [11].

3 Methodologies

The SAR image contains two-dimensional information of both the amplitudes and the phases of the returns from targets within the imaging area. The amplitude stands for the reflectivity of the scene while the phase is a term proportional to the sensor-to-target distance and records possible surface movements. Two SAR satellite images can be combined to generate an interferogram. In particular, the phase difference of two images with the same viewpoint can accurately measure any shifts of the returned phase. From this, the earth's surface movement towards or away from the satellite is computed. Moving along their orbit, the SAR sensor scan simultaneously acquires the investigated scene. This process is called single-pass interferometry. Another one criterion is that it can look at the target at different time, i.e., a single antenna is available and the satellite overpasses the investigated area twice. It is referred to as repeat-pass interferometry. In the repeat-pass configuration, the temporal baseline is the time difference between the acquisitions. This will be useful for DInSAR process. The contribution of this paper is developing a model for DInSAR process.

3.1 Baseline Estimation

The baseline is an important parameter to choose the interferometric image pair. Here is the procedure to estimate the baseline value. At first, we interpolated the two orbits corresponding to the two images using Lagrange interpolation. The baseline is calculated by using Eq. (1) as described below:

$$B = \frac{\lambda R \tan(\theta)}{2R_r} \tag{1}$$

where λ is the wavelength, R is the range distance, R_r is the pixel spacing in range, and θ is the incidence angle.

3.2 Spectral Shift Filtering

It is based on a time-frequency analysis of the signals that are acquired. Only the energy contained in frequencies that were acquired in both scenes provides information for the coherence image. The non-overlapping spectral areas are not correlated, that is, they are not coherent one. Hence, they produce a source of noise in the final coherence image. Thus, a filtering is needed to preserve the common overlapping frequency bands. Spectral shift filter gives solution to this problem. This phenomenon of non-overlapping spectral bands takes place both in range and azimuth. But the origins are completely different and, as a matter of fact, filtering is performed independently in each one of the dimensions. The shift depends on parameters such as the looking angle and the local slope of the terrain. Standard range filtering approximates the shift as constant within the image. In order to perform such correction, a simulated phase based on external digital elevation model is required.

3.3 Co-registration

Co-registration is the process of superimposing two or more SAR images in the slant range geometry. External DEM is needed for co-registration process. The co-registration process is of two types: a coarse process and a fine co-registration process. The fine co-registration process is calculated at sub-pixel accuracy by taking advantage of the complex image information. Basically, the fine co-registration is done by the evaluation of cross-correlation measurements based on the complex input data. But coarse co-registration is done at pixel level. Most of the time, fine co-registration is preferred.

3.4 Coherence Estimation and Interferogram Generation

Coherence is estimated for individual component; it determines the size of the interferogram also. The interferogram is the combination of the signals S1 and S2 received at antennas 1 and 2, respectively. It is described in Eqs. (2) and (3)

$$S_1 = A_1 e^{-j\frac{4\pi r_1}{\lambda}} \tag{2}$$

$$S_2 = A_2 e^{-j\frac{4\pi r_2}{\lambda}} \tag{3}$$

The interferogram is calculated by using two signals received, and it is described in Eq. (4)

$$S_1 \cdot S_2 = A_1 A_2 e^{-j\frac{4\pi(r_1 - r_2)}{\lambda}} \tag{4}$$

where S_1 and S_2 are the complex images, and A_1 and A_2 are the amplitudes of the complex images.

The phase difference of the image φ_{int} is given in Eq. (5). It is due to atmospheric disturbances, mode of acquisition, etc.

$$\varphi_{\text{int}} = \varphi_f + \varphi_{topo} + \varphi_{displ} + \varphi_{atm} + \varphi_{err} \tag{5}$$

where φ_f is the flat earth component, the topographic phase is φ_{topo}, the displacement phase is φ_{displ}, the atmospheric term φ_{atm}, and the error phase φ_{err}. Except for φ_{err} and the φ_f, each term contains information relevant to specific issues. The φ_{disp} is the phase component accounting for the satellite-to-target distance change ΔR, and it is shown in Eq. (6).

$$\varphi_{displ} = \frac{4\pi}{\lambda} \Delta R \tag{6}$$

The similarity of the two radar signals can be measured by calculating the interferometric coherence, which is the normalized complex correlation

$$\gamma = \frac{\langle S_1 S_2 \rangle}{\sqrt{< |S_1 {*} S_1| > < |S_1 {*} S_2| >}} \tag{7}$$

where γ varies between zero in the case of decorrelation and one if the two signals are perfectly correlated; it is described in Eq. (7) given above. This phase is given modulo 2π and represented in the image by fringes.

3.5 Synthesize the Interferogram from DEM

Synthetic interferogram from an SRTM DEM is created. The synthetic interferogram is used in the subsequent interferometry process to eliminate the topographic effects in the interferogram. In particular for DInSAR, the resolution of the DEM is similar to the resolution of the input SAR images. The topographic effect to be removed significantly varies with the scene content and the perpendicular interferometric baseline value. The higher the resolution of the reference DEM is, the better the orbit correction will be.

3.6 Removal of Atmospheric Effects and Differential Interferogram

The obtained phase calculated in interferogram is affected due to earth's curvature. It is removed by using synthesized interferogram from digital elevation model. This provides fringes which are related to the change in displacement, elevation, noise, and atmospheric effects. The flat phase can be estimated in the Eq. (8)

$$\Delta\varphi flat = 4\pi/\lambda[B \sin(\theta - \alpha) - B \sin(\theta o - \alpha)] \tag{8}$$

where θo is the look angle to each point in the image. Then, assume local height as zero.

3.7 Phase Unwrapping

The range of interferogram phases is from –π to π. This phase unwrapping is used to add the correct multiple of phases which resolves the 2π ambiguity. It is used to reconstruct the absolute interferometric phase from its wrapped values in the differential interferogram. The absolute interferometric phase is nothing but the surface motion component towards the line-of-sight direction and is shown in Eq. (9).

$$\varphi wrap = \varphi u\text{unwrap} + 2n\pi \tag{9}$$

where $\varphi wrap$ is the wrapped phase, $\varphi unwrap$ is the unwrapped phase, and n is a positive integer.

3.8 Displacement Monitoring

The floating value of the image represents the line-of-sight motion. Pixel values in these images are directly derived from the unwrapped differential interferogram and represent displacement values in meters. Vertical motion is derived from line of sight and displacement from phase unwrapping. For horizontal displacement, local incidence angle is also taken for consideration. The displacement is calculated as in Eq. (10).

$$\text{Displacement} = \lambda/4\pi\,[\text{ph} - (\text{H}_{\text{res}} * \text{R}_1) \tag{10}$$

where λ is the wavelength, H_{res} is the topographic height, and R_1 is the ratio between baseline and topographic values. Ph is the phase value from the result of phase unwrapping algorithm.

4 Implementation

The DInSAR model for displacement monitoring for volcano eruption in Hawaii Island is shown in Fig. 1. The image pair is filtered using spectral shift filter for both range and azimuth shifts to remove topographic effect. Azimuth spectral filtering for stripmap eliminates the two parallel bands which do not overlap. As a

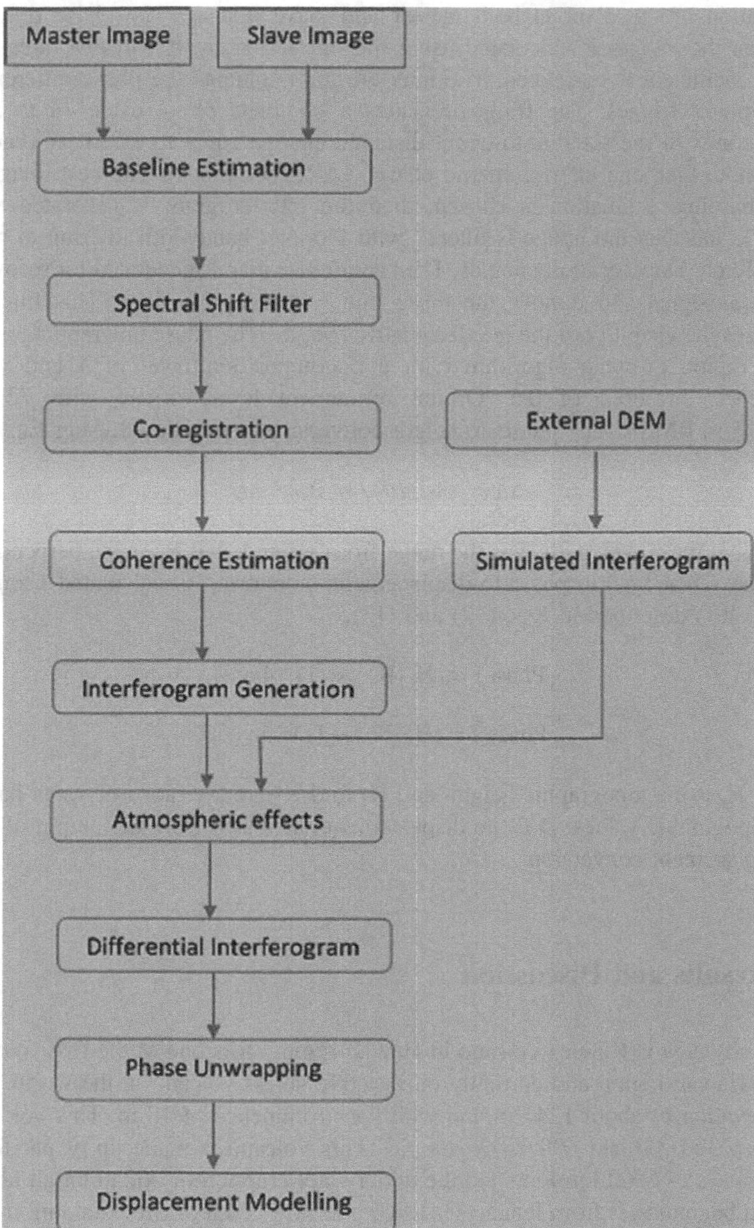

Fig. 1 Architecture diagram of the proposed method

consequence, a unique reference of Doppler centroid for each range is needed. Such values will be provided as a polynomial. This is actually representing the center of the images. Then, the images are superimposed on one another for calculating the correlation of the pixel in both master and slave images. The phase difference between the images is calculated using the interferogram. It will give how much phase coefficient is correlated. The interferogram contains the phase difference in the form of fringes. The fringe indicates a complete phase cycle $-\pi$ to π and corresponds to the satellite-to-target distance change equal to λ\2. It is generated from more than one interferometric phase. SARScape model: The best image pair from baseline estimation is chosen, then the interferogram is generated as per Eq. (7), and then the image is filtered with Doppler bandwidth filtering to reduce the different Doppler squint angles. Then coherence map is generated for recovering the radar signal and remove the noise due with adaptive filter. This filter also increases the visibility of the interferometric fringes. The phase unwrapping is done using region growing algorithm with a decomposition level of 5 and with a coherence threshold of 0.5. Orbital refinement is performed with 72 GCP with <5 m RMSE. The phase–to-height conversion is performed using Eq. (11).

$$\Delta h = [\lambda \mathrm{S} \sin(\theta)/4\pi \, \mathrm{B_P}] * \Delta\varphi \tag{11}$$

where λ is the wavelength, S is the range from ground, and B_P is the perpendicular baseline. Then finally, phase-to-displacement monitoring is calculated using step model. It is described in Eqs. (12) and (13).

$$\mathrm{Phase1} = (\mathrm{H_r} * \mathrm{K_1}) + (\mathrm{D} * 4\pi/\lambda) \tag{12}$$

$$\mathrm{Phase\,2} = (\mathrm{H_r} * \mathrm{K_2}) + (\mathrm{D} * 4\pi/\lambda) \tag{13}$$

where H_r is the topographic height, and K_1 and K_2 are the ratios between baseline and topographic values. D is the displacement calculated from the output of phase to displacement conversion.

5 Results and Discussion

The study area is Kilauea volcano in Hawaii region. It is one of the five volcanoes in the Hawaii region and currently most active shield volcano in the world. It has the elevation of about 1247 m and with the prominence of 18 m. This volcano is between 300,000 and 600,000 years old. This volcano is made up of pacific tectonic plates of 6000 km long and the underwater eruption of Alkali basalt lava. Its current beginning is from January 3, 1983, and flows eastern rift extending from 11 to 12 km to the sea. The highest point center latitude and longitude is 19°25′16″N and 155°17′12″W. It is the fault of unknown depth moving vertically at an average of 0.1 to 0.8 inch per year. The test data site is TerraSAR-x, a two-pass image with the slant range of 587.7 and 588.16 km, respectively. The input master and slave SAR image is shown in Figs. 2 and 3. The interferogram generated from the proposed model is shown in Fig. 4, and the interferogram generated from the

Fig. 2 Master image

Fig. 3 Slave image

existing model is shown in Fig. 5. Figures 6 and 7 shown below are the differential interferograms calculated from proposed model and existing model.

The coherence values are extending from 0 to 1, that is 0 corresponds to the area where deformation occurred. Table 1 shown below gives the details of horizontal displacement map. Displacements 1 and 2 represent the displacement in horizontal direction calculated by the proposed method and existing algorithm. From the overall results, the displacement is 0.02 inch for the temporal period of 176 days. The vertical displacement is shown in Fig. 8. From this profile, the orange color line, i.e., displacements 1 and 2, indicates vertical displacement from the proposed method and the existing Model.

The vertical displacement map shows that the displacement varies from 0.01 to 0.02 inch for 176 temporal days. The validation is performed by calculating the

Fig. 4 Int1

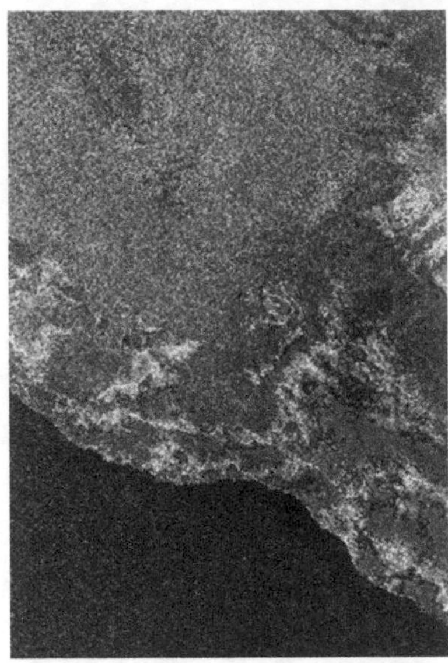

Fig. 5 Int2

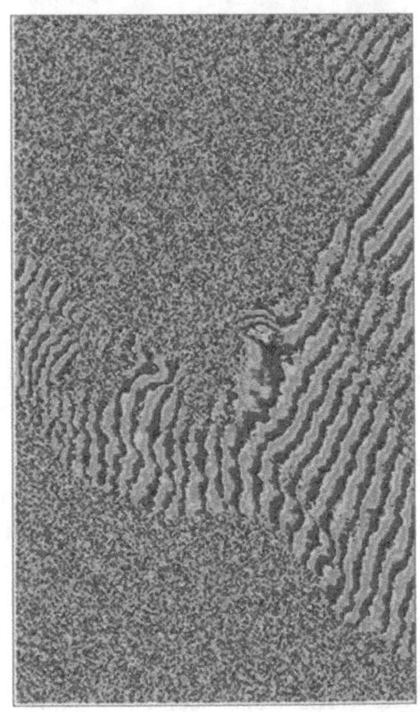

Fig. 6 Differential Int1

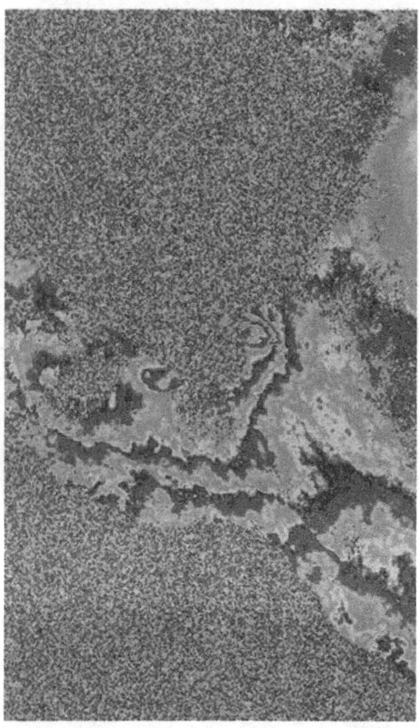

Fig. 7 Differential Int2

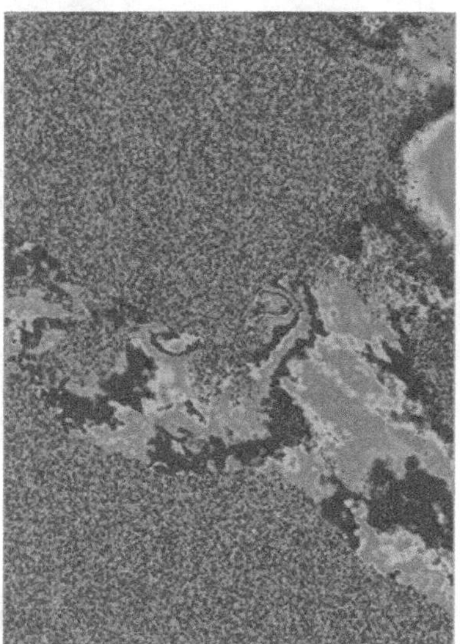

relative error between the results from the proposed method to results from existing software. The relative error is the difference between the results of proposed model and the result from software tool to the results from software tool. The observed

Table 1 Horizontal displacement map

Latitude	Longitude	Displacement 1	Displacement 2
19.2547 N	155.141 W	0.0008	0.0003
19.26192 N	155.135 W	0.0168	0.0158
19.24199 N	155.1448 W	0.0221	0.0159
19.2411 N	155.1434 W	0.0296	0.0258

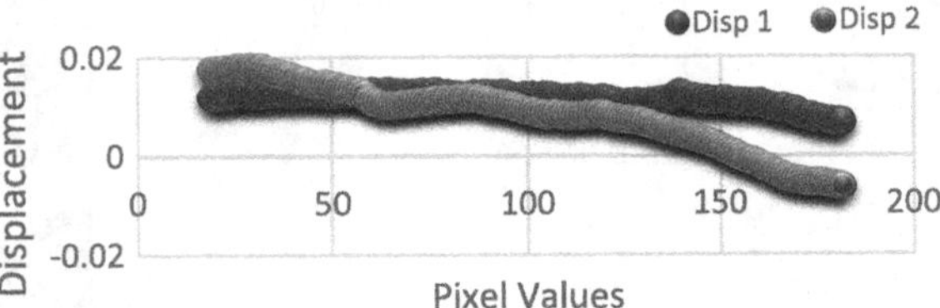

Fig. 8 Vertical displacement map

relative error obtained is 0.03. From the results, it can be shown that the results of the proposed method are closely related to the results from existing tool.

6 Conclusion and Future Work

In this research work, a DInSAR model is developed for displacement monitoring of environmental hazards such as volcano. The high-resolution TerraSAR-X image pairs are taken, and then interferometric coherence is calculated. This will give difference in the phase value. Then, the atmospheric effects and noise are removed using filtering method. After that phase is unwrapped using minimum cost flow method. Finally, displacement is calculated. Then, the displacement calculated from the proposed model is validated with the results from SARscape model. The results show that the proposed model is closely related with the existing tool, and the observed relative error is 0.03. The improvement of effective algorithm for calculating displacement is considered the future scope of the research.

References

1. Aster, R.C., R.P. Meyer, G. De Natale, A. Zollo, M. Martini, E. Del Pezzo, R. Scarpa and G. Iannaccone, Seismic Investigation of the Campi Flegrei: A summary and synthesis of results, *Volcanic* Seismology, IAVCEI Proceedings, 462–483, 1992.
2. Briole, P., D. Massonnet, C. Delacourt, Post-eruptive deformation associated with the 1986-87 and 1989 lava flows of Etna detected by radar interferometry, *Geophysics. Res. Lett.*, *24*, 37–40, 1997.

3. Dvorak, J.J. and G. Berrino, Recent Ground Movement and Seismic Activity in Campi Flegrei, Southern Italy: Episodic Growth of a Resurgent Dome, *J. Geophysics. Res.*, *96*, B2, 2309–2323, 1991.
4. Lanari, R., P. Lundgren, and E. Sansosti, Dynamic Deformation of Etna Volcano observed by Satellite Radar Interferometry, Geophysics. Res. Lett., 25, 1541–1544, 1998.
5. Agustan, Kimata, F., Pamitro, Y.E. and Abidin, H.Z., 2012, Understanding the 2007–2008 eruption of Anak Krakatau Volcano by combining remote sensing technique and seismic data. International Journal of Applied Earth Observation and Geoformation, Volume 14, Issue 1, February 2012, Pages 73–82.
6. Hill, D. P., F. Pollitz, and C. Newhall, 2002, Earthquake-volcano interactions, Phys. Today, 55(11), 41–47.
7. Kriswati, E., Agustan, Loeqman A., Meilano, I., Pamitro, Y.E., Yunazwardi, M., Nurhayati, S., 2011, Inflation of Indonesian Volcanoes before the Eruption, Proceeding of 1st International Seminar of Environmental Geoscience in Asia.
8. Shimada, M., 1999, Verification processor for SAR calibration and interferometry, Adv. Space Res. Vol. 23, No. 8, pp. 1477–1486.
9. FornaroG., G. Franceschetrti., Lanari, D, Rossia, and M. Tesauro, interferometric SAR phase unwrapping via the finite element method, IEEE Proc. Radar, Sonar, Navigation, 144, 266–274, 1997.
10. Lanari, R., Lundgren, P., Manzo, M. & Casu, F., 2004a. Satellite radar interferometry time series analysis of surface deformations for Los Angeles, California, *Geophys. Res. Lett.,* 31.
11. Lanari, R., Mora, O., Manunta, M., Mallorqui, J., Berardino, P. & Sansosti, E., 2004b. A small-baseline approach for investigating deformations on full-resolution differential SAR interferograms, *IEEE Trans. Geosci. Remote Sens.,* 42(7), 1377–1386.

Effective Printed Tamil Text Segmentation and Recognition Using Bayesian Classifier

S. Manisha and T. Sree Sharmila

Abstract Text segmentation and recognition of Indian languages have gained a lot of research interest in the recent years. The existence of a huge number of symbols and varying characteristics in these languages makes segmentation and extraction of text a challenging task. The Tamil language has a wide variety of the literature, and printed text is available in various forms such as newspaper, books, and magazines. In this paper, extraction of printed Tamil text from an image is done irrespective of the characteristics of the text such as font style, color, and size. The proposed work uses scanned printed Tamil text as the input image. This input image is binarized since text is always available in the foreground, and histograms can be used to segment them into lines and words. The morphological operator, dilation, is used to remove outliers such as dots and commas present in an underlying object and segment the printed text into words to facilitate text detection. Further, each character is identified using bounding box technique. Classification of Tamil letters is done by extracting features such as gradient information and curvature-based information obtained from grayscale and binary images. These features are trained, and characters are classified using Bayesian classifier. The recognized characters are documented as text using Unicode format. The performance of the approach is evaluated using precision, recall, and F-measure.

Keywords Binarization · Bounding box · Character recognition · Classification · Dilation · Segmentation · Tamil text detection

S. Manisha (✉)
Department of CSE, SSN College of Engineering, Chennai, India
e-mail: manishas@ssn.edu.in

T. Sree Sharmila
Department of IT, SSN College of Engineering, Chennai, India
e-mail: sreesharmilat@ssn.edu.in

H.S. Behera and D.P. Mohapatra (eds.), *Computational Intelligence in Data Mining*, Advances in Intelligent Systems and Computing 556,
DOI 10.1007/978-981-10-3874-7_69

1 Introduction

In recent years, character recognition has gained significant interest, and most of the work done is in the English language, and further has been extended to Arabic, Bangla, Urdu, Chinese, Japanese, Devanagari, and other Indian languages [1, 2]. In the last decade, recognizing Tamil characters has increased in popularity [3, 4]. The characteristics of the Tamil script are generally curve shaped and contains 247 letters, comprising of 12 vowels (known as uyir ezhuthukkal), 18 consonants (known as mei ezhuthukkal), 216 composite letters (known as uyir mei ezhuthukkal) derived from vowels and consonants, and a special character (known as ayutha ezhuthu) [5]. A universal character encoding system (Unicode) is available for written characters that can be used for documenting text of any language for the purpose of character recognition. The Unicode is designed for all the Tamil characters available. In this paper, the interest is to segment, extract, and recognize the printed Tamil text [6].

Tamil text extraction and detection is one of the major research issues in document analysis. Mallikarjun et al. use distance-based histogram analysis technique to compute the width, height, gap, and center distance of the extracted connected components [4]. Various histogram features such as gradient vector directions and level of luminance of a grayscale image are used to identify the features of the text [2]. The straightness of the vertical and horizontal lines at various points is used for discriminating the handwritten characters from the printed characters. This can be done using various classifiers such as neural networks, where a multilayer perceptron can be used for horizontal-, vertical-, and slope-based features for classification of text characters [3]. Pal et al. [1] have developed an algorithm for identifying two major Indian scripts: Devanagari [6, 7] and Bangla [8, 9] texts using projection profile and statistical features using decision tree classifier. For separation of text at word level [10] and character level, hidden Markov models (HMMs) use projection profile for identifying the statistical features. To maintain the structural details of the character, [11] uses Gabor filter, such that it does not affect the edges of the image. These filters are also useful for the characters that are similar to Arabic characters [12]. Other classifiers such as SVM (support vector machines) [11] and k-nearest neighbor (k-NN) [13] are also useful for classifying the characters based on the features extracted. The neural network concept was initially used to predict handwritten numerals.

The other techniques such as zoning- and projection-based features use the concept of centroids for classification and recognition of each character in the text [14]. It is also assumed that SVM produces better classification results for handwritten Tamil text documents at a document level [12]. Here, two stages are used for unsupervised classification. The first level uses clustering method to recognize the handwritten documents from the printed text. The next level is to use a supervised classifier for the purpose of recognizing those characters. Pal et al. use the other solution to recognize the characters based on direction features [8]. The images are separated using bounding box, and each character is segmented into different blocks and is downsampled using the Gaussian filter [12]. The Unicode refers to base that emphasizes the principle that each character in a particular language has a 16-bit unique code. These are simple to parse and process as they have a well-defined semantics [15].

2 Proposed Work

The major steps involved in character recognition include the following: preprocessing, text extraction and segmentation, feature selection, text classification, and character recognition as shown in Fig. 1.

2.1 *Preprocessing*

Initially, the noise is removed using median filter, and binarization technique is done using Otsu's threshold selection method [13] to divide the input image into foreground and background. The main objective of binarization technique is to extract text in an image, and the accuracy of binarization affects the accuracy of

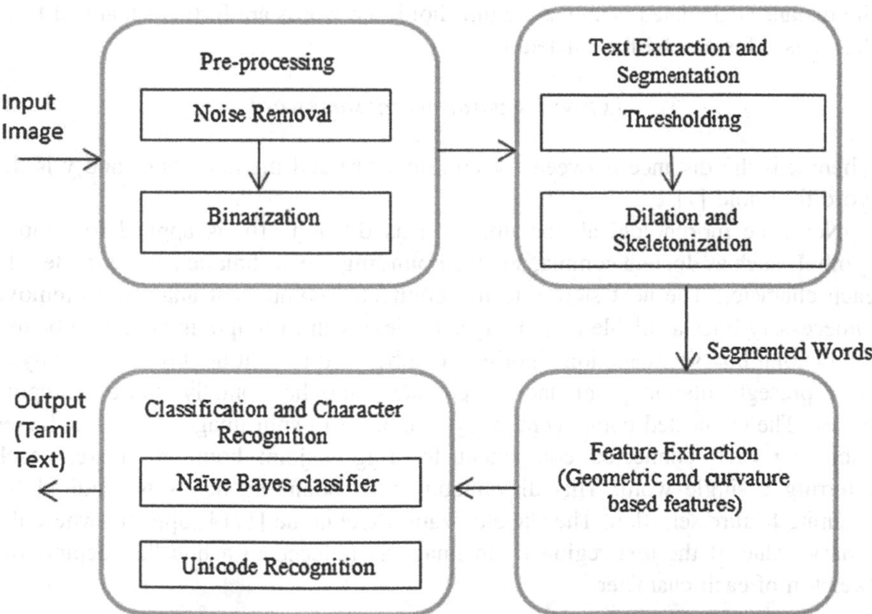

Fig. 1 Architecture of printed Tamil text segmentation and recognition

segmentation. It decreases the presence of unwanted data and preserves only the desired data in document images, thus reducing the complexity of text extraction [11].

2.2 Text Extraction and Segmentation

In an image, since text is present as foreground, the vertical histogram analysis can be performed to identify the number of lines available to separate the lines in a paragraph. On applying horizontal-based histogram analysis, the words are separated using space-based technique. The lines and words are separated using the gap between each line and a word. Two reference lines such as upper and lower lines are taken, and the distance between the middle region and the reference region is calculated, and if the distance is more than the threshold, the lines are separated. The same can be mathematically formulated as

$$(a > b \,\&\&\, c > d)\ ?\ \textit{separate line}: \textit{same line} \tag{1}$$

where a is the distance between upper reference matra and the mid-reference matra, b is the upper threshold, c is the distance between lower reference matra and the mid-reference matra, and d is the lower threshold [2]. Similarly, the starting and the ending points of the text are taken as references, and distance is calculated. If the distance calculated is above the threshold, the words are further separated from the lines. This can be formulated as

$$(x > y)\ ?\ \textit{separate word}: \textit{same word} \tag{2}$$

where x is the distance between the current word and the next word and y is the word threshold [1].

Next, the morphological operator such as dilation [16] is applied to remove symbols such as dot and comma, and the bounding box technique is used to identify each character. The next step is to use connected component analysis to remove unnecessary lines available in an image. To deal with overlapping bounding boxes, the disjoint box segmentation algorithm can be used to split the boxes vertically at each presegmentation point and is separated into horizontally non-overlapping boxes. The connected component analysis is used to split image and detect boxes enclosing each connected component forming disjoint bounding boxes, each referring a single word. This disjointedness of bounding boxes is required for accurate feature selection. The skeletonization technique [17] is applied, where the binary value of the text region in an image is reduced to a line that depicts the skeleton of each character.

2.3 Feature Selection

The segmented words are now interpolated to same size using bilinear interpolation technique with height and width as 32 × 32 pixels. Since each character can be extracted using bounding box technique, character-level features are selected. Features such as height, width, horizontal and vertical lines, segment concavity with respect to centroids, and curvature information are extracted. The height and the width of the characters can be identified using vertical and horizontal boundaries of the character, respectively. The Sobel mask can be used for extracting horizontal, vertical, and curved lines [18] as shown in Fig. 2.

Two small and large thresholds for horizontal, vertical, positive, and negative slope lines are set. Using these masks slopes, horizontal and vertical gradients in a character are identified. The curvature information can be extracted from a 5 × 5 mask that is shown in Fig. 3 and is used for classification of characters.

Fig. 2 Masks used for identifying lines **a** horizontal, **b** vertical, **c** positive slope, and **d** negative slope lines [18]

(a)

-1	-1	-1
2	2	2
-1	-1	-1

(b)

-1	2	-1
-1	2	-1
-1	2	-1

(c)

-1	-1	2
-1	2	-1
2	-1	-1

(d)

2	-1	-1
-1	2	-1
-1	-1	2

Fig. 3 Mask used for identifying curved lines in a text [18]

2	2	2	2	2
2	-1	-1	-1	2
2	-1	-1	-1	2
2	-1	-1	-1	2
2	2	2	2	2

2.4 Classification and Recognition

In this paper, naïve Bayes classifier is used to classify and recognize each character. The features of all the 247 characters in the Tamil language along with the special characters are trained to the classifier. Using Bayes theorem, a probability is calculated for each class and is stored as the universal probability of that class [19].

$$p(A/B) = \{p(A) * p(B/A)\}/p(B) \tag{3}$$

In the testing phase, a character is given as input, and probability is calculated for each input character based on the features extracted. The obtained probability is compared with the probability of the class, and the character is classified to the closest class. Once the character is classified, the universally available Unicode is used for recognition [15]. Here, each character has a 16-bit code and has well-defined semantics. Unicode of each character is mapped to its corresponding letters available based on these semantics using rule-based classification and is printed using the Unicode font to achieve character recognition.

3 Results and Discussions

In this paper, images for segmentation and recognition are taken from World Languages dataset available as an open source [5]. Recognition is done at image level and word level. At image level, 550 images are taken and are recognized. Similarly, at word level, 20,000 words are available in the chosen dataset. The recognized words are printed using Arial Unicode Microsoft font style. A sample text is given in Fig. 4.

In the preprocessing step, the noise present in the image is removed using median filter, and the result after binarization is displayed in Fig. 5.

The result of applying bounding boxes on each character of the binarized image is given in Fig. 6.

The overlap present in the bounding boxes is removed, and words are segmented in Fig. 7.

Once the text given is segmented into individual words, it is important to choose the correct features that uniquely identify a particular character and distinguish itself from the rest of the characters. Thus, each of the values of the features is unique to a character. Features of a particular character அ are listed in Table 1.

இருள்நீங்கி இன்பம் பயக்கும் மருள்நீங்கி
மாசறு காட்சி யவர்க்கு.

Fig. 4 Sample printed Tamil text

இருள்நீங்கி இன்பம் பயக்கும் மருள்நீங்கி
மாசறு காட்சி யவர்க்கு.

Fig. 5 Binarized image

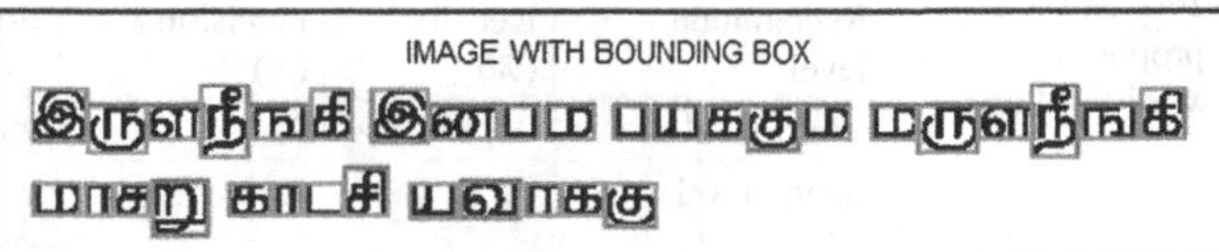

Fig. 6 Characters separated using bounding box

Fig. 7 Segmented Tamil text

Table 1 Values of features extracted for the character அ [18]

Features	Values
No. of short horizontal lines	0
No. of short vertical lines	0
No. of long horizontal lines	1
No. of vertical arcs	1
No. of circles	1
No. of horizontal arcs	1
Height	23
Width	20

Each segmented word is given to the Bayesian classifier, and the characters are recognized. The noise from the image is removed and binarized, and bounding boxes are used to identify every character. The overlap in the bounding boxes is removed, and the text is segmented at word level. The features of every character are extracted from these segmented words and are trained in the naïve Bayes classifier, and are used to recognize the text at image level without segmentation of words. The result of recognized characters at image level and word level is given in Fig. 8.

The metrics such as precision, recall, F-measure, and misclassified rate are calculated using the following equations:

இருள்நீங்கி இன்பம் பயக்கும் மருள்நீங்கி
மாசறு காட்சி யவர்க்கு.

Fig. 8 Text recognized at word level and image level

Table 2 Performance measures of proposed approach at word level and image level

Recognition level	Recall (%)	Precision (%)	F-measure (%)
Word level	95.3	97.4	96.3
Image level	92.8	93.2	92.9

$$Recall\,(R) = CRW/ANW \tag{4}$$

$$Precision\,(P) = CRW * (CRW + FRW) \tag{5}$$

$$F\,measure = (2 * P * R)/(P + R) \tag{6}$$

where *CRW* is the correctly recognized word, *FRW* is the falsely recognized word, and *ANW* is the actual number of words in dataset (counted manually). The precision and recall of the proposed method are tabulated in Table 2 at word level and image level.

From above Table 2, it is evident that the level of accuracy increases at word level is higher when compared to that of the image level. Thus, this emphasizes the need for segmentation of text into words.

The major disadvantage of the proposed system is that it does not support similar looking characters. This is because the features of the characters are closer to each other such that they are classified as a same character. The possible solution to overcome misclassification is to identify features that have major difference in the values between each character and are used for training the classifier.

4 Conclusion

In this paper, median filter is used to remove noise, and binarization is using Otsu's thresholding to separate text from background. Using space detection techniques, horizontal histogram analysis and vertical histogram analysis are done to segment lines and words from the printed text paragraphs. Line-, circle-, and curvature-based features are extracted for each character and are used for training the naïve Bayes classifier based on probability for each class for classification. The classified text characters are recognized and are printed into a text file using the Unicode technique. Performance of the proposed approach is evaluated and compared between the word level and image level, and it is evident that segmentation of words provides a better level of accuracy compared to that of the image level. Distinguishing between similar looking characters can be concentrated in the future.

References

1. Ayan Kumar Bhunia, Ayan Das, Partha Pratim Roy, UmapadaPal, "A Comparative Study of Features for Handwritten Bangla Text Recognition", International Conference on Document Analysis and Recognition (ICDAR), pp 636–640, 2015.
2. S.M. Shyni, M. Antony Robert Raj and S. Abirami, "Offline Tamil Handwritten Character Recognition Using Sub Line Direction and Bounding Box Techniques", Indian Journal of Science and Technology, Vol 8(S7), pp 110–116, 2015.
3. M. Antony Robert Raj, Dr. S. Abirami, "A Survey on Tamil Handwritten Character Recognition using OCR Techniques", CCSEA, SEA, CLOUD, DKMP, CS & IT 05, pp. 115–127, 2012.
4. Dr C.P. Sumathi, S. Karpagavalli, "Techniques and methodologies for Recognition of Tamil Typewritten and Handwritten Characters: A Survey", International Journal of Computer Science & Engineering Survey, Vol 3 (6), pp 23–35, 2012.
5. Mallikarjun Hangarge, K.C. Santosh, Srikanth Doddamani, Rajmohan Pardeshi, "Statistical Texture Features based Handwritten and Printed Text Classification in South Indian Documents", In proceedings of ICECIT, pp 215–221, 2012.
6. Jomy John, Pramod K.V., Kannan Balakrishnan, "Handwritten Character Recognition of South Indian Scripts: A Review", National Conference on Indian Language Computing, Kochi, pp 1–6, 2011.
7. U. Pal, T. Wakabayashi, F. Kimura, "Comparative Study of Devnagari Handwritten Character Recognition using Different Feature and Classifiers", International Conference on Document Analysis and Recognition, pp 1111–1115, 2009.
8. LTG (Language Technologies Group), "Optical Character Recognition for Printed Kannada Text Documents", SERC, IISc Bangalore, 2003.
9. U. Pal and B. B. Choudhuri, "A Complete Printed Bangla OCR System", Pattern Recognition. Vol 31 (5), pp 531–549, 1997.
10. U. Patil, M. Begum, "Word level handwritten and printed text separation based on shape features", International Journal of Emerging Technology and Advanced Engineering Vol. 2 (4), pp 590–594, 2012.
11. Kefali A, Sari, Sellami M, "Evaluation of binarization techniques for old Arabic document images", MISC 2010, Algeria, pp. 88–99, 2010.
12. Seethalakshmi R., Sree Ranjani T.R., Balachandar T., "Optical Character Recognition for Printed Tamil text using Unicode", Journal of Zhejiang University Science, 6A (11), pp 1297–1305, 2005.
13. Otsu. N, "A threshold selection method from gray level histograms", IEEE Trans. Systems, Man and Cybernetics, vol. 9, pp. 62–66, 1979.
14. Trier. O.D, Jain. A.K and Taxt. J, "Feature extraction methods for character recognition - A survey", Pattern Recognition, vol. 29, no. 4, pp. 641–662, 1996.
15. R. Indra Gandhi, Dr. K. Iyakutti, "An Attempt to Recognize Handwritten Tamil Character Using Kohonen SOM", International Journal of Advanced Networking and Applications, Vol 1 (3), pp 188–192, 2009.
16. Aparna K G and A G Ramakrishnan, "A Complete Tamil Optical Character Recognition System", white paper pages 11, 2000.

17. Siromoney et al., "Computer recognition of printed Tamil character", Pattern Recognition, Vol. 10, pp 243–247, 1978.
18. Palaiahnakote Shivakumara, Rushi Padhuman Sreedhar, Trung Quy Phan, Shijian Lu, "Multi-oriented Video Scene Text Detection Through Bayesian Classification and Boundary Growing", IEEE Transactions on Circuits and systems for Video Technology, Vol. 22 (8), pp 1227–1235, 2012.
19. Mohamed Ben Halima, Hichem Karray and Adel M. Alimi, "Arabic Text Recognition in Video Sequences", pp 603–608, 2010.

Retrieval of Homogeneous Images Using Appropriate Color Space Selection

L.K. Pavithra and T. Sree Sharmila

Abstract In this paper, convenient color space selection issue in content-based image retrieval system for low-level feature mining is addressed by the exploration of color edge histogram feature extraction on HSV, YIQ, YUV, and YCbCr color spaces. Moreover, Haar wavelet transform is applied to reduce feature vector count for the beneficial to speed up the retrieval process, and then semantic retrieval is obtained via similarity metric. Retrieval accuracy of each color space is analyzed through the parameters such as precision, recall, and response time of the system. Experimental results show that HSV color space-based retrieval system averagely gives 5%, 18.3%, and 26% high retrieval than the YIQ, YUV, and YCbCr color spaces, respectively.

Keywords Content-based image retrieval • Feature extraction • HSV • YCbCr • YIQ • YUV

1 Introduction

In recent years, there is a rapid increase in digital image collections. Large collection of image database is used in different applications such as data mining, medical, satellite imaging, crime prevention, and military. [1]. Two kinds of image search are available—text-based search and visual content (image)-based search. In typical text-based image search, keywords are created in non-automatic manner, which would often lead to implausible retrieval. To overcome these flaws, content-based image retrieval (CBIR) system was introduced [2]. This search system can automatically grasp more key information than text-based search and provide high relevant response compared to typical search, and it mainly depends

L.K. Pavithra (✉) · T. Sree Sharmila
SSN College of Engineering, Chennai, India
e-mail: pavithralki@ssn.edu.in

T. Sree Sharmila
e-mail: sreesharmilat@ssn.edu.in

H.S. Behera and D.P. Mohapatra (eds.), *Computational Intelligence in Data Mining*, Advances in Intelligent Systems and Computing 556,
DOI 10.1007/978-981-10-3874-7_70

on low-level features (color, texture, shape, and spatial relation,) extracted from images. Color is the most prominent low-level content descriptor, which is used to calculate the feature information of the image [3]. Color space states a way to specify, create, and visualize color [4]. Color space selection is the familiar controversy in color feature extraction [5]. The most commonly used color spaces are RGB, HSV, YIQ, YUV, and YCbCr. RGB color space is the origin of other color models, since all other models were derived from RGB [6]. It is an additive color model of three primary colors: red (R), green (G), and blue (B). The main disadvantage of using RGB is that, processing real-world images with RGB color space will not give coherent outcome because of non-uniform perceptual quantization in color space. It creates large gap between the distance function defined in RGB space and human perception theory [7]. On account of this, color space transformation has been encouraged. HSV color space mainly related to human visual system. Another name of the HSV color space is perceptual color space. HSV stands for hue, saturation, and value. Hue and saturation are related to human visual skills. Value holds the decoupled data in the form of intensity about color. The conversion of HSV into RGB is defined in [6].

YIQ model was developed by the American National Television System Committee (NTSC) mainly for transmission purpose. Y contains luminance information, and I and Q define the in-phase and quadrature modulation components. These two components mainly hold the color information. YIQ is a device-dependent color space. The transformation from YIQ into RGB is given in [8].

YUV and YIQ are the television (TV) color spaces. Y holds the luminance data, U and V carry the chrominance information. U has blue chrominance details, while V band carries red chrominance information. Linear transformation from YUV into RGB is defined in [9]. YCbCr color model is obtained from the YUV model. Here, Y is the luminance information channel. Cb and Cr are the blue-difference and red-difference chroma channels [10]. YCbCr is mainly used for compression and speeding up the system performance in field programming gate array (FPGA) system. This transformation is essential in high-definition television (HDTV), digital coding of TV pictures, video digital libraries, etc [11]. At first, color histogram is being used to represent the color attribute of an image by its count. Well-established color histogram with Euclidean distance is the strategy for color similarity measure [12]. However, the location of the color pixel is not present in histogram and lacks in response similar to human perception, and handling of large database is inadequate. In [13], quantization is done in RGB color space to extract the color correlogram features, and it illustrates the spatial relationship between the pairs of color change within the desired distance. It gives consistent response even if the viewing position of the object changes.

Aleksandra et al. [7] proposed a quantized color code book system for color matching and meaningful image retrieval. Smith and Chang [14] suggested HSV nonlinear transformation with RGB space because of its easily reversible property.

It is applicable for color extraction in region of image using multiple thresholds. In [15], RGB image is converted to HSV, L * a * b * color spaces with distinct quantization level, and then histogram is plotted by utilizing quantization information. Histogram intersection metric is used for similarity evaluation. Octree data structure [16] is proposed to cluster different color features, such as dominant color and color histogram to produce meaningful image retrieval. Wavelet compression technique [17] is used to compress image data for efficient transmission. Haar wavelet [18] transform was applied to image in HSV color space in different quantization levels to obtain fast and efficient matching system. Edge features and color features [19] are extracted from YCbCr color space, and Manhattan distance metric is applied to retrieve similar images. Above studies shows the problem associated with color space selection and color feature extraction.

Our proposed work addresses the color space selection issues by analyzing retrieval accuracy of CBIR system on different color spaces through features mined from the image. In our work, color images used in multimedia image database [20] (Wang's database) are taken for the implementation and analysis purpose.

2 Proposed Work

Proposed work contributes the well-known CBIR system by addressing the color space selection issue by comparing the performance of the retrieval with distinct color space. The standard flow of the CBIR system is given in Fig. 1. The back end of the retrieval system computes color and edge features from the image database and they are accumulated in the feature database, and the front end of the CBIR system can get query image as input from the user. In the intermediate phase, query image feature vectors are extracted as same as done in the back end process and gets compared with feature database. Then, similar feature-corresponding images are retrieved from image database and displayed to the user.

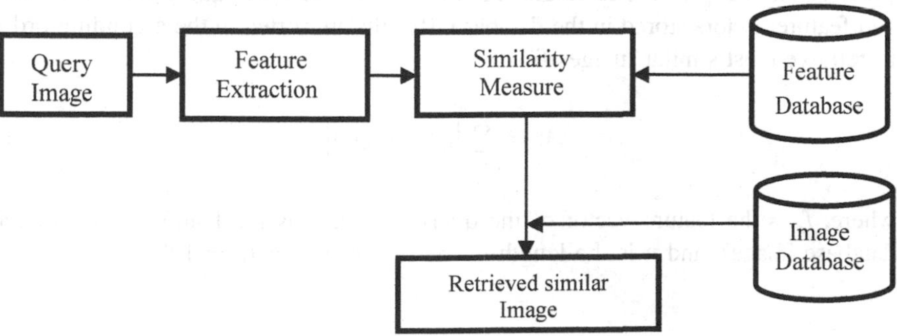

Fig. 1 Standard CBIR system

2.1 Color Space Selection

The proposed system selects HSV, YIQ, YUV, and YCbCr color spaces for color feature extraction, since these color spaces are having both color (chrominance) and luminance information. Then their performances are compared and analyzed. In the first step, back end of the CBIR system takes images from the database, after that RGB to corresponding color space conversion is done. Then, each channel is separately processed in further steps.

2.2 Edge Feature Extraction

CBIR system with either edge feature or color feature will not produce suitable results [19]. So the proposed system extracts edge feature [21] along with the color feature to obtain the adequate match. According to human perception theory, human eye is more sensitive to luminance channel than chrominance. Hence, the canny edge operator is applied to luminance channel (Y) of each color model. In HSV, canny edge detection is done in value channel. Other chrominance channels are remains the same. Then these edges extracted from HSV, YIQ, YUV, and YCbCr models images are transformed back to RGB space. After that histogram is computed with 256 bins for R, G, and B channels separately. At the end of this step, each channel has 256 feature vectors.

2.3 Feature Reduction

Front end of the CBIR system will get requested input from the user in the form of query image, and subsequently same feature extraction mechanism is carried out in that image. Toward the final step, 128 feature vectors are extracted for the input image. This feature vectors are given to Manhattan distance [19] similarity metric (SM) given in Eq. (1). It calculates the distance between the query feature vectors and feature vectors stored in the database. Results are sorted in the ascending order to retrieve most similar images first.

$$SM = \sum_{i=1}^{n} \left| f_q(i) - f_d(i) \right| \tag{1}$$

where, f_q is the feature vector of the query image, f_d is the feature vector of the database images, and n is the length of the feature vector (n = 128).

3 Experimental Results

Color and edge features are extracted from query image and compared with features in the database. Then, large number of similar images are retrieved. The performance of the proposed system on different color spaces is measured with well-known factors such as precision (2), recall (3) [18, 19], and retrieval time.

$$\text{Precision} = \frac{\text{No. of relevant images retrieved}}{\text{Total no. of images retrieved}} \tag{2}$$

$$\text{Recall} = \frac{\text{No. of relevant images retrieved}}{\text{Total no. of relevant images in the database}} \tag{3}$$

Wang's database has 1000 images that are classified into 10 groups; name and label values of the each group in this database are given as follows: African tribes—1, Sea—2, Buildings—3, Bus—4, Dinosaurs—5, Elephants—6, Flowers—7, Horse—8, Mountains—9, and Food—10. Retrieval results of distinct color spaces are analyzed on the three class images (Dinosaurs, Rose, and Horse) by randomly selected 5 images per class, and for 'Rose' query image retrieval results are shown in Figs. 2, 3, 4, and 5. Figure 2 shows the top 15 proximity results of the requested 'Rose' image on the HSV color space and they belong to the same 'Rose' class. Miscellaneous retrieval is obtained in YIQ color space for the same 'Rose' query image which is shown in Fig. 3, whereas in Fig. 4 YUV color space gives the highly matched 15 homogeneity retrieval result of the requested image. Figure 5

Fig. 2 Top most 15 retrieval results of query image Rose in HSV color space

Fig. 3 Top most 15 retrieval results of query image Rose in YIQ color space

Fig. 4 Top most 15 retrieval results of query image Rose in YUV color space

exhibits the YCbCr color space retrieval results which produce 6 dissimilar images among the topmost 15 nearest retrieval results.

Performance of the image retrieval system on different color space is evaluated for a number of retrievals equal to 50 which is shown in Table 1. The precision and recall for the 'Dinosaur' image are high in YIQ and YUV color space compared to

Fig. 5 Top most 15 retrieval results of query image Rose in YCbCr color space

Table 1 Performance comparison on HSV, YIQ, YUV, and YCbCr color spaces

Class name	Color space											
	HSV			YIQ			YUV			YCbCr		
	P (50)	R (50)	T	P (50)	R (50)	T	P (50)	R (50)	T	P (50)	R (50)	T
Dinosaurs	84	42	51	100	50	31	100	50	35	33	16	31
Rose	87	43.5	50	48	24	31	71	35.5	33	63	31	31
Horse	84	42	51	52	26	31	69	34.5	34	81	40	32
Average	85	42.5	50.7	66.7	33.3	31	80	40	34	59	29	31.3

P—Precision, R—Recall, and T—Time in seconds

other color spaces. HSV color model offers the second highest evaluation results for 'Dinosaur' test image. The second test image 'Rose' yields the highest precision and recall in HSV color space. For the test image 'Rose', YIQ-based image retrieval system gives negligible performance that is listed in Table 1, whereas HSV, YUV, and YCbCr based retrieval gives acceptable performance. Moreover, retrieval time of the CBIR system on HSV, YIQ, YUV, and YCbCr color space is shown in Table 1. Amount of retrieval time taken by HSV is relatively higher than the other color spaces.

4 Conclusion

In this paper, color edge histogram feature extraction is explored on HSV, YIQ, YUV, and YCbCr color spaces. Moreover, retrieval accuracy of each color space is evaluated using the precision, recall, and retrieval time of the CBIR system. Performance of each color space is exhaustively analyzed through altering the query image of the user. Even though retrieval time of the HSV color space is more than other color spaces, it provides consistent retrieval and approximately (in terms of precision) 5, 18.3, and 26% high accuracy with respect to the YIQ, YUV, and YCbCr color spaces. Future scope of this work is to extract additional low-level features from the image and give more adequate retrieval in minimal amount of time.

References

1. Singha, M., Hemachandran, K.: Content Based Image Retrieval using Color and Texture. Signal and Image Processing: An International Journal (SIPIJ). 3(1), (2012) 39–57.
2. Rui, Y., Huang, T.S.: Image Retrieval: Current Techniques, Promising Directions, and Open Issues. Journal of Visual Communication and Image Representation. 10, (1999) 39–62.
3. Swain, M.J, Ballard, D.H.: Color indexing. International Journal of Computer Vision. 7(1), (1991) 11–32.
4. Ford, A., Roberts, A.: Colour Space Conversions. (1998).
5. Wan, X., Kuo, C.-C.: Color distribution analysis and quantization for image retrieval. SPIE Storage and Retrieval for Image and Video Databases IV 2670. 8, (1996) 9–16.
6. Liu, Y., Zhang, Y., Zhang, C.: A fast algorithm for YCbCr to perception color model conversion based on fixed-point DSP. Multimedia Tools Appl. 74, (2015) 6041–6067.
7. Mojsilovic, A., Hu, J., Soljanin, E.: Extraction of Perceptually Important Colors and Similarity Measurement for Image Matching, Retrieval, and Analysis. IEEE Transactions on Image Processing. 11(11), (2002) 1238–1248.
8. Martinez-Alajarin, J., Luis-Delgado, J. D., Tomas-Balibrea, L.M.: Automatic System for Quality-Based Classification of Marble Textures. IEEE Transactions on Systems, Man, and Cybernetics, Part C: Applications and Reviews. 35(4), (2005) 488–497.
9. Tkalcic, M., Tasic, J.F.: Color spaces: perceptual, historical and applicational background EUROCON 2003. Computer as a Tool The IEEE Region 8. 1, (2003) 304–308.
10. Liu, Y., Zhang, Y., Zhang, C.: A fast algorithm for YCbCr to perception color model conversion based on fixed-point DSP. Multimedia Tools Applications. 74, (2015) 6041–6067.
11. Sapkal, A., Munot, M., Joshi, M.: R'G'B' to Y'CbCr Color Space Conversion Using FPGA. The IET International Conference on wireless, Mobile and Multimedia Networks. (2008) 255–258.
12. Wang, X.Y., Wu1, J.F., Yang, H.Y.: Robust image retrieval based on color histogram of local feature regions. Multimedia Tools Appl. 49 (2), (2010) 323–345.
13. Huang, J., Ravi Kumar, S., Mitra, M., Zhu, W.J., Zabih, R.: Image indexing using color correlograms. IEEE Conference on Computer Vision and Pattern Recognition. (1997) 762–768.
14. Smith, J.R., Chang, S.F.: Single Color Extraction and Image Query. Proceedings of the IEEE International Conference on Image Processing. 3, (1995) 528–531.

15. Singha, M., Hemachandran, K.: Performance analysis of Color Spaces in Image Retrieval. Assam University Journal of Science & Technology, Physical Sciences and Technology. 7(2), (2011) 94–104.
16. Wan, X., Kuo, J.C-C.: A New Approach to Image Retrieval with Hierarchical Color Clustering. IEEE Transactions on Circuits and Systems for Video Technology. 8(5), (1998) 628–643.
17. Tsai, M.J.: Very Low Bit Rate Color Image Compression By Using Stack-Run-End Coding. IEEE Transactions on Consumer Electronics. 46(2), (2000) 368–374.
18. Singha, M., Hemachandran, K., Paul, A.: Content-based image retrieval using the combination of the fast wavelet transformation and the colour histogram. IET Image Processing. 6 (9), (2012) 1221–1226.
19. Agarwal, S., Verma, A.K., Dixit, N.: Content Based Image Retrieval using Color Edge Detection and Discrete Wavelet Transform. IEEE International Conference on Issues and Challenges in Intelligent Computing Techniques (ICICT). (2014) 368–372.
20. Wang, J.Z., "Wang Database," [Online], Available at: http://wang.ist.psu.edu/.
21. Canny, J.: A Computational Approach to Edge Detection. IEEE Transactions on Pattern Analysis and Machine Intelligence. 8(6), (1986) 679–698.

Stylometry Detection Using Deep Learning

K. Surendran, O.P. Harilal, P. Hrudya, Prabaharan Poornachandran and N.K. Suchetha

Abstract Author profiling is one of the active researches in the field of data mining. Rather than only concentrated on the syntactic as well as stylometric features, this paper describes about more relevant features which will profile the authors more accurately. Readability metrics, vocabulary richness, and emotional status are the features which are taken into consideration. Age and gender are detected as the metrics for author profiling. Stylometry is defined by using deep learning algorithm. This approach has attained an accuracy of 97.7% for gender and 90.1% for age prediction.

Keywords Readability metrics · Vocabulary richness · Stylometry · CNN

1 Introduction

Author profiling, comes under the field of Computational Linguistics, is the process of extracting relevant information such as age, gender, language, and mood of the author from a text, which in great way will help to conduct interesting behavioral analyses. It has large scale usage in many fields such as advertising and low enforcement [1]. For example, nowadays most of the e-commerce businesses revolve around related algorithms for providing their users great buying experience.

K. Surendran (✉) · O.P. Harilal · P. Hrudya · P. Poornachandran · N.K. Suchetha
Amrita Center for Cyber Security Systems & Networks, Amrita School of Engineering, Amrita Vishwa Vidyapeetham, Amrita University, Amritapuri, India
e-mail: surendrank@am.amrita.edu

O.P. Harilal
e-mail: harilalop@am.amrita.edu

P. Hrudya
e-mail: hrudyap@am.amrita.edu

P. Poornachandran
e-mail: praba@am.amrita.edu

N.K. Suchetha
e-mail: suchetha.nk1988@gmail.com

H.S. Behera and D.P. Mohapatra (eds.), *Computational Intelligence in Data Mining*, Advances in Intelligent Systems and Computing 556,
DOI 10.1007/978-981-10-3874-7_71

The common approach followed by most of the researchers for profiling authors is by extracting relevant features from the texts written and mapping them to a model [2, 3]. Here, the foremost step is selecting the features. Extracting author information from any social media such as Twitter and Facebook is one of the challenging tasks. The profile we need to monitor might be private and will be able to access only some text data. In such situations, predicting the age and gender from those limited text data has substantial importance.

In this paper, Convolutional Neural Network (CNN), a deep learning approach is used for age and gender estimation. The proposed system is able to predict age and gender of a text with high accuracy. Overfitting problem overcomes by deep neural networks with their large number hyper parameters.

This paper is organized as follows. Section 2 presents Related Works. In Sect. 3, we provide the details of Stylometry Construction followed by Stylometry formation and Empirical analysis, respectively, in Sects. 4 and 5.

2 Related Work

Author profiling is one of the active research area which mainly consists of extracting age and gender. One of the substantial works on classifying user attributes such as age, gender, personalization, and recommendation has done by Rao, Delip, et al. [4]. They have proposed a stacked SVM-based approach for the classification. Cheng, Na, Rajarathnam Chandramouli, and K.P. Subbalakshmi [5] proposed the work for gender detection from text which is helpful for forensics. They have come up with considerable amount of features for detecting the gender of the author from any piece of text. Nguyen, Dong-Phuong, et al. [6] explored the hardness of age prediction from tweets. They have found out that older the twitter user, younger they are. Murugaboopathy, G., et al. [7] also did an experiment to detect gender from a text. They have achieved the prediction by using structural, stylometric as well as gender-specific features on Support Vector Machine learning algorithm. Rangel, Francisco, and Paolo Rosso [8] identified the age and gender of the user by based on the language usage using SVM classifier. Gender identification alone is detected by Deitrick, William, et al. using stylometric features. Another gender prediction task from email stream is done by Deitrick, William, et al. using neural network. Gender prediction using N-gram character with streaming algorithm is also there [9]. One of the gender prediction work is done on music metadata [10] also. Using speech and acoustic signals Banchhor, Sumit Kumar, and S.K. Dekate [11] identified the gender of the author. Using Facebook comments with different classifiers, Talebi, Masoud, and Cemal Kose [12] identified gender and age of the user. Using gradient features of the text [13], profiling the author has been achieved. Also inferring the age alone from the text is done in [14, 15]. Some other works are also depicts the interest in the area of author profiling [16]. Relevant work based on women characteristics [17–19] substantial work has done in this area.

3 Stylometry Constructions

Besides widely used word-based and character-based n-gram features, syntactic-, semantic-, readability-based, and person's mood-based features are proposed in order to enhance the accuracy. Figure 1 shows important features which are considered to find out the age of authors. Figure 2 shows the significant feature to detect the gender of authors.

3.1 Character Level Features

Unigram frequency is one of the features in this classification issue. Other important features are:

- Average number of characters per word: This feature will quantify the average occurrence of characters in a word for a particular category.
- Average number of special characters: It consists of average occurrences of special characters.
- Average Number of repetitive special characters: Even though this feature comes under special characters, it has significant role in gender identification. Popular repeated characters are exclamation (e.g., wow!!!!), ellipse (e.g., hope u r fine....), and question marks (e.g., u there????).

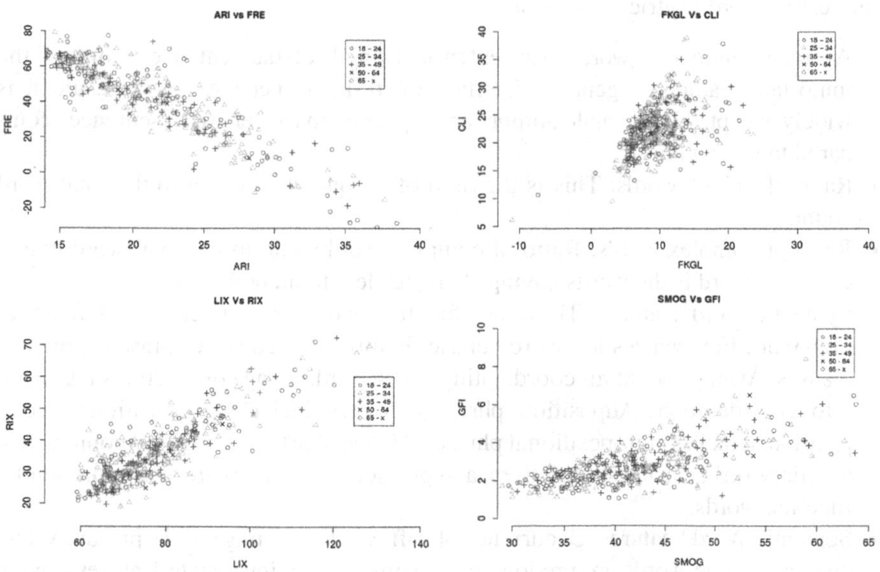

Fig. 1 Age specific features

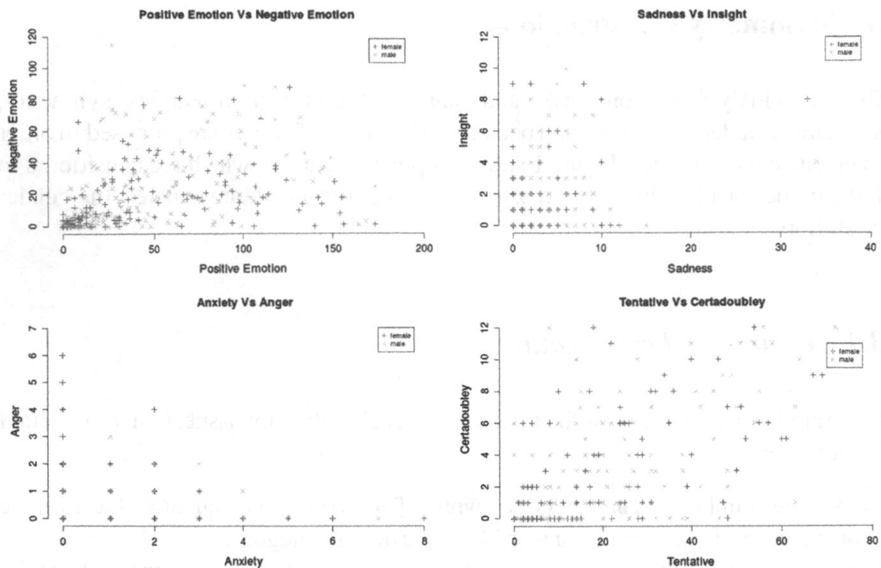

Fig. 2 Gender specific features

3.2 *Word Level Features*

N-gram is one of the popular features which are considered. Other than this features, interesting word-centric features are:

- Average Number of words per sentence: Length of the sentence is one of the important features in gender identification task. As per the survey result, it is widely accepted that female authors are supposed to use lengthier sentences compared to male authors.
- Ratio of unique words: This is the ratio of unique word count to the total word count.
- Ratio of complex words: Ratio of complex word count to the total word count. Complex word is the words having character length more than 6.
- Syntactic word features: These are function words. Conjunction and Interjection Word Frequencies are word centric features that connect phrases, words or clauses. Average count for coordinating and subordinating conjunctions has taken into consideration. Adposition phrase category includes prepositional, post-positional, and circum-positional phrases. Usage of article words, pronoun words, auxiliary verb, particles, expletives, and pro-sentence counts are also comes under function words.
- Semantic Word Features: Occurrence of soft words, greeting words, profanity, and gender-specific words are predominant features. Emotional state features are the words which will determine the mood of a person. Negative (No, Never, etc.), negative emotional (hurt, shame, etc.), positive (Yes, always, etc.), positive emotional

(Love, sweet, etc.), tentative (guess, maybe, etc.), certainty (Always, of course, etc.), stopping (block, constrain, etc.), agreeing (ok, sure, etc.), anxiety words (nervous, fearful, etc.) are some of the words which shows the mental status of the author.

3.3 Sentence Level Features

Sentence richness is one of the important features. Occurrence of proper sentence is taken into consideration.

3.4 Vocabulary Richness

This feature will give an idea about the vocabulary richness of a text. Important features are: Hapax legomenon or Hapax is a word that occurs once in a text. Frequency of occurrence of those words will be calculated. Like this, dis legomenon (words coming twice in the text), tris legomenon (words coming thrice in the text), and tetrakis legomenon (words coming quadruple times in the text) are also calculated.

3.5 Readability Metrics

The purpose of readability metrics is to assess the receptiveness of a piece of text in a scientific way. The readability measures which are taken into consideration are:

- Automated Readability Index (ARI): This feature will provide grade value corresponding to different age category. For example, grade value of 8 corresponds to a teenager.

$$ARI = 4.71(\frac{Letters}{Words}) + 0.5(\frac{Words}{Sentences}) - 21.43 \quad (1)$$

- Flesch Reading Ease (FRE): Is one of the oldest readability formulas. Score range is in between zero to hundred and 60–70 score corresponds to good text.

$$RE = 206.835 - 1.015(\frac{Words}{Sentences}) - 84.6(\frac{Syllables}{Words}) \quad (2)$$

- Flesch Kincaid Grade Level (FKGL): This is one of the grading formulas for text.

$$FKGL = [0.39(\frac{Words}{Sentences})] + 11.8(\frac{Syllables}{Words})] - 15.59 \quad (3)$$

- Simple Measure of Gobbledygook (SMOG): This will give an idea about the person's educational level.

$$SMOG = 1.0430\sqrt{PlySyllables * \frac{30}{Sentences}} + 3.1291 \quad (4)$$

- Gunning fog index (GOI): This value range is from 6 to 17 and it can classify the authors from a range of sixth standard to college graduate.

$$GOI = [0.4(\frac{Words}{Sentences}) + 100(\frac{ComplexWords}{Words})] \quad (5)$$

- Coleman Liau Index (CLI): One of the readability tests to measure the understandability of text.

$$CLI = [0.0588(\frac{Letters}{Words}) + 0.296(\frac{Sentences}{Words})] - 15.8 \quad (6)$$

- Lycée International Xavier (LIX): It is a readability measure based on long words.

$$LIX = (\frac{Words}{Periods}) + (\frac{LongWords * 100}{Words}) \quad (7)$$

- Anderson's Readability Index (RIX): One of the easiest readability tests.

$$RIX = (\frac{LongWords}{Sentences}) \quad (8)$$

- Spache Readability Formula (SRF): This formula will grade the children up to 4th grade.

$$SRF = 0.121 * AvgSentenceLength + 0.082 * Unfamiliarwords + 0.659 \quad (9)$$

- Dale- Chall (DC): This formula will grade the children above 4th grade. This is an advanced way to improve Flesch Reading Ease formula.

$$DC = 0.1579(\frac{DifficultWords}{Words} * 100) + 0.0496(\frac{Words}{Sentences}) \quad (10)$$

- Linear Write (LW): Is a formula to score the easiness of text. Its base value to check the easiness is 20.
- Raygor estimate graph (REG): It is calculated based on the average number of characters and sentences per 100 words. After extracting the features, classifier module will train the system using deep Learning algorithm.

4 Stylometry Formation

Methodologies used to construct the stylometry of a particular author groups are explained in this module. Model construction with different algorithms gives an idea about the accuracy differences. Tested algorithms for stylometry construction are Naïve Bayes, Decision tree, Random Forest, and Deep Learning. For age calculation, considered age groups are:

$$18\text{–}24, 25\text{–}34, 35\text{–}49, 50\text{–}64, 65\text{–}x$$

and for gender classes are male and female.

4.1 Network Architecture

Convolutional Neural Network(CNN) is a deep learning algorithm which is used to accomplish the task of both age and gender prediction. CNNs composed of several layers of convolutions and will be having nonlinear activation functions. Mostly used activation functions are ReLU and tanh. Instead of giving each input neurons to the next layer, CNNs connect the convolutions to the next layer. In each layer, pooling or subsampling will be taken place. Pooling is like a filter inorder to reduce the computational complexity of the model net. MAXPOOL with 2×2 is the pooling layer used in this work inorder to prevent overfitting. The input matrix which is feeding as input neurons to CNNs is taken as feature matrix. Each row corresponds to each user having the aggregated feature set of 50. Feature considered are all words. These words are converted to vectors in feature collection or stylometry construction module. The network architecture will have input neurons, convolution, RELU, POOL, and FC.

- Input neurons: Input neurons are having 50 features having 2 lakh rows.
- Convolution layer: This layer will calculate the convolution by dot product. We have used 12 filters for the same. If the input neuron is N × N matrix with filter, then the output will be (N-n+1) × (N-n+1). The filter ϕ is having the size n × n. The convolution is as follows:

$$x_{ij}^{\ell} = \sum_{a=0}^{n-1} \sum_{b=0}^{n-1} \phi_{ab} y_{(i+a)(j+b)}^{\ell-1} \tag{11}$$

- RELU: Activation function used is Rectified Linear Unit (ReLU) which is shown as:

$$f(x) = \ln(1 + e^{x}), \tag{12}$$

- POOL: The next step is down sampling/Pooling operation. Here, the input neuron is an N × N matrix with sample space as M × M and the resulting output will become N × M matrix.
- FC: The final step is Fully Connected (FC) layer. This layer will predict the class.

5 Empirical Analysis

This section describes the different evaluation results. Tested the dataset set with different machine learning algorithms such as Naïve Bayes, Decision tree, Random forest, SVM, and CNN. Naïve Bayes algorithms performed the least on the data sets; 63.7413% for gender and 19.6% for age prediction. Tables 1 and 2 shows the classifier accuracy of gender and age, respectively. Deep learning algorithm performed better with an accuracy of 97.7% for gender and 90.1% for age prediction.

In this work, we have come up with different feature set for both age and gender prediction having considerably good accuracy on online social media data. We have trained our data on Twitter data, but the same feature set will provide good accuracy for other social media data such as Facebook chat, Blogs, Emails, and News. Twitter data can be in any language, but our model we have trained mainly for English corpus data. We have tested our training set in Deep learning4J and Tensorflow, machine learning libraries. System attained good accuracy using Convolutional Neural Network with multiple hidden layers.

Table 1 Accuracy for gender

Algorithm	Prefix (%)
Convolutional Neural Network	97.7
SVM	80.3
Random Forest	74.36
Decision Tree	63.0485
Naive Bayes	63.7413

Table 2 Accuracy for age

Algorithm	Prefix (%)
Convolutional Neural Network	90.1
SVM	55.6
Random Forest	43.88
Decision Tree	29.7921
Naive Bayes	19.6305

Acknowledgements We are grateful to the Amrita University for providing us the entire infrastructure to accomplish our task. We would like to thank all Cyber Security Department staff members for their help and support to complete this project successfully.

References

1. Arroju, Mounica, Aftab Hassan, and Golnoosh Farnadi. "Age, gender and personality recognition using tweets in a multilingual setting." Sort 100 (2014): 250.
2. Deitrick, William, et al. "Gender identification on Twitter using the modified balanced winnow." (2012).
3. Peersman, Claudia, Walter Daelemans, and Leona Van Vaerenbergh. "Predicting age and gender in online social networks." Proceedings of the 3rd international workshop on Search and mining user-generated contents. ACM, 2011.
4. Rao, Delip, et al. "Classifying latent user attributes in twitter." Proceedings of the 2nd international workshop on Search and mining user-generated contents. ACM, 2010.
5. Cheng, Na, Rajarathnam Chandramouli, and K. P. Subbalakshmi. "Author gender identification from text." Digital Investigation 8.1 (2011): 78–88.
6. Nguyen, Dong-Phuong, et al. "Why gender and age prediction from tweets hard: Lessons from a crowdsourcing experiment." Association for Computational Linguistics, 2014.
7. Murugaboopathy, G., et al. "Appropriate Gender Identification from the Text."
8. Rangel, Francisco, and Paolo Rosso. "Use of language and author profiling: Identification of gender and age." Natural Language Processing and Cognitive Science 177 (2013).
9. Miller, Zachary, Brian Dickinson, and Wei Hu. "Gender prediction on Twitter using stream algorithms with N-gram character features." (2012).
10. Wu, Ming-Ju, Jyh-Shing Roger Jang, and Chun-Hung Lu. "Gender Identification and Age Estimation of Users Based on Music Metadata." ISMIR. 2014.
11. Banchhor, Sumit Kumar, and S. K. Dekate. "Comparison of text-dependent method for Gender Identification."
12. Talebi, Masoud, and Cemal Kose. "Identifying gender, age and Education level by analyzing comments on Facebook." Signal Processing and Communications Applications Conference (SIU), 2013 21st. IEEE, 2013.
13. Bouadjenek, Nesrine, Hassiba Nemmour, and Youcef Chibani. "Age, gender and handedness prediction from handwriting using gradient features." Document Analysis and Recognition (ICDAR), 2015 13th International Conference on. IEEE, 2015.
14. Moseley, Nathaniel, Cecilia Ovesdotter Alm, and Manjeet Rege. "Toward inferring the age of Twitter users with their use of nonstandard abbreviations and lexicon." Information Reuse and Integration (IRI), 2014 IEEE 15th International Conference on. IEEE, 2014.
15. Nguyen, Dong-Phuong, et al. "How old do you think I am? A study of language and age in Twitter." (2013).
16. Maarten, et al. "Developing age and gender predictive lexica over social media." (2014).
17. Burger, John D., et al. "Discriminating gender on Twitter." Proceedings of the Conference on Empirical Methods in Natural Language Processing. Association for Computational Linguistics, 2011.
18. Ratcliffe, Krista. Rhetorical listening: Identification, gender, whiteness. SIU Press, 2005.
19. Schmader, Toni. "Gender identification moderates stereotype threat effects on women's math performance." Journal of Experimental Social Psychology 38.2 (2002): 194–201.

SVPWM-Based DTC Controller for Brushless DC Motor

G.T. Chandra Sekhar, Budi Srinivasa Rao and Krishna Mohan Tatikonda

Abstract Brushless DC motor is one of the growing electrical drives in present days, because of their higher efficiency, high power density, easy maintenance and control, and high torque to inertia ratio. In this paper, a sensorless space vector modulation-based direct torque and indirect flux control of BLDC has been investigated. There are several methods that are projected for BLDC to gain better torque and current control, i.e., with minimum torque and current pulsations. The proposed sensorless method is similar to the usual direct torque control method which is utilized for sinusoidal alternating current motors so that it controls toque directly and stator flux indirectly by varying direct axis current. And the electric rotor position can be found by using winding inductance and stationary reference frame stator fluxes and currents. In this paper, space vector modulation technique is utilized which permits the regulation of varying signals than that of PWM and vector control techniques. The validity of the projected sensorless three-phase conduction direct torque control of BLDC motor drive method is established in the Simulink, and the results are observed.

Keywords Direct torque and indirect flux controller · BLDC · SVM technique

G.T. Chandra Sekhar (✉) · B.S. Rao
Electrical and Electronics Engineering, Sri Sivani College of Engineering,
Chilakapalem, Srikakulam, AP, India
e-mail: gtchsekhar@gmail.com

B.S. Rao
e-mail: bsreee2013@gmail.com

K.M. Tatikonda
Electrical and Electronics Engineering,
Andhra Loyola Institute of Engineering & Technology, Vijayawada, India
e-mail: t.krishnamohan02@gmail.com

H.S. Behera and D.P. Mohapatra (eds.), *Computational Intelligence in Data Mining*, Advances in Intelligent Systems and Computing 556,
DOI 10.1007/978-981-10-3874-7_72

1 Introduction

For applications such as high accuracy, high efficiency, and high power density [1], brushless DC motors are the better choice in present days. Usually, BLDC motor is accounted as high pursuance motor which is efficient in generating more amounts of torque over wide speed ranges. BLDC motors are inside out of common DC motors, and they exhibit the same torque–speed characteristics [2]. The main difference lies in usage of brushes. Like DC motor, brushless DC motor does not have brushes so that they are electronically commutated. Commutation is nothing but changing the motor phase currents at desired time to create rotational torque.

For brushless DC motors with trapezoidal back emf [3] obtaining low-frequency ripple free torque, instantaneous torque and flux are major considerations. So in order to obtain the control on flux and torque, there are different methods that are stated for sensorless control of BLDC and they are as follows:

1. Measurement of back emf and
2. Freewheeling current detection method.

The above stated methods have their own advantages and disadvantages, and moreover, the newer techniques that are evolved made them a bit effective less as some of the techniques need hardware equipment for sensing purpose.

This paper presents a simple position-sensorless direct torque and indirect flux control of BLDC motor, similar to the normal DTC structure used for sinusoidal pulsating direct current motors where torque and flux are regulated at the same time. This process provides features of conventional DTC [4] such as fast torque response compared to vector control and position-sensorless drive. The electrical rotor position is famous by calculating winding inductance and stationary reference framework stator flux linkages and currents [5].

2 Modeling of Brushless DC Motor

BLDC motor is one of the classifications in long-lasting magnet synchronous motors [5]. As its name indicates as synchronous electric motor, the magnetic field created by both the stator and rotor rotates with the same frequency. Thus, BLDC motors do not experience any “slip” which is normally noticed in introduction motors. To achieve proper commutation, PMDCM uses mechanical commutators and brushes. But in case of BLDC motor, it uses Hall effect sensors [6] in place of mechanical commutators and brushes. So BLDC is said to be electronically commutated. Brushless DC motor is just inside out of DC motor. The stator of BLDC motor contains winding coils and the rotor with permanent magnets. The stator develops the magnetic field to make the rotor to rotate. The Hall effect sensors that are placed 120 electrical degrees apart detect the rotor position so as to make proper commutation sequence [7]. Therefore, BLDC motors replace the coils with permanent magnets in armature so it does not require any brushes and commutators as shown in Fig. 1.

And the schematic diagram for brushless DC motor is shown in Fig. 2 [8]. The mathematical modeling for BLDC drive is obtained by considering the following considerations:

1. It has three symmetrical windings.
2. It has no magnetic saturation.
3. It neglects the hysteresis and eddy current losses.
4. It Ignores the mutual inductance.
5. And it neglects the armature reaction.

The mathematical modeling is obtained by considering the KVL equations for Fig. 2 [9].

$$\begin{aligned} V_a &= i_a r_a + L\frac{di_a}{dt} + e_a \\ V_b &= i_b r_b + L\frac{di_b}{dt} + e_b \\ V_c &= i_c r_c + L\frac{di_c}{dt} + e_c \end{aligned} \tag{1}$$

For solving these equations, in this paper, we have used a concept of line-to-line Park's transformation technique. This line-to-line Park's transformation converts the three-phase voltages into two-phase coordinators, which is expressed as follows:

$$\begin{bmatrix} Vab \\ Vca \end{bmatrix} = \begin{bmatrix} -\frac{1}{3} & -\frac{1}{3} \\ \frac{\sqrt{3}}{3} & -\frac{\sqrt{3}}{3} \end{bmatrix} \begin{bmatrix} Va \\ Vb \\ Vc \end{bmatrix} \tag{2}$$

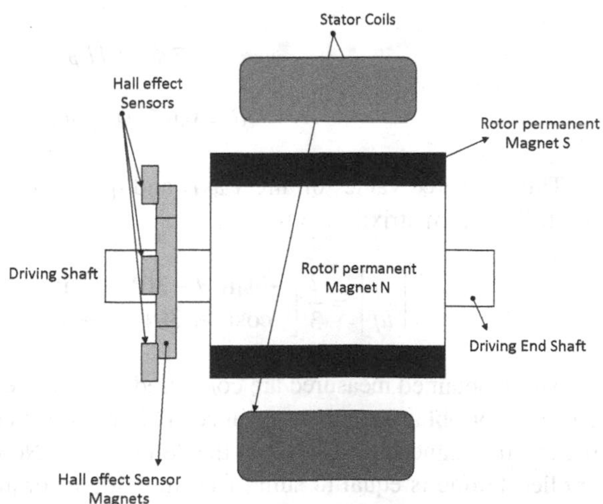

Fig. 1 Architecture of BLDC motor

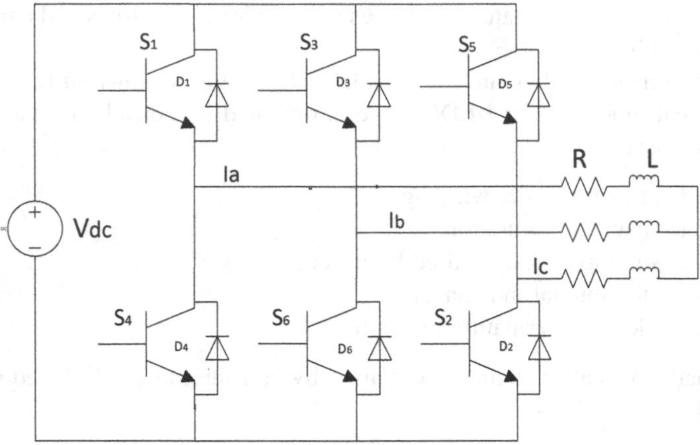

Fig. 2 Basic schematic diagram for BLDC

The matrix coordinates obtained from the above line-to-line Park's transformation are transformed into orthogonal matrix coordinates (α, β).

Similarly, like voltage, the three-phase currents also transformed into two-phase orthogonal matrix. These two-phase currents (Iα, Iβ) and voltage (Vα, Vβ) are used for calculating the flux linkages (ψα, ψβ) from the expression described as follows [10]:

$$\begin{aligned}\psi_{\alpha} &= \frac{1}{L_{\alpha}}(V_{\alpha} - i_{\alpha} r_{a}) \\ \psi_{\beta} &= \frac{1}{L_{\beta}}(V_{\beta} - i_{\beta} r_{a})\end{aligned} \tag{3}$$

And from this equation, the phase angle is calculated as follows:

$$\psi = \psi_{\alpha} + j\psi_{\beta} \tag{4}$$

$$\theta = \tan^{-1}(\psi_{\beta}/\psi_{\alpha}) \tag{5}$$

The measured values of direct axis and quadrature axis currents are obtained by the following matrix:

$$\begin{bmatrix} i_d \\ iq \end{bmatrix} = \frac{2}{3}\begin{bmatrix} -\sin(\theta - 30) & \sin(\theta + 30) \\ \cos(\theta + 30) & -\cos(\theta - 30) \end{bmatrix}\begin{bmatrix} i_{\alpha} \\ i_{\beta} \end{bmatrix} \tag{6}$$

These obtained measured are compared with reference direct and quadrature axis currents for obtaining error tolerance. The reference current signals are obtained by the electromagnetic torque. From the definition of Newton's law of motion, the total applied torque is equal to sum of all individual torques across each element.

$$T_e = T_m + J\frac{dw_m}{dt} + Bw_m \tag{7}$$

The electromagnetic torque generated by a brushless DC motor is expressed as follows [11]:

$$T_e = \frac{e_a i_a + e_b i_b + e_c i_c}{w_m} \tag{8}$$

From the above two equations, the electromagnetic torque developed by a BLDC motor at any instant is as follows:

$$T_e = \frac{2e_p i_p}{w_m} \tag{9}$$

where e_p is called phase back emf and i_p is a nonzero phase current.

The back emf for a BLDC motor is given as follows:

$$e_p = kw_m \tag{10}$$

The error difference obtained from comparison of the currents is given to SVM controller for obtaining the gate pulses to the three-phase inverter.

3 Space Vector Modulation Technique

It is a different approach for getting gate triggering signals instead of general pulse width modulation technique which is based on the space vectors generated [12] by the system two-phase vector components: α- and β-axis. Figure 3 shows the space vector representation of the adjacent vectors S1 and S2 with 8 space vector switching pattern positions of inverter as shown in figure.

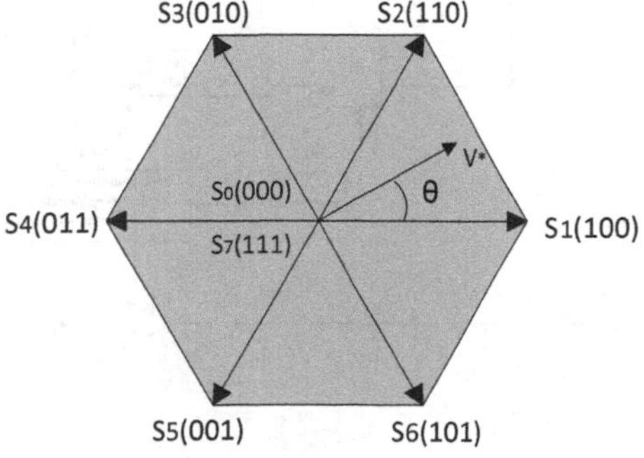

Fig. 3 Space vector modulation technique

Generally, the space vector modulation technique is one of the most popular methods in pulse width modulation techniques fed for the three-phase voltage source inverters. By using space vector modulation, the harmonic content in both output voltage and output currents is reduced. The space vector modulation technique is used in this paper for creating the reference vectors generated by modulating the switching time sequence of space vectors in each of six sectors as shown in Fig. 3 [13]. From Fig. 3, six switching sectors are used for inversion purpose and two sectors are behaved like null vectors.

$$V * Tz = S1 * T1 + S2 * T2 + S0 * (T0/2) + S7 * (T0/2) \tag{11}$$

4 Principle of Operation of Space Vector Modulation Scheme for BLDC Drive

The basic control block diagram shows the implementation of the direct torque control-based space vector modulation technique [14], which is shown in Fig. 4. With this proposed control technique, first, the values for estimated torque and flux linkages are determined from the actual three-phase component currents and the three-phase stator voltages. For doing these calculations, we considered the two-phase rotational orthogonal matrix vectors. And after the determination of estimated torque and flux linkages, these estimated values are used for generating triggering sequences. Two proportional integral controllers are used to regulate the current errors [15]. The gate switching signals for the inverter are obtained from the voltage vectors which are obtained from controlling and comparison of actual phase

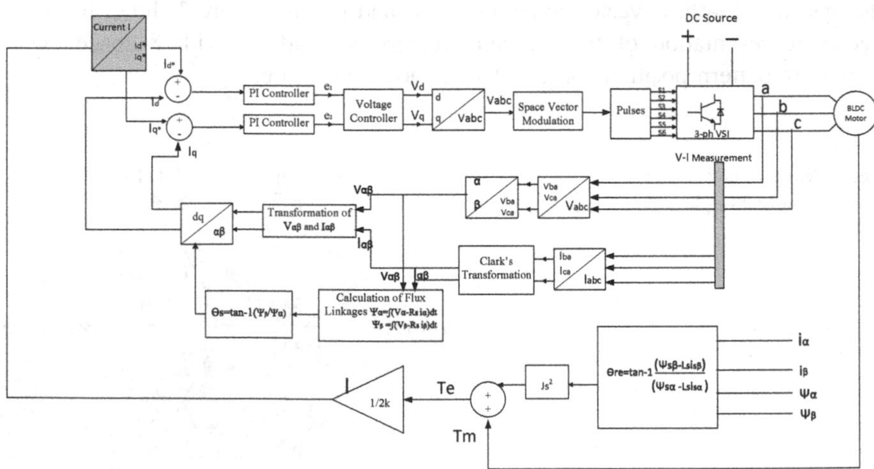

Fig. 4 Control diagram of DTC-SVM technique

values of voltage and current vectors. The complete block diagram for the SVM-based DTC controller is shown in Fig. 4.

5 Selection of Electric Rotor Position

The electric rotor position θre, which is required in torque estimation, can be found using the equation.

$$\theta_{re} = \tan^{-1}\left(\frac{\psi_{s\beta} - L_s i_{s\beta}}{\psi_{s\alpha} - L_s i_{s\alpha}}\right) \quad (12)$$

And the value of θreS is used in calculation of electromagnetic torque Te [16].

6 Simulation Diagram and Results

The experimental setup for DTC-SVM-based BLDC drive is done in MATLAB/Simulink model. Switching pulses for the three-phase inverter are obtained from the switching table which decides the pulses from the error signals of stator currents. The absolute value of current is estimated from the estimated torque which is derived from the mechanical modeling and motor parameters such as phase voltage and phase currents (Figs. 5, 6, 7, 8, 9, and 10).

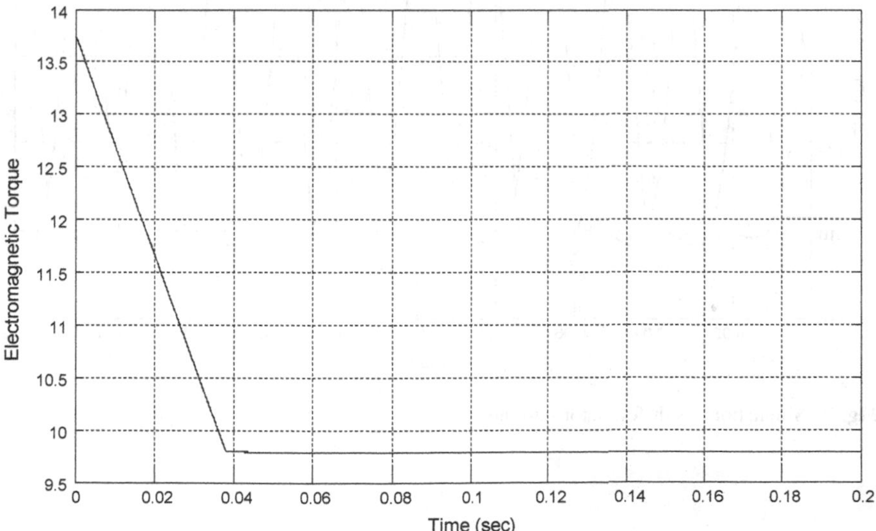

Fig. 5 Simulation result for electromagnetic torque at Tm = 10.5 N-m

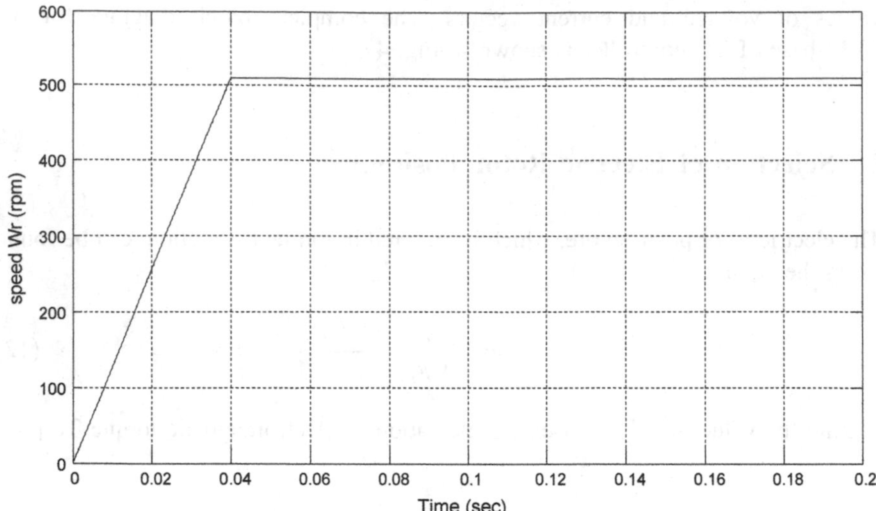

Fig. 6 Simulation result for speed

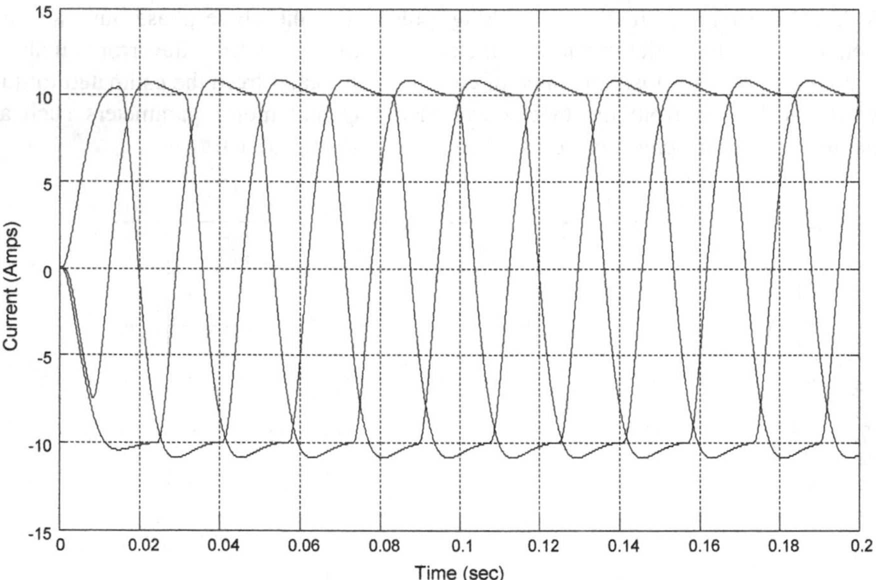

Fig. 7 Simulation result for stator currents

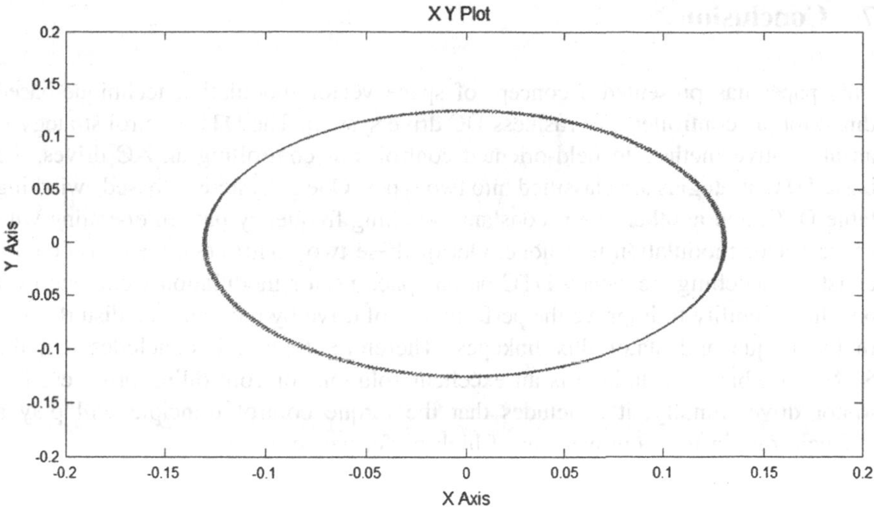

Fig. 8 Simulated indirectly controlled flux linkage when Ids is zero under 10.5 N-m load torque

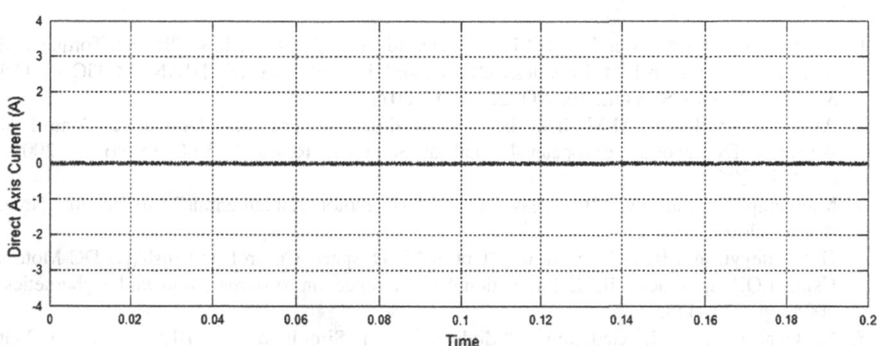

Fig. 9 Simulation result for stator direct axis current

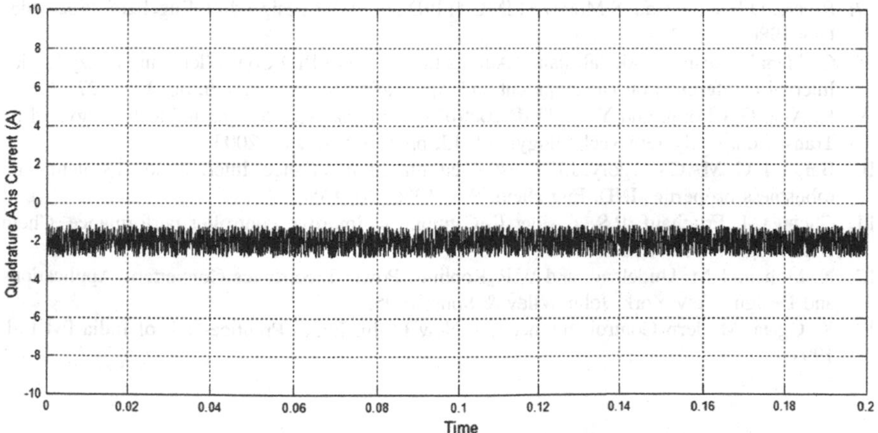

Fig. 10 Simulation result for quadrature axis current

7 Conclusion

This paper has presented a concept of space vector modulation technique-based direct torque controller for brushless DC drive system. The DTC control strategy is an alternative method to field-oriented control. For controlling an AC drives, the basic DTC strategies are classified into two types: One is hysteresis-based switching table DTC, and another one is constant switching frequency pattern operating with space vector modulation technique. Out of these two controllers, we considered a constant switching frequency DTC-based space vector modulation technique as it has the capability to improve the performance of drive by reducing the disturbances in the torque and stator flux linkages. Therefore, finally, it concludes that the SVM-DTC-based technique is an excellent solution for controlling brushless DC motor drive. Finally, it concludes that the torque control principle will play a strategic role in the improvement of high-performance drives.

References

1. Salih Baris Ozturk, Member, IEEE, and Hamid A. Toliyat, Fellow, "Direct Torque and Indirect Flux Control of Brushless DC Motor" in IEEE/ASME TRANSACTIONS ON MECHATRONICS, VOL. 16, NO. 2, APRIL 2011.
2. Atef Saleh Othman Al-Mashak- beh "Proportional Integral and Derivative Control of Brushless DC Motor" European Journal of Scientific Research Vol. 35 No. 2 (2009), pp. 198– 203.
3. Microchip Technology, "Brushless DC (BLDC) motor fundamentals", application Note, AN885, 2003.
4. Gwo-Rueyyu and Rey-Chue Hwang "Optimal PID Speed Control of Brushless DC Motors Using LQR approach" IEEE International Conference on systems, Man and Cybernetics, 2004, pp. 473–478.
5. C. Gencer and M. Gedikpinar "Modelling and Simulation of BLDCM using Mat lab/Simulink" Journal of Applied Sciences 6(3):688–691, 2006.
6. Allan R. Hambley, "Electrical Engineering Principles and Application", Prentice Hall, New Jersey 1997.
7. Rivera, D.E.skogestad, S.Morari M.IMC 4: PID controller design. Ind. Engchem. Processdes. Dev 1986, 25,252.
8. Gaddam Mallesham, AkulaRajani, "Automatic Tuning of PID Controller using Fuzzy Logic", Internal Conference on Development and Application Systems, 2006, pp. 120–127.
9. K. Ang, G. Chong, and Y. Li, "PID control system analysis, design, and technology," IEEE Trans. Control System Technology, vol. 13, pp 559–576, July 2005.
10. Bergh, L.G. MAC Gregory. J.F. constrained minimum variance- Internal model structure and robustness properties.IND. Eng chem. Res. 1986, 26, 1558.
11. Chein, L-L Fruehauf, P.S consider IMC tuning to improve controller performance. Chem. Eng. Prog 1990, 86, 33.
12. N. Mohan, T.M. Undeland, and W.P. Robbins, Power Electronics Converters, Applications, and Design, New York: John Wiley & Sons, 1995.
13. K. Ogata, Modern Control Engineering, New Delhi, India: Prentice-Hall of India Pvt Ltd., 1991.

14. Performance evaluation of BLDC motor with conventional PI and fuzzy speed controller, Singh M Electr. Eng. Dept., Delhi Technol. Univ., New Delhi, India; Garg.
15. Direct torque control of brushless DC motor drives with reduced starting current using fuzzy logic controller. Parhizkar N. Dept. of Electr. Eng., Islamic Azad Univ., Fars, Iran; Shafiei, M., Kouhshahi, M.B.
16. Combined Flux Observer with Signal Injection Enhancement for Wide Speed Range Sensorless Direct Torque Control of IPMSM Drives, Andreescu, G.; Univ. Politeh. Of Timisoara, Timisoara; Pitic, C.I.; Blaabjerg, F.; Boldea, I.K. Padiyar, Power System Dynamics: Stability & Control. New York: Wiley, 1996.

Implementation of IoT-Based Smart Video Surveillance System

Sonali P. Gulve, Suchitra A. Khoje and Prajakta Pardeshi

Abstract Smart video surveillance is a IOT-based application as it uses Internet for various purposes. The proposed system intimates about the presence of any person in the premises, also providing more security by recording the activity of that person. While leaving the premises, user activates the system by entering password. System working starts with detection of motion refining to human detection followed by counting human in the room and human presence also gets notified to neighbor by turning on alarm. In addition, notification about the same is send to user through SMS and e-mail. The proposed system's hardware implementation is supported by Raspberry Pi and Arduino board; on the other hand, software is given by OpenCV (for video surveillance) and GSM module (for SMS alert and e-mail notification). Apart from security aspect, system is intelligent enough to optimize power consumption wastage if user forgets to switch off any electronic appliances by customizing coding with specific appliances.

Keyword IOT (Internet of Things)

1 Introduction

In the present world, situation security assumes a vital part. Numerous individuals utilize distinctive sorts of security system to keep their property from unapproved person's entry. Security system helps individuals to feel somewhat safe while they have to travel or avoid their home for work. A large number of the security system works just inside a specific territory limit [1], for instance, CCTV, as a person need

S.P. Gulve (✉) · S.A. Khoje · P. Pardeshi
MAEER's MITCOE, Kothrud, Pune (MH), India
e-mail: sonaligulve@gmail.com

S.A. Khoje
e-mail: suchitra.khoje@mitcoe.edu.in

P. Pardeshi
e-mail: prajakta.pardeshi@mitcoe.edu.in

H.S. Behera and D.P. Mohapatra (eds.), *Computational Intelligence in Data Mining*, Advances in Intelligent Systems and Computing 556,
DOI 10.1007/978-981-10-3874-7_73

to see camera footage from control room. The current security systems against robbery are entirely costly as a certain measure of cash must be paid to administration supplier to store the recorded video despite the fact that there is no human movement is recognized. The solution for this problem is an intelligent surveillance system that can start recording video only after a human motion is detected. This eventually minimizes the required storage space and makes system cost-effective.

The proposed framework gives more security with the assistance of Web at less expensive cost and requires less storage space. In literature [2 and 3], researchers have proposed various methods for people counting. In literature [4–10], researchers have proposed many image processing methods/algorithms for human counting which are prone to problems such as occlusion or shadow and overlapping. To address these problems at some extent, Rossi and Bozzoli [4] and Sexton et al. [5] proposed a technique in which the position of camera is vertical as for the plane of the floor. In literature [11], researchers proposed an improved adaptive background mixture model for real-time tracking with shadow detection. The proposed framework gives a smart security system which gives home security with SMS and e-mail notice about the unapproved people nearness, programmed human checking and switching off all the appliances which consumes more power by customizing coding with particular appliances. Proposed system performs various tasks such as motion detection, human detection and counting, alarm activation, SMS notification through GSM and Internet Twilio account, and e-mail notification.

To improve the system performance, two boards are used—Raspberry Pi and Arduino. Raspberry Pi works in surveillance mode and Arduino works in normal mode. Arduino verifies the password and allows Raspberry Pi to start the surveillance mode. Once the password is verified, Arduino turns off all the electrical appliances by customizing coding with specific appliances. Raspberry Pi performs various tasks in surveillance mode such as motion and human detection, human counting, sending SMS, and e-mail notification to user after human detection. After human detection, Raspberry Pi sends command to Arduino for sending SMS to user by communicating with GSM module. By default, system remains in normal mode.

As the user enters correct password, system starts working in surveillance mode. In surveillance mode, Raspberry Pi detects human motion and counts number of people in a room. The location of a camera is at the entrance of a room. The human count is implemented by background subtraction [2] method in OpenCV. If any human is detected in surveillance mode, then using the GSM module and Twilio account message is sent to the owner of the house. The highlights of proposed system are as follows:

(1) The proposed framework includes people counting, and two notices are sent to client by SMS: One SMS is sent through GSM and one SMS is sent through Twilio trial account with the assistance of Web. The recorded video is sent as a e-mail to client. At the point when there is no individual in the premises, the framework works in ordinary mode.

(2) Raspberry Pi detects motion and human presence and it counts number of humans in a room. As the system detects human presence, immediately a SMS notification is sent to the user. The system also sends the recorded video to users mail id. As a human is detected, GSM module gets instruction from Arduino regarding SMS notification. Another SMS notification is sent through Internet Twilio trial account. The alarm is turned on as human presence is detected.
(3) The proposed system also provides a facility to control electrical appliances by turning them off. The proposed system offers few advantages such as-
    (i) Less memory storage space is used for recording video as system start recording the video only after motion is detected.
    (ii) Recorded video is e-mail to user so that the user can inspect it later.
    (iii) User gets noticed (SMS and Email) just after human detection, so that he can take necessary actions immediately.

2 Working Principle

The proposed framework is initiated by entering right password. The movement recognition algorithm is actualized to distinguish the moving items and human count in room is done by utilizing OpenCV. After the password verification system starts working in surveillance mode and all the electrical appliances are turned off appliances by customizing coding with specific appliances. If motion is detected, the system checks for human detection. As system detects human presence, a notification through SMS is sent and alarm is turned on. The activity of that human is recorded and e-mailed to user.

If the video consists of less than or equal to 100 moving frames, the video is immediately sent to user, and if the video exceeds 100 moving frames, then the video of those moving frames will be sent in the next e-mail. Motion is detected by background subtraction method MOG2 algorithm. In the event if the movement is identified, then human discovery is executed by HAAR cascade classifier. The proposed framework is sufficiently keen to identify human movement, checks number of individuals, and informs client by sending SMS, and e-mails the recorded video to client. Figure 1 demonstrates the block diagram of proposed framework. Step by step working process is as follows:

(1) Start process by entering correct password. System goes in surveillance mode as Arduino allows Raspberry pi to turn on the camera and all electrical appliances are turned off.
(2) Wait for motion detection—Confirm the human detection, people counting mode and send SMS to the owner, Alarm is turned ON.

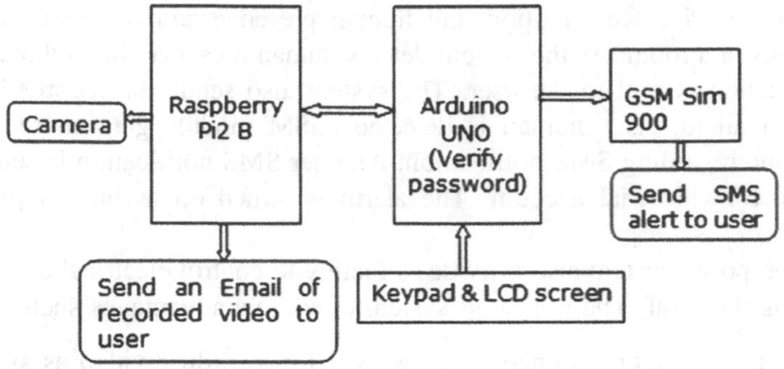

Fig. 1 Block diagram

(3) Security mode is ON—Record video as security system is broken and e-mail that recorded video to user.
(4) Enter password again to make system work in normal mode.

3 System Architecture

3.1 Elements of the System

Raspberry Pi2 is the primary handling unit. OpenCV is build and introduced on it for image processing. Arduino is subprocessing unit which is in charge for initialization of fundamental handling unit after getting password from client. Python is utilized for interfacing between Raspberry Pi2 and Arduino, and Python is additionally utilized for sending SMS through Web; furthermore, it is utilized to send e-mail to the client/user.

3.2 Hardware and Software Design

In addition to Raspberry Pi, Arduino and GSM module are the principle equipments utilized for the framework. The GSM module (needs a SIM card to work) is associated with Arduino by USB to serial converter. A LCD screen is utilized to show the entered secret word. Two relays of 12 volts are utilized which are associated between raspberry pi and Arduino. One relay is in charge of activation and deactivation of fundamental procedure, i.e., Raspberry Pi. Other relay gives a control to turn off/on electrical appliances. Figure 2 demonstrates the algorithm utilized for the proposed framework.

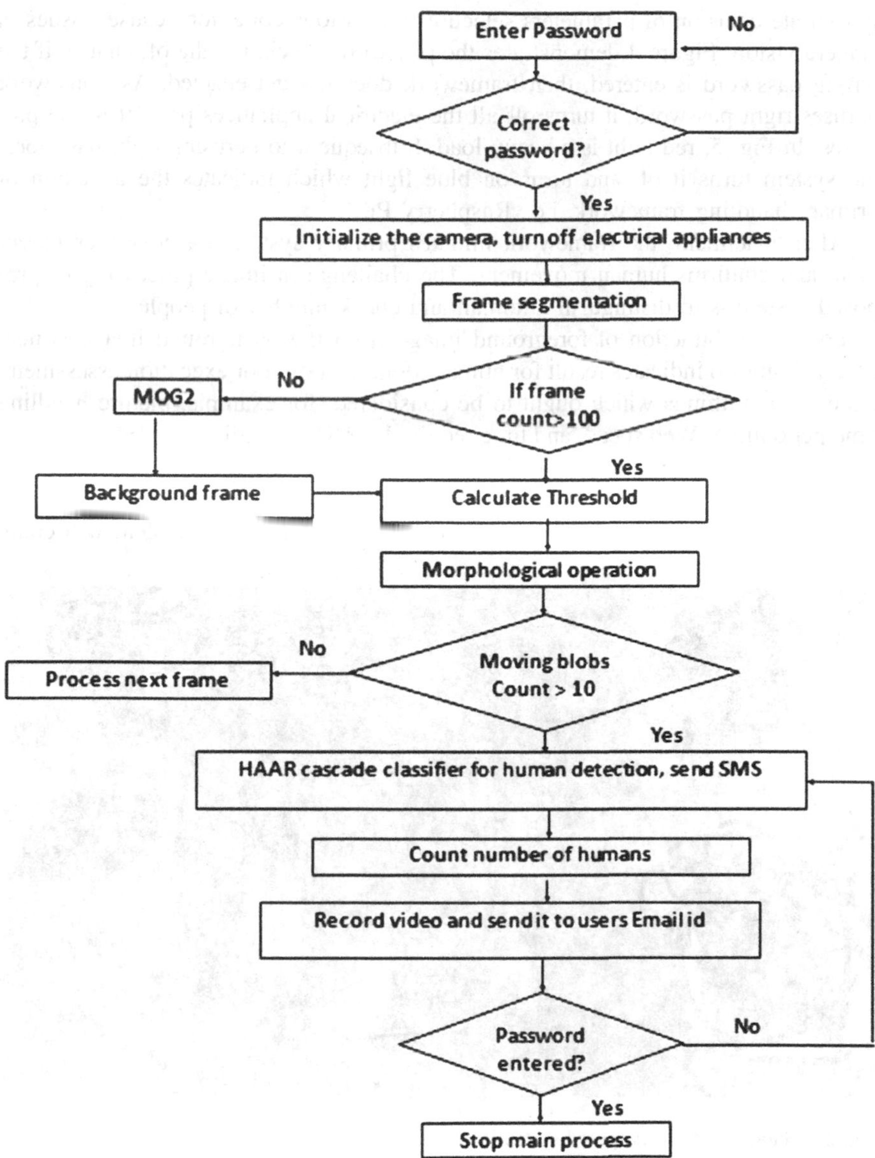

Fig. 2 Algorithm

4 Results

Figure 3 demonstrates the pictorial view of the proposed framework. For the proposed framework, different components should be dealt with, for example, distinctive lighting condition intensely aggravates the nature of a camera pictures.

Inaccurate decision of parameters selection for various conditions causes issues in camera vision. Figure 4 demonstrates the password check. On the off chance if the wrong password is entered, then framework does not get enacted. As framework peruses right password, it turns off all the electrical appliances present in the premises. In Fig. 5, red light is a heavy load. Subsequent to perusing right password, the system turns it off and turns on blue light which indicates the actuation of primary handling framework, i.e., Raspberry Pi.

After enactment, the framework initiates primary system that identifies movement and confirms human movement. The challenge in image processing of proposed system is to distinguish a human and check number of people.

For this, subtraction of foreground image from the background image is necessary. Figure 6 indicates result for human identification. For execution assessment, there are few things which ought to be considered, for example, picture handling time per outline, Web speed, and time required for SMS sending by GSM. Figure 7 indicates result for SMS and e-mail notification.

The recorded video for security mode is in JPEG format as it can be played using any standard video player. As human motion is detected, main system will count

Fig. 3 Pictorial view of proposed system

Fig. 4 Password verification

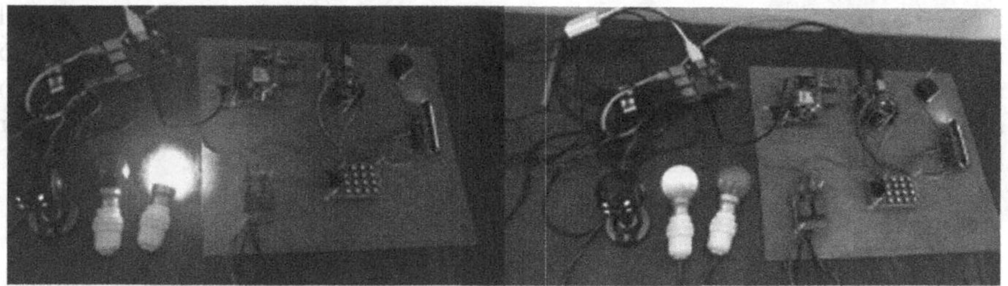

Fig. 5 Controlling electrical appliances

Fig. 6 Human detection

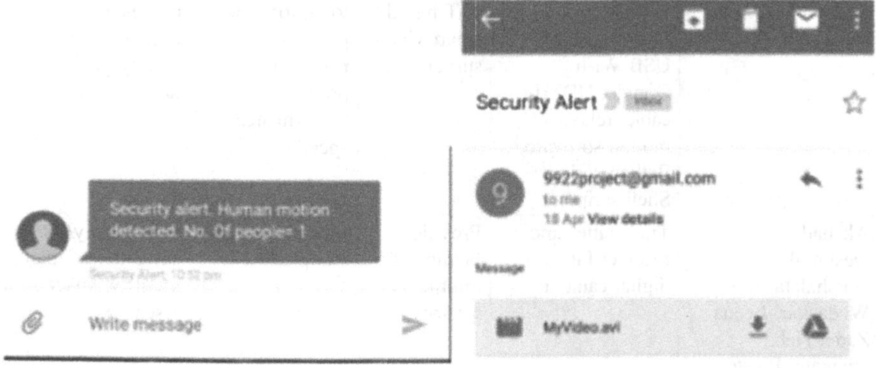

Fig. 7 SMS and e-mail notification

number of people and a SMS will be sent to user regarding security alert which will notify user about unauthorized persons' presence. GSM module will send a SMS to user. A video will be recorded and saved on SD card as well as will be e-mailed to user. Figures 6 and 7 show e-mail and SMS notification results send to user after human detection. SMS sending is done through GSM and Internet.

4.1 Observations and Comparison

Depending on the experiment performed on the proposed system, Table 1 shows the observation that is made and Table 2 shows comparison of existing system.

Table 1 Observation table

Parameters	Results
Camera speed	30 frames per second
Motion detection speed	3.1–9 s as it checks for 100 moving frames with great accuracy
Human detection speed	2–5 s
Distance between camera and human (for human detection)	10–15 ft (depending on .xml file)
Time required to send SMS	20–30 s (depends on Internet speed and Web site load) for 50 Mbps LAN connection

Table 2 Comparison between existing systems

Reference	Hardware and software	Advantages	Disadvantages	Application
Khot Harish S, Gote Swati R, Khatal Sonali B, Pandarge Sangmesh [12]	IBM Smart Surveillance Engine (SSE), camera, Ethernet	Provides front-end video analysis capabilities	Online video streaming which requires more Internet data usage	Home/office surveillance
U. Ramakrishna, N. Swathi [13]	Raspberry Pi, USB Web Camera, GSM, USB Wi-fi Dongle, HDMI cable, relay, motion software, Python scripts, Shell script	Provides IOT-based smart video surveillance	Use of low-processing power chips may result in poor performance speed	Industries, offices/home, military areas, elderly person falling sick
Akshada Deshmukh, Harshalata Wadaskar, Leena Zade, Neha Dhakate, Preetee Karmore [14]	Transmitter and receiver kit, digital camera	Provides facility of multilevel security	More hardware complexity	Office/army/home surveillance, bank security, space research

5 Conclusion

The proposed framework is cheaper in cost as it requires less storage space and no individual to monitor persistently from control room. In the proposed framework, two hardware boards are utilized to enhance the execution of the framework. The proposed system also provides facility of instantaneous alert to user so action can be taken immediately. The proposed system can be implemented at high-alert places such as banks, industry, or any other places where this type of security is required.

The future thought is to attempt and add some more elements to the framework like face recognition for user for activation and deactivation of the system and mobile-based home automation framework which will permit client to control the proposed framework through mobile. As Internet assumes an imperative part in the proposed framework, utilization of 3G/4G network would be suggested for better execution.

References

1. Md. Syadus Sefat, Abdullah Al Mamun Khan, Md. Shahjahan "Implementation of vision based intelligent home automation and security system", 3rd INTERNATIONAL CONFERENCE ON INFORMATICS, ELECTRONICS & VISION 2014.
2. F. Bartolini, V. Cappellini and A. Mecocci, Counting people getting in and out of a bus by real time image-sequence processing, Image and Vision Computing, vol. 12, no. 1, Jan. 1994, pp. 36–41.
3. A. Albiol, I. Mora and V. Naranjo, Real-time high density people counter using morphological tools, IEEETrans. Intelligent Transportation Systems, vol. 2, no. 4, Dec. 2001, pp. 204–218.
4. M. Rossi and A. Bozzoli, Tracking and counting moving people, IEEE International Conference on Image Processing (ICIP), vol. 3, 1994, pp. 212–216.
5. G. Sexton, X. Zhang and G. Redpath, Advances in automatic counting of pedestrians, 1995 European Convention on Security and Detection (ECSD95), 1995, pp. 106–110.
6. K. Terada, D. Yoshida, S. Oe and J. Yamaguchi, A method of counting the passing people by using the stereo images, IEEE International Conference on Image Processing (ICIP), vol. 2, 1999, pp. 338–342.
7. O. Masoud and N. P. Papanikolopoulos, A novel method for tracking and counting pedestrians in real-time using a single camera, IEEE Trans. Vehicular Technology, vol. 50, no. 5, Sep. 2001, pp. 1267–1278.
8. J. W. Kim, K. S. Choi, B. D. Choi and S. J. Ko, Real-time vision-based people counting system for security door, International Technical Conference on Circuits/Systems Computers and Communications, 2002, pp. 1416–1419.
9. J. Bescos, J. M. Menendez and N. Garcia, DCT based segmentation applied to a scalable zenithal people counter, IEEE International Conference on Image Processing (ICIP), vol. 3, 2003, pp. 1005–1008.
10. T. H. Chen and C. W. Hsu, An automatic bi-Directional passing-people counting method based on color-image processing, 37th IEEE International Carnahan Conference on Security Technology, Taiwan, Oct. 2003, pp. 200–207.
11. KaewTraKulPong P, Bowden R. An improved adaptive background mixture model for real-time tracking with shadow detection Proceedings 2nd European Workshop on Advanced

Video Based Surveillance Systems (AVBS 2001), Kingston, UK, September 2001 3rd INTERNATIONAL CONFERENCE ON INFORMATICS, ELECTRONICS & VISION.
12. Khot Harish S, Gote Swati R, Khatal Sonali B, Pandarge Sangmesh Smart Video Surveillance, IJ EERT Volume 3, Issue 1, January 2015, PP 109–112 ISSN 2349-4395 (Print) & ISSN 2349-4409 (Online).
13. U. RAMAKRISHNA, N. SWATHI: Design and Implementation of an IoT Based Smart Security Surveillance System, IJSETR ISSN 2319–8885 Vol. 05, Issue. 04, February-2016, Pages: 0697–0702.
14. Akshada Deshmukh, 2 Harshalata Wadaskar, 3 Leena Zade, 4Neha Dhakate, 5 Preetee Karmore: Webcam Based Intelligent Surveillance System, IJES Vol. 2, 8 March 2013, Pp 38–42 Issn(e): 2278-4721, Issn(p):2319-6483.

Context-Aware Recommendations Using Differential Context Weighting and Metaheuristics

Kunal Gusain and Aditya Gupta

Abstract Context plays a paramount role in language and conversations, and since their incorporation into traditional recommendation engines, which made use of just the user and item details, an effective method to utilize them in the best possible manner is of great importance. In this paper, we propose a novel approach to handle the sparsity of contextual data, their increasing dimensionality, and the development of an effective model for a context-aware recommender system (CARS). We further go on given relevance, in the form of assigning weights even to the individual attributes of each context. Differential context weighting (DCW) is used as the rating model to obtain the desired ratings. Optimization of weights required for DCW is done through metaheuristic techniques, and toward this, we have further gone on to experimentally compare two of the most popular ones, namely particle swarm optimization (PSO) and the firefly algorithm (FA). Recommendations using the optimal one were then obtained.

Keywords Recommender system · Context · Context-aware recommendation · Differential context · Metaheuristics · Particle swarm optimization · Firefly algorithm

1 Introduction

Recommender systems continue to be one of the most researched fields in recent times and find extensive usage in e-commerce; personalized marketing; or recommending products, places, music among other things. An Internet user is quite likely to come across a recommendation service, sometimes without even being

K. Gusain (✉) · A. Gupta
Computer Science Department, Bharati Vidyapeeth's College of Engineering,
Guru Gobind Singh Indraprastha University, Delhi, India
e-mail: kunalgusain1995@gmail.com

A. Gupta
e-mail: adityag95@gmail.com

H.S. Behera and D.P. Mohapatra (eds.), *Computational Intelligence in Data Mining*, Advances in Intelligent Systems and Computing 556,
DOI 10.1007/978-981-10-3874-7_74

explicitly aware of it. Whereas one cannot deny the advantage a good and timely recommendation offers to the user, we do, however, have to accept that a huge margin of improvement still remains in their working. Usage of a high number of, and forever evolving, artificial intelligence, and machine learning techniques, makes sure that there is a boundless potential in terms of their optimization. Traditional recommendation engines were used to working on just two factors, which are the users and the items. More recently, there have been efforts to include 'contexts' in the recommendation process as well [1], enabling us to give even more specific and improved suggestions. These recommendation engines are called context-aware recommender systems or simply CARS. A context can be as simple as the users' location, day, time, or it could be their mood and even their companions. Both implicit and explicit contexts are relevant, and as contexts do, they have the ability to change over time for the same user. Context-aware recommender systems (CARS) incorporate them in the hope to provide efficient recommendations, and toward this extent, a number of models have been put forward. There are two primary ways in which this process of incorporation takes place, namely the filtering and the modeling approach as given by Adomavicius et al. [2]. The filtering approach is further subdivided into two approaches, depending on the time filtering is done, namely pre-filtering and post-filtering approaches. However, in both of these the contexts are not incorporated directly instead used as a filter; thus, the second approach of modeling is preferred since it directly utilizes contexts to give recommendations. Once the approach has been finalized, we now have to choose a rating model to find the desired ratings.

One rating model that could be used in CARS is the differential context relaxation (DCR) model [3, 4]. In DCR, a subset of features is selected that are then used for giving recommendations. What happens as a result is that not all features are effectively incorporated and instead the suggestions are just based on a small subset of, sometimes even arbitrary constraints. Due to the lack of an effective model, it was perhaps necessary at the beginning to tackle the problem of sparsity, but now, the more nuanced and accurate approach call differential context weighting (DCW) has stormed the marketplace of ideas and has been shown to be a substantial improvement over DCR. This pioneering work was also done by a team of the foremost researchers and authorities in the field of contextual recommendations [5]. DCW started giving weights to each of the features, and then, these weighted contexts were used for the evaluation. So unlike in the relaxation technique of leaving behind some contexts, now all the features could be included and the end recommendation was as a result more potent. The process of assigning weights further needs an optimization step, wherein we decide how and what are the weights that are going to be assigned to the features. This optimization can be done by applying metaheuristic techniques.

Metaheuristics [6, 7] are being increasingly used and researched in optimization-related methodologies, primarily due to their ability to reduce the time in computation and also since they are independent of gradient-based analysis; the caveat is that a globally optimal solution is not necessarily always found. Our focus is on nature-based metaheuristic techniques, namely particle swarm optimization

(PSO) and firefly algorithms (FA). PSO, like other swarm intelligence techniques, is inspired by insects, birds, or animals' behavior in nature and how they swarm [8, 9]. A population-based optimization, where each particle has a defined position in the space of particles, and depending upon the velocity, which is the speed and direction, their convergence can be evaluated. Firefly algorithms, unlike PSO which used bees and birds as a reference, uses fireflies and their organization as a metric [6, 10, 11], with the idea being the correlation between the brightness/glow of the fireflies to the values of the objective function. In our paper, we have compared and analyzed both these methods across a number of metrics to find out which would be more suitable for optimization in DCW and eventually better recommendations in CARS.

The objective of this paper is to find an efficient and economic approach of giving context-aware recommendations. Toward this extent, we have used DCW as a rating model. We have further enhanced and improved upon the existing methodology of using contexts by giving relevance in the form of weights to the individual attributes of all the features. Let us say that if the choice of companion is a feature, then apart from giving weights to the entire metric, individual attributes of a companion, such as family, alone, significant other and son, have also been assigned weights. This has helped to make the recommendations even more potent, and to the best of our knowledge, this method has not been implemented before. The data we use, which are of movie ratings, are vast and have a plethora of features to choose upon from. Due to the paucity of good contextual data available, we have used one of the most extensive data available, and the permission for the same has also been obtained. We have further gone on to compare two of the most usable, popular, and efficient optimization techniques that exist, PSO and FA, and compared them on a number of metrics to find which of them is more suitable to our problem. PSO was found to be the better alternate and was thus used to obtain desired recommendations.

2 Differential Context Weighting (DCW)

Differential context relaxation (DCR) [3] works on the principle of assigning binary values to each context, i.e., 0 or 1. The value is of '1' when the context is present, and the value is '0' when the context is absent. Unlike DCR, differential context weighting (DCW) works on the principle of assigning some weights to the given contexts instead of just assigning them values of '0' or '1' [5]. These weights have the range of all real numbers between '0' and '1'. These weights can be seen as the amount of contribution each context makes to the final rating. If a context has more weight, then that means it contributes more to the actual rating as compared to the context with low weights. The similarity metric which we use for calculating the similarity between the two users, given the weights for each context is weighted, is Jaccard similarity function. The key parameters for the weighted Jaccard similarity function are σ sigma and 'c, d,' where sigma is the vector containing weights for all the contexts, whereas c and d are the given contexts for the user. The weighted Jaccard similarity, which is used to calculate the similarity between all the users, is given by Eq. (1):

$$J(c,d,\sigma)=\frac{\sum_{f\in c\cap d}\sigma_f}{\sum_{f\in c\cup d}\sigma_f} \tag{1}$$

An example of the dataset used for movie recommendations is given in Table 1. A significant change that we have proposed is the assignment of weights to the attributes of the individual context. This makes the dataset even sparser. This can be done by first converting the data in Table 1. To the bit matrix like the one shown in Table 2.

The DCW model is given in Eq. (2):

$$P_{a,i,\sigma}=\overline{\rho}(a,\sigma)+\frac{\sum_{u\in N_{a,\sigma}}(\rho(u,i,\sigma)-\overline{\rho}(u,\sigma))*sim_w(a,u,\sigma)}{\sum_{u\in N_{a,\sigma}} sim_w(a,u,\sigma)} \tag{2}$$

'P' is the predicted rating, 'a' is the user, 'i' is the item for which the rating 'r' has been assigned with ρ being the rating function and $\overline{\rho}(a,\sigma)$ is the baseline function. The equation mainly consists of 4 parts:

1. Selection of neighbors,
2. Contribution by neighbors,
3. User baseline, and
4. Similarity between users.

Selection of neighbors involves selecting all the users from the data who have rated the given item 'i' in the given context 'c'. It is also possible that a neighbor 'u' might have rated the movie differently in different context, so to tackle this problem, we take maximally similar context of the neighbor and use it for the rating. The equation for selection of neighbors is given in (3).

$$N_{a,\sigma}=\{u\colon \max_{r_{u,i,d}}(J(c,d,\sigma)\} \tag{3}$$

Table 1 Contextual ratings of users for movie 'i'

User	Day type	Season	Location	Mood	Rating
U1	Working day	Spring	Home	Positive	5
U2	Holiday	Spring	Friend's house	Neutral	3
U3	Weekend	Autumn	Home	Positive	4
U1	Holiday	Autumn	Friend's house	Positive	?

Table 2 Bit matrix of Table 1

User	Day type working day	Day type holiday	Day type weekend	Season spring	Season autumn	Location home	Friend's home	Mood positive	Mood neutral	Rating
U1	1	0	0	1	0	1	0	1	0	5
U2	0	1	0	1	0	0	1	0	1	3
U3	0	0	1	0	1	1	0	1	0	4
U1	0	1	0	0	1	0	1	1	0	?

Neighbor contribution is the process of calculating the contribution by the user to the rating of the user. Sometimes, it is possible that the neighbor has rated the same item in multiple contexts differently. So we take the weighted average of the rating given by that neighbor in all the contexts as given by Eq. (4).

$$\rho(u, i, \sigma) = \frac{\sum_{r_{u,i,d}} r_{u,i,d} * J(c, d, \sigma)}{\sum_{r_{u,i,d}} J(c, d, \sigma)} \tag{4}$$

Equation (5) shows the formula of overall average for all those items that were given ratings for similar contexts, 'I_u' depicts the set of all the items that were rated by our user 'u'. User baseline is the average rating given by that user in that similar context.

$$\overline{\rho}(u, \sigma) = \frac{\sum_{i \in I_u} \rho(u, i, \sigma)}{|I_u|} \tag{5}$$

3 Metaheuristics

Recommendation engines, like life, require continuous concerted efforts to arrive at a feasible and desired result. Thus, the heuristic approach is built into the fundamental structure of recommendation systems, and this trial and error as well as the continuous improvement structure of theirs makes metaheuristics indispensable in optimization techniques. A number of metaheuristic algorithms are available [12], and their efficiency for solving problems helps them stay relevant in optimization domains. Out of all the population-based methods or evolutionary algorithms, particle swarm optimization (PSO) and firefly algorithm (FA) are the most relevant ones pertaining to our problem set.

3.1 Particle Swarm Optimization (PSO)

This swarm intelligence technique takes inspiration from the behavior of swarming creatures such as bees and organizational patterns of birds. PSO [9] works by identifying a set of target points, where all particles are plotted in the target space. Once all the target points are obtained, a concerted effort is undertaken in which at each successive step PSO tries to improve the given solution, in an effort to find the value which is closest to the target value. Ascribing velocity and direction to particles, it primarily has three components, namely momentum, cognitive, and the social components. The first one is used to describe the earlier velocity, due to which the particle was able to reach the current position. Cognition is that property of the particle which makes it want to return to the best possible position/scenario it had come across during its travails to the current position. Finally, the third and the social component, another beautiful feature of this algorithm, describes the tendency of the particle to attach itself or move toward the best possible position in its neighboring target space. PSO has a few clear and distinct advantages, which makes it one of the more used and researched optimization algorithms. First is its tendency to be aware and knowledgeable about its environment and the population of particles in the target space that it is working in. Secondly, it has a fast convergence characteristic, which makes it time efficient. Finally, PSO has shown to be highly efficacious in obtaining results in static as well as dynamic search spaces.

3.2 Firefly Algorithm (FA)

Firefly algorithm (FA) [6, 10, 11], another of the metaheuristic approaches, can be utilized in solving optimization problems. FA is a swarm intelligence technique; FA works by studying the behavioral pattern of fireflies and how they mate with each other. Glowing in the dark, these bioluminescent creatures make use of their light to attract mates to breed. In the search space, one firefly is used to represent a single candidate. Using the brightness as a metric, these fireflies are then evaluated on their ability to absorb and emit light, as well as their inherent attractiveness. Light intensity is the direct measurement of how attractive a firefly is. These characteristics of FA are possible because of certain natural features which make them rather unique and interesting: First of all, fireflies are unisex in gender and attract others only by the strength or intensity of their glow. Secondly, it also has the ability to move randomly in search of a brighter prospect. Lastly, the degree of the glow of a firefly decreases in proportion to the decreasing distance between them, as the two creatures move toward one another. One of the basic features of any metaheuristic algorithms is its ability to randomize the paths occasionally, such that it allows us to avoid local optimums. In FA as well, the randomizing of path factor plays a similarly crucial role. After each move of the firefly, its brightness function is updated by evaluating it at this recent position. This brightness function is nothing but a representative manifestation of the objective function. At each successive move, the function is evaluated, and if it is found to be better, it is updated at this new location. We set limits to the possible movements, by deciding for how many iterations it can be run, or we can also have a cutoff value, which once reached results in the termination of the algorithm. A number of variations and hybrids have been proposed over time, and many trade-offs have also been discussed; however, the basic idea remains same that of models being made based on the behaviors of bioluminescent fireflies, their mating habits, and their attractiveness to each other being a characteristic of the degree of their brightness. All these qualities have made them a fascinating phenomenon and their effectiveness has caused us to study it as a problem-solving alternate.

4 Experimental Setup

We use the dataset of a movies recommendation engine. With more than a thousand movies, the data have around fourteen different contexts such as the users age, their sex, city, country, time of the day, location, weather, companion, mood, emotional state, and interaction (whether it is the first interaction with a movie or nth time). The attributes of these contexts are also quite rich like the emotional state has sad, happy, scared, surprised, angry, disgusted as its attributes. Since in our research proposal, we want to treat each attribute as equally important and contributing feature like individual contexts; thus, we have a total of seventy-four features, for which we need to calculate the recommendation on. This LDOS—CoMoDa, movie dataset is a sparse one. We tested our results or obtained root-mean-squared value (RMSE) for over a thousand entries.

Pertaining to our first proposal, which was to assign weights even to the individual attributes of the contexts in the dataset. Such that fourteen or so contexts now turned into over seventy-four features. This though made the data sparse, it also made the recommendations in the end more potent with a reduced RMSE, since now even more user- and context-specific recommendations were being obtained. Now we have the newly expanded seventy-four features, which need to be optimized and assigned weights before being entered into our rating model. Since there are a number of optimization techniques available, our research has tried to narrow down and find the most effective technique that could be used for the aforementioned purpose. We use the extremely popular and widely used PSO as the first technique and FA as the second one since both of these offer the most relevant methodologies that can be applied to the said problem. The results of the optimization techniques would be the optimized weights for each context, to be entered into the rating model (DCW here).

The first metric for comparison is the number of iterations versus completion time (in seconds), given in Fig. 1. Number of iterations for both the algorithms, PSO and FA, were compared over a wide margin to gauge their behavior and offer a succinct analysis. We observed that with an increase in the number of iterations, there was a gradual but sure increase in the completion times of both the algorithms. However, both also showed some erratic behavior with the iteration increase, which is characteristic of all metaheuristic algorithms due to their inherent random functions and thus does not significantly deter our prognosis. As is evident from Fig. 1, PSO is significantly better than FA for our movie recommendation data.

The second metric we used to compare accuracy is RMSE value. The graph for the comparison is given in Fig. 2. This again showed PSO at an advantage with significantly less RMSE for the given data, and PSO had 1.3079 whereas FA had 1.8288. Thus, the two metrics employed by us, helped to evaluate and find out that PSO is markedly better than its alternate FA.

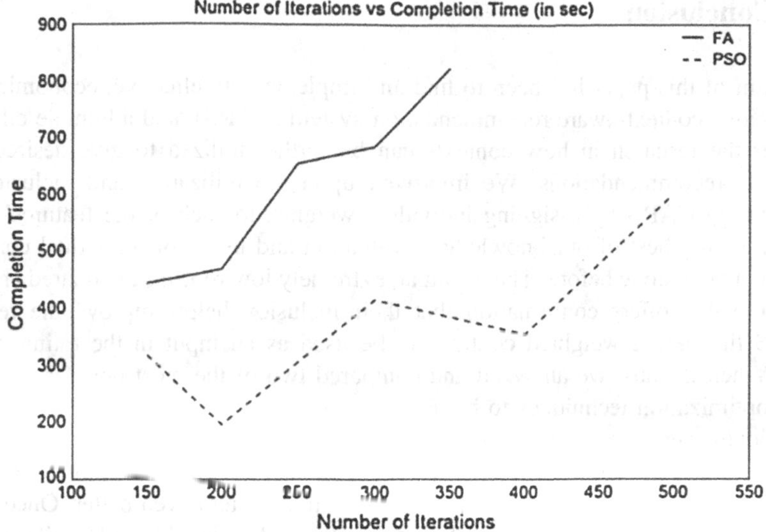

Fig. 1 Number of iterations versus completion time graph for PSO and FA

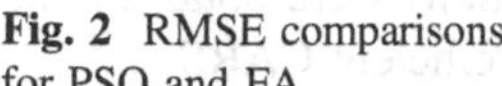

Fig. 2 RMSE comparisons for PSO and FA

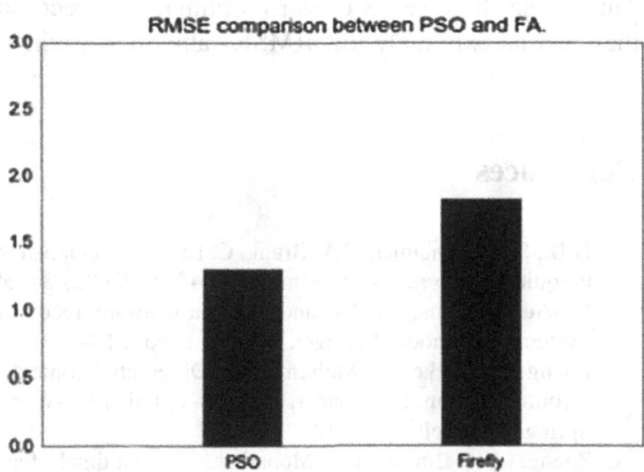

The final step is the implementation of the rating model using the weighted contexts obtained from the metaheuristic optimization techniques used above. The model we use is the differential context weighting or the DCW model. For finding the similarity between the users in DCW, we use the weighted Jaccard similarity function given in Eq. (1). Also, the ratings were calculated by using the DCW equation given in (2). Thus, we were able to successfully implement a context-aware recommendation engine incorporating our proposed suggestions.

5 Conclusion

The aim of this paper has been to find and implement an effective, economic, and efficacious context-aware recommendation system (CARS), and a humble effort to further the research in how contexts can be further utilized to give desired and accurate recommendations. We improved upon the utilization and inclusion of contexts in CARS by assigning individual weights, to each of the features' attributes. To the best of our knowledge, evaluation and usage of individual contexts have not been done before. The eventual, extremely low root-mean-squared error or RMSE value offers confirmation that their inclusion helps improve the results. CARS first needs weighted contexts to be used as an input in the rating model (DCW here). Thus, we analyzed and compared two of the most popular and relevant optimization techniques to be used to provide weights to the contexts. These techniques were the metaheuristic, PSO, and FA. Experimental study proved that PSO was far more suited for contextual recommendations. It outperformed FA by a fair estimate and helped make the recommendation system even better. Once these values from metaheuristic approaches were obtained, using Jaccard similarity, the similarity between the different users was found. Here, we take DCW as our rating model. DCW also been shown to be superior to other rating models as well [3, 4, 13]. Thus using these series of steps, optimal final recommendations were obtained and they had an extremely low RMSE value, thus giving us an efficient CARS.

References

1. D.B., Schilit Theimer, C.A. Brunk, C. Evans, B. Gladish, M. Pazzani, Adaptive interfaces for ubiquitous web access, Comm. ACM 45 (5) (2002) 34–38.
2. G. Adomavicius, T. Alexander, Context-aware recommender systems, in Recommender Systems Handbook, Springer, US, 2011, pp. 217–253.
3. Zheng, Y., Burke, R., Mobasher, B.: Differential context relaxation for context-aware travel recommendation. In: Huemer, C., Lops, P. (eds.) EC-Web 2012. LNBIP, vol. 123, pp. 88–99. Springer, Heidelberg (2012).
4. Zheng, Y., Burke, R., Mobasher, B.: Optimal feature selection for context-aware recommendation using differential relaxation. In: ACM RecSys 2012, Proceedings of the 4th International Workshop on Context-Aware Recommender Systems (CARS 2012). ACM (2012).
5. Zheng, Yong, Robin Burke, and Bamshad Mobasher. "Recommendation with differential context weighting." International Conference on User Modeling, Adaptation, and Personalization. Springer Berlin Heidelberg, 2013.
6. X. S. Yang, "Nature-Inspired Metaheuristic Algorithms", Luniver Press, 2008.
7. Christian Blum, Maria Jos´e Blesa Aguilera, Andrea Roli, Michael Sampels, Hybrid Metaheuristics, An Emerging Approach to Optimization, Springer, 2008.
8. Xiang-yin Meng, Yu-long Hu, Yuan-hang Hou, Wen-quan Wang, The Analysis of Chaotic Particle Swarm Optimization and the Application in Preliminary Design of Ship", International Conference on Mechatronics and Automation, August, 2010.
9. J. Kennedy, R. C. Eberhart, "Particle swarm optimization", IEEE International Conference on Neural Networks, Piscataway, NJ., pp. 942–1948, 1995.

10. Sh. M. Farahani, A. A. Abshouri, B. Nasiri, and M. R. Meybodi, "A Gaussian Firefly Algorithm", International Journal of Machine Learning and Computing, Vol. 1, No. December 2011.
11. Xin-She Yang, Chaos-Enhanced Firefly Algorithm with Automatic Parameter Tuning, International Journal of Swarm Intelligence Research, December 2011.
12. P. M. Pardalos and H. E. Romeijn, Eds., Handbook of Global Optimization, vol. 2, Kluwer Academic Publishers, 2002.
13. Zheng, Yong, Bamshad Mobasher, and Robin Burke. "Integrating context similarity with sparse linear recommendation model." International Conference on User Modeling, Adaptation, and Personalization. Springer International Publishing, 2015.

A Multi-clustering Approach to Achieve Energy Efficiency Using Mobile Sink in WSN

Samaleswari Pr. Nayak, S.C. Rai and Sipali Pradhan

Abstract A wireless sensor network (WSN) consists of a large number of interconnected sensors, which provides unique features to visualize the real world scenario. It explores many opportunities in the field of research due to its wide range of applications in current fields that require survey and periodic monitoring which is inevitable in our daily life. However, the main limitations of such sensors are their resource constrained nature, mainly to conserve battery power for extending the network lifetime. We have proposed an algorithm for energy efficiency in WSN in which mobile sink node is used to operate the routing process considering the shortest path between multiple unequal clusters with reduced energy. This model also ensures non-occurrences of energy hole problems within the network area.

Keywords WSN · CR · MAAM · Mobile sink

1 Introduction

A WSN is a collection of sensors, which are constrained with limited battery power and are capable of performing communication in wireless medium. When multiple sensors are deployed in an area, all of them are expected to sense different types of events for collecting the sensed data from the network [1]. The received data is required to be further transmitted to the neighboring nodes. In wireless environment, data transmission consumes highest amount of energy over other tasks [2].

S.Pr. Nayak (✉)
Department of CSE, Silicon Institute of Technology, Bhubaneswar, India
e-mail: samaleswari.nayak@gmail.com

S.C. Rai
Department of IT, Silicon Institute of Technology, Bhubaneswar, India

S. Pradhan
Department of Computer Science, North Orissa University, Baripada, India

H.S. Behera and D.P. Mohapatra (eds.), *Computational Intelligence in Data Mining*, Advances in Intelligent Systems and Computing 556,
DOI 10.1007/978-981-10-3874-7_75

Every data packet consists of a field that stores the energy and node position by considering the maximum and minimum remaining battery. During data aggregation after receiving all the data from the network, sink gets to know the position of all sensor nodes. Here, while making decisions for the movement, sink chooses the node with the maximum residual battery [3]. Although optimization of energy consumption in areas of a network can expand the lifetime of the network to a certain amount of time, the balancing of energy for entire network cannot be found. Hence we propose an algorithm to increase the lifetime of WSN using sink mobility and multiple clustering concepts [4].

2 Related Work

2.1 Moving Scheme for Mobile Sinks in Wireless Sensor Network

For data collection process in WSN, if complete aggregated data from total network will forward to a mobile sink through multiple hops, then maximum energy will be consumed by the node which is more nearer to the sink for forwarding the data to other nodes or sink, which may create a situation where entire network will fail [4]. Introduction of mobile sink concept may enhance the network lifetime through less energy consumption of individual node. The author has advised to implement two different schemes for implementing sink mobility. Firstly, without having the entire idea about the topology of network the sink can take decision for moving. Secondly, movement decision can be taken by considering different energy distribution among all sensors [4].

The node with highest residual energy is allowed to invest energy for directing data for rest of the nodes to balance the energy distribution among different sensors within its communication range which then becomes the next destination from the nodes that are within its sensing range of two hops and closer to the target node than the destination itself and carry forward in the same way as the decided node [4]. If the devices within two hops are away from the targeted node than the sink, then it will precede the destination. While the sink enters the predefined range of communication of the target, the same principle is applied as that in the single step moving idea to get the position to move around the destination [5].

2.2 A Novel Energy-Efficient Routing Algorithm for Wireless Sensor Network with Sink Mobility

A number of applications of WSN are now an important part of our day-to-day life out of which in maximum situations human intervention [6] is not possible. So the sensors with limited battery power are a great threat to prolong the life of the network. Thus, maximum research work is going on to find different solutions to provide a better idea for maximization of network life. As more energy being consumed during communication process, the author in this article has given an idea to do some clustering, and by using the concept of REQ and REPLY, the sink will become mobile to collect data from each node.

2.3 An Energy Aware Source Routing with Disjoint Multipath Selection for Energy-Efficient Multihop Wireless Adhoc Networks (EASR)

In general, the nodes more closely to the sink consume maximum battery power for extra processing of data. A moving idea proposed by the author in energy-aware sink relocation [7] being followed for mobile sinks in WSNs. Through this paper, they have incorporated the principle of energy-aware transmission range adjustment to tune the communication range of each sensor according to its remaining battery power. The authors have proposed two components i.e., the energy-aware communication range adjustment and the relocation of sink mechanism.

For relocation, it determines four positions of the moving destination along with the calculation of the weight value of each which is evaluated by the path capacity of neighboring nodes for each destination [5]. Then, the EASR method [7] is used to move the sink to the correct position with the maximum value among the four selected positions by following better routing paths to forward the message to increase the lifetime of the network. Also the data transmission is based on overhearing ratio. But if multiple request reply will appear with less overhearing ration, then EASR protocol will fail to communicate. As well as if always a single path will be followed for transmission, then the nodes present in that path will die soon because of maximum energy consumption [8].

2.4 Energy Efficient Zone Divided and Energy Balanced Clustering Routing Protocol (EEZECR)

In this work, it is being considered as the network is divided into multiple zones [9] which is being simplified the design mechanism of the network. It has followed heterogeneity clustering method to solve the hot spot problem. The zone division approach has given improvement in maximization in network lifetime over the clustering approach of LEACH [10]. It has better load balancing in the network due to double cluster head approach which is in result reduces the load from the cluster heads. But the problem in this approach is that the CH collects all data and then sends to the next cluster head of nearest cluster. But collecting data reduces huge energy from the cluster head [11] and also the energy is being reduced from head while sending data to the BS.

3 Proposed Model: MAAM

3.1 Multiple Cluster Formation and Selection of Cluster Representative (CR) and Formation of Chain

In our model, multi-clustered approach with adaptive mobile sink (MAAM), at first the nodes are deployed in the wide range of field. The deployed sensors are divided into suitable number of clusters by using K-medoids algorithm. It first calculates the initial centroid requirements and then gives better, effective, and stable cluster [12]. It also takes less execution time because it eliminates unnecessary distance computation by using previous iteration. It generates stable clusters which improves accuracy.

After the formation of clusters, cluster representative (CR) is chosen for each cluster after every round of data sending, and correspondingly, the chain is being formed. A node is selected randomly from the live nodes for each round. If N is the number of nodes, then by using random function, a node is selected as cluster representative for *i*th round. Randomly selecting representative node also gives benefit likely for those nodes that will die at random locations [13], thus provides robust network. Whenever a node dies the chain reconstruction starts from the extreme nodes from the considered periphery using the greedy approach [14].

3.2 Sending of Data Through Nodes and Aggregation by Cluster Representative

After creation of clusters, data transmission begins as per the fixed schedule. If we assume that each node always has information to forward, then they send as per their given transmission time to the CR [15]. Data aggregation is applied at each CR [16]. For data aggregation, simple model is used which simply reduces the redundant data where if there are x nodes in a cluster and each of the node sends a packet of same length to CR, then at CR there are x packets each of same length. After data aggregation, cluster head produces packets of same length [17]. So the number of packets in the output is a function of number of packets in the input.

4 Energy Consumption Model Using Mobile Sink

The algorithm provides the brief idea of the proposed model.

arr_ch: array containing coordinates of all cluster heads
arr_rem: array containing those cluster heads which doesn't satisfy the conditions.
d_thres: threshold distance

Move_Sink (arr_ch)

1. Find centroid C of all the cluster heads (i.e., arr_ch)
2. Move sink to C
3. if(dist(arr_ch, c) < d_thres)
4. send data to sink
5. else
6. arr_rem = arr_ch
7. Move_Sink(arr_rem)
8. end

The model for receiving and transmitting a k bit message is used where energy consumption occurs during communication, aggregation, storage, and forwarding to the next node.

So we can represent the total energy consumption for communication as

$$E_{tx}(k, d) = k * E_{elec} + k * E_{efs} * d^2, \text{ for } d < d_0 \tag{1}$$

$$E_{tx}(k, d) = k * E_{elec} + k * E_{efs} * d^4, \text{ for } d > d_0 \tag{2}$$

Here, d_0 represents the threshold value based upon which the free space communication channel model or multi-path communication channel model is used.

Also to receive the message, the node consumes

$$E_{rx}(k) = k * E_{elec} \tag{3}$$

where E_{elec} notifies the energy consumption of transmits or receives 1-bit message and E_{efs} denotes the energy consumption for traveling 1-bit message in free space.

Similarly to merge n number of different messages, the total energy consumption will be

$$E_{dx}(k) = n * k * E_{da} \tag{4}$$

Here, E_{da} notifies the energy consumption for merging 1-bit message.

In our proposed model, each node will consume energy based upon the minimum distance from neighboring node or directly from sink.

So the energy consumption by any node for communication can be stated as

$$E_{tx}(k, d) = \min\left(k * E_{elec} + k * E_{efs} * d_{n2n}^2, k * E_{elec} + k * E_{efs} * d_{n2s}^2\right) \quad (5)$$

where d_{n2n} represents distance between node to node and d_{n2s} represents node to sink as sink moves nearer to the node to collect the data. So the energy consumption will be less in this proposed model as distance for transmission will decrease due to the sink movement.

All the cluster representatives (CR) are being identified by sink. Now to find where the sink should move is solved by using the concept of centroid. The centroid of CR is found and sink is placed at that obtained location. Placing sink at the centroid reduces the distance between sink and node and hence helps in transferring the data efficiently. Then, it is checked for each of the CR in the network that whether the distance between corresponding CR and sink is less than the threshold distance. If it is, then considered CR sends the data to the sink otherwise the CR will send the data to the nearby CR. Again, this method is repeated for those CR which have yet not sent their data to the sink.

5 Experimental Results

For the simulation work, we have used MATLAB with network size is 1000 m × 1000 m area. Here, the position of source is fixed but the destination node i.e., the sink node is mobile. The total area is divided into five different unequal clusters using k-medoid clustering algorithm. In Fig. 1, the result has been shown that the proposed model provides best result for five numbers of clusters. Initially, the networks with six clusters have shown better result but by increasing the number of rounds the numbers of dead nodes are also more. But the model shows best result

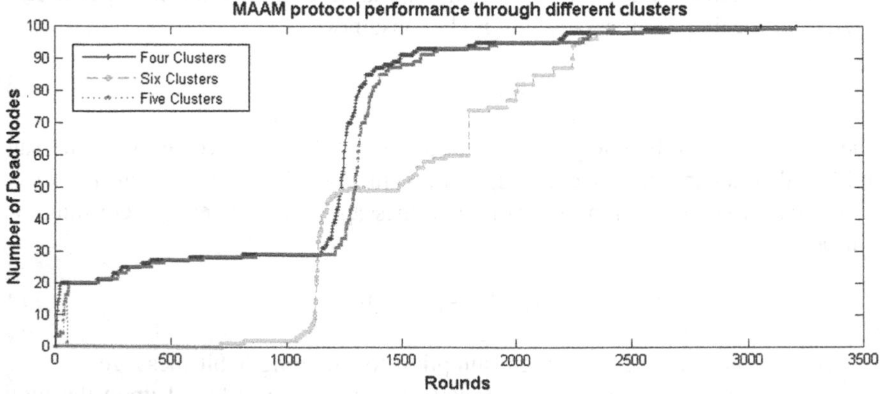

Fig. 1 Performance comparison of MAAM using different clusters

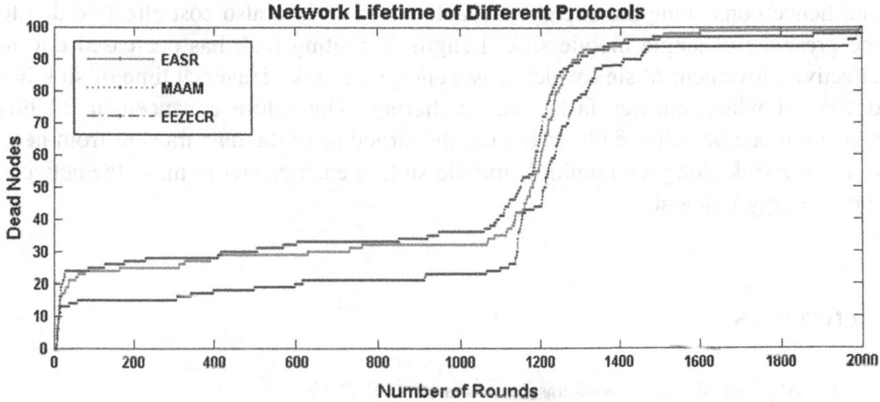

Fig. 2 Network lifetime comparison between MAAM, EASR, and EEZECR protocols

during five total numbers of clusters. The proposed model is also showing better result for four clusters but while the number of rounds will increase the performance degrades on later stage.

From the comparison graph as given in Fig. 2, it is inferred that the proposed model is better than the EASR and EEZECR models. EASR model provides better result than other two models but when the network size will be more it failed to support. But our proposed model is showing better result for large networks where EEZECR protocol is also providing better result than EASR but for larger network the nodes dies rapidly than MAAM. We have experimented by considering 100 homogeneous nodes deployed randomly in the network. As per the result of dead nodes about 30%, EASR has shown better result than other two. Similarly for 60%–80% dead nodes, EASR has given better performance but between EEZECR and MAAM, MAAM has shown approximately better result. But while the network size becomes large, the performance of both EASR and EEZECR has degraded, and MAAM has shown continuous better result with different clusters. So the lifetime of network using MAAM model is more than other two compared models.

6 Conclusion

Extending network lifetime through optimal energy consumption is a challenging task in WSN. Our proposed sink mobility assisted algorithm provides a mechanism to carry out the sink mobile so as to reduce the total number of hop counts to the destination which ensures reduction in consumption of battery power. It provides a stable, better accuracy cluster with significantly less time in comparison with the existing models like EASR and EEZECR. Congestion in “Cluster Representative” has also become reduced which ensures faster message relaying. Data aggregation model reduces redundancy in packets by sending data through active nodes only and hence conserving the energy of inactive nodes. It is also cost effective due to the presence of single mobile sink. Length of routing path has decreased due to effective movement of sink which saves energy of sink. Traversal time of sink has decreased which ensures faster data gathering. The future enhancement of this algorithm can be achieved by changing the structure of the data transfer from head to mobile sink along with multiple mobile sink in each cluster to make the network more energy efficient.

References

1. I.F. Akyildiz, W. Su, Y. Sankarasubramaniam and E. Cayirci, A Survey on Sensor Networks, IEEE Communications Magazine, vol. 40, no. 8, August (2002) 102–114.
2. Y. Gu, D. Bozdag, E. Ekici, F. Ozguner, and C. G. Lee, A Network Lifetime Enhancement Method for Sink Relocation and its Analysis in Wireless Sensors, Oct (2013) 386–395.
3. Kim, Haeyong, Yongho Seok, Nakjung Choi, Yanghee Choi, and Taekyoung Kwon, Optimal Multi-Sink Positioning and Energy-Efficient Routing in Wireless Sensor Networks, In International Conference on Information Networking, Springer Berlin Heidelberg (2005) 264–274.
4. Bi, Yanzhong, Jianwei Niu, Limin Sun, Wei Huangfu, and Yi Sun, Moving Schemes for Mobile Sinks in Wireless Sensor Networks, International Performance, Computing, and Communications Conference IEEE (2007) 101–108.
5. Lindsey, Stephanie, and Cauligi S. Raghavendra, PEGASIS: Power-Efficient Gathering in Sensor Information Systems, Aerospace Conference Proceedings IEEE, (2002) 1125–1130.
6. Jose, Deepa V., and G. Sadashivappa, A Novel Energy Efficient Routing Algorithm For Wireless Sensor Networks Using Sink Mobility, International Journal of Wireless & Mobile Networks (2014) 15–20.
7. Piyush Gupta, P. R. Kumar, Critical Power for Asymptotic Connectivity in Wireless Network, Stochastic Analysis, Control, Optimization and Applications, Birkhauser, Boston, (1999) 547–566.
8. Wang, Chu-Fu, Jau-Der Shih, Bo-Han Pan, and Tin-Yu Wu, A Network Lifetime Enhancement Method For Sink Relocation and Its Analysis in Wireless Sensor Networks, IEEE sensors journal, (2014) 1932–1943.
9. Hwang, Do-Youn, Eui-Hyeok Kwon, and Jae-Sung Lim, EASR: An Energy Aware Source Routing With Disjoint Multipath Selection for Energy-Efficient Multihop Wireless adhoc Networks, In International Conference on Research in Networking, Springer Berlin Heidelberg, (2006) 41–50.
10. Verma, Sandeep, and Kanika Sharma, Energy Efficient Zone Divided and Energy Balanced Clustering Routing Protocol (EEZECR) in Wireless Sensor Network, Circuits and Systems: An International Journal (CSIJ), January 2014.
11. Xiangning F, Yulin S, Improvement on LEACH Protocol of Wireless Sensor Network, International Conference on Sensor Technologies and Applications IEEE (2007) 260–264.
12. Taruna, S., Rekha Kumawat, and G. N. Purohit, Multi-hop Clustering Protocol Using Gateway Nodes in Wireless Sensor Network, International Journal of Wireless & Mobile Networks (2012) 169–176.
13. Samaleswari P. Nayak, Kasturi Dhal, S. C. Rai, and Sateesh K. Pradhan., TIME: Supporting Topology Independent Mobility With Energy Efficient Routing in WSNs, 1st International Conference on Next Generation Computing Technologies (NGCT), IEEE (2015) 350–355.

14. Raval, Gaurang, and Madhuri Bhavsar, Improving Energy Estimation Based Clustering With Energy Threshold for Wireless Sensor Networks, International Journal of Computer Applications (2015) 113–119.
15. Nam, Choon Sung, Young Shin Han, and Dong Ryeol Shin, Multi-hop Routing-Based Optimization of The Number of Cluster-Heads in Wireless Sensor Networks, (2011) 2875–2884.
16. Samaleswari P. Nayak, S. C. Rai, and Sateesh K. Pradhan, MERA: A Multi-clustered Energy Efficient Routing Algorithm in WSN, 14th International Conference on Information Technology (ICIT), IEEE, (2015) 37–42.
17. Kumar, Surender, Manish Prateek, N. J. Ahuja, and Bharat Bhushan, MEECDA: Multihop Energy Efficient Clustering and Data Aggregation Protocol For HWSN, (2014).

Selection of Similarity Function for Context-Aware Recommendation Systems

Aditya Gupta and Kunal Gusain

Abstract Earlier recommendation engines used to work on just the user and item details; however, more recently, users' specific contexts have found an equally significant place as a metric in finding recommendations. The addition of contexts to the mix makes for more personalized suggestions and search for a truly efficient context-aware recommendation system (CARS) continues. Differential context weighting (DCW)-based CARS needs to compute similarities between different users to give recommendations. Our objective is to analyze, compare, and contrast various similarity functions, not only to find the best-suited one but also to implement an efficacious and economic CARS. To optimize the weights in DCW, we use a metaheuristic approach.

Keywords Recommender system • Context • Context-aware recommendation • Similarity functions • Differential context weighting

1 Introduction

Recommendations guide many choices in our daily lives, and thus, the process of recommending is one which merits time and attention. Traditionally, recommendation engines were built using the features of users × items. Over time, a third input called 'context' has also come into the picture, which is nothing but the users' specific data about the choice they made [1]. For a particular user, it could be the time recommendation was made in, the day or even the weather. Contexts could be collected either implicitly by the Web site/app/device or explicitly by asking for users' input. Simple collaborative filtering techniques were not able to handle this

A. Gupta (✉) · K. Gusain
Computer Science Department, Bharati Vidyapeeth's College of Engineering,
Guru Gobind Singh Indraprastha University, Delhi, India
e-mail: adityag95@gmail.com

K. Gusain
e-mail: kunalgusain1995@gmail.com

H.S. Behera and D.P. Mohapatra (eds.), *Computational Intelligence in Data Mining*, Advances in Intelligent Systems and Computing 556,
DOI 10.1007/978-981-10-3874-7_76

increased dimension, users × items × contexts. The increased metric was not the only issue since contexts could be different for the same user at different times, and the collection of all the contexts was not possible, the data often obtained was sparse. Even in dense data, we needed a metric to find ways to implement a successful context-aware recommendation system (CARS). The pioneering work into CARS was first done by Adomavicius et al. [2] since then there has been no looking back.

There are primarily two problems with using the contextual dataset: firstly, the selection of contexts that will be used; secondly, increasing sparsity and dimensionality in the data. Zheng et al. have worked and compared many methodologies which can be used to solve the above problems, such as context-aware matrix factorization (CAMF) [3], contextual sparse linear method (CSLIM) [4], and differential context relaxation (DCR) [5, 6]; Unger et al. have done significant work using autoencoders to tackle the problem of dimensionality [7]. More recently, an improved variant of DCR, which is the differential context weighting (DCW), has also been proposed [8]. Context-aware recommendation system (CARS) although significantly better than traditional engines is still a relatively recent entrant and thus a fascinating research topic; and much more work is needed to be done to arrive at a truly efficient CARS. Out of all the rating models, DCR has proved to be quite efficient; however, its property of taking into account only selective contexts or 'relaxing' of contexts gave rise to DCW, which now gave weights to all the contexts [8]. In this paper, we analyze the working of DCW with the aim to improve its ratings and further the research in CARS. Depending upon different similarity functions, substantially different results can be obtained, and thus, various functions need to be analyzed to arrive at the best possible one. To the best of our knowledge, no work has been done in this regard.

The objective of this paper was to study, compare, and contrast various similarity functions with the aim of finding the best one that can be employed in DCW to obtain the recommendations. The most prominent and relevant functions such as Jaccard, Euclidean, cosine, Manhattan, Dice, and Minkowski are hereby studied. We further found the root-mean-squared error (RMSE) feature and entered the rich contextual dataset available, which is the LDOS-CoMoDa Movie Dataset. To optimize the weights required for DCW, we use the swarm-based metaheuristic [9, 10] technique called particle swarm optimization (PSO) [11, 12]. Experimental results are studied, and recommendations using the best similarity metric are obtained.

2 Methodology

2.1 Rating Model

To find the recommendations, we need to assign ratings based on the data, and this is the core of any CARS. We employ differential context weighting (DCW) [8] to find the ratings because out of the filtering and modeling techniques [2], giving weights to all the contexts improves the recommendation and makes sure that all contexts play a role in finding of the ratings, thus increasing the relevance of the CARS. In DCW, we assign weights to all the features depending upon their importance or relevance in the data. Weights are between numeric 0 and 1 and depend upon the contribution of that feature. Higher weight simply means that the contribution of that feature is more than the one with a comparatively lower weight. Now, we use the similarity function to find the similarities between different users based on their weighted contexts as given by Eq. (1). Here, 'σ' denotes the weights, and 'c' and 'd' are the contexts for the user.

$$sim_w(c, d, \sigma) \tag{1}$$

As in [8], the DCW rating model; if 'P' depicts the predicted rating and user is 'a', then the rating assigned by the user to an item 'i' is shown in Eq. (2). Rating is given by 'r', and 'ρ' is the rating function.

$$P_{a,i,\sigma} = \overline{\rho}(a, \sigma) + \frac{\sum_{u \in N_{a,\sigma}} (\rho(u, i, \sigma) - \overline{\rho}(u, \sigma)) * sim_w(a, u, \sigma)}{\sum_{u \in N_{a,\sigma}} sim_w(a, u, \sigma)} \tag{2}$$

To start, first, we must find the neighbors ('N') that have rated the item 'i' with the context 'c'. Since for the same user, the contexts change over a period of time, thus giving rise to multiple ratings over different contexts, we fuse the maximally similar ones to their neighbors. The formula for selection is given in Eq. (3).

$$N_{a,\sigma} = \{u\colon \max_{r_{u,i,d}} sim_w(c, d, \sigma)\} \tag{3}$$

Once the neighbors have been selected, we must find their individual contributions. For this, we have given weights to all the contexts and thus use the mathematical average of all the ratings given by that neighbor (Eq. (4)).

$$\rho(u, i, \sigma) = \frac{\sum_{r_{u,i,d}} r_{u,i,d} * J(c, d, \sigma)}{\sum_{r_{u,i,d}} J(c, d, \sigma)} \tag{4}$$

DCW rating model is implemented using the equations above.

2.2 *Similarity Function*

Similarity function is the metric used to calculate the similarity between two objects. They usually take two objects as input and give a large positive value for the objects with a high degree of similarity, either negative value or zero for the objects with a low degree of similarity. It can also be defined as the distance between the dimensions or features representing the objects. The similarity functions find applications in recommender systems, clustering, sequence alignment, etc. Similarity is generally measured in the range from 0 to 1. Similarity equals '1' if two objects are identical and '0' if two objects are completely different. There are various similarity functions which are being used these days; few of them which we considered for our experiment are as follows:

1. Euclidean distance,
2. Manhattan similarity,
3. Minkowski similarity,
4. Cosine similarity,
5. Jaccard similarity, and
6. Dice similarity.

All the similarity functions are represented in their modified version, which incorporates the weighted contexts. These weights are shown by 'σ' in the formulas below.

Euclidean Distance

Euclidean distance is described as the distance between two objects in the Euclidean space. Euclidean distance works well when data is dense and continuous. It can also be referred as the length of the path connecting two objects. This distance is calculated using Pythagoras theorem. The function for Euclidean distance is given in Eq. (5).

$$sim(c, d, \sigma) = \left(\sqrt{\sum_{i=1}^{k} \sigma_i * (c_i - d_i)^2} \right)^{-1} \tag{5}$$

Manhattan Distance

Manhattan distance is a metric to calculate the absolute sum of difference between the coordinates of two objects. In this, we are tasked with finding the sum of absolute difference between the coordinates of objects under review. The function of Manhattan distance is given in Eq. (6).

$$sim(c, d, \sigma) = \left(\sum_{i=1}^{n} \sigma[i] * |c[i] - d[i]| \right)^{-1} \tag{6}$$

Minkowski Distance

Minkowski distance can be considered as the generalized metric form of Manhattan distance and Euclidean distance. The function of Minkowski distance is given in Eq. (7).

$$sim(c, d, \sigma) = \left(\sum_{i=1}^{n} \sigma[i] * |c[i] - d[i]|^p \right)^{\frac{1}{p}} \tag{7}$$

The most important parameter in the above equation is p. If p = 1, then the equation behaves as Manhattan distance; if p = 2, then the Equation behaves as Euclidean distance. Similarly, when p equals infinity, then the resulting equation is called Chebyshev distance.

Cosine Similarity
Cosine similarity is the measure of similarity between two vectors which finds the normalized dot product of two vectors. By trying to calculate the cosine similarity, we are effectively trying to figure out the cosine of the angle between two objects. If two vectors have the same orientation, then their cosine similarity is 1; if two vectors are at the right angle, then their cosine similarity is 0. The cosine similarity is generally used in positive vector space, where the result is bounded by [0, 1]. The function cosine similarity is given in Eq. (8).

$$sim(c, d, \sigma) = \cos(\varphi) = \frac{(\sigma \cdot c) \cdot (\sigma \cdot d)}{|\sigma \cdot c||\sigma \cdot d|} \tag{8}$$

Jaccard Similarity
Jaccard similarity or Jaccard index is a similarity measure which is used to calculate the similarity between two finite sample sets. It can be defined as the cardinality of the intersection of two sets divided by the cardinality of the union of two sets. The function Jaccard similarity is given in Eq. (9).

$$sim(c, d, \sigma) = \frac{\sigma \cdot |x \cap y|}{\sigma \cdot |x \cup y|} \tag{9}$$

Dice Similarity
Dice similarity is also known by several other names such as Sorensen index and Dice coefficient. It is used to calculate the similarity between two samples. Dice similarity is very similar to Jaccard similarity. It also does not satisfy the triangle inequality property just like Jaccard index. Dice similarity works well with heterogeneous data and gives less weight to outliers, hence reducing the effect of outliers on the final output. The function Dice similarity is given in Eq. (10).

$$sim(c, d, \sigma) = \frac{2 * \sigma \cdot |x \cap y|}{|x \cdot \sigma^2| + |y \cdot \sigma|^2} \tag{10}$$

2.3 *Optimization*

Once the weighted values have been obtained, DCW finally has to optimize these weights to find the ratings. The optimization technique we employ for this step is particle swarm optimization (PSO). This is a swarm intelligence methodology which is based on the behavior of birds and bees in nature [11, 12]. A nature-based metaheuristic technique—the idea is that just like in nature bees swarm together, similarly, we can make the particles swarm together for optimization—is proposed. All the particles are plotted in what is called the target space, and then, each particle is allowed to swarm or move forward in the space to find the best possible location it can find. Each particle has a definite velocity; that is, it has a defined speed and direction. In PSO, each particle has a fascinating characteristic called the social component of that particle, due to which it tends to attach itself to the best location in its neighborhood. As the particle moves through target space looking for best possible value, and if passes an optimal value, and then cannot find a value better than this one, it can also come back to this previous location. It is working in a step-by-step characteristic feature of any metaheuristic algorithm. PSO is a knowledgeable algorithm, in the sense that at each particular location, all the particles are aware of their environments and have knowledge regarding the path they have taken, their direction, and what speed are they moving by. It has also been shown to have a high convergence rate and is a highly optimal technique for optimization problems like ours in DCW.

3 Results

The dataset used for comparison of similarity function is LDOS-CoMoDa Movies Dataset with data of over thousand movies and fourteen different contexts such as time, day type, season, and location. Once the features have been assigned weights, we use all the six similarity functions to find and compare the necessary similarities. We first calculate the weights that are to be assigned to each context using particle swarm optimization. After calculating the weights for all the contexts, the next step is to run DCW with different similarity functions on our dataset. The accuracy of the similarity functions is measured using root-mean-squared value (RMSE). We calculated the RMSE for all the similarity twice, first time on the test data and the second time on the overall complete data. The results for both are shown in Table 1 and graphically depicted in Figs. 1 and 2, respectively.

Figure 1 shows that Dice similarity performs slightly better than the other 5 functions, and Minkowski similarity function performs the worst.

Table 1 Similarity functions with their respective RMSE values

Similarity functions	RMSE values for test data	RMSE values for entire data
Dice	1.0654	0.9695
Jaccard	1.0523	0.9614
Cosine	1.0539	1.5894
Manhattan	1.0374	0.9369
Minkowski	1.1903	0.9981
Euclidean	1.1912	0.9897

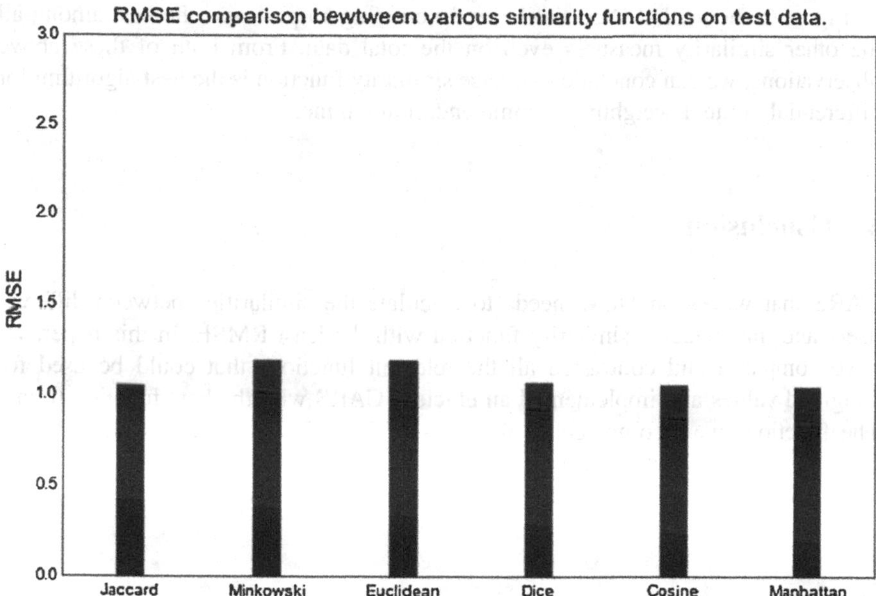

Fig. 1 RMSE comparison between various similarity functions on test data

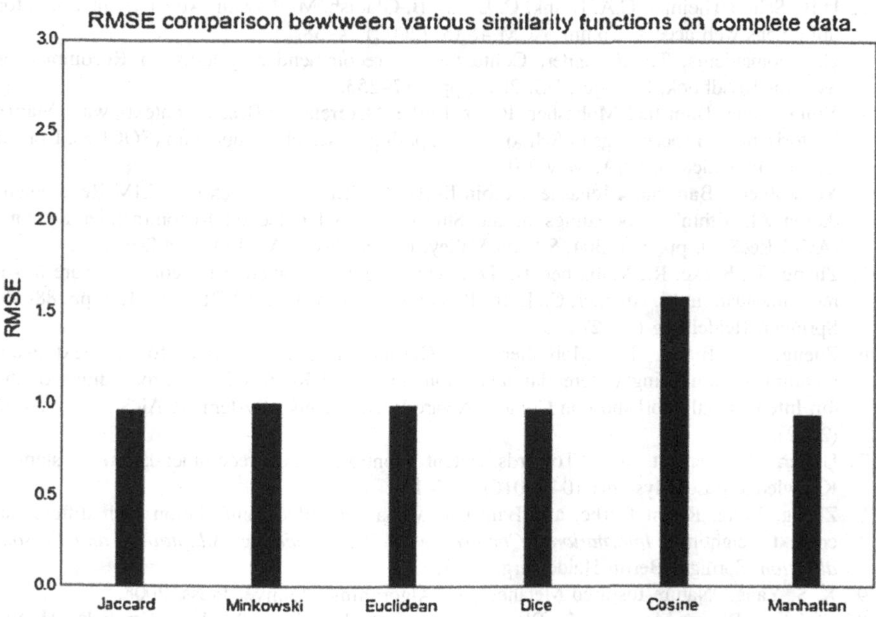

Fig. 2 RMSE comparison between various similarity functions on the entire data

Figure 2 shows that Dice similarity function again performs the best among all the other similarity measures even on the total data. From both of these above observations, we can conclude that Dice similarity function is the best algorithm for differential context weighting recommendation engine.

4 Conclusion

CARS that works on DCW needs to calculate the similarities between different users and thus needs a similarity function with the least RMSE. In this paper, we have compared and contrasted all the relevant functions that could be used for weighted values and implemented an efficient CARS with the best function found. The functions were compared on the basis of RMSE, and it was found that Dice similarity function outperforms all its counterparts. Thus, it is our recommendation that Dice similarity function should be used for DCW-based CARS to obtain the best result.

References

1. D.B., Schilit Theimer, C.A. Brunk, C. Evans, B. Gladish, M. Pazzani, Adaptive interfaces for ubiquitous web access, Comm. ACM 45 (5) (2002) 34–38.
2. G. Adomavicius, T. Alexander, Context-aware recommender systems, in Recommender Systems Handbook, Springer, US, 2011, pp. 217–253.
3. Yong Zheng, Bamshad Mobasher, Robin Burke. "Correlation-Based Context-aware Matrix Factorization", Proceedings of School of Computing Research Symposium (SOCRS), DePaul University, Chicago, USA, May 2015.
4. Yong Zheng, Bamshad Mobasher, Robin Burke. "CSLIM: A Contextual SLIM Recommendation Algorithm", Proceedings of the 8th ACM Conference on Recommender Systems (ACM RecSys), pp. 301–304, Silicon Valley, Foster City, CA, USA, Oct 2014.
5. Zheng, Y., Burke, R., Mobasher, B.: Differential context relaxation for context-aware travel recommendation. In: Huemer, C., Lops, P. (eds.) EC-Web 2012. LNBIP, vol. 123, pp. 88–99. Springer, Heidelberg (2012).
6. Zheng, Y., Burke, R., Mobasher, B.: Optimal feature selection for context-aware recommendation using differential relaxation. In: ACM RecSys 2012, Proceedings of the 4th International Workshop on Context-Aware Recommender Systems (CARS 2012). ACM (2012).
7. Unger, Moshe, et al. "Towards latent context-aware recommendation systems." Knowledge-Based Systems 104 (2016): 165–178.
8. Zheng, Yong, Robin Burke, and Bamshad Mobasher. "Recommendation with differential context weighting." *International Conference on User Modeling, Adaptation, and Personalization*. Springer Berlin Heidelberg, 2013.
9. X. S. Yang, "Nature-Inspired Metaheuristic Algorithms", Luniver Press, 2008.
10. Christian Blum, Maria Jos´e Blesa Aguilera, Andrea Roli, Michael Sampels, Hybrid Metaheuristics, An Emerging Approach to Optimization, Springer, 2008.

11. Xiang-yin Meng, Yu-long Hu, Yuan-hang Hou, Wen-quan Wang, The Analysis of Chaotic Particle Swarm Optimization and the Application in Preliminary Design of Ship", International Conference on Mechatronics and Automation, August, 2010.
12. J. Kennedy, R. C. Eberhart, "Particle swarm optimization", IEEE International Conference on Neural Networks, Piscataway, NJ., pp. 942–1948, 1995.

Medical Dataset Classification Using k-NN and Genetic Algorithm

Santosh Kumar and G. Sahoo

Abstract This paper proposes a hybrid technique that applies artificial bee colony (ABC) algorithm for the feature selection and combined k-nearest neighbor (k-NN) with genetic algorithm (GA) used for effective classification. The aim of this paper was to select the finest features including the elimination of the insignificant features of the datasets that severely affect the classification accuracy. The proposed approach used in heart disease and diabetes diagnosis, which has higher impact rate on reducing quality of life throughout the world, is developed. The datasets including heart disease, diabetes, and hepatitis are taken from UCI repository and evaluated by the proposed technique. The classification accuracy is achieved by 10-fold cross-validation. Experimental results show the higher accuracy of our proposed algorithm compared to other existing systems.

Keywords Artificial bee colony (ABC) · k-nearest neighbor (k-NN) · Genetic algorithm (GA) · Heart disease

1 Introduction

Data mining is the method of digging out significant information and knowledge from huge amount of data. It provides popular technique commonly known as classification, association, and clustering involved in knowledge extraction [1]. In classification, class attributes are selected based on their predefined class and finally class attributes involved in construction of classifier. In medical domain, quick and cost-effective treatment are one of the major challenges to provide quality of services like

S. Kumar (✉) · G. Sahoo
Department of Computer Science & Engineering, Birla Institute of Technology,
Mesra, Ranchi 835215, Jharkhand, India
e-mail: san77j@gmail.com

G. Sahoo
e-mail: gsahoo@bitmesra.ac.in

H.S. Behera and D.P. Mohapatra (eds.), *Computational Intelligence in Data Mining*, Advances in Intelligent Systems and Computing 556,
DOI 10.1007/978-981-10-3874-7_77

diagnose patients correctly. Many healthcare organizations need a decision support system to identify the disease on time in cost-effective manner.

k-nearest neighbor is a distinctively popular and real classification technique that classifies the training samples themselves. One of the best commonly used simple classification techniques is the nearest neighbor (NN) method that classifies a new case into the class of its nearest neighbor case [2].

Genetic algorithm is the branch of evolutionary computation that is formulated on the basis of Charles Darwin's principles of evolution [3]. Genetic algorithms and other evolution algorithms [4, 5] have been utilized in various complex optimization and simulation problems because of their powerful search and optimization capabilities.

Genetic algorithms have been used with various machine learning methods to optimize weighting properties of the method. Since our research is based on the nearest neighbor and genetic algorithm, and its implication in classification process, we concentrate on related works where GAs have been applied only with the *k*-nearest neighbor method.

1.1 Related Work

The proposed typical model has been verified on medical datasets which is affected more lives in our society and is commonly seen these days. We have tested the developed system on heart disease, diabetes, and hepatitis datasets.

Kelly and Davis [6] combined the GA with a weighted *k*-nearest neighbor (*wk*-NN) method in the algorithm called GA-WKNN in order to find a single attribute weight vector that would improve the classification results of the *wk*-NN. A similar kind of approach was used in [7] where GA was combined with the *wk*-NN and a parallel processing environment in order to optimize classification of large datasets. In both studies, a set of real-valued weights for attributes to discriminate all classes of data were achieved as a result after GA runs. The study of Hussein et al. [8] showed that GA can be applied successfully in setting a real-valued weight set for 1-NN classifier, but the improvement of accuracy happened at the cost of raised processing time. Results showed that GA methods combining the *wk*-NN outperformed the basic *k*-NN [6–8]. However, a single set of weights for all classes is not always the best solution because attributes have a different effect on classes. Therefore, solutions for searching for a weight for each class and attribute have been developed.

These medical data have been our test data in our previous researches [9]. Deekshatulu also applied genetic search for the attribute selection and GA with *k*-nearest neighbor method to perform classification for better accuracy [10]. Uzer et al. [11] applied ABCFS method for the feature selection and SVM algorithm for hepatitis and diabetes disease classification with 94.92% accuracy. Aibinu et al. [12] classify the diabetes disease by NN method with 81.28% classification accuracy.

In this innovative work, we have offered a hybrid classification algorithm which combines k-NN and GA to classify the heart disease, diabetes, and hepatitis for the diagnosis purpose.

This paper is structured as follows: In Sect. 2, we cover the feature selection method, basics of k-NN algorithm, genetic algorithm, and the performance evaluation strategies. Section 3 explained the proposed algorithm, and its methodology and outcomes are discussed in Sect. 4 with final remarks in Sect. 5.

2 Basic Concepts

2.1 Feature Selection

A feature selection method reduces the dimensionality of the original feature space $[Y_1, Y_2, \ldots, Y_n]$ to a lower-dimensional space by selecting a small significant feature set.

$$\begin{bmatrix} Y_1 \\ Y_2 \\ Y_3 \\ \vdots \\ Y_n \end{bmatrix} \rightarrow \begin{bmatrix} Yi_1 \\ Yi_2 \\ Yi_3 \\ \vdots \\ Yi_m \end{bmatrix} \quad m < n \tag{1}$$

The outcomes in both reduced computational time and increased classification accuracy. The three most popular feature selection techniques are categorized as follows: filter, wrapper, and embedded methods [13]. A commonly applied wrapper method is ABC algorithm [14], as an optimization technique, which simulates the intelligent foraging habits of honey bees.

Artificial Bee Colony. (ABC) is the newest among the nature-motivated algorithm defined by Karaboga [15] in 2005, inspired by the intelligent habits of honey bees. Although the performance of different optimization algorithm is dependent on applications, some recent works demonstrate that the artificial bee colony is more fast than either GA or particle swarm optimization (PSO) solving certain problems [16]. Additionally, ABC has demonstrated an ability to attack problems with a lot of variables (high-dimensional problems) [17]. ABC feature selection pseudocode is listed in Fig. 1.

2.2 k-Nearest Neighbor Classifier

k-nearest neighbor [18] classification algorithm is as follows: when a test sample (unknown sample) is given, firstly search the pattern space to find out the k training

```
Begin
  InitPopulation()
  While remain iterations do
    Select sites for the local search
    Recruit bees for the selected sites and to evaluate fitness
    Select the bee with the best fitness
    Assign the remaining bees to looking for randomly
    Evaluate the fitness of remaining bees
    UpdateOptimum()
  EndWhile
  Return BestSolution
End
```

Fig. 1 ABC pseudocode

samples (known samples) which are closest to the test samples, namely *k*-nearest neighbors, and then count the selected *k*-nearest neighbors.

Euclidean distance is used to calculate the distance between the test sample and all the training samples. The formula is

$$distance(X, Y) = \sqrt{\sum_{i=1}^{N} (x_i - y_i)^2} \tag{2}$$

where *X* is a test sample and *Y* is a training sample. The following steps show the functionality of the algorithm:

1. *k* value is evaluated based on iterative experiment and best result is chosen and stored in sample space.
2. Distance is calculated for this step by Euclidean and Manhattan distances.
3. Chosen *k* value is sorted in increasing order by its distance and minimum *k* distance is finally taken.
4. Existing classes of *k* nearest data are identified.
5. In the last and final step, identified *k* classes which have maximum class ratio are taken.

k-NN has encountered many drawbacks during its application such as low efficiency and dependency on the selection of good values of *k*.

2.3 Genetic Algorithm

The elementary idea of the genetic algorithm is the following: In the beginning, a population of individuals is formed either randomly or with information about the application domain. Traditionally, a binary representation of the individuals has been

Table 1 Genetic algorithm parameter

Genetic algorithm	Parameters values
Crossover rate	0.6
Mutation rate	0.033
Population size	21
Generation	20
Elitism	Yes

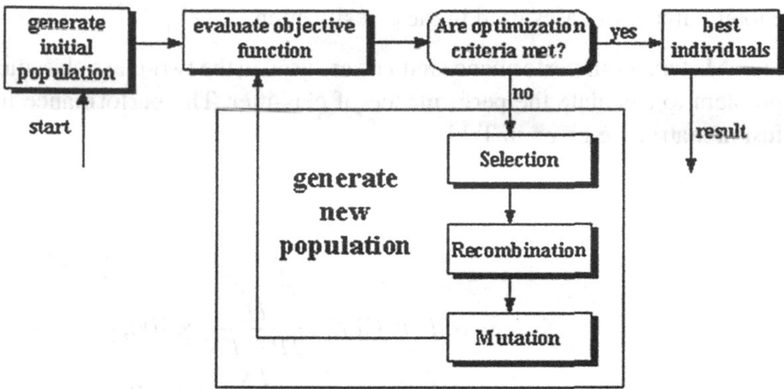

Fig. 2 Structure of a simple genetic algorithm

used, but in multidimensional and numerical problems real-valued representation are nowadays used [4].

In respective generation, the individuals of the population are calculated with an objective evaluation function, thus giving the individual its fitness rate. A selection method is used to find the fittest individuals for a new population. Some individuals of the new population go through the reproduction by means of crossover and mutation. The parameters used in the GA runs are described in Table 1, and construction of GA is shown in Fig. 2.

2.4 Performance Evaluation

We have applied four criteria for performance analysis on UCI datasets [19]. These criteria are 10-fold cross-validation analysis of specificity and sensitivity, classification accuracy, and confusion matrix.

Classification Accuracy. In our experiments, the classification accuracies showed the percentage of all correctly classified cases within the dataset are the following:

$$Accuracy = 100\frac{t_{pos}}{n_{cases}}\% \tag{3}$$

Table 2 Confusion matrix

Predicted class		
Actual class	Positive	Negative
Positive	TP	FN
Negative	FP	TN

where t_{pos} was the total number of cases correctly classified within classes and n_{cases} was the total number of cases used in the classification.

Confusion Matrix. Four performance indices are used in the two-class classification-based problem to calculate the performance of classifier. The performance indices of confusion matrix are given in Table 2.

Analysis of Sensitivity and Specificity. The following formulas (4) were applied to calculate sensitivity, specificity, positive predictive value, and negative predictive value in percentage.

$$\begin{aligned} Sensitivity\ (\%) &= \frac{TP}{TP+FN} \times 100, \\ Specificity\ (\%) &= \frac{TN}{TN+FP} \times 100, \\ Positive\ predictive\ value\ (\%) &= \frac{TP}{TP+FP} \times 100, \\ Negative\ predictive\ value\ (\%) &= \frac{TN}{TN+FN} \times 100. \end{aligned} \tag{4}$$

***k*-Fold Cross-validation**. *k*-fold cross-validation is used for the test outcomes to be more valuable [20]. In *k*-fold cross-validation original samples are divided into test and training datasets as *k* subsamples and $k-1$ subsamples, respectively. A process is repeated *k* times with each of *k* subsamples, and average value of resulting *k* gives test accuracy of the certain classifier.

3 Proposed Method

Poorly distinguished features of the dataset affect the classification accuracy and decrease the performance of the system in terms of computational speed. In this proposed system, less distinctive features are finally eliminated by the feature selection algorithm. The compactness of dataset improves the performance of classifier and increases the system rating too.

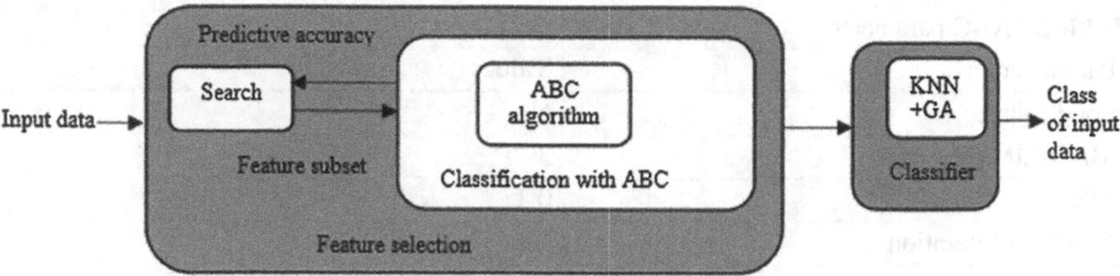

Fig. 3 Block diagram of the proposed method

Proposed algorithm

Step 1) load the data set
Step 2) Apply Artificial bee colony (ABC) algorithm on the data set
Step 3) attributes are ranked based on their value
Step 4) select the subset of higher ranked attributes
Step 5) Apply (KNN+GA) on the subset of attributes that maximizes classification accuracy
Step 6) calculate accuracy of the classifier, which measures the ability of the classifier to correctly classify unknown sample.

Fig. 4 Proposed algorithm

Our proposed method has categorized in two parts that is depicted in Fig. 3. In the first part, classification with ABC is used for feature selection, and hence, effective method for feature selection was developed. At the second phase, *k*-NN+GA classifier is applied on the reduced dataset to calculate the accuracy rate. The accuracy rate and consistency was cross-determined by 10-fold cross-validation process.

Proposed algorithm is shown in Fig. 4, and steps 1–4 deal with feature selection and their individual ranking. Step 5 is applied to construct the classifier, and step 6 records the accuracy of the classifier. In our study, ABC and *k*-NN+GA systems are deployed to solve the classification problem of medical datasets such as heart disease dataset, diabetes dataset, and hepatitis dataset.

3.1 Feature Selection with ABC

In this system, best feature subset is identified by ABC classification and is given in Fig. 1. Datasets divided into train data and test data have taken from the whole datasets. The train data contain 75% instances taken from whole datasets and 25% in case of test data.

ABC algorithm is applied on each of the classes of the train dataset, which has been modified by classification. The best feature vectors are identified as a food that is representing the classes of training set. The best chosen features' class is representing the test dataset that is based on food value accuracy. The ABC parameter which

Table 3 ABC parameter

Parameters	Values
Food source	N
MAX LIMIT	3
MR	0.1
Number of iteration	100

Table 4 List of datasets

Databases	Number of classes	Sample	Number of features	Number of selected features
Hepatitis	2	155	19	11
Diabetes	2	768	8	6
Heart disease	2	303	13	7

Table 5 *k*-NN parameter

Parameter	Values
k	1, 2, 3, …, N
Crossvalidate	True
Debug	True
Distance weighting	k = 1
Mean squared	True
No normalization	False

Table 6 Performance analysis of the classification for the medical datasets

Performance criteria	Hepatitis dataset	Diabetes dataset	Heart disease dataset
Classification accuracy (%)	95.94	80.20	98.84
Sensitivity (%)	98.14	89.92	98.22
Specificity (%)	92.24	59.92	68.14
Positive predictive value (%)	97.93	90.62	98.20
Negative predictive value (%)	92.24	76.40	94.12

is used for experimental work has listed in Table 3. The datasets including heart disease, diabetes, and hepatitis are used for evaluating *k*-NN+GA performance, and their features are listed in Table 4.

Table 7 Accuracy assessment with other algorithm

Dataset name	k-NN+GA (our proposed method)	NN+PCA	GA+NN	GA+SVM
Hepatitis	95.94	94.05	93.7	92.59
Diabetics	80.20	80.08	83.3	82.50
Heart disease	98.84	98.14	97.70	96.80

Table 8 Accuracy evaluation with other existing techniques

Author	Techniques	Accuracy (%)	Our study
Uzer (2013) [hepatitis] [11]	ABCFS+SVM	94.92	95.94
Aibinu (2011) [diabetes] [12]	NN	81.28	82.20
Deekshatulu (2013) [heart] [10]	k-NN+GA	95.73	98.84

3.2 k-Nearest Neighbor and GA Parameters

The 10-fold cross-validation of the classifier shows its reliability and the experimental parameters used by training process has been listed in Tables 1 and 5.

4 Result and Discussion

ABC and *k*-NN+GA methods are developed to test on hepatitis, heart disease, and diabetes datasets are defined in Fig. 4. The test result shows the classification accuracy measured by 10-fold cross-validation. The experimental details and performance of the proposed algorithm are listed in Table 6.

The result shows that our developed system works well in terms of accuracy and correctness rate evaluated by 10-fold cross-validation. The comparison is made on 3 datasets, namely hepatitis, heart disease, and diabetic datasets; other algorithms are listed in Table 7. The evaluation analysis of our results with literature-reviewed techniques is listed in Table 8.

5 Conclusion

This study was designed to diagnose the heart and diabetic diseases. In our work, ABC algorithm has been deployed for the feature selection. In feature selection method, redundant and unimportant features are eliminated from training datasets that is reflected in our classification accuracy. We have applied k-nearest neighbor combined with the genetic algorithm on datasets that are subjected to feature selection for the classification. Each dataset is classified using the k-NN+GA classifier and the 10-fold cross-validation is conducted for the performance analysis of classifier. The outcomes compared with the listed literature review articles showed that our proposed system accuracy is 95.94%, 80.20%, and 98.84% for hepatitis, diabetes, and heart dataset, respectively. The experimental outcomes show the effectiveness of our proposed methods and their high accuracy than some existing methods makes it very promising for medical practitioner.

References

1. Han, Jiawei, Jian Pei, and Micheline Kamber.: Data mining: concepts and techniques. Elsevier, (2011)
2. T. M. Cover and P. E. Hart.: Nearest neighbor pattern classification. IEEE Transactions on Information Theory, vol. 13, no. 1, pp. 2127 (1967)
3. Goldberg, David E.: Genetic algorithms. Pearson Education India (2006)
4. Z. Michalewicz.: Genetic Algorithms + Data Structures = Evolution Programs. Springer, Berlin, Germany (1992)
5. A. E. Eiben and J. E. Smith.: Introduction to Evolutionary Computing. Springer, Berlin, Germany (2003)
6. Kelly Jr, James D., and Lawrence Davis.: A Hybrid Genetic Algorithm for Classification. IJCAI, vol. 91, pp. 645–650 (1991)
7. Punch III, William F., Erik D. Goodman, et al.: Further Research on Feature Selection and Classification Using Genetic Algorithms. ICGA, pp. 557–564 (1993)
8. Hussein, Faten, Nawwaf Kharma, and Rabab Ward.: Genetic algorithms for feature selection and weighting, a review and study. Document Analysis and Recognition, 2001. Proceedings. Sixth International Conference, pp. 1240–1244. IEEE (2001)
9. Kumar, Santosh, and G. Sahoo.: Classification of Heart Disease Using Nave Bayes and Genetic Algorithm. Computational Intelligence in Data Mining-Volume 2, pp. 269–282. Springer India (2015)
10. Deekshatulu, B. L., and Priti Chandra.: Classification of heart disease using k-nearest neighbor and genetic algorithm. Procedia Technology 10, pp. 85–94 (2013)
11. Uzer, Mustafa Serter, Nihat Yilmaz, and Onur Inan.: Feature selection method based on artificial bee colony algorithm and support vector machines for medical datasets classification. The Scientific World Journal 2013 (2013)
12. Aibinu, A. M. et al.: A novel signal diagnosis technique using pseudo complex-valued autoregressive technique. Expert Systems with Applications, 38(8), 9063–9069 (2011)
13. Saeys, Y., Inza, I., and Larranaga, P.: A review of feature selection techniques in bioinformatics. Bioinformatics, 23(19), pp. 2507–2517 (2007)
14. Tan, Feng et al.: Improving feature subset selection using a genetic algorithm for microarray gene expression data. IEEE International Conference on Evolutionary Computation, pp. 2529–2534 (2006)
15. Karaboga, Dervis.: An idea based on honey bee swarm for numerical optimization. Vol. 200. Technical report-tr06, Erciyes University, engineering faculty, computer engineering department (2005)
16. Karaboga, Dervis, and Bahriye Basturk.: A powerful and efficient algorithm for numerical function optimization: artificial bee colony (ABC) algorithm. Journal of global optimization 39, no. 3, pp. 459–471 (2007)

17. Akay, Bahriye, and Dervis Karaboga.: Parameter tuning for the artificial bee colony algorithm. International Conference on Computational Collective Intelligence, pp. 608–619. Springer Berlin Heidelberg (2009)
18. Bishop, Christopher M.: Pattern recognition. Machine Learning 128 (2006)
19. Blake, C., E. Keogh, and C. J. Merz.: UCI repository of machine learning databases. Irvine, CA: Department of Information and Computer Science, University of California (1998)
20. Franois, Damien et al.: Resampling methods for parameter-free and robust feature selection with mutual information. Neurocomputing 70, no. 7, pp. 1276–1288 (2007)

Analysis of Static Power System Security with Support Vector Machine

B. Seshasai, A. Santhi, Ch Jagan Mohana Rao, B. Manmadha Rao and G.T. Chandra Sekhar

Abstract Security analysis is the task of evaluating security and reliability limits of the power system, up to what level the system is secure. Power system security is divided into four classes, namely secure, critically secure, insecure, and highly insecure, depending on the value of security index. A multi-class support vector machine (SVM) classifier algorithm is used, in this paper, to categorize the patterns. These patterns are generated at different generating and loading conditions for IEEE 6 bus, IEEE 14 bus, and New England 39 bus systems by Newton–Raphson load flow method for line outage contingencies. The main target is to give a forewarning or hint to the system operator at security level which helps to actuate requisite regulating actions at the suitable time, to put a stop to the system collapse.

Keywords Security analysis · Support vector machine (SVM) · Classifier · Newton–Raphson method · Contingency

---

B. Seshasai · A. Santhi · C.J.M. Rao · B.M. Rao · G.T. Chandra Sekhar (✉)
Sri Sivani College of Engineering, Chilakapalem, Srikakulam, AP, India
e-mail: gtchsekhar@gmail.com

B. Seshasai
e-mail: bseshasai211@gmail.com

A. Santhi
e-mail: santhi.dunna@gmail.com

C.J.M. Rao
e-mail: ch.jagan211@gmail.com

B.M. Rao
e-mail: lovelymanu207@gmail.com

H.S. Behera and D.P. Mohapatra (eds.), *Computational Intelligence in Data Mining*, Advances in Intelligent Systems and Computing 556,
DOI 10.1007/978-981-10-3874-7_78

1 Introduction

Electricity plays a vital role in one of the quick developing countries like India. Nowadays, power system security is the prime electrical issue all over the world. Serious perturbations can cause the power system to an unpleasant state. Realizing the power system limits or boundaries may help the security-level system operator to take obligatory action preventing the system collapse. Hence, potent control of power systems [1] is needed for expeditious security assessment [2–4]. Static security is the capacity or ability of the power system to withstand unexpected contingencies to reach steady-state operating point without transgressing the constraints of power system [5, 6]. A powerful tool is, therefore, required to classify and assess the power system security level with high accuracy in minimum time. Patterns are generated for IEEE 6 bus, IEEE 14 bus, and New England 39 bus systems. It is necessary to run load flow of each system to generate these patterns. After running load flow, the maximum MVA limits of each line and the maximum and minimum voltage limits of each bus of each system are calculated. Maximum MVA limit will be the maximum value of the thermal limit which the line could withstand. Support vector machine (SVM) is a tool, which is being used as a solution for pattern classification problem [5]. Today, pattern classification is a procuring progress in troubleshooting power system [6, 7]. The classification function is designed depending on the training set.

2 Power System Security Assessment

Security analysis is the testing process of power system safety limits, up to what extent it is secure. In this paper, the main concentration is on static security assessment (SSA). The security evaluation process includes set of contingencies, namely generator/line outage, system phase faults, and sudden changes in the load. The line outage contingency is taken in this paper for testing the system security. Power system security is divided into four classes, viz. fully secure, partially secure, insecure, and terribly insecure, based on the value of security index [2]. The static security level is defined based on the computation of a term called Static Security Index (SSI).

The Static Security Index (SSI) is computed by evaluating the Line Overload Index (LOI) for each branch and Voltage Deviation Index (VDI) for each bus. The equations of LOI, VDI, and SSI are shown below which are represented as (1), (2), and (3), respectively:

$$LOI_{km} = \begin{cases} \frac{S_{km} - S_{km}^{\max}}{S_{km}} * 100, & if \quad S_{km} > S_{km}^{\max} \\ 0, & if \quad S_{km} < S_{km}^{\max} \end{cases} \tag{1}$$

where S_{km} is the MVA flow representation of branch km and $S^{max}{}_{km}$ is the representation of mega volt ampere flow that the branch km could withstand.

$$VDI_k = \begin{cases} \frac{|V_k^{min}| - |V_k|}{|V_k^{min}|} * 100, & if \ |V_k| < |V_k^{min}| \\ \frac{|V_k| - |V_k^{max}|}{|V_k^{max}|} * 100, & if \ |V_k| > |V_k^{max}| \\ 0, & Otherwise \end{cases} \quad (2)$$

where V_k^{min} and V_k^{max} are the minimum and maximum voltage limits, respectively, and V_k is the bus voltage magnitude of bus k.

$$SSI = \frac{W_1 \sum_{i=1}^{N_l+N_t} LOI_i + W_2 \sum_{i=1}^{N_b} VDI_i}{N_l + N_t + N_b} \quad (3)$$

where the number of transmission lines, number of transformers, and no of buses are denoted by N_l, N_t, and N_b, respectively.

3 Proposed System

The proposed system consists of two stages:

(A) static security assessment [7] and (B) SVM-based pattern classification [6] as shown in Fig. 1.

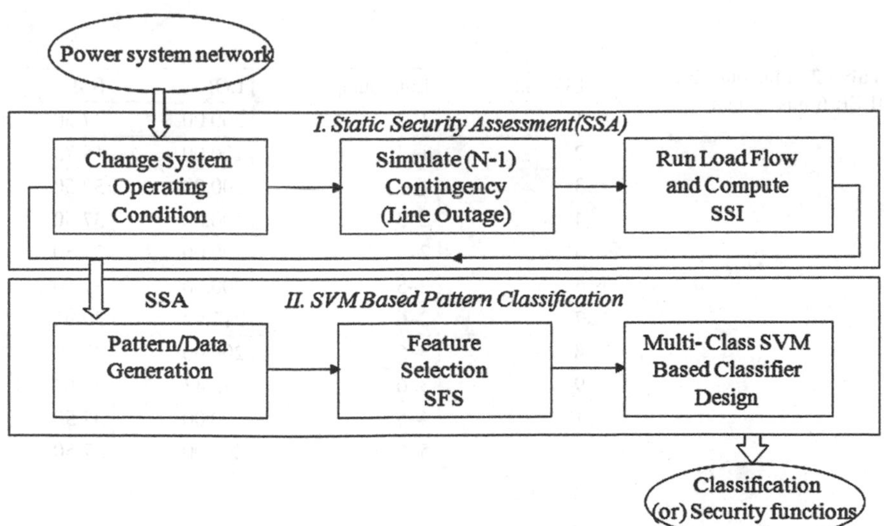

Fig. 1 SVM classifier design for static security assessment

A. **Static Security Assessment**

Static security is the capacity to reach a steady-state operating point which is capable of withstanding the system operating constraints following a contingency. Feature selection (FS) is the selection process of a subset of pattern attributes called *'features,'* removing the unwanted and unrelated variables for building robust learning models [2]. The feature selection block reduces the huge dimension of pattern vector. The classifier design block uses a suitable learning algorithm and develops the classifier model based on a training data set [8]. In this stage, a power system network is taken and the operating conditions are changed to test the system. Each system consisting N lines is applied to a line outage contingency, hence named as N-1 line outage contingency. Numbers of contingency conditions have been applied to Newton–Raphson load flow method to obtain Static Security Index (SSI) at each condition. The class labels have been categorized depending on different values of SSI which is tabulated in Table 1. Line outage of each line for an IEEE 6 bus system is shown in Table 2.

It is clear from the above table that the 6 bus system is much affected for a 3–6 line outage contingency. Number of lines in a 6 bus system is taken as 11. Similarly, the line outage contingency is applied to the IEEE 14 bus system to know for

Table 1 Class labels for static security assessment

Class	SSA
Class A: static fully secure	SSI = 0
Class B: static partially secure	SSI > 0 & SSI ≤ 5
Class C: static insecure	SSI > 5 & SSI ≤ 15
Class D: static terribly insecure	SSI > 15

Table 2 Line outage for IEEE 6 bus system

Line. no	Line outage	LOI	SSI
1	1–2	200.00	37.50
2	1–4	100.00	18.75
3	1–5	200.00	37.50
4	2–3	200.00	37.50
5	2–4	200.00	37.50
6	2–5	200.00	37.50
7	2–6	212.92	39.92
8	3–5	200.00	37.50
9	3–6	243.43	45.64
10	4–5	200.00	37.50
11	5–6	200.00	37.50

which line outage contingency the system is mostly affected. The numbers of lines in 14 bus system are more, and hence, the line outage that affects the system is directly given as in Table 3.

B. SVM-based Pattern classification

Patterns are generated for IEEE 6 bus and IEEE 14 bus systems by running at different line outage contingencies. To generate these patterns, Newton–Raphson load flow methods are applied to each system. Voltage deviation of each bus and thermal violation of each branch are calculated to find SSI. These generated patters are classified by using SVM classifier.

Observations

1. SSI is high, when outage of line consisting generators.
2. SSI is less, when outage of line consisting high load.
3. Overload of the line reduces by increasing the number of lines.

The single-line diagram of IEEE 6 bus system is shown in Fig. 2 with the respected generated patterns. These patterns are categorized according to the class labels of SSI as shown in Table 4.

Table 3 Line outage for IEEE 14 bus system

Line. no	Line outage	LOI	SSI
7	6–12	142.3362	31.82

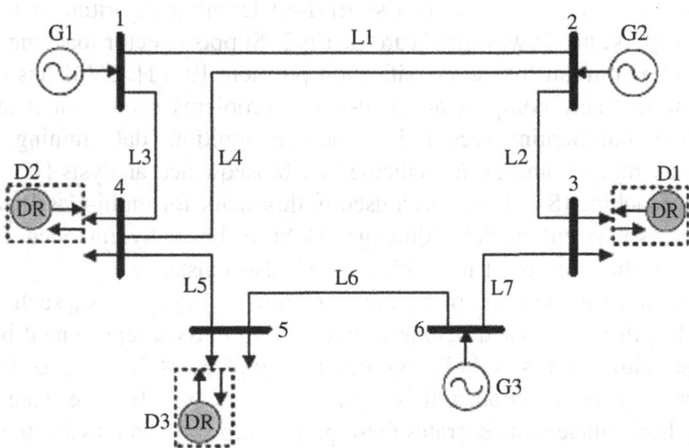

Fig. 2 Single-line diagram of IEEE-6 bus system

Table 4 SSI for IEEE—6 bus system

Class	VDI	LOI	SSI
Class A	0	0	0
Class A	0	0	0
Class A	0	0	0
Class A	0	0	0
Class A	0	0	0
Class B	5.4127	0	0.6388
Class B	8.6318	0	1.0155
Class B	0	17.9273	3.1636
Class B	0	12.6149	2.2262
Class B	0	28.1761	4.9722
Class C	0	31.2023	5.5063
Class C	0.2267	66.0729	12.417
Class C	6.5271	34.5286	14.0689
Class C	2.8216	82.1016	7.9073
Class C	0	100	13.6497
Class D	0	147.9216	27.7353
Class D	0.019	100	20.0025
Class D	0	138.3153	25.9341
Class D	0	100	20
Class D	0	100	17.6471

4 SVM Overview

Support vector machine (SVM) is a supervised learning algorithm developed by Vladimir Vapnik, and it was first heard in 1992. Support vector machine (SVM) is used to find a solution for the classification problem [9–11]. SVM has successful applications in many complex and real-world problems such as text and image classification, handwriting recognition, face recognition, data mining, bioinformatics, stock market, and even medicine and biosequence analysis [12, 13]. Support vector machine (SVM) has been used in this work for multi-classification task in security assessment model. Although SVM is basically intended for binary classification, the concept of multi-class SVM also exists.

Consider an example of set of n points (vectors): $x_1, x_2, \ldots x_n$ such that x_i is a vector of length m, and each belong to one of two classes represented by +1 and −1. So the training set is $(x_1, y1), (x_2, y2), (x_n, y_n), x_i \in R^m, y_i \in \{+1, -1\}$. Hence, the decision function will be $f(x) = \text{sign}\ (w.\ x + b)$. We want to find a separating hyperplane that separates these points into the two classes: the positives (class +1) and the negatives (class −1). (Assuming that they are linearly separable.) The separating hyperplane for the above considered input and output patterns is shown in Fig. 3. But there are many possibilities for such hyperplanes, which is shown in Fig. 4. The selection of optimal hyperplane by using SVM is also shown in Fig. 5.

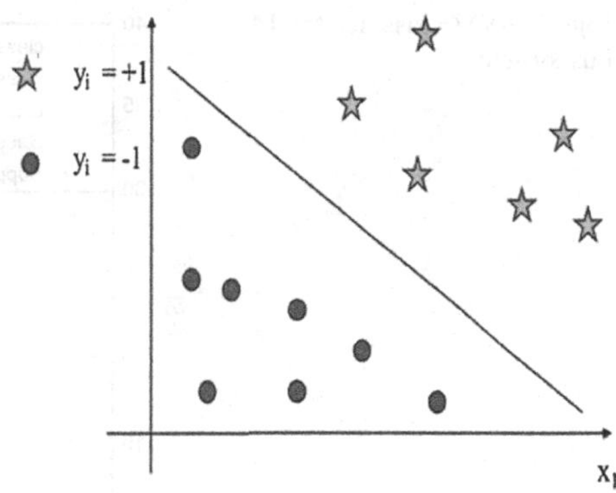

Fig. 3 Hyperplane for linearly separable patterns

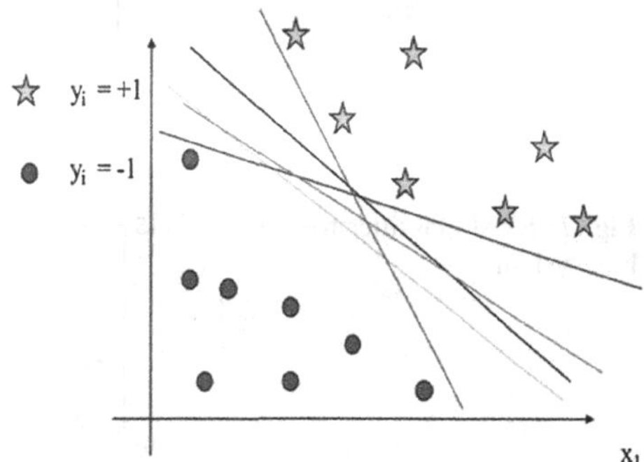

Fig. 4 All possible hyperplanes

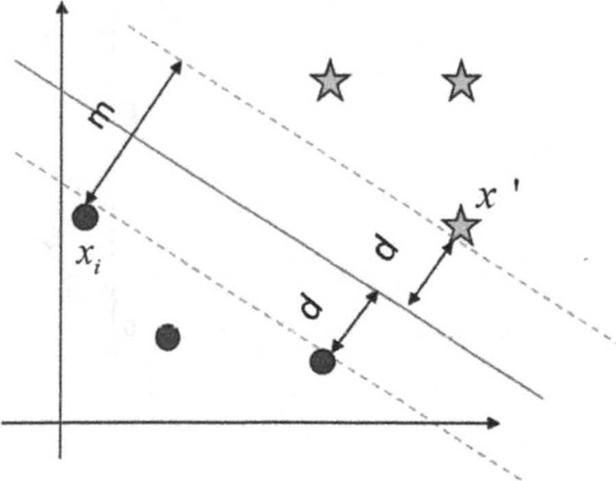

Fig. 5 Choosing optimal hyperplane

5 Results and Discussion

The generated patterns are classified by using SVM classifier. The SVM classifier classifies two classes at a time. The decision surface or hyperplane is obtained by using SVM algorithm. The software used for this algorithm is MATLAB R2009a in

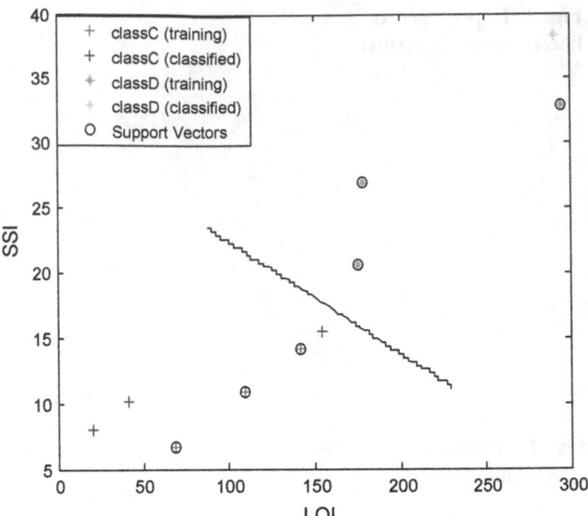

Fig. 6 SVM classifier for 14 bus system

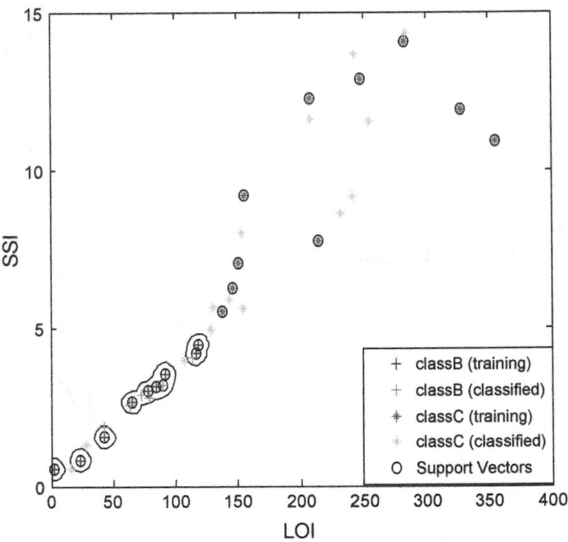

Fig. 7 SVM classifier for 39 bus system

Windows XP operating system. The results obtained for 14 bus and 39 bus system are shown in Figs. 6 and 7, respectively.

Decision tree: A tree test function is used in support vector machine algorithm, due to which an optimal decision tree is automatically generated as shown in Fig. 8.

3-Dimensional hyperplane:

SVM is capable of obtaining 3D hyperplane to classify the patterns. The decision surface in a two-dimensional axis is a plane, while it is a surface in a

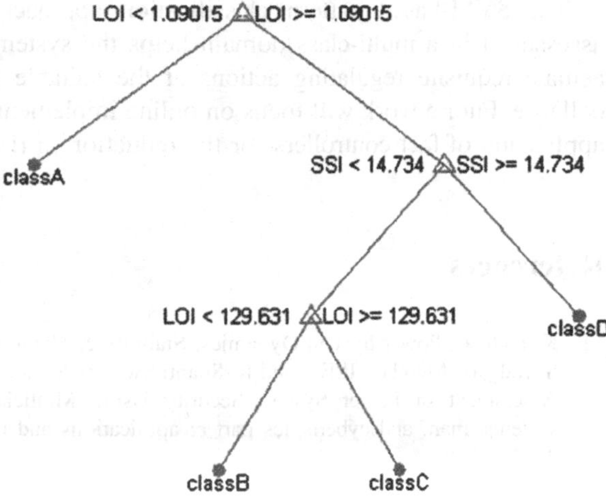

Fig. 8 Decision tree for New England 39 bus system

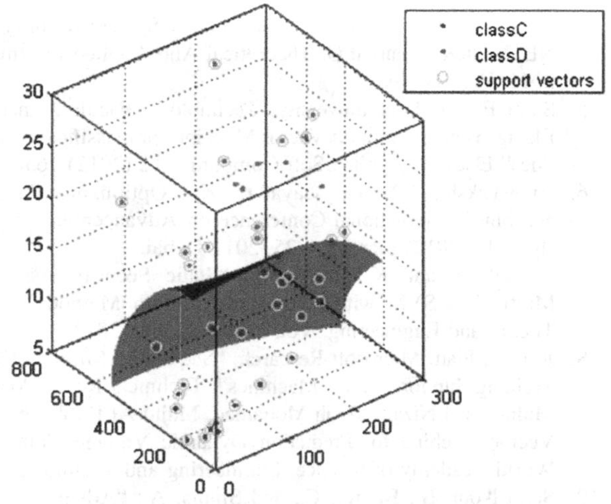

Fig. 9 3D SVM classifier for New England 39 bus system

three-dimensional axis which is shown in Fig. 9. All the patterns under the surface are classified as static insecure, while all the patterns above the surface are static highly insecure cases.

6 Conclusion and Future Scope

Support vector machine (SVM) is the leading solution for the classification problem. SVM technique is much effective when compared to other techniques in neural networks. It is clear that very low misclassification rate and high accuracy are obtained.

The SVM-based pattern classification approach for power system security assessment in a multi-class domain helps the system operator at security level to actuate requisite regulating actions at the suitable time and to stop the system collapse. Future work will focus on online implementation of proposed system and application of fact controllers for the reduction in risk, due to security.

References

1. K. Padiyar, Power System Dynamics: Stability & Control. New York: Wiley, 1996.
2. S. Kalyani, Member, IEEE, and K. Shanti Swarup, Senior Member, IEEE; "Classification and Assessment of Power System Security Using Multiclass SVM" IEEE Transactions on systems, man, and cybernetics part c: applications and reviews, vol. 41, no. 5, September 2011.
3. Craig A. Jensen, Mohamed A. El-Sharkawi, and Robert J Marks, "Power System Security Assessment Using Neural Networks: Feature Selection Using Fisher Discrimination" IEEE Transactions on Power Systems, Vol. 16, No. 4, November 2001.
4. J. Jasni, M.Z.A Ab Kadir, "Static Power System Security Assessment Via Artificial Neural NETWORK", Journal Of Theoretical And Applied Information Technology 30th September 2011. Vol. 31 No. 2.
5. Sami Ekisci, Firat university, Technology Faculty, Energy Systems Engineering, 23119, Elazig, Turkey "Support Vector Machine for classification and locating faults on transmission lines" Elsevier, Applied Soft Computing 12 (2012) 1650–1658.
6. Vijayalakshmi .N and Gayathri .K, "Optimizing Power Flow Using Support Vector Machine", International Conference on Advancements in Electrical and Power Engineering (ICAEPE'2012) March 24–25, 2012 Dubai.
7. S. Kalyani and K. S. Swarup, "Static Security Assessment in Power Systems Using Multi-Class SVM with Parameter Selection Methods", International Journal of Computer Theory and Engineering, Vol. 5, No. 3, June 2013.
8. John C. Platt, Microsoft Research, "Sequential Minimal Optimization: A Fast Algorithm for Training Support Vector Machines", Technical Report MSR-TR-98-14 April 21, 1998.
9. Muhammad Nizam, Azah Mohamed, Majid Al-Dabbagh, and Aini Hussain, "Using Support Vector Machine for Prediction Dynamic Voltage Collapse in an Actual Power System", World Academy of Science, Engineering and Technology 41 2008.
10. Scho lkopf, B., Burges, C., and Smola, A.: "Advances in kernel methods—support vector learning" (MIT Press, Cambridge, MA, 1999).
11. R. SwarnaLatha, Ch. Sai Babu, K. DurgaSyam Prasad Department of Electrical and Electronics Engineering, Jawaharlal Nehru Technological University, Kakinada, Andhra Pradesh, INDIA, " Detection & Analysis of Power Quality Disturbances using Wavelet Transforms and SVM", Vol 02, Issue 02; August-December 2011 International Research Journal of Signal Processing.
12. S. R. Samantaray, P .K. Dash, G. Panda National institute of Technology, Rourkela, India College of Engineering, Bhubaneswar, India, "Fault Classification and Ground detection using Support Vector Machine", Digital Object Identifier. doi:10.1109/TENCON.2006. 344216.
13. Dahai You, Ke Wang, Lei Ye, Junchun Wu, Ruoyin Huang, "Transient stability assessment of power system using support vector machine with generator combinatorial trajectories inputs", Elsevier, Electrical Power and Energy Systems 44 (2013) 318–325.

Credit Card Fraud Detection Using a Neuro-Fuzzy Expert System

Tanmay Kumar Behera and Suvasini Panigrahi

Abstract In this paper, a two-stage neuro-fuzzy expert system has been proposed for credit card fraud detection. An incoming transaction is initially processed by a pattern-matching system in the first stage. This component comprises of a fuzzy clustering module and an address-matching module, and each of them assigns a score to the transaction based on its extent of deviation. A fuzzy inference system computes a suspicious score by combining these score values and accordingly classifies the transaction as genuine, suspicious, or fraudulent. Once a transaction is detected as suspicious, a neural network trained with history transactions is employed in the second stage to verify whether it was an actual fraudulent action or an occasional deviation by the legitimate user. The effectiveness of the proposed system has been verified by conducting experiments and comparative analysis with other systems.

Keywords Credit card · Fraud detection · Fuzzy clustering · Fuzzy expert system · Neural network

1 Introduction

The use of credit card as a mode of payment for online as well as daily purchases has increased in the last few decades. As a consequence, associated credit card fraud has also proliferated. Fraud in credit card is practiced with an intension of acquiring goods without paying for it. This kind of illegitimate actions on credit cards is done in two possible ways: by using stolen credit cards physically (physical

T.K. Behera (✉) · S. Panigrahi
Department of Computer Science and Engineering & IT,
Veer Surendra Sai University of Technology, Burla, Sambalpur 768017,
Odisha, India
e-mail: tanmay.vssut@gmail.com

S. Panigrahi
e-mail: spanigrahi_cse@vssut.ac.in

H.S. Behera and D.P. Mohapatra (eds.), *Computational Intelligence in Data Mining*, Advances in Intelligent Systems and Computing 556,
DOI 10.1007/978-981-10-3874-7_79

fraud) and by exploiting the card details without the knowledge of the genuine cardholder (virtual fraud) via online transactions.

The loss because of credit card fraud is found to be very high in the past few years.

Statistics depicts that online banking has been growing rapidly in these years. 2014 Global Consumer Fraud Survey which entitled "Global Consumers: Losing Confidence in the Battle against Fraud" states that of all cardholders, 27% of them have experienced card fraud in the past five years among which 14% have undergone fraud multiple times [1]. This resulted in loss of cardholders' confidence on financial institutions for providing them protection against fraudulent activities. The survey report of 2012 states that fraud due to credit card is 27% only in India which is increased to 32% within two years of span [2]. A survey of fraud cases across 160 business organizations reveals that the virtual fraud is twelve times higher than the physical fraud [3]. Thus, virtual fraud has become a greater issue to the economic system. Moreover, this kind of fraudulent activities may remain undetected for longer period of time as the actual cardholder is not aware of it. Hence, this issue needs to be addressed in the strongest possible manner.

Various prevention methodologies such as such as credit card authorization, address verification system (AVS), and rule-based detection have been employed to deal with this problem by various organizations. However, every time the fraudsters come up with new ideas and tricks to break the prevention methodologies. Once the prevention measure fails, so as to preserve the viability of the payment system, there is a requirement of developing an effective credit card fraud detection system (CCFDS).

2 Related Work

In detecting credit card fraud, various techniques have been used including neural network (NN), genetic algorithm, data mining, game-theoretic approach, support vector machines, and meta-learning.

Artificial neural network (ANN) is the first technique which was employed in credit card fraud detection (CCFD). A feasibility study has been performed by Ghosh and Reilly for Mellon Bank to test the potency of the ANN in CCFD and reached at a reduction of 20–40% losses due to fraud [4]. An NN-based data mining technique known as CARDWATCH has been proposed by Aleskerov et al. [5]. An online CCFDS has been presented by Dorronsoro et al. based on a neural classifier where Fisher's discriminant analysis is employed to distinguish the fraudulent activities from the normal ones [6].

Another approach presented by Liu and Li suggests a technique for CCFD using game theory [7]. Vasta et al. proposed a model which models the interaction between an attacker and the FDS as a repeated game in which each is trying to maximize the payoffs [8].

Quah and Sriganesh proposed a framework that can be used for detecting fraud in real time [9]. It uses self-organizing map (SOM) for outlier analysis of each customer individually, and then, the classification algorithm is applied to determine whether a transaction is fallacious or legitimate. Panigrahi et al. have proposed a CCFDS which is based on combining evidence from multiple sources by using Dempster–Shafer theory [10]. The transactions found to be suspicious are further examined by applying Bayesian learning.

The fraud detection problem can be envisioned as a problem of data mining, where goal is to determine whether an operation of transaction is a genuine one or a fraudulent one. Chan et al. took huge number of transactions and divided them into smaller subsets, and then, distributed data mining is applied for building prototypes of users' behaviors [11]. Chiu and Tsai have also applied data mining concept in their work where Web service for data exchange was taken into consideration [12].

The prime concern in such real-life problem is that the fraction of fraudulent transactions is comparatively less in numbers and hence requires finding out a rare occasion from a huge set of genuine transactions. This may result in the causation of false alarms in many cases and thus requires minimization. In case of failure in detecting a fraud case, there is a straight loss to the company; moreover, the follow-up actions taken in addressing false alerts are high-priced too. In addition, the change of behavior of cardholders and fraudsters is essential to be captured by the CCFDS for reducing the misclassification. Thus, our objective is to design an adaptive CCFDS, which detects the unlawful activities efficaciously while reducing the rate of false alarms by learning the spending patterns of the card users with time.

The remaining part of the paper has been arranged as follows: Sect. 3 presents the proposed work in detail along with the block diagram depicting the flow of events in the proposed system. The next section discusses the experimental findings of the approach. The paper is then concluded along with directions for future research in Sect. 5.

3 Proposed Work

In this work, we have proposed a *neuro-fuzzy expert system* for credit card fraud detection (NFES_CCFD) which ingrates evidences obtained from two distinct sources based on different transaction attributes for analyzing the deviation of user's behavior from his normal spending profile. Furthermore, learning mechanism based on NN is used to affirm the suspicious cases. The proposed FDS is divided into the following four components:

3.1 Credit card validation component (CCVC);
3.2 Input pattern-matching component (IPMC);
3.3 Fuzzy rule combiner component (FRCC); and
3.4 Neural network based learning component (NNLC).

We have used two preset threshold values lower threshold (Φ_{Lth}) and upper threshold (Φ_{Uth}) for classifying the transactions as genuine or fraudulent. The undecided transactions are allowed to execute but tagged as suspicious for further verification by the learning component. The threshold values are determined experimentally. The methodology of fraud detection along with the system components has been depicted in Fig. 1.

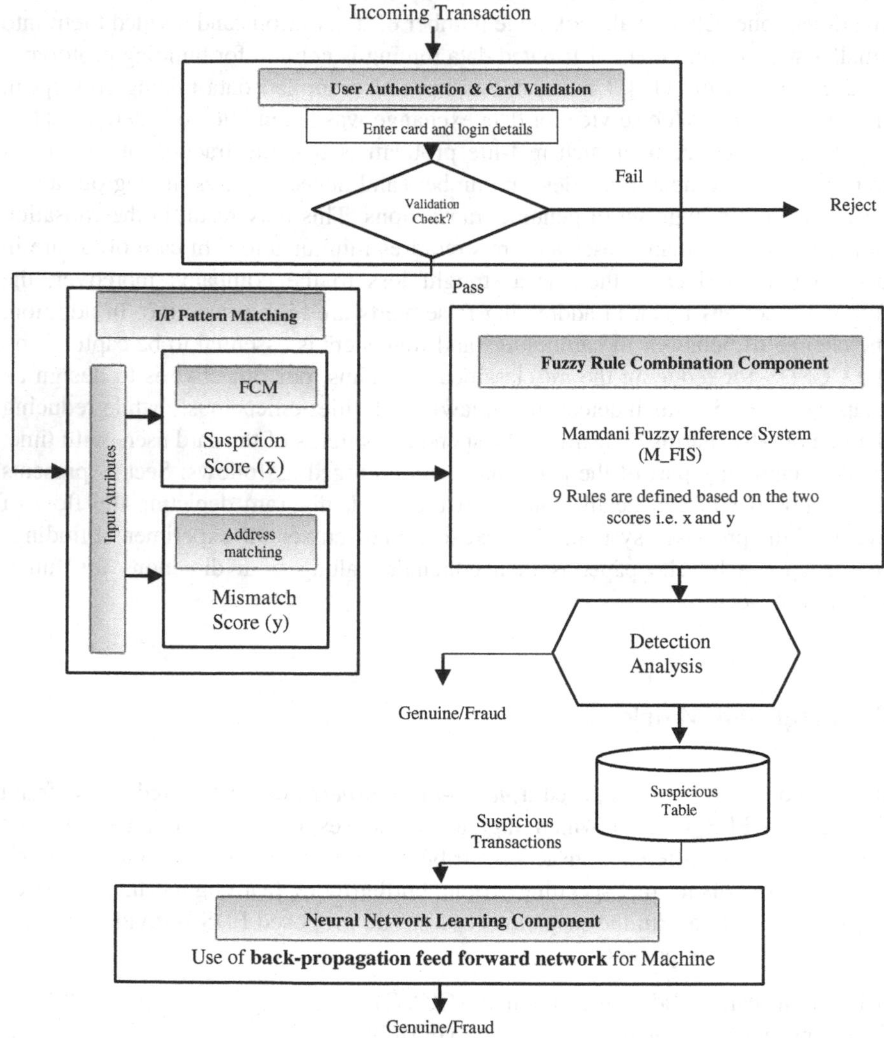

Fig. 1 Methodology of the proposed CCFDS

3.1 Credit Card Validation Component (CCVC)

For every incoming transaction on a specific card, initially, the basic checks such as verification of the details of credit card, e.g., pin of the card, amount of the transaction within the limit of the card, and expiry date of the card, are inspected and verified in this component.

3.2 Input Pattern-Matching Component (IPMC)

Once the basic checks are cleared, then the spending patterns in the transaction are analyzed by passing it through the Fuzzy C-means (FCM) clustering block and address-matching block. FCM is a soft clustering technique which overcomes the disadvantages of the hard clustering algorithms by allowing individual data point to belong more than one cluster. Thus, it is widely accepted in this kind of real-life applications [13].

Before testing an incoming transaction, the FCM is first applied on the historical transactions for forming clusters that build the user profiles as concerning to the attributes *items purchased, amount of purchase, and location of the transaction.* If a transaction made on a card C_k is T_{Ck} and d_{ik} is the Euclidean distance that is measured from cluster head p_i to the object point x_k of the transaction T_{Ck}, then a suspicion score can be assigned to the transaction, which is expressed as follows:

$$S_{score} = d_{ik} = \|x_k - p_i\| \tag{1}$$

where S_{score} is the degree of deviation from usual profile.

Similarly, the other block checks the *billing address* and the *shipping address* of the transaction. If both the addresses are found to be same, then it can be considered as a legitimate transaction with very high probability as a fraudster would normally never sent the purchased item to the genuine cardholder's billing address. However, if there is any mismatch found in the addresses, then the transaction can be considered as suspicious and a score is assigned in between the lower threshold and upper threshold [10]. In this case, we cannot declare the transaction as certainly fraudulent because a genuine user may gift an item to someone else, where the shipping address will be different from that of the billing address.

3.3 Fuzzy Rule Combiner Component (FRCC)

The fuzzy rule combiner is the component that determines whether a transaction is intrusive or normal based on the input scores from the FCM (x) and the address mismatch block (y). For each input transaction, it employs two-input single-output

Mamdani fuzzy inference system (FIS) to combine two scores x and y for determining the output z [14]. The Mamdani FIS uses simple inference procedure to predict the output. Based on the input membership functions (MFs), the FRCC uses the following nine fuzzy if-then rules:

- Rule 1: if *x* is *low_x* and *y* is *low_y* then *z* is *genuine*.
- Rule 2: if *x* is *low_x* and *y* is *medium_y* then *z* is *suspicious*.
- Rule 3: if *x* is *low_x* and *y* is *high_y* then *z* is *suspicious*.
- Rule 4: if *x* is *medium_x* and *y* is *low_y* then *z* is *suspicious*.
- Rule 5: if *x* is *medium_x* and *y* is *medium_y* then *z* is *suspicious*.
- Rule 6: if *x* is *medium_x* and *y* is *high_y* then *z* is *suspicious*.
- Rule 7: if *x* is *high_x* and *y* is *low_y* then *z* is *suspicious*.
- Rule 8: if *x* is *high_x* and *y* is *medium_y* then *z* is *suspicious*.
- Rule 9: if *x* is *high_x* and *y* is *high_y* then *z* is *fraudulent*.

In the current work, we have used the Mamdani FIS having the *max-min composition* for fuzzification and the *centroid of area* method for defuzzification. Three-output fuzzy sets, namely *genuine, suspicious,* and *fraudulent,* are defined. The genuine transactions are allowed, whereas the fraudulent transactions are simply blocked and confirmation is obtained from the genuine cardholder. However, the suspicious transactions are maintained in a table, namely *suspicious table* for further processing by the learning component.

3.4 Neural Network Learning Component (NNLC)

The suspicious transactions are examined and classified by applying the feed forward neural networks (FFNNs) with back-propagation (BP). In the proposed method, we have used a supervised NN for learning based on the Bayesian Regulation (BR) back-propagation technique [15]. This learning algorithm is preferred as it has set a benchmark against other back-propagation algorithms due to its higher accuracy of prediction.

In the current work, the attributes those have been considered are *items purchased*, *time of purchase,* and the *time gap* between two consecutive transactions on the same card for learning purpose. The fraud cases obtained from the dataset are statistically analyzed based on these attribute measures to find out the interrelation between input data and values for some crucial factors for realizing several fraud patterns. This information about fraud is then repeatedly fed to the FFNN for training. Based on the resemblance with the trained patterns, finally a distrustful transaction is classified as fraudulent or legitimate. The learning module is added to FDS for reducing the fraction of wrong classifications.

4 Simulation and Results

The proposed CCFDS is being evaluated with large-scale data, and the performance is measured on various performance metrics.

4.1 Simulation

Synthetic datasets developed with the simulator by Panigrahi et al. [10] have been applied for analyzing the proposed CCFDS because of the non-availability of the real-life credit card datasets or any other benchmark datasets. Large-scale datasets have been developed on which the performance of the proposed system is being tested. Gaussian distributions and Markov modulated poisson processes are used in the simulator for generating the synthetic transactions which reflects the characteristics of legitimate users as well as fraudsters. The design of the simulator is being done in such a way that it can capture various events of the real life which are normally found in credit card payment processing system.

4.2 Results

The experiments are performed by inputting the inputs to IPMC component, where the FCM module computes a score (x) based on the Euclidian distance that is being measured for each of the object points existing in the clusters. In the address-matching module, if there is a mismatch of the billing address and shipping address, then a score in between the thresholds is assigned, and if no deviation occurs, then a low value around 0.2 is assigned as discussed earlier in Sect. 3.2. The threshold values are set experimentally as lower threshold (Φ_{Lth}) = 0.22 and upper threshold (Φ_{Uth}) = 0.78. Based on the threshold values, the scores are divided as low, medium, and high. Nine fuzzy if-then rules have been defined in the FRCC block, where an incoming transaction is classified as genuine, suspicious, or a fraudulent one. The suspicious ones are directed toward a *suspicious_table* for further analysis.

The suspicious transactions are fed to the NN where the BR back-propagation algorithm is employed for learning. In this work, we have used 5-hidden layers to train the network with the datasets. As the hidden layer numbers increase, better results are drawn, but it also increases the computation time. The whole dataset is being split into three groups −70% for training, 15% for validation, and 15% for testing.

The receiver operating characteristic (ROC) curve is plotted in Fig. 2, which is used as a metric to check the quality of the classifier. The curve can be produced by plotting the true positive rate (Sensitivity) against the false positive rate

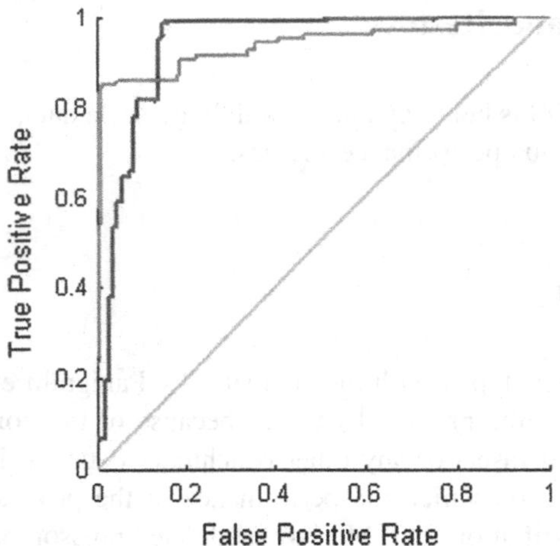

Fig. 2 ROC curve

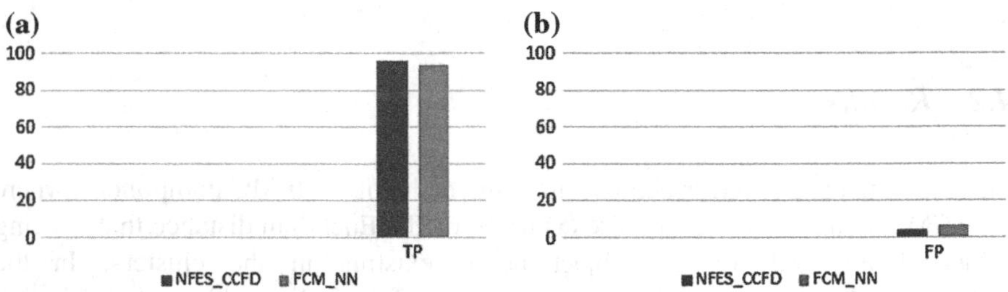

Fig. 3 **a** NFES_CCFD versus FCM_NN in terms of TP, **b** NFES_CCFD versus FCM_NN in terms of FP

(1-Specificity). Points in the upper left corner with 100% sensitivity and 100% specificity represent an ideal ROC. The performance of the proposed FDS can be apparent from the arch in Fig. 2 that it is perfect with regard to ROC.

We have analyzed the performance of the proposed NFES_CCFD system with the variation of the number of genuine and fraudulent transactions. For further verifying our results, we have compared our results with one of our previous work that applies a combination of fuzzy clustering and neural network (FCM-NN) [16]. It is clearly apparent from Fig. 3 that NFES_CCFD performs better than that of the FCM-NN having higher accuracy and lowered misclassification.

5 Conclusions

In this research, we have presented a novel scheme for CCFD by combining a rule-based fuzzy inference system and a learning component that uses back-propagation neural network. We have tested the proposed system by carrying out experiments using stochastic models. Based on the results obtained, it is inferred that incorporation of neural network along with fuzzy inferencing is appropriate in addressing this sort of real-world issues.

References

1. Inscoe, S. W.: Global Consumers: Losing Confidence in the Battle against Fraud. ACI Universal Payments, June, 2014, www.aciworldwide.com.
2. Credit Card Fraud Statistics-2013, <http://www.cardhub.com/edu/creditfraud-Statistics>, 5 Feb, 2015.
3. Online Fraud is twelve times higher than offline fraud, <http://sellitontheweb.com/ezine/ne-ws034.shtml>, 20, June, 2007.
4. Ghosh, S., Reilly, D.L.: Credit card fraud detection with a neural-network. In: Proceedings of the Annual International Conference on System Science, pp. 621–630 (1994).
5. Aleskerov, E., Freisleben, B., Rao, B.: CARDWATCH: a neural-network based database mining system for credit card fraud detection. In: Proceedings of the Computational Intelligence for Financial Engineering, pp. 220–226 (1997).
6. Dorronsoro, J.R., Ginel, F., Sanchez, C., Cruz, C.S.: Neural fraud detection in credit card operations. IEEE Transactions on Neural Networks, pp. 827–834 (1997).
7. Liu, P., Li, L.: A Game-Theoretic Approach for Attack Prediction. Technical Report, PSU-S2-2002-01, Penn State University (2002).
8. Vatsa, V., Sural, S., Majumdar, A.K.: A game-theoretic approach to credit card fraud detection. In: Proceedings of the International Conference on Information Systems Security, Lecture Notes in Computer Science, vol. 3803. pp. 263–276 (2005).
9. Quah, J. T. S., Srinagesh, M.: Real-time credit fraud detection using computational intelligence. Expert Systems with Applications, 35, pp. 1721–1732 (2008).
10. Panigrahi, S., Kundu, A., Sural, S., Majumdar, A.: Credit card fraud detection a fusion approach using Dempster–Shafer theory and bayesian learning. Information Fusion, pp. 354–363 (2009).
11. Chan, P.K., Fan, W., Prodromidis, A.L., Stolfo, S.J.: Distributed data-mining in credit card fraud detection. In: Proceedings of the IEEE Intelligent Systems, pp. 67–74 (1999).
12. Chiu, C., Tsai, C.: A web services-based collaborative scheme for credit card fraud detection. In: Proceedings of the IEEE International Conference on e-Technology, e-Commerce and e-Service, pp. 177–181 (2004).
13. Nayak, J., Naik, B., Behera, H. S.: Fuzzy C-Means (FCM) Clustering Algorithm: A Decade Review from 2000 to 2014. In Computational Intelligence in Data Mining-Vol. 2, Springer India, pp. 133–149 (2015).
14. Mamdani, E.H.: Applications of fuzzy logic to approximate reasoning using linguistic synthesis. IEEE Transactions on Computers, Vol. 26, No. 12, pp. 1182–1191 (1977).
15. MacKay.: Neural Computation: Bayesian Regulation NN algorithm, Vol. 4, No. 3, pp. 415–447 (1992).
16. Behera, T. K., Panigrahi, S.: Credit Card Fraud Detection: A Hybrid Approach Using Fuzzy Clustering & Neural Network. 2015 Second International Conference on Advances in Computing and Communication Engineering (ICACCE-15), IEEE, pp. 494–499 (2015).

Author Index

H.S. Behera and D.P. Mohapatra (eds.), *Computational Intelligence in Data Mining*, Advances in Intelligent Systems and Computing 556,
DOI 10.1007/978-981-10-3874-7

Printed in the United States
By Bookmasters